Data Networks, IP and the Internet

Data Networks, IP and the Internet

Protocols, Design and Operation

Martin P. Clark
Telecommunications Consultant, Germany

WILEY

Other Wiley Editorial Offices

John Wiley & Sons Inc., 111 River Street, Hoboken, NJ 07030, USA

Jossey-Bass, 989 Market Street, San Francisco, CA 94103-1741, USA

Wiley-VCH Verlag GmbH, Boschstr. 12, D-69469 Weinheim, Germany

John Wiley & Sons Australia Ltd, 33 Park Road, Milton, Queensland 4064, Australia

John Wiley & Sons (Asia) Pte Ltd, 2 Clementi Loop #02-01, Jin Xing Distripark, Singapore 129809

John Wiley & Sons Canada Ltd, 22 Worcester Road, Etobicoke, Ontario, Canada M9W 1L1

Wiley also publishes its books in a variety of electronic formats. Some content that appears
in print may not be available in electronic books.

Library of Congress Cataloging-in-Publication Data

Clark, Martin P.
  Data networks, IP, and the Internet : networks, protocols, design, and operation / Martin
P. Clark.
      p. cm.
  Includes bibliographical references and index.
  ISBN 0-470-84856-1
    1. Computer networks. 2. TCP/IP (Computer network protocol) 3. Internet. I. Title.

TK5105.5 .C545 2003
004.6–dc21
                                                        2002191041

British Library Cataloguing in Publication Data

A catalogue record for this book is available from the British Library

ISBN 0-470-84856-1

Typeset in 9.5/11pt Times by Laserwords Private Limited, Chennai, India

Für Ruth,

in Erinnerung an Wade

in dessen Gesellschaft dieses Buch entstanden ist.

Contents

Preface

The business world relies increasingly upon data communications, and modern data networks are based mainly on the Internet or at least on the IP (Internet Protocol). But despite these facts, many people remain baffled by IP and multiprotocol data networks. How do all the protocols fit together? How do I build a network? And what sort of problems should I expect? This book is intended for experienced network designers and practitioners, as well as for the networking newcomer and student alike: it is intended to provide an explanation of the complex jargon of networking: putting the plethora of 'protocols' into context and providing a quick and easy handbook for continuing reference.

Even among experienced telecommunications and data-networking professionals, there is confusion about how data network components and protocols work and how they affect the performance of computer applications. I have myself bought many books about the Internet, about IP and about multiprotocol networks, but found many of them 'written in code'. Some have the appearance of computer programmes, while others perversely require that you understand the subject before you read them!

Putting the pieces of knowledge and the various components of a network together — working out how computers communicate — can be a painstaking task requiring either broad experience or the study of a library full of books. The experience has spurred me to write my own book and handy reference and this is it. My goal was a text in plain language, building slowly upon a solid understanding of the principles — introducing a newcomer slowly and methodically to the concepts and familiarising him or her with the language of data communications (the unavoidable 'jargon') — but always relating new topics back to the fundamentals:

- relating to the real and tangible;

- sharing experiences and real examples;

- not only covering the theoretical 'concepts'; but also

- providing practical tips for building and operating modern data networks.

The book covers all the main problems faced by data network designers and operators: network architecture and topology, network access means, which protocol to use, routing policies, redundancy, security, firewalls, distributed computer applications, network service applications, quality of service, etc.

The book is liberally illustrated and written in simple language. It starts by explaining the basic principles of packet-data networking and of layered protocols upon which all modern data communications are based. It then goes on to explain the many detailed terms relevant to modern IP networks and the Internet. My goal was that readers who only wanted to 'dip in' to have a single topic explained should go away satisfied — able to build on any previous knowledge of a given subject.

The extensive set of annexes and the glossary of terms are intended to assist the practising engineer — providing a single reference point for information about interfaces, protocol field names and formats, RFCs (Internet specifications) and acronyms (the diagrams and some of the appendices are also available for download at: http://www.wiley.co.uk/clarkdata/). With so many acronyms and other terms, protocols, code-fields, and technical configuration information to remember, it is impossible to expect to keep all the details 'in your head'! And to distinguish where jargon and other special 'telecommunications vocabulary' is being used in the main text, I have highlighted terms as they are being defined by using *italics*.

The book is intended to provide a complete foundation textbook and reference of modern data networking — and I hope it will find a valued position on your bookshelf. Should you have any suggestions for improvement, please let me know!

Martin Clark

Acknowledgements

No book about the Internet can fail to recognise the enormous contribution which has been made to the development of the Internet by the Internet Engineering Task Force (IETF) and its parent organisation, the Internet Society. Very many clever and inspired people have contributed to the process and all those RFC (request for comments) documents — unfortunately far too many to allow individual recognition.

I would also like to thank the following organisations for contribution of illustrations and granting of copyright permission for publication:

- Apple Computer;

- Black Box Corporation and Black Box Deutschland GmbH;

- France Télécom;

- IBM;

- International Telecommunications Union;

- Microsoft Corporation/Waggener Edstrom;

- RS Components Ltd.

The media departments of each of the organisations were both kind and helpful in processing my requests, and I would like to thank them for their prompt replies. The experience leads me in particular to recommend the online IBM archive (www.ibm.com/ibm/history) as well as the cabling and component suppliers: Black Box Corporation, Black Box Deutschland GmbH and RS Components Ltd.

The copyright extracts drawn from ITU-T recommendations were chosen by the author, but reproduced with the prior authorisation of the ITU. All are labelled with their source accordingly. The full texts of all ITU copyright material may be obtained from the ITU Sales and Marketing Division, Place des Nations, CH-1211 Geneva 20, Switzerland, Telephone: +41 22 730 6141 (English) / +41 22 730 6142 (French) / +41 22 730 6143 (Spanish), Telex: 421 000 uit ch, Fax: +41 22 730 5194, email: Sales@itu.int or Internet: www.itu.int/publications.

Finally I would like to thank my 'personal assistants' — who assisted in wading through the voluminous drafts and made suggestions for improvement:

- my brother, Andrew Clark;

- my close friend and data networking colleague, Hubert Gärtner;

- Jon Crowcroft of Cambridge University — who spent many hours patiently reviewing the manuscript and explaining to me a number of valuable suggestions;

- Susan Dunsmore (the poor copy editor)—who had to struggle to correct all the italics, 'rectos' and 'decrements' — and not only that, but also had to make up for what my English grammar teacher failed to drill into me at school;

- the production and editorial staff at John Wiley — Zoë Pinnock, Sarah Hinton and Mark Hammond.

Martin Clark

Foreword

Before we start in earnest, there are three things I would like you, the reader, to keep in mind:

1. The first part of the book (Chapters 1–3) covers the general principles of data communications. This part is intended to introduce the concepts to data communications newcomers. Chapters 4–15 build on this foundation to describe in detail the IP (Internet protocol) suite of data communications protocols and networking procedures.

2. Terms highlighted in *italics* on their first occurrence are all telecommunications vocabulary or 'jargon' being used with their strict 'telecommunications meaning' rather than their meaning in common english parlance.

3. Although the book is structured in a way intended to ease a reader working from 'cover to cover', you should not feel obliged to read it all. The extensive index, glossary and other appendices are intended to allow you to find the meaning of individual terms, protocols and other codes quickly.

1

The Internet, Email, Ebusiness and the Worldwide Web (www)

Nowadays every self-respecting person (particularly if a grandparent!) has a personal email address. And many modern companies have encompassed ebusiness. They have prestigious Internet 'domain names' (advertised with modern lower case company names) and run Worldwide Web (www) sites for advertising and order-taking. What has stirred this revolution? The Internet. But when, why and how did data networking and interworking start? And how did the Internet evolve? Where will it lead? And what does all that frightful jargon mean? (What are the acronyms and the protocols?). In this chapter we shall find out. We shall talk about the emergence of computer networking, the Worldwide Web (www), about ISPs (Internet service providers) and about where the Internet started — in the US Defense Department during the 1970s. We discuss the significance of the Internet Protocol (IP) today, and where it will lead. And most important of all — we start 'unravelling' the jargon.

1.1   In the beginning — ARPANET

The beginnings of the Internet are to be found in the *ARPANET*, the *advanced research project agency network*. This was a US government-backed research project, which initially sought to create a network for resource-sharing between American universities. The initial tender for a 4-node network connecting UCLA (University of California, Los Angeles), UCSB (University of California, Santa Barbara), SRI (Stanford Research Institute) and the University of Utah took place in 1968, and was won by BBN (Bolt, Beranek and Newman). The network nodes were called Internet message processors (IMPs), and end-user computing devices were connected to these nodes by a protocol called *1822* (1822 because the *Internet engineering note (IEN)* number 1822 defined the protocol). Subsequently, the agency was increasingly funded by the US military, and consequently, from 1972, was renamed DARPA (Defense Advanced Research Project Agency).

These beginnings have had a huge influence on the subsequent development of computer data networking and the emergence of the Internet as we know it today. BBN became a leading manufacturer of *packet switching* equipment. A series of *protocols* developed which are sometimes loosely referred to either as *TCP/IP (transmission control protocol/Internet protocol)* or as *IP (Internet protocol)*. Correctly they are called the 'IP-protocol suite'. They are defined in documents called *RFCs (request for comment)* generated under the auspices

Data Networks, IP and the Internet: Protocols, Design and Operation   Martin P. Clark
© 2003 John Wiley & Sons, Ltd   ISBN: 0-470-84856-1

of the *Internet Engineering Task Force (IETF)*. The current most-widely used version of the Internet protocol (IP)—version 4 or IPv4—is defined in RFC 791. The current version of TCP (transmission control protocol) is defined in RFC 793.

1.2   The emergence of layered protocols for data communication

In parallel with the development of the ARPANET, a number of standardised layered protocol 'stacks' and protocol suites for simplifying and standardising the communication between computer equipment were being developed independently by various different computer and telecommunications equipment manufacturers. Most of these protocols were 'proprietary'. In other words, the protocols were based on the manufacturers' own specifications and documentation, which were kept out of the public domain. Many manufacturers believed at the time that 'proprietary' protocols gave both a 'competitive advantage' and 'locked' customers into using their own particular brand of computer hardware. But the principles of the various schemes were all similar, and the ideas generated by the various groups of developers helped in the development of the standardised protocols which came later.

All data communications protocols are based upon packet switching, a form of electronic inter-computer communication advanced by Leonard Kleinrock of MIT (Massachusetts Institute of Technology—and later of UCLA—University of California in Los Angeles) in his paper 'Information flow in large communication networks' (July 1961). The term packet switching itself was coined by Donald Davies of the UK's National Physical Laboratory (NPL) in 1966.

Packet switching is analogous to sending letters through the post—the data content, analogous to a page of a letter is the *user content* (or *payload*) of a standard *packet*. The user content is packed in the *packet* or *frame* (analogous to an 'envelope') and labelled with the destination *address*. When the size of a single packet is too small for the message as a whole, then the message can be split up and sent as a sequence of numbered packets, sent one after another (see Figure 1.1). The networking nodes (which in different types of data networks have different names: routers, switches, bridges, terminal controllers, cluster controllers, front-end

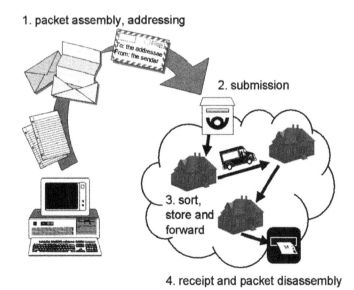

Figure 1.1   Post Office analogy illustrating the principles of packet switching.

processors, etc.) all essentially work like a postal sorting office. They read the 'address' on each packet (without looking at the contents) and then *forward* the packet to the appropriate next node nearer the destination.

The best-known, most successful and widely used of the 1970s generation of packet-switching protocols were:

- *SNA (systems network architecture)* — the networking protocols used for interconnecting IBM (International Business Machines) computers;

- *DECnet* — the networking protocols used for interconnecting computers of the Digital Equipment Corporation (DEC);

- *X.25* (ITU-T recommendation X.25) and its partner protocol, X.75. This was the first attempt, coordinated by the International Telecommunications Union standardisation sector (ITU-T), to create a 'standard' protocol — intended to enable computers made by different manufacturers to communicate with one another — so-called *open systems interconnection (OSI)*.

1.3 SNA (systems network architecture)

The *systems network architecture (SNA)* was announced by IBM in 1974 as a standardised communications architecture for interconnecting all the different types of IBM computer hardware. Before 1974, transferring data or computer programs from one computer to another could be a time-consuming job, sometimes requiring significant manual re-formatting, and often requiring the transport of large volumes of punched cards or tapes. Initially, relatively few IBM computers were capable of supporting SNA, but by 1977 the capabilities of the third generation of SNA (SNA-3) included:

- *communication controllers* (otherwise called *FEPs* or *front end processors*) — hardware which could be added to *mainframe* computers for taking over communication with remote devices;

- *terminal controllers* (otherwise called *cluster controllers*) — by means of which, end-user terminals (*teletypes* or computer *VDUs, video display units*) could be connected to a remote host computer;

- the possibility to connect remote terminal controllers to the mainframe/communication controller site using either leaselines or dial-in lines;

- the possibility of multi-host networks (terminals connected to multiple mainframe computers — e.g., for bookkeeping, order-taking, personnel, etc. — by means of a single communications network).

Figure 1.2 illustrates the main elements of a typical SNA network, showing the typical star topology. Point-to-point lines across the *wide area network (WAN)* connect the front end processor (FEP or communications controller) at the enterprise computer centre to the terminals in headquarters and remote operations offices. The lines used could be either *leaselines*, point-to-point X.25 (i.e., *packet-switched*) connections, *frame relay* connections or dial-up lines.

During the 1980s and 1990s, SNA-based networks were widely deployed by companies which used IBM mainframe computers. At the time, IBM mainframes were the workhorse of the computing industry. The mainframes of the IBM S/360, S/370 and S/390 architectures became well known, as did the components of the SNA networks used to support them:

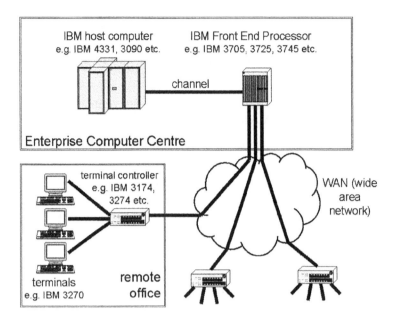

Figure 1.2   A typical SNA network interconnecting IBM computer hardware.

- Front end processor (FEP or communication controller) hardware: IBM 3705, IBM 3725, IBM 3720, IBM 3745;

- Cluster controller hardware: IBM 3174, IBM 3274, IBM 4702, IBM 8100;

- *VTAM (virtual telecommunication access method)* software used as the mainframe communications software;

- *CICS (communication information control system)* mainframe management software;

- *NCP (network control program)* front end processor communications control software;

- *NPSI (NCP-packet switching interface)* mainframe/FEP software for use in conjunction with X.25-based packet-switched WAN data networks;

- *TSO (time sharing option)* software allowing mainframe resources to be shared by many users;

- *NetView* mainframe software for network monitoring and management;

- *APPN (advanced peer-to-peer networking)* used in IBM AS-400 networks;

- *ESCON (enterprise system connection)*: a high-speed 'channel' connection interface between mainframe and front-end processor;

- *Token ring local area network (LAN)*.

Due to the huge popularity of IBM mainframe computers, the success of SNA was assured. But the fact that SNA was not a public standard made it difficult to integrate other manufacturers' network and computer hardware into an IBM computer network. IBM introduced products intended to allow the integration of public standard data networking protocols such as X.25 and Frame Relay, but it was not until the explosion in numbers of *PCs (personal computers)* and *LANs (local area networks)* in the late 1980s and 1990s that IBM lost its leading role in the data networking market, despite its initial dominance of the personal computer market. LANs and PC-networking heralded the Internet protocol (IP), *routers* and a new 'master' of data networking — Cisco Systems.

1.4 DECnet

The Digital Equipment Corporation (DEC) was another leading manufacturer of mainframes and computer equipment in the 1980s and 1990s. It was the leading force in the development of *mini-computers, workstations* and *servers* and an internationally recognised brand until it was subsumed within COMPAQ (which in turn was swallowed by Hewlett Packard). DEC brought the first successful minicomputer (the PDP-8) to the market in 1965.

Like IBM, DEC built up an impressive laboratory and development staff. The main philosophy was that software should be 'portable' between the various different sizes of DEC hardware platforms and DEC became a prime mover in the development of 'open' and public communications standards.

DECnet was the architecture, hardware and software needed for networking DEC computers. Although some of the architecture remained proprietary, DEC tended to incorporate public standards into DECnet as soon as they became available, thereby promoting 'open' interconnectivity with other manufacturers' devices. The technical legacy of DEC lives on — their very high performance *alpha servers* became the basis of the server range of COMPAQ. In addition, perhaps the oldest and best-known Internet search engine, Alta Vista, was originally established by DEC. Unfortunately, however, the commercial management of DEC did not match its technical prowess. The company overstretched its financial resources, largely through over-aggressive sales, and was taken over by COMPAQ in 1998 (and subsequently subsumed by Hewlett Packard in 2002).

1.5 Other mainframe computer manufacturers

In the 1970s and 1980s, there were a number of large computer mainframe manufacturers — Amdahl, Bull, Burroughs, DEC, Honeywell, IBM, Rockwell, Sperry, Sun Microsystems, UNIVAC, Wang, etc. Each had a proprietary networking and operating system architecture, or in the case of Amdahl and Wang, positioned their products as low cost alternatives to IBM hardware. Where these companies have survived, they have been largely 'reincarnated' as service, maintenance, support and application development companies. Typically they sell other people's computer and networking hardware and specialise in *system integration*, software development and support. Burroughs, Sperry and UNIVAC, for example, all became part of the computer services company known today as UNISYS.

1.6 X.25 (ITU-T recommendation X.25)

ITU-T's recommendation X.25 defines a standard interface for connecting computer equipment to a *packet switched data network* (see Figure 1.3). The development of the X.25-interface and the related packet-switched protocols heralded the appearance of *public data networks (PDN)*. Public data networks were meant to provide a cost-effective alternative for networking enterprise computer centres and their remote terminals.

By using a public data network, the line lengths needed for dedicated enterprise-network connections could be much shorter. No longer need a dedicated line stretch from the remote site all the way to the enterprise computer centre as in Figure 1.2. Instead a short connection to the nearest *PSE (packet switch exchange)* was adequate. In this way, the long distance

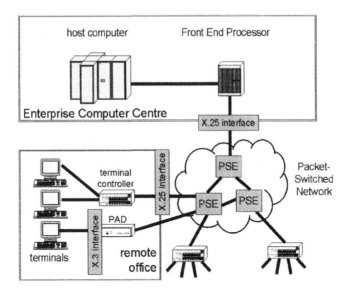

Figure 1.3   A typical public packet-switched network.

lines between PSEs in the *wide area network* and the costs associated with them were shared between different networks and users (see Figure 1.3). Overall network costs can thus be reduced by using public data networks (assuming that the tariffs are reasonable!). In addition, it may be possible to get away with fewer ports and connection lines. In the example of Figure 1.3, a single line connects the front end processor (FEP) to the network where in Figure 1.2. three ports at the central site had been necessary.

The X.25-version of packet switching, like SNA, DECnet and other proprietary data networking architectures, was initially focused on the needs of connecting remote terminals to a central computer centre in enterprise computing networks. In commercial terms, however, it lacked the success which it deserved. Though popular in some countries in Europe, X.25 was largely ignored in the USA. The X.25 standard (issued in 1976) had arrived late in comparison with SNA (1974) and did not warrant a change-over. On an economic comparison, it was often as cheap to take a *leaseline* and use SNA than it was to use a public X.25 network to connect the same remote site. As a result, enterprise computing agencies did not rush to X.25 and the computer manufacturers did not make much effort to support it. The IBM solution for X.25 using *NPSI (NCP-packet switching interface)*, for example, always lacked the performance of the equivalent SNA connection. Only in those countries where leaselines were expensively priced (e.g., Germany) did X.25 have real success. In Germany, the Datex-P packet-switched public data network of the Deutsche Bundespost was one of the most successful X.25 networks.

In the case where a remote dumb terminal is to be connected to a computer across a public data network, a PAD may be used. A *PAD (packet assembler/disassembler)* is a standard device, defined by the packet-switching standards in ITU-T recommendation X.3. Its function is to convert the keystrokes of a simple terminal into packets which may be forwarded by means of a packet-switched network to a remote computer. A number of different parameters are defined in X.3 which define the precise functioning of the PAD. The parameters define the linespeed to be used, the content of each packet and the packet *flow control*. Typically the PAD would be adjusted to forward complete commands to the central computer. Thus a number of keystrokes, as making up a series of command words, would first be collected by the PAD, and only forwarded in a single packet once the human user typed the <return>

key. But by setting the PAD parameters differently it is possible to forward each keystroke individually, or all the keystrokes typed within a given time period.

Flow control is used in an X.25 packet-switched network to regulate the sending of data and to eliminate errors which creep in during transmission. In simple terms, flow control is conducted by waiting for the acknowledgement by the receiver of the receipt of the previous packet before sending another one. This ensures that messages are received and do not get out of order. An adaptation of this method is sometimes used for terminal-to-computer communication: the user's terminal does not display the character actually typed on the keyboard at the time of typing, but instead only displays the *echo* of each character. The echo is the same character sent back by the computer as an acknowledgement of its receipt. Using echo as a form of flow control prompts the human user to re-type a character in the case that the computer did not receive it (otherwise the character will not appear on the user's terminal screen).

Despite its relatively poor commercial success, X.25 left a valuable legacy. X.25 created huge interest in the development of further public standards which would permit *open systems interconnection (OSI)*. Computer users were no longer satisfied with using computers merely for simplifying departmental calculations and record-keeping. Instead they were impatient to link various computer systems together (e.g., the computer running the 'salary' programme to that running the 'bookkeeping' and that carrying the 'personnel records'). This required that all the different types of manufacturers' computers be interconnected with one another using 'open' rather than proprietary interconnection standards — open systems interconnection. Soon, the interconnection of all the company-internal computers was not sufficient either: company staff also wanted to swap information electronically with both customers and suppliers. Users were demanding *electronic data interchange (EDI)*. The rapid development of both OSI and EDI are both important legacies of X.25. But in addition, much of the basic vocabulary and concepts of packet switching were established in the X.25 recommendations.

Understanding the basic problems to be overcome in order to realise both open systems interconnection (OSI) and electronic data interchange (EDI) is key to understanding the challenge of *internetworking*. But before discussing these subjects, it will be valuable to discuss in detail some of the basic components of a data network and the jargon which goes to describe them. We next introduce *DTEs (data terminal equipment)*, *DCEs (data circuit-terminating equipment)*, *protocols*, *UNIs (user-network interfaces)* and *NNIs (network-network interfaces)*. . .

1.7 DTE (data terminal equipment), DCE (data circuit-terminating equipment), line interfaces and protocols

A simple *wide area* (i.e., long distance) data communications link is illustrated in Figure 1.4. The link connects a PC or computer terminal on the left of the diagram to a mainframe computer on the right. The long-distance network which actually carries the connection is shown as a 'cloud' (in line with modern convention in the networking industry). It is not clear exactly clear what is in the 'cloud' in terms of either technology, the line types and interfaces, or the topology. This is often the case, and as we shall see, need not always concern us. What is more important are the interfaces used at each end of the network to connect the computing equipment. These interfaces are defined in terms of the *DTE (data terminal equipment)*, the *DCE (data circuit-terminating equipment)* and the *protocols* used.

You will note that the communicating devices (both the PC and the mainframe computer) are DTE (data terminal equipment) in the jargon. The 'T' in DTE does not necessarily refer to a computer device (computer terminal) with a screen and a keyboard (though this is one example of a DTE). The DTE could be any piece of computer equipment connected to a data

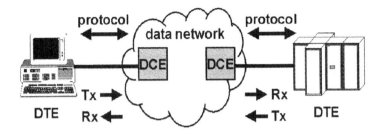

DCE = data circuit terminating equipment
DTE = data terminal equipment

Figure 1.4   Explanation of the terms DTE, DCE and protocol.

network. In contrast to the DTE, the DCE is the 'start of the long-distance network' (the first piece of equipment in the long-distance network to which the DTE is connected—i.e., the DTE's 'direct communications partner').

If only a short cable were to be used to connect the two DTEs in Figure 1.4, then the two devices could be directly connected to one another, without requiring the DCEs or the long-distance network. But whenever the distance between the DTEs is more than a few metres (up to a maximum of 100 m, depending upon the DTE interface used), then a long distance communication method is required. In simple terms, the DCE is an 'adaption device' designed to extend the short range (i.e., *local*) communication capabilities of DTE into a format suitable for long distance (i.e., *wide area*) data networking. A number of standardised DTE/DCE interfaces have been developed over the years which allow all sorts of different DCEs and *wide area network (WAN)* types to be used to interconnect DTE, without the DTE having to be adapted to cope with the particular WAN technology being used to transport its data.

The cable connection and the type of plug and socket used for a particular DTE/DCE connection may be one of many different types (e.g., *twisted pair cable, UTP (unshielded twisted pair), STP (shielded twisted pair), category 5 cable (Cat 5), category 7 cable (Cat 7),* coaxial cable, optical fibre, wireless, etc.). But all DTE/DCE interfaces have one thing in common—there is always a *transmit (Tx)* data path and a *receive (Rx)* data path. At least four wires are used at the interface, one 'pair' for the transmit path and one 'pair' for the receive path. But in some older DTE/DCE interface designs, multiple cable leads and multi-pin cable connectors are used.

DTE/DCE interface specifications are suitable for short-range connection of a DTE to a DCE[1] (typical maximum cabling distance 25 m or 100 m). Such specifications reflect the fact that the DTE is the 'end user equipment' and that the DCE has the main role of 'long distance communication'. The three main elements which characterise all *DTE/DCE* interfaces are that:

- The DCE provides for the signal transmission and receipt across the long distance line (wide area network), supplying power to the line as necessary;

- The DCE controls the speed and timing (so-called *synchronisation*) of the communication taking place between DTE and DCE. It does this in accordance with the constraints of

---

[1] Though intended for DTE-to-DCE connection, DTE/DCE interfaces may also be used (with slight modification, as we shall see in Chapter 3) to directly interconnect DTEs.

the wide area network or long distance connection. The DCE determines how many data characters may be sent per second by the DTE and exactly when the start of each character shall be. This is important for correct interpretation of what is sent. The DTE cannot be allowed to send at a rate faster than that which the network can cope with receiving and transporting!

- The <u>DTE sends</u> data to the network <u>on the path labelled 'Tx' and receives on the path</u> <u>labelled 'Rx'</u>, while the <u>DCE receives on the 'Tx' path and transmits on the 'Rx' path</u>. No communication would be possible if both DCE and DTE 'spoke' to each other's 'mouths' instead of to their respective 'ears'!

The terms DTE and DCE represent only a particular function of a piece of computer equipment or data networking equipment. The device itself may not be called either a *DTE* or a *DCE*. Thus, for example, the personal computer in Figure 1.4 is undertaking the function of *DTE*. But a PC is not normally called a 'DTE'. The DTE function is only one function undertaken by the PC.

Like the DTE, the DCE may take a number of different physical forms. Examples of DCEs are *modems, network terminating (NT)* equipment, *CSUs (channel service units)* and *DSUs (digital service units)*. The DCE is usually located near the DTE.

The physical and electrical interface between a DTE and a DCE may take a number of different technical forms. As an example, a typical computer (DTE) to modem (DCE) connection uses a 'serial cable' interface connecting the male, 25-pin D-socket (ISO 2110) on the DTE (i.e., the computer) to the equivalent female socket on the DCE (modem). This DTE/DCE interface is referred to as a *serial interface* or referred to by one of the specifications which define it: ITU-T recommendation *V*.24 or EIA *RS-232*. The interface specification sets out which control signals may be sent from DTE to DCE, how the timing and synchronising shall be carried out and which leads (and socket 'pins') shall be used for 'Tx' and 'Rx'.

In addition to a standardised physical and electrical interface, a protocol is also necessary to ensure orderly communication between DTE and DCE. The protocol sets out the etiquette and language of conversation. Only speak when asked, don't speak when you're being talked to, speak in the right language, etc. Understanding the plethora of different protocols is critical to understanding the Internet, and we shall spend much of our time talking about protocols during the course of this book.

Why are there so many protocols? Because most of them have been designed to undertake a very specific function, something like 'identify yourself' or 'send a report'. If you need to 'identify yourself' before 'sending a report' two different protocols may need to be used.

Line interfaces

Before we leave Figure 1.4, you may have noticed that our discussion has not concerned itself at all with what you might think is the most important part of the communication — conveying the data through the data network from one DCE to the other. Surprising as it may seem, this may not concern us. The realisation of the network itself has been left to the network operator and/or the data network equipment manufacturer! As long as the network transports our data quickly and error-free between the correct two end-points why should we care about the exact topology and technology inside the network? Maybe the internal protocols and *line interfaces*[2] of the network are not standardised! But why should this concern us? If there is a problem in the network what will we do other than report the problem and demand that the network operator sort it out?

---

[2] See Chapter 3.

The first data networks comprised a number of different 'switches', all of which were supplied by the same manufacturer. There are significant advantages to a single source of supply of switches, commercial buying-power being perhaps the most important. In addition, a single source of supply guarantees that devices will interwork without difficulty, and that advanced 'proprietary' techniques may be used for both the transport of data between the different switches and for *network management*.

Using a specific manufacturer's proprietary transport techniques can be advantageous, because at any one time the agreed public data networking standards are some way behind the capabilities of the most modern technology. A proprietary technique may offer benefits of cost, efficiency, better performance or afford capabilities not yet possible with standardised techniques. Thus, for example, proprietary versions of IP *tag-switching* appeared before a standardised version (called MPLS — multiprotocol label switching) became available. MPLS we shall meet in Chapter 7.

The advantage of having network equipment and *network management system* supplied by a single manufacturer is that it is easy to correlate information and to coordinate configuration changes across the whole network. For example, it is relatively easy to change the physical location of a given data network address or destination from one switch to another and to adjust all the network configuration data appropriately. In addition, any complaints about poor network quality can be investigated relatively easily.

1.8   UNI (user-network interface), NNI (network-network interface) and INI (inter-network interface)

The initial priority of interface standardisation in data networks was to create a means for connecting another manufacturer's computer (or DTE — data terminal equipment) to an existing data network (at a DCE — data circuit-terminating equipment) using a protocol or suite of protocols. The combination of a DTE, DCE and relevant protocol specification describes a type of interface sometimes called a *user-network interface (UNI)*. For some types of networks (e.g., X.25, *frame relay* and *ATM — asynchronous transfer mode*), a single document (the UNI specification) replaces separate specifications of DTE, DCE and protocol. A *user-network interface (UNI)* is illustrated in Figure 1.5. Despite the fact that the term *UNI* is not generally used in Internet protocol suite specifications, it is wise to be familiar with the term, since it is used widely in data networking documentation. We explain them briefly here.

A UNI (user-network interface) is typically an asymmetric interface, by means of which an end-user equipment (or DTE) is connected to a wide area network (WAN). The point of connection to the WAN may go by one of a number of different names (e.g., DCE — data circuit terminating equipment, modem, switch, router, etc.), but all have one thing in common — the network side of the connection (the DCE or equivalent) usually has the upper hand in controlling the interface.

As well as UNIs (user-network interfaces), there are also *NNIs (network-network interfaces or network-node interfaces)* and *INIs (inter-network interfaces)*. An NNI specification defines the interface and protocols to be used between two subnetworks of a given operator's network. Within each of the individual subnetworks of a large network, a single manufacturer's equipment and the associated proprietary techniques of data transport and network management may be used. The NNI allows the subnetwork (which may comprise only a single *node*) to be inter-linked with other subnetworks as shown in Figure 1.5.

Unlike the UNI, the NNI is usually a more 'symmetrical' interface. In other words, most of the rights and responsibilities of the subnetworks (or single nodes) on each side of the interface are identical (e.g., management control, monitoring, etc.). Since the basic physical and electrical interface technology used for some NNIs was adapted from technology originally

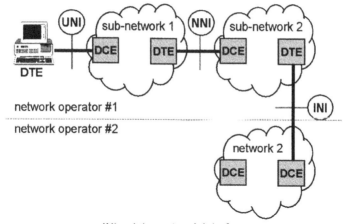

INI = inter-network interface
NNI = network-node interface or network-network interface
UNI = user-network interface

Figure 1.5   UNI, NNI and INI type interfaces between end devices and data networks.

designed for UNI interface, it is often the case that one of the networks may be required to act as DCE, while the other acts as DTE. Symmetry is achieved simply by allowing both ends to assume either the DCE or the DTE role — as they see fit for a particular purpose. Some physical NNI interfaces are truly symmetric.

The third main type of interface is called the *INI (inter-network interface)* or *ICI (inter-carrier interface)*. This is the type of interface used between networks under different ownership, i.e., those administered by different operators. Most INI interfaces are based upon standard NNI interfaces. The main difference is that an INI is a 'less trusted' interface than an NNI so that certain security and other precautions need to be made. An operator is likely to accept signals sent from one subnetwork to another across an NNI for control or reconfiguration of one his subnetworks, but is less likely to allow third party operators to undertake such control of his network by means of an INI. In a similar way, information received from an INI (e.g., for performance management or accounting) may need to be treated with more suspicion than equivalent information generated within another of the operator's subnetworks and conveyed by means of an NNI.

1.9   Open systems interconnection (OSI)

In the early days of computing, the different computer manufacturers developed widely diverse hardware, operating systems and application software. The different strengths and weaknesses of individual computer types made them more suited to some *applications* (i.e., uses) than others. As a result, enterprise customers began to 'collect' different manufacturers' hardware for different departmental functions (e.g., for bookkeeping, personnel records, order-taking, stock-keeping, etc.).

The business efficiency benefits of each departmental computer system quickly justified the individual investments and brought quick economic payback. But the demands on computers and computer manufacturers quickly moved on, as company IT (information technology) departments sought to interconnect their various systems rather than have to manually re-type output from one computer to become input for another. As a result, there was pressure to develop a standard means for representing computer information (called *data*) so that it could be understood by any computer. Similarly, there was a need for a standard means of electronic conveyance between systems. These were the first standards making up what we now refer to as *open systems interconnection (OSI)* standards.

It is useful to assess some of the problems which have had to be overcome, for this gives an invaluable insight into how a data network operates and the reasons for the apparently bewildering complexity. In particular, we shall discuss the *layered* functions which make up the OSI (open systems interconnection) model.

When we talk as humans, we conform to a strict etiquette of conversation without even realising it. We make sure that the listener is in range before we start talking. We know who we want to talk to and check it is the right person in front of us before we start talking. We make sure they are awake, paying attention, listening to us, not talking to or looking at someone else. We know which language they speak, or ask them in clear, slow language at the start. We change language if necessary. We talk slowly and clearly and keep to a simple vocabulary if necessary. While we talk, we watch their faces to check they have heard and understood. We repeat things as necessary. We ask questions and we elaborate some points to avoid misunderstanding. When we are finished we say 'goodbye' and turn away. We know to 'hang up' the telephone afterwards (if necessary), thereby ensuring that the next caller can reach us.

Computers and data networks are complex, because they are not capable of thinking for themselves. Every situation which might possibly arise has to have been thought about and a suitable action must be programmed into it in advance. Computers have no 'common sense' unless we programme it into them. If one computer tries to 'talk' to another, it needs to check that the second computer is 'listening'. It needs to check it is talking to the right piece of equipment within the second computer. (We might send a command 'shut down', intending that a given 'window' on the screen of the second computer should receive the command and that the 'window' should thus close. But if instead the receiving computer directs the command to the power supply, the whole PC would shut down instead.)

When a computer starts 'talking', it has to 'speak' in a 'language' which the listening computer can understand, and must use an agreed set of alphabetic characters. When 'talking' it has to check that the listener has heard correctly and understood. And when talked to itself, it may be appropriate to stop 'talking' for a while in order to concentrate on 'listening' or to wait for a reply. Finally, when the communication *session* is over, it is proper formally to close the conversation. The 'listener' need no longer pay attention, and the 'talker' may turn attention to a third party.

The list of potential problems and situations to be considered by designers of data networks is a long one. Here are a few examples:

- Different types of computer, using different operating systems and programming languages wish to 'talk' to one another;

- Data information is to be shared by different types of application (e.g., bookkeeping and order-taking programs), which use information records stored in different data formats;

- Different character representations are used by the different systems;

- It is not known whether the computer we wish to send data to or receive data from is active and ready to communicate with us;

- To 'reach' the destination device we must 'transit' several intermediate networks of different types;

- There are many different physical, electrical and mechanical (i.e., plug/socket) interfaces.

The OSI model

The open systems interconnection (OSI) model, first formalised as a standard by ISO (International Organization for Standardization) in 1983 subdivided the various data communications functions into seven interacting but independent *layers*. The idea was to create a modular structure, allowing different standard functions to be combined in a flexible manner to allow any two systems to communicate with one another. Although the model no longer covers all the functions of data networks which have come to be needed, the idea of 'layered' protocols and *protocol stacks* has come to be a cornerstone of modern data communications. It is thus useful to explain the basics of the model and the jargon which it lays down.

To understand the OSI model, let us start with an analogy, drawn from a simple exchange of ideas in the form of a dialogue between two people as illustrated in Figure 1.6. The speaker has to convert his ideas into words; a translation may then be necessary into the grammar and syntax of a foreign language which can be understood by the listener; the words are then converted into sound by nerve signals and appropriate muscular responses in the mouth and throat. The listener, meanwhile, is busy converting the sound back into the original idea. While this is going on, the speaker needs to make sure in one way or another that the listener has received the information, and has understood it. If there is a breakdown in any of these activities, there can be no certainty that the original idea has been correctly conveyed between the two parties.

Note that each function in our example is independent of every other function. It is not necessary to repeat the language translation if the receiver did not hear the message — a request (prompt) to replay a tape of the correctly translated message would be sufficient. The specialist translator could be getting on with the next job as long as the less-skilled tape operator was on hand. We thus have a layered series of functions. The idea starts at the top of the talker's *stack* of functions, and is converted by each function in the stack, until at the bottom it turns up in a soundwave form. A reverse conversion stack, used by the listener, re-converts the soundwaves back into the idea.

Each function in the protocol stack of the speaker has an exactly corresponding, or so-called *peer* function in the protocol stack of the listener. The functions at the same layer in the two stacks correspond to such an extent, that if we could conduct a direct *peer-to-peer* interaction then we would actually be unaware of how the functions of the lower layers protocols had

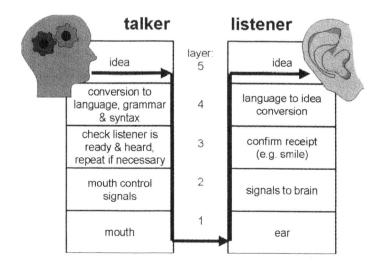

Figure 1.6   A layered protocol model for simple conversation.

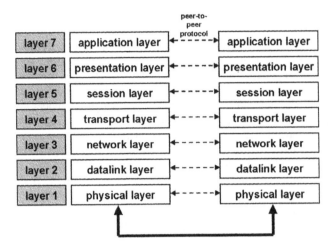

Figure 1.7   The Open Systems Interconnection (OSI) model.

been undertaken. Let us, for example, replace layers 1 and 2 by using a telex machine instead. The speaker still needs to think up the idea, correct the grammar and see to the language translation, but now instead of being aimed at mouth muscles and soundwaves, finger muscles and telex equipment do the rest (provided the listener also has a telex machine, of course!). We cannot, however, simply replace only the speaker's layer-1 function (the mouth), if we do not carry out simultaneous *peer protocol* changes on the listener's side because an ear cannot pick up a telex message!

As long as the layers interact in a peer-to-peer manner, and as long as the interface between the function of one layer and its immediate higher and lower layers is unaffected, then it is unimportant how the function of that individual *protocol layer* is carried out. This is the principle of the open systems interconnection (OSI) model and all *layered* data communications protocols.

The OSI model sub-divides the function of data communication into seven layered and peer-to-peer sub-functions, as shown in Figure 1.7. Respectively from layer 7 to layer 1 these are called; the *application layer*, the *presentation layer*, the *session layer*, the *transport layer*, the *network layer*, the *data link layer* and the *physical layer*. Each layer of the OSI model relies upon the *service* of the layer beneath it. Thus the transport layer (layer 4) relies upon the network service which is provided by the stack of layers 1–3 beneath it. Similarly the transport layer provides a transport service to the session layer, and so on. The functions of the individual layers of the OSI model are defined more fully in ISO standards (ISO 7498), and in ITU-T's X.200 series of recommendations. In a nutshell, they are as follows.

Application layer (Layer 7)

This layer provides communications functions services to suit all conceivable types of data transfer, control signals and responses between *cooperating* computers. A wide range of *application layer protocols* have been defined to accommodate all sorts of different computer equipment types, activities, controls and other applications. These are usually defined in a modular fashion, the simplest common functions being termed *application service elements (ASEs)*, which are sometimes grouped in specific functional combinations to form *application entities (AEs)* — standardised communications functions which may be directly integrated into computer programs. These communications functions or protocols have the appearance of computer programming commands (e.g., get, put, open, close etc.). The protocol sets out how the command or action can be invoked by a given computer programme (or *application*) and the sequence of actions which will result in the peer computer (i.e., the computer at the remote end of the communication link). By standardising the protocol, we allow computers to 'talk' and 'control' one another without misuse or misinterpretation of requests or commands.

Presentation layer (Layer 6)

The presentation layer is responsible for making sure that the data format of the application layer command is appropriate for the recipient. The *presentation layer protocol* tells the recipient in which language, syntax and character set the *application layer* command is in (in other words, which particular application layer *protocol* is in use). If necessary, the presentation layer can undertake a format conversion.

The binary digits (called *bits*) in which information is stored as *data* within computers are usually grouped in 8-bit patterns called *bytes*. Computers use different codes (of either one or two bytes in length) to represent the different alphanumeric characters. The most commonly used standard codes are called *ASCII (American standard code for information interchange)*, *unicode* and *EBCDIC (extended binary coded decimal interchange code)*. Standardisation of the codes for representing alphanumeric characters was obviously one of the first fundamental developments in allowing inter-computer communication.

Session layer (Layer 5)

A *session* between two computers is equivalent to a conversation between two humans, and there are strict rules to be observed. When established for a session of communication, the two devices at each end of the communication medium must conduct their 'conversation' in an orderly manner. They must listen when spoken to, repeat as necessary, and answer questions properly. The *session protocol* regulates the 'conversation' and thus includes commands such as start, suspend, resume and finish, but does not include the actual 'content' of the communication.

The session protocol is rather like a tennis umpire. He or she cannot always tell how hard the ball has been hit, or whether there is any spin on it, but he/she knows who has to hit the ball next and whose turn it is to serve, and he/she can advise on the rules when there is an error, in order that the game can continue. The session protocol negotiates for an appropriate type of session to meet the communication need, and then it manages the session.

A session may be established between any two computer applications which need to communicate with one another. In this sense the application may be a 'window' on the computer screen or an action or process being undertaken by a computer. Since more than one 'window' may be active at a time, or more than one 'task' may be running on the computer, it may be that multiple 'windows' and 'tasks' are intercommunicating with one another by means of different sessions. During such times, it is important that the various communications sessions are not confused with one another, since all of them may be sharing the same communications *medium* (i.e., all may be taking place on the same 'line').

Transport layer (Layer 4)

The *transport service* provided by the *transport layer protocol* provides for the end-to-end data relaying service needed for a communication session. The transport layer itself establishes a transport connection between the two end-user devices (e.g., 'windows' or 'tasks') by selecting

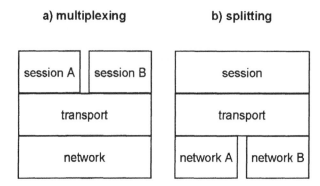

Figure 1.8   Protocol multiplexing and splitting.

and setting up the network connection that best matches the session requirements in terms of destination, quality of service, data unit size, flow control, and error correction needs. If more than one network is available (e.g., leaseline, packet-switched network, telephone network, router network, etc.), the transport layer chooses between them.

An important capability of the *transport protocol* is its ability to set up *reliable* connections in cases even when multiple networks need to be traversed in succession (e.g., a connection travels from LAN (local area network) via a wide area network to a second LAN). The IP-related protocol *TCP (transmission control protocol)* is an example of a transport layer protocol, and it is this single capability of TCP combined with IP (TCP/IP), that has made the IP-suite of protocols so widely used and accepted.

The transport layer supplies the network addresses needed by the network layer for correct delivery of the message. The network address may be unknown by the computer application using the connection. The *mapping* function provided by the transport layer, in converting *transport addresses* (provided by the session layer to identify the destination) into network-recognisable addresses (e.g., telephone numbers) shows how independent the separate layers can be: the conveyance medium could be changed and the session, presentation and application protocols could be quite unaware of it.

The transport protocol is also capable of some important *multiplexing* and *splitting* functions (Figure 1.8). In its multiplexing mode the *transport protocol* is capable of supporting a number of different sessions over the same connection, rather like playing two games of tennis on the same court. Humans would get confused about which ball to play, but the transport protocol makes sure that computers do nothing of the kind.

Two sessions from a mainframe computer to a PC (personal computer) in a remote branch site of a large shopping chain might be used simultaneously to control the building security system and (separately) to communicate customer sales. Different software programmes (for 'security' and for 'sales') in both the mainframe computer and in the PC could thus share the same telecommunications line without confusion.

Conversely, the *splitting* capability of the transport protocol allows (in theory) one *session* to be conducted over a number of parallel network communication paths, like getting different people to transport the various volumes of an encyclopaedia from one place to another.

The transport protocol also caters for the end-to-end conveyance, segmenting or concatenating (stringing together) the data as the network requires.

Network layer (Layer 3)

The *network layer* sets up and manages an end-to-end connection across a single real network, determining which permutation of individual links need be used and ensuring the correct transfer of information across the single network (e.g., LAN or wide area network). Examples of layer-3 *network protocols* are IP and X.25.

Datalink layer (Layer 2)

The datalink layer operates on an individual link or subnetwork part of a connection, managing the transmission of the data across a particular physical connection or subnetwork (e.g., LAN — local area network) so that the individual bits are conveyed over that link without error. ISO's standard datalink protocol, specified in ISO 3309, is called *high level data link control (HDLC)*. Its functions are to:

- *synchronise* the transmitter and receiver (i.e., the link end devices);

- control the *flow* of data bits;

- detect and correct *errors* of data caused in transmission;

- enable multiplexing of several *logical channels* over the same physical connection.

Typical commands used in datalink control protocols are thus *ACK (acknowledge), EOT (end of transmission)*, etc. Another example of a 'link' protocol is the IEEE 802.2 *logical link control (LLC)* protocol used in Ethernet and Token Ring LANs (local area networks).

Physical layer (Layer 1)

The *physical layer* is concerned with the *medium* itself. It defines the precise electrical, interface and other aspects particular to the particular communications *medium*. Example *physical media* in this regard are:

- the cable of a DTE/DCE interface as defined by EIA RS-232 or ITU-T recommendations: X.21, V.35, V.36 or X.21bis (V.24/V.28);

- a 10 Mbit/s *ethernet LAN* based on twisted pair (the so-called *10baseT* medium);

- a 4 Mbit/s or 16 Mbit/s Token ring LAN using Twinax (i.e., 2 x coaxial cable);

- a digital leaseline (e.g., conforming to ITU-T recommendation I.430 or G.703);

- a high speed digital connection conforming to one of the SONET (synchronous optical network) or SDH (synchronous digital hierarchy) standards (e.g., STM-1, STM-4, STM-16, OC3, OC12, STS3 etc.);

- a fibre optic cable;

- a radio link.

1.10   EDI (electronic data interchange)

By the 1980s, companies had managed to interconnect their different department computer systems for book-keeping, order-taking, salaries, personnel, etc., and the focus of development turned towards sharing computer data directly with both suppliers and customers. Why take an order over the telephone when the customer can submit it directly by computer — eliminating both the effort of taking down the order and the possibility of making a mistake in doing so?

In particular, large retail organisations and the car manufacturers jumped on the bandwagon of *EDI (electronic data interchange)*.

The challenge of electronic data interchange (EDI) between different organisations is considerably greater than the difficulties of 'mere' interconnection of different computers as originally addressed by OSI. When data is transferred only from one machine to another within the same organisation, then that organisation may decide in isolation which information should be transferred, in which format and how the information should be interpreted by the receiving machine. But when data is moved from one organisation to another, at least three more problems arise in addition to those of interconnection:

- The content and meaning of the various information fields transferred must be standardised (e.g., order number format and length, address fields, name fields, product codes and names, etc.).

- There needs to be a means of reliable transfer from one computer to the other which allows the sending computer to send its information independently of whether the receiving computer is currently ready to receive it. In other words, the 'network' needs to cater for *store-and-retrieve* communication between computers (comparable with having a postbox at the post office for incoming mail which allows you to pick up your mail at a time convenient to you as the receiver).

- There needs to be a way of confirming correct receipt.

Various new standardisation initiatives emerged to support EDI, among the first of which were:

- The standardisation of *bar codes* and unique product identification codes for a wide range of grocery and other retail products was undertaken. The industry-wide standard codes provided the basis for the 'just-in-time' re-stocking of supermarket and retail outlet shelves on an almost daily basis by means of EDI.

- The major car manufacturers demanded EDI capability from their component suppliers, so that they could benefit from lower stock levels and the associated cost benefits of 'just-in-time' ordering. Car products and components became standardised too.

- The banking industry developed *EFTPOS (electronic funds transfer at the point-of-sale)* for ensuring that your credit card could be directly debited while you stood at the till.

All of the above are examples of EDI, and whole data networking companies emerged specialising in the needs of a particular industry sector, with a secure network serving the particular 'community of interest'. Thus, for example, the ODETTE network provided for EDI between European car manufacturers. TRADERNET was the EDI network for UK retailers. SWIFT is the clearing network of the banks and SITA was the network organisation set up as a cooperative venture of the airlines for ticket reservations and flight operations. Subsequently, some of these networks and companies have been subsumed into other organisations, but they were important steps along the road to modern *ebusiness* (electronic business).

The *store-and-retrieve* methods used for EDI include *email* and the ITU's *message handling system (MHS)* [as defined in ITU-T recommendation X.400]. Both are *application layer* protocols which cater for the *store-and-retrieve* method of information transport, as well as the confirmation of reply.

1.11 CompuServe, prestel, minitel, BTx (Bildschirmtext) and teletex

The idea of equipping customers with computer terminals, so that they could log-in to a company's computers and make direct enquiries about the prices and availability of products and services emerged in the later 1970s. Equipping the customer with the terminal improved the level of customer service which could be offered, while simultaneously reducing the manpower required for order-taking. Since the customer was unlikely to put a second terminal on his desk (i.e., a competitor's terminal), it also meant reduced competition.

The travel industry rapidly reorganised its order-taking procedures to encompass the use of computer terminals by customers. There was soon a computer terminal at every airport check-in desk and even some large travel agents. Other travel agents, meanwhile, continued to struggle making phone calls to overloaded customer service agent centres. For a real revolution, all the travel agents needed a terminal and an affordable means of network access. It came with the launch of the first dial-up information service networks, which appeared in the late 1970s and early 1980s. The first information services were the *Prestel* service of the British Post Office (BPO) and the *CompuServe* information service in the USA (1979). Both were spurred by the modem developments being made at the time by the Hayes company (the Hayes 300 bit/s *modem* appeared in 1977).

The Prestel service followed the invention by the BPO laboratories of a simple terminal device incorporating a modem and a keyboard, which could be used in conjunction with a standard TV set as a 'computer terminal' screen. It spurred a new round of activity in the ITU-T modem standardisation committees — as the V.21, V.22 and V.23 modems appeared. And it became the impetus for the new range of *teletex* services which were to be standardised by ITU-T. The *facsimile* service appeared at almost the same time and also saw rapid growth in popularity, so that the two together — *teletex* and facsimile tolled the death knell for *telex* — the previous form of text and data communication which had developed from the *telegraph*.

Other public telephone companies rapidly moved to introduce their own versions of teletex. France Télécom introduced the world-renowned *minitel* service (Figure 1.9) in 1981 and Germany's Deutsche Bundespost introduced *Bildschirmtext* (later called *BtX* and *T-Online*

Figure 1.9   France Télécom's first minitel terminal (1981) [reproduced courtesy of France Télécom].

classic). But while none of these services were truly profitable businesses, they nonetheless were an important development towards what we today call the *Worldwide Web (www)*. They demonstrated that there was huge potential for greatly increased usage of the public telephone networks for access to data information services.

1.12   The role of UNIX in the development of the Internet

In 1969, the *UNIX* computer operating system was developed by Ken Thompson of AT&T Bell Laboratories. It has turned out to be one of the most powerful and widely accepted computer operating systems for computer and telephone exchange systems requiring *multi-tasking* and *multi-user* capabilities. Standard UNIX commands allow for access to computer files, programs, storage and other resources. Encouraged by the hardware volumes purchased by AT&T (American Telegraph and Telephone company), UNIX was quickly adopted by many computer manufacturers as their standard operating system, so that computer programs and other applications written for UNIX could easily be *ported* (i.e., moved with only very few changes) from one computer system to another.

Most importantly for the development of the Internet, one of the participants in the ARPANET, the University of California in Berkeley, at the request of DARPA, wrote an extension to UNIX to incorporate the newly developed TCP/IP protocols. This version of UNIX was called UNIX 4.2BSD (Berkeley System Distribution). It was immediately used in the ARPANET and was released to the public domain in 1983. It opened the door for rapid further development of applications for *file transfer* between computers and for a more-widely standardised form of *email*. The embedding of TCP/IP within UNIX also made UNIX servers the natural choice of hardware for web servers, which would appear later.

1.13   The appearance of the PC (personal computer)

Ted Hoff at Intel invented the microprocessor in 1971. At the same time, IBM invented the floppy disk as a convenient, small and cheap means of storing computer data. Now, using a single processor chip, complemented by a few memory chips and input/output devices, it was possible to create a working micro-computer. The first commercially available computer kit (the MITS Altair) duly appeared in 1975, and the Commodore PET computer was the hit of 1977. A period of intense further development of the microprocessor chip took place at Intel. The 8086 chip was released in 1979 and the 8088 in 1980.

Based on the Intel 8088 microprocessor, the *IBM PC (personal computer)* appeared in August 1981 (Figure 1.10). This set the standard for PCs as we know them today. The IBM PC incorporated the *DOS (disk operating system)* software developed by the Micro-Soft company (later renamed Microsoft) which had been set up by Bill Gates and Paul Allen in 1975. By 1983, a new version of the IBM PC, the IBM PC XT, included a hard disk for storage of data.

Apple Computer, founded by Steve Jobs and Steve Wozniak in 1976, introduced the *Macintosh* computer in 1984 (Figure 1.11). It revolutionised personal computing with the *graphical user interface (GUI)*, the use of a mouse to 'point and click' and the opening of different 'windows' for different tasks. Microsoft quickly reacted by introducing a new operating system software, *Microsoft Windows*, in 1985. The 'look and feel' of Microsoft *Windows* were so similar to the Macintosh operating system that it led Apple Computer to file a lawsuit.

1.14   Local area networks (LANs)

The PC took the business world by storm. Word processing programmes and spreadsheet programmes made life easier for office staff, and meant that their managers could reduce the

Figure 1.10   The first IBM PC (IBM 5150 : 1981) [reproduced courtesy of IBM].

Figure 1.11   The first 128k Apple Macintosh computer (1984) [reproduced courtesy of Apple Computer, Inc].

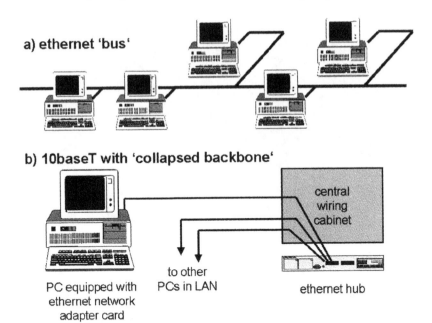

a) ethernet 'bus'

b) 10baseT with 'collapsed backbone'

central
wiring
cabinet

to other
PCs in LAN

ethernet hub

PC equipped with
ethernet network
adapter card

Figure 1.12   Bus and 10baseT 'collapsed backbone' alternative structures for ethernet LANs.

secretarial staff. And as quickly as the use of PCs grew, so did the need for networking them all together. Company staff wanted to be able to share data easily, and to be able to securely store data.

The foundation stone for *LANs (local area networks)* was laid by the Xerox company, at its *Palo Alto Research Centre (PARC)* in 1973. Robert Metcalfe and David Boggs invented the principles of the *ethernet LAN* and published them in 1976. Initially *ethernet* was based on coaxial cable interconnecting all the PCs together in a *bus* structure or *backbone* cable network simply linking all the PCs together in a chain (Figure 1.12a). But as *structured office* cabling based upon *twisted pair* cabling became popular, the most popular form of ethernet emerged — 10baseT. Using a 10 Mbit/s adapter card in each, all the office PCs could be connected in a star-fashion (a so-called *collapsed backbone*) over twisted pair cabling to the central wiring cabinet, where a LAN hub is used to connect all the PCs together into an ethernet LAN (Figure 1.12b).

The 3Com company introduced the first 10 Mbit/s ethernet LAN adapter card in 1981. The official *link layer* and *physical layer* protocol standards were standardised in the renowned IEEE 802 standards in 1982. (The *link layer control (LLC)* is IEEE 802.2 and the ethernet physical layer is defined in IEEE 802.3).

Meanwhile at IBM, there was also work going on to develop the *token ring LAN* as specified in IEEE 802.5. This work culminated in the introduction of the 4 Mbit/s token ring LAN in 1985 and the 16 Mbit/s version in 1988. But while some experts claimed that token ring LAN had higher performance and reliability than equivalent ethernet LANs, token ring lost out commercially because of its later introduction and the higher costs of the adapter cards. Nonetheless, IBM's work on LANs was important because of its development of *NETBIOS (network basic input/output system)*. *NETBIOS* provides a 'layer' of software to link a network adapter operating software to a particular PC hardware and computer operating system. It extends the basic operating system with transport layer capabilities for inter-application communications and data transfer.

1.15   LAN servers, bridges, gateways and routers

With LANs came *servers, bridges, gateways* and *routers*. Initially, the servers had the primary function of being 'administration' terminals for managing the LAN itself, and for being shared *file and print servers*. As a *file server*, the server PC provided a central resource for storing, backing-up and sharing files on behalf of PC users in the LAN. As a *print server*, the server took over the job of queueing print jobs (correctly called *spooling*) so that a shared printer within the LAN could print each in turn.

Initially, many servers were normal PCs, but the higher storage and greater performance requirements of the servers quickly led to the use of much more powerful, specially developed hardware and software. The Novell *Netware* software, for example, became a popular LAN operating system — the *de facto* standard. It was introduced in 1983 and was hugely successful until it was supplanted by the introduction of Microsoft's WindowsNT in 1993.

Bridges, gateways and routers are all types of hardware which can be added into LANs (see Figure 1.13) to provide for interconnection with other local area networks (LANs) and wide area networks (WANs). Bridges are special telecommunications equipment introduced into LANs to allow the LAN 'boundary' to be extended by connecting two separate LANs together. A bridge, in effect, makes two separate LANs operate as if they were a single LAN.

Gateways are typically PCs within the LAN which are equipped with relevant software and network adapter hardware to allow the LAN to be connected to an existing *mainframe* computer network. The most common forms of gateways performed some kind of *terminal emulation*: in effect, allowing a PC within the LAN to appear to a remote mainframe computer as if it were one of its standard terminals (so-called *dumb terminal*). This allowed the new world of PC users to replace the old world of mainframe terminal users, without loss of their mainframe applications. The most commonly used forms of gateway and terminal emulation software were 3270-emulation (for IBM-mainframe connection of LANs) and VT-100-emulation (for connection of LANs to DEC mainframe and minicomputers).

As the number of PCs in companies grew, so the number of LANs and *LAN segments* (LAN subnetworks) grew, and it became unwieldy to operate all the individual segments as single LAN using bridges. Instead, *routers* appeared. Rather than 'flooding' the data around the whole LAN in an attempt to find the correct destination as bridges do, routers are more

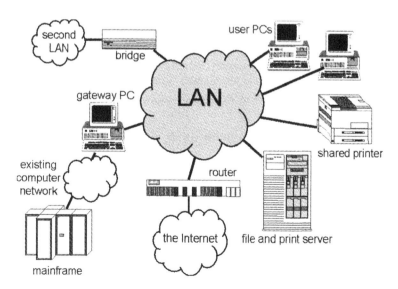

Figure 1.13   Shared LAN devices.

careful in keeping the amount of data to be sent to a minimum and they select carefully the best 'route' to the destination from routing tables. Routers are now the most commonly used type of equipment used in a LAN to provide 'access' to the Internet.

1.16   Why did IP win through as the standard for 'open' communication?

Internet protocol (IP) has become the predominant means of modern data communication. In doing so, it had to fight off strong competition from alternative internationally standardised technologies. Why did the IP protocol suite win through? Technically, because it offers a reliable means of transmitting various different types of network as data makes its way to the destination. But probably more important was the simple fact of its embedding into the major computer operating systems — UNIX, WindowsNT and Windows95.

1.17   The development and documentation of IP (Internet protocol) and the Internet

The large public data network which we know today as the *Internet* has its routes in the ARPANET, the conception of which was set out in the request for proposals of summer 1968 for the original four node network. But the protocols in their current form first started to take shape in 1974 when Vinton Cerf and Robert Kahn published the basic principles of TCP/IP. Further important 'landmarks' in the history of the evolution of the Internet were the adoption of the *domain name system (DNS)* in 1983, the establishment of the *IETF (Internet Engineering Task Force)* in 1989 and the formal re-naming of the ARPANET AS the *Internet* in 1990.

The IETF is the standards body of the Internet, and the standards themselves are documented in documents called *RFCs (request for comments)*. The name reflects the somewhat 'informal' manner in which Internet standards have been evolved — by sharing ideas and documents over the network itself. All the current standards are publicly available via the Internet at www.rfc-editor.org. If you do refer to the standards, you may notice that many of the most important ones were written by Jon Postel. He was the Deputy Internet Architect and RFC editor during the 1980s and 1990s.

Following the renaming of the Internet in 1990, a number of large US-based research and educational networks were inter-linked with the original ARPANET, with the effect of greatly extending the network's reach. Networks connected at this time included the US National Research and Education Network (NREN), the US National Science Foundation (NSF) network, the NASA (National Aeronautical and Space Administration) and the US Department of Education network.

1.18   Electronic mail and the domain name system (DNS)

The earliest forms of electronic mail *(email)* appeared around 1971. By 1972, email was available on ARPANET. But the early forms of email were restricted to simple text messaging between different terminal users of the same mainframe and its associated data network. It was not until the late 1980s and early 1990s that email began to take off as a major means of inter-company and finally, private communications. The critical developments took place in the mid-1980s with the development of the *domain name system (DNS)* and *SMTP (simple mail transfer protocol)*. A lot of work was also undertaken under the auspices of ITU-T[3] in

---

[3] ITU-T stands for International Telecommunications Union — standardization sector. Until 1989, ITU-T was previously called CCITT — International Telegraph and Telephone Consultative Committee.

developing their recommendation X.400 for a *message handling system (MHS)* based on a 'post office' system of *storage-and-retrieval*. However, apart from some of the ideas which were assumed into the IP-based standards, X.400-based systems have largely disappeared.

The domain name system (DNS) established the familiar email addresses with the @-symbols. The *domain* is the part of the *email address* which appears after the @-symbol. The *domain name* identifies the *post office* where the users incoming email is stored in his *postbox*. The name or number which appears before the @-symbol identifies the individual user within this post office *domain*[4]. The use of a *domain name* means that the user does not have to remember the 12-digit Internet address of the relevant post office (e.g., 255.255.164.101).

The domain name is usually split into at least two parts, separated by a full stop or 'dot'. The first part (before the 'dot') is the name part (typically a company name). The second part (after the 'dot') is the domain name type (e.g., 'com' = commercial, 'edu' = education, 'gov' = government, 'net' = network, 'org' = organization, 'tv' = television etc.). Thus a well-known domain name is @microsoft.com. 'microsoft' tells you the name of the company, and '.com' tells you that this company has *registered* its domain name on the 'com' *root server*, which is intended for 'commercial' entities. Microsoft could have chosen to register itself as '.net'. There are no particular rules, but the idea is that the name should be easy to remember or to guess if you don't know it beforehand.

The domain name system (DNS) allows the relevant post office server to be found using a number of sequential steps. First, an enquiry to the relevant *root server* (e.g., the '.com' server) is made. The root server is requested to provide the relevant Internet address of the 'microsoft.com' *domain name server*. A subsequent enquiry (in this case to the microsoft.com domain name server) reveals the actual network location of the relevant domain *post office* (i.e., email server and software). The mail can then be appropriately transmitted to the incoming mail postbox of the intended recipient, by using the Internet address which results from the above described address *resolution* process. Once the Internet address is known by a sending post office, it can be *cached* (i.e., stored), to prevent the network address (i.e., *IP address*) from having to be *resolved* from the *domain name address* every time a mail needs to be sent.

There are a number of different domain name types and each has a corresponding *root server*. The main ones used originally had the 3-letter suffixes .com, .edu, .gov, .net, .org. But as demand for registering domains has grown, it has become necessary to add further root servers. The two-letter suffixes (e.g., .au (Australia), .de (Germany), .es (Spain) .fr (France), .jp (Japan), .nz (New Zealand), .uk (United Kingdom) and .us (United States)) are national root directories corresponding to the ISO standard two-letter country codes defined in ISO 3166[5]. These root directories are administered by national organisations, some of whom choose (e.g. the United Kingdom) to subdivide into the com/edu/gov/net/org categories by making three-part domain names, e.g., @microsoft.co.uk. The UK has chosen to shorten the initial three letter-symbols to two-letter ones (e.g., 'com' shortened to 'co' etc.). Other countries, meanwhile, use the full three letters, e.g., www.environment.gov.au. But some countries (e.g., France and Germany) have elected to simplify domain names by dispensing with the com/edu/gov/net/org sub-categories. Thus common domain names used in France and Germany are of the format: @microsoft.fr and @microsoft.de.

The user name part of the email address (appearing before the @-symbol) may include the 26-letter roman alphabetic and/or arabic numeral characters plus the 'dot'(.) character. Usually the post office (or email) administrator of the company likes to choose a standard format for all the users within the domain (e.g., bill.gates or bgates etc.), so that a typical email address

---

[4] A *domain* is defined to be 'the part of a computer network in which the data processing resources are under common control' (e.g., a particular enterprise's network). A domain name server undertakes the task of *resolving* the 'easily rememberable' *domain names* of devices to the exact port addresses (IP addresses) where they are connected to the network.

[5] See Appendix 3.

might appear: bill.gates@microsoft.com. Incidentally, you may have noticed that some people have their email addresses printed on their business cards including capital or upper-case letters, e.g., Bill.Gates@Microsoft.com. This is acceptable, but unimportant. This is simply the same address as bill.gates@microsoft.com, since Internet addresses are case-insensitive.

1.19   html, WindowsNT and the Worldwide Web

The *Worldwide Web* (*www* or simply *web*) was an invention of the 1990s. The first version of *html (hypertext markup language)* was drafted in 1990, primarily for allowing scientific papers to be published in a formatted page layout and subsequently cross-referenced and researched by other remote users. Html 1.0 included six levels of heading, character attributes, quotations, source code listings, list and hyperlinks to other documents and images. A real boost came in 1993 when the *Mosaic* graphical web Browser appeared (developed by Marc Andreessen and Eric Bina). The *Navigator* browser of the Netscape Communications Corporation followed in 1994.

The WindowsNT (NT stands for 'new technology') operating system introduced by Microsoft in 1993 greatly improved the inter-networking capabilities of LANs by making their servers easily reachable using the Internet protocol. Within two years the computing world was speaking IP. The critical technology of the Worldwide Web (www): *hypertext markup language* version 2 (*html/2.0*) and the *hypertext transfer protocol (http/1.0)* was developed by Tim Berners-Lee of CERN (European Organisation for Nuclear Research) in 1995. Immediately, Microsoft's Internet Explorer was released, and included with the Windows95 operating system for PCs.

The most widely used initial version of html, version 3.2, was developed by cooperation between IBM, Microsoft, Netscape, Novell, SoftQuad, Spyglass and Sun Microsystems. Html version 3.2 included tables, *applets* (mini-programs running within a webpage), text around images, super and subscripts, frames, 'ActiveX' controls, etc. Following this, the web was ready to boom. WindowsNT gained powerful capabilities for acting as a *web server* (i.e., a computer, connected to the Internet, where the *web pages* of a given web domain are stored). Sun Microsystems adapted its *Java* programming language to enable it to be *plugged in* to browsers and web pages as applets. Such applets allow a web-page to be 'interactive' and alive with animation.

1.20   Internet addresses and domain names

In the Internet world there are two three types of *addresses* which are important to distinguish between. The three are:

- **Internet addresses**. These have the form of four numbers, each between 1 and 255, separated by 'dots'. (e.g., 255.255.164.101) and identify end-points in the Internet network itself. An Internet address is necessary for all communication across the Internet but may be invisible to the end-user, having been *resolved* by enquiry to a root server or domain name server as we discussed earlier in the chapter.

- **Worldwide Web addresses**. These have the domain name form http://www.company-name.com. They identify world wide *websites*, characterised by a domain name and the individual *web pages* which users of the Worldwide Web (www) may *browse*. In order for the user's browser to contact the appropriate server, the domain name must first be *resolved* into the Internet address of the server by enquiries made to the *root server* and *domain name server*.

- **Email addresses**. These have the form user.name@companyname.com and allow an email to be delivered to the correct post office and ultimately the correct *incoming mailbox* of the intended recipient. Like the corresponding worldwide address, the Internet address of the recipient's post office must first be resolved by querying the relevant root server and domain name server. For this reason the addresses are sometimes known as email *alias* addresses.

The administration and registration of both Internet addresses and domain names are carried out by a number of international and national organisations. The best-known of these and their web addresses are as follows:

- InterNIC (Internet Network Information Centre)   www.internic.net
- Internet Assigned Numbers Authority (IANA)   www.iana.net
- RIPE (Reseaux IP Européens)   www.ripe.net
- ARIN (American Registry for Internet Numbers)   www.arin.net
- APNIC (Asia Pacific Network Information Centre)   www.apnic.net

1.21   What are ISPs (Internet service providers) and IAPs (Internet access providers)?

An *Internet service provider (ISP)* is a company which provides its customers with an account for accessing the Internet. The best-known and most used ISPs are AOL (America Online), BTInternet (British Telecom), CompuServe, T-Online (Deutsche Telekom) and wanadoo.fr (France Telecom). An ISP typically distributes CD-ROMs (Compact Disk — Read Only Memory) with browser and other Internet software. When you subscribe to a given ISP's service, you typically are directed to his *home page* each time you log on.

An *Internet access provider (IAP)* is a company which provides the network service (usually a dial-up service, but it can be a *leaseline* or a *cable modem* or a *DSL (digital subscriber line)*) for connecting the Internet user (often called a 'surfer') to the server of his ISP, where he or she gains access for browsing the Internet. Many ISPs operate their own user access networks, so that ISP and IAP are the same company. But in other cases the IAP is subcontracted by the ISP. Thus, for example UUNET (part of WorldCom) provides the IAP service for AOL. Similarly, in Germany, Deutsche Telekom is the IAP for T-Online (Deutsche Telekom's ISP subsidiary).

1.22   The emergence of ebusiness

With the Worldwide Web (www) in existence and a plethora of ISPs (Internet service providers) to promote it, the use of the Internet boomed from 1995. Companies turned their attention to how they could streamline their business, or launch into new forms of *ebusiness* (electronic business — conducted via the Worldwide Web). Cisco, one of the leading manufacturers of Internet networking equipment, takes most of its orders via the web! Meanwhile the marketing machine got to work. Companies changed their names to include the @-symbol, and to use only lower-case characters in keeping with the *egeneration*. Educational bodies introduced courses via the Internet as a basis for *elearning* and governments launched into *egovernment*. Where will it lead? Who knows?

2

Fundamentals of Data Communication and Packet Switching

'Data', a plural noun, is the term used to describe information which is stored in and processed by computers. In this chapter we how such data (computer text or graphics) are represented electronically and explain the basic physical principles and practicalities of telecommunications line transmission. We explain binary code, ASCII, EBCDIC, pixels and graphics arrays, computer-to-network interfaces, digital transmission, modems, synchronisation, the basics of packet switching and the measures necessary to avoid data communications errors.

2.1 The binary code

Binary code is the means used by computers to represent numbers. Normally, we as humans quote numbers in *decimal* (or *ten-state*) code. A single digit in decimal code may represent any of ten different *unit* values, from nought to nine, and is written as one of the figures 0, 1, 2, 3, 4, 5, 6, 7, 8, 9. Numbers greater than nine are represented by two or more digits: twenty, for example, is represented by two digits, 20, the first '2' indicating the number of 'tens', so that 'twice times ten' must be added to '0' units, making twenty in all. In a three digit decimal number, such as 235, the first digit indicates the number of 'hundreds' (or 'ten times tens'), the second digit the number of 'tens' and the third digit, the number of 'units'. The principle extends to numbers of greater value, comprising four or indeed many more digits.

Consider now another means of representing numbers using only a *two-state* or *binary* code system. In such a system a single digit is restricted to one of two values, either zero or one. How then are values of 2 or more to be represented? The answer, as in the decimal case, is to use more digits. 'Two' itself is represented as the two digits one-zero, or '10'. In the binary code scheme, therefore, '10' does not mean 'ten', but 'two'. The rationale for this is similar to the rationale of the decimal number system with which we are all familiar.

In decimals the number one thousand three hundred and forty-five is written '1345'. The rationale is

$$(1 \times 10^3) + (3 \times 10^2) + (4 \times 10) + 5$$

Data Networks, IP and the Internet: Protocols, Design and Operation   Martin P. Clark
© 2003 John Wiley & Sons, Ltd   ISBN: 0-470-84856-1

the same number in binary requires many more digits, as follows.

$$1345 \text{ (decimal)} = 10101000001 \text{ (binary)}$$

$$1345 \text{ (decimal)} = 10101000001 \text{ (binary)}$$

	(binary)		(decimal)	
$=$	1	$\times 2^{10}$		1024
$+$	0	$\times 2^{9}$	$+$	0
$+$	1	$\times 2^{8}$	$+$	256
$+$	0	$\times 2^{7}$	$+$	0
$+$	1	$\times 2^{6}$	$+$	64
$+$	0	$\times 2^{5}$	$+$	0
$+$	0	$\times 2^{4}$	$+$	0
$+$	0	$\times 2^{3}$	$+$	0
$+$	0	$\times 2^{2}$	$+$	0
$+$	0	$\times 2$	$+$	0
$+$	1		$+$	1
			$=$	1345

Any number may be represented in the binary code system, just as any number can be represented in decimal.

All numbers, when expressed in binary consist only of 0s and 1s, arranged as a series of Binary digITS (in the jargon: *bits*). The *string* of bits of a binary number are usually suffixed with a 'B', to denote a binary number. This prevents any confusion that the number might be a decimal one. Thus 41 (forty-one in decimal) is written '101001B'.

2.2   Electrical or optical representation and storage of binary code numbers

The advantage of the binary code system is the ease with which binary numbers can be represented electrically. Since each digit, or bit, of a binary number may only be either 0 or 1, the entire number can easily be transmitted as a series of 'off' or 'on' (sometimes also called *space* and *mark*) pulses of electricity. Thus forty-one (101001B) could be represented as 'on-off-on-off-off-on', or 'mark-space-mark-space-space-mark'. The number could be conveyed between two people over quite a distance, transmitting by flashing a torch, either on or off, say every half second, and receiving using binoculars. Figure 2.1 illustrates this simple binary communication system in which two binary digits (or bits) are conveyed every second. The speed at which the binary code number, or other information can be conveyed is called the *information conveyance rate* (or more briefly the *information rate*). In this example the rate is 2 bits per second, usually shortened to 2 bit/s.

transmitter   receiver

Figure 2.1   A simple binary communication system.

Figure 2.1 illustrates a simple means of transmitting numbers, or other binary coded data by a series of 'on' or 'off' electrical or optical states. In fact, the figure illustrates the basic principle of modern optical fibre transmission.

As well as providing a means for bit transmission across the communications *medium*, a telecommunications system also usually provides for temporary data storage. At the sending end the data has to be stored prior to transmission, and at the receiving end data may have to 'wait' momentarily before the final receiving device or computer program is ready to accept it.

2.3 Using the binary code to represent textual information

The letters of the alphabet can be stored and transmitted over binary coded communications systems in the same way as numbers, provided they have first been *binary-encoded*. There have been four main binary coding systems for alphabetic text. In chronological order these are the *Morse code*, the *Baudot code* (used in telex, and also known as *international alphabet IA2*), *EBCDIC (extended binary coded decimal interchange code), and ASCII (American (national) standard code for information interchange*, also known as *international alphabet IA5*).

2.4 ASCII (American standard code for information interchange)

As a 7-bit binary code for computer characters, *ASCII (American standard code for information interchange)* [pronounced 'Askey'] was invented in 1963 and is the most important code. The original code (Tables 2.1 and 2.2) encompassed not only the alphabetic and numeric characters (which had previously also been catered for by the Morse code and the Baudot code but

Table 2.1 The original 7-bit ASCII code (International alphabet IA5)

HEX CODE YX			X	0	1	2	3	4	5	6	7	8	9	A	B	C	D	E	F
	BIT		4	0	0	0	0	0	0	0	0	1	1	1	1	1	1	1	1
			3	0	0	0	0	1	1	1	1	0	0	0	0	1	1	1	1
			2	0	0	1	1	0	0	1	1	0	0	1	1	0	0	1	1
Y			1	0	1	0	1	0	1	0	1	0	1	0	1	0	1	0	1
	8	7	6	5																
0	0	0	0	0	0 NUL	1 SOH	2 STX	3 ETX	4 EOT	5 ENQ	6 ACK	7 BEL	8 BS	9 HT	10 LF	11 VT	12 FF	13 CR	14 SO	15 SI
1	0	0	0	1	16 DLE	17 DC1	18 DC2	19 DC3	20 DC4	21 NAK	22 SYN	23 ETB	24 CAN	25 EM	26 SUB	27 ESC	28 FS	29 GS	30 RS	31 US
2	0	0	1	0	32 spce	33 !	34 "	35 #	36 $	37 %	38 &	39 '	40 (| 41)	42 *	43 +	44 ,	45 -	46 .	47 /
3	0	0	1	1	48 0	49 1	50 2	51 3	52 4	53 5	54 6	55 7	56 8	57 9	58 :	59 ;	60 <	61 =	62 >	63 ?
4	0	1	0	0	64 @	65 A	66 B	67 C	68 D	69 E	70 F	71 G	72 H	73 I	74 J	75 K	76 L	77 M	78 N	79 O
5	0	1	0	1	80 P	81 Q	82 R	83 S	84 T	85 U	86 V	87 W	88 X	89 Y	90 Z	91 [	92 \	93]	94 ^	95 _
6	0	1	1	0	96 `	97 a	98 b	99 c	100 d	101 e	102 f	103 g	104 h	105 i	106 j	107 k	108 l	109 m	110 n	111 o
7	0	1	1	1	112 p	113 q	114 r	115 s	116 t	117 u	118 v	119 w	120 x	121 y	122 z	123 {	124 |	125 }	126 ~	127 DEL

Table 2.2 ASCII control characters

ASCII character	Meaning
ACK	Acknowledgement
BEL	Bell
BS	Backspace
CAN	Cancel
CR	Carriage Return
DC1	Device Control 1
DC2	Device Control 2
DC3	Device Control 3
DC4	Device Control 4
DEL	Delete
DLE	Data Link Escape
EM	End of Medium
ENQ	Enquiry
EOT	End of Transmission
ESC	Escape
ETB	End of Transmission Block
ETX	End of Text
FF	Form Feed
FS	File Separator
GS	Group Separator
HT	Horizontal Tab
LF	Line Feed
NAK	Negative Acknowledgement
NUL	Null
RS	Record Separator
SI	Shift In
SO	Shift Out
SOH	Start of Header
STX	Start of Text
SUB	Substitute Character
SYN	Synchronisation character
US	Unit Separator
VT	Vertical Tab

also a range of new *control characters* as needed to govern the flow of data in and around the computers.

An adapted 8-bit version of the code, developed by IBM and sometimes called *extended ASCII* is now standard in computer systems, the most commonly used code used is the 8-bit ASCII code corresponding to the DOS (disk operating system) *code page 437*. This is the default character set loaded for use in standard PC keyboards, unless an alternative national character set is loaded by means of re-setting to a different *code page*.

The extended 8-bit ASCII code (international alphabet IA5 and code page 437) is illustrated in Table 2.3. The 'coded bits' representing a particular *character*, number or control signal are numbered 1 through 8 respectively (top left-hand corner of Table 2.3). These represent the *least (LSB)* (number 1) through *most significant bits (MSB)* (number 8) respectively. Each alphanumeric *character*, however, is usually written most significant bit (i.e., bit number 8) first. Thus the letter C is written '01000011'. But to confuse matters further, the least significant bit is transmitted first. Thus the order of transmission for the letter 'C' is '11000010' and for the word 'ASCII' is as shown in Figure 2.2.

Table 2.3 Extended ASCII code (as developed for the IBM PC; code page 437)

HEX CODE YX				X	0	1	2	3	4	5	6	7	8	9	A	B	C	D	E	F
			4		0	0	0	0	0	0	0	0	1	1	1	1	1	1	1	1
	BITS		3		0	0	0	0	1	1	1	1	0	0	0	0	1	1	1	1
			2		0	0	1	1	0	0	1	1	0	0	1	1	0	0	1	1
Y	8	7	6	1 / 5	0	1	0	1	0	1	0	1	0	1	0	1	0	1	0	1
0	0	0	0	0	0	1 ☺	2 ●	3 ♥	4 ♦	5 ♣	6 ♠	7 •	8 ◘	9 ○	10 ◙	11 ♂	12 ♀	13 ♪	14 ♫	15 ☼
1	0	0	0	1	16 ►	17 ◄	18 ↕	19 ‼	20 ¶	21 §	22 ▬	23 ↨	24 ↑	25 ↓	26 →	27 ←	28 ∟	29 ↔	30 ▲	31 ▼
2	0	0	1	0	32 spce	33 !	34 "	35 #	36 $	37 %	38 &	39 '	40 (	41)	42 *	43 +	44 ,	45 -	46 .	47 /
3	0	0	1	1	48 0	49 1	50 2	51 3	52 4	53 5	54 6	55 7	56 8	57 9	58 :	59 ;	60 <	61 =	62 >	63 ?
4	0	1	0	0	64 @	65 A	66 B	67 C	68 D	69 E	70 F	71 G	72 H	73 I	74 J	75 K	76 L	77 M	78 N	79 O
5	0	1	0	1	80 P	81 Q	82 R	83 S	84 T	85 U	86 V	87 W	88 X	89 Y	90 Z	91 [	92 \	93]	94 ^	95 _
6	0	1	1	0	96 `	97 a	98 b	99 c	100 d	101 e	102 f	103 g	104 h	105 i	106 j	107 k	108 l	109 m	110 n	111 o
7	0	1	1	1	112 p	113 q	114 r	115 s	116 t	117 u	118 v	119 w	120 x	121 y	122 z	123 {	124 |	125 }	126 ~	127 ⌂
8	1	0	0	0	128 Ç	129 ü	130 é	131 â	132 ä	133 à	134 å	135 ç	136 ê	137 ë	138 è	139 ï	140 î	141 ì	142 Ä	143 Å
9	1	0	0	1	144 É	145 æ	146 Æ	147 ô	148 ö	149 ò	150 û	151 ù	152 ÿ	153 Ö	154 Ü	155 ¢	156 £	157 ¥	158 Pts	159 ƒ
A	1	0	1	0	160 á	161 í	162 ó	163 ú	164 ñ	165 Ñ	166 ª	167 º	168 ¿	169 ⌐	170 ¬	171 ½	172 ¼	173 ¡	174 «	175 »
B	1	0	1	1	176 ░	177 ▒	178 ▓	179 │	180 ┤	181 ╡	182 ╢	183 ╖	184 ╕	185 ╣	186 ║	187 ╗	188 ╝	189 ╜	190 ╛	191 ┐
C	1	1	0	0	192 └	193 ┴	194 ┬	195 ├	196 ─	197 ┼	198 ╞	199 ╟	200 ╚	201 ╔	202 ╩	203 ╦	204 ╠	205 ═	206 ╬	207 ╧
D	1	1	0	1	208 ╨	209 ╤	210 ╥	211 ╙	212 ╘	213 ╒	214 ╓	215 ╫	216 ╪	217 ┘	218 ┌	219 █	220 ▄	221 ▌	222 ▐	223 ▀
E	1	1	1	0	224 α	225 β	226 Γ	227 π	228 Σ	229 σ	230 µ	231 τ	232 Φ	233 θ	234 Ω	235 δ	236 ∞	237 ø	238 ε	239 ∩
F	1	1	1	1	240 ≡	241 ±	242 ≥	243 ≤	244 ⌠	245 ⌡	246 ÷	247 ≈	248 °	249 •	250 ·	251 √	252 ⁿ	253 ²	254 ■	255 blnk

CODED CHARACTER

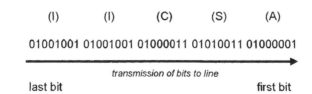

(I)   (I)   (C)   (S)   (A)

01001001 01001001 01000011 01010011 01000001

→ *transmission of bits to line*

last bit   first bit

Figure 2.2 Transmission of bits to line, most significant bit first.

Table 2.3 shows the 256 characters of the standard set, listing the *binary* value (the 'coded bits') used to represent each *character*. The decimal value of each character (in the top left-hand corner of each box), appears for reference as well, since some protocol specification documents refer to it rather than the binary value.

The *hexadecimal* values shown in Table 2.3 have the same numerical value as the *binary* and *decimal* equivalents in the same 'cell' of the table, but are simply expressed in base 16 (as opposed to base 2 for binary or base 10 for decimal). Computer and data communications specialists often prefer to talk in terms of hexadecimal values, because they are easier to remember and much shorter than the binary number equivalent. It is also very easy to convert from a binary number to its equivalent hexadecimal value and vice versa[1]. Hexadecimal notation is used widely in Internet protocol suite specifications. Hexadecimal number values are usually preceded with '0x' or suffixed with an 'H'. Thus the binary value 0100 1111B (decimal value 79 and ASCII character 'O') can be written as a hexadecimal value either as 0x4F or as 4F H.

It was convenient to extend the original ASCII code (which used a standard length of 7 bits) to 8 bits (of the *extended ASCII* code) for three reasons: First, it allowed additional characters to be incorporated corresponding to the various European language variations on the roman character set (e.g., ä, å, â, α, ç, é, è, ñ, ø, ö, ü, ß, etc). Second, it also allowed the addition of a number of graphical characters to control the formatting of characters (e.g., 'bold', 'italics', 'underlined', 'fontstyle', etc.) as well as enabling simple tables to be created. Third, it is convenient for computer designers and programmers to work with the standard character length of *1 byte* (8 bits). The memory and processing capabilities of a computer are designed and expressed in terms of bytes, Megabytes (1 Mbyte = 1024 bytes = 8192 bits) and Gigabytes (1 Gigabyte = 1024 × 1024 bytes = 8 388 608 bits).

As well as the 8-bit IBM PC-version of ASCII, there are a number of other 8-bit *extended ASCII* codes including the text/html code (ISO 8859-1) which we shall encounter in Chapter 11 (Table 11.7) and the Microsoft Windows Latin-1 character set (code page 1252) detailed in Appendix 1. The different character sets are adapted for slightly different purposes. In addition, a 16-bit (2 byte) character set which is based on ASCII but correctly called *unicode* has also been developed as a general purpose character set. *Unicode* allows computer programs to represent all the possible worldwide characters without having to change between code sets (Arabic, Chinese, Greek, Hebrew, Roman, Russian, Japanese, etc.) and is sometimes used in multilingual computer programs. But under normal circumstances, an 8-bit version of ASCII is used instead of unicode in order to keep the storage needed for each character down to only 8-bits.

2.5   EBCDIC and extended forms of ASCII

EBCDIC (extended binary coded decimal interchange code) (pronounced ebb-si-dick) is an alternative 8-bit scheme for encoding characters, and is the main character set used by IBM mainframe computers. The 8-bit EBCDIC code existed before the ASCII code was extended to 8 bits and afforded the early IBM mainframes the scope of 128 extra control characters.

---

[1] In the hexadecimal (or base 16) numbering scheme the digits have values equivalent to the decimal values 0–15. These digits are given the signs 0, 1, 2, 3, 4, 5, 6, 7, 8, 9, A, B, C, D, E and F respectively. Thus 'A' represents the decimal value 'ten', 'B' represents 'eleven' and so on, up to 'F' which represents 'fifteen'. The conversion of hexadecimal digits into binary values is easy, since each digit in turn can be progressively replaced by four binary digits. Thus the value 0H (hexadecimal) = 0000B (binary), 1H = 0001B, 2 H = 0010B, 3H = 0011B, 4H = 0100B, 5H = 0101B, 6H = 0110B, 7H = 0111B, 8H = 1000B, 9H = 1001B, AH = 1010B, BH = 1011B, CH = 1100B, DH = 1101B, EH = 1110B and FH = 1111B. Surprisingly, perhaps, even multiple digit hex (hexadecimal) numbers can be converted easily to binary. So, for example the hex value 9F is equivalent to the binary number 1001 1111.

In the case that an IBM mainframe receives an ASCII-coded file from another computer, the character set needs to be converted using a translation program. This is the *presentation layer* functionality we discussed in Chapter 1.

2.6   Use of the binary code to convey graphical images

Besides representing numerical and alphabetical (or textual) characters, the *binary code* is also used to transmit pictorial and graphical images as well as complex formatting information. This information is typically processed by the computer *graphics card*.

Pictures are sent as binary information by sending an 8-bit number (a value between 1 and 256) to represent the particular colour and shade (from a choice of 256) of a miniscule dot, making up a part of the picture. The picture itself is an *array* of dots. Typically a *video graphics array (VGA)* card supports an array of 640 dots width and 480 dots high (a so-called *resolution* of 640×480^2 picture elements or *pixels*). Put all the coloured dots together again in the right pattern (like an impressionist painting) and the picture reappears. This is the principle on which computer images are communicated. Send a series of pictures, one after the other at a rate of 25 Hz (25 picture frames per second) and you have video signal.

Figure 2.3 illustrates the principle of creating a computer graphic. In our example, the letters 'VGA' appear in a graphic array of 25 dots width and 16 dots height. Each dot (correctly called a pixel — a picture element) in our example is either the colour 'black' or 'white'. We could thus represent each pixel with a single bit, value 1 for black and 0 for white. For each of the 400 pixels in turn, we code and send a single bit, according to the colour of the picture we wish to send. We start at the top left-hand corner and scan across each line in turn. So that in binary (usually called *digital* code) our picture becomes:

```
00000 00000 00000 00000 00000
00000 00000 00000 00000 00000
01000 00100 11111 00000 10000
01000 00101 00000 10000 10000
01000 00101 00000 10000 10000
00100 01001 00000 00001 01000
00100 01001 00000 00001 01000
00100 01001 00000 00001 01000
00010 10001 00000 00010 00100
00010 10001 00111 10011 11100
00010 10001 00000 10010 00100
00001 00001 00000 10100 00010
00001 00001 00000 10100 00010
00001 00000 11111 00100 00010
00000 00000 00000 00000 00000
00000 00000 00000 00000 00000
```

In the binary code format the image is much harder to pick out than on the 'screen' of Figure 2.3! In fact, it would have been even harder if we were not to have typed the bits in the convenient array fashion, but instead had printed them as a continuous line of 400 characters — which is as they appear on the 'transmission line' between the PC graphics card and the screen.

On reflection, it might seem rather strange, given our previous discussion about the ASCII code — in which the three letters 'VGA' could be represented by 24 bits (01010110 01000111

---

[2] Alternative, commonly used VGA picture sizes are 800×600 pixels and 1024×768 pixels.

Figure 2.3 An example of a video graphic array (VGA) comprising 25 × 16 picture elements (pixels).

01000001), that we should now find ourselves requiring 400 bits to represent the same three letters! How have we come to need an extra 376 bits, you might ask? The reason is that the form in which we wish to present the three letters (on a video screen, to a computer user) requires the 400-bit format. The conversion of the 24-bit format into the 400-bit format is an example of the OSI model *presentation layer* function, as discussed in Chapter 1. Such a conversion is typically part of the function carried out by the computer graphics card. But in reality, most modern computer graphics images are not a mere 'black and white' image of only 400 bits for a 25 × 16 pixel matrix. Instead the colour of each pixel is usually represented by between 1 and 4 bytes of code (representing between 256 and 43 million colours) and most computer screens nowadays are *arrays* of either 640 × 480 pixels, 800 × 600 pixels or 1024 × 768 pixels, so that a single 'screenshot image' may correspond to 3 Mbytes (1024 × 768 × 4 = 3.1 million bits) of data.

2.7 Decoding binary messages — the need for synchronisation and for avoiding errors

Next, we consider the challenge posed by the decoding of a binary message at the receiver end of a connection, and also the problems caused by *errors* introduced during the communication.

Let us consider sending a short message, containing the single word 'Greeting'. Coding each of the letters into ASCII using the table of Table 2.3 we derive a bit sequence as follows, where the right-hand bit should be sent first:

s	g	n	i	t	e	e	r	G
01110011	01100111	01101110	01101001	01110100	01100101	01100101	01110010	01100111
last bit to be sent								first bit to be sent

All well and good: easy to transmit, receive and decode back to the original message. But what happens if the receiver incorrectly interprets the 'idle' signal which is sent on the line (value '0') prior to the first '1' bit of the real message as a leading bit of value '0'? In this case, an extra '0' appears at the right-hand end of our *bit string* and all the other bit values are shifted one position to the left. The pattern decoded by the receiver will be as below. Now the meaning of the decoded message is gibberish! [decoded message: ╬ Σ╩╚╩ Φ╥■╬ µ]

last bit received   first bit received

11100110   11001110   11011100   11010010   11101000   11001010   11001010   11100100   11001110

[last bit lost]   [extra '0' assumed]

μ   ╫   ■   π   Φ   ╨   ╨   Σ   ╫

Our example illustrates perfectly the need for maintaining *synchronisation* between the transmitter and the receiver, in order that both take the same bit as the first of each *byte*. We shall return to the various methods of ensuring synchronism later in the chapter. But first, let us also consider the effect of errors.

Errors are bits which have changed their value during conveyance across a network. They may be caused by a large number of different reasons, some of which we shall consider later in the chapter. If the three underlined errors below occur in the original code for 'Greetings', then the received message is corrupted. Unfortunately, the result may not obviously be corrupted gibberish, but may instead appear to be a 'valid' message. In this example, rather than pleasing our recipient with 'Greetings', we end up insulting him with the message 'Greedy~gs'!

s   g   y   d   e   e   r   G

01110011   01100111   01111110   01111001   01100100   01100101   01100101   01110010   01100111

last bit to be received   errors underlined   first bit to be received

There is a clear need to minimise errors. We do this by ensuring that the quality of the transmission lines we use is very high. The quality we measure in terms of the *bit error ratio (BER)*. In our example we had three bit errors in a total of $9 \times 8 = 72$ bits, a bit error ratio (BER) of 4%. This would be an unacceptably high BER for a modern data network, most of which operate in the range BER $= 10^{-7}$ to 10^{-9} (1 error in 10 million or 1000 million bits sent). In addition to using very high quality lines, data protocols also usually include means for *detecting* and *correcting* errors. These methods are called *error detection* or *error correction* codes. The simple fact is that we cannot afford any corruptions in our data!

2.8 Digital transmission

We have learned how we can code textual, graphic, video and other types of computer *data* into binary code — in particular into 8-bit *blocks* of binary code which we call *bytes*. And we have seen how we can convey this data across a communications *medium* by means of *digital transmission* — essentially turning the electricity or light on the line either 'on' (to represent binary value '1') or 'off' (to represent binary value '0'). Digital transmission media which operate according to this basic principle include:

- the *serial ports* of computers and the *local* connection lines connected to them;

- *local area networks (LANs)*;

- digital *leaselines* (including all *PDH (plesiochronous digital hierarchy), SDH (synchronous digital hierarchy)* and *SONET (synchronous optical network)* type lines... e.g., lines conforming to RS-232, V.24, X.21, G.703, '64 kbit/s', '128 kbit/s', 'E1', 'T1', 'E3', 'T3', 'STM-1', 'OC-3', etc);

- digital radio links;

- digital satellite connections;

- point-to-point fibre optic transmission.

In reality, however, the transmission on digital line systems is rarely a simple two-state 'on-off' process. For the purpose of line synchronisation and error avoidance it is instead normal to use a *line code*. We shall discuss shortly what a line code is and how it works, but beforehand we need to understand signal *modulation*. *Modulation* is the technical term used to describe how real transmission *media* can be made to carry data and other *digital* signals. The discussion will help us to understand some of the causes of bit errors.

2.9   Modulation of digital information over analogue media using a modem

A *modem* is a device which can be connected to the *serial port* of a computer, to convert the *digital* data information arising within the computer into a form suitable to be conveyed across an *analogue* telecommunications medium (a line or network such as the telephone network). The word *modem* is a derivation from the two words *MODulator* and *DEModulator*. The first modem was invented in 1956 by AT&T Bell Laboratories.

Three basic data modulation techniques are used in modems for converting digital computer data into a form suitable for carriage across various types of *media*. There are also more sophisticated versions of each modulation type and even hybrid versions, combining the various techniques.

Amplitude modulation (AM)

Modems employing *amplitude modulation (AM)* alter the amplitude of the *carrier* signal between a set value and zero (effectively 'on' and 'off') according to the respective value '1' or '0' of the modulating bit stream. This form of digital modulation is correctly called *on-off-keying (OOK)*. OOK was the technique which was used in the earliest modems and is also widely used in modern optical fibre transmission — where a laser or LED (light-emitting diode) light source is switched 'on' and 'off'. Figure 2.4 illustrates an example of OOK in which the *carrier* signal (of frequency f_1) is simply switched on and off. Alternatively, two different, non-zero values of amplitude may be used to represent '1' and '0', as in the case of Figure 2.5c.

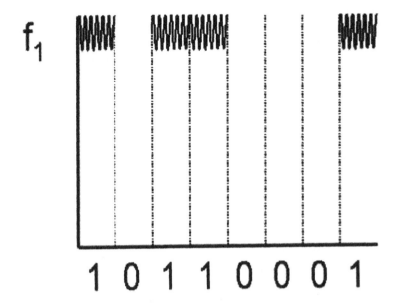

Figure 2.4   On-off keying form of amplitude modulation.

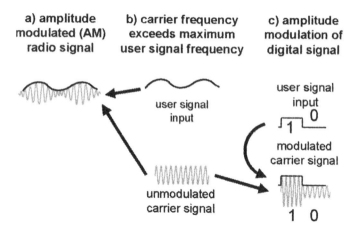

Figure 2.5   Amplitude modulation (AM).

For amplitude modulation to work correctly, the frequency of the *carrier signal* must be much higher than the highest frequency of the *information signal*. (This ensures that the 'peaks and troughs' of the carrier signal (Figure 2.5a) are more frequent than the 'peaks and troughs' of the information signal (Figure 2.5b), thus enabling the carrier signal to 'track' and record even the fastest changes in the information signal. The carrier signal is the high pitched tone which computer users will be familiar with listening to, when they make a 'dial-up' connection from their PC and their modem 'synchronises' with the partner device at the other end of the line.

Amplitude modulation (AM) is carried out simply by using the information signal (i.e., user data) to control the power of a carrier signal amplifier.

Frequency modulation (FM)

In *frequency modulation* (Figure 2.6a), it is the frequency of the *carrier signal* that is altered to reflect the value '1' or '0' of the modulating bit stream (the *information signal*). The amplitude and phase of the carrier signal are otherwise unaffected by the modulation process.

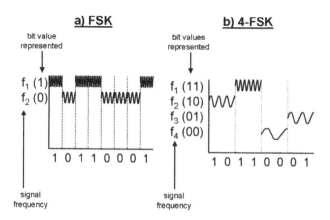

Figure 2.6   Frequency modulation (FM) and frequency shift keying (FSK).

If the number of bits transmitted per second is low, then the signal emitted by a frequency modulated modem is heard as a 'warbling' sound, alternating between two frequencies of tone. Modems using frequency modulation for coding of digital signals are more commonly called *FSK*, or *frequency shift key* modems.

A common form of *FSK* modem uses four different frequencies (or tones), two for the transmit direction and two for the receive. The use of four separate frequencies allows simultaneous sending and receiving (i.e., *duplex* transmission) of data by a modem, using a single bi-directional medium (i.e., using a 'two-wire' circuit rather than a 'four-wire' circuit comprising separate circuits or 'pairs' for 'transmit' and 'receive' directions).

A further form of FSK, called 4-FSK also uses four frequencies, but all four in each direction of transmission. The use of four frequencies allows one of four different 'two-bit combinations' to be transmitted. Thus a single pulse of tone conveys 2 *bits* of information (Figure 2.6b). In this case the *information rate* (the number of bits carried per second — in the example of Figure 2.6b the rate is 2 bits/s) is higher than the *Baud rate* (the number of changes in carrier signal tone per second (in our example, the Baud rate is 1 *Baud*). This is an example of *multilevel transmission*, which we shall return to later.

Phase modulation

In *phase modulation* (Figure 2.7), the carrier signal is *advanced* or *retarded* in its *phase* cycle by the modulating bit stream (the information signal). The frequency and amplitude of the carrier signal remain unchanged. At the beginning of each new bit, the signal will either retain its phase or change its phase. In the example of Figure 2.7, the initial signal phase represents value '1' and the change of phase by 180° represents next bit '0'. In the third bit period the value to be transmitted is '1' and does not therefore require a phase change. This is often confusing to newcomers of phase modulation, since as a result the <u>absolute phase</u> of the signal in both time period 2 and 3 is the same, even though it represents different bit values of the second and third bits. It is important to remember that the coding of digital signals using *phase modulation* (often called *phase shift keying* or *PSK*) is <u>conducted by comparing the signal phase</u> in one time period to that in the previous period. It is <u>not the absolute value</u> of the signal phase that is important in phase modulation, <u>rather the phase change</u> that occurs at the beginning of each time period.

Figure 2.8 illustrates an advanced form of phase shift keying called 4-PSK or *quarternary phase shift keying (QPSK)*. Just as in Figure 2.7, it is the phase change at the beginning of

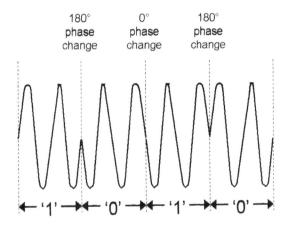

Figure 2.7   Phase modulation (PM) or phase shift keying (PSK).

4-PSK (QPSK)

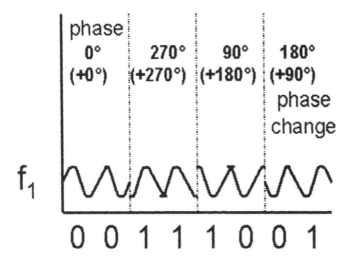

Figure 2.8   4-PSK or QPSK (quarternary phase shift keying).

each bit and not the absolute phase which counts. Like the example of 4-FSK we discussed in Figure 2.6, QPSK is an example of *multilevel transmission*, which we discuss next.

High bit rate modems and 'multilevel transmission'

The transmission of high bit rates can be achieved by modems in one of two ways. One is to modulate the carrier signal at a rate equal to the high bit rate of the modulating signal. This creates a signal with a high *Baud rate*. (The rate (or frequency) at which we fluctuate the carrier signal is called the Baud rate.) The difficulty lies in designing a modem capable of responding to the line signal changes at the high Baud rate. Fortunately an alternative method is available in which the Baud rate is lower than the bit rate of the modulating bit stream (the so-called *information rate*). The lower Baud rate is achieved by encoding a number of consecutive bits from the modulating stream to be represented by a single line signal state. The method is called *multilevel transmission*, and is most easily explained using diagrams. Figures 2.7 and 2.9 both illustrate a bit stream of 2 bits per second (2 bit/s) being carried respectively by 4-FSK and 4-PSK modems, both of which use four different line signal states. Both modems are able to carry the 2 bits/s information rate at a Baud rate of only 1 per second (1 Baud).

The modem used in Figure 2.6 achieves a lower Baud rate than the bit rate of the data transmitted by using each of the line signal frequencies f1, f2, f3 and f4 to represent two consecutive bits rather than just one. The modem used in Figure 2.8 achieves the same end-result using four different possible phase changes. In both cases it means that the modulated signal 'sent to line' is always slightly delayed relative to the original source data signal.

There is a signal delay associated with *multilevel transmission*. In the examples of Figures 2.7 and 2.9, the delay is at least 1 bit duration, since the first of each 'pair' of bits cannot be sent to line until its 'partner' bit arrives. But the benefit of the technique is that the receiving modem will have twice as much time to detect and interpret each bit of the received datastream. *Multi-level transmission* is invariably used in the design of very high bit rate modems.

Modem 'constellations'

Modem constellation diagrams assist in the explanation of more complex *amplitude* and *phase-shift-keyed (PSK)* modems. Figure 2.9a illustrates a *modem constellation* diagram composed of four dots. In fact, it is the *constellation diagram* for a 4-PSK modem with <u>absolute</u> signal phase values of +45°, +135°, +225° and +315°. Each dot on the diagram represents the relative phase and amplitude of one of the four allowed line signals generated by the modem. The distance of the dot from the origin of the diagram axes represents the amplitude of the signal, and the angle subtended between the X-axis and a line from the 'point of origin' of the diagram represents the signal phase. In our example, each of the four optional signal states have the same amplitude (i.e., all the points are the same distance from the point of origin).

Figures 2.9b and 2.9c illustrate signals of different *signal phase*. Figure 2.9b shows a signal of 0° phase: the signal starts at zero amplitude and increases to maximum amplitude.

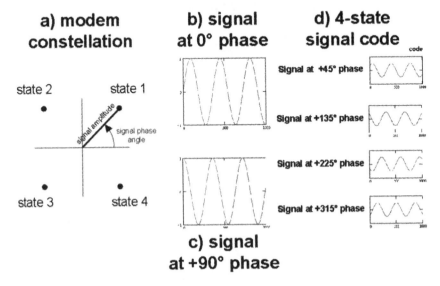

Figure 2.9 Modem constellation diagram of an example 4-PSK modem.

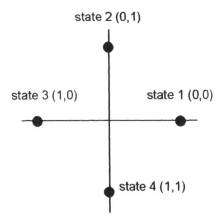

Figure 2.10   Modem constellation diagram for the 4-PSK modem also shown in Figure 2.8.

Figure 2.9c, by contrast, shows a signal of 90° phase, which commences further on in the cycle (in fact, at the 90° phase angle of the cycle). The signal starts at maximum amplitude but otherwise follows a similar pattern to Figure 2.9b. Signal phases, for any phase angle between 0° and 360° could similarly be drawn. Returning to the signals represented by the constellation of Figure 2.9a we can now draw each of them, as shown in Figure 2.9d. The phase changes (which may be signalled using the modem illustrated in Figure 2.9) are 0°, +90°, +180° and +270°. But it is not the same modem as that illustrated in Figure 2.8, because the <u>absolute</u> signal phases are not the same. For comparison, the constellation diagram of the 4-PSK modem of Figure 2.8 is shown in Figure 2.10.

Quadrature amplitude modulation (QAM)

We are now ready to discuss a complicated but common modem modulation technique known as *quadrature amplitude modulation (QAM)*. QAM is a technique using a complex hybrid of phase (or *quadrature*) as well as *amplitude modulation*, hence the name. Figure 2.11 shows an eight-state form of QAM (8-QAM) in which each line signal state represents a 3-bit signal (values nought to seven in binary can be represented with 3 bits). The eight signal states are a combination of four different relative phases and two different amplitude levels. The table in Figure 2.11a relates the individual 3-bit patterns to the particular phases and amplitudes of the signals that represent them. The fourth column of the table illustrates the actual line signal pattern that would result if we sent the signals in the table consecutively as shown. Figure 2.11b shows the *constellation* of this particular modem.

  To finish off the subject of modem constellations, Figure 2.12 presents, without discussion, the constellation patterns of a couple of very sophisticated modems, specified by ITU-T recommendations V.22 bis and V.32. As in Figure 2.11, the constellation pattern would allow the interested reader to work out the respective 16 and 32 line signal states. Finally, Table 2.4 lists some of the common modem types and their uses. When reading the table, bear in mind that *synchronous* and *asynchronous operation* is to be discussed later in the chapter, and that *half duplex* means that 2-way transmission is possible but only one direction of transmission is possible at any particular instant of time. This differs from *simplex* operation where only one-way transmission is possible.

a) bit combinations & signal attributes

Bit combination	Signal amplitude	Phase shift	Typical signal
000	Low	0°	
001	Low	+90°	
010	Low	+180°	
011	Low	+270°	
100	High	+0°	
101	High	+90°	
110	High	+180°	
111	High	+270°	

Note: Each signal in the fourth column of the table in Figure 2.11a is shown phase state relative to the signal in the box above immediately above it. Remember that it is the <u>phase change</u> and <u>not the absolute phase</u> that is important.

Figure 2.11   Constellation diagram of an example 8-QAM modem.

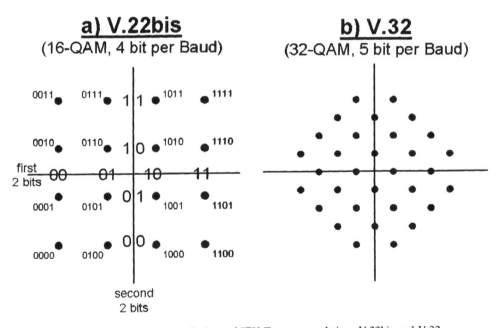

Figure 2.12   Modem constellations of ITU-T recommendations V.22bis and V.32.

2.10   Detection and demodulation — errors and eye patterns

No matter which transmission *medium* and *modulation* scheme is used, some kind of *detector* or *demodulator* is necessary as a receiver at the destination end of a communication link.

Table 2.4 Common modem types and related functions

Modem type (ITU-T recommendation)	Modulation type	Bit speed (bit/s)	Synchronous (S) or asynchronous (A) operation	Full or half duplex	Circuit type required
V.21	FSK	up to 300	A	Full	2 w telephone line
V.22	PSK	1200	S or A	Full	2 w telephone line
V.22 bis	QAM	2400	S or A	Full	2 w telephone line
V.23	FSK	600/1200	A	Full	4 w leaseline
				Half	2 w telephone line
V.26	4-PSK	2400	S	Full	4 w leaseline
V.26 bis	4-PSK	2400/1200	S	Half	2 w telephone line
V.27	8-PSK	4800	S	Full	2 w leaseline
V.27 ter	8-PSK	4800	S	Half	2 w leaseline
V.29	16-QAM	9600	S	Half	4 w leaseline
V.32	QAM	up to 9600	S or A	Full	2 w telephone line
V.32 bis	QAM	up to 14 400	S or A	Full	2 w telephone line
V.33	QAM	14 400	S or A	Full	4 w leaseline
V.34	QAM	28 800	S or A	Full	2 w telephone line
V.35	AM	48 000	Wideband	Full	Groupband leaseline
V.36	AM	48 000	Wideband	Full	Groupband leaseline
V.37	AM	72 000	Wideband	Full	Groupband leaseline
V.42	—	—	Convert synchronous to asynchronous format	—	Error correcting protocol

(continued overleaf)

Table 2.4 (continued)

Modem type (ITU-T recommendation)	Modulation type	Bit speed (bit/s)	Synchronous (S) or asynchronous (A) operation	Full or half duplex	Circuit type required
V.42 bis	—	up to 30 kbit/s in association with V.32 modem	—	—	Data compression technique
V.43	—	—	—	—	Data flow control
V.44	—	—	—	—	Data compression procedure
V.54	—	—	—	—	Loop test device and procedure for modems
V.61	QAM	4800 to 14 400	Voice plus data modem	Full	2 w telephone line
V.90	QAM	56 000 up/ 33 600 down	Digital and analogue modem pair	Full duplex but asymmetric rates upstream and downstream	2 w telephone line
V.91	QAM	64 000	Digital modem	Full	4 w digital leaseline
V.92	QAM	64 000	Enhancements to V.90	Full	2 w telephone line
V.130	—	—	—	—	ISDN terminal adapter—DCE for ISDN
V.300	—	128 000 or 144 000	DCE for digital leaseline network	Full	4 w digital leaselines

For on-off keying (OOK) or similar techniques the term *detector* is most commonly used, otherwise the device at the end of an *analogue* medium is called a *demodulator*. Both have to perform a similar 'decision-making function' based on each bit received and some kind of *threshold* criterion.

As you might imagine, the nice square-shaped signal initially transmitted across a digital transmission line (Figure 2.13a) gets rather degraded during its journey. By the time it reaches the receiver, having been subjected to *attenuation, distortion* and electrical *noise*, the signal (Figure 2.13b) may no longer be anything like the 'square' form original. The job of the *detector* is to determine, for each individual *bit* whether the received value is meant to represent binary value '0' or a binary value '1'. It does this using a threshold *decision criterion*. In the example of Figure 2.13b the original signal values were amplitude '0' and amplitude '1', and the threshold criterion used at the detector is amplitude value '0.5'. If, at the relevant point in time corresponding to a particular bit (the so-called *decision point*), the amplitude of the received signal is less than 0.5, then this is *detected* as binary value '0'. Otherwise amplitude values higher than 0.5 are *detected* as binary value '1'. In the corresponding case of an analogue medium, a *threshold criterion* based on the nearest *state point* in the modem constellation pattern is used (see Figure 2.12).

At the *decision point* corresponding to bit 9 of Figure 2.13b, the received signal has been distorted to such an extent that the detected value is '0', rather than the original value '1' sent. This results in a *bit error* in the received signal. The cause of the bit error is a combination of attenuation and distortion on the line. It might equally have been due to signal interference caused by electrical noise. But attenuation, distortion and noise are not the only causes of bit errors...

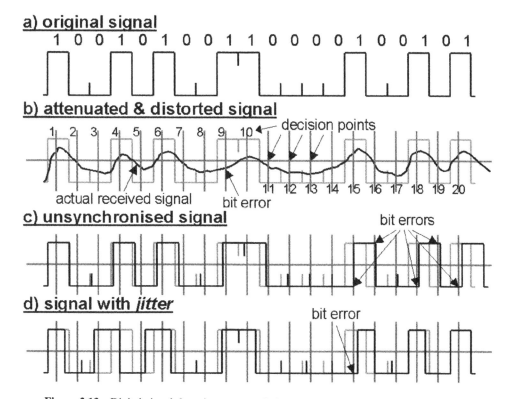

Figure 2.13   Digital signal detection, causes of signal degradation and received bit errors.

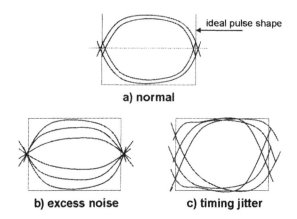

Figure 2.14   Eye diagrams.

Figure 2.13c illustrates how bit errors arise if the transmitter and receiver are not exactly *synchronised* with one another. In our example, the transmitting bit rate is very slightly too low, compared with the rate of detection. Thus the transmitted signal (shown as the square pattern) is a slightly 'stretched' version of the intended signal. (The intended signal was that of Figure 2.13a.) By the time of the detector's fifteenth *decision point*, the 'stretched' signal has *slipped* by a whole bit period — causing the fourteenth bit (value '0') to be detected twice. This misleads the receiver into creating a new extra bit, also value '0' between the original 14th and 15th bits. Such a *slip* would lead to the type of message corruption problem we discussed earlier (the 'Greedy~gs'/'Greetings' problem discussed in § 2.7).

Jitter can also lead to bit errors, as Figure 2.13d illustrates. Jitter is the term used to describe signals transmitted with unequal time duration (i.e., some pulses are longer than others). Jitter usually arises because of the poor quality of the transmitting device, and is usually resolved at the equipment design stage, through choice of good quality components, and tested as part of equipment *homologation* or *approvals* testing.

When trying to trace the source of bit errors, technicians often use oscilloscopes or other test equipment either to plot the *constellation diagram* (Figures 2.11 and 2.13) or to plot the *eye diagram* of a received signal (Figure 2.14). In an 'ideal' constellation diagram plot, each of the 'dots' representing the *state points* is sharply defined. But where noise or other disturbances are present on the line, the 'dot' gets spread over a larger area and appears 'fuzzy'.

An *eye diagram*, when displayed on an oscilloscope or other test equipment shows the actual pulse shapes of the bits being received. There are various characteristic shapes of the eye diagram (Figure 2.14) which help to pin down the various different sources of bit errors.

2.11   Reducing errors — regeneration, error detection and correction

One of the simplest precautions used to eliminate errors during transmission is the technique of *regeneration* (Figure 2.15). A number of *regenerators* are placed at regular intervals along the length of a digital transmission line. Each regenerator detects the signal it receives and retransmits it as a sharp square wave signal. In this way, a perfect square signal is *regenerated* at regular points along the route. The distance between regenerators is chosen, so that comparatively little distortion of the digital pulses of the signal can take place during the relatively short hop between each pair of regenerators. In this way, major signal distortion is prevented, and so the chance of bit errors is drastically reduced.

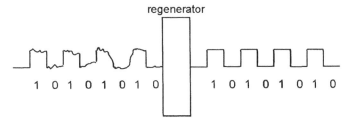

Figure 2.15   The principle of regeneration.

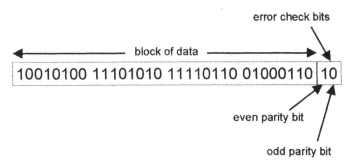

Note: As an aside, the example of Figure2.16 is some what unusual,since it would not be normal to send two simple parity bits, since the second parity bit does not provide much more information about errors beyond that already provided by the first.

Figure 2.16   The principle of error detection.

In addition to regeneration, which is generally used on all digital line systems, it is common in data communication also to apply *error detection* and/or *error correction*. Error detection is carried out by splitting the data to be sent up into a number of *blocks*. For each *block* of data sent, a number of *error check bits* are appended to the back end of the *block* (Figure 2.16).

The error *check bits* (also sometimes called the *FCS* or *frame check sequence*) are used to verify a successful and error-free transmission of the data block to which they relate. In the example in Figure 2.16, the data block is 4 bytes long (32 bits) and there are two error check bits, including an *even parity bit* and an *odd parity bit*. The even parity bit is set so that the total number of bits of binary value '1' within the data block <u>and</u> the parity bit (33 bits in total) is *even*. Since there are 17 bits of value '1' in the data block, then the even parity bit is set to value '1' to make the total number of '1' values 18 and thus even. Similarly, the odd parity bit is set to make the total number of '1' values odd. If on receipt of the new extended block, the total number of bits set at binary value '1' does not correspond with the indication of the parity bits, then there is an error in the message.

Sophisticated error detection codes even allow for the *correction* of errors. *Forward error correction (FEC)* codes allow an error (or a number of errors) in the data block to be corrected by the receiver without further communication with the transmitter. Alternatively, without using forward error correction, the detection of an error could simply lead to the *re-transmission* of the affected data block. Different data protocols either use *parity checks*, forward error correction (FEC), data *retransmission* or no error correction as their means of eliminating bit errors.

Error detection and correction have nowadays become a sophisticated business. Standard means of error detection and correction are provided by *cyclic redundancy checks (CRCs), Hamming codes, Reed-Solomon* and *Viterbi codes*. These methods use extra bits and complex mathematical algorithms to dramatically reduce the probability of errors surviving detection and correction. They therefore greatly increase the dependability of transmission systems. The most common detection and correction technique used by data protocols intended for terrestrial networks are *cyclic redundancy check codes*, and we discuss these next.

Cyclic redundancy check (CRC) codes

Error detection and correction codes can be divided into *cyclic* and *non-cyclic* codes. Cyclic codes are a special type of error correcting *block code*, in which each of the valid *codewords*[3] are a simple lateral shift of one another.

To illustrate a cyclic code, let us consider a (10, 3) code (the codewords are 10 bits long in total, comprising 7 bits of user data and 3 error check bits). If we assume that the following is a valid codeword of our code:

$$c = (1\ 0\ 1\ 0\ 1\ 0\ 0\ 1\ 0\ 0)$$

then $c = (0\ 1\ 0\ 1\ 0\ 0\ 1\ 0\ 0\ 1)$ must also be a valid codeword, if the code is cyclic. (All bits are shifted one position to the left, with the extreme left-hand bit of the original codeword reinserted on the right-hand end. The other 8 codewords of this CRC are:

$c = (1\ 0\ 1\ 0\ 0\ 1\ 0\ 0\ 1\ 0)$ $c = (0\ 1\ 0\ 0\ 1\ 0\ 1\ 0\ 1\ 0)$
$c = (0\ 1\ 0\ 0\ 1\ 0\ 0\ 1\ 0\ 1)$ $c = (1\ 0\ 0\ 1\ 0\ 1\ 0\ 1\ 0\ 0)$
$c = (1\ 0\ 0\ 1\ 0\ 0\ 1\ 0\ 1\ 0)$ $c = (0\ 0\ 1\ 0\ 1\ 0\ 1\ 0\ 0\ 1)$
$c = (0\ 0\ 1\ 0\ 0\ 1\ 0\ 1\ 0\ 1)$ $c = (0\ 1\ 0\ 1\ 0\ 1\ 0\ 0\ 1\ 0)$

Cyclic codes are capable of correcting larger numbers of errors within the data block than are non-cyclic codes. A *CRC-n* is an n-bit *cyclic redundancy check* code. The value of the code is set by first multiplying the data block by a given *multiplier*, then dividing (in binary, or *modulo 2*) the result of the multiplication by a *generator polynomial*. The *remainder* is used to set the value of the CRC field. A number of common CRC codes are shown in Table 2.5.

Table 2.5 Common cyclic redundancy check (CRC) codes

CRC-n type	Field multiplier	Generator polynomial
CRC-1		(simple parity check)
CRC-3	x^3 (in other words: 1000 Binary)	$x^3 + x + 1$(1011 B)
CRC-4	x^4 (10 000 B)	$x^4 + x + 1$ (10 011 B)
CRC-5	x^5 (100 000 B)	$x^5 + x^4 + x^2 + 1$(110 101 B)
CRC-6	x^6 (1 000 000 B)	$x^6 + x + 1$ (1 000 011 B)
CRC-7	x^7 (10 000 000 B)	$x^7 + x^3 + 1$ (10 001 001 B)
CRC-10	x^{10} (10 000 000 000 B)	$x^{10} + x^9 + x^5 + x^4 + x + 1$ (11 000 110 011 B)
CRC-16	$x^{16} + x^{15} + x^{14} + x^{13} + x^{12} + x^{11} + x^{10} + x^9 + x^8 + x^7 + x^6 + x^5 + x^4 + x^3 + x^2 + x + 1$ (11 111 111 111 111 111 B)	$x^{16} + x^{12} + x^5 + 1$ (10 001 000 000 100 001 B)

---

[3] The **codeword** is a fancy name for the resulting pattern of bits which is transmitted after adding the error check bits. The codeword thus comprises both the original *user data block* (the first 32 bits of Figure 2.16) and the *error check bits* (the last 2 bits). One talks of (x, y) codewords, where x is the total number of bits in the codeword (in the case of Figure 2.16, x = 34) and y is the number of error check bits (y = 2 in Figure 2.16). Thus, the example of Figure 2.16 is a (34,2) codeword.

Other well-known cyclic redundancy check codes include *BCH (Bose-Chaudhuri-Hocquen-hem)* and *Reed-Solomon* codes.

2.12   Synchronisation

The successful transmission of data depends not only on the accurate coding of the transmitted signal (e.g., using the ASCII code, as we discussed earlier), but also on the ability of the receiving device to decode the signal correctly. This calls for accurate *synchronisation* of the receiver with the transmitter, so that the beginning and end of all received bits occur at regular and predictable intervals. For the purpose of synchronisation a highly accurate clock must be used in both transmitter and receiver.

It is usual for the receiver to *sample* the communication line at a rate much faster than that of the incoming data, thus ensuring a rapid detection of any change in line signal state, as Figure 2.17c shows. Theoretically it is only necessary to sample the incoming data at a rate equal to the nominal bit rate of the signal, but this runs a risk of data corruption. If we chose to sample near the beginning or end of each bit (Figures 2.18a and 2.18b) we might lose or duplicate data as the result of a slight fluctuation in the time duration of individual bits. Much faster sampling ensures rapid detection of the start of each '0' to '1' or '1' to '0' transition. In this case, the signal <u>transitions</u> are interpreted as bits, and the exact clock rate of the transmitter can be determined.

Variations in the clock rate arising from different time durations of individual received bits come about because signals are liable to encounter time shifts during transmission, which may or may not be the same for all bits within the message. These variations are usually random and they combine to create an effect known as jitter. As we showed in Figure 2.13d, jitter can lead to bit errors.

The purpose of synchronisation is to remove all short-, medium- and long-term time effects of timing or clocking differences between the transmitter and the receiver. Short-term variations are called jitter. Long term variations are instead called *wander*.

In the short term, synchronisation between transmitter and receiver takes place at a bit level, by *bit synchronisation*. This keeps the transmitting and receiving clocks in step, so that bits start and giving pulses of constant duration. (Recall the errors which occurred as a result of incorrect pulse durations in Figures 2.14c and 2.14d.) Medium-term *character* or *word synchronisation* prevents confusion between the last few bits of one character and the first few bits of the next. If we interpret the bits wrongly, we end up with the wrong characters, as we found out in our earlier example, when the message 'Greetings' was turned into the gibberish '⊣⊢ ∑⊓⊔ ⏀⊓∎⊓⊢ μ'. Finally there is *frame synchronisation*, which ensures data reliability and *integrity* over longer time periods.

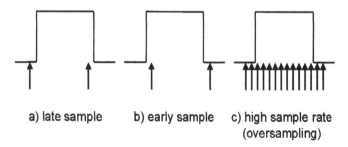

a) late sample   b) early sample   c) high sample rate
(oversampling)

Figure 2.17   Effect of sampling rate.

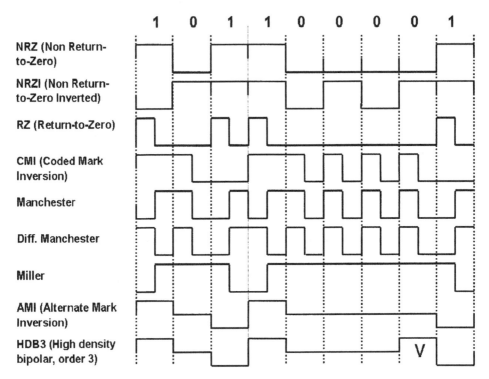

Figure 2.18   Commonly used line codes for digital line systems V = Violation.

Bit synchronisation using a line code

The sequence of ones and zeros (*marks* and *spaces*) making up a digital signal is not usually sent directly to line, but is first arranged according to a *line code*. The *line code* serves for the purpose of bit synchronisation.

One of the biggest problems to be overcome by *bit synchronisation* is that if either a long string of 0's or 1's were sent to line consecutively, then the line would appear to be either permanently 'on' or permanently 'off' — effectively a *direct current (DC)* condition is transmitted to line. This is not advisable for two reasons. First, the power requirement is increased and the attenuation is greater for *direct current (DC)* as opposed to *alternating current (AC)*. Second, it may be unclear how many bits of value '0' have been sent consecutively. It may even be unclear to the receiver whether the line is actually still 'alive'. The problem gets worse as the number of consecutive 0's or 1's increases. Line codes therefore seek to ensure that a minimum frequency of line state changes is maintained.

Figure 2.18 illustrates some of the most commonly used line codes. Generally they all seek to eliminate long sequences of 1's or 0's, and try to be *balanced* codes, i.e., producing a net zero direct current voltage. Thus, for example, the three-state codes AMI and HDB3 try to negate positive pulses with negative ones. This reduces the problems of transmitting power across the line.

The simplest *line code* illustrated in Figure 2.18 is a *non-return to zero (NRZ)* code in which '1' = 'on' and '0' = 'off'. This is perhaps the easiest to understand. All our previous examples of bit patterns were, in effect, shown in a non-return to zero (NRZ) line code format.

In *NRZI (non-return-to-zero inverted)* it is the presence or absence of a *transition* (a transition is a change of line state, either from '1' to '0' or from '0' to '1') which represents a '0' or a '1'. Such a code is technically quite simple (even if confusing to work out) and may be advantageous where the line spends much of its time in an 'idle' mode in which a string of '0s' or '1s' would otherwise be sent. Such is the case, for example, between an *asynchronous terminal* and a *mainframe* computer. NRZI is used widely by the IBM company for such connections.

A *return-to-zero (RZ)* code works in a similar manner to *NRZ*, except that *marks* return to zero midway through the bit period, and not at the end of the bit. Such coding has the advantage of lower required power and constant *mark* pulse length in comparison with basic NRZ. The length of the pulse relative to the total bit period is known as the *duty cycle*. Synchronisation and timing adjustment can thus be achieved without affecting the mark pulse duration.

A variation of the NRZ and RZ codes is the *CMI (coded mark inversion)* code recommended by ITU-T. In *CMI*, a '0' is represented by the two signal amplitudes A1, A2 which are transmitted consecutively, each for half the bit duration. '1s' are sent as full bit duration pulses of one of the two line signal amplitudes, the amplitude alternating between A1 and A2 between consecutive marks.

In the *Manchester code*, a higher pulse density helps to maintain synchronisation between the two communicating devices. Here the transition from high-to-low represents a '1' and the reverse transition (from low-to-high) a '0'. The Manchester code is used in *ethernet LANs*.

In the *differential Manchester code* a voltage transition at the bit start point is generated whenever a binary '0' is transmitted but remains the same for binary '1'. The IEEE 802.5 specification of the *token ring* LAN demands differential Manchester coding.

In the *Miller code*, a transition either low-to-high or high-to-low represents a '1'. No transition means a '0'.

The *AMI (alternate mark inversion)* and *HDB3 (high density bipolar)* codes defined by ITU-T (recommendation G.703) are both three-state, rather than simple two-state (on/off) codes. In these codes, the two extreme states (if you like, '+' and '−') are used to represent marks (value '1') and the mid-state (if you like, value '0') is used to represent *spaces* (value '0'). The three states are often realised as 'positive' and 'negative values', with a mid-value of '0'. Or in the case of optical fibres, where light is used, the three states could be 'off', 'low intensity' and 'high intensity'. In both AMI and HDB3 line codes, alternative marks are sent as positive and negative pulses. Alternating the polarity of the pulses helps to prevent direct current being transmitted to line. (In a two-state code, a string of marks would have the effect of sending a steady 'on' value to line.)

The HDB3 code (used widely in Europe and on international transmission systems) is an extended form of AMI in which the number of consecutive zeros that may be sent to line is limited to 3. Limiting the number of consecutive zeros bring two benefits: first, a null signal is avoided, and second, a minimum *mark density* can be maintained (even during idle conditions such as pauses in speech). A high mark density aids the regenerator timing and synchronisation.

In HDB3, the fourth zero in a string of four is *marked* (i.e., forcibly set to 1) but this is done in such a way that the 'zero' value of the original signal may be recovered at the receiving end. The recovery is achieved by marking fourth zeros in *violation*, that is to say in the same polarity as the previous 'mark', rather than in opposite polarity mark (opposite polarity of consecutive marks being the normal procedure).

Other line codes used on WAN lines (particularly in North America) include B8ZS (bipolar 8 zero substitution) and *ZBTSI (zero byte time slot interchange)*.

Character synchronisation — synchronous and asynchronous data transfer

Character synchronisation ensures that the receiver knows which is the first bit of each character code pattern. Misplacing the first bit can change the interpretation of the character. (In our earlier example, the message 'Greetings' became '⌐ ∑⅃⅃L Φ╥■⌐ μ').

Let us consider the following sequence of 9 received bits:

(last bit received)   0 0 1 1 0 0 1 1 0   (first bit received)

As a 'raw stream' of received bits, it is difficult to determine which 8 bits (when grouped together) represent an ASCII character. If we assume that the first 8 bits represent a character, then the value is '01100110' and the character decoded is 'f' (see Table 2.3) On the other hand, if the first bit shown is the last bit of the previous character, then the code is '00110011' and the decoded character is '3'. So how do we determine whether 'f' or '3' is meant? The answer is by means of *character synchronisation*.

Character synchronisation (or *byte synchronisation*) can be achieved using either *asynchronous mode transmission* or *synchronous mode transmission*.

Asynchronous transmission

In *asynchronous data transfer* each data character (represented, say, by an 8-bit 'byte') is preceded by a few additional bits, which are sent to mark (or *delineate*) the start of the 8-bit string to the receiver. This assures that character synchronisation of the transmitting and receiving devices is maintained.

When a character (consisting of 8-bits) is ready to be sent, the transmitter precedes the 8-bit pattern with an extra *start bit* (value '0'), then it sends the 8-bits, and finally it suffixes the pattern with two 'stop bits', both set to '1'.[4] The total pattern appears as in Figure 2.19, where the user's eight bit pattern 00110011 is being sent.

In *asynchronous* transmission, the line is not usually in constant use and the spacing of characters need not be regular. The idle period between character patterns (the *quiescent period*) is filled by a string of 1's which serve to 'exercise' the line. The receiver can recognise the start of a new character by the presence of the *start bit* transition (from state '1' to state '0'). The following 8-bits then represent the character pattern and are followed by the two *stop bits* (Figure 2.19).

The advantage of asynchronous transmission lies in its simplicity. The start and stop bits sent between characters help to maintain synchronisation without requiring very accurate clock hardware in either the transmitter or the receiver. As a result, asynchronous devices

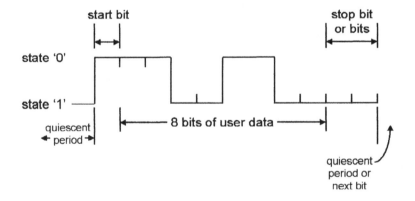

Figure 2.19   Asynchronous data transfer.

---

[4] Usually nowadays, only one stop bit is used. This reduces the overall number of bits which need to be sent to line to convey the same information by 9%.

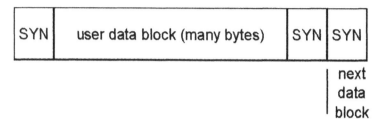

Figure 2.20 Synchronous data transfer.

can be made quite simply and cheaply. Asynchronous transmission is widely used between computer terminals and the computers themselves because of the simplicity and cheapness of terminal design. Given that human operators type at indeterminate speeds and sometimes leave long pauses between characters, asynchronous transmission is ideally suited to this use. The disadvantage of asynchronous transmission lies in its relatively inefficient use of the available bit speed. As we can see from Figure 2.19, out of 11 bits sent along the line, only 8 (i.e., 73%) represent useful information.

Synchronous transmission

In *synchronous mode transmission*, data characters (usually a fixed number of *bits*, or one or more *bytes*) are transmitted at a regular periodic rate.

Synchronous data transfer is the most commonly used mode in modern data communications. In synchronous data transfer, the data transmitted and received must be clocked at a steady rate. A highly accurate clock is used at both ends, and a separate circuit may be used to transmit the timing between the two. Provided all the data bit patterns are of an equal length, the start of each is known to follow immediately the previous character. The advantage of synchronous transmission is that much greater line efficiency is achieved (since no start and stop bits need be sent for each character). The disadvantage is that the complexity of the clocking hardware increases the cost as compared with asynchronous transmission equipment.

Byte synchronisation is established at the very beginning of the transmission or after a disturbance or line break using a special *synchronisation (SYN)* pattern, and only minor adjustments are needed thereafter. Usually an entire block of user information is sent between the synchronisation (SYN) patterns, as Figure 2.20 shows. The SYN byte shown in Figure 2.20 is a particular bit pattern, used to distinguish it from other user data.

2.13 Packet switching, protocols and statistical multiplexing

The need for packet switching

Until the 1970s, wide area networks (i.e., nationwide or international networks spanning long distances) were predominantly *circuit-switched* networks. Circuit switching is the technology of telephone communication in which a circuit of a fixed bandwidth is 'permanently' allocated to the communicants for the entire duration of a conversation or other communication. (In telephone networks the 'permanently' allocated circuit bandwidth is 3.1 kHz. In modern digital telephone networks (called *ISDN—integrated services digital network*) the bandwidth is 64 kbit/s). But while such networks can be used for the carriage of data, they are not ideally suited to data communication.

The main limitation of circuit-switched networks when used for data transport is their inability to provide variable bandwidth connections. When only a narrow bandwidth (or low bit-rate) is required compared with that of the standard circuit bandwidth, then the circuit is used inefficiently (under-utilised). Conversely, when short bursts of much higher bandwidth are required (for example, when a large computer file is to be sent from one computer to another or downloaded from the Internet), there may be considerable data transmission delays, since the circuit is unable to carry all the data quickly. A more efficient means of data conveyance, *packet switching*, emerged in the 1970s. *Packet switching* has become the basis of most modern data communications, including the Internet protocol (IP), X.25 'packet-switched' networks, *frame relay* and *local area networks (LANs)*.

Packets and packet formats

Packet switching is so-called because the user's overall message is broken up into a number of smaller *packets*, each of which is sent separately. Packets are carried from node to node across the network in much the same way in which packages make their way one post office to the next in a postal delivery network (recall Figure 1.1 of Chapter 1). To ensure that the data packets reach the correct destinations, each packet of data must be labelled with the address of its intended destination. In addition, a number of fields of 'control information' are added (like the stickers on a parcel 'for internal post office use'). For example, each packet of data can be protected against errors by adding a *frame check sequence (FCS)* of error check bits. A SYN byte in the packet header also serves for the purpose of synchronisation. Figure 2.21 illustrates the typical format of a data packet, showing not only the FCS and SYN fields, but also some of the other control fields which are added to the *user data* (or *payload*).

The *flag* delimits the packet from the previous packet and provides for synchronisation. In conjunction with the *packet length* field, it prepares the receiver for the receipt of the packet, enabling the receiver to determine when the frame check sequence (FCS) for detecting errors in the packet will start.

The *destination address* tells the network to which destination port the packet must be delivered. The *source address* identifies the originator of the packet. This information is important in order that the sender can be informed if the packet cannot be delivered. It

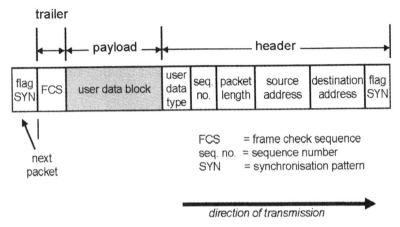

Figure 2.21   Typical data packet format.

allows the recipient easily to reply to any messages and in some cases is used to assure the packet preferential treatment or priority during its journey across the network (sender-oriented priority).

The *sequence number* enables longer messages which have been split up into multiple *packets* to be re-assembled by the receiver in the correct order. The *user data type* field lets the receiver know which type and format of information are held in the *payload* or user data field.

All of the 'control field' information added to the original user data is correctly called *protocol control information (PCI)*. The addition of PCI to the user data enables the sending and receiving end devices (as well as the network nodes along the way) to communicate with, and control, one another. The manner in which they do so is set out in the relevant *protocol*. Thus a protocol is a process by which nodes within a network can communicate with, and control the actions of, one another for the purpose of carriage of *user data*. The end user (a person or computer application) is unaware of the *protocol control information (PCI)*, except for the fact that the capacity of the network is reduced by the extra PCI data which must be carried. Because of this, the extra data (i.e., the PCI added over and above the user data) is usually referred to as the *overhead*.

Packet switching

Packet-switches are complicated devices with a large data storage capacity and powerful communications software, known by various names: *packet switch exchange (PSE)* for X.25 networks, *frame relay node, ATM (asynchronous transfer mode) switch, router* to name a few.

The principle of packet switching is illustrated in Figure 2.22. Each packet is routed across the network to its indicated destination address according to the most efficient path available at the time. In theory, this means that individual packets making up a single message, may take different paths through the network. (This is called *datagram* relaying.)

In contrast to *datagram* relaying, in *path-oriented* routing all the packets belonging to a given 'connection' take the same route. This greatly reduces the problem of sorting out

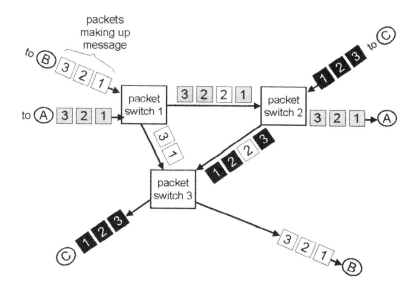

Figure 2.22 The principle of packet switching.

mis-sequenced packets (i.e., those out of order) on receipt, and provides for more consistent propagation delays. On failure or loss of a particular link in the path, an alternative path is usually selected without having to clear the 'call' and without losing packets. In practice, *path-oriented* routing is mostly used.

Thanks to the flexible routing made possible by both *datagram relaying* and *path-oriented forwarding*, a variable amount of bandwidth can be made available between the two end-points, and network link utilisation is optimised.

Packets are routed across the individual paths within the network according to the connection quality needs of the communicants, the prevailing network traffic conditions, the link error reliability, and the shortest path to the destination. The route chosen for any particular connection is controlled by the layer 3 (network layer) protocol (e.g., the *Internet protocol — IP*) and the *routing protocol*.

Packet switching gives good end-to-end reliability. With well designed switches and networks it is possible to bypass network failures (even during the progress of a 'call'). Packet switching is also efficient in its use of network links and resources — sharing them between a number of calls thereby increasing their utilisation.

Virtual circuits and logical channels

None of the physical connections between nodes (i.e., links) in a packet-switched network are dedicated to the carriage of any single message. Instead, the use of the links is shared. The individual packets of a single message are jumbled up with packets from other messages (see Figure 2.22). For the end-users, however, the effect is nonetheless as if a 'permanent' channel existed between the two ends (as shown in Figure 2.23).[5] Each 'channel' is known as a *logical channel, virtual channel* or *virtual circuit*.

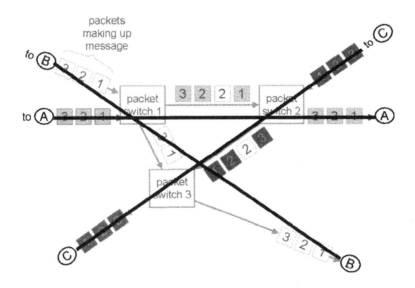

Figure 2.23   *Virtual circuits* created by packet switching.

---

[5] Although there are only three actual circuits between the packet exchanges in Figure 2.23, we could continue to add virtual circuits across this network without limitation. This is a key feature of packet switching: the number of virtual circuits can greatly exceed the number of physical circuits. The only risk is that should they all try to send data at once, then the network would snarl up in congestion.

Statistical multiplexing

Packet switching relies upon statistical multiplexing. Imagine that we *statistically multiplexed* two telephone conversations. Then we would suppress the silent periods in one conversation and insert words from other conversations in the resulting gaps. This principle is at the heart of packet switching and modern data communications. Separate packets (originating from different data users) are sent across a shared network, one after another.

The major benefit of statistical multiplexing is that the useful carrying capacity of the line is maximised by avoiding the unnecessary transmission of redundant information (i.e., pauses). But in addition, since the full speed of the line is available for each individual connection or carriage of data information, the transmission time (*propagation time*) may be reduced for short 'bursts' of communication. In stark contrast to *circuit-switched* networks, in which capacity shortfalls between two nodes in a network are alleviated by adding more circuits, it is generally better in packet-switched networks simply to upgrade the speed of the (single) line between each pair of nodes.

The technique of statistical multiplexing is illustrated in Figure 2.24.[6] Three separate users (represented by sources A, B and C) are communicating over the same transmission line, sharing the resources by means of statistical multiplexing. The three separate source circuits are fed into a statistical multiplexor (a packet-switch), which is connected by a single line to the demultiplexor (a second packet switch) at the receiving end.

The statistical multiplexor sends whatever it receives from any of the source channels directly onto the transmission line, or stores the signal (in a *buffer*) prior to sending, if the line is temporarily being used by one of the other data sources. Once the line is free, the *buffer* is emptied by transmitting its contents onto the line. And why is it called *statistical*

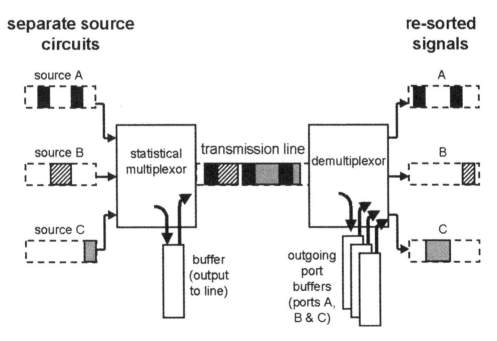

Figure 2.24 The principle of statistical multiplexing.

---

[6] For *duplex* communication, a similar arrangement, using a second line for the receive channel, but with multiplexor on the right and demultiplexor on the left, will also be necessary but this is not shown.

multiplexing? Simply because it relies on the statistical unlikelihood of all three sources wanting to send information simultaneously. The *buffer* serves to resolve the 'collisions' which would otherwise occur on the rare occasions during which more than one source is active at a time. The most important of the signals is sent first, while the less important signals are temporarily *queued* in the buffer.

Statistical multiplexing systems cannot cope with long periods of simultaneous transmission by two or more sources. During periods of prolonged simultaneous transmission, the *buffer* progressively fills up and starts to overflow (like a bucket overflowing).

Surprisingly, the loss of small amounts of information may not be catastrophic. The various protocols are designed to help prioritise the discarding of information from the 'bucket', should it become necessary. Subsequently, other protocols can detect the loss of information and arrange for *retransmission* of the information from the source.

In order to guard against the possibility of undue information loss, the system, line or network needs to be planned so that the sum of the *average* throughputs of each of the source channels is less than the maximum bit rate of the transmission line (whereby the *average rate* should be measured over a relatively short period of time). In other words, the line bit rate must exceed the sum (in bits/second) of the average source output rates (A + B + C). In practice, the line bit rate should be at least 1.5 to 2 times the sum of the average source output rates, otherwise frequent periods of congestion will be experienced.

Congestion in a circuit-switched network (like the telephone network) manifests itself in the rejection of new calls. New callers do not 'get through' but instead receive a 'network busy' tone, and are advised to 'try again later'. By contrast, congestion in a statistically multiplexed (i.e., packet-switched) data network manifests itself as propagation delay. The delays are caused by the mounting number of packets being stored in the sending *buffers*. The delays affect all the connected users. Users establishing new 'connections' are not rejected, the service simply gets slower for everyone else.

Locating the congested points in a network and understanding the causes of delay are some of the hardest challenges faced by data network operators! The control of network delay and the prudent adding of capacity to a data network to overcome congestion are highly skilled tasks requiring considerable experience![7]

2.14   Symmetrical and asymmetrical communication: full duplex and all that!

The terms *simplex, half duplex* and *full duplex* as well as the terms *symmetric, asymmetric, upstream, downstream, broadcast, multicast* all describe different modes of communication and the capabilities of networking devices. It is important to know what all the terms mean, so here goes:

Simplex

Simplex communication is one-way communication, allowing a single listener to hear a single speaker. A typical simplex system has a single transmitter, a single transmission *medium* and a single receiver (Figure 2.25a).

Half duplex

Half duplex communication makes two-way communication possible, but of the two communicating parties, only one may 'speak' at a time. Not even 'butting in' is possible. A typical

---

[7] See Chapter 14.

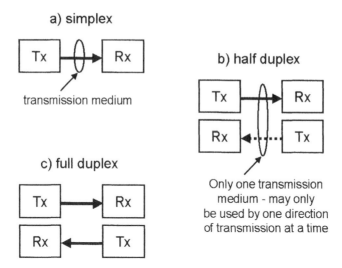

Figure 2.25   Simplex, half duplex and full duplex communication.

half duplex system comprises two transmitters and two receivers (one transmitter and one receiver at either end of the link) but only one transmission medium (Figure 2.25b). A good example of a half duplex system is a 'push to use' walkie-talkie radio. The speaker 'takes the line' by pressing the button and says 'over' when he or she is finished. In the early days of modem development, it was difficult to separate 'transmit' from 'receive' signals on a single pair of wires (2-wire connection), so that many early modems were either *half duplex* for use on 2-wire connections, or *full duplex*, for use on 4-wire connections (one pair each for 'transmit' and 'receive' directions).

There are rarely two large files being communicated in opposite directions between a particular pair of computers across a data network, so that most data communication is largely 'half duplex' in character. Nonetheless, it is inconvenient if no communication is available in the reverse direction, since the reverse channel provides a useful means of 'butting in' and of data *flow control*. Most modern transmission systems are therefore *full duplex* systems.

Full duplex

Full duplex communication systems allow 2-way communication without restriction. Either or both parties may speak at any time, 'butt in' or speak at the same time. Most communication line systems are nowadays full duplex systems, using either two separate transmission media (e.g., a 4-wire communication line as in Figure 2.25c) or a single medium (such as a 2-wire communication line), but in this case a 'shared medium' technique is used to keep 'transmit' and 'receive' directions of transmission strictly apart.

Symmetric and asymmetric communication: upstream and downstream

While the communication link itself may allow full duplex communication, the actual communication taking place may require unequal bit rates and volumes of data to be moved in the two different directions. When the volume of data transported in both directions is approximately equal, then we talk of *symmetric* communication. When greatly different volumes are transported in the two directions, we speak of *asymmetric* communication.

Many Internet users receive far more data from the network (e.g., webpages, software downloads, etc.) than they send (typically only an occasional email, or the keyboard commands requesting the downloads). As a result, a number of *asymmetric* connection line technologies have appeared in recent years, which although full duplex (since communication is possible in both directions at the same time), have different bit rate capacities in the two directions of communication. A typical *ADSL (asymmetric digital subscriber line)* connection used for Internet access offers 768 kbit/s in the *downstream* direction (i.e., from network to Internet end-user) while only offering a maximum of 128 kbit/s in the *upstream* direction (from the end-user sending into the network).

Broadcasting and multicasting

Much of our discussion so far has been centred upon achieving data communication on a *point-to-point* basis between two fixed end-points, and many transmission media and data networks are optimised for such communication. Having said this, there are some applications which require that the same information is sent to multiple receivers at the same time (so-called *broadcast* applications). Some transmission media (e.g., radio) are ideally suited to such broadcasting, but even 'terrestrial line' networks can be made to appear like broadcast networks by distributing the same message progressively throughout all the 'branches' of the network. This is called *multicasting*. It is common to use such broadcasting techniques, for example, for 'advertising' network status, topology or other changes to all the devices within a network.

2.15   Serial and parallel communication

All communication links and interfaces are designed either for *parallel data transmission* or for *serial data transmission*. The latter is the 'norm' in *wide area networks (WANs)* — networks spanning long distances, whole nations or even the globe. The two different methods of transmission differ in the way each 8-bit data pattern (i.e., each character) is conveyed.

Internally, computers operate using the parallel transmission method, employing eight *parallel* circuits to carry one bit of information each. Thus during one time interval all 8 bits of the data pattern are conveyed. The advantage of this method is the increased computer processing speed which is made possible. The disadvantage is that it requires eight circuits instead of one. Parallel data transmission is illustrated in Figure 2.26a, where the pattern '10101110' is being conveyed over the 8 parallel wires of a computer's *data bus*.[8] Examples of parallel interfaces are those specified by ITU-T recommendations V.19 and V.20.

Serial transmission requires only one transmission circuit, and so is far more cost effective for data transmisssion on long links between computers. The data on the computer's parallel *bus* is converted into a serial format simply by 'reading' each line of the bus in turn. The same pattern '10101110' is being transmitted in a serial manner in Figure 2.26b. Note that the Baud rate (as we discussed earlier) needed for serial transmission is much higher than for the equivalent parallel transmission interface.

2.16   The problem of long lines — the need to observe the maximum line length

During the course of this chapter we have covered the fundamental manner in which data can be coded in binary code in *bits* and then transported across a telecommunications line. We

---

[8] Nowadays 32-bit data *buses* are commonplace and even more bits lead to even higher processing speeds.

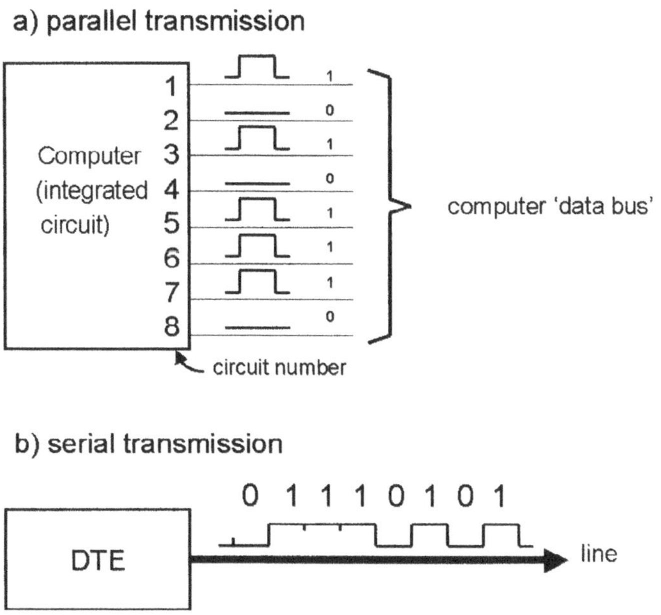
a) parallel transmission
Computer (integrated circuit)
1
2
3
4
5
6
7
8
1
0
1
0
1
1
1
0
computer 'data bus'
circuit number
b) serial transmission
DTE
0 1 1 1 0 1 0 1
line

Figure 2.26 Parallel and serial transmission.

have also how *protocols* are necessary to set out an etiquette of communication. But before we go on to discuss how the protocols themselves work, it is worth considering the 'physical' constraints which protocol developers are up against: what they must consider and incorporate in their protocol designs. In particular, it is important to recognise the influence of the line length. The best way of conducting the communication, and thus the best type of protocol to use, depends greatly upon the line length.

For many data networks and protocols a maximum line length, bit rate and other parameters are specified (or is not specified, but is nonetheless implicit in the design of the protocol!). It is important to be aware of these limitations when using the particular type of interface, line or protocol. Here's why.

Attenuation and distortion

One of the problems of long lines is the degradation caused by attenuation and distortion, and this largely determines the maximum allowed line and link lengths possible in a given network and using a given protocol. But in addition to attenuation and distortion, there are some far less obvious problems which also limit line length. These are caused by the 'storage medium' and 'transient' effects of a long line, as we shall see next.

The effect of propagation delays on how protocols function

The electrical pulses which represent the individual *bits* in a data signal typically travel across a network at a speed of around 10 million metres per second (10^8 m/s). This is around one third of the speed of light. The delay (as compared to the speed of light) is caused by the various buffers and electronic components along the way. If we assume quite a modest *bit rate* of 2 Mbit/s (2×10^6 bit/s) is used to convey a signal along a given transmission line, we can actually work out the length of each pulse!

$$\text{Bit length} = [\text{speed in m/s}]/[\text{bit rate}] = 10^8/(2 \times 10^6) = 50 \text{ m}$$

The first time I contemplated the length of a *bit* in metres, I was surprised by the result. After all, one tends to assume when considering electrical circuits that the transmission is instant: 'the light goes on as I flick the switch!' But this is far from true when we are dealing with high speed communication lines! If our 2 Mbit/s line above is 1000 km long (quite possible within a nationwide network, never mind an international connection!), then there are actually 20 000 bits (or 2500 *bytes*) in transit on the line at any one time [1000 000 m/50 m]. Say 'stop' and you'll still receive at least all the 20 000 bits already on their way! Actually many more: because the 'stop' message will take some time to get to the receiver (this message is itself in a queue of at least 20 000 bits heading in the other direction). 20 000 bits will be received before the 'stop' message is acted upon by the transmitter, by which time a further 20 000 bits are already on their way. So after saying 'stop' be prepared to receive a further 40 000 bits (5000 bytes).

For the 5000 bytes you need a *buffer* unless of course this data is simply thrown away! But throw the data away at your peril! For if you throw it all away, you'll have to ask for it to be sent again, once you're ready to receive it. That will mean another 5000 bytes (of 'idle line information') for the dustbin — wasted while the message requesting the retransmission gets through to the other end. Plus, of course, the 5000 bytes will have to be sent again. That's 10 000 bytes of information we didn't need to send over the line! We'll have to be careful about all this waste, if the maximum capacity of the line is to be used efficiently!

A similar flow control problem to that discussed above is illustrated as a signal flow diagram in Figure 2.27. Protocol interchanges are frequently illustrated in this manner, so it is worth becoming familiar with how to read the chart. The left-hand vertical line represents the 'A-end' of a point-to-point connection, the right hand vertical line the 'B-end'. The top of each line represents the beginning of the illustrated time period, and the 'time axis' progresses towards the bottom of the page. The diagonal lines represent individual bits sent from one end of the line to the other. (The bits progress along the line from one side of the diagram to the other, while time elapses down the page — hence the diagonal line.)

At the beginning of the time period shown, a 'request' signal is generated at the A-end of the line, to request data to be returned by the B-end. Since the request message is more than one bit long, a certain amount of time expires (shown as T_R on the diagram) to transmit all of the message onto the line. Once on the line, the message propagates along the line towards the B-end (represented by the diagonal lines for the 'first' and 'last' bits of the request message). After a period T_P (called the *propagation time*), the first bit of the request reaches the B-end. The final bit of the request arrives a short time, T_R, later. Now the message can be interpreted by the B-end computer. We assume in Figure 2.27 that it reacts instantaneously and starts to send the first packet of the requested data. (Normally a further processing delay would be encountered here.)

Further diagonal lines represent the first and last bits of the returned data packet as they make their way back to the A-end. After a total elapsed time T_C the first packet of data has completely arrived at A.

The protocol shown in Figure 2.27 requires that the A-end *acknowledges* receipt of the first packet of data before the B-end is allowed to send the second packet. The acknowledgement of the first packet and the transmission of the second packet is shown in the diagram.

Now let us consider the relative efficiency of our usage of the line. The total duration of a 'cycle' of sending a request (or acknowledgement) and receiving a packet of data is T_C, where:

$$T_C = (T_P + T_R) + (T_P + T_D) = 2 \ T_P + T_R + \ T_D$$

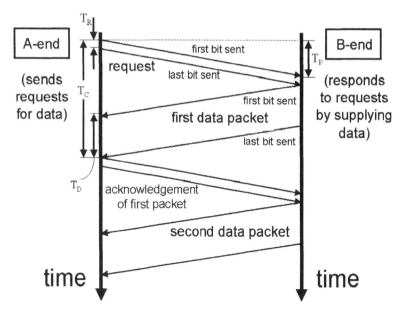

Figure 2.27   A protocol signal flow diagram.

The proportion of the time for which the two communication paths (A-to-B and B-to-A) are in use reflects the efficiency of usage of each:

$$\text{Usage efficiency (A-to-B)} = T_R/T_C = T_R/(2\ T_P + T_R + T_D)$$

$$\text{Usage efficiency (B-to-A)} = T_D/T_C = T_D/(2\ T_P + T_R + T_D)$$

Let us assume that we keep our request message very short; we assume it can be transmitted onto the line instantly (e.g., because of a very high line bit rate). Then we set $T_R = 0$. This is usually a reasonable assumption. Now, when the line is very short we can also ignore the time required for an individual bit to propagate along the line, by setting $T_P = 0$. In this case, we discover that:

For very short lines:

$$\text{Usage efficiency (A-to-B: request direction)} = 0\%$$

$$\text{Usage efficiency (B-to-A: data direction)} = 100\%$$

No surprises here perhaps. Data is being downloaded nearly all the time (100% line usage B-to-A), and our acknowledgements are a help in making sure that the A-end is always ready to receive the next packet. All hunky dory! But look at what happens when the propagation time T_P becomes significant. In particular, let us return to our previous example of a 1000 km line of 2 Mbit/s bit rate, and assume *packet lengths* of 256 bytes (2048 bits). In this case $T_P = 10^6$ m/$[10^8$ m/s$] = 0.01$ s and $T_D = 2048$ bits/2 000 000 bits/s $= 0.001$ s. And what are the line usage efficiencies?

Example long line of 1000 km and 256 byte packet size:

$$\text{Usage efficiency (A-to-B: request direction)} = 0\%$$

$$\text{Usage efficiency (B-to-A: data direction)} = 4.8\%$$

What? 4.8% line usage efficiency! A shock perhaps? Let it be a warning to all communications programmers and protocol designers! *Acknowledgements* used with short packet sizes can lead to very low line utilisation on long, high bit rate lines!

The effect of delays on 'collision' protocols

In some *shared medium* networks (e.g., radio networks and ethernet LAN networks), users are allowed to transmit signals onto the *medium* provided they first check that the medium is free (i.e., idle). In such networks, the protocol has to be able to deal with collisions of packets sent by different sources.

Because of the propagation delay incurred by the signal sent by a first transmitter, a second remote device may start transmitting on the already 'busy' line, unaware that the transmission of the first device has not yet reached it. The danger is in the length of the line! The longer the line, or the greater the size of the *shared medium* network (e.g., LAN — local area network), so the greater the chance of a collision. The more collisions that occur, the less efficient is the use of the medium.

If the network is too big, the bit rate is too high (making the pulses too short) or the line is too long for the protocol in use, then the efficiency of the network may drop dramatically!

MORAL: Know the limitations of the network and protocols in use.

3

Basic Data Networks
and Protocols

Modern data communications networks are built up in a 'modular' fashion — using standardised network components, interfaces and protocols based on digital line transmission, packet switching and layered communications protocols. In this chapter we present the basic components of a data network, and explain in detail the 'networking' or lower-layer protocols (protocol layers 1–3) which make them work. We shall explain physical and electrical interfaces and connectors, as well as physical, datalink, network, transport and higher-layer protocols: everything that goes to ensure efficient propagation across a network. Complex though it may seem, these are only the network basics!

3.1 The basic components of a data network

The basic components of a data network are illustrated in Figure 3.1. A personal computer (PC) and a mainframe computer — both examples of *DTE (data terminal equipment)* — are connected to the network (shown as a 'cloud') by means of DCE *(data circuit-terminating equipment)*. The DCE is a device which provides the 'starting point of the long-distance network' and is a device installed either near to the DTE, or sometimes is incorporated into it. The network of Figure 3.1 (providing the connection between the DCEs) comprises a meshed network of 4 *nodes*.[1]

The devices (DTEs, DCEs and nodes) in Figure 3.1 are interconnected by means of standardised communications interfaces, and every possible physical, electrical and protocol aspect of the interconnection is precisely defined.

There are different classes of general interfaces describing different types of connections, as we learned in Chapter 1. *User-network interfaces (UNIs)* are used to connect end devices (DTEs) to the network and are generally 'asymmetric' — the balance of control of the interface being with the network. *Network-node interfaces (NNIs)*, meanwhile, are more 'symmetric' arrangements intended to be used to interconnect nodes of equal importance in the main *backbone* part of the network.

Both UNI and NNI specifications define in precise detail the physical and electrical aspects of the interconnection, as well as the *lower-layer* or *networking protocols* (layers 1–3 of

---

[1] A *node* is any kind of switch-type device, upon which numerous communications links and other connections converge. A node might technically be a *switch*, a *router*, a *LAN hub*, a *multiplexor* or some other kind of exchange. In this example, as often, the exact technical nature of the node does not concern us.

Data Networks, IP and the Internet: Protocols, Design and Operation   Martin P. Clark
© 2003 John Wiley & Sons, Ltd   ISBN: 0-470-84856-1

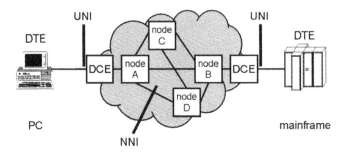

Figure 3.1   The components and interfaces making up a simple data network.

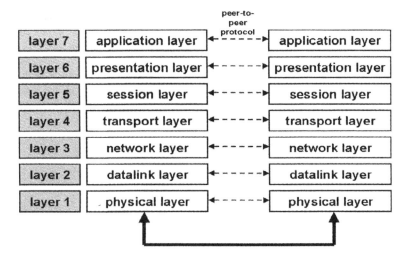

Figure 3.2   The protocol layers of the OSI (Open Systems Interconnection) model.

Figure 3.2) which must be used at the interfaces. It is the detailed functions of these layers (layer 1 — *physical layer*; layer 2 — *datalink layer* and layer 3 — *network layer*) which will be the main subject of discussion in this chapter. We start with an overview of the various layer roles, before we go on later to review the details.

Physical layer (layer 1)

The physical layer (layer 1) interface and protocol specification define how the physical *medium* (e.g. the copper cable, optical fibre or radio channel) is to be used to transport the *bitstream* which comprises a digital signal: the precise shape and electrical or optical characteristics of the signal, the bit rate and clocking, as well as the physical connectors to be used.

Different physical layer specifications correspond to different physical *media* and *modulation* or *line coding* schemes. Thus one physical layer specification might cover the use of a *twisted-pair* copper cable to carry a signal of a given bit rate, while another physical layer specification may define the use of light pulses of a given laser light wavelength — to be sent over an optical fibre medium.

While one physical layer (i.e., the medium and its related modulation, control and line coding — the *protocol*) may be used as an alternative for another, media and protocols within the physical layer are not interchangeable. Thus, for example, it may be possible to use <u>either</u> a twisted-pair line <u>or</u> a fibre-optic-based communications system to carry a given signal, but it is ludicrous to consider using an optical transmitter and detector (intended for use with a fibre cable) in conjunction with a copper cable!

The *physical layer* interface usually provides for a simple point-to-point connection or *link* between two end-points. Alternatively, the physical layer of a *shared medium* (e.g., radio) provides for the 'broadcast' of a signal from any of a number of devices sharing the communications *medium* to all of the others. In the case of a *shared medium*, all but the intended recipient(s) will ignore the transmission. The most common shared medium systems used for data communications are LANs (local area networks) and wireless data networks.

Physical layer interfaces specifically designed for long-distance lines or NNIs (network-node interfaces) are usually designed for long-distance, permanently established point-to-point links. Such *line interfaces* use a minimum number of cable leads for the connection.

Each link of an end-to-end connection across a network may use a different physical medium and thus a different physical layer. Thus, for example, the path across the network from DTE-to-DTE in Figure 3.1 (a minimum of five links) might traverse different fibre, copper and radio communications media along the way. Each node will perform the necessary physical layer adaptations and signal conversions to achieve this.

In the case of physical layer interfaces intended for use as UNIs (user-network interfaces), the role of a special 'network terminating' or 'adaption' device is defined: the *DCE — data-circuit terminating equipment*. As we shall learn, the DCE undertakes a physical layer conversion of the short-range communications port capabilities of the DTE into a *line interface* format suitable for long-distance communication. The role of the DCE is to do the following:

- convert the physical interface emanating from the *DTE (data terminal equipment)* into a *line interface* format suitable for long-distance transmission, and provide for digital/analogue signal conversion if necessary;

- provide for network *termination*, being a source of power for the line and network as necessary;

- <u>forward</u> data received from the DTE to the network;

- <u>deliver</u> data received from the network to the DTE;

- *clock* and *bit-synchronise*[2] the data transmission of the DTE during the *data transfer* phase;

- set up the physical connection forming the medium and clear it as required and/or requested by the DTE (This may be necessary where the physical *link* is actually a dial-up connection across a telephone network, or a temporary radio path).

Datalink layer (layer 2)

The *datalink layer* and *datalink protocol* are concerned with the formatting of the 'raw' stream of bits carried across a link by the physical layer into a byte-synchronised and structured form recognisable as *blocks* or *frames* of data. Apart from this, the *datalink layer* is also concerned with *data flow control*. The sending and receiving devices at either end of the physical link need to be coordinated to make sure that they 'listen' when 'spoken to' and only 'speak' when it is 'their turn'. A *frame check sequence (FCS)* may be applied by the *datalink* protocol in

---

[2] See Chapter 2.

the case where the physical medium is likely to be prone to a high level of bit errors. We discussed all this in Chapter 2.

Together, a datalink layer protocol, a physical medium and a physical layer protocol are all that is needed for point-to-point transport of *blocks* or frames of data. A network layer (layer 3) protocol only becomes necessary when a *network* (i.e., a meshed topology of numerous links) needs to be traversed.

While various different physical layer media and protocols may be used with any given datalink layer, not all datalink protocols are suitable for all physical media and physical protocols. MORAL: you can mix and match to some degree — but not all combinations of physical layer (layer 1) and datalink layer (layer 2) protocols are viable!

Network layer (layer 3)

The network layer (layer 3) protocol is used to combine a number of separate individual *datalinks* into a 'chain' — thereby forming a complete path across a network comprising multiple nodes (as, for example, in Figure 3.1).

The *network protocol* is concerned with *packets* and *packet switching*. As we learned in chapters 1 and 2, *packets* of data are analogous to the parcels and letters of a postal service. The frames of data carried by the datalink layer (layer 2), meanwhile, are analogous to the postal delivery vans which run from one postal depot or sorting office to the next. In just the same way in which postal delivery vans are loaded before a 'run' from one depot to the next, and then completely emptied and 're-sorted' before packing into another van for the next 'leg' of the journey, so the frames of the datalink layer are filled and emptied of their packet contents before and after each datalink of the path through the network. The frames of the datalink protocol only carry the data for a single leg of its journey. The packets of the network protocol, on the other hand, are transferred from one end of the complete path to the other.

Network layer protocols have a number of responsibilities to fulfil:

- identification of destination *network address*;

- *routing* or *switching* of packets through the individual network nodes and links to the destination;

- *statistical multiplexing* of the data supplied by different users for carriage across the network;

- *end-to-end data flow control*: the flow control conducted by layer 2 protocols only ensure that the receiving data buffers on each individual link can receive more data. The layer 3 protocol has the more onerous task of trying to ensure a smooth flow of data across all the datalinks or subnetworks taken together. Uneven 'hold-ups' along the route need to be avoided;

- correct *re-sequencing* of packets (should they get out of order having travelled different routes on their way to the destination);

- error correction or transmission re-request on a network end-to-end basis.

Now, let's get down to details.

3.2   Layer 1 — physical layer interface: DTE/DCE, line interfaces and protocols

The physical layer interface defines the manner in which the particular *medium* (e.g., twisted pair cable, coaxial cable, fibre line or radio link) should be coded to carry the basic digital information. At a minimum, the physical interface specification needs to define the precise nature of the medium (e.g., wire grade, impedance etc); the exact electrical (or equivalent) signals which are to be used on the line and the details of the *line code* which shall be used for *bit synchronisation* as discussed in Chapter 2. Many modern physical interface specifications also stipulate the precise mechanical connectors (i.e., plugs and sockets) which should be used for the interface, but this is not always defined. As well as the basic electrical (radio or optical) interface, the physical layer specification defines control procedures (the *physical layer protocol*) which allows one or both of the devices at either end of the line (i.e., the physical *medium*) to control the line itself.

There are three main types of physical interface to be distinguished from one another:

- DTE-to-DCE interfaces. These are the asymmetric point-to-point *user-network interfaces (UNIs)* used typically to connect end devices (e.g., computers) to *modems, line terminating units (LTU), channel service units (CSU), data service units (DSUs), network terminating units (NT or NTU)*. All the latter are examples of *data circuit terminating equipment (DCE)*;

- Line interfaces or *trunk* interfaces (symmetrical point-to-point *line interfaces*)—most NNIs and some UNIs are of this type;

- *Shared medium* interfaces (point-to-multipoint interfaces)—usually used as UNIs. [The most commonly used *shared media* are *local area networks (LANs)*. LANS are widely used to connect end-users PCs to internal office data networks].

In general terms, the UNI (user-network interface) always employs a DTE-to-DCE type interface, while trunk interfaces tend to use symmetrical, higher bit rate NNI (network-node interface).

Figure 3.3 shows a network in which three routers and two DTEs are interconnected. Both DTEs are connected to the network by means of DTE/DCE interfaces. Routers C and B are also connected to the line which interconnects them by means of DTE/DCE (e.g., V.24 or RS-232) interfaces. Routers A & B and A & C, meanwhile, are interconnected by means of direct trunk interfaces (DCE/DCE[3]). In these cases, the DCE function is included within the router itself.

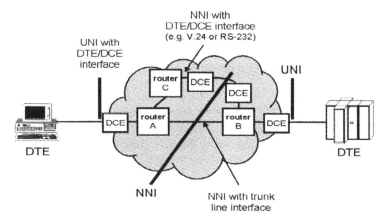

Figure 3.3   DTE/DCE Interface and Trunk or line interface.

---

[3] The 'DCE/DCE interface': tempting as it may be, it is not correct to call the long-distance connection between DCEs a 'DCE/DCE interface'. Instead, one should refer to the *line interface* or *network interface*. The term 'DCE/DCE interface' is reserved for the case in which the UNI interface (normally used to connect a DTE to a DCE) is used to connect a second DCE instead. A cross-cable is used for this purpose, as we discover in Figure 3.8b.

DTE-to-DCE interfaces

DCEs (data circuit-terminating equipment) form the 'end-point' of a network or long-distance telecommunications line, allowing point-to-point communication between remote computers and terminals (which are generically called *DTE* or *data terminal equipment*).

The first DCEs were modems and were designed to be used at either end of a *datalink* created by means of a dial-up telephone connection. The modem (like all other types of DCE) had to provide an interface for connecting to the *serial* data communications port of the computer, but also be capable of *setting up, receiving, clearing* and controlling telephone connections.

As we have already discussed, the basic functions of all types of DCE are to:

- convert the physical interface emanating from the DTE (data terminal equipment) into a *line interface* format suitable for long-distance transmission, and provide for digital/analogue signal conversion if necessary;

- provide for network *termination* of the long-distance line, being a source of power for the line and network as necessary;

- <u>forward</u> data received from the DTE to the network;

- <u>deliver</u> data received from the network to the DTE;

- clock and bit-synchronise[4] the data transmission of the DTE during the *data transfer* phase;

- set up the physical connection forming the medium and clear it as required and/or requested by the DTE. This may be necessary where the physical *link* is actually a dial-up connection across a telephone network, or a temporary radio path.

Some DCEs are controlled by the associated DTE, others act autonomously on behalf of the DTE. The user's data and the control signals (*functions*) are conveyed from DTE to DCE or DCE to DTE by means of dedicated control leads defined as part of the DTE–DCE interface. Detailed *procedures* define how the functions are used and how the states of the DTE and DCE can be changed from *idle*, through *ready* to the *data transfer* phase and afterwards arrange for *clearing* of the connection.

Typical physical layer interface specifications for a DTE-to-DCE interface comprise up to four different components, including a definition of:

- the physical connector;

- the electrical interface;

- the controls and functions for establishing the link: changing from one state (e.g., idle) to another (e.g., ready or data transfer) or vice versa;

- the procedures (i.e., sequences of commands) which define the use of the controls. (The controls and procedures form together form the physical layer protocol.)

The most commonly used DTE/DCE interfaces are illustrated in Figure 3.4. There are two main categories of DTE/DCE interfaces: X.21-type interfaces (for digital lines) and X.21$_{bis}$-type interfaces (used in *modems* for analogue lines). The other V-series and X-series recommendations listed in Figure 3.4 define individual aspects of specific interfaces.

---

[4] See Chapter 2.

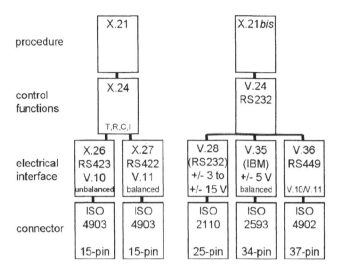

Note: You may be wondering why all the specifications and standards defining DTE/DCE interfaces have such different designations. This is because different naming standards are used by the different standards-publishing organisations. ISO (International Organization for Standardization) uses a simple numbering scheme (but I haven't yet worked out the logic behind the individual numbers). ITU-T (International Telecommunications Union—Standardization sector) issues *recommendations* in various series. The X-series defines 'interfaces and procedures intended for use in general data communications'. The V-series defines' data communications over the telephone network' (e.g. modems). Sometimes the same recommendation is issued with a recommendation number in both series (e.g.V.10/X.26).The nomenclature RS, mean while, stands for *recommended standard*. This designation is used by the United States EIA/TIA (Electronic Industries Alliance or Association/ Telecommunications Industries Association).

Figure 3.4 Standards, specifications and ITU-T recommendations defining DTE/DCE interfaces.

DCEs intended for use between DTEs and analogue *wide area network (WAN)* lines are called *modems*. Such modems conform at their DTE/DCE interface with ITU-T recommendations X.21$_{bis}$ and V.24/V.28. X.21$_{bis}$ sets out the entire framework of the DTE-to-DCE interface used by the modem. V.24 (and RS-232 as well) set out the signals and circuits (together with their names and numbers) used at the interface. But which signal is sent on exactly which wire and via which pin of the connector is defined by either V.28 (RS-232), V.35 or V.36 accordingly.

There are five DTE-to-DCE interfaces in common usage. These (as shown in Figure 3.4) are:

- V.24/V.28 (25-pin DB-25 plug) is the most common interface used between computers and analogue modems. In Europe this interface is simply referred to as 'V.24' and in North America as RS-232.

- V.36 (37-pin plug), usually referred to as RS-449) is the most commonly used interface in North America, the UK and France for high bit rate DCEs and those used to connect digital lines. Very confusingly, many people refer to 'V.35' even though they really mean 'V.36'.

- V.35 is a similar interface to V.36 but with a different connector. Its usage is restricted to certain types of IBM computers and networking equipment.

- X.21 (V.11) is the main interface used in Germany, Austria, Switzerland for interfacing digital line DCEs and high bit rate lines to DTEs. It is often referred to simply as 'X.21' without specifying V.11.

- X.21 (V.10) is the version of the X.21 interface designed for use in conjunction with coaxial cables. It is also simply referred to as 'X.21' without specifying V.10.

Unfortunately, newcomers to DTE/DCE standards can easily be confused by the various specifications, since lazy experts often do not define in full the interfaces they wish to refer to. It is commonplace, for example, to refer only to 'V.24' when in fact the complete V.24/V.28 interface is meant. The V.24/V.28 (or RS-232) interface uses the 25-pin DB-25 connector (ISO 2110) commonly seen on older modems. Similarly, when referring to DCEs used on digital line circuits, people often speak of 'X.21' without saying whether the interface is V.10 (for *unbalanced circuits*, i.e., *coaxial* cables) or V.11 (for *balanced circuits*, i.e., *twisted pair cable*).

Network synchronisation

One of the major functions of the DCE (data circuit terminating equipment) is to ensure that *the* DTE (data terminal equipment) transmits its data in a manner bit-synchronised with the network (i.e., at precisely the right bit rate and at the correct interval in time).

In a public digital transmission network, the network operator uses an extremely accurate *master clock* (typically a caesium clock or the extremely accurate clock signal of the satellite *global positioning system*) to synchronise his or her entire network. This ensures that *slip, jitter, wander* and other undesirable effects (as discussed in chapter 2) do not affect signals as they move from one node to the next through the network.

The clock signal is sent to all devices within the network, by means of a hierarchical *synchronisation plan* (Figure 3.5). Each node in the network is configured to receive a *primary clock signal* and a *secondary* (or *back-up*) *clock signal*. When the *primary source* fails, the node reverts to the secondary. The clock is propagated as far as the DCEs, and from each DCE is passed on to the corresponding DTE, thus ensuring that all the DTEs maintain an accurate and network-compatible transmitting bit rate.

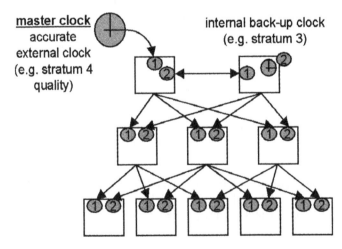

Figure 3.5   Hierarchical network synchronisation plan with nominated primary and secondary sources DCEs and DTE/DCE Interfaces intended for use with analogue lines (Modems).

In the case of analogue transmission lines, the DCE cannot rely on the network to provide accurate clocking information, since the bit rate of the signal received from the remote end is not accurately regulated by the *PTO's* (*public telecommunications organisation* or *operator*) analogue network. For this reason, an internal clock is needed within the DCE in order to maintain an accurate transmitting bit rate.

DTE/DCE control signals

DTE/DCE interfaces were originally multi-lead interfaces, with separate wires or *circuits* dedicated for each control action. The control signals defined by ITU-T recommendation V.24 and EIA RS-232 are the most widely used. These are listed in Table 3.1.

DTE/DCE electrical interface

As an example of a standardised physical layer electrical interface, Figure 3.6 illustrates the 'negative logic' voltages defined by RS-232 and V.28 for use on each of the DTE/DCE circuits defined in Table 3.1. For a *mark* (binary value '1'), the transmitter should output a voltage between -5 V and -15 V. The receiver, meanwhile, interprets any received voltage between -3 V and -15 V as a mark. Similarly, when transmitting a *space* (binary value '0'), the

Table 3.1 DTE/DCE control signals defined by EIA RS-232/ITU-T recommendation V.24

Signal code	Signal meaning	DB-25 Connector pin number (ISO 2110)	Db-9 pin number (EIA 562)	ITU-T Rec. V.24 signal name	EIA signal name	Signal meaning for binary value '1' or 'mark'
CTS	Clear-to-send	5	8	106	CB	From DCE to DTE.. 'I am ready when you are'
DCD	data carrier detect	8	1	109	CF	From DCE to DTE.. 'I am receiving your carrier signal'
DSR	data set ready	6 (not always used)	6	107	CC	From DCE (the 'data set' to DTE).. 'I am ready to send data'
DTR	data terminal ready	20	4	108.2	CD	From DTE to DCE.. 'I am ready to communicate'
GND	Ground	7	5	102	AB	Signal ground—voltage reference value
RC	Receiver clock	17 (little used)	—	115	DD	Receive clock signal (from DCE to DTE)
RI	Ring indicator	22	9	125	CE	From DCE to DTE.. 'I have an incoming call for you'
RTS	Request-to-send	4	7	105	CA	From DTE to DCE.. 'please send my data'
RxD	Receive data	3	2	104	BB	DTE receives data from DCE on this pin
TC	Transmitter clock	15 (little used)	—	114	DB	Transmit clock signal (From DCE to DTE)
TxD	Transmit data	2	3	103	BA	DTE transmits data to DCE on this pin

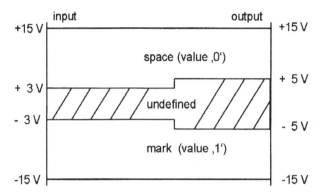

Figure 3.6   Defined output (send) and input (detect) voltages of RS-232 and V.28 circuits.

output voltage shall be between +5 V and +15 V, and any received voltage between +3 V and +15 V shall be interpreted as a *space*. This is a *non-return-to-zero /NRZ)*[5] code. The wider voltage span used by the receiver rules out the possibility of marks or spaces being 'lost' on the line due to signal attenuation. Note that the maximum cabling length of a V.28/RS-232-based interface is around 15 to 25 m.

DTE/DCE connectors and cables

The familiar 25-pin *DB-25* connector (ISO 2110) used on modems is illustrated in Figure 3.7a. The DTE normally has a *male* connector (i.e., pins exposed in the socket) while the DCE has a *female* connector (socket with holes to receive pins). The DTE-to-DCE connection cable comprises a cable with a *female plug* at one end (for the DTE) and a *male plug* at the other

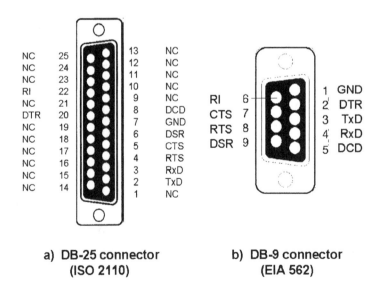

a) DB-25 connector   **b) DB-9 connector**
(ISO 2110)   **(EIA 562)**

Figure 3.7   Connectors used for V.24/V.28 or RS-232 interface.

---

[5] See Chapter 2.

(for the DCE). This ensures, in particular, that the *TxD (transmit data) and RxD (receive data)* leads are correctly wired. (Recall from Chapter 2: the DTE transmits data on 'TxD' for the DCE to receive. The DTE receives data from the DCE on lead 'RxD'.) The cable simply connects pin 1 at the DTE to pin 1 at the DCE, pin 2 to pin 2, etc. (Figure 3.8a).

Some personal computers still present a 25-pin male connector as their means of connecting to an external modem or similar device, but more common nowadays is the 9-pin *DB-9* connector (EIA 562) as the *serial interface* port. The nine most critical leads of the V.24-defined interface can simply be transposed into this smaller connector (Figure 3.7b) and therefore save valuable space on the back of the *laptop* computer.

As time has moved on, computer users have not only wanted to have smaller serial port connectors on their PCs, they have also been innovative in finding ways of interconnecting their computers directly over the serial port. You can even directly connect two DTEs together by using a special *null modem* cable (Figure 3.8b). This is designed to 'trick' each of the computers into thinking it is communicating with a DCE. In a null modem cable, the TxD and RxD leads are 'crossed' since both devices expect to send on TxD and receive on RxD. MORAL: be very careful with computer cables, socket adapters and *gender changers* (male to female plug or socket adapters). Just because you can make the plug fit into the socket doesn't mean it's the right cable for the purpose! Don't be surprised if the system doesn't work!

Hayes AT-command set

Early modem development concentrated on the need to adapt the telephone network to enable the carriage of data communication. The earliest modem was developed by a telephone company (AT&T) and was designed, like telephone networks of the day (date: 1956), in a hardware-oriented fashion. There were no microprocessors at the time. From this heritage resulted the relatively complex hardware-oriented interface V.24, with its multiple pin plugs, cables and sockets. But as time progressed, modems became more sophisticated, introducing the potential to control the modem by means of software commands. The most commonly

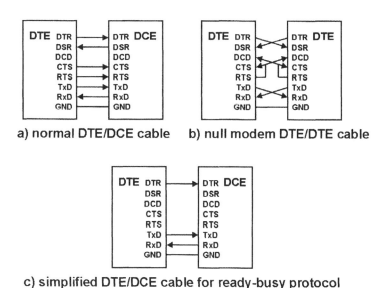

Figure 3.8   DTE/DCE and null modem cable types.

used command set used for this purpose in the *Hayes AT command set*, developed by Hayes Microcomputer corporation in the 1970s.

AT is short for *attention*. These are the two ASCII characters which precede each of the commands sent from a computer to a *Hayes compatible modem* (most modems are nowadays). Common commands are:

- ATDT (followed by a number) = dial this number using tone dialling

- ATDP (followed by a number) = dial this number using pulse dialling

- [X1] = dial immediately

- [X3] = wait for dial tone

- [W] = wait for second dial tone (e.g., when dialling via a *PBX, private branch exchange*)

- ATA = manual answer incoming calls

- ATH = hang up

- ATS0 = 0 = disable auto answer

- ATZ = modem reset

The modem responds by acknowledging the commands with the response 'OK'. As well as the above commands, there are a number of signals which allow the internal configuration of the modem to be changed (e.g., change bit rate, timeouts, etc.). Rather than requiring dedicated leads, as in the V.24 interface, AT commands are sent directly over the TxD and RxD leads prior to establishment of a connection. (They therefore cannot be confused with user data.)

Since the AT command set provides a powerful means of controlling the modem, the need for the multiple pins of the V.24 interface has reduced. Hence the typical use of seven leads (Figure 3.8a). But when using the *ready-busy-protocol*, only four are necessary (Figure 3.8c).

An alternative scheme to the Hayes AT command set is offered by ITU-T recommendation V.25$_{bis}$.

DTE/DCE interfaces for digital line bit rates exceeding 64 kbit/s — V.35/V.36 and X.21

DCEs (data circuit terminating equipment) used for the termination of digital leaselines (bit rates from 64 kbit/s to 45 Mbit/s or even higher) go by a variety of names:

- *CSU (channel service unit)*;

- *DSU (data service unit)*;

- *LTU (line terminating unit)*;

- *NTU (network terminating unit)*.

Such devices have to cope with the special demands of high bit rate signals — in particular their sensitivity to electromagnetic interference.

In the UK, France and North America, the V.36 interface is most commonly used to connect DTEs to DCEs designed to terminate lines of bit rates between 64 kbit/s and 1.544 Mbit/s

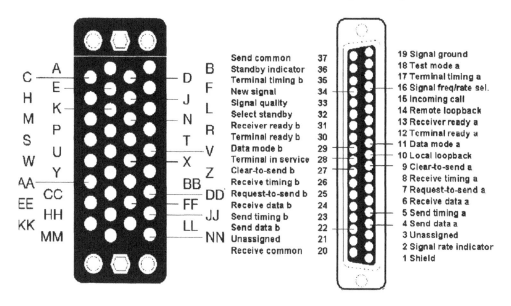

Send common	37
Standby indicator	36
Terminal timing b	35
New signal	34
Signal quality	33
Select standby	32
Receiver ready b	31
Terminal ready b	30
Data mode b	29
Terminal in service	28
Clear-to-send b	27
Receive timing b	26
Request-to-send b	25
Receive data b	24
Send timing b	23
Send data b	22
Unassigned	21
Receive common	20

19	Signal ground
18	Test mode a
17	Terminal timing a
16	Signal freq/rate sel.
15	Incoming call
14	Remote loopback
13	Receiver ready a
12	Terminal ready a
11	Data mode a
10	Local loopback
9	Clear-to-send a
8	Receive timing a
7	Request-to-send a
6	Receive data a
5	Send timing a
4	Send data a
3	Unassigned
2	Signal rate indicator
1	Shield

a) V.35
(connector: M/50 [ISO 2593])

b) RS-449 (V.35/V.36)
(connector: ISO 4902 [DB37])

Figure 3.9   Usual connectors for V.35 and V.36 (RS-449) interfaces.

(North America) or 2.048 Mbit/s (Europe). The V.36 interfaces is well suited for cases in which the DCE is connected to the main network by means of coaxial cable. Coaxial cables were initially used to carry digital leaselines from customer premises to the nearest telephone exchange building (coaxial cable *local loops*).

The connectors and pin layouts of the V.35 and RS-449 (V.35/V.36) interfaces are illustrated in Figure 3.9. Note that many more circuits and pins are needed than in the case of V.24 or RS-232, because each of the main signals (transmit, receive, etc.) is allocated a <u>pair</u> of wires ('a' and 'b' leads) rather than using only the 'a' lead and sharing a *common* 'b-lead' or *ground*. This precaution helps prevent possible interference between the signals, and is necessary to support the higher bit rates which V.35 and RS-449 interfaces are designed for.

In Germany and other continental European countries, the X.21 (V.11) interface is normally used for the DTE/DCE interface of digital leaselines. The emergence of X.21 and V.11 was driven by the use of standard telephone cabling (two *twisted pair* cables—i.e., 4-wire) for the *local loop* section of digital leaselines (from customer premises to the nearest telephone exchange). X.21 was also a natural choice because, at the time of conception of X.21, Deutsche Telekom (then called Deutsche Bundespost) needed a control mechanism capable for setting up dial-connections across their (now extinct) Datex-L *public circuit-switched data network (PSDN)*.

X.21 was developed later than V.24 and RS-232 with the specific objective of catering for higher bit rate lines, which became possible with the advent of digital telephone leaselines. X.21 line bit rates are usually an integral multiple of the standard digital telephone channel bandwidth (64 kbit/s)—hence the so-called n*64 kbit/s rates: 128 kbit/s, 192 kbit/s, 256 kbit/s, 384 kbit/s, 512 kbit/s, 768 kbit/s, 1024 kbit/s, 1536 kbit/s, 2048 kbit/s.

Together with the related recommendations X.24, V.10 (also known as X.26) and V.11 (also known as X.27), X.21 sets out an interface requiring far fewer interconnection circuits than necessary for the earlier V.24 and RS-232 interfaces. In addition, X.21 was specifically targetted on *synchronous communication.*[6]

X.21 uses separate pairs of 'a' and 'b' leads for the main signal circuits, rather than employing a *common* signal (or *ground*). This helps prevent interference between signals and allows for much greater DTE-DCE cabling distances of up to 100 m (the maximum cabling length had previously been limited by V.24/V.28 to around only 15–25 m).

Instead of a large number of separate circuits for the individual control signals (as in V.24–Table 3.1), X.21 uses the *control lead* (from DTE to DCE) and the *indication lead* (from DCE to DTE) in a way which enables multiple control and indication messages to be sent using the transmit (T) and receive (R) pairs, thus greatly reducing the number of wires required for DTE/DCE cables and the number of pins and sockets needed in X.21 connectors (Figure 3.10). A simple logic achieves this:

- When the *control leads (C)* are set 'on', then the data transmitted on the *transmit leads (T)* is to be interpreted by the DCE as a control message and acted on accordingly.

- When the *indicate leads (I)* are set by the DCE to 'on', then the DCE is ready to receive data. Otherwise, when 'off', the data sent by the DCE to the DTE on the *receive leads (R)* may be either *information* being sent to the DTE or the *idle* state (a string of '1's is sent in this case).

The control signals and states defined by X.21 are listed in Table 3.2. DTE and DCE negotiate their way from the idle state through *selection* and *connection* to the *data transfer* state. Following the end of the communication session, the *clearing* procedure returns the line back to the idle state.

Subrate multiplexing

When bit speeds below the basic digital *channel* bit rate of 64 kbit/s are required by the DTE, then the DCE may have to carry out an additional function: breaking down the line

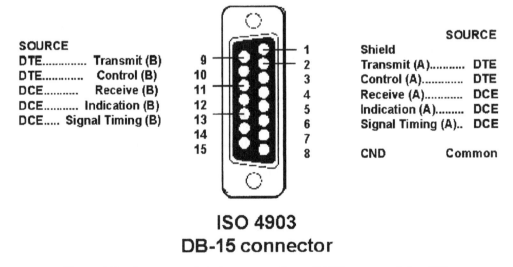

Figure 3.10   Connector and pin-layout according to ITU-T recommendation X.21.

---

[6] See Chapter 2.

Table 3.2 Communication states: ITU-T recommendation X.21

State number	State name	T (transmit)	C (control)	R (receive)	I (indication)
1	Ready	1	Off	1	Off
2	Call request	0	On	1	Off
3	Proceed-to-select request (i.e., request dial)	0	On	+	Off
4	Selection signal sequence (i.e., number to be dialled)	ASCII [7 bit] (IA5)	On	+	Off
5	DTE waiting	1	On	+	Off
6	DCE waiting	1	On	SYN	Off
7	Call progress signal	1	On	ASCII [7 bit] (IA5)	Off
8	Incoming call	1	Off	Bell	Off
9	Call accepted	1	On	Bell	Off
10	Call information (from DCE)	1	On	ASCII [7 bit] (IA5)	Off
11	Connection in progress	1	On	1	Off
12	Ready for data (i.e., to communicate)	1	On	1	On
13	Data transfer (i.e., communication)	Data	On	Data	On
13S	Send data	Data	On	1	Off
13R	Receive data	1	Off	Data	On
14	DTE controlled not ready, DCE ready	01	Off	1	Off
15	Call collision	0	On	Bell	Off
16	DTE clear request	0	Off	X(any signal)	X(any signal)
17	DCE clear confirmation	0	Off	0	Off
18	DTE ready, DCE not ready	1	Off	0	Off
19	DCE clear indication	X(any signal)	X(any signal)	0	Off
20	DTE clear indication	0	Off	0	Off
21	DCE ready	0	Off	1	Off
22	DTE uncontrolled not ready (fault), DCE not ready	0	Off	0	Off
23	DTE controlled not ready, DCE not ready	01	Off	0	Off
24	DTE uncontrolled fault, DCE ready	0	Off	1	Off
25	DTE provided information	ASCII [7 bit] (IA5)	Off	1	On

bandwidth of 64 kbit/s into a number of lower bit rate channels. This process is called *sub-rate multiplexing* (or *terminal adaption*). *Terminal adaption* is defined in ITU-T recommendation V.110. The process can derive one or a number of lower bit rate channel bit rates from a 64 kbit/s channel: 600 bit/s, 1200 bit/s, 2.4 kbit/s, 4.8 kbit/s, 7.2 kbit/s, 9.6 kbit/s, 12 kbit/s, 14.4 kbit/s, 19.2 kbit/s, etc.

International leaseline connections

Anyone who has ordered an international *leaseline* or other *wide area network (WAN)* connection (e.g., a *frame relay* connection) may have had cause to worry about whether an RS-232 DTE in the USA can be connected to a V.24/V.28 DTE in Europe or whether a V.35 DTE in the USA could be connected to an X.21 DTE in Europe. The answer is 'YES'. You order the international digital leaseline with RS-232 interface in the USA and V.24/V.28 interface in Europe (or V.35 in USA and X.21 in Europe as appropriate). The fact that the two DTE-DCE connections have different formats is immaterial.

Line interfaces

During the previous discussion about the DTE-to-DCE interface, you might well (with justification) have been wondering about what happens on the network side of the DCE at its *network* or *line interface*. A line interface is required for long-distance transport of data (anything more than the few metres of cable possible with DTE/DCE interfaces, right up to many thousand kilometres). As well as being used on the 'line' or 'network side' of DCE devices, *line interfaces* are also used for *trunk* connections between nodes in a wide area network (e.g., between the nodes of Figure 3.1). In this case, the DCE is dispensed with and the node equipment (e.g., a router or a switch) is equipped directly with a *line interface card*, to which the trunk line is directly connected.

For obvious reasons of economy, line interfaces are usually designed to use the equivalent of only two pairs of wires (one pair for transmitting and one pair for receiving) rather than the multiple leads used by DTE/DCE interfaces.

Where two devices in a network (DCEs, nodes, etc.) are interconnected by means of a line interface, the connections are always made from the Tx (transmit) port of the sending device to the Rx port of the receiver and vice versa (Figure 3.11). Thus the trunk port cards of routers or other node equipment are interconnected in this way: output A-to-input B and input A-to-output B.

Why bother with the DCE and the DTE/DCE interface at all?

In the case where a DCE is used, a device external to the DTE (the DCE) has to perform a conversion of the physical layer interface from the DTE/DCE format used by the DTE (e.g., V.24/V.28, X.21, RS-232, RS-449 etc.) into a *line interface* format required by the wide area network (WAN). The line interface of a DCE may conform either to an analogue modem modulation format (e.g., V.23 or V.32) or to a digital line transmission format (e.g., G.703, PDH, SDH, SONET).

So, if the DCE only performs a physical layer conversion, why bother with it at all? Why not instead let the DTE output its data directly in the line interface format? Answer: Because by using the DTE/DCE interface, the DTE need only implement this relatively simple interface and can be isolated from network technology changes. The bit rate of the long-distance line and the modulation technique can be upgraded without affecting the DTE (i.e., the user's computer). The output speed of the DTE can be increased simply by increasing the clock speed delivered to it by the DCE across the DTE/DCE interface.

Digital line interface types — digital transmission in wide area networks (WANs)

Wide area network (WAN) lines used in data networks are typically lines leased from public telephone companies and carried across their digital transmission networks. Since these

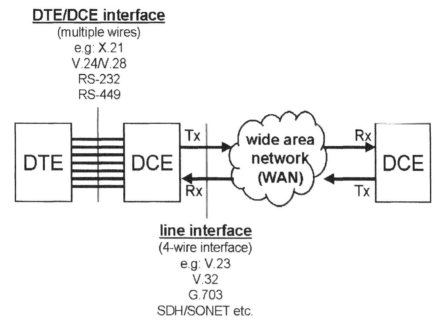

Figure 3.11 DCE (data circuit-terminating equipment) connections: the DTE/DCE interface and line interface.

networks were developed mainly for the initial purpose of telephone communication, the transmission systems which they use, and the bit rates which they offer as possible DCE line interfaces (see Figure 3.11) are geared around those employed for telephone (i.e., voice) communication.

In contrast to DTE/DCE interfaces, line interfaces usually comprise a maximum 4 wires (or equivalent—(two twisted pairs, two coaxial cables or two fibres, etc.) rather than multi-wire interfaces. A further difference is that, unlike DTE/DCE interfaces, there is no opportunity for the physical layer of a line interface to be 'not ready'. Bits are permanently flowing in both directions, if only a string of '0's or '1's is being sent as an *idle pattern*.

The first digital transmission systems appeared as early as 1939, when *pulse code modulation (PCM)* was invented as a means of converting analogue signals such as speech into a *digital* format. But widespread use of PCM and digital transmission systems began only in the 1960s.

All digital transmission systems are based on a basic telephone channel of 64 kbit/s bit rate. This is the bit rate corresponding to a 'basic' analogue telephone channel of 4 kHz bandwidth. The rate of 64 kbit/s comes about because during the process of converting an analogue telephone channel to a *pulse code modulated (PCM)* signal, the signal must be *sampled* at a frequency at least twice that of the bandwidth (i.e., at 2×4 kHz $= 8$ kHz—or every 125 μs). At each *sample point*, the amplitude of the original analogue signal is replaced by a numerical value between -128 and $+127$ (8 bits) corresponding to the analogue signal *amplitude*. The 'basic' telephone channel (of 4 kHz when in *analogue* form) thus has a *digital* bit rate of 8 kHz x 8 bits per sample $= 64$ kbit/s. This bit rate forms the 'basic building block' of all *higher order* digital transmission systems, and is referred to as a *64 kbit/s channel*, a *B-channel (bearer-channel)* or DS_0-channel *(digital signal, hierarchy level 0)*.

Higher order digital *multiplexing* systems operate at rates which are an exact integral multiple of the 64 kbit/s basic channel rate. Three basic types of digital transmission are used in public telecommunications networks and these systems provide the transport for digital leaselines *plesiochronous digital hierarchy (PDH)*; *synchronous digital hierarchy (SDH)* and *SONET (synchronous optical network)*. The various different hierarchy levels and bit rates have various names (e.g., DS1, DS3, T1, T3, E1, E3, STM-1, STS-3, OC3 etc.) but all are based on the same principles of *multiplexing*. We discuss multiplexing first, then each of the specific digital line transmission types in turn: PDH, SDH and SONET.

Multiplexing

The suffix 'plex' in communications technology is used to describe the use of a single medium or channel to carry multiple signals simultaneously (from the latin *perplexus*: to entwine). Thus *simplex* is the carriage of a single signal, while *duplex* is the carriage of two separate signals (in opposite directions) to allow two-way communication. Logically, *multiplex* is the carriage of multiple signals on a signal cable or other medium. *Multiplexing* is the process by which the multiple individual signals are combined together and prepared for transport. *Demultiplexing* reverses the multiplexing process at the receiver end of the connection.

In the case of digital multiplexing, a number of digital telephone channels (of 64 kbit/s bit rate) are combined to share a higher order digital bit rate (e.g., 32 telephone channels × 64 kbit/s may be multiplexed together to create a higher order bit rate of 2.048 Mbit/s).

The multiplexing process works by taking 1 byte (corresponding to a single amplitude value of a PCM [pulse code modulation] signal or one ASCII character) in turn from each of the channels to be multiplexed. The process can be imagined to be like a 'rotating selector' as shown in Figure 3.12.

In the example of Figure 3.12, 32 channels of 64 kbit/s are being multiplexed, taking one byte in turn from each channel: starting with channel 0 and working through to channel 31 before sending the next byte from channel 0. Since each digital telephone channel transmits one byte of data (one signal amplitude value) every 125 µs (corresponding to 8000 Hz), the higher order bit rate must be able to carry 32 bytes within the same time period.

In other words, the bit rate must be 32 times higher. The resulting higher order signal has a *byte-interleaved* pattern and a bit rate of 2.048 Mbit/s.

The *demultiplexing* process merely reverses the multiplexing process — 'unweaving' the individual signals from one another to recover the original thirty-two 64 kbit/s signals. Each of the individual *tributary* channels of a higher order digital bit rate are carried *transparently*, i.e., without affecting one another and without being altered in any way. The recovered signals (once demultiplexed) are identical to the input bit streams.

The same principles (as described above) are the basis of all digital multiplexing and explain why higher order digital line bit rates are always an exact multiple of the basic telephone channel bit rate of 64 kbit/s. But for purposes of datacommunication, it is often more appropriate to use the higher order bit rates (e.g., 1.544 Mbit/s [T1], 2.048 Mbit/s [E1], 34 Mbit/s [E3], 45 Mbit/s [T3], 155 Mbit/s, etc.) directly for high speed data transport. Such higher bit rate signals can easily be carried by standard public telecommunications networks.

Plesiochronous digital hierarchy (PDH)

Until the 1990s the backbone transmission networks of digital telephone networks were based upon a technology called *PDH (plesiochronous digital hierarchy)*. Three different PDH hierarchies evolved in different regions of the world, as summarised in Figure 3.13. They share three common attributes:

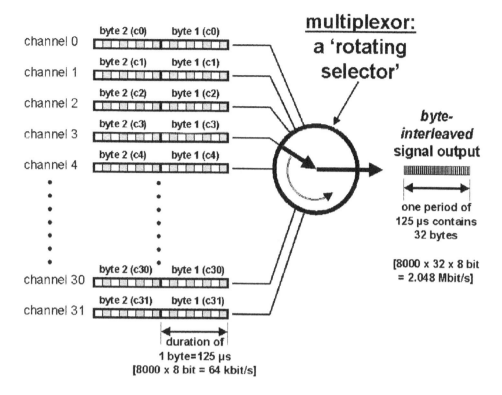

Figure 3.12   The principle of digital multiplexing.

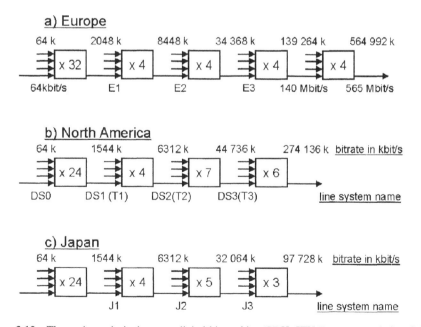

Figure 3.13   The various plesiochronous digital hierarchies (PDH; ITU-T recommendation G.571).

- they are all based on the needs of telephone networks, i.e., offering integral multiples of 64 kbit/s channels;

- they require <u>multiple multiplexing stages</u> to reach the higher bit rates, and are therefore difficult to manage and to measure and monitor performance, and relatively expensive to operate;

- they are basically incompatible with one another.

Each individual transmission line within a PDH network runs *plesiochronously*. This means that it runs at a clock speed which is nominally identical to all the other line systems in the same operator's network but is not locked *synchronously* in step (it is said to be *free-running*). This results in certain practical problems. Over a relatively long period of time (say, one day) one line system may deliver two or three bits more or less than another. If the system running slightly faster is delivering bits for the second (slightly slower) system, then a problem arises with the accumulating extra bits. Eventually, the number of accumulated bits become too great for the storage available for them, and some must be thrown away. The occurrence is termed *slip* and can result in *bit errors* when transmitting data (as we saw in Chapter 2 — Figure 2.14c). But slip is not so damaging to speech connections. The problem of slip is kept under control in PDH transmission by adding extra bits, called *framing* bits (also called *bit-stuffing* or *justification*).

The addition of the framing bits mean that the nominal line rate is slightly higher than the sum of the user signals to be carried. The framing bits allow the two end systems have to communicate with one another, slowing up or slowing down as necessary to keep better in step with one another. Not only this: whenever the normal user bits are not sufficient to carry the 'accumulating' bits, we can borrow some of the framing bits. The extra framing bits account for the difference, for example, between 4×2048(E1 bit rate) = 8192 kbit/s and the actual E2 bit rate (8448 kbit/s) — see Figure 3.13. Framing bits also account for the difference between 24×64 kbit/s = 1.536 Mbit/s and the actual *T1*-line system bit rate of 1.544 Mbit/s.

Extra framing bits are added at each stage of the PDH multiplexing process. In consequence, the efficiency of higher order PDH line systems (e.g., 139 264 kbit/s — usually termed 140 Mbit/s systems) are relatively low (91%). But more critical still, the framing bits added at each stage make it very difficult to break out a single 64 kbit/s channel or a 1.5 Mbit/s or 2 Mbit/s *tributary* from a higher order line system at an intermediate point without complete demultiplexing. In the example of Figure 3.14, six multiplexors are required to 'split out' a single E1 tributary from the 140 Mbit/s line at the intermediate point B. This makes PDH networks expensive, rather inflexible and difficult to manage.

The electrical interface used for all the various bit rates of the PDH is defined in ITU-T recommendation G.703. Basically, it is a four wire interface, with one pair each used for transmit and receive directions of transmission. Unlike DTE/DCE interfaces, there are no additional wires or circuits for control messages. These messages (such as are necessary) are carried within the *framing bits*. The line codes employed by ITU-T recommendation G.703 are *HDB3 (high density bipolar, order 3)* in the European PDH hierarchy *and AMI (alternate mark inversion)* in the North American and Japanese hierarchies. These were illustrated in Chapter 2 — Figure 2.18. Recommendations G.704, G.732 and G.751 set out the allowed bit rates and the correct signal framing of the higher order bit rates.

Structured and unstructured formats of PDH (plesiochronous digital hierarchy)

When you come to order a WAN (wide area network) connection for your data you may need to specify not only the user *bit rate* which you require, but also whether the line should be *structured* or *unstructured*.

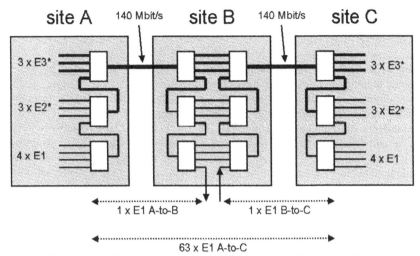

Note:* these E3 and E2 tributaries must also be demulitplexed to E1 rate

Figure 3.14 Breaking-out a 2 Mbit/s tributary from a 140 Mbit/s PDH line system.

Structured or *unstructured* refers to whether the framing has been added to the bitstream. The two different formats are incompatible, but both formats are not always available.

The unstructured format is used when the full line bit rate is required (e.g., an E1-line of 2048 kbit/s or a T1-line of 1544 kbit/s). Since the framing (and thus the *synchronisation* timing) has not been 'dictated' by the network of the telecommunications carrier, the framing bits are available to be set and used by you, the 'end user'. Thus your devices at either end of the point-to-point connection may *stuff, justify* and *synchronise* with one another directly, as discussed in the previous section.

The structured format is that in which the framing has already been added by the public telecommunications network carrier. In this case, *timeslot 0* (corresponding to the first 64 kbit/s channel within the frame) has been set by the network (for *synchronisation* purposes), and is not available for framing by the end user devices. Some carriers only offer the structured format (with maximum bit rates of 1920 kbit/s or 1984 kbit/s (in the case of E1 lines) or 1536 kbit/s (in the case of T1 and J1 lines)) instead of the full line bit rates of 2048 kbit/s and 1544 kbit/s respectively. The structured format is also used in the case that *fractional-E1* or *fractional-T1* bit rates arc requested. *Fractional* bit rates are bit rate multiples of 64 kbit/s (i.e., n × 64 kbit/s) where n is an integer value between 1 and 31). For these bit rates, the carrier may choose to provide a *local loop* connection (i.e., the wire from your premises to the nearest telephone exchange building) based on a full E1 (2.048 Mbit/s) or T1 (1.544 Mbit/s) transmission system. By structuring the signal, the exact n × 64 kbit/s signal can be carried. The excess bit rate (making up the full 1544 kbit/s or 2 Mbit/s line signal) is simply thrown away.

Any *DCE (data circuit terminating equipment)* needed to *terminate* the carrier's PDH line in your premises, and convert to the physical interface (e.g., X.21 or V.35/V.36) for connecting the *data terminal equipment (DTE)* is usually provided and maintained by the carrier. However, in some countries, regulation requires that the customer be able to rent or purchase DCE, like other *customer premises equipment (CPE)*, from a range of competing retailers. DCE intended to terminate digital leaselines go by a variety of names: *CSU, channel service unit*; *DSU, data service unit*; *LTU, line terminating unit*; *NTU, network terminating unit*, etc.

Synchronous digital hierarchy (SDH)

SDH, in contrast to PDH, requires the exact synchronisation of all the links and devices within a network. It uses a multiplexing technique which was specifically designed to allow for the *adding* and/or *dropping* of the individual *tributaries* within a high speed bit rate and for ring network topologies with circuit *protection switching*. Thus, for example, a single *add and drop multiplexor (ADM)* is used to break out a single 2 Mbit/s *tributary* from an *STM-1 (synchronous transport module)* of 155 520 kbit/s (Figure 3.15).

The *containers* (i.e., available bit rates) of the *synchronous digital hierarchy (SDH)* were designed to correspond to the bit rates of the various PDH hierarchies, as illustrated in Figure 3.16 and Table 3.3. *Containers* are multiplexed together by means of *virtual containers* (abbreviated to *VCs* but not to be confused with *virtual circuits* which we will discuss elsewhere in this book), *tributary units (TU), tributary unit groups (TUG), administrative units (AU)* and finally, *administrative unit groups (AUG)* to form *synchronous transport modules (STM)*. Synchronous transport modules (STM-1, STM-4, STM16, STM-64, etc.) are carried by SDH line systems.

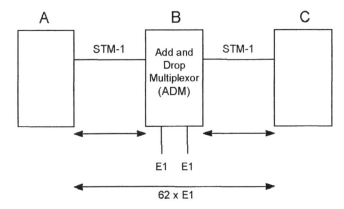

Figure 3.15   Add/drop multiplexor (ADM) used to break-out 2 Mbit/s (E1) from a 155 Mbit/s (STM-1) line.

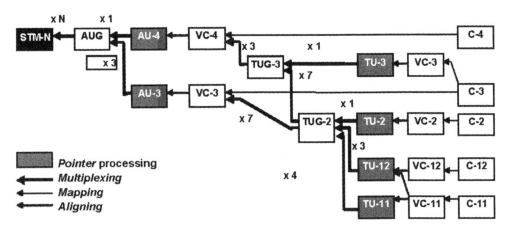

Figure 3.16   Synchronous Digital Hierarchy (SDH) Multiplexing Structure (ITU-T /G.709) [reproduced courtesy of ITU].

Table 3.3   SDH (Synchronous digital hierarchy) and SONET (synchronous optical network) hierarchies

North American SONET	Carried Bit rate/Mbit/s	SDH
VT 1.5	1.544	VC-11
VT 2.0	2.048	VC-12
VT 3.0	3.172	—
VT 6.0	6.312	VC-21
—	8.448	VC-22
—	34.368	VC-31
	44.736	VC-32
—	149.76	VC-4
STS-1 (OC-1)	51.84	—
STS-3 (OC-3)	155.52	STM-1
STS-6 (OC-6)	311.04	—
STS-9 (OC-9)	466.56	—
STS-12 (OC-12)	622.08	STM-4
STS-18 (OC-18)	933.14	—
STS-24 (OC-24)	1244.16	—
STS-36 (OC-36)	1866.24	—
STS-48 (OC-48)	2488.32	STM-16
STS-96 (OC-96)	4976.64	—
STS-192 (OC-192)	9953.29	STM-64

The basic building block of the SDH hierarchy is the *administrative unit group (AUG)*. An AUG comprises one AU-4 or three AU-3s. The AU-4 is the simplest form of AUG, and for this reason we use it to explain the various terminology of SDH (*containers, virtual containers, mapping, aligning, tributary units, multiplexing, tributary unit groups*).

The container comprises sufficient bits to carry a full frame (i.e., one cycle) of user information of a given bit rate. In the case of *container 4 (C-4)* this is a field of 260×9 bytes (i.e., 18720 bits), as illustrated in Figure 3.17a. In common with PDH (plesiochronous digital hierarchy) bit rates, the *frame repetition rate* (i.e., number of cycles per second) in SDH is 8000 Hz. Thus a *C4-container* can carry a maximum user throughput rate (*information payload*) of 149.76 Mbit/s (18720 bits/frame × 8000 frames/second). The payload is available to be used as a point-to-point bandwidth.

To a C-4 container is added a *path overhead (POH)* of 9 bytes (72 bits). This makes a virtual container (VC). The process of adding the POH is called *mapping*. The POH information is communicated between the point of *assembly* (i.e., entry to the SDH network) and the point of *disassembly* (i.e., exit from the SDH network). It enables the management of the SDH transmission system and the monitoring of the overall network performance.

The virtual container is *aligned* within an *administrative unit (AU)* (this is the key to synchronisation). Any spare bits within the AU are filled with a defined filler pattern called *fixed stuff*. In addition, a *pointer* field of 9 bytes (72 bits) is added (Figure 3.17b). The pointers (3 bytes for each VC — up to three VCs in total, thus 9 bytes maximum) indicate the exact position of the virtual container(s) within the AU frame. Thus in our example case, the AU-4 contains one 3-byte pointer indicating the position of the VC-4. The remaining 6 bytes of pointers are filled with an idle pattern. One AU-4 (or three *multiplexed* AU-3s containing three pointers for the three VC-3s) forms an AUG. Alternatively, the AUG (AU-4) could be used to transport a complete 140 Mbit/s PDH line system, with *fixed stuff* padding out the unused bytes in each C-4 container.

To a single AUG is added 9×8 bytes (576 bits) of *section overhead (SOH)*. This makes a single *STM-1 frame* (of 19 440 bits). The SOH is added to provide for *block framing* and

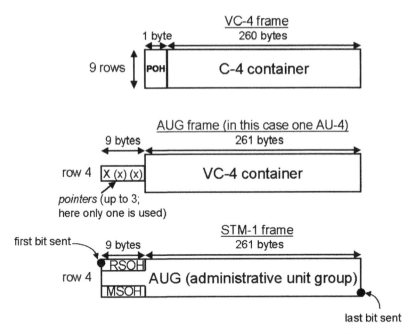

Figure 3.17   Basic structure of an STM-1 (SDH) or STS-3 (SONET) Frame.

for the maintenance and performance information carried on a transmission line *section* basis. (A *section* is an administratively defined point-to-point connection in the network, typically an SDH-system between two major exchange sites, between two intermediate multiplexors or simply between two regenerators.) The SOH is split into 3 bytes of *RSOH (regenerator section overhead)* and 5 bytes of *MSOH (multiplex section overhead)* as illustrated in Figure 3.17c. The RSOH is carried between, and interpreted by, SDH line system *regenerators*. The MSOH is carried between, and interpreted by the devices assembling and disassembling the AUGs (i.e., the line systems of higher order bit rate). The MOH ensures integrity of the AUG.

Since the *frame repetition rate* of an STM-1 frame is 8000 Hz, the total line speed is 155.52 Mbit/s (19440 × 8000). Figure 3.17 shows the gradual build-up of a C-4 container into an STM-1 frame. The diagram conforms with the conventional diagrammatic representation of the STM-1 frame as a matrix of 270 columns by 9 rows of bytes. The transmission of bytes, as defined by ITU-T standards, starts at the top left-hand corner, working along each row from left to right in turn, from top to bottom. The structure is defined in ITU-T recommendations G.707, G.708 and G.709.

Tributary unit groups (TUGs) and *tributary units (TUs)* provide for the breakdown of the VC-4 or VC-3 payload into lower speed tributaries, suitable for carriage of PDHs T1, T3, E1 or E3 line rates (1.544 Mbit/s, 44.736 Mbit/s, 2.048 Mbit/s or 34.368 Mbit/s).

For higher bit rates (above the STM-1 rate of 155 Mbit/s), the SDH line systems also support 'power-of-four' (i.e., 1, 4, 16, etc.) multiples of AUGs. AUGs are multiplexed together with a proportionately increased *section overhead*, to make larger STM frames. Thus an STM-4 frame (4 AUGs) has a frame size of 77 760 bits, and a line rate of 622.08 Mbit/s. An STM-16 frame (16 AUGs) has a frame size of 311 040 bits, and a line rate of 2488.32 Mbit/s (2.5 Gbit/s).

SONET (synchronous optical network)

SONET (synchronous optical network) is the name of the North American variant of SDH. It is virtually identical to SDH, but the terminology differs. The SONET equivalent of an

SDH synchronous transfer module (STM) has one of two names: either *optical carrier (OC)* or *synchronous transport system (STS)*. The SONET equivalent of an SDH virtual container (VC) is called a *virtual tributary (VT)*. Some SDH STMs and VCs correspond exactly with SONET STS and VT equivalents. Some do not. Table 3.3 presents a comparison of the two hierarchies, presenting the various system bit rates, as well as the subrates offered by the various *tributaries* (*virtual tributary* or virtual container).

Physical media used for SDH and SONET

SDH transmission signals, i.e., those conforming to the framing requirements of ITU-T recommendations G.707, G.708 and G.709, may be transmitted via either an electrical (copper wire) or an optical (glass fibre) *physical interface*. The electrical interface is defined by ITU-T recommendation G.703 (as with PDH). Otherwise a number of different optical fibre types may be used, as set out by ITU-T recommendation G.957. The individual fibre types are defined by ITU-T recommendations G.651 to G.655 (Table 3.4). When using the optical interface

Table 3.4 Classification of optical fibre interfaces for SDH equipment

	Single mode fibre (SMF)		Multi-mode fibre (MMF)	
FibreCore diameter	8 μm	10 μm	50 μm	62.5 μm
Fibre mantle diameter	125 μm	125 μm	125 μm	125 μm
Glass cladding diameter	250 μm	250 μm	250 μm	250 μm
Wavelength & attenuation				
850 nm	Not used	Not used	■ 3.0 dB/km ■	■ 3.7 dB/km ■
1310 nm	Not used	■ <1.0 dB/km ■	■ 2.0 dB/km ■	■ 1.0 dB/km ■
1550 nm	■ 0.3 dB/km ■	■ <0.5 dB/km ■	Not used	Not used
Relevant specifications	ITU-T rec. G.654 (loss minimised, 1550 nm only)	ITU-T rec. G.652	ITU-T rec. G.651	
	ITU-T rec. G.653 (restricted use at 1310 nm)	ITU-T rec. G.655		
Typical range	Long range	Metropolitan	Metropolitan and short range	Metropolitan and short range
	>60 km	<60 km		
Optical system designations	ZX	LX, LH (1300 nm — long wave)	SX (850 nm — short wave)	SX (850 nm — Short wave)
			LX, LH (1300 nm — long wave)	LX, LH (1300 nm — long wave)
	STM-L (long range)	STM-S (short range)	STM-S (short range)	STM-S (short range)
		STM-I (intra-office)	STM-I (intra-office)	STM-I (intra-office)

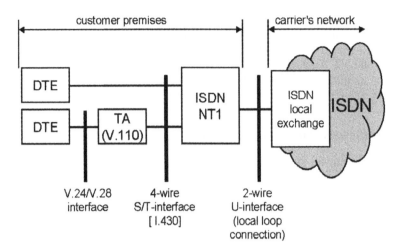

Figure 3.18 ISDN S/T-interface (physical interface: ITU-T recommendation I.430), ISDN U-interface and V.110.

a number of different physical connectors may be used, but the ST-connector (Figure 3.18) is the one most used by public telecommunications carriers. Taken together, the SDH (or SONET) framing recommendations, the optical or electrical interface specification and the physical connector form a physical layer interface (layer 1).

ISDN line interfaces

ISDN (integrated services digital network) was invented in the early 1970s as the digital version of the telephone network. As its name suggests, it was originally intended to be an ideal network technology for integrating telephone and data communications in a single network. Unfortunately, it was not well suited to data transport, and lost out rapidly to the *packet switched networks* which were being developed at the same time. Nonetheless, certain aspects of ISDN have retained significance for data networking:

- *I.430* (ITU-T recommendation): the physical layer interface used in ISDN;

- *BRI (basic rate interface)* and *PRI (primary rate interface)*: ISDN interfaces used for 'dial-up' connections.

ITU-T recommendation I.430 defines a *physical layer* interface (the ISDN *S/T interface*) capable of duplex transmission (equivalent to 4-wire transmission) at a rate of 192 kbit/s. Its charm lies in the fact that the associated ISDN *U-interface* (see Figure 3.18) requires only a single twisted pair of wires for use in the *local loop* (connecting the end-user's premises to the carrier's nearest telephone exchange building). By using the U-interface and I.430, a single pair of wires previously used to connect an analogue public telephone can be converted to ISDN and the data throughput of the line is correspondingly greatly increased.

When using the ISDN U-interface, a device called a *network termination type 1 (NT1)* is used instead of a DCE. Meanwhile, the DCE/DTE interface is replaced by the I.430 interface (the *ISDN S/T-interface* — Figure 3.18). Optionally, an additional piece of equipment, called a *terminal adapter (TA)*, conforming to ITU-T recommendation V.110 can be used to provide a 'modem-like' DTE/DCE interface for low bit rate lines conforming to V.24/V.28

(e.g., 600 bit/s, 1200 bit/s, 2.4 kbit/s, 4.8 kbit/s, 7.2 kbit/s, 9.6 kbit/s, 12 kbit/s, 14.4 kbit/s, 19.2 kbit/s, etc.).

The I.430 interface is used for the ISDN basic rate interface (BRI). It uses *AMI* line coding (*alternate mark inversion* — as we saw in Chapter 2 — Figure 2.18).

ISDN is widely used in data networks for providing 'dial-up' connections and temporary point-to-point interconnection of data devices. *BRI (basic rate interface)* is widely used, for example, for dial-up access from end user PCs (personal computers) to the Internet. Temporary connections can be useful as back-up trunk connections between nodes in a data network during periods when the 'normal' link has failed. Alternatively, they can be used to provide for extra capacity during periods of network congestion.

Physical layer cabling and connectors used in data networks

Appendix 7 illustrates (for reference purposes) a number of the commonest cabling and connector types used in data communications networks. Meanwhile, Figure 3.19 illustrates the most commonly used connector type employed in internal office data networks — the RJ-45 plug and socket. The connector is a rectangular shape, 11.6 mm wide, with 8 contacts. It should not be confused with the similar connectors RJ-11 (4 or 6 contacts) or RJ-12 (6 contacts) which are also widely used in telecommunications. RJ-11 is typically used for connecting the telephone handset to the base of a standard telephone; RJ-11 and RJ-12 are also commonly used as the PC 'modem' lead connector (from a PC with an internal modem to the telephone network socket).

Be extremely careful when using RJ-45 connectors and patch cables! All the cables look similar, but the *pin-out* configurations of the individual cable circuit connections vary widely: there are contradictory standards for the pin-outs. Some of the more common pin-out configurations are listed in Table 3.5.

Normal RJ-45 patch cables are wired with 8-core cable (all 8 circuits wired, pin one to pin one, pin two to pin two, etc.), using either *UTP (unshielded twisted pair)* or *STP (shielded twisted pair)* cable. But like the *null modem cable* of Figure 3.8, there are a large variety of speciality *crossover cables* for various purposes, e.g., for direct PC-to-PC connection (a DTE-to-DTE connection). Always check the suitability of the cable you are using!

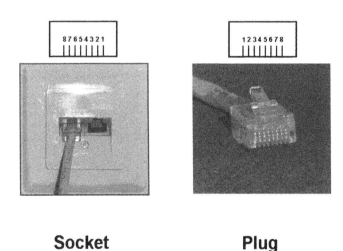

Socket   Plug

Figure 3.19   RJ-45 connector (8-pole socket and plug).

Table 3.5 Common RJ-45 patch cables and pin-out configurations

	Pin 1	Pin 2	Pin 3	Pin 4	Pin 5	Pin 6	Pin 7	Pin 8
RJ-45 pin colours (EIA/TIA 568B)	White/Orange	Orange	White/Green	Blue	White/Blue	Green	White/Brown	Brown
	Pair 2 ←	→ Pair 2	Pair 3 ←	Pair 1 ← ←←←	→→→ → Pair 1	→ Pair 3	Pair 4 →	← Pair 4
Uses:								
Standard ethernet patch cable	DTE receive + (hub transmit)	DTE receive- (hub transmit)	DTE transmit + (hub receive)	Not used	Not used	DTE transmit- (hub receive)	Not used	Not used
2-wire telephone line (DTE-to-WAN)	Not used	Not used	Not used	+	−	Not used	Not used	Not used
4-wire telephone line (DTE-to-WAN)	DTE receive +	DTE receive -	Not used	DTE transmit+	DTE transmit-	Not used	Not used	Not used
V.24/V.28 or RS-232	RTS (cct 4)	DTR (cct 20)	TXD (cct 2)	GND (cct 7)	GND (cct 7)	RXD (cct 3)	DSR/CD (ccts 6/8)	CTS (cct 5)
RJ-45 pin colours (EIA/TIA 568A)	White/Green	Green	White/Orange	Blue	White/Blue	Orange	White/Brown	Brown
	Pair 3 ←	→ Pair 3	Pair 2 ←	Pair 1 ← ←←←	→→→ → Pair 1	→ Pair 2	Pair 4 ←	→ Pair 4
Older RJ45 colouring	Blue	Orange	Black	Red	Green	Yellow	Brown	Grey

Other commonly used connector types

The pin-outs and connector forms of some other common PC communication ports are presented in Figures 3.20 and 3.21. The normal DB-25 PC parallel port shown in Figure 3.20a is used to extend the computer's internal IEC (International Electrotechnical Commission) bus

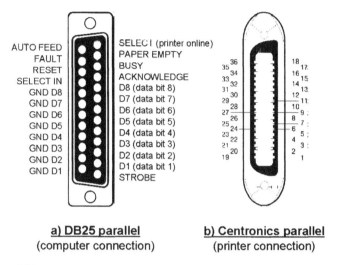

AUTO FEED	SELECI (printer online)
FAULT	PAPER EMPTY
RESET	BUSY
SELECT IN	ACKNOWLEDGE
GND D8	D8 (data bit 8)
GND D7	D7 (data bit 7)
GND D6	D6 (data bit 6)
GND D5	D5 (data bit 5)
GND D4	D4 (data bit 4)
GND D3	D3 (data bit 3)
GND D2	D2 (data bit 2)
GND D1	D1 (data bit 1)
	STROBE

a) DB25 parallel
(computer connection)

b) Centronics parallel
(printer connection)

Figure 3.20   PC parallel port connection to local external devices (e.g., printer).

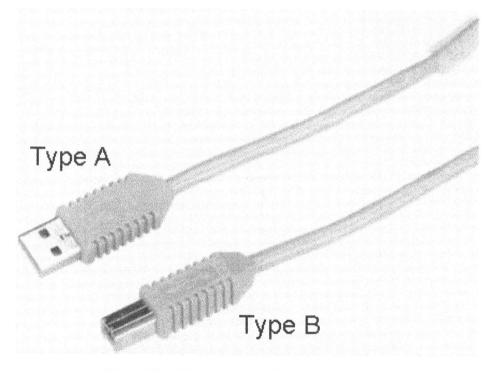

Type A

Type B

Figure 3.21   USB (universal serial bus) interface connector.

(specification: IEC 7210 or IEEE 488.2) to locally connected peripheral devices like print-ers, scanners and other input/output devices. The corresponding socket on the printer usually employs the centronics parallel connector shown in Figure 3.20b.

Figure 3.21 illustrates the *USB (universal serial bus)* now commonly used as a local serial bus for connecting multiple local external devices to a PC. The connectors used are either of type A or of type B. USB is beginning to take over from the *parallel port* connection (Figure 3.20), the SCSI *(small computer system interface)*, and the normal *serial port* as the 'preferred' method of connection of devices to a PC.

The latest version of the USB (universal serial bus) interface — USB 2.0 — defines three types of interfaces, depending upon the speed required: low speed (up to 1.5 Mbit/s); medium speed (up to 12 Mbit/s) and high speed (up to 480 Mbit/s).

3.3   Layer 2 — data link layer

Physical layer interfaces carry digital bitstreams but are not able to control the flow of data across a link or network. A number of additional mechanisms are necessary: to set up the correct type of connection to the right destination port, to indicate the coding of the data, to control the message flow between the end devices, to provide *statistical multiplexing* and to provide for *error correction* of incomprehensible information. These functions are the responsibility of *layer 2 (datalink)* and *layer 3 (network)* protocols.

The datalink (layer 2) protocol formats the raw stream of bits carried by the physical layer into data frames and controls the flow of these frames from one node to the next (i.e., across a single *datalink, shared medium* or subnetwork — e.g., a LAN).

To support the protocol functions, the *datalink (layer 2) protocol* adds to each block of data a *frame header* and *trailer* for *byte* and *character synchronisation*, for *flow control* of the data transmission and (optionally) for basic bit error detection and correction (Figure 3.22).

The *network (layer 3) protocol* is concerned with the communication across the series of individual *datalinks* which make up the entire end-to-end *network* connection. Layer 3 protocols regulate the process of *statistical multiplexing*, which as we learned in Chapter 2, makes for efficient data networks, by allowing multiple users to share the bandwidth available. The network (layer 3) protocol establishes the connection, identifies the correct destination port, supports statistical multiplexing, collects packets together as they reach their destination (maybe via different routes), and sorts them back into order using the *sequence numbers* or requests re-transmission of any missing ones. The network protocol, like the datalink protocol, may also be concerned with *data flow control*, but in this case on an end-to-end basis rather than on the link-by-link flow control performed by the datalink layer.

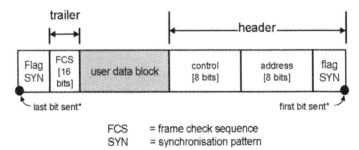

Figure 3.22   The HDLC (higher level datalink control) frame structure — a typical layer 2 protocol.

Layer 2 data flow control — initially called 'handshaking'

The term *handshaking* is nowadays used by protocol designers to describe a range of different procedures, particularly the type of procedures carried out when networking devices speak to one another for the first time — to 'introduce themselves' to one another and indicate what they are capable of. But originally *handshaking* was the term used to describe the early data flow control procedures used to regulate the speed of data transfer across point-to-point data communications lines. From these early handshaking procedures, today's datalink (layer 2) protocols developed. Perhaps the best known of these simple software handshake procedures are *XON/XOFF* and *ETX/ACK*.

In the *XON/XOFF* handshake protocol, the sender is only allowed to send data having received the XON signal (transmission on — ASCII character 11H) from the receiver. Should the receiver return XOFF (transmission off — ASCII character 13H), then the sender is prohibited from sending further data until once more permitted to do so by receipt of a renewed XON signal. The simple procedure ensures that the receiver is always ready for any data sent: in particular, that the receiving data buffer is not full. The charm of XON/XOFF protocol is its simplicity, and this technique is the basis of many modern protocol flow control procedures.

A slightly more complex handshake procedure is offered by the *ETX/ACK* protocol. In this protocol, the sender sends each *byte* or *block* of data and appends the ETX character (ASCII 03H) as the 'end text' character. The sender is then required to wait before sending the next *byte* or *block* until the receiver responds with the ACK (acknowledge) character (ASCII 06H) signalling its readiness to receive again.

Like XON/XOFF, the ETX/ACK protocol performs the important job of *data flow control* — ensuring that the sender only sends data when the receiver is ready to receive it. But while ETX/ACK is extremely effective in ensuring the *acknowledgement* of data receipt, it has a number of weaknesses if each data block transmitted has to be individually acknowledged:

- A lot of time is wasted while the sender waits for the next ACK. This severely limits the effective bit rate of the line — maybe to a level as low as 5–15% of the real line bit rate (we discussed this effect in Chapter 2 — Figure 2.27);

- The network efficiency is further reduced by the need to carry the ACK characters themselves.

Next in the history of data flow control came the development of the *window* (or *packet window*). When a *window* is employed (usually on an end-to-end basis as part of a layer 3 (network) protocol), each of the sent packets is given a *sequence number*. The receiver is then only required occasionally to send an ACK, confirming which sequence number is expected next, and thus by deduction which frames have already been correctly received. The sender is allowed to continue sending data, as long as the number of packets not yet acknowledged (ACK) by the receiver is less than the *window size*. This significantly increases the efficiency of the ETX/ACK protocol (as compared to XON/XOFF), as long as the window size is not set too small. On the other hand, if the value is set too big, the receiver may be swamped with data it cannot handle before it has had a chance to say 'stop'.

In the early days of digital networks, *datalink* and *network* protocols had to deal with a relatively high proportion of bit errors. Until the mid-1980s, digital networks typically operated with a *bit error ratio (BER)* of around 10^{-5} (i.e., 1 bit error in every 100 000 bits (12 500 bytes) sent). Since a typical data block was between 256 and 4096 bytes in size, this meant that between 2% and 33% of data blocks (layer 2 frames or layer 3 packets) contained an error — an unacceptably high proportion. As a result, very robust error correction methods had to be built into protocols and *retransmission* of corrupted packets had to be undertaken.

Nowadays, digital networks have much higher quality. *Bit error ratios (BER)* rarely exceed 10^{-9}, or in the case of fibre networks, 10^{-12}. Much fewer blocks are corrupted, and heavyweight error correction is superfluous. Meanwhile, the prices of the electronic storage components required to create large receive *buffers* have dropped significantly. As a result, the problem of regulating receive buffer overflows has decreased and the need for performing error correction has reduced. In consequence, the emphasis of layer 2 protocol design today has moved to ensuring maximum data flow efficiency over very high bit rate lines.

Layer 2 (datalink) protocol functions and packet format (HDLC, higher layer datalink control)

The bitstream carried by the *physical layer (layer 1)* is divided up into *blocks* of data (called frames) by the headers and trailers of the *layer 2 (datalink) protocol*. The user data is sandwiched between a *header* and a *trailer*. Figure 3.22 illustrates one of the most widely used layer 2 protocols, *HDLC (higher level datalink control)*. HDLC is used for *synchronous* data transfer across point-to-point data lines. The headers and trailer of HDLC carry the information the protocol needs (the *protocol control information (PCI)*) to do its work, in order of transmission:

Flag

The *flag* indicates the start and end of each frame. A single flag separates each frame from the next (HDLC uses *synchronous transmission*). The flag is coded with the *synchronisation byte (SYN)* used for byte and character synchronisation as described in Chapter 2. SYN is a 1 byte field with the value '0111 1110'.

Since the flag is the means used to determine the total length of the frame, it is important that the same pattern '0111 1110' does not appear in any of the other fields of the frame, otherwise the end of the frame would be incorrectly determined. This is done by a process called *bit stuffing*.[7] The process of bit stuffing entails inserting an extra bit of binary value '0' after any sequence of five simultaneous '1's. The extra '0' bit is removed by the HDLC receiver, by simply converting any received patterns of '111110' into '11111'.

Address

The *address* is the *layer 2 address*. This is not the same as (and should not be confused with) the *network address* used by layer 3 protocols to identify the destination end-point in the network (a telephone number or an IP address is a layer 3 address). The layer 2 address field labels a frame as either a *command* or a *response* and indicates the DTE or DCE as the intended destination. When identified as a command frame, the address identifies the DCE or DTE of a point-to-point link to which the command is being sent. Similarly, in a response frame (in which a given DTE or DCE provides information in response to a command) the

---

[7] Do not get confused between bit stuffing carried out by a layer 2 protocol such as HDLC and the coding changes (*violations*) performed by the *line codes* of the *physical layer (layer 1)* (as discussed in chapter 2). The two mechanisms work entirely independently of one another. ITU-T refers to the layer 2 bit stuffing process as *adding transparency*).

address identifies the DCE or DTE sending the response. Except in the case of shared media (LANs), it is unusual in the case of a point-to-point datalink that more than a few addresses are actually in use.[8] *LANS (local area networks)* use this field in an adapted form of HDLC (called *LLC — link layer control*) to identify the intended destination device within the LAN as discussed in Chapter 4.

A *datalink* or medium using HDLC must always have a designated *primary station* (usually the DCE) and at least one *secondary station* (usually the DTE). The primary station controls the link. The secondary station may only send when the primary station grants permission. In the case of a *LAN* or a *bus*-configured network access line (such as the *S/T interface* of *basic rate ISDN, BRI*), there are multiple stations.

In the *balanced mode* of HDLC, each of the stations connected to the *datalink* or *shared medium* acts simultaneously as both a primary and a secondary station, thereby giving them all equal priority and eliminating the single controller. *The LAPB (link access protocol balanced)* used in ISDN *BRI (basic rate interface)* is a variation of the balanced form of HDLC.

Control field

The *control field* controls the flow of data across the *datalink* and normally comprises 8 bits. This field sets the 'mode' of transfer to be used. It arranges for *data flow control* by determining the 'readiness' of the link and organises the *acknowledgment* or *retransmission* of blocks, frames or packets. It also *disconnects* the link and generally provides information about the line status. In some layer 2 protocols, additional controls are provided, e.g., for authorisation of the user before use of the link or for secure transfer of data (e.g., by encryption). As necessary, the control field may be extended by an integral multiple of 8 further bits to accommodate such controls.

The control field may contain one of three different types of frame, called *information frames (I-frames), supervisory frames (S-frames)* or *unnumbered frames (U-frames)*. In reality, the messages are all of the same basic format: the different names merely serve to categorise the different types of controls. All I-frames begin with bit 1 = '0', all S-frames with the first two bits '10' and all U-frames with the bit values '11', as Table 3.6 illustrates.

HDLC offers only three *commands* (in *U-frame* format): *SNRM, SABM* and *disconnect*. These set the configuration of the link. The *response* messages (also in U-frame format) merely acknowledge these commands. Normal data frames are sent as *I-frames*. When a datalink is running smoothly, the frames received at one end of the link contain the *receive sequence number (N(R))* to acknowledge the receipt of all the frames sent — up to and including the N(R)−1th frame). The values used by N(R) correspond with the *send sequence numbers [N(S)]* which appear in the *I-frames* sent in the opposite direction. The transmission of N(S) and N(R) *sequence numbers* in this way allows for full data flow control, but simultaneously also for *full duplex* operation of the datalink. Only when there is a 'hiccup' in the data flow,

---

[8] The normal address field of 8 bits allows 256 different addresses for devices sharing the physical medium (i.e., for multipoint networks), although 2 of these addresses are *reserved*. These are '1111 1111' which is the *all stations* (or *broadcast address*) and '0000 0000' which is the *no stations address* (ignored by all stations, but used for network testing purposes). A frame with the *all stations* address is used when sending the same command to all stations (e.g., a flow control command: 'stop sending'). Meanwhile, a frame with a no stations address in a ring topology network will progress around the ring and return (or *loopback*) without being removed by a receiving station, provided the ring is not broken by a failure. If necessary, the HDLC address field can be extended by any integral number of a further 8 bits. Use of the longer address space is made in the case of LANs, where unique numbers are allocated to individual *network access cards* (e.g., a PC ethernet card) at the time of their manufacture. These addresses are 48-bits in length and allocated by IEEE. Sometimes they are referred to as the *IEEE-address*, but also as the *MAC (media-specific access control)-name* or *MAC-address*.

Table 3.6 HDLC control field signals and coding (bit 1 is the least significant bit — the first bit sent)

Frame type	Signal type	Signal purpose	C F 1	o i 2	n e 3	t l 4	r d 5	o- 6	l 7	- - 8	Remarks
I-frame	I	*Sequence numbers* of sent frames N(S) and next frame number expected to be received N(R), effectively an acknowledgement of all previously received frames	0	N	(S	)	P/F	N	(R	)	P = Poll bit F = Final bit
S-frame	RR	Receive ready and acknowledgement of all frames up to and including N(R)-1th	1	0	0	0	P/F	N	(R	)	P = Poll bit F = Final bit
	RNR	Receive not ready, but acknowledgement of all frames up to and including N(R)-1th	1	0	1	0	P/F	N	(R	)	P = Poll bit F = Final bit
	REJ	Reject, requests retransmission of all frames from and including N(R)th	1	0	0	1	P/F	N	(R	)	P = Poll bit F = Final bit
	SREJ	Selective Reject, requests retransmission only of the N(R)th	1	0	1	1	P/F	N	(R	)	P = Poll bit F = Final bit
U-frame	SNRM	Set normal response mode; secondary stations may only send frames when *polled* by the primary station by setting P(poll bit) = '1'	1	1	0	0	P	0	0	1	*P* = Poll bit
	SABM	Set asynchronous balanced mode; secondary stations may send at any time	1	1	1	1	P	1	0	0	P = Poll bit
	DISC	Disconnect	1	1	0	0	P	0	1	0	P = Poll bit
	UA	Unnumbered acknowledgement; acknowledges having received a U-frame with a 'set' command	1	1	0	0	F	1	1	0	P = Poll bit
	DM	Disconnected mode; notification of a protocol error	1	1	1	1	F	0	0	0	F = Final bit

do *supervisory* messages (*s-frames*) need to be used to sort things out (calling a temporary halt in communication or requesting the retransmission of a particular numbered frame).

The *poll bit (P)* is set to '1' when one station wishes to solicit (i.e., *poll*) a response from the other. *The final bit (F)* is set to '1' to indicate the corresponding response to the poll.

In the normal HDLC frame, the control field has a length of 8 bits (1 byte), and the sequence numbers are three bits in length or *modulo 8*. In other words the count goes from 0 to 7 and then restarts at 0 again — counting continuously. In order for the acknowledgement process to work correctly, the layer 2 *window size* must be set at 7 or less, thereby limiting the sender to sending no more than 7 unacknowledged frames. If it were to send eight or more frames without first receiving an acknowledgement, then different unacknowledged frames would share the same sequence number. It would then be impossible to tell which of the two frames was being acknowledged, or from which of the two frames retransmission should occur. Since small window sizes (e.g., 7) can lead to very inefficient usage of high bit rate lines (as we saw in Chapter 2 — Figure 2.28), HDLC is sometimes used instead in a *modulo 128* mode. In this case, the *control field* is extended to 16 bits, thus enabling both N(S) and N(R) fields to be extended to 7 bits each. The extended N(S) and N(R) fields allow a *sequence number* count from 0 to 127, and thus a *window size* of up to 127 unacknowledged frames.

The *commands* and *responses* carried by the HDLC *control field* are communicated directly between layer 2 protocol handlers. The physical layer hardware and software are oblivious to the layer 2 communication. This is *layered* and *peer-to-peer communication*!

User data field

The user data is the information itself. This field may contain only real end-user information but more likely is that it contains the *PCI (protocol control information)* of a higher layer protocol (e.g., PCI for one or more of the layers 3, 4, 5, 6, 7) in addition to the real user information. This will become clearer in Figure 3.28.

Frame check sequence (FCS)

The frame check sequence (FCS) is normally a 16-bit code used for detection of errors in received frames. A *cyclic redundancy check (CRC)*[9] is used to set and check the value which appears in this field.

Least significant bit (LSB) sent first

The order of transmission to line of an HDLC frame is always 'flag-address-control-information-FCS-flag' and each of the individual frames is always sent *least significant bit (LSB)* first.

Well-known layer 2 protocols

Apart from HDLC, other well-known and widely used layer 2 protocols include: *SDLC (synchronous datalink control)* which is part of IBM's *SNA (systems network architecture)* (it was the protocol from which HDLC was developed); *LLC (logical link control — IEEE 802.2)*[10] as used in LANs, *LAPB (link access protocol balanced)* used in X.25 packet networks, *SLIP (serial line internet protocol)* and *PPP (point-to-point protocol)*[11] — part of the IP (Internet Protocol) suite. All of these are similar to, or based upon, HDLC.

Media access control (MAC)

In some layer 2 protocols, the various functions are split into further sub-layers. In the case of the datalink layer (IEEE 802.2) used in LANs, for example, this is comprised not only

---

[9] See chapter 2.
[10] See chapter 4.
[11] See chapter 8.

of the *logical link control (LLC)* but also of the *media access control (MAC)* layer. The approximate 'split' of responsibilities is that the LLC assumes responsibility for the 'higher level' functions of control, supervision, user data transfer and frame check sequencing, while the MAC takes over medium-specific tasks related to the interface with the physical layer. Thus the prime tasks undertaken by the MAC-layer are those of addressing and coordination with the physical layer.

The first MACs appeared during the emergence of LANs (local area networks), when HDLC had to be adapted to cope with multiple devices communicating using a shared medium. This required a more powerful capability for addressing than the point-to-point datalink networks which had previously existed. The emergence of MAC-protocols also lead to the standardisation by IEEE of the 48-bit address format and numbering plan hardcoded into network adapter cards at their time of manufacture. These addresses are variously referred to as *MAC-addresses, MAC-names* or *IEEE-addresses*. We shall return to the subject of the local area network MAC protocol and addresses in more detail in Chapter 4.

As a quick-reference for those who need to know, since it is mathematically complicated, the value set in the *FCS (frame check sequence)* field of HDLC is the *ones complement* (a mathematical term meaning that the '0's of the binary number are changed to '1's and vice-versa) of the sum (calculated in binary) of:

1) the remainder of $x^k(x^{15} + x^{14} + x^{13} + x^{12} + x^{11} + x^{10} + x^9 + x^8 + x^7 + x^6 + x^5 + x^4 + x^3 + x^2 + x + 1)$ divided (modulo 2) by the *generator polynomial* $x^{16}x^{12}x^5 + 1$. k is the number of bits in the frame existing between the final bit of the opening flag and the first bit of the FCS, excluding *stuffing bits* (synchronous transmission) or start and stop bits (asynchronous transmission);

The expression '$x^{15} + x^{14} + x^{13} + x^{12} + x^{11} + x^{10} + x^9 + x^8 + x^7 + x^6 + x^5 + x^4 + x^3 + x^2 + x + 1$' means the binary number '1111 1111 1111 1111' and '$x^{16} + x^{12} + x^5 + 1$' means the binary number '1 0001 0000 0010 0001' etc. x^k means a binary number starting with a '1' and followed by k bits of value '0'. In the general case of any CRC (cyclic redundancy check) the value to be divided by the *generator polynomial* always takes the same form, only the length of the field to be *coded* (i.e., the value of k) and the length of the *codeword* (i.e., the number of '1's in the second number — in our case $15 + 1 = 16$) changes].

and:

2) the remainder of the division (modulo 2) by the generator polynomial $x^{16} + x^{12} + x^5 + 1$ of the product (i.e., multiplication) of x^{16} by the content of the frame (i.e., the data to be coded). The frame content in this sense is all the bits between the final bit of the opening flag and the first bit of the FCS, excluding stuffing bits (synchronous transmission) or start and stop bits (asynchronous transmission).

Usually a register is available in the transmitter for calculating the remainder value described in (1) above. This register is preset to all '1's and then modified according to the result of the generator polynomial division. The *ones complement* of the resulting remainder is the 16-bit FCS.

At the receiver, the incoming *codeword* (i.e., the entire frame including address, control, information and FCS fields, but excluding stuffing bits) is multiplied by x^{16} (i.e., shift register of 16 places to the left, filling from the right (least significant bit) from the 'remainder register' which is initially pre-set to all '1's) and then divided (modulo 2) by the generator polynomial $x^{16} + x^{12} + x^5 + 1$. The final content of the 'remainder register' is '0001 1101 0000 1111' if there were no transmission errors.

3.4 Layer 3—network layer and network layer addresses

The *network* (or *layer 3*) *protocol* is responsible for managing the statistical multiplexing, end-to-end carriage of information and other functions which combine and coordinate a number of datalinks to produce a *network service*. The distinguishing feature of a network (layer 3) protocol is its capability to 'find', *route* to and/or *switch* to remote communications partners who are not reachable via a single *datalink* (such as a single point-to-point link or a LAN, local area network). Thus in communicating between one LAN and another, or between any two devices across a *wide area network (WAN)*, a *network protocol* has to be used.

When invoking a layer 3 protocol, an end-user or 'higher layer application' (i.e., computer program) must specify the *network address* (equivalent to a telephone number) with which communication is to be established. This address identifies the destination port of the network. Network layer (layer 3) protocols have a number of responsibilities to fulfil:

- identification of destination network address;

- *routing* or *switching* of packets across the individual network nodes and links to the destination;

- *statistical multiplexing* of the data supplied by different users for carriage across the network;

- *end-to-end data flow control*: the flow control conducted by layer 2 protocols only ensure that the receiving data buffers on each individual link can receive more data. The layer 3 protocol has the more onerous task of trying to ensure a smooth flow of data across all the datalinks or subnetworks taken together. Uneven 'hold-ups' along the route need to be avoided;

- correct *re-sequencing* of packets (should they get out of order having travelled via different routes on their way to the destination); and

- error correction or transmission re-request on a network end-to-end basis.

As with layer 1 and layer 2 protocols, there are a number of alternative layer 3 protocols available for use in different types of network and there are marked differences between how they operate. It is, for example, important to understand the differences between *connection-oriented* and *connectionless* network protocols, and this we shall study next.

Connection-oriented network service (CONS) and connectionless network service (CLNS)

Telephone networks, circuit-switched data networks, X.25- and frame relay-based packet-switched networks, as well as the cell-switched networks of *ATM (asynchronous transfer mode)* are examples of *connection-oriented* switching or *connection-oriented network service (CONS)*. Under connection-oriented switching, a *circuit, virtual circuit (VC), connection* or *virtual connection (VC)* is established between sender and receiver before information is conveyed. Thus a telephone *connection* is first established by dialling the telephone number before the conversation takes place. This ensures the readiness of the receiver to receive information before it is sent, there is no point in talking if nobody is listening.

In contrast, a *connectionless-network service (CLNS)* such as that provided by the *Internet protocol (IP)* allows *messages* (i.e., *packets*) to be despatched, without even checking the validity of the address. Thus, for example, the postal service is analogous to a *connectionless* service. The sender posts the entire message (envelope and contents) into the post box and forgets about it. Sometime later, the receiver receives the message—delivered through his letter box (or alternatively: it gets lost somewhere or for some other reason cannot be delivered!).

The main advantage of a connectionless-network service is that the sender need not wait for the receiver to be ready and the network need not be encumbered with the extra effort of setting up a connection. The message, called a *packet* or *datagram* is simply despatched. Since a single packet or datagram (for example, containing a single data file) represents the entire content of many data communications 'messages', it is easier to treat the datagram like a postal telegram and *route* it from node to node along its path to the destination based upon the network address appearing in the header (the equivalent of the telegram's envelope). The disadvantage is that the sender gets no clear guarantee or confirmation of message delivery. The sender is left in doubt: Did the receiver get the message? — Were they simply too lazy to reply? Or did the receiver not get the message? Was the address not written correctly?

Theoretically, the multiple packets making up a large message, when sent in a connectionless manner, may take different paths through the network to the destination. This would make it very difficult for 'eavesdroppers' to intercept the entire message, so making the communication more 'secure'. This idea appealed to the military users who were some of the first users of IP (Internet protocol). On the other hand, the different routes of the different packets usually lead to different propagation delays and so, in turn, can lead to complications in the communication between the two end-points. In practice, this is overcome by the use of *deterministic* or *path-oriented routing* (all packets are routed along the same path). But where this is used (nearly always), the 'security' benefit is lost.

One of the important distinctions between connection-oriented and connectionless networks is the type of device used as a network node in the respective networks. In connection-oriented networks, it is normal to refer to the network nodes as *switches* and to the function they carry out as *switching*. A switch *switches* a *connection*. In contrast, connectionless networks have no connections to be switched. The nodes of a connectionless network, usually called *routers*, merely route and *forward* datagrams (like the 'sorters' in a postal network).

Connection-oriented and connectionless networks have very different strengths and weaknesses. And since we shall concentrate in this book most on the connectionless ways of IP, let us take just a few moments to consider the comparative strengths of connection-oriented switching — and I make no apologies for doing so!. While some readers may consider X.25, *frame relay, ATM (asynchronous transfer mode)* and other connection-oriented protocols as only of 'historic' significance (i.e., overtaken and replaced by IP-suite protocols), this ignores their continuing widespread use — including in the backbone of some IP-networks! Not only this, but the principles (and even the details!) of the protocol operation of connection-oriented packet-switching (as developed for X.25, frame relay and ATM) are nowadays being adopted into the IP-suite protocols. The following brief discussion of connection-oriented packet switching and X.25 will help us later to understand the motivation for using a connection-oriented transport layer protocol in the IP suite: *TCP (transmission control protocol)*. It will also lay the foundations for understanding *MPLS (multiprotocol label-switching)*. Both TCP and MPLS are discussed in more detail in Chapter 7.

Circuit-switched, X.25-packet, frame relay and ATM networks are connection-oriented data networks!

The distinguishing property of a *circuit-switched* network is the existence throughout the communication phase of the *call*, of an unbroken physical and electrical *connection* between origin and destination points. The connection is established at *call set-up* and *cleared* after the call. The connection may offer either one direction (simplex) or two direction (duplex) use. Telephone networks are said to be *circuit-switched* networks and are also *connection-oriented*.

Conversely, although X-25-[12] and *frame relay*-based[13] *packet-switched networks* are also *connection-oriented*, an entire and exclusive physical connection from origin to destination will not generally be established at any time during communication. This may seem confusing at first, but is important to understand, since it is at the root of how a *connection-oriented* data networking protocol works. The important point about *connection-oriented network services (CONS)* is that a connection *set-up* phase confirms the readiness of the receiver to receive information and determines the route through the network which will be used to carry the packets <u>before</u> data transfer commences. The *connection* which results is actually termed a *logical channel, virtual circuit (VC)* or *virtual channel*, since though it appears to the two end-users as though a dedicated ('virtual') path exists, the physical connection is actually shared with other users. Bandwidth is only consumed when an actual packet needs to be conveyed.

Historically, *circuit-switched* networks (providing connections for users' exclusive use) were considered necessary when very rapid or instantaneous interaction was required (as is the case with speech or live video). Conversely, *packet-switched* networks (both connection-oriented ones such as *X.25* and frame relay, as well as *connectionless* networks such as those based on IP) are more efficient when instantaneous reaction is not required, but when very low 'corruption' of data is paramount.

Cell-switching (as used in *ATM—asynchronous transfer mode*)[14] is a specialised connection-oriented form of packet switching in which the packet lengths are standardised at a fixed length. Cell switching (also called *cell relay switching*) was developed by ITU-T as the basis of the *broadband integrated services digital network (B-ISDN)* and *ATM (asynchronous transfer mode)*. It is intended to carry high bit rate mixed voice, data and video (so-called *multimedia*) signals.

Logical channels, relaying and tag- or label-switching

A special mechanism is usually built into *connection-oriented* network protocols to cater efficiently for the needs of *statistical multiplexing*. The packets of the various different end-users and their connections have to be 'labelled' in an efficient manner in order that they do not get mixed up with one another en route. In the case of the X.25-packet switching protocol, this is done by labelling each packet sent during the data transfer phase of a particular connection with the same *logical channel number (LCN)*. In the case of other protocols, a range of other names are used to describe the 'label', but all have a similar function. Example label names are:

- *logical channel number (LCN)* (X.25 layer 3 protocol);

- *data link connection identifier (DLCI)* (frame relay protocol);

- *virtual path identifier(VPI)/virtual channel identifier(VCI)* (ATM, asynchronous transfer mode);

- *tag* (IP *tag-switching*);

- *label* (*MPLS, multi-protocol label switching*).

It is important to understand how *logical channels* (and their above-listed equivalents) are used and thus how *label switching* works, since it is used in the design of connection-oriented data switches and routers to improve their data throughput capacity. The *logical channel number* or *label* is usually allocated only for the duration of a *connection* at the time of the

---

[12] See Appendix 8.
[13] See Appendix 9.
[14] See Appendix 10.

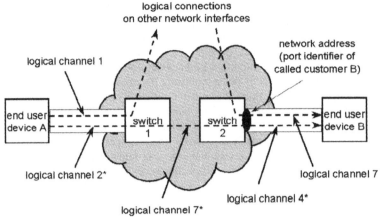

Figure 3.23   Layer 3 addresses: Logical channels and network addresses.

connection- or *call-set-up*. It provides for fast identification (and thus for fast-switching) of packets belonging to a given connection.

Let us study the connection from end-user device A to end-user device B in Figure 3.23. The connection traverses two switches (Switch 1 and Switch 2). There are thus three separate links or 'hops' along the route. Each of these links or *hops* represents a *network interface* (e.g., UNI, NNI, etc). For each network interface, a *logical channel number (LCN)* or other-named *label* is allocated and 'tagged' to each of the packets traversing the connection. Thus the end-to-end connection of our example comprises three separate logical channels (one for each of the *network interfaces*), concatenated end-to-end. On the first (end-user device A to Switch 1) logical channel 2 is allocated. On the inter-switch trunk, logical channel number 7 is allocated, and from Switch 2 to end-user device B, logical channel 4 is allocated.

During the data transfer phase (of the *call* or *connection*), the switches revert to providing a *switching* or *relaying* service as follows. Switch 1 *switches* or *relays* packets with the 'logical channel number 2' (from end-user device A) to the 'logical channel 7' on the trunk to Switch 2. Similarly, Switch 2 only needs to switch or relay 'logical channel 7' on the trunk to 'logical channel 4' on the end-user connection. This *switching* or *relaying* can be performed by specialised hardware. This has significant advantages for the design and performance of the switch hardware.

We shall assume that in the example of Figure 3.23 there are a maximum of 8 logical channels on each of the three data links, each represented by a logical channel number between 0 and 7, and represented therefore by a three- logical channel number (LCN). The *relaying procedure* need only study this relatively short address and be aware of the pre-determined 'next hop' and 'next logical channel' of the connection to which it must switch. For this purpose very fast hardware and software can be designed. This specialised hardware and software can perform a lot faster by restricting itself to *relaying* than rather than having to perform a 'look-up' of the [60 or more bit-] destination *network address* for each packet to be forwarded. The importance of this relaying technique is underlined in its recent adoption as part of the modern IP (Internet Protocol)-based techniques called *tag switching, MPLS (multi-protocol label switching)* and *IP version 6 (IPv6)*. (More on this subject in Chapter 7.)

The logical channel numbers are allocated at the time of connection set-up. End-user device A may have selected the preferred logical channel number '2' on the first link. Switch 1, meanwhile, allocated logical channel number '7' on the middle link of the connection, and

Switch 2 allocated logical channel 4 on the last link. The basis of the choice of logical channel might have been simply the 'next free one available' or 'the one best-suited to the type of connection requested by the caller'. Usually, only a limited number of logical channels are available at each *network interface*. This depends upon the number of labels made available in the protocol and reflects the limitations posed by the hardware design of the network switches or end-devices. There is a limited amount of storage capacity and number of different data buffers which can be assured to serve multiple simultaneous connections!

The *network address*[15] of the destination needs to be identified to each of the two switches in our example of Figure 3.23 only during the connection *set-up* phase. The network address allows each switch or router to select the most appropriate 'next hop' of the connection. After the decision, a logical channel number (or label) for the next hop is allocated and the connection is switched through. The network address is *signalled* to switches or routers by means of the layer 3 (network) protocol, using a call or *connection* set-up procedure. However, once the data transfer phase of the connection commences, the network address is no longer required and the subsequent user data packets 'transferred' do not include it. Instead, the packets are merely labelled with the (much shorter) logical channel numbers (LCN) or other label equivalent.

Layer 3 (network) protocol control information (PCI) and frame structure

The *protocol control information (PCI)* necessary for, and the procedures undertaken by *connection-oriented* packet networks differ greatly from those of *connectionless* networks. In particular, in connectionless routing, each packet sent from the source to the destination must be labelled with the full network address of the destination and must be *routed* and *forwarded* separately by intermediate *routers* according to this address.

Next, we shall use the X.25 layer 3 packet-format here to illustrate the functions of a *connection-oriented* protocol. Afterwards we will compare how this differs with the format and function of a *connectionless* protocol, using IP (Internet protocol) as our main example. IP itself will be covered in detail in Chapter 5.

X.25-layer 3 packet format and protocol

The X.25-protocol is a network (layer 3) protocol used between peer devices at the UNI (user-network interface) of a public packet-switched data network (often called an *X.25 network* or *packet-switched network*). Imagine that the DTE of Figure 3.1 is 'talking' to the network (DCE) using the X.25 layer 3 protocol to tell the first node in the network how to handle the X.25 calls. The protocol handlers in the DTE and in the first switch node in the packet network are *peer* partners, there being a *network interface* between them.

Being a *connection-oriented* protocol, X.25 (ITU-T recommendation X.25) defines distinct procedures for *signalling* and *data transfer* during the various phases of the *call*, including:

- *call request* and *connection set-up*;

- *data transfer*, including flow control;

---

[15] The network address (the layer 3 address) is the unique address identifying the network port to which the destination device — end-user device B (see Figure 3.23) — is attached. (It should not be confused with the layer 2 and MAC-addresses we spoke of earlier). The network address is usually a lengthy number. International telephone numbers, for example, may be up to 15 digits long. Coding each decimal digit as a 4-bit *binary coded decimal (BCD)* number, the network address is a total of 60 bits long. Internet addresses (which are examples of *network addresses*) usually comprise a 32-bit main address and up to 32 bits of subaddress (64 bits in total). And the ever increasing demand for more numbers means that we can only expect network addresses to get even longer than 64 bits (8 bytes). IPv6 address, as we shall see in Chapter 5, are 128 bits long!

- *supervision*, including retransmission, interrupt, reset, restart, registration and diagnostics; and

- *call clearing* (once communication is finished).

Before being able to communicate across a packet-switched-network, an X.25 DTE first has to signal its desire for a connection to be set-up. For this purpose the DTE generates an X.25 *call request packet* which includes all the information needed by the network 'control point' to set up the connection (including *called address* (the B-end destination), *calling address* (the A-end origin of the call), features, *facilities* and network services needed for the connection and any related *call user data*. The DTE selects the preferred logical channel number (LCN) it would like to use for the call, *signalling* this in the call request packet (Figure 3.24). This logical channel is then put in the *DTE-waiting* state, while the network node (i.e., DCE) decides what to do next.

During the DTE-waiting period, the network node uses its internal *routing table* to determine the best route to the indicated *called address*, and continues the call set-up procedure by negotiating with nodes further along the connection. When all the nodes have mutually 'agreed', the connection of the various links and nodes to the destination can be established. At this point, the DCE sends a *call accept* packet back to the DTE and data transfer can commence.

During *data transfer*, a much simpler packet can be sent (Figure 3.25), including only the *general format identifier* (bits Q, D and *modulo*), the logical channel number (LCN), comprising the 4-bit *logical channel group number (LCGN)* and the 8-bit logical channel number (LCN), the user data and the packet type, which during data transfer contains nothing more than the send and receive packet sequence numbers, P(S) and P(R), as used for data flow control. The *Q-bit (qualifier bit)* distinguishes user data packets from control packets.

You may wonder why there is a similar data flow control mechanism (using sequence numbers) to that implemented in layer 2 protocols. Why do we need a second one, you may ask? This lies in the use and meaning of the D-bit (*delivery confirmation bit*) in the first

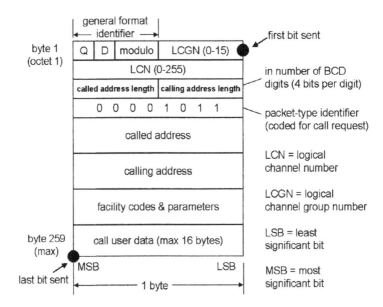

Figure 3.24   X.25 call request packet format (call set-up phase).

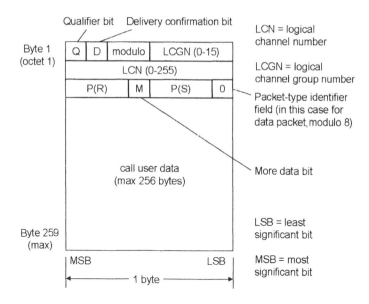

Figure 3.25   X.25 Data packet format (data transfer phase).

byte of the packet.[16] Set at D = '1', this bit indicates that the local DTE wishes to receive a confirmation from the remote DTE of the receipt of a given packet on an 'end-to-end' basis. In other words, the layer 3 flow control mechanism is acknowledging packets received across the entire end-to-end connection across the network.

More details of the X.25 protocol and packet format appear in Appendix 8.

Connectionless layer 3 protocols — the internet protocol (IP)

The principle difference between connectionless networks and connection-oriented ones is that a connectionless network has to handle and forward each packet of data (each *message*) separately. There is no 'stream' of messages passing between the two same end-points making up a *virtual circuit*. As far as a connectionless network is concerned, no two messages are related, each individual packet must have its destination network address inspected and each must be *routed* individually. Consequently, the nodes in connectionless networks are usually called *routers*. Both the Internet and *IP (Internet protocol)* are based on connectionless networking.

A layer 3 (network) protocol intended to support connectionless network service (e.g., IP) differs from a layer 3 protocol such as X.25 intended to support connection-oriented switching. Instead of there being a distinct two-phase establishment of communication (first, connection set-up and then an optimised data transfer phase using logical channels), there is only one *routing* function and a single packet-format. If you like, each packet has to be built like a *call request packet* of X.25, but with the entire user data (a much larger field) already attached to it. There is no *confirmation* or *acceptance* of the connection, merely a forwarding of each

---

[16] The process of confirmation is occurring by means of a 'cascaded' process across each of the interfaces of the network, not necessarily using identical packet sequence numbers at each of the interfaces along the way. Remembering the delay problems we discussed in conjunction with Figure 2.28, you can imagine how careful we have to be in setting the layer 3 window size, for now the unacknowledged packets are spread all along the route.

packet to the *next hop*. Many more details of the exact IP (Internet protocol) packet formats and protocol procedures follow in Chapter 5.

Route selection and routing tables

No matter whether a connection-oriented or a connectionless network protocol is in use in a data network, each network node relies on an internal *routing table* to determine the most appropriate *next hop* (i.e., next link and next node) towards the destination, as identified by the destination network address. Figure 3.26 illustrates an example network in which 5-digit network addresses are in use. Thus all addresses in the range 001XX (e.g., 00146) are located on node A. Similarly the ports with addresses in the range 091XX are connected to node F. The diagram shows possible routing tables (for use in nodes A and B) for each of the six available number ranges. For each number range, the routing table provides first, second and third choice routes.[17] The first choice route will always be selected, provided that it is both available (i.e., in operation rather than in a faulty state) and not already over-loaded with traffic. Should the first choice route be unavailable or congested, then the second choice route will be taken. And if this in turn is also unavailable or congested, then the third choice route will be used. Only if all three routes are busy or unavailable will the connection set-up fail.

The process of looking an address up in a routing table involves comparing the first few digits of the destination *network address* with the entries in the *routing table*. The routing table entries are normally listed in ascending numerical order (as in Figure 3.26). The routing table entry which provides the nearest match (to the most number of leading digits) is used to determine the appropriate route (i.e., next hop).

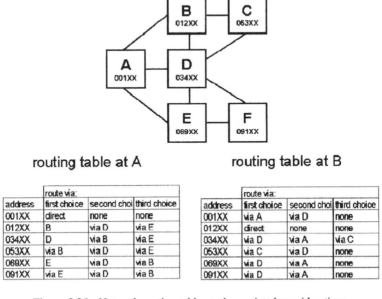

routing table at A

address	route via: first choice	second choi	third choice
001XX	direct	none	none
012XX	B	via D	via E
034XX	D	via B	via E
053XX	via B	via D	via E
069XX	E	via D	via B
091XX	via E	via D	via B

routing table at B

address	route via: first choice	second choi	third choice
001XX	via A	via D	none
012XX	direct	none	none
034XX	via D	via A	via C
053XX	via C	via D	none
069XX	via D	via A	none
091XX	via D	via A	none

Figure 3.26   Network routing tables and associated considerations.

---

[17] In many *routers* only a first choice route is saved in the *routing table*. The best second choice route is calculated only following the failure of the first choice route.

The three greatest challenges faced when creating routing tables are:

- determining the full list of *reachable addresses* and keeping this permanently up-to-date;

- creating a routing algorithm (calculation procedure) for ensuring the most efficient overall use of the network when working out the best routes to each individual destination; and

- avoiding network instability, which might occur if different nodes chose routes uncoordinated with their neighbours. Examples of instabilities are *circular routing* and *route flapping*.

Figure 3.26 illustrates how easy it is to introduce the possibility of *circular routing*. Consider a call being made to the network address range 069XX located on node E. The routing table at B for reaching node E is: first choice via D, second choice via A. There could also have been a third choice via C (in case the link BD was congested). Now let us consider routes from node A to node E: first choice direct, second choice via D, third choice via B. This all seems to make sense! But what happens in the case that node D fails? First: calls from B to E are routed via A to E. But maybe link AE becomes congested as a result, causing A to select its third choice route to E — via B! Now *circular routing* occurs: B will select A as the next hop to reach E, but node A will decide subsequently to route the call back to B. B, of course, sends back to A and so on ad infinitum, until either the caller 'hangs up' or until either link AE or node D recovers. In the meantime, the AB link rapidly congests, as the call spirals between the two nodes, locking up further capacity.

Route flapping can occur in the case where routing tables are automatically updated. Automatic adjustments of routing might be undertaken in an attempt to adapt the network routing according to current overall traffic load. Let us assume that particularly heavy traffic is affecting the route originating at node B and destined for node E. Let us further assume, that the first and second route choices selected at node B for the address range 069XX are automatically swapped around (to be via A second choice via D) in the case that the link BD exceeds a certain traffic congestion level. What might happen in practice? Answer: all the traffic from B to E is first routed via D and loaded onto link BD. The traffic congestion threshold is reached on this link, so the first choice of node B for address range 069XX is changed to be 'via node A'. Immediately all the traffic routes via node A, so link BD becomes virtually free again — and the first choice route reverts back to be via link D. Subsequently, the route continues switching from 'via D' to 'via A' ad infinitum (until the overall traffic drops). This is called route flapping. It is dangerous because it leads to very unpredictable traffic flows across the individual links of the network.

In public telephone networks, it is normal to establish the routing tables manually — working out the network address ranges which need to be *reachable* and determining the best outgoing route to reach them. Data networks, on the other hand, have traditionally been designed to work the routing tables out for themselves. This requires an automatic mechanism allowing network nodes to inform one another of the exact whereabouts of each destination network range or even individual network address. This is the domain of *routing protocols* and *routing algorithms*, examples of which are *RIP (routing information protocol), OSPF (open shortest path first)* and *BGP (border gateway protocol)*. We shall detail the precise functioning of these protocols and the operational network management problems which they bring in Chapter 6.

3.5   Layer 4—transport layer protocol

The transport layer (layer 4) protocol is responsible for providing network-independent communication services between computer applications running in end-user terminals. In theory, data communication (i.e., *transport* of data) may have to traverse different *networks* along the way. The transport layer in this case has to ensure the coordination and control of all the various networks and arrange communication end-to-end. The transport layer provides what is correctly called the *transport service.*

The most commonly used *transport protocols* nowadays are *TCP (transmission control protocol)* and *UDP (user datagram protocol).* Both of these protocols are normally used in conjunction with IP (Internet protocol)[18] based networks. TCP is said to provide a *reliable* transport service, and works as a *connection-oriented transport service (COTS).* The reliable and connection-oriented nature of TCP/IP gives it similar properties to an X.25 network. Note how X.25 did not need a transport layer to achieve this — users of X.25 typically use a *null layer* at the transport layer. UDP, meanwhile, is unreliable and is based on a connectionless *transport service.* Unlike TCP, UDP is unable to guarantee delivery of the message, but UDP is efficient in the use of the network for lower priority and short messages, since it requires much less *protocol control information (PCI).* In other words, it uses a shorter packet *header.* One often speaks of UDP being a 'best effort' protocol.

Peer-to-peer communication at the transport layer usually takes place between peer partner software *applications* running in the two end-user computing devices. At the originating end, the transport layer organises a 'connection' on behalf of the computer application, thereby isolating the application completely from the constraints of the network or networks over which the data will be carried. The transport layer need only be informed by the application of the destination network address and software application (i.e., the *peer partner* — for example, the IP address of the destination device to which data is to be transported).

As necessary, the transport layer may *split* up a long stream of data into separate packets or individual messages and organise for these to be transported over one or more different networks and paths to the destination (Figure 3.27). This is important when using IP networks, since each packet is carried at the IP level as if it were a completely separate message. At the destination, the separate packets are once again reassembled into the correct data sequence. For this, the transport layer protocol uses *data sequence numbers* just like layer 2 and layer 3 protocols. (Yet more sequence numbers, I here you say!) These are necessary for the reassembly process — helping ensure that packets which travelled via different networks are reassembled in order. In addition, the transport layer (as also at layers 2 and 3) provides for flow control and error correction on an application-to-application basis.

You may be wondering why each protocol layer appears to have yet another provision for sequence numbers, flow control and error correction. The simple answer is that each progressively higher layer may have to coordinate multiple networks or datalinks of the layer beneath it, and has a responsibility to 'bring this all together' as a single, coordinated *service.*

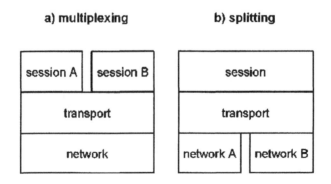

Figure 3.27   Transport layer functions of multiplexing and splitting.

---

[18] TCP is also well suited to use with frame relay.

The sequence numbers of the transport layer, for example, control the flow of data directly between the end applications. These numbers are *synchronised* on an end-to-end basis, making the coordination much simpler.

Why not simply throw away the flow control and error correction at the lower protocol levels? To some extent, more modern protocols (e.g., *frame relay*) have much weaker or no flow control and error correction procedures at the lower protocol layers, in order to reduce the total overhead of the *protocol control information*. When necessary, packet corruptions or losses can be recovered by having them retransmitted from their source. But in the past, when bit errors during transmission were more common, it made no sense to 'start all over again' from the start of the connection when half the journey (i.e., some of the data links) had been traversed without errors. Recovering the error at a midway point and 'carrying on' is sometimes far more efficient than taking the risk that on the next attempt from the beginning you do not even get as far, but meanwhile increase the load on the network.

Consider a simple connection comprising two data links, the first of which runs at a faster bit rate than the second. Then at the datalink level, the first link can deliver data into the intermediate node faster than the intermediate node is able to forward it. This is the reason why the intermediate node needs its datalink (i.e., layer 2) flow control. It ensures, in the short term, that the buffer does not overflow. Meanwhile the data flow controls of the higher layers (3 and 4) regulate the flow end-to-end across the intermediate networks and complete connection. If, for example, the receiving end-user device can only accept data at a relatively slow rate, then it is worth trying to match the original sending rate to the same speed. If more data is sent into the network than can get past a given 'bottleneck', or more data is sent than the receiver can accept, then the data only serves to clog up all the data buffers along the way, like a queue of cars at every traffic lights because everyone left for the same holiday destination at the same time. The solution? To spread the departures over a longer period of time.

Another important capability of the transport layer protocol is its ability to *multiplex* different *sessions* across the same transport 'connection'. The different possible sessions correspond to different functions which the end-user computers may be coordinating between one another. Different sessions are identified with different *port numbers* corresponding to the different activities. Different port numbers are allocated within TCP and UDP, for example, for the following different *session, presentation* and *application* services:

- *FTP (file transfer protocol)* — an *application layer protocol* which transfers data files between computers;

- *Telnet* — an network application allowing a PC to log-on to a remote mainframe computer via the Internet;

- *SMTP (simple mail transfer protocol)* — used for email service;

- *HTTP (hypertext transfer protocol)* — used for the Worldwide Web;

- *DNS (domain name system)* — used as a 'directory' for Worldwide Web addresses in the form www.company.com

- *SNMP (simple network management protocol)* — used for network management communication and control of the devices;

- *POP3 (post office protocol)* and *IMAP (interactive mail access protocol)*–used for email service.

In addition to TCP and UDP, there are a number of other widely used transport protocols, but we will not discuss these in detail (More about transport protocols and application *port numbers* in Chapter 7!):

- *TLI (transport layer interface)*, as defined in 1986 as part of the open systems interconnection (OSI) work and defined in ISO 8072 and ITU-T recommendation X.214;

- *Transport protocol* (associated with TLI above and defined by ISO 8073). This defines different service levels, specifically TP1, TP2, TP3, TP4 and TP5, of which TP4 (class 4) is the most widely used. TP4 stands for OSI transport protocol class 4 (error detection and recovery class). TP4 is the OSI 'equivalent' to TCP. It was developed by the US National Bureau of Standards and forms part of the US *government OSI Profile (GOSIP)* architecture;

- *Windows95 transport protocol/NDIS (network driver interface specification)* — this allows multiple higher layer protocol stacks to be *multiplexed* across a network;

- *NetBEUI (NetBIOS extended user interface)* — this provides a standard transport layer frame format for conveyance of NetBIOS (network basic input output system), as used to control sessions in LANs.

3.6 Layers 5–7 — higher layer protocols

The *higher layer* protocols of layer 5 (*session layer*), layer 6 (*presentation layer*) and layer 7 (*application layer*), like the transport layer (layer 4) are resident in the computer end-devices which communicate end-to-end across the network. The *peer partners* are computer software programs at the *operating system* and *application* program level. Because such programs are often delivered as a complete 'suite' of software by a given software or operating system manufacturer (e.g., Microsoft, Sun Microsystems, etc.), the standardisation of software interfaces at these *higher layers* has come to follow the 'proprietary' and 'de facto' standards established by the computing industry rather than the original strict layers envisaged by OSI. The number of 'layers' of protocols and software and the boundaries between layers can thus become somewhat blurred. Thus, for example, a single software may spread across all the layers 5–7 (or even 4–7). Put simply, only a few still comply rigidly to the OSI model and standards. Nonetheless, it is instructive to consider examples of commonly used software which provides the functionality of these layers.

Session layer

The *session layer* is an end-user computer program responsible for *binding* (setting up and managing communications) and *unbinding* logical links between two application programs (e.g., computer-to-computer, computer-to-printer, etc.). Put simply, the session layer manages the different computer hardware resources which are interconnected by means of a network. It is thus strongly linked to the computer *operating system (OS)* software. The session layer provides 'network gateway capabilities' which link the 'computer world' to the 'telecommunications world'.

Above the session layer, computers are largely concerned only with datafiles (in different formats) and 'real' application software. These are the concerns respectively of the *presentation* and application layers. In this world of programming and computing hardware, software designers, network administrators and users prefer to refer to devices with names or acronyms such as 'LPT1', 'local printer', 'group-printer', 'peter-pc', etc. Referring to the devices like this means you don't have to change the software each time the network gets reconfigured, somebody moves or the network addresses change. But it does mean that a central network resource management is needed, linking computer names to network addresses, binding and unbinding these devices to their addresses and setting-up sessions when they wish to communicate with one another. This is the job of the session layer.

The session layer has to be able to cope with recovering such problems as the remote computer being switched off. Having initiated a session, the session layer has to be able to check that it remains 'alive'. The software commands issued to the session layer and used to invoke and control session activities are 'input' and 'output' commands which will be familiar to computer programmers:

accept	Msg.Call	Read	Sendto
bind	Msg.Listen	recv	setsocket
close	Msg.Receive	select	shutdown
connect	Msg.Send	semdmsg	socket
listen	Msg.Session.Status	send	—

Typical and well-known examples of software providing session services and protocols[19] are:

- *Berkeley socket service* or *socket interface* (this was the first consistent application programming interface, used in UNIX and TCP/IP networking environments since the early 1980s);

- *NetBIOS (network basic input output system)* — provides for session and transport services between PCs and servers in LANs. It was originally developed for IBM's LAN manager software, but subsequently became the basis of Windows NT and Windows 2000 networking, which is nowadays the most commonly used LAN management software.

- *RPC (remote procedure call)* — used for the initial UNIX *client server* networking;

- *Secure sockets layer (SSL)* — a layer introduced between session and transport layers to provide for extra authentication and encryption of communications between a web user (client) and a web server;

- *Winsock* (Windows socket interface);

- *SIP* (session initiation protocol) for VOIP (voice over IP).

A session *binds* two *logical units (LUs)* by arranging for the interconnection of the two relevant network ports (identified by their *network addresses* and transport layer *ports*). Session multiplexing allows a single device (e.g., a *server*) to 'speak' with multiple remote devices (e.g., *clients*) at once.

The term *bind* originates from *Berkeley Internet Name Daemon*. This was originally a software and a related file which provided a link between the logical names known by computer programs and Internet addresses. In a UNIX system, the IP address information related to computer and device *usernames* is held in a file called 'etc/hosts'. The Windows equivalent is called *WINS (Windows Internet name server)*. It deals with name registration, queries and 'releases'.

As the Internet rapidly became more complex, it became impractical to store all possible remote device names, web addresses (e.g., www.company.com) and related network addresses in a local file called /etc/hosts and to keep this file up-to-date. Instead, *root directories* and the *domain name system (DNS)*[20] were developed.

---

[19] See Chapter 10.
[20] See Chapter 11.

Session (layer 5) and transport layer (layer 4) addresses

Before we leave the subject of the session layer, we should briefly explain the two terms: *session target* and *socket number*. The session target (a term being increasingly used) is the (layer 5) destination 'address' of a session in its initiation phase (it is a *logical* username or equivalent). The *socket address*, meanwhile, is a layer 4 (transport layer) address, usually quoted as a combination of the IP *network address* (layer 3) and the TCP *port number* (layer 4).

Presentation layer

The *presentation layer* (layer 6) is concerned with the *format*[21] and *syntax* of data communicated between applications. Since most computers uses ASCII format for the coding of basic characters, presentation conversion does not have to be carried out at this level. On the other hand, software commands and datafile formats can vary widely, and need to be standardised or converted for the purpose of open communication. Control codes, special graphics, graphical user interfaces, browsers and character sets work in this layer. The different applications and associated datafile (presentation) formats which serve them are the domain of *object-oriented* programming.

Typical examples of presentation layer related services and formats are:

- *Graphical user interface (GUI)* — a presentation format conversion which typically converts communicated characters into 'graphics' — typically screen 'windows' or *browsers*;

- *XDR (eXternal data representation* — RFCs 1014 and 1832) is a language used to describe different data formats. It was developed by Sun Microsystems to facilitate the interchange of files between different types of computer;

- *ASN.1 (abstract syntax notation 1)* — is an ITU-T and ISO-defined language somewhat similar to XDR — and also used to describe different data formats. ASN.1 and the SMI (structure of management information) subsets of it, have become one of the standard means of defining application layer protocols, as we shall discover in Chapters 9 and 10;

- *html (hypertext markup language* — RFC 1866) — provides a character set (ISO 8859) and a method of punctuating text with html *tags*. The tags indicate how the text is to be presented to the human user receiving the file (e.g., bold, italics, background colour, blinking, position, size, etc.). html also provides for the possibility of hyperlinking text.[22]

- *MIME (multipurpose Internet mail extensions)* — a standard format for including non-text information in Internet email which enables the attachment of multimedia data files;

- *Network virtual terminal (NVT)* — this is a process and format makes remote terminals with different characteristics (e.g., a PC, an appleMac, a 'dumb terminal', etc.) all appear the same to a server or host computer;

- *PostScript* — a widely used 'print' file format used to send data to computer printers;

- *Server message block (SMB)* and the similar *CIFS (common Internet file system)* — provides a standard control block structure for server communications in Microsoft Networks, including, among other things, file and printer management and queuing. SMB was the original version, CIFS is the name of the latest version;

---

[21] See Chapter 2.
[22] See Chapter 11.

- *Connection-oriented presentation service (COPS)* and protocol (ISO 8823 and ITU-recommendations X.216 and X.226).

Application layer

Application layer (layer 7) protocols perform input and output routines (or sub-programs) on behalf of more complicated end-user application programmes. Invoking an application layer protocol is typically done using modern *object-oriented* computer commands and file structures that C, C++ and JAVA programmers are familiar with. Typical examples of application layer services and protocols are:

- Berkeley remote commands (UNIX);

- *FTAM (file transfer, access and management)* — the OSI protocol suite for transferring files between computers;

- *FTP (file transfer protocol)*;

- *HTTP (hypertext transfer protocol)*;

- *NFS (network file system)* — developed by Sun Microsystems for networked sharing of files between UNIX computers and servers;

- *POP3 (post office protocol)* and *SMTP (simple mail transfer protocol)* — widely used by modern email systems;

- *Telnet* (remote log-in protocol);

- *X.400* or *message handling service (MHS)* — the OSI version of email service.

3.7 Protocol stacks and nested protocol control information (PCI)

Modern telecommunications protocols, as we have seen, are designed as layered *peer-to-peer* protocols. Each *layer* performs a separate distinct function which is coordinated by the *peer* partners (the two protocol handlers which deal with the particular function or layer at the two ends of the connection). Each layer adds its own *protocol control information (PCI)* to the original user data. PCI identifies the *address, socket* or *target* of the *peer partner* and allows for such functions as sequence numbering, data flow control, error correction, port or logical channel identification, information regarding the data format, and so on.

In reality, the communication starts at the *application layer* of the sender. The *user data* (i.e., the user message or data) is packed into the *user data field* of the *application layer frame*. In the jargon, the user data field is called the application *SDU (service data unit)*. The SDU is passed to the application layer to make use of the *application service*. The point of handover is called the *service access point (SAP)* of the *application layer*. Protocols identify the SAP with some kind of address. The address is called the *SAPI (service access point identifier)*.

In order to invoke the *application service* at the application *SAP* (Figure 3.29) a number of commands have to be issued, to tell the application layer protocol exactly what it has to do. These commands are part of the protocol and are called *primitives*. Once the application SDU has been received, the application layer adds the application layer *protocol control information (PCI)* — as if it were 'talking' directly to the application layer protocol handler in the receiver (i.e., in *peer-to-peer communication*). The combination of the user data (the application layer SDU) and the application layer PCI (added as *headers* and *trailers*) is an *application layer frame*, but also called the application layer *protocol data unit (PDU)*.

Though the sending application layer has the impression it is talking directly *peer-to-peer* with the application layer at the receiver, in reality, the application layer PDU has to be processed by all the other protocol layers. Under a new name (*presentation layer SDU*), the (unchanged) application layer PDU is passed to the *presentation layer* at the *presentation-SAP*, using presentation layer primitives as appropriate — to 'order' the correct *presentation service*. The entire application layer frame becomes the 'user data' as far as the *presentation layer* is concerned, and to it are added the headers and trailers of the presentation layer *protocol control information (PCI)* to form the presentation layer PDU.

In turn, the *presentation layer PDU* becomes the *session layer SDU*; passed over at the session layer SAP, using the session layer primitives. And so on. Incidentally, the transport layer SAPI is the *port number* or *socket number* we talked of earlier, while the network layer *SAPI* is the *network address* and the datalink layer *SAPI* is the MAC-address.

The actual frame transmitted over the physical medium (the end result of all the individual layer protocol activities) thus has a structure which appears like a set of 'nested' frames, the layer 7 frame is the user information of layer 6. The layer 6 frame is the user information of layer 5, and so on, with the frame itself finally delimited by the layer 2 (datalink) frame and flags (Figure 3.28).

The *headers* and *trailers* comprising the *protocol control information (PCI)* of each *protocol layer* are stripped from the received message by the corresponding protocol handler (the *peer* partner) in the distant end device, and the *user data field* information (i.e., the remainder of the message) is passed to the next higher protocol layer, and finally to the 'end-user'. In other words, the actual communication is first <u>down</u> the *protocol stack* of the sender and then <u>up</u> the *protocol stack* at the receiver (Figure 3.29), though each individual protocol layer believes itself to be in direct *peer-to-peer* communication with its remote *peer partner*.

If you consider the actual data frame sent over the medium, it comprises a flag, followed by a large number of PCI fields (for all the various layers), the user data and then all the various further PCI fields in the trailer. This is when you really begin to question whether all the different fields are necessary (must I really send multiple sequence numbers?). The more PCI

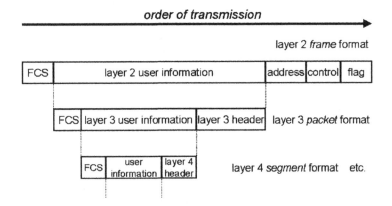

FCS = frame check sequence

Note: An FCS (frame check sequence) is not always used at each of the protocol layers. Whether one is used depends upon the design of the protocol—and in particular upon the reliability of the network for which the protocol is intended. In addition, the terminology FCS (frame check sequence) is not used by all protocols. The IP-suite of protocols, as we will discover later has a checksum file instead (see Chapter5). The checksum algorithm used is much simpler than the FCS of HDLC (higher level data link control), which we presented earlier in this chapter.

Figure 3.28   'Nested' frame formats of layer protocols.

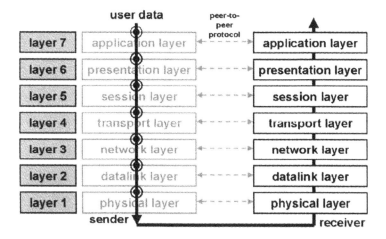

⊙ = SAP = service access point

Figure 3.29   Processing and transfer of user data by layered protocols.

fields I can do without, the lower is the total *overhead* — and the more efficient is my usage of the network capacity. Unsurprisingly, the relative efficiency of different alternative protocol suites is a hotly and continuously debated subject. And protocol designers are continually developing protocols as the needs of the users change and the reliability of networks improve.

In my experience, different protocols designed for similar purposes are rarely significantly more or less efficient than their competitors, as their designers would like to have us believe. Quite simply, they have to add similar amounts of overhead to carry out the required functions, even if the functions are performed slightly differently. As the Germans would say: 'We can all only cook with water', i.e., you can't break the laws of physics!

3.8   Real networks and protocol stack representations

It is often helpful, in attempting to understand protocols and their relationships with one another to illustrate the *protocol stacks* of the various devices. Books about data communications and protocols are littered with protocol stacks (as indeed this one is). A diagram of the protocol stack helps to clarify the roles played by different pieces of networking and computer hardware and software. Figure 3.30, for example, illustrates a typical office LAN (local area network), comprised of end-user devices, *structured cabling*, a LAN *hub* and a *bridge* or *router*. Note how the end-user PC of Figure 3.30 takes part in all the protocol layers (including layers 4–7 which are not illustrated). The LAN devices (the hub, LAN switch and bridge), meanwhile, only have functionality corresponding to layers 1 and 2 at most (the LAN *medium* and *datalink* layer protocol, *LLC—logical link control*.

3.9   Protocol encapsulation

When new protocols are devised, it is often useful to be able to use older (so-called *legacy*) networks to transport them, or conversely, to be able to carry the old protocols across the new network. A 'quick-and-dirty' method of achieving this is by means of *protocol encapsulation*. In purist terms, protocol encapsulation relies on 'cheating' — taking the *PDU* (*protocol data unit*—i.e., the complete frame including real user data as well as headers and trailers) of a

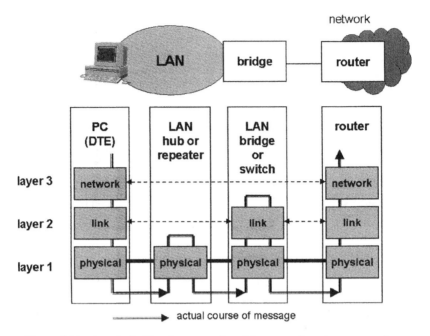

Figure 3.30   A typical office data network and its protocol stack representation.

low level protocol and transmitting it across another network by pretending it is 'user data' of a higher level protocol of the second protocol suite.

By using *protocol encapsulation* you could mimic a simple layer 2 datalink between two devices, even if the actual communication takes place across an IP-based router network. How? By delivering the layer 2 frame of the datalink to the transport layer of the TCP/IP network as if it were 'user data'.

Protocol encapsulation (also called *tunnelling*[23]) often proves a useful way of interconnecting and integrating networks of different vintages and originally meant for different purposes. It is useful to remember that, however many protocol layers are placed 'underneath' the *encapsulated* protocol, the resulting protocol *service* will remain the same. For example, *PPPoE — PPP over ethernet —* provides for a *datalink layer service — PPP (point-to-point protocol)* service, but carried over an ethernet LAN. Similarly, encapsulating a *datalink* protocol (layer 2) into a layer 4 protocol (such as TCP/IP) will leave the end-to-end service unchanged — as a layer 2 service.

Be careful when using protocol encapsulation! Protocols which have been encapsulated do not always behave quite the way they would when left to themselves. The extra delays caused by the *encapsulation* and the transport can lead to serious problems, if not with the encapsulated protocol itself, then with the application which is being run across it.

3.10   Control and management protocols

When I think back to my first encounters with layered protocols and my attempts to understand *multiprotocol* networks, I can recall a series of long mental struggles: trying to relate the different protocols onto the 7-layer OSI model and trying to figure out the roles of individual

---

[23] See Chapter 13.

protocols in coordinating the overall activities of the network. One of my biggest breakthroughs was to realise that many of the protocols used in networks are not intended to carry messages between the A-party and B-party of a connection, but instead for control and management of the network itself.

Figure 3.31 illustrates a *protocol reference model (PRM)* developed by ITU-T to illustrate how different protocol stacks on the *user plane (u-plane)*, the *control plane (c-plane)* and the *management plane (m-plane)* combine with one another. So far in this chapter, we have concerned ourselves mainly with the protocols used on the *user plane*. These are the protocols used directly in the transfer of user data from end-user DTE to another.

The control plane is used to describe protocols which communicate information from an end-user DTE to the (control part) of a node, or between nodes within the network — for the purpose of conveying information needed for establishing, controlling and clearing connections (Figure 3.31a).

Management plane protocols, meanwhile, are used only within the network itself. They might carry status information regarding network or end-user equipment performance to a human operator at a network management centre. Alternatively, they may be used to deliver controls and commands as required to reconfigure the network. Management plane protocols are used, among other things, to allow nodes in a network to mutually update one another's routing tables. Routing protocols (as we will discuss in Chapter 6) come in this category.

It is important to recognise that entirely different protocol stacks may be used for the different *user plane*, control plane and management plane functions (Figure 3.31), though usually the same lower layers (layers 1–3) are used (e.g., IP and ethernet or HDLC).

For a user plane application, the two end-points of the communication are the A-end and B-end of the connection. For a control plane application, the communication end-points are the *control* part of a DTE and at the control part of a network node. Sometimes the

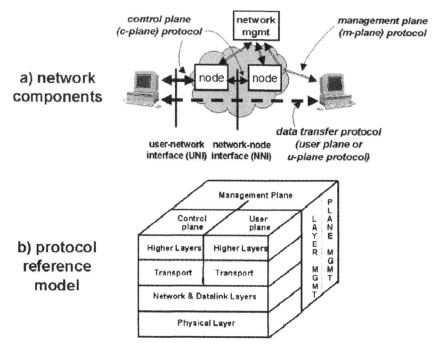

Figure 3.31 Protocol reference model: user, control and management planes [Figure 3.31b reproduced courtesy of ITU].

control protocol procedure is referred to as *signalling*. Between the DTE and node *control points* (or *signalling points*), a second connection may exist (over and above that used for the A-to-B user plane connection). The control plane connections are often dedicated virtual channels with pre-determined logical channel numbers (sometimes called *signalling virtual channels*). Similarly, pre-configured virtual channels may be used for different pre-determined *management plane* activities.

Unfortunately many books, and especially marketing brochures do not refer to the protocol stacks of different network functions accurately and compare protocols of different layers and functions, and even on different *planes* with one another. This confused me for a long time, and I have therefore tried hard in this chapter to navigate you around the main pitfalls. I hope I have succeeded, for a thorough understanding of the basics of layered protocols will make the rest of the book (and networking life as a whole) much easier to follow and understand.

3.11   Propagation effects affecting protocol choice and network design and operation

It might seem an obvious statement to make, but not all protocols are the same and not all protocols are equally well suited for a particular purpose or application. Obvious, but often forgotten, ignored or simply unknown! If a computer program or application is to run well, then the programmer should consider precisely how the protocols he or she selects operate. Similarly, network equipment designers need to consider the specific needs of each protocol and how they are to interact with one another. Unfortunately, too many programmers spend too little effort here, with the result that many applications run much slower than they need to. It's not always the case that the network is overloaded: far too often the application itself is poorly designed or is expecting too much from the network. The best solution to a slow application is not always simply to upgrade the network bandwidth or bit rate!

We discussed the effect on protocol delays of the indiscriminate use of data *acknowledgement* on a *datalink* we discussed in Figure 2.27 of Chapter 2. Here, we shall consider similar propagation delay problems and constraints as arise from packet buffering, error correction and relaying between multiple links and nodes of a network path.

Figure 3.32 illustrates how the number of 'hops' in a connection (the *hop count*) has a significant impact on the end-to-end propagation delay across the communications path [even

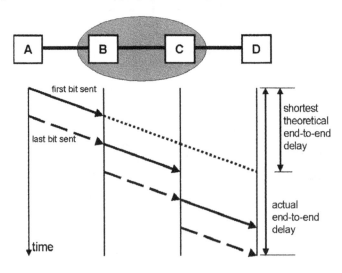

before considering the additional 'processing time' required by the nodes themselves]. Each packet (or frame) must be fully buffered at each intermediate station before it can be forwarded; and the packet length governs the period of time required to 'read' the entire packet into the *buffer* (this is the time which expires between the receipt of the first and last bits of the packet). Including the 'reaction and handling time' of each of the intermediate and end-nodes, the overall delay between the 'first bit sent' and the 'last bit received' (assuming that all the links have the same bit rate) is something like:

End-to-end propagation delay $= T_p + (h + 1)t_N + h(L_p/B) + Q$

Where: B = line bit rate in bit/s
h = hop count
L_p = packet length in bits (including protocol control information)
Q = sum of intermediate node *queueing* delays
t_N = node reaction and packet handling delay
T_p = minimum theoretical propagation delay = connection length/speed of propagation

---

The end-to-end propagation delay is thus very sensitive to the *packet length* and to the hop count, as well as to the *bit rate* of the line! Merely upgrading the bit rate on one of the links along the route will not have much effect, unless that link is overloaded — in other words, is adding significantly to the overall *queueing* delay, (the queueing is that caused by packets having to wait in a *queue* at intermediate nodes for use of the outgoing line).

Perhaps we could save some of the delay at the intermediate nodes by forwarding each bit immediately as it arrives (i.e., the first bit is sent on before the last bit arrives)? Some hardware designs and specific protocols do offer scope for tricks like this, but in many cases two constraints prevent simultaneous onward relaying:

- the need to set and check the error correction code (frame check sequence);

- the different bit rates of lines entering and leaving intermediate nodes (if the outgoing line is faster than the incoming line there is a problem, because the outgoing line expects to send bits faster than it is receiving them).

4

Local Area Networks (LANs)

LANs emerged in the late 1980s as the most important means of conveying data between different computers and computer peripheral devices (printer, file server, electronic mail server, fax gateway, host gateway, computer printer, scanner, etc.) within a single office, office building, or small campus. They were originally designed as *shared media* (layer 2 or *datalink* communications media) and are ideally suited for relatively short distance, high speed data transport and have thus become the foundation for modern 'electronic offices' — interconnecting workstations, word processors, shared printers, file servers, email systems, web servers and so on. We shall explain the various types of LAN and explain how they work but we shall primarily be concerned with the *ethernet* LAN in its various forms — 10baseT, 100baseT (fast ethernet) and Gigabit ethernet, for this has become the predominant standard for PC and server-based networking.

4.1   The different LAN topologies and standards

The different types of LAN may be characterised by their distinctive topologies, but all comprise a single *shared medium* transmission path interconnecting all the data terminal devices, together with appropriate *protocols* (called the *logical link control* and the *medium access control (MAC)*) to enable data transfer. The three most common LAN topologies are the *star, ring* and *bus* topologies. These are illustrated in Figure 4.1.

The original *IEEE 802.3 (ISO 8802.3)* standard defines a physical layer protocol called *CSMA/CD (carrier sense multiple access with collision detection)* which is usually used with a bus topology and referred to as *ethernet*. *IEEE 802.4* (ISO 8802.4) defines an alternative physical layer (layer-1) protocol for a *token bus*, but is also suitable for a star topology. *IEEE 802.5* defines a layer 1 protocol suitable for use on a *token ring* topology. Meanwhile, *IEEE 802.2 (ISO 8802.2)* defines the *logical link control (LLC)* protocol (equivalent to the layer 2 or *datalink layer*) which is used with any of the above physical layer protocols. LLC provides for the transfer of information in the form of data *frames* between any two devices connected to the LAN. The information to be transported (i.e., information frame or packet) is submitted to the *logical link control (LLC)* layer together with the address of the device to which it is to be transmitted. Much like HDLC[1] (from which it was developed), LLC assures successful transfer, error detection, retransmit etc. Figure 4.2 shows the various protocols and their relationship to the layers of the OSI model. The main difference between the different

---

[1] See Chapter 3.

Data Networks, IP and the Internet: Protocols, Design and Operation   Martin P. Clark
© 2003 John Wiley & Sons, Ltd   ISBN: 0-470-84856-1

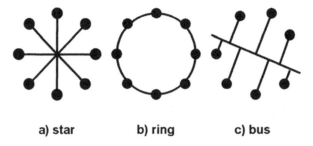

a) star   b) ring   c) bus

Figure 4.1   Alternative LAN topologies.

datalink layer (OSI layer 2)	logical link control (LLC)	IEEE 802.2		
	medium access control (MAC)	IEEE 802.3	IEEE 802.4	IEEE 802.5
physical layer (OSI layer 1)	physical protocol layer (PHY)			
	physical medium dependent (PMD)	ethernet	token bus	token ring

Figure 4.2   IEEE standards for LANs.

LAN technologies lies in how they prevent more than one terminal using the bus or ring at the same time.

4.2   Ethernet (CSMA/CD; IEEE 802.3)

Ethernet is based upon *CSMA/CD (carrier sense multiple access with collision detection)*. It is a *contention protocol*. On a CSMA/CD LAN the terminals do not request permission from a central controller before transmitting data on the *shared medium*: instead they *contend* for its use. Before transmitting a packet of data, a sending terminal 'listens' to check whether the *medium* is already in use, and if so, it waits before transmitting its data. Even when it starts to send data, it has to continue checking the path to make sure that no other stations have started sending data at the same time. If the sending terminal's output does not match that which it is simultaneously monitoring on the transmission path, then there must have been a *collision* (with one or more packets sent by other terminals). Following a collision, a *back-off* procedure has to be followed before re-attempting to resend data.

To receive data, the medium access control (MAC) software in each terminal monitors the transmission path, decoding the destination address of each *block* or frame of data passing to find out whether it is the intended destination of the data. If it is, the data is read and decoded, if not, the data is ignored.

Ethernet was originally a proprietary LAN standard (predating the IEEE 802.3 standard) developed by the Xerox corporation of USA at PARC (Palo Alto Research Centre). The original design was based upon a length of coaxial cable, with 'tee-offs' to individual work stations, with a maximum of around 500 stations. It could also be achieved using FM (frequency modulation) radio. The idea was to simplify the cabling needs of offices in which many terminals or personal computers were in use. Simply by laying a single coaxial cable along each of the corridors and connecting all the cables together, a bus could be created over which all the office computer devices could intercommunicate. Each time a new device was installed, a new tee-off could be installed from the corridor into the particular office where the device was situated (Figure 4.3a). Meanwhile, no new cabling needed to be installed along the corridor, so saving space in the conduits and averting the constant removal and replacement of the ceiling tiles when adding users to the network.

The technology of the initial coaxial cable form of *ethernet* (called 10base5) developed rapidly. First, a clever range of connectors for the tee-off points were developed, which enabled new devices to be connected very quickly without first severing the main coaxial cable bus. Another special connector, called the *MAU (medium attachment unit)*, converted the coaxial cable format of the main LAN bus to a convenient DB-15 (male) connector interface (called *AUI, attachment unit interface*) which could easily be connected to a special ethernet LAN *network interface card (NIC)* incorporated in a PC. These developments considerably reduced the time needed for new installations and reduced the disturbance to existing users. *Thin-ethernet (cheapernet* or *10base2)* also appeared. This allowed the use of narrower gauge coaxial cable as the main bus in smaller networks, and helped to reduce the installation costs.

As the number of computer devices in offices increased, multiple ethernets and increased flexibility became necessary, and this prompted the development of office *structured cabling*, using *twisted pair* telephone cabling laid to *wiring cabinets* in a star configuration, and the appearance of *collapsed backbone* LANs based on this cabling. The ethernet *10baseT* standard was born (Figure 4.3b).

The original coaxial cable realisation of ethernet was a single cable bus (Figure 4.3a), usually installed in the cable conduits and office corridor ceiling voids. The main network element was called the *backbone*. Tee-offs into individual offices were installed as needed, either by teeing directly into the main bus, or by using pre-installed sockets and connectors. A *baluns*, usually built into the socket in the end location, provided for correct impedance matching (50 Ω), independent of whether or not an end-user device was connected into the socket.

a) ethernet as coaxial cable bus

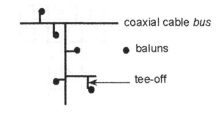

b) ethernet as structured twisted pair cabling

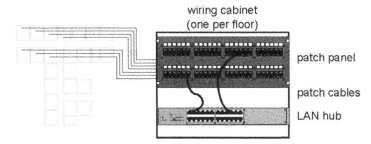

Figure 4.3 Typical coaxial cable and twisted pair wiring configurations for Ethernet.

When installed as part of a *structured cabling scheme* (nowadays the most common realisation of ethernet — Figure 3b), *twisted pair* cabling (e.g., *category 5 – Cat5 –* cabling) provides for the transmission medium. Multiple twisted pair cables are usually installed in each individual office and near each individual desk during office renovation, and wired back to a wiring cabinet, of which there is usually one per floor (office storey), installed in a small equipment room or cupboard. Usually next to the wiring cabinet and in the same rack, a LAN *hub* (or alternatively nowadays, a *LAN switch*) is installed. A LAN hub replaces the coaxial cable *backbone*, so that this arrangement used to be referred to as a *collapsed backbone* topology. The bus topology still exists, but now only within the hub itself, which provides for the interconnection of all the devices forming the LAN, ensuring physical connection and appropriate electrical impedance matching.

Should new PCs or other end-user devices need to be added to the LAN, then they are equipped with a *network interface card* (*NIC*, also called an *ethernet adapter* card) and subsequently plugged into the relevant office socket (typically RJ-45 format). The relevant connection is then *patched* through at the wiring cabinet to the LAN hub, adding a new port card (if necessary) at the hub (Figure 4.3b). Should any of the devices need to be moved from one office to another this can be achieved by re-*patching* at the wiring cabinet. The *adds and changes* are thus far less disruptive both to other LAN users and office furnishings. In the *structured cabling* scheme, the baluns is no longer needed, since the hub provides for this function.

Ethernet LAN components are relatively cheap. The topology is easy to realise and manage, and is relatively resilient to failures. Though it was disputed for a long while among theoreticians whether ethernet was the best LAN topology, it was incorporated into all sorts of different types of devices and became the *de facto* standard. As a result, ethernet has become the predominant type of LAN. The fact that any station may use the transmission path, as long as it was previously idle, means that fairly good use can be made of the LAN, even at times when some destinations are unavailable because of a transmission path break. Such resilience to cable breaks is not enjoyed by some other LAN types employing more sophisticated data transmission.

During the surge in the number of LANs during the late 1980s, there was a hot theoretical debate about the relative strengths of different types of LANs. IBM, in particular, were keen to promote their *token ring* LAN. Theoreticians suggested that the random collisions of a large number of competing devices all trying to communicate over the same CDMA/CD (ethernet) LAN would lead to rapid network performance degradation under heavy load as compared with token ring, in which transmissions are only permitted from stations who have received a *token* permitting them to send. In practice, however, the traffic in a LAN is rarely completely 'random', since most users communicate with the various main central *server* devices within the network. As a result the cheaper ethernet LAN technology has outlived token ring, the use of which was largely restricted to providing networks for the IBM AS400 mid-range computer.

4.3   Ethernet LAN standards (IEEE 802.3 and 802.2)

The original coaxial cable bus-form of ethernet has largely disappeared, and the most common forms used today are:

- *10baseT* (standard 10 Mbit/s ethernet based on *twisted pair* copper cables (two pairs));

- *100baseT* (the generic name for *fast ethernet* of 100 Mbit/s);

- *10/100baseT* (an alternative generic name for fast ethernet, reflecting the *autosensing* capability and backwards-compatibility with 10baseT);

- *100baseFX* (a specific realisation of fast ethernet using a pair of fibre optic cables);

- *100baseTX* (a specific (and the most common) realisation of fast ethernet using category 5 twisted pair copper cables (two pairs));

- *100baseT4* (a specific realisation of fast ethernet using <u>four pairs</u> of category 3, 4 or 5 twisted pair copper cables. This form is very robust against noise); and

- *Gigabit ethernet (1000baseX)*, of which there are five different versions: 1000baseLX for 1300 nm *long wavelength* fibre transmission; 1000baseSX for 850 nm *short wavelength* fibre transmission; 1000baseZX for 1550 nm wavelength fibre; 1000baseCX for 150 Ω twinaxial copper cable or 1000baseT for 4-pair unshielded twisted pair category 5 cable and RJ-45 connectors.

The *datalink* and *physical layer* standards which define the modern form of 10baseT ethernet are defined in IEEE 802.2 and 802.3 standards (ISO 8802.2 and ISO 8802.3). We review these briefly next, before returning to a review and comparison of the different *physical layer* standards employed by fast ethernet and Gigabit ethernet. Note that the modern standards differ slightly from the original version of ethernet (called *ethernet II* which was developed by Digital Equipment Corporation, Intel and Xerox (DIX)).

4.4   Ethernet LAN datalink layer protocols — LLC and MAC

The datalink layer functions of an ethernet LAN are split up into two sublayers, comprising the *logical link control (LLC)* protocol (as defined in IEEE 802.2) and the MAC (medium access control) layer (as defined in IEEE 802.3 in the case of ethernet). The LLC takes on the slightly 'higher' functions of 'logical' dataflow control between the end-point DTEs (LAN stations), while the MAC takes care of LAN station addressing and functions which have to be adapted according to the characteristics of the physical medium.

Slightly different standards, protocol layers and sublayers are defined for the different flavours of ethernet (10baseX, fast ethernet (100baseX) and Gigabit ethernet (1000baseX), as illustrated in Figure 4.4.

When combined, the LLC and MAC create a frame (Figure 4.5) with a similar frame format to HDLC (higher level datalink control). LLC performs similar functions to HDLC (from which it was derived):

- layer 2 *addressing (MAC-ad*dress) to identify source and destination stations on the LAN;

- *length indicator* to signal the overall length of the frame;

- the *control field* used for layer 2 sequence numbers and data *flow control* between the DTEs on the LAN;

- the *frame check sequence (FCS)* used for error detection and correction.

The *preamble* field is used to establish frame and bit timing (*bit synchronisation*[2]) between the receiving and destination stations. Such preamble patterns are commonly used in protocols designed for networks in which a single medium is contended for (e.g., in digital mobile telephone networks as well!). The problem is that there is not a continuous stream of data being transmitted on the medium, and each new transmission is generated by a different transmitting device. Therefore we need a little time to 'warm up' the network and synchronise

---

[2] See Chapter 2.

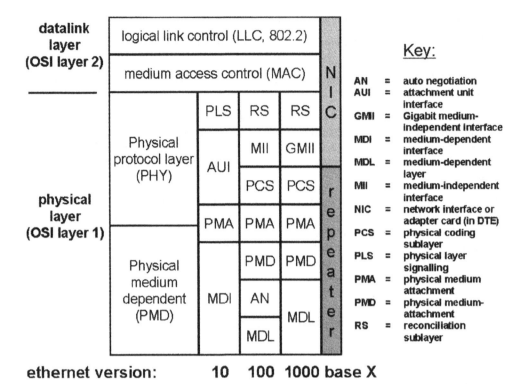

Figure 4.4   Ethernet (10baseX), fast ethernet (100baseX) and Gigabit ethernet (1000baseX) standards and layers.

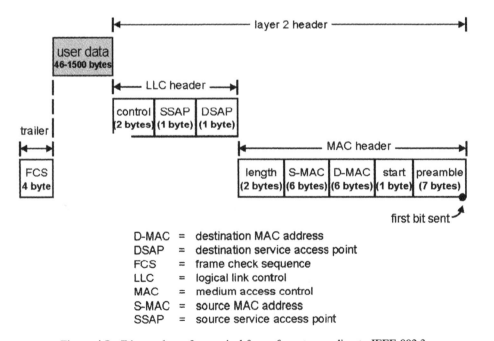

Figure 4.5   Ethernet layer 2 canonical frame format according to IEEE 802.3.

the two ends. If some of the preamble goes missing while the receiver 'adjusts its hearing aid to the right volume', this doesn't matter. The pattern itself is 7 identical bytes with value '0101 0101' [55H].

The *start* byte is the equivalent of the HDLC SYN byte: it provides for character- or *byte-synchronisation*[3] and delimitation of the frame. The bit pattern used, however, is slightly different. The start of frame (SOF) or *start of frame delimiter (SFD)* is set to the binary value '1101 0101' [C5] (but is transmitted, as is the case with all ethernet header fields, in the *canonical format*[4] (i.e., *least significant bit* first).)

The MAC address is the *datalink address* which identifies the stations within the LAN. The addresses themselves have a 6-byte (48-bit) format and are typically programmed into LAN *network interface cards (NICs)* at their time of manufacture. The destination MAC address identifies the intended destination of the packet while the source MAC address identifies the originating station. The first 3 bytes (24 bits) of the address value (last three bytes sent) form the *organizational unique identifier (OUI)*, some well-known examples of which are listed in Table 4.1. The first bit of the OUI (last bit sent) is the I/G bit. When set to '1' it indicates that the MAC address is a *group address*. Otherwise it is an *individual address*. The second bit (second-to-last bit sent) is the U/L bit. When set to '1' it means that the address is *locally* administered (i.e., configured by the LAN administrator). But more usual nowadays are ethernet *NIC (network interface cards)* with their addresses already pre-configured (with *IEEE unique identifiers*) and 'burned in' to the card. In this case, the U/L bit is set to '0' (meaning that the address is *universally* recognised). The last three bytes of the address are set uniquely by the manufacturer.

The *length* field describes the total length of the frame fields which follow, but precede the frame check sequence. This includes the LLC header and the *transmission unit* (or *user data*). Together these make up the LLC PDU (protocol data unit). The minimum length is 64 bytes. The maximum length is 1518 bytes. Since the LLC header has a length of 18 bytes, this means that the minimum transmission unit length is 46 bytes. In cases where the actual user data is less than this minimum length it is filled out with *padding*. Meanwhile the *maximum transmission unit (MTU)* is 1500 bytes. When the maximum size is exceeded, multiple frames need to be sent.

The *service address point (SAP)* header fields are used to help direct the other protocol fields in the header to their correct process or application (i.e., to the *application* or *higher*

Table 4.1   IEEE Unique identifiers (48-bit) used in the MAC-address field: well known examples of the organisation unique identifier (OUI)

Organisation	Organisational unique identifier (OUI) value in hexadecimal (3 bytes total)
Apple computer	08-00-07
Cisco	00-00-0C
Compaq (ex-Digital)	08-00-2B and 00-00-F8
Hewlett Packard (HP)	08-00-09
IANA (Internet Assigned Numbers Authority)	00-00-5E
IANA multicast (RFC 1054)	01-00-5E
IBM	10-00-5A
Novell	00-00-1B

---

[3] See Chapter 2.
[4] Though the ethernet header is always sent in canonical format, the user data field may be transmitted either in *canonical* format (like X.25) or in *non-canonical* format like IP (Internet protocols are always sent *most significant bit* first).

layer protocol using the LLC datalink *service*). The SAP field thus allows multiple protocols (*multiprotocols*) to share the same LAN.

The seventh bit (but transmitted second) is the *U/L bit* (= *universally/locally*) of a *destination SAP (DSAP)*. It is set to '1' for universally assigned protocol addresses; '0' indicates that the protocol addresses are locally assigned. The eighth bit (but transmitted first as the *least significant bit*) is the *I / G* bit *(individual / group bit)*. When when set to '1' I/G indicates that a group address is in use, otherwise value '0' indicates an individual address. In the source SAP field, the *least significant bit* indicates whether the control field is a *command* or a *response*.

'Globally assigned' SAP values identify different *higher layer* protocols in the *user data* field. Commonly used values are 04, 08 and 0C (or 05, 09 and 0D as *group addresses*) for IBM SNA (systems network architecture) over ethernet. 7E identifies ISO 8208 (X.25 over ethernet), F0 is NetBEUI, E0 is 'Novell' and 42 identifies IEEE 802.1D transparent bridging (about which we shall talk later).

The SAP-field value 'AA' indicates an extended SAP address. In this case, a slightly amended LLC frame format is used to accommodate the longer SAP address. This format is called the *SNAP (subnetwork access protocol)* format (Figure 4.6). In effect *SNAP* is itself a 'higher protocol sublayer' inserted above the LLC SAP. The additional fields in the SNAP-frame format allow a much wider range of SAP addresses to be allocated (i.e., many more higher layer protocols to be used in association with ethernet) than the 255 limit of a standard 1-byte SAP field. This gives the potential for network and computer equipment manufacturers to develop their own 'proprietary' networking and routing protocols and still have them carried by *multiprotocol* ethernet LANs. Table 4.2 lists some of the better known SNAP protocol types.

The *control field* of LLC contains *I-frames (information frames), s-frames (supervisory frames)* and *u-frames (unnumbered frames)* which serve the same purpose of sequence

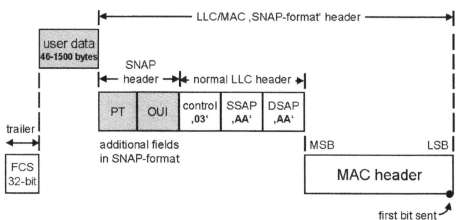

DSAP = destination service access point: set to value ‚AA‘
FCS = frame check sequence
LLC = logical link control
MAC = medium access control
OUI = organisational unique identifier
PT   = protocol type
SNAP = subnetwork address protocol
SSAP = source service access point: set to value ‚AA‘

Figure 4.6   IEEE 802.2 SNAP(subnetwork address protocol)-format of LLC.

Table 4.2 Protocol types supported by ethernet SNAP-format

Protocol	Protocol type (PT) value in hexadecimal (2 bytes total)
Address resolution protocol (ARP)	08-06
Appletalk	80-9B
Appletalk ARP (address resolution protocol)	80-F3
DECnet maintenance operations protocol (MOP)	60-01
DECnet local area transport (LAT) protocol	60-04
DECnet routing	60-03
IBM SNA over ethernet	80-D5
IEEE 802.1Q (VLANs, virtual LANs)	81-00
Internet protocol version 4 (IP v4)	08-00
Internet protocol version 6 (IP v6)	86-DD
Novell IPX (Internetwork packet exchange) protocol	81-37
Reverse address resolution protocol (RARP)	80-35
Simple network management protocol (SNMP)	81-4C
Xerox network system (XNS)	06-00

numbering and flow control as the equivalent frames in *HDLC (higher level datalink control* — see Chapter 3).

The *frame check sequence (FCS)* used by ethernet is a 4 byte (32 bit) *cyclic redundancy check (CRC)* code.

The term canonical format used to describe the layer 2 format used in 802.3-based ethernet refers to the fact that each of the bytes is sent least significant bit (LSB) first. The canonical format is sometimes also referred to as the *little end-ian* or *hexadecimal representation* format.

In contrast to the canonical format, the non-canonical format refers to the transmission of the bytes *most significant bit (MSB)* first. The non-canonical format is used for the frame headers in token ring LANs (IEEE 802.5) and for the *user data* field of all link-layer protocols (including ethernet) when sending IP-related protocols. The non-canonical format is also known as the *big end-ian* or *bit-reversed representation*. Naturally, data which traverses both a canonical and non-canonical network along the course of a 'connection' must be converted in format at each of the network boundaries.

Depending upon the *service* needs of the higher layer protocol being carried by LLC, there are two different variant forms of the protocol, called LLC type 1 (LLC1) and LLC type 2 (LLC2).

LLC1 uses only *u-frames (unnumbered frames)* with a 1 byte *control field*. It is suited to providing an unacknowledged *connectionless* service, as is suited to the carriage of IP (Internet protocol). In the LLC1 format, there are only three different types of frame: *UI (unnumbered information*: control field value = 03H); *XID (exchange information*: control field value = AF or BF) and *test frames* (control field value = E3 or F3).

In contrast to LLC1, LLC2 provides a *connection mode service* very similar to HDLC. It is based on a 2-byte control field (1 byte for u-frames) as illustrated in Figure 4.7 and Table 4.3. This format is used for *connection-oriented* network, transport or session protocols such as IBM's *SNA (systems network architecture)*, X.25 over ethernet or NetBEUI. Under LLC2, frames need to be *acknowledged*.

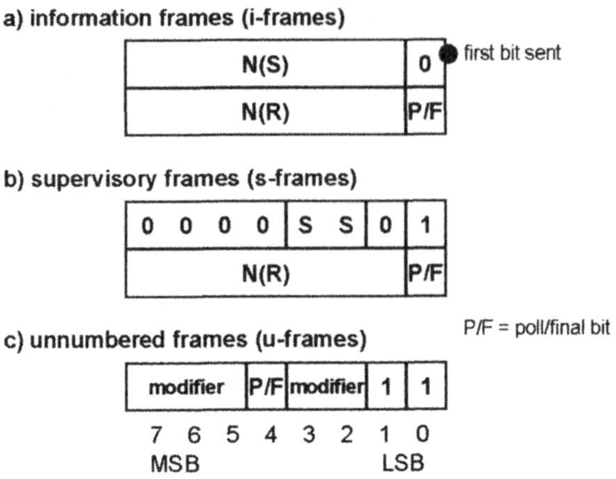

a) information frames (i-frames)

N(S)	0	● first bit sent
N(R)	P/F	

b) supervisory frames (s-frames)

0	0	0	0	S	S	0	1
N(R)					P/F		

c) unnumbered frames (u-frames)   P/F = poll/final bit

modifier	P/F	modifier	1	1

```
7  6  5  4  3  2  1  0
MSB                LSB
```

Figure 4.7 LLC control field (canonical) format (for coding see Table 4.3).

Table 4.3 LLC type 2 (LLC2) control field signals and coding

Frame type	Signal type	Signal purpose	Control field coding MSB LSB	Remarks
I-frame	I	*Sequence numbers* of sent frames N(S) and next frame number expected to be received N(R), effectively an acknowledgement of all previously received frames	1st byte value: 7 bit N(S), then 0 (sent 0 then N(S), LSB first 2nd byte value: 7 bit N(R) then Poll bit	P = Poll bit
S-frame	RR	Receive ready and acknowledgement of all frames up to and including N(R)-1th	1st byte value: '0000 0001' 2nd byte value: 7 bit N(R), then P/F bit	P = Poll bit F = Final bit
	RNR	Receive not ready, but acknowledgement of all frames up to and including N(R)-1th	1st byte value: '0000 0101' 2nd byte value: 7 bit N(R), then P/F bit	P = Poll bit F = Final bit
	REJ	Reject, requests retransmission of all frames from and including N(R)th	1st byte value: '0000 1001' 2nd byte value: 7 bit N(R), then P/F bit	P = Poll bit F = Final bit
U-frame	SABME	Set asynchronous balanced mode; secondary stations may send at any time	1st byte value: '011P 1111'	P = Poll bit
	DISC	Disconnect	1st byte value: '010P 0011'	P = Poll bit

Table 4.3   (*continued*)

Frame type	Signal type	Signal purpose	Control field coding		Remarks
			MSB	LSB	
	UA	Unnumbered acknowledgement; acknowledges having received a U-frame with a 'set' command	1st byte value: '011F 0011'		F = Final bit
	DM	Disconnected mode; notification of a protocol error	1st byte value: '000F 0011'		F = Final bit
	FRMR	Frame reject	1st byte value: '100F 0111'		F = Final bit

4.5   Ethernet physical layer — basic functions of the physical layer signalling (PLS)

The basic mode of operation of the *physical layer* of *ethernet* is reflected in the original name used for the IEEE 802.3 standard: CSMA/CD (*carrier sense multiple access* with *collision detection*). It is said to be a *contention protocol*. The various functions carried out are divided into sublayers within layer 1 (physical layer) as illustrated in Figure 4.4.

The basic principle is that *stations* should always listen on the bus for transmissions. Each transmission 'heard' on the bus must have its destination MAC address inspected to check whether it matches the station's own address. If the two addresses match (that in the 'heard' frame and that of the station itself), then the station reads the frame into its receive buffer. Otherwise the frame is ignored.

When a station wishes to send data, it must wait until the bus (Figure 4.1c or Figure 4.3a) is free. It may then transmit a frame onto the bus. While doing so, the transmitter must check that the frame does not *collide* with a frame sent almost simultaneously from another station. The possibility of *collision* arises because all the stations *contend* equally for the use of the bus. To check for a collision, a station compares the actual signal received on the bus with the signal it is transmitting. If the two differ, then the station knows that there has been a collision, and stops transmitting immediately (it is said to *back-off*). It then sends a *jam signal* of at least 32 bits in length, which all stations will receive and interpret as a signal to cease any concurrent transmission attempts. An algorithm called *truncated binary exponential back-off* ensures that different waiting times (chosen at random using a *back-off algorithm*) are adopted by the individual stations before the renewed transmission of the frame is attempted.

In order that collisions can be reliably detected, transmissions are required to last a minimum duration of one *slot* (MAC slot) period of 512 bit periods (or 4096 bit periods in the case of 1000baseX). This sets the minimum transmission unit length of 64 bytes we encountered earlier when discussing the layer 2 (LLC/MAC) frame format. (The actual minimum frame size including all the preamble, MAC-header and FCS is 72 bytes). Between slots or frames sent, an *interframe gap (IFG)* of 96 bit periods ensures that other stations get a chance to 'butt in' and thus also get use of the bus. Otherwise one station could dominate the use of the LAN.

Among other parameters (see Table 4.4), a maximum *round-trip delay* is defined for different types of ethernet LAN. This value limits the maximum physical size of the LAN (i.e., the maximum length of the bus). If the round trip delay were too long, then a reflection of the

Table 4.4   Ethernet physical layer signalling parameters and constraints

	10baseT (ethernet)	100baseX (fast ethernet)	1000baseX (gigabit ethernet)
Bit duration	100 ns	10 ns	1 ns
Slot duration	512 bits	512 bits	4096 bits
Minimum frame size	64 bytes (72 bytes with all PCI)	64 bytes (72 bytes with all PCI)	64 bytes
Max. Round trip delay	51.2 μs	5.12 μs	512 ns
Maximum frame size	1500 (MTU only) (1530 with all PCI)	1500 (MTU only) (1530 with all PCI)	1500 (MTU only) (1530 with all PCI)
Interframe gap (IFG)	9.6 μs	0.96 μs	0.96 μs
Maximum number of stations	100	100	100
Usual physical medium	2-pair Cat 5 copper cable	Cat 5 copper cable (2-pair for 100baseTX or 4-pair for 100baseT4) or multimode fibre (100baseFX)	Multimode fibre or monomode fibre or 150 ohm twinax or cat 5 cable(1000baseT — IEEE 802.3ab)
Maximum hub-to-station or point-to-point length	100 m	100 m (unshielded twisted pair copper cable) 400 m (half duplex fibre) 2 km (full duplex fibre)	320 m (half duplex [HDX] fibre) 3 km (full duplex [FDX] monomode fibre) 200 m copper cable
Maximum collision domain size	500 m	250 m	25 m (1000baseCX) 200 m (1000baseT) 320 m (half duplex [HDX] fibre)

transmitting station's own signal from the most remote end of the LAN would be interpreted by that station as a signal different to that which it was transmitting, and the collision back-off procedure would be commenced. Another effect of too long a round-trip delay is that remote stations would not start to receive a frame of the minimum size until after the sender had completed sending it. In other words, it would not be possible to detect and avert a collision fast enough: the sender of the packet would not repeat the transmission in the case of a collision (since it did not detect the collision).

The *maximum transmission unit (MTU)* size of 1500 bytes (giving a total frame size including preamble of 1530 bytes — see Figure 4.5) is stipulated to ensure that a given station does not dominate the use of the LAN; and to assist the *network interface card (NIC)* designers determine the appropriate size of data buffers.

4.6   Ethernet hubs (half duplex repeaters)

Originally, ethernet was conceived as a bus topology (Figure 4.1 c and Figure 4.3a), but with the emergence of office structured cabling systems based on twisted pair cabling (either *unshielded twisted pair, UTP* or *shielded twisted pair (STP)* and often to *Category 5 (Cat.*

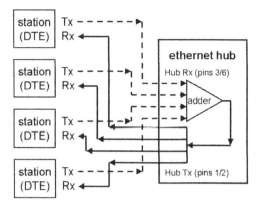

Figure 4.8 Broadcasting function of an ethernet hub (half-duplex repeater, HDR).

5) cable specification), the use of ethernet reverted almost exclusively to a star topology (Figure 4.3b) or point-to-point usage (for direct connection of two intercommunicating devices). The new star topology was heralded by the appearance of the ethernet hub.

An ethernet hub (also called a *half duplex repeater, HDR*) is a device designed to be placed at the 'star point' of a star topology. It gives each connected device the impression it is connected to a shared bus. Connections of each of the stations of the LAN (i.e., the DTEs) are connected to the hub by means of two twisted pairs of wires (usually pins 1;2 and 3;6 of an RJ-45 connector). The hub receives data from each station on RJ-45 pins 3 and 6 (Tx) and transmits data to each station on pins 1 and 2 (Rx).[5] The main action of the hub is simply to 'add' all the signals it receives from stations together and *broadcast* the same signal back to all stations (on the Rx leads). This is illustrated in Figure 4.8. In this way, each station receives all the signals transmitted by other stations, as if they had all shared the same bus. Even the transmitting station receives a copy back from the hub of the original signal it sent. This allows it to check for collisions.

In addition to signal *broadcasting*, the hub acts to *repeat* the signals received. In other words, it cleans up the line coding (*regenerates* the signal) and strengthens it as necessary.

The fact that at any one time the same signal appears on at least one of the receive leads (Tx) and all of the transmit (Rx) leads of the hub ports restricts the operation of a normal ethernet LAN to *half duplex* operation. This is the so-called *shared ports* configuration. On the other hand, it is possible (using a so-called *cross-cable* or *cross-over cable*, which switches pins 1 and 2 at one end of the cable to pins 3 and 6 at the other and vice-versa) to connect two ethernet DTEs directly to one another without using a hub. In this case a *point-to-point* configuration results, which optionally can be used in a *full duplex mode* if the collision detection is disabled (defined by IEEE 802.3x). When used in the full duplex mode, LLC uses the X-ON X-OFF protocol for data *flow control*. Receipt of the X-OFF character (ASCII 13H) means that the receive buffer is full and sending should stop until a subsequent X-ON character (ASCII 11H) is received.

Nowadays, it is becoming more common to use *LAN switches* rather than *LAN hubs* at the 'star point' of the LAN topology. This enables the use of full duplex transmission. In addition, it is an important means of reducing the broadcast traffic on a LAN. For as the traffic levels in a LAN grows (which is inevitable as the number of stations and overall usage grows), the level of broadcast traffic (i.e., the fact that each message has to be sent to each of the stations)

---

[5] Tx and Rx leads are named relative to the DTE — so they appear to be 'the wrong way round' at the hub.

turns out to be a real limitation on the overall traffic capacity of the LAN. More on LAN switches later in the chapter.

4.7   Alternative physical layers — ethernet, fast ethernet and gigabit ethernet

So that we don't constantly have to keep referring back to it, Figure 4.9 is a repeat of Figure 4.4. It illustrates the various physical layer sublayers defined by the IEEE 802.3 suite of standards. Here it shall serve to help us explain the functions of the various sublayers and also compare the slightly different realisations of ethernet (10baseT), fast ethernet (100baseX) and Gigabit ethernet (1000baseX).

Ethernet physical layer (IEEE 802.3 10baset:IEEE 802.3i)

The basic IEEE 802.3 standards for the ethernet *physical layer* used in 10baseT evolved directly from the original standards intended for coaxial cable-based networks (10base5 for *thicknet* and 10base2 for *thinnet* ethernet). Some of the sublayers therefore reflect a split of the functionality of the physical layer which is more apparent when considering 10base5 and 10base2 networks than it is in a modern 10baseT ethernet. Let us consider each of the sublayers in turn, starting at the 'top'.

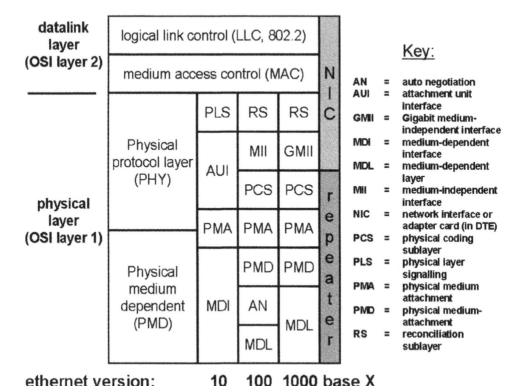

Figure 4.9   Ethernet protocols and sublayers.

The *physical layer signalling (PLS)* sublayer controls the *carrier sensing* and reacts to *collision detection* as we described earlier. The *AUI (attachment unit interface)* passes signals from the PLS to the *PMA (physical medium attachment)*. In practice, AUI is a cable with DB-15 plugs and sockets. Such connectors (labelled AUI) are still found on older networking equipment and ethernet *network interface cards (NIC)*. AUI also defines the coding of the physical layer signal to be *Manchester coding* (see Chapter 2–Figure 2.18).

The *physical medium attachment (PMA)* is achieved using a device known as a *medium attachment unit (MAU)*. It is the PMA which is responsible for the actual detection of collisions, notifying them by means of the AUI to the PLS for action. It also regulates when transmissions may be sent onto the medium, but otherwise merely forwards the already *line-coded* signal, adapting it for the actual type of coaxial cable or other medium in use. In the later days of coaxial cable networks, the MAU was a connection device which could be incorporated into the cable network itself by means of a BNC socket. Alternatively it was sometimes incorporated into wall sockets (behind which was the coaxial cabling of the bus). The *medium dependent interface (MDI)* in this case is the specification of one of the different allowed 50 Ω coaxial cable types (thicknet, thinnet) and the associated *BNC (bayonet connector)* connectors.

In modern 10baseT ethernet, it is normal for the AUI and MAU functionality to be combined into the network interface card (NIC). This is both cheaper and reduces the possible sources of failures. The standard interface format from the NIC is nowadays an RJ-45 socket. An RJ-45 category 5 patch cable is used to connect the NIC to a similar socket on the LAN hub. Should, however, a modern NIC be required to be connected to an older coaxial cable ethernet or standard AUI (DB-15 connector) then a *transceiver* is used to do the conversion. This is a small device with a single RJ-45 socket on one side, and an AUI interface (DB-15 connector or BNC connector) on the other — for direct connection to the coaxial cable ethernet backbone.

Fast ethernet physical layer (100baseT, IEEE 802.3u)

The *active hubs* (*switches*) used in fast ethernet networks make it possible to combine the different types of fast ethernet and older 10baseT ethernet devices in a single network. Such *backward-compatibility* was given high importance by fast ethernet designers. Naturally therefore, the combined *10/100baseT* hubs have to cope with much more than the *passive* hubs of simple 10baseT networks. First of all, each of the ports of a 10/100baseT *active hub* may be running at different speeds (either 10 Mbit/s or 100 Mbit/s). Second, each port speed may first have to be either *autosensed* (the hub adjusts to the speed of the device) or *auto-negotiated* between the devices (which discover which particular fast ethernet technology is in use (TX, T4, etc).

There were new considerations to be taken care of in the specifications of fast ethernet:

- The higher bit rate demanded faster interfaces between the protocol layers and on the physical medium.

- Backward compatibility was essential, in order that older DTE devices with existing ethernet cards could coexist with newer *fast ethernet* devices in the same LAN.

- Full duplex operation was defined, since *point-to-point* connections of fast ethernet were envisaged as backbone links between different ethernets.

At the time of introduction of *fast ethernet*, modifications had to be made to the ethernet MAC and physical layers to take account of the much higher speed of transmission. Fast ethernet uses technology for high speed physical data transfer which came from FDDI (fibre distributed data interface).[6] The physical layer technology of FDDI was simply adapted to interface the existing ethernet PLS/MAC interface. At the same time, the old sublayers (reflecting the coaxial cable heritage of basic ethernet) were dispensed with, and a new sublayer model was born (see Figure 4.9).

Backward compatibility with the 10baseT version of ethernet demanded a similar MAC-frame structure (Figure 4.5 and Table 4.4) to that of 10 Mbit/s ethernet and a similar serial interface (1-bit at a time) from the MAC layer. This was achieved with a new *reconciliation sublayer (RS)*.

The *MII (medium independent interface)* of fast ethernet replaces the AUI of 10 Mbit/s. The MII is the interface which carries the *physical layer signalling (PLS)* to the *physical medium* itself and provides a standard medium-independent interface for this interface. It comprises 4 separate transmit and receive wires, each operating at 25 Mbit/s (for fast ethernet) or at 2.5 Mbit/s (for 10 Mbit/s ethernet). This enables four bits at a time (a *nibble*) to be carried by the MII at a lower *baud rate*. As discussed in chapter 2, such *multilevel transmission* allows a higher bit rate at a lower *baud rate* by sending multiple bits at once using a single *symbol*. Instead of using *Manchester coding*, the MII uses *NRZ (non return-to-zero)* coding. The *reconciliation sublayer (RS)* converts to the 'old' one-bit-at-a-time interface required by the MAC. Both MII and RS also are adapted to support *full duplex* operation. If present as a cable interface (like AUI), the MII cable may be a maximum of 0.5 m long and has 18 pins (4 transmit, 4 receive, 10 control).

A new layer, the *physical coding sublayer (PCS)* performs similar functions to that of the 10 Mbit/s *physical layer signalling (PLS)* sublayer. Meanwhile, the *physical medium attachment (PMA)* sublayer converts the standard internal format of MII (a parallel format like that shown in Figure 2.26) into one of the different physical layer alternatives, the connectors, pin layouts, line codes and physical properties of which are defined by the remaining layers: *PMD (physical medium dependent)* and *MDL (medium dependent layer)*.

There are several alternative physical forms of fast ethernet, the most important of which are:

- *100baseTX*, which uses 2-pair category 5 cabling like 10baseT, *MLT-3 (multi-level transmission 3)* line coding (a multiple state higher level transmission technique requiring only 32 MHz bandwidth: category 5 *unshielded twisted pair [UTP]* cable can only provide 100 MHz of bandwidth);

- *100baseFX*, which uses a single fibre pair, together with a combination of *4B/5B block code* and *non-return-to-zero-inverted (NRZI)* line code;

- *100baseT4*, which uses 4-pair (2 pairs for transmit and 2 pairs for receive) category 3, 4 or 5 cabling and 8B/6T line coding (This technique overcomes the limited bandwidth of the UTP cable by using multiple cable pairs for the transmission); and

- *100baseT2*, which is a half-duplex version of 100baseT similar to 100baseT2 but requiring only 2 pairs.

An optional *auto-negotiation* sublayer may be included in the case of the most common *100baseTX* version of fast ethernet. This standard, like 10baseT ethernet supports hub-to-end-station cabling distances of up to 100 m and allows a fast ethernet hub station to negotiate with a newly connected station to determine the optimal speed and technology of transmission. Auto-negotiation is conducted by means of a *fast link pulse (FLP)* exchanged between the two devices when they are first switched on or connected. Alternatively, some hubs also support *autosensing*. Autosensing allows older 10baseT ethernet devices to be incorporated into a 100baseTX fast ethernet network. The combination of devices is possible because the hub

---

[6] Described later in this chapter.

is capable of autosensing the actual maximum bit rate of the end station and adjusting its mode of operation accordingly. When provided with autosensing, the 100baseTX form of fast ethernet is often referred to as 10/100baseT. This is becoming the most common form of fast ethernet for connecting end stations to the hub.

Most fast ethernets and 10/100baseT networks are operated as a switched star topology rather than as a *collision domain*. This is because an active hub in a collision domain has to 'throttle back' the rate of data transmission from the devices connected to the higher speed ports, so that the frames can be broadcast even to the slower speed ports. By doing so, the 100 Mbit/s ports are effectively reduced to 10 Mbit/s throughput — even when communicating with another 100 Mbit/s port. Collision domains make sense only in LANs which comprise exclusively either 10 Mbit/s or 100 Mbit/s devices, otherwise switches and point-to-point connection of the end-devices makes more sense.

The 100baseFX version of fast ethernet is most commonly found as a point-to-point full duplex technology for interconnecting different LANs on large campus sites using optical fibre cables.

The most commonly used connectors for fast ethernet are RJ-45 (for copper cable) and the *MIC (medium interface connector)* or *SC-connector* (for fibre cables).

Gigabit ethernet physical layer (1000baseX, IEEE 802.3x, 802.3z and 802.3ab)

The protocol sublayer model of Gigabit ethernet (1000baseX) is similar to that of fast ethernet (100baseX), as is clear from Figure 4.9. The most striking immediate differences are the dispensing with the *auto negotiation* (which maybe can return once the 1000baseT standard stabilises) and the change of the MII to the GMII interface.

The *GMII (Gigabit medium-independent interface)* interface uses two 8-wire transmission paths for both the transmit and receive directions and is capable of transferring an entire byte all at once. It employs 8B/10B line coding. Naturally, the *RS (reconciliation sublayer)* for Gigabit ethernet is also slightly different from that of fast ethernet, but the same basic functions of converting the 1-bit-at-a-time MAC layer format to the 8-bit *parallel* format of GMII remain. Unlike the MII and AUI, GMII is not an inter-device physical interface but an electronic component level interface within a Gigabit ethernet network interface card.

As with fast ethernet, the *PMA (physical medium attachment)* sublayer of Gigabit ethernet converts the internal parallel format of GMII into the precise format used on the physical medium, and presents it to the lower *physical medium dependent (PMD)* and *medium-dependent layer (MDL)* for adaption for the medium itself. As with fast ethernet, there are a number of different optional alternative physical realisations of Gigabit ethernet.

The fibre means of physical transmission for Gigabit ethernet uses serial line coding techniques based on the *fibre channel* physical transmission (standards which pre-dated Gigabit ethernet). These are based on 8B/10B *block coding* and sent in *NRZI (non-return-to-zero inverted)* line code. The copper cable alternatives meanwhile use high bandwidth coaxial cable (1000baseCX) or multilevel transmission and multiple transmission leads (1000baseT):

- *1000baseLX*, for a pair of multimode or monomode cables operating at 1300 nm wavelength;

- *1000baseSX*, for a pair of multimode or monomode cables operating at 850 nm wavelength;

- *1000baseZX*, for a pair of multimode or monomode cables operating at 1550 nm wavelength;

- *1000baseCX*, for shielded balanced copper cable (twinax cable and connectors); and

- *1000baseT* (IEEE 802.3ab) for 4-pair unshielded twisted pair category 5 cabling employing the *dual duplex* transmission mode and PAM5 (5-state pulse amplitude modulation) coding (4D-PAM5).

All of the Gigabit ethernet interfaces are complex and expensive. Given this, and the fact that many existing computers may be unable to handle the bit rates it makes possible, its use is likely to be restricted to backbone and carrier networks for some time.

A note on fast ethernet and gigabit ethernet block codes and line codes

We discussed in Chapter 2 some of the most important line codes which are used to carry digital data streams over a physical medium. We covered in detail the line codes NRZ (non-return-to-zero), NRZI (non-return-to-zero inverted) and Manchester coding. But you may now be wondering about how the other *line codes* and *block codes* we have encountered in fast ethernet and Gigabit ethernet operate:

- 4B/5B

- 8B/10B

- 8B/6T

- MLT-3 (multilevel transmission-3)

- 4D-PAM5 (pulse amplitude modulation-5)

For detailed documentation of the coding, you will need to refer to a databook or the relevant specification, but it is nonetheless important to understand the principles:

4B/5B and 8B/10B are so-called *block codes* and not *line codes*. The *block code* 4B/5B converts 4 *bits* of the original data stream (4B) into 5 bits of block code (5B) which are actually transmitted on the line. Such block codes are used widely in conjunction with optical line systems. They take advantage of the high bandwidth capability of optical fibre transmission to increase the reliability of the system in terms of its synchronization and resistance to errors. Thus a 4B/5B code increases the bit rate which needs to be carried by the line by 20%. But by so doing, there is potential to ensure that repeat patterns of bits (e.g., 00000 or 11111) do not lead to constant 'on' or 'off' signals on the line. The optical line coding itself can only be two-state 'on' or 'off', and typically the NRZI line code is employed to transfer the 4B/5B or 8B/10B block-coded signal.

The 8B/10B code is also a *block code* which converts an entire byte of 8 bits into a block *codeword* of 10 bit length.

8B/6T is a *line code*. It converts an 8-bit (8B) signal pattern into 6 digits of *ternary* (i.e., 3-state) code (6T). An 8-bit pattern may have any one of 256 different binary values, as we learned in Chapter 2. But a ternary (i.e., 3-state code, e.g., using three signal values, correctly called *symbols*, $+$, 0 and $-$) can represent $3^6 = 729$ different values using only 6 digits. By using this line code we have 'spare states' [$729 - 256 = 473$ of them!] which we can use to build in a capability equivalent to *block coding*, thereby making our signal less susceptible to noise from other disturbing sources of *electromagnetic interference (EMI)*. Meanwhile, the fact that the line only has to carry 6 *tertiary* symbols rather than 8 *bits* means we have reduced the *baud rate*. We can now get away with using a lower bandwidth medium! 8B/6T is a specific form of 3-state multilevel transmission (MLT-3). Other well-known MLT-3 codes are 4B/3T (used across the ISDN U-interface), MMS43 and that defined by ANSI X3T9.5 (used for 100 Mbit/s transmission over twisted pair cable).

4D-PAM5 is a 5-state *pulse amplitude modulation* transmission which allows an 8-bit signal to be *block-coded* into 4 line-coded digits (each represented by one of the 5 different symbols of the code, each symbol in turn being represented by a different *line code* pulse amplitude allowed by the code).

4.8   LAN segments and repeaters — extending the size of a single collision domain

Figure 4.10 illustrates three typical ethernet networks, showing how the number of devices within a single ethernet LAN (in the jargon called a single collision domain) can be increased by *cascading* end devices and/or hubs to create the necessary number of ports, provided the cabling length does not exceed the maximum specified (Table 4.5).

The maximum geographic dimension[7] of an ethernet LAN is determined as the product of the maximum allowed cabling length within a *segment* multiplied by the maximum allowed number of segments. A segment is a zone or subnetwork part of a collision domain. A segment of a coaxial cable LAN (10base5 or 10base2) is that part of an ethernet LAN within which end-users devices can intercommunicate with one another without passing a repeater. The boundary of two different adjoining segments is thus established by means of a repeater.

In the days of coaxial cable-based ethernet (10base5 and 10base2), a repeater was necessary to amplify the signals on the ethernet bus, in order that the signal would reach all segments of the bus with sufficient strength to be correctly received by all connected stations (see Figure 4.10c). One could determine how many segments a particular LAN comprised by adding up the number of repeaters and adding one. Thus a network with no repeaters comprised one segment. A LAN comprised of a 'chain' of subnetworks and two repeaters has three segments, and so on.

In 10baseT networks, by comparison, repeaters tend to be referred to as hubs or switches, and all the end-user ports attached to a single hub are considered to be in the same segment. A

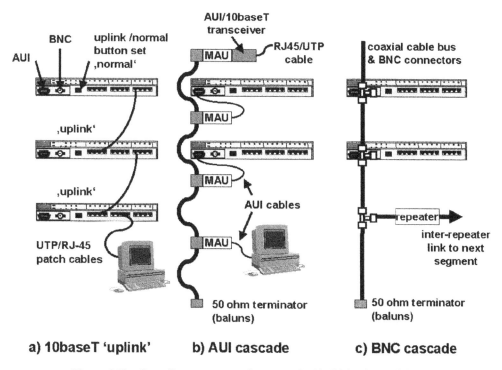

a) 10baseT 'uplink'   b) AUI cascade   c) BNC cascade

Figure 4.10   Cascading segments and repeaters in 10Mbit/s ethernet LANs.

---

[7] The maximum geographic dimension of a LAN is limited, as we saw earlier, by the maximum allowed round-trip delay time.

Table 4.5 Ethernet network size limitations

	Thicknet (10base5)	Thinnet (10base2)	10BaseT	100BaseTX	100BaseFX	100BaseT4	1000BaseT	1000BaseLX	1000BaseSX
Cable type	10 mm coax	RG-58	2-pair Cat 5	2-pair Cat 5	2x MMF	4-pair Cat 5	4-pair Cat 5	2xSMF /MMF	2x MMF
Maximum number of stations	1024	30	100						
Minimum distance between stations	2.5 m	0.5 m	N/A	N/A	N/A	N/A	N/A	N/A	N/A
Maximum segment or station-to-hub connection length	500 m	185 m	100 m	100 m	412 m (HDX) 2 km (FDX)	100 m	200 m	320 m (HDX) 440 m (multi-mode FDX) 3 km (monomode FDX)	260 m (multi-mode HDX) 320 m (monomode HDX) 550 m (monomode FDX)
Maximum number of segments	5	5	5	Class 1: max 1 repeater Class 2: Max 2 repeater 250 m	Class 1: max 1 repeater Class 2: Max 2 repeater 1 km	Class 1: max 1 repeater Class 2: Max 2 repeater 250 m	Max 1 repeater per domain	Max 1 repeater per domain	Max 1 repeater per domain
Maximum size of collision domain	2500 m	925 m	500 m				200 m	320 m	320 m

hub is said to provide a single *shared port* (equivalent to the bus of a coaxial cable segment). Thus in a 10baseT network, the number of segments within the network as a whole is equal to the number of hubs. 10baseT hubs (also called *shared port hubs* or *half duplex repeaters*) can be *cascaded* together as shown in Figure 4.10a to create a larger single collision domain.

When cascading 10baseT hubs, the ports of the different hubs are usually connected together using a normal (i.e., parallel) UTP (unshielded twisted pair) patch cable as shown in Figure 4.10a. A special port on the hub is usually provided as an *uplink* port. The only difference between an uplink port and a normal port is that the pins allocated in the RJ45 connector for transmission (pins 3 and 6) and for receive paths (pins 1 and 2) have been crossed over. If this *cross-over* were not undertaken, then both hubs would be trying to transmit on the same pair of wires and both to receive on the other pair of wires. Alternatively two normal ports can be interconnected using a *cross-over cable* — as discussed in §4.6. Cascading of 10baseT hubs can also be effected using the AUI ports (as shown in Figure 4.10b) or using a coaxial cable segment (Figure 4.10c). In this way, different vintages of 10 Mbit/s ethernets can be combined together into a single LAN (single collision domain).

No matter how the various segments of a single 10 Mbit/s collision domain are interconnected, the worst case path between two end-stations may not pass through more than five segments. This does not mean that the maximum number of segments within a single LAN (single collision domain) is five. It means that apart from the origin and destination segments, the path should not have to traverse more than three other segments. Thus in a purely 10baseT network, for example, the longest allowable path is a cascade of 5 hubs from one end-station to the other. Meanwhile, in a purely coaxial cable environment, the path may not pass more than 4 repeaters.

In fast ethernet and Gigabit ethernet one talks of *half duplex repeaters (HDR)* for *shared port* networks (i.e., HDRs are used in *collision domain* networks) and *full duplex repeaters (FDR)* for point-to-point or switched network connections. Because of the much higher speeds, fast and Gigabit ethernet networks are more sensitive to propagation delays, so that much fewer repeaters are allowed on the maximum path length, particularly when operating as collision domain networks (i.e., with half duplex repeaters). In contrast, *full duplex (FDX)* connections can generally be longer than *half duplex (HDX)* ones (see Table 4.5) because of the less stringent timing and length constraints. (Collisions do not have to be detected within the slot duration in the case of full duplex.)

4.9   LAN switches — extending coverage and managing traffic in LAN networks

The popularity of LANs in the working environment grew rapidly in the 1980s and 1990s. But while a 10 Mbit/s *shared medium* might suffice among a small number of users, sooner or later the limits of a single LAN are reached, and one of the following problems arises:

- user traffic demand exceeds the aggregate 10 Mbit/s total capacity of the LAN;

- the required geographical coverage exceeds the maximum cabling lengths of a single *collision domain*;

- the number of users exceeds the maximum allowed in a collision domain or the number of physical ports available on the ethernet hubs.

For all three of these problems there is a simple solution: split the single LAN (single collision domain) into two or more smaller, interconnected LANs. In this way the above problems are solved, but a new one arises: how can I interconnect the new smaller LANs to one another, so that the stations and their end-users can continue to intercommunicate with one another? The subject of LAN interconnection we shall return to later in the chapter. In the meantime we discuss how one of the above problems (that of meeting traffic demand) may nowadays be most easily be solved by the use of a *LAN switch* or *ethernet switch*.

In the shared port or single collision domain configuration of an ethernet LAN, any two of the end-user stations may communicate with one another at up to 10 Mbit/s half duplex (i.e., they may only communicate in one direction at a time). At first glance, the 10 Mbit/s bit rate may appear to offer fantastically fast data transfer — and it does, provided only a few users share the LAN. But when the number of stations gets nearer the limit of 100 allowed by 10baseT ethernet LANs, and all the users are active at once (as they typically are in an office network), the situation looks quite different: if each user wants to send and receive data in equal volumes, then the equivalent bit rate available for each is the equivalent of 50 kbit/s 'full duplex'. This is only the equivalent of a dial-up ISDN line, and as any frequent Internet 'surfer' will tell you: it can take a frustatingly long time to download large datafiles at this speed. Worse still, the theoretical maximum aggregate throughput capacity of an ethernet LAN comes nowhere near the nominal 10 Mbit/s transmission rate of the bus itself. Particularly at high traffic loadings, much of the bus capacity is lost due to collisions of the packets.

Full- and limited-availability (full-mesh and partial-mesh) switches

In contrast to a *shared medium*, which only allows one of the end stations to transmit at any one time, a switch allows multiple paths between different end-stations to be established at the same time. This multiplies accordingly the capacity of the LAN. Figure 4.11a illustrates a *full availability* 6-port switch configuration. The full availability (i.e., full-mesh and *non-blocking*) switch matrix of Figure 4.11a allows all six of the stations to be simultaneously communicating: A with C; B with E and D with F. In this configuration, the maximum

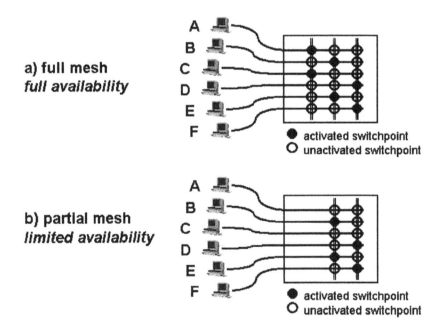

Figure 4.11   Full availability (full-mesh) and limited availability (partial mesh) switches.

throughput of the network has been multiplied to 3×10 Mbit/s $= 30$ Mbit/s, reflecting the three simultaneous paths which may be established across the *backplane* of the switch. The switch of Figure 4.11a is termed a full availability matrix, since all of the ports can be configured to communicate at once (assuming that the desired destination port is free).

Figure 4.11b shows a *limited availability* switch matrix. In a limited availability (or *partial mesh*) matrix, not all of the ports can communicate at once, since insufficient paths are available. The number of available paths (in our example, two) is less than half the number of ports (in our case 6 ports/2 $= 3$ paths are required for *full availability*). When all the available paths of a limited availability switch are already in use, then any attempts to set up further communications will be *blocked*, irrespective of whether the destination port is free or busy. Thus, for example, it is not currently possible to establish communication between the free ports A and C in Figure 4.11b, since all the available switch paths are already in use.

How an ethernet switch operates — the creation of the source address table (SAT)

Like most other modern data networking components, ethernet switches are designed to be able to administer themselves. When you first plug-in the end-user equipment (i.e., the DTE or station) to an ethernet switch (which mostly nowadays are based upon 10/100baseT), it starts the initial period of auto-sensing or *auto-negotiation* (also called *NWAY*) on each port. By so doing, the switch is able to configure each of its ports to the correct bit rate, to half-duplex (HDX) or full duplex (FDX) and (as appropriate) to the correct ethernet or fast ethernet technology (10baseT, 100baseTX, 100baseT4 or 100baseT2). Now the switch is ready to support communication between the different ports.

Since the switch initially knows none of the MAC addresses of the attached DTE devices, it starts operating in the same manner in which a hub would operate. Any packets received from any of the attached devices are simply *flooded* to all of the attached ports. In this way, the switch can be sure that the packet will reach the correct MAC destination (provided of course that the destination device is connected). But meanwhile, the switch learns from its experience. It learns where the MAC address of the device which originated the packet is. How? Because each packet contains a MAC header, which contains both the destination and the *source* MAC address. In this way, the switch is able to relate the source MAC address to the port where the packet originated. When subsequently a packet is sent to the switch with this address as its *destination* MAC-address, the switch knows already which port the packet must be forwarded to.

Over time, the switch is able to build a complete *source address table (SAT)* of all the MAC addresses of devices connected to it and their respective port numbers. In other words, the source address table (SAT) is determined by observing source addresses, and all entries in the table are the MAC addresses of DTEs connected to local switch ports. With the SAT to hand, the switch no longer needs to *flood* all packets to each of the ports, but instead, can direct them only to the relevant port. As a result, multiple simultaneous paths can be established between different pairs of ports, as we saw in Figure 4.11.

Apart from the benefit of increasing network traffic capacity, an ethernet *switch* also offers more data security than an ethernet hub, since the packets are switched directly between only the relevant two communicating ports and not usually broadcast or flooded to all the devices in the LAN.

The individual ports of an ethernet LAN switch may be used either to connect individual DTEs (data terminal equipment), such as a single PC, to connect whole collision domains, or to connect other switches. It does not matter, as far as the operation of the switch is concerned, whether one or more individual MAC-source addresses are assigned to each port.

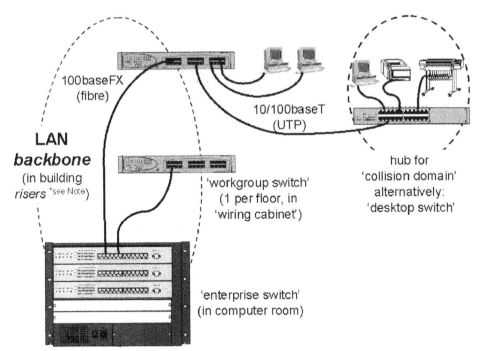

Note: *Risers are inter-floor channels designed into modern office buildings by architects to provide a passageway or conduit for inter-floor telecommunications and electrical cabling, as well as piping and other building services.

Figure 4.12   Typical office LAN based on ethernet switches.

Thus switches can be used either as the 'star-point' of individual LANs, or for interconnecting different switched *subnetworks* or collision domains (i.e., subnetworks using *LAN hubs*). Figure 4.12 illustrates the huge flexibility of modern *switched LANs.*, showing a typical office LAN comprising a backbone network with an *enterprise switch* at the centre of it and fast ethernet (100baseFX) fibre cabling trunking it to separate *workgroup switches* on each floor. The workgroup switches are normally placed beside the patch panel in the wiring cabinet of the office floor, from where the UTP (unshielded twisted pair) cabling (typically Cat 5) leads out to the sockets installed in each office.

End-user PCs are typically connected directly to the switch, though it is also possible to install small hubs or desktop switches to create small collision domains for interconnected local printers, scanners and other peripheral equipment. The maximum number of DTEs (i.e., end-stations) a switch port can support is known as the *end-station density (ESD)* or *port address support (PAS)*. The value of the ESD or PAS is set by the switch designer, when deciding how much capacity to build-into the SAT memory space. Differing values of ESD or PAS are the main distinguishers between *enterprise, workgroup* and *desktop* switches.

Because a LAN switch is able to 'learn' about the devices connected to it and the network around it, it is easy to add further devices at will to the switch later. In addition, it is possible to unplug devices, or move them from one location to another (and thus from one switch port to another). Thus ethernet switches are 'plug-and-play' devices — you connect them together to configure the network topology you want, and they self-administer themselves to make it work.

The source address table (SAT) can be kept up-to-date even when devices are unplugged from a switch by associating an *ageing time* (typically 300 seconds) with each of the entries in the table. If no traffic is received from a particular address in the SAT within the ageing time window, then the entry for this address is deleted from the SAT. A 300 second ageing time may seem a little short, and some devices allow the human LAN administrator to adjust the value, but the ageing time should not be set too long. Long ageing times hamper the ease with which devices can be moved around an office (i.e., from one switch port to another) and lead to the potential of misdirected traffic. Users complain of problems, and the human administrator cannot figure out what is going wrong—all the wire connections appear to be alright, but the communication doesn't work! *Ageing* is a useful way of ensuring that routing tables automatically remain updated, no matter how the network topology changes! But ageing times should not be set too short, for this reduces the efficiency of the network with the extra volume of packet flooding and slows up the switch by keeping it permanently busy updating SAT entries.

Mirrored ports — for measuring traffic in switched ethernet lans

Before LAN switches emerged, human *LAN administrators* used to monitor the traffic volumes and flows in LANs by the use of LAN *probes* or *sniffers*. These are devices which, in the case of ethernet collision domains, monitor the source and destination addresses of all the packets broadcast over the ethernet *bus*. Once switches appeared, most of the packets within the LAN were no longer broadcast or flooded to all the ports. As a result, it became difficult to monitor either the traffic in the LAN as a whole or the traffic flows to and from a given DTE (data terminal equipment). For this reason, switches usually offer a *mirrored port* facility. When a port has been mirrored a second port is set up in parallel with the first. It receives all the traffic that the first port receives, and any packets originated by it are handled as if they had been originated by the first. The mirror port is useful for the connection of a *probe* or *sniffer* for analysing the traffic to the mirrored port.

4.10   Other types of LAN (token ring and token bus)

There have been a number of different LAN technologies developed over the years, all of which are tending to be replaced by ethernet, fast ethernet or Gigabit ethernet technology. Three other technologies (*token ring LAN, FDDI [fibre distributed data interface]* and *token bus*) were, like ethernet, made into official IEEE 802-series standards and are still to be found deployed in corporate networks. For this reason, they deserve mention here. Given the large number of IBM computers deployed with token ring LAN networks, token ring LANs may live on for a while yet. Indeed, there is still a level of ongoing standardisation effort looking to upgrade token ring to encompass 100 Mbit/s and Gigabit versions.

Wireless LANs

In recent years, wireless LANs (using radio transmission) have become popular. ETSI (European Telecommunications Standards Institute) developed a system called HIPERLAN (HIgh PErformance Radio LAN), but the most popular version looks likely to be that based on the IEEE 802.11 standard. IEEE 802.11 *wireless LANs (WLANs)* are, in effect, wireless versions of ethernet LANs—as explained in Appendix 6.

Token ring LAN (IEEE 802.5)

The *token ring LAN* standard (defined by IEEE 802.5-series standards) employs a *token* (passed between each of the terminals connected to the ring topology (Figure 4.1) in turn) to assign the 'right to transmit data' on the LAN. The manner in which the token is passed is as follows: the token itself is used to carry the packet of data. The transmitting terminal sets the token's *flag*, putting the destination address in the *header* to indicate that the token is full. The token is then passed around the ring from one terminal to the next. Each terminal checks whether the data is intended for it, and passes it on. Sooner or later the token reaches the destination terminal where the data is read. Receipt of the data is confirmed to the transmitter by changing a bit value in the token's flag. When the token gets back to the transmitting terminal, the terminal is obliged to empty the token and pass it to the next terminal in the ring.

One of the beneficial features of IEEE 802.5 MAC protocol is its ability to establish priorities among the ring terminals. This it does through a set of priority indicators in the token. As the token is passed around the ring, any terminal may request its use on the next pass by putting a *request* of a given priority in the reservation field. Provided no other station makes a higher priority request, then access to the token is given next time around. The reservation field therefore gives a means of determining demand on the LAN at any moment by counting the number of requests in the flag. In addition, the system of prioritisation ensures that terminals with the highest pre-assigned authority have the first turn. High speed operation of certain pre-determined, time-critical devices is likely to be crucial to the operation of the network as a whole, but they are unlikely to need the token on every pass, so that lower priority terminals get a chance to use the ring when the higher priority stations are not active.

Token ring was developed by IBM, and is most common in office installations where large IBM mainframe and mid-range computers (particularly AS400) are in use. The original form required specialised cabling (IBM type 1) and operated at 4 Mbit/s. The idea was that a single cable loop could be laid through all the offices on a floor or in a building and devices added on demand. To avoid the disturbances and complications which might arise when connecting new devices to the ring (any break in the ring renders the LAN inoperative), IBM developed a sophisticated cabling system, including the various IBM special cables. The cable loop was pre-fitted with a number of sockets at all possible user device locations. The sockets ensured that when no device was connected, the ring was through-connected. But on plugging in a new device, the ring is diverted through that device (Figure 4.13). The *baluns* (special socket) for early token ring networks thus catered not only for correct impedance matching, but also for the ring continuity.

Token ring network interface cards (NICs) in the individual end user computer devices connected to token ring LANs also have to be designed in such a way as to ensure ring

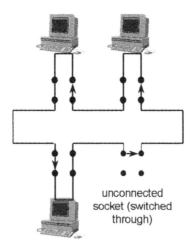

Figure 4.13   Socket design in Token Ring LANs to ensure ring continuity.

continuity in the case that the device is switched off. Thus the card reverts to a 'switched-through' state when no power is applied, so that even though the end device itself plays no active part in token-passing while switched off, the tokens nonetheless still have a complete ring available.

The further development of the token ring technology (mainly by IBM) has brought about the capability of use of twisted pair cabling, and the emergence of a 16 Mbit/s as well as the original 4 Mbit/s version. In the 16 Mbit/s version (IEEE 802.5f and 802.5n), higher quality cabling (typically *category 5* cable) is required. There is also a 100 Mbit/s version (IEEE 802.5t).

Token ring LAN hubs have also developed alongside ethernet hubs, and allow for similar collapsed backbone topologies in conjunction with structured cabling systems. The ring topology is collapsed into the hub itself, while two sets of wires to each individual user station allows for the extension of the ring to each user device. The switch-through function previously performed by the socket is undertaken at the hub, thus reducing the complexity and cost of individual sockets, so that standard category 5 structured cabling and the associated RJ-45 sockets may be used. The *token ring LAN* may differ from the *ethernet LAN* only in the port cards used within the hub and the LAN cards used in the individual PCs. Otherwise cabling, wiring cabinet and LAN hub unit may be identical. Indeed, in some companies, ethernet and token ring LANs exist alongside one another, without the user being aware to which type of LAN he or she is connected.

Token ring LANs, like ethernet LANs, are common in office environments, linking personal computers for the purpose of data file transfer, electronic messaging, mainframe computer interaction or file sharing. Some LAN administrators used to be enthusiastic about whether ethernet or token ring offered the best solution, but in reality, for most office users, there was little to choose between them. Token ring LANs perform better than ethernets at near full capacity or during overload but can be more difficult and costly to install — especially when only a small number of users are involved. In most cases, the choice between ethernet and token ring comes down to the recommendation of a user's computer supplier, since hardware and software of a particular computer type may have been developed with one or other type of LAN in mind. Thus token ring remains the recommendation of the IBM company, while in all other environments, ethernet has gained the upper hand.

Should the ring of a token ring LAN be broken (i.e., lose its continuity), then the LAN ceases to operate, since the token can no longer be returned to its sender and passed on, as required by the protocol. This made it unpopular with some LAN administrators. To provide for a ring *continuity check*, the layer 2 addressing scheme (i.e., the MAC addressing scheme) of token ring allocates a special *loopback address* (also called the *no station address*). When the no station address (a string of all 0s) is set as the MAC address, then each terminal is expected to ignore the packet and token, merely forwarding it around the ring. Provided the ring is complete, the packet and token return around the ring (i.e., are *looped back*) to their origin. In contrast to the *no station address* (binary string of all 0's), the *broadcast address* is set as a binary string of all 1s). When the broadcast address is set in the token, each of the stations in the terminal will receive the same broadcast message. Such messages can be useful for simultaneous reconfiguration of the LAN.

FDDI (fibre distributed data interface)

The *fibre distributed data interface (FDDI)* is a 100 Mbit/s token ring network defined by ANSI (American National Standards Institute) X3. FDDI is a *metropolitan area network (MAN)* technology which can be used to interconnect LANs over an area spanning up to 100 km, allowing high speed data transfer. Originally conceived as a high-speed link for the needs of broadband terminal devices, FDDI was most used as an optimum 'backbone' transmission system for campus-wide wiring schemes, especially where network management and fault recovery were required. In particular, FDDI became popular in association with the very first optical fibre building cabling schemes, since it provided one of the first means to connect LANs on different floors of a building or in different buildings on a campus via optical fibre. Due to its expensive nature and the rapid development of alternative technologies (including *ATM — asynchronous transfer mode*; and later, 100 Mbit/s *fast ethernet*), FDDI fell into decline, no longer being recommended or further developed by most LAN and computer manufacturers. Nonetheless, some of the physical layer standards developed for FDDI live on as fast ethernet.

A second generation version of FDDI, FDDI-2 was developed to include a capability similar to *circuit-switching* to allow voice and video to be carried reliably in addition to packet data. But these capabilities were never widely used. Nor was the copper cable version: *CDDI (copper distributed data interface).*

The FDDI standard is basically a *physical layer* and *MAC (medium access control)* standard, defined in four parts, and to be used in conjunction with the standard logical link control (LLC) defined by IEEE 802.2:

- *media access control (MAC)* defines the rules for token passing and packet framing;

- *physical layer protocol (PHY)* defines the data encoding and decoding;

- *physical media dependent (PMD)* defines drivers for the fibre optic components; and

- *station management (SMT)* defines a multi-layered network management scheme which controls MAC, PHY and PMD.

The ring of an FDDI is composed of dual optical fibres interconnecting all stations. The dual ring allows for fault recovery even if a link is broken by reversion to a single ring, as Figure 4.14a shows. The fault need only be recognised by the *CMTs (connection management mechanisms)* of the station immediately on either side of the break. To all other stations the ring will appear still to be in its normal contra-rotating state (Figure 4.14b).

When configured as a ring, each of the stations is said to be in *dual-attached* connection. Alternatively, a fibre star connection can be formed using *single-attached stations* with a

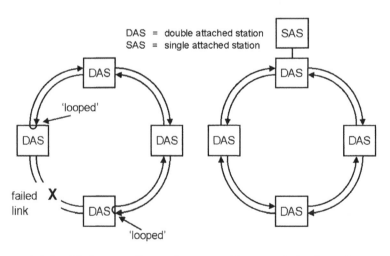

Figure 4.14   Fibre distributed data interface (FDDI).

multiport concentrator at the hub (itself a double attached station — Figure 4.14b). *Single-attached stations (SASs)* do not share the same capability for fault recovery as *dual-attached stations (DASs)* on a dual ring.

FDDI-2, the second generation of FDDI has a maximum ring length of 100 km and a capability to support around 500 stations including telephone and packet data terminals. The FDDI-2 ring is controlled by one of the stations, called the *cycle master*. The cycle master maintains a rigid structure of cycles (which are like packets or data slots) on the ring. Within each cycle a certain bandwidth is reserved for circuit switched traffic (e.g., voice and data). This guarantees bandwidth for established connections and ensures adequate delay performance. Remaining bandwidth within the cycle is available for packet data use.

The voice and video carriage capability of FDDI-2 is possible because of its interworking with the *integrated voice data (IVD)* LAN standard defined in IEEE 802.9.

Fibre channel

Another alternative medium for high speed data transfer based on a switched point-to-point LAN topology is the fibre channel (FC) — as standardised by ANSI (American National Standards Institute) and the FCA (Fibre Channel Association).

Switched multimegabit digital service (SMDS)/DQDB (dual queue dual bus) MANs

SMDS (switched multimegabit digital service) networks are *metropolitan area networks (MANs)* which conform to IEEE 802.6 and use a protocol called *distributed queue dual bus (DQDB)*. DQDB was co-developed by Telecom Australia, the University of Western Australia and their joint company, QPSX communications limited. It was designed to provide a basis for initial broadband *metropolitan area* interconnection of networks (like a LAN, but on a larger geographical scale), suitable for simultaneous transmission of not only data, but also voice and video signals. As a public data communications service, the switched multimegabit digital service became available in the United States in 1991. Like FDDI, the technology was too expensive, and it has fallen into disuse, but it also established some important principles for later communications and protocol design, which are worthy of discussion.

The DQDB protocol uses two *slotted* buses of bit rates up to 155 Mbit/s for transporting *segments* of information (the DQDB name for the basic unit of *user data* carried by the network) between communicating broadband devices. *Segments* are 48 byte frames of user data information.

Figure 4.15 illustrates the structure of a switched multimegabit digital service (SMDS) network using the DQDB protocol. Two unidirectional high speed buses run out from *master* and *slave frame generators* at opposite ends of the 'ribbon' topology. Each of the devices (*nodes*) connected to the network are connected to both buses for sending and receiving data.

The role of the *frame generators* is to structure the bit stream carried along the buses into fixed length, 57 byte, *slots*. Slots are really data *frames*, but of a fixed length. They are filled by nodes wishing to send user information and are then carried downstream along the bus. The relevant receiving node reads information out of the slot being sent to it, but does not delete the slot contents. The slot thus remains on the bus, travelling further downstream until it falls off the end.

When a node wishes to send information it may do so in the first available empty slot, but in doing so must follow the procedure set out in the *medium access control (MAC)* protocol. The MAC protocol is intended to ensure a fair use of the available bandwidth of the buses between all the devices wishing to send information.

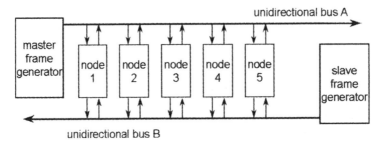

unidirectional bus A

unidirectional bus B

Figure 4.15   Bus structure of DQDB.

Before sending information, a sending node must know the relative position of the receiving node on the bus. It then sends a request in the opposite direction to request bandwidth. For example, say node 2 of Figure 4.15 wished to transmit to node 5, then it would send a request on bus B. This advises the *upstream* nodes of bus A (i.e., node 1 in our case) that node 2 requires capacity on bus A. Node 2 must then wait until all other previously *pending requests* from other *downstream* nodes on bus A have been cleared. Once these are cleared, it may send in any free slot, and may continue to fill slots until a further slot request appears from a *downstream* node (i.e., node 3, 4 or 5).

It is a simple and yet very effective medium access control. Requests for use of bus A are sent on bus B. Meanwhile the use of bus B is governed by the requests on bus A. The control of the use of the network is decentralised, so that each node may independently determine when it may transmit information, but must be capable of keeping track of the pending requests.

When a node is not communicating on one of the buses (say, bus A), it monitors the requests for use of the bus, keeping a running total of the outstanding requests using its *request counter*. Each time a request passes on bus B, the request counter is incremented, and when a free slot goes by on bus A, the counter is decremented. In this way it can keep track of whether a free slot on bus A is *available* to it or not. The request counter is never decremented to a value less than zero.

Each time a node has a segment it wishes to send on bus A, it generates a *waiting counter*. The initial value copied into the waiting counter is that currently held in the request counter. The waiting counter is decremented each time a free slot passes on bus A until the value reaches '0', when the segment may be sent in the next free slot.

When transmitted onto one of the buses the 48 byte segment of user information is supplemented with a 4 byte *segment header*, a 1 byte *access control field* and a 4 byte *slot header* as shown in Figure 4.16, so that the total length of a slot is 57 bytes. The frame structure of DQDB, and the functioning of the protocol is designed to make it compatible with *asynchronous transfer mode (ATM)*,[8] an important WAN technology. The 53-byte *cell* comprised within each slot (Figure 4.16) is equivalent to the ATM cell.

The DQDB *slot header* carries a 2 byte *delimiter field* and 2 bytes of control information used by the physical layer for the layer management protocol. The *access control field* may be written-to by any of the nodes on the bus. This is the field in which the slot requests are transmitted. The *segment header* carries a 20-bit *virtual channel identifier*, like the logical channel number of X.25. This identifies the cells corresponding to a particular connection to the appropriate receiving node.

Data blocks to be carried by DQDB are formatted in the standard LLC (layer 2 protocol) manner of frame *header*, followed by *user data block* and the frame *trailer* (Figure 4.17). The frame header contains the address of the originating and destination nodes. The user data block

---

[8] See Appendix 10.

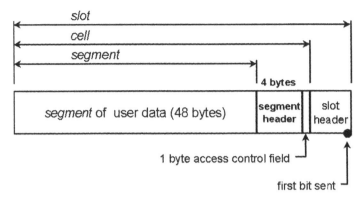

Figure 4.16   Slot and segment structure of DQDB.

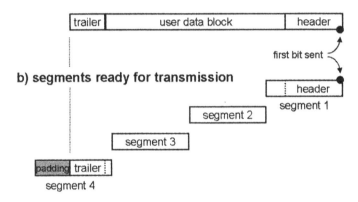

Figure 4.17   Segmentation and reassembly (SAR) of a data block for transmission using DQDB or ATM.

is the data frame to be carried which may be up to 9188 bytes in length (192 segments), and the trailer includes the *frame check sequence*. But before being transmitted across the physical medium, the SMDS protocols break down these data blocks (LLC frames) into individual *segments* (i.e., 48 byte chunks), each of which is formatted as *slot* for transmission. This process is called *segmentation*. If necessary, the last segment is filled with *padding* (Figure 4.17). At the receiving end, the slots are *reassembled* into the original LLC frame. The protocol sublayer which performs these functions is called the *segmentation and reassembly* sublayer. It is common to both SMDS and ATM. The *ATM adaptation layer* (*AAL*—the data frame carrying variants are AAL 3/4 and AAL5) operates in a similar manner.[9]

Segmentation and reassembly (SAR) of data blocks into fixed length cells or slots for transmission across a physical medium is one of the methods frequently chosen by protocol designers attempting to build networks suitable for both real-time signals like voice and video as well as carriage of data files. The use of slots allows the transmission of a large data frame to be interrupted temporarily when a high priority signal (such as real-time voice or video) needs to be carried. By so doing, we minimise the possibility of unacceptably long or variable

---

[9] See Appendix 10.

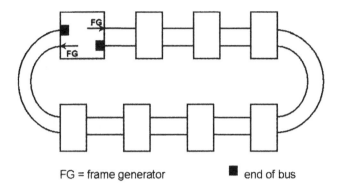

FG = frame generator   ■ end of bus

Figure 4.18   DQDB or SMDS configured in a *looped bus* topology.

propagation delays. Such delays lead to perceptible quality impairment of voice and video connections.

As well as in a straight forward bus configuration, SMDS (switched multimegabit digital service) networks which use the DQDB protocol may also be configured in a *looped-bus* topology. In this case the bus is looped so that the two frame generators of Figure 4.15 are contained in the same node. This node also contains two *end of bus* devices as shown in Figure 4.18. In real terms, the network is still two independent buses, but there may be a practical advantage in not needing two separate *frame generator* nodes.

Token bus LAN (IEEE 802.4)

A *token bus* LAN (like the *token ring LAN*) also controls the transmission of data onto the transmission path by the use of a single token. Only the terminal with the token may transmit packets onto the bus. When a terminal has the token it sends any data frames it has ready, and then passes the token on to the next terminal. To check that its successor has received the token correctly the terminal makes sure that the successor is transmitting data. If not, the successor is assumed to be on a failed part of the network, and in order to prevent 'lock-up' of the LAN, the original terminal creates a new successor by generating a new token. Transmission faults in the LAN bus can therefore be circumvented to some extent. However, those parts of the LAN that are isolated from the token remain cut off.

Token bus networks were never commonly used in office environments — where ethernet and token ring networks predominate. Token bus networks were most common in manufacturing premises — often operating as broadband (high speed) networks for the tooling and control of complex robotic machines.

4.11   LAN operating software and LAN servers

So far we have talked about the physical structure of LANs, and the logical procedures used to convey data across them. This alone, however, is not a sufficient basis for creation of an office LAN. In addition, a *LAN operating system* (software) is required. At the start, a number of different manufacturers offered alternative proprietary systems. Over time, the number of systems reduced to four: *Novell Netware, IBM LAN Manager/NetBIOS, Appletalk* and *Microsoft Networking* (the LAN server software being *Windows for workgroups, WindowsNT* or *Windows 2000*).

LAN operating systems provide for the software *sockets* (i.e., interface) between normal computer operating software (e.g., *Microsoft DOS, Windows, Windows95, Apple Macintosh,* etc.) and the new functions made possible by LAN networks (e.g., file server, host gateway, fax server, common printer, etc.). In addition, they provide for easy LAN administration as well as for the management of the user directories and other network resources.

The early LAN operating systems were installed to provide for LAN *file servers* and *print servers*. LAN *servers* are typically powerful and expensive computers, capable of faster processing and additional functions useful to the workgroup as a whole. Such servers are connected to the LAN, and usually remain in operation 24 hours per day. A print server allows a PC on the LAN to choose between any of the printers connected to the LAN rather than being limited to the one directly connected to his computer, and *spools* (i.e., queues) the *print jobs* to the printer. Sometimes an end-user might choose the fast black and white laser printer, while on other occasions maybe the colour printer is more appropriate. Meanwhile, a common file server allows LAN users to share a common data filing system. A file server is usually a computer with a large amount of storage capability which may be rapidly accessed and easily *backed up* by specialist computer staff on a once per day or once per week basis. It provides for secure storage of information and easy sharing of information basis on a *workgroup* or defined *closed user group* basis.

LAN operating systems are closely linked with network protocols, and most of them have now standardised on the IP protocol suite at their core. But manufacturers still tend to add extra 'proprietary' protocols to provide additional facilities. Thus, for example, *Novell Net-Ware (net*work operating system soft *ware*) uses the proprietary Novell *IPX (internetwork packet exchange)* network protocol and its *Novell directory service (NDS)* as the basis of its 'network' services. Meanwhile *Appletalk* is a suite of operating software and protocols for LANs interconnecting *Apple* computers. The most widely used LAN operating system (used mainly for authentication and authorisation of users at their time of log-in, as well as for file server, printer server, directory and LAN administration services) is *WindowsNT* and its successors, *Windows2000* and *WindowsXP*. The protocol stack, which is based on IP (Internet protocol) is called *Microsoft Networking*. It includes *the Windows Internet Name Server (WINS)* and the Windows *Active Directory Service (ADS)*. Another important proprietary LAN operating system for UNIX-based networks was Sun Microsystems' combination of the *network information service (NIS)* and *NFS (network file system)*.

4.12   Interconnection of LANs — bridges, switches, VLANs, routers and gateways

The interconnection of numerous LANs, perhaps of different types, or the connection of a LAN to a mainframe computer or other external network or device requires the use of *bridges, switches, routers* or *gateways*. We discuss these next.

LAN bridges

Bridges were the first type of devices used widely for interconnecting different LANs to make them appear as if they were a single LAN. A bridge is an intelligent hardware connected to the LAN, which examines the address in the *LLC (logical link control)* header of each data *frame* originated on the local LAN. With the knowledge of the local *source addresses* (source MAC addresses), the bridge creates a *source address table (SAT)*, also known as a *bridge address table, BAT*. As long as all frames detected in the LAN have destination addresses which are held in the SAT, then the bridge 'knows' that they will successfully reach their destination

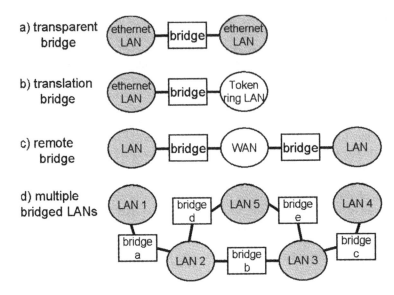

Figure 4.19   Different types of LAN bridges.

within the local LAN. But when a frame with a destination address not known in the SAT crops up, the bridge has to act.

Frames with unknown destination addresses are copied and passed across the bridge connection to the second LAN (Figure 4.19a. By forwarding frames with unknown addresses to all the stations on both LANs, the frame is bound to reach its destination (or so the theory goes).

The bridge of Figure 4.19a is a *local bridge*, since it is designed to connect LANs within the same building. It is also a *transparent bridge* because the LAN protocols used on either side of the bridge are both the same (both are ethernet LANs). In cases where the two LANs use different layer 2 protocols (e.g., in Figure 4.19b, which connects an ethernet LAN (and thus an ethernet MAC) to a token ring LAN (token ring MAC)), a *translation bridge* is required to translate the *MAC* and *physical layer*. The translation bridge of Figure 4.19b is also a local bridge, since both LANs are located in the same building.

In the early days of LANs, since most of the traffic was 'local' (e.g., from PC to local printer or local server, with only occasional network access to remote devices), the *filtering* process undertaken by LAN bridges was quite effective. Only a small proportion of packets would be *bridged*, the remainder would be *filtered* out. In this way, a bridge connection with only limited bit rate capacity (e.g., a 64 kbit/s leaseline) might not become overloaded with traffic. This brought the potential of economic *wide area* interconnection of LANS.

Remote bridges (Figure 4.19c) are intended to be used in pairs, to connect different LANs located in two different geographic locations, which are to be interconnected by means of a *wide area network (WAN)* connection (e.g., a leaseline or frame relay connection).

As office networks became ever more complex, the number of bridges used in a given network increased, so that typical networks comprised multiple individual LANs and bridges like those shown in Figure 4.19d. Even in this example, packets originated in LAN1 will find their way to destination addresses in LAN 4, since all the bridges along the way will forward the 'unknown' destination address. But meanwhile, an undesired effect occurs, which jams the network with traffic. . .

The packet originates in LAN 1 and is forwarded by bridge a to LAN 2. The packet is then forwarded by both bridges b and d to LANs 3 and 5 respectively. The packet arriving in LAN 3 is forwarded by both of bridges c and e to LANs 4 and 5 respectively. By this stage, at least one copy of the packet reaches its destination in LAN 4. But unfortunately, two unneeded copies are still underway (currently both are in LAN 5). These copies continue to circulate around the loop LAN 2-LAN 3-LAN 5 and continue to generate further copies until eventually the network seizes up in congestion (this effect is known as a *broadcast storm*). This problem is overcome by one of two techniques:

- *spanning tree protocol (STP)*; or,

- *source route bridging (SRB)*.

The *spanning tree protocol (STP)* was defined in 1990 and documented in IEEE 802.1d. It is designed to resolve the *circular routing* and packet *replication* problems we discussed above.

When using STP, a *bridge protocol* is conducted between all the bridges in a given network, who exchange *bridge protocol data units (BPDUs)*. The BPDUs are contained within normal layer 2 packets, for which the destination address is a *multicast address* understood by all bridges. (Packets bearing a *multicast address* are flooded through the network in a similar manner to packets with the broadcast address. The difference is that multicast packets are received by <u>multiple</u>, but <u>not all</u> stations in the network.)

The first step in establishing a *spanning tree* is the *election* of a *root bridge*. The bridge *elected* to be the *root bridge* is that with the lowest *bridge identity (BID)*. The bridge identity comprises a 2-byte *bridge priority* value which can be set by the network administrator (0-65535; default value 32768) and the bridge's 6-byte MAC address.

Once the root bridge has been *elected* (by default the one with the lowest MAC address), the other bridges then calculate their respective shortest *path cost*[10] to the root bridge. BPDUs are generated by the root bridge and *multicast* to the other bridges. For each link along the way, the *path cost* value held in the packet is incremented according to the bandwidth of the link. Each bridge is therefore able to determine (from the content of the incoming packet's path cost value and the incoming port on which it arrives — called the *root port*) the best path to the root bridge. This is the port and path with the minimum path cost. By means of BPDUs, the root bridge is advised of the path for direction of incoming traffic from the root bridge.

In LANs in which there is more than one bridge, the two bridges determine the one with the lowest path cost to the root bridge and declare this the *designated port* for incoming traffic from the root bridge. The spanning tree (routes to and from each bridge to the *root bridge*) is now complete! The spanning tree serves to ensure that a single non-duplicated route (via the root bridge) is available from all possible source LANs to all possible destinations.

Bridges in LANs which use the *spanning tree protocol (STP)* work in much the same way as transparent bridges. They continue to 'learn' with each received data frame behind which ports given MAC addresses can be reached, and build this knowledge into their source address table (SAT). Thus, over time, bridge e of Figure 4.19d learns that addresses in LAN 4 should only be forwarded towards LAN 3 and not towards LAN 5. They learn as a result of packets originated in LAN 4. Only when a packet originates in a LAN where the bridge does not already 'know' the destination address is the spanning tree necessary. In this case, the packet is directed only in the direction of the root bridge. If none of the bridges along the path know the destination address, then the packet is forwarded from the *source bridge* via the *root port* to the *root bridge* and thereafter to the *designated port* of the destination LAN.[11] If however, any of the bridges along the path from the source bridge to the root bridge 'know' the destination, then the frame will instead be directed along this more direct route, provided the route is not *blocked*.

---

[10] STP is a so-called *distance vector protocol (DVP)*, more about which we shall learn in Chapter 6.

[11] Given the prominent position of the root bridge, it is likely to 'know' most addresses available in the network and have them stored in its SAT. However, if necessary, the *root bridge* can arrange to flood frames to all attached LANS in a non-duplicating manner.

It is important to note that the route determined by the spanning tree is not necessarily the best or shortest route between source and destination. But it is a guarantee that the packet will not be circular routed or greatly replicated. One use of the route determined by the spanning tree (via the root bridge) helps each of the bridges 'learn' a shorter route for the next time the same destination is addressed. The *spanning tree protocol (STP)* is widely used in the ethernet LAN environment.

In the case of a change of topology within a bridged-LAN network, non-root bridges may generate BPDUs called *topology change notifications (TCN BPDU)*. These messages are used to convey status information about the bridge and its ports. Ports may be in any of the following states:

- *disabled* by the human network administrator;

- *blocking* user data, but receiving BPDUs;

- *listening* to and sending BPDUs; undertaking the *convergence* process to develop the spanning tree;

- *learning* about the network and building bridge tables, but not sending user data;

- *forwarding* data on this port.

The alternative to the spanning tree protocol (STP) for avoiding circular routing in bridged LAN networks is *source route bridging (SRB)*. This is the technique developed by IBM for use with token ring LAN networks. In source route bridging (SRB), the source determines the entire route across the network to the destination and includes this information within a special control field in the token ring header called the *routing information field (RIF)*. The route is determined by a *path discovery method*, which works by broadcasting packets from the destination, and recording the paths they take in reaching a given source. The shortest route from the destination to the source is assumed to be the best route from source to destination! So the route taken by the path discovery packet is recorded in the routing table of the source device (a DTE or *source route bridge*). Other bridges will only forward packets when there is routing information in the *routing information field* and the *RI (routing information)* bit is enabled.

Just as with **switch** source address tables (SATs), the entries in *bridge address tables (BATs*, but also called *SATs)* are *aged*. In other words, table entries and BPDU information which remains unused for a period longer than the *ageing time* are deleted. This ensures that the information does not get out of date, should the network be reconfigured. New information and routes have to be re-learnt to replace the deleted information.

The main advantage of the use of bridges for interconnecting LANs is their relative simplicity and cheapness. There is usually very little which needs to be (and which can be) configured.

Encapsulation bridges

In many office networks, different technologies are used for the end-user LANs (usually ethernet) and the backbone (the network for interconnecting the individual LANs). For example, in the past it was common to use FDDI as a high speed backbone technology for fibre connections between LANs in different buildings on a campus or on different storeys of a building. Instead of the *WAN* connection of Figure 4.19c, imagine a MAN (metropolitan area network) connection using FDDI technology. In this case, the source and destination LANs are both of the same type, even though the intermediate network is not. This is a typical case for the employment of *protocol encapsulation* (as we explained in Chapter 3). An FDDI/ethernet *encapsulation bridge* simply carries the entire ethernet frame from the source ethernet LAN to the destination LAN, wrapping it up in an FDDI *user data* field while it is carried by the FDDI network. *Encapsulation* is thus equivalent to *bridging*.

Use of LAN switches for interconnecting LANs

While bridges initially provided a relatively cheap means of interconnecting a small number of simple LANs, they were not suited to large, complex office networks, since they result in very complicated topologies which are extremely difficult to manage. Most large modern office data networks are based on ethernet switches rather than on ethernet hubs and bridges. This makes for much easier traffic and resource management, as we explained earlier in this chapter.

Switches can be interconnected to one another by means of simple interconnection of their ports (Figure 4.20a). But since switches use the same spanning tree protocol (STP) as bridges for directing traffic between the different LANs and LAN segments, they are also subject to its constraints. Spanning tree protocol (IEEE 802.1d) normally allows only one link between two LANs to be active at any one time. Second and further links may exist, but will normally remain in the *blocked* state until the first link fails. For this reason, some switches offer a *port trunking* capability which enables multiple port trunks to be grouped together to create the effect of a much higher bit rate single connection. This is termed *link aggregation* (Figure 4.20b). The *lead port* in the group takes part in the process of the spanning tree *convergence*. The other ports merely derive their configuration from the lead port.

VLANs (virtual LANs) and virtual bridged LANs (IEEE 802.1p and IEEE 802.1q)

With the emergence of switches *VLANs (virtual LANs* or *virtual-bridged LANs)* appeared as well. *VLANs* are networks of devices connected together as if they were in the same LAN collision domain (or, more correctly: *broadcast domain*), even though the devices may be physically connected to different devices.

In *switched* ethernet networks, individual end-user devices (e.g., PCs) are typically each allocated a separate *port*, in which case there are no *collisions* as such and no collision domains. Nonetheless, there is still the potential to *broadcast* or *multicast* the same message towards all the ports in the equivalent LAN (e.g., all the ports on the switch). For this reason one refers instead to a broadcast domain. Thus all collision domains are broadcast domains but

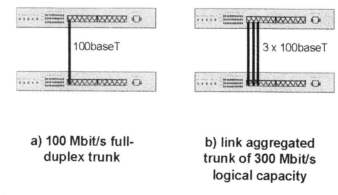

a) 100 Mbit/s full-
duplex trunk

b) link aggregated
trunk of 300 Mbit/s
logical capacity

Figure 4.20   Interconnecting switches: port trunking and link aggregation.

not all broadcast domains are collision domains. Broadcasts and *multicasts*, as we have seen earlier in the chapter, may be used when the LAN ports are flooded with copies of a given data frame, for which there is no entry in the switch SAT (source address table). Broadcasts or multicasts may also be used to update network devices (e.g., bridges) with network status or topology changes (e.g., using bridge protocol data units, BPDUs). Alternatively a given user in a LAN may wish to send the same information simultaneously to multiple (multicast) or all other (broadcast) users.

What characterises a VLAN is that all the devices within a VLAN shared the same broadcast domain, even though they may not all be connected to the same switch. VLANs are typically created to reflect the organisational or department structure of a company, even though the individual user-PCs may be in different locations and connected to different ethernet switches (Figure 4.21).

There are four main reasons why a network administrator may wish to configure his office network as a series of VLANs:

- to subdivide the network into VLANs reflecting the organisational structure of the company. This would give all the users equal access to a common set of shared resources.

- to provide for secure segmentation of the network into distinct subnetworks of *closed user groups*. This provides for protection of resources — not allowing access by outside users.

- to reduce the *broadcast traffic* on the LAN backbone. As we have seen, broadcast traffic can severely reduce the overall traffic capacity of a an office LAN network. By using VLANs, broadcast and multicast traffic still reaches all the relevant destinations within the VLAN, without loading the rest of the LAN ports with unnecessary traffic.

- to prioritise the use of the LAN or router backbone network, giving higher priority to certain users, as defined to belong to a particular VLAN.

The membership of a VLAN may be defined according to one of the following three schemes:

- *port-based VLANs(Class 1 VLANs)* — the individual switch ports (and all connected devices) belong to the VLAN;

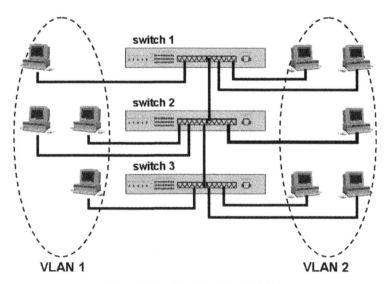

Figure 4.21   Virtual LANs (VLANs).

- *MAC address-based VLANs (Class 2 VLANs)* — individual device MAC addresses are defined in the VLAN configuration server to belong to one or more VLANs. The advantage of basing the VLAN on MAC-addresses is that no network reconfiguration is necessary when a particular device is moved from one port to another;

- *upper layer protocol (ULP)-based VLANs (Class 3 VLANs)* — define VLAN members as all endpoints *(service access points — SAPs)* of a particular protocol type (e.g., for conveniently splitting *multiprotocols* on the same network, such as *IP, IPX, Appletalk*, etc.).

Individual switch ports, MAC addresses or protocols may be configured to belong to multiple VLANs. The configuration itself is carried out manually by the network administrator. But once the membership of the VLAN has been defined and the *VLAN identifier (VID)* has been allocated, the network itself automatically takes over the necessary VLAN network functions.

The IEEE 802.1p and IEEE 802.1q standards lay out the functions and operation of VLANs. Three-layered functions are defined:

- the *configuration layer* is a signalling mechanism for informing all relevant switches about VLAN membership and update information. It provides a communication link between each of the switches and the central VLAN configuration server, using the *generic attribute registration protocol (GARP)* and the *generic VLAN registration protocol (GVRP)*;

- the *distribution/resolution layer* performs the function of switching outgoing and incoming packets at the LAN to and from their relevant *VIDs (VLAN identifiers)* during data transfer; and

- the *mapping layer* associates tags to the layer 2 (i.e., LLC-SNAP header[12]). The *tag* contains the *VID (VLAN identifier)* as well as the *user priority field (UPF*, as defined by IEEE 802.1p). The VID value limits the broadcasting of the packet to the VLAN broadcast domain (i.e., restricting its receipt to same VLAN members only at the destination switch). This creates the *virtual bridge* from VLAN source to VLAN destination switch. The UPF, meanwhile, indicates the priority level to be applied to the packet as it makes its way from the VLAN source to VLAN destination switch across the backbone.

VLANs may be created by *bridging* either local area networks (LANs) comprising only switches and other LAN components or alternatively can bridge IP-based wide area networks (WANs).

Routers

Routers are even more intelligent devices than bridges or switches. They are designed to 'learn' the topology of complicated *wide area* (i.e., layer 3) networks (even ones which are constantly growing or changing) and accordingly *route* frames or packets across them to the destination indicated by the *network address* (e.g., IP address) in the header. Routers learn about network changes through experience using *routing protocols* to 'learn' the best routes to all possible *reachable destinations.* In this way, communication is possible even across very complicated and cumbersome networks which have been built by different parties and simply connected together.

Many LAN protocols (e.g., *Novell's IPX, Appletalk*, etc.) may be *routed* in their *native* (i.e., raw) form, but it is nowadays normal instead to use the *Internet protocol (IP)* [a *layer 3* or *network protocol*] as the main protocol for interconnecting complicated LAN networks. IP

---

[12] See Chapter 3.

a) 'normal' SNA-network linking terminal to mainframe

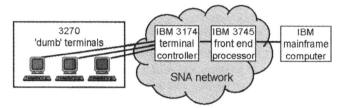

b) access to mainframe from LAN using 3270 gateway

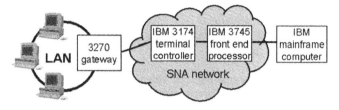

Figure 4.22   Replacement of dumb terminals using a LAN and 3270 gateway.

uses a worldwide standardised addressing scheme (*Internet addressing*) to identify end user stations uniquely. This provides for the ability to connect all LANs worldwide into a single common network, the Internet, thus extending the information sharing and electronic mail capabilities of single LANs to the world computer community as a whole.

The problem with networks of multiple routers (including the Internet itself) is that the individual routes through the network are difficult to monitor, manage and keep up-to-date. It is difficult to know which networks are being transmitted along the way, so that optimal network loading and the security of information is difficult to achieve. For this reason, many different routing and network management protocols have been developed as part of the IP suite, as we shall discover in later chapters of this book.

Gateways

A *gateway* provides for interpretation and conversion of protocols. A *LAN gateway* provides access for a LAN user to an external service, such as a mainframe computer. Typically a gateway consists of a Personal Computer (PC) equipped with appropriate network hardware to reach the external device and running gateway software. An example of a widely used LAN gateway is a *3270-* or *SNA-gateway*. *IBM 3270* is the communication protocol used between the host computer and the terminal of an IBM mainframe. The 3270 protocol allows the computer to interpret keyboard interaction at the 3270-terminal and control the exact image appearing on the terminal screen, without there having to be 'intelligence' in the terminal. Thus 3270 type terminals are sometimes described as *dumb terminals*. The conversion necessary to make an intelligent terminal (such as a personal computer) appear to talk to a host computer like a dumb terminal is carried out by *terminal emulation* or *3270 emulation* software. Where this software resides in a LAN, it is called a *3270 gateway, SDLC (synchronous datalink control)-gateway* or *SNA (systems network architecture)-gateway* (Figure 4.22).

5

WANs, Routers and the Internet Protocol (IP)

The Internet Protocol (IP), as the centrepiece of the Internet and IP-suite of protocols, has established itself as the most widely-used data networking protocol. It provides an interface function on which the Internet—the 'network of networks' is based. It is used not only in end-devices which access the Internet (so-called hosts) but also between the nodes (referred to as routers) of wide area data networks. This chapter describes how wide area networks (WANs) can be built using routers and the Internet Protocol. It explains how the routers work and the aspects of the Internet Protocol (including the IP addressing scheme) which make router functions possible. We will discuss in detail the two most important variations of the Internet protocol—version 4 (IPv4) and version 6 (IPv6)—and we shall cover the functions of an IP-based router, as well as the various types of network communication made possible by IP—unicasting, broadcasting, multicasting and anycasting.

5.1 WANs (wide area networks), routers, Internet protocol (IP) and IP addresses

The *Internet protocol (IP)* was originally conceived for *internetworking*—connecting different types of data *networks* together into a single network. But while this single network has come to be known under a single name—the *Internet*—it is made up of a wide range of networks of different technologies. In short, it is the 'network of networks'. Figure 5.1 illustrates a typical data network based on the Internet protocol (IP). Figure 5.1 shows a network between two *hosts* and comprising six routers, two different LANs, and two different WAN technologies. Incidentally, there are no strict rules for the designations LAN, MAN and WAN: their commonly understood meanings are:

- *LANs (local area networks)* are typically in-building or on-campus networks with maximum cabling lengths around 100–500 m. Typical technology is ethernet or token ring, though some in-building LANs of the past were based on *FDDI* and *ATM*;

- *MANs (metropolitan area networks)* are typically networks capable of covering an area requiring cabling distances up to about 20 km. Example technologies are *Gigabit ethernet* (based on fibre), *FDDI* and *SMDS*;[1]

---

[1] See Chapter 4.

Data Networks, IP and the Internet: Protocols, Design and Operation   Martin P. Clark
© 2003 John Wiley & Sons, Ltd   ISBN: 0-470-84856-1

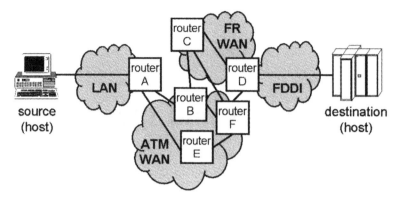

Figure 5.1   A typical IP-based network: the 'network of networks'.

- *WANs (wide area networks)* are network technologies without distance constraints. This phrase is generally used to describe networks based on IP-backbone router networks, leaselines, frame relay, ATM, X.25 or similar technology.

Communication in IP data networks takes place between hosts, across a network made up of *nodes* (called *routers* or *switches*). This is the terminology we shall use in this book. But during the course of the development of technical standards for the Internet (spanning more than 30 years since 1969), different terminology has been used. The original ARPA standards referred to hosts and to the nodes as *Internet message processors (IMPs)*; the original version of the Internet protocol had two variants, called simply IMP-IMP and Host-IMP. During the 1970s and early 1980s, the nodes were sometimes referred to as *packet switched exchanges (PSEs)*, but also as *gateways*. As we shall discover, some of the routing protocols used between routers are still referred to as *interior* or *exterior gateway protocols*. The Organization for International Standardization (ISO), meanwhile, has issued a number of protocol specifications which refer instead to the hosts as *end systems (ES)* and the nodes as *intermediate systems (IS)*. We shall encounter this terminology particularly in the chapter on routing protocols which comes later. In the meantime, we shall refer to hosts and routers.

Working together, the six routers and four networks of Figure 5.1 are required to transfer messages between the two hosts (i.e. communication endpoints — PCs, servers or computer applications) which are illustrated. The routers of Figure 5.1 act like the sorting offices of a postal network — they read the addresses of each packet received in turn (the address is indicated on the *packet header*). According to the address, individual packets are *routed* (or *forwarded*) to a next router (nearer their destination) by way of the *next hop* of their *path*. The Internet protocol (IP) provides the packet header which is critical to this process.

There are two main functions which the Internet protocol has to fulfil. These are:

- *routing* based upon a *unique network addressing* scheme (*IP address*) recognised by all networks; and

- *fragmentation* of packets as necessary, if unduly long packets exceed the size allowed (the *maximum transmission unit — MTU*) length of a particular network technology (*medium*) along the route.

The unique address is provided by means of the *Internet address* or IP address, which we shall discuss in detail later in the chapter. Originally the IP address (as supported in IP version

4 — IPv4) was a 32-bit address — written in the familiar form of four decimal numbers (value 0–255) separated by 'dots', e.g.,

$$173.65.8.0.$$

But the demand for IP addresses was so great that *IP version 6 (IPv6)* was designed to accommodate a much greater address range — of 128 bits, to be written in the form:

$$\text{xxxx : xxxx : xxxx : xxxx : xxxx : xxxx : xxxx : xxxx}$$

Each x represents a hexadecimal digit — one of the 15 possible values: 0, 1, 2, 3, 4, 5, 6, 7, 8, 9, A, B, C, D, E or F, each of which is equivalent to 4 bits, as we saw in Chapter 2.

Fragmentation of packets is necessary when long packets exceed the maximum size allowed by the *maximum transmission unit — MTU —* of the next transit network on the *path* to the destination. We have seen[2] how ethernet LANs support an MTU of 1500 bytes, but we shall discover that some other media can support only much shorter message transmission units. IP packets in general may be up to 65 535 bytes (or octets) long. But if an IP *packet* of 65 535 bytes needs to *transit* an ethernet LAN as the next hop of the path to a given destination, then the 65 535-byte packet would need to be *fragmented* (i.e. cut up) into smaller fragments of a maximum of 1500 bytes in length each (about 45 fragments are needed). Later in the journey, the *fragments* are *reassembled* into the original packet. The *reassembly* may take place either at the ultimate destination or at another gateway router (i.e. at the entry point to another transit network capable of carrying a larger MTU).

The routers of Figure 5.1 provide gateway functionality for interconnecting the different types of networks. At a basic level, they may have to convert between different types of *physical layer* (layer one) and *datalink* layer (layer two) interfaces (as we learned about in Chapter 3). But there is much more to a router than this.

5.2   Main functions of routers

Since the main purpose of a router is to forward IP packets (also called IP *datagrams*) according to the IP address held in the datagram *header*, it is reasonable to conclude that the required functionality of routers can be largely determined from the specification for the Internet protocol and its various options (IPv4 is defined in RFC 791; IPv6 is defined in RFC 2460). This is true. However in addition, in 1995 the *Internet Engineering Task Force (IETF)* issued RFC 1812 as a supplement to RFC 791 to formally define the requirements of IPv4 routers. RFC 1812 sets out in detail the appropriate implementation in IPv4 routers of the various Internet protocol functions and fields. From it we are able to determine the minimum set of basic functions which should be undertaken by a router. Routers are required to support the following functions:

- network *interfacing* (including fragmentation as necessary);
- packet *forwarding* according to a routing table;
- *routing table* creation and updating;
- Internet protocol (IP) processing;
- network congestion and admission control;

---

[2] See Chapter 4.

- network security and access control; and

- network configuration, monitoring and administration.

The following sections provide more details of each function.

Network interfacing

A router provides for:

- the *internetworking* and interfacing of networks of different types (e.g., interfacing the *physical* and *datalink layers* of a LAN to a WAN);

- the adaptation of data-packet or frame formats at network boundaries as necessary, e.g., conversion from *canonical* to *non-canonical* data transmission formats[3] or vice versa;

- the *resolution* and *translation* of addresses as necessary;

- fragmentation and reassembly of large packets as necessary;

- the support of *PPP (point-to-point protocol)* as a standard IP-suite layer 2 (*datalink protocol*) interface for general purpose, point-to-point serial lines. PPP[4] is similar to HDLC[5].

Packet forwarding according to a routing table

A router supports the following packet forwarding and routing capabilities:

- the ability to *forward* packets towards their *ultimate destination* based upon the indicated *destination IP-address*;

- the creation and maintenance of a *routing table* for the purpose of determining the *next hop* of the packet's path to its destination;

- the *routing table* recognises all standard *broadcast* (e.g., 255.255.255.255), *multicast* and *unicast* addresses (explained later) and has a *default route* for all possible addresses (e.g., 'if I don't recognise this address, I'll pass the packet to another default router, which will know how to deal with it'). The default route is usually indicated in the routing table with the notation '0.0.0.0'. Each '0' value means 'any value between 1 and 255'.[6] Thus 0.0.0.0 matches 'all possible IP-addresses'.

- the *destination IP address* is compared with all the entries in the routing table, to determine the *routing table* entry which matches the greatest number of leading digits (i.e., the *longest match*). This is the entry with the greatest number of consecutive matching digits (reading the number from left to right — Figure 5.2). Thus the destination IP address of Figure 5.2 (173.65.8.1) generates the following 'longest matches' in each of the routing tables:

  - at router A: the longest match is '173.0.0.0' and the chosen next hop is via router B

  - at router B: the longest match is '173.65.0.0' and the chosen next hop is via router C

---

[3] See Chapter 3.
[4] See Chapter 8.
[5] See Chapter 3.
[6] The value '0' in a routing table is taken to mean 'any allowable value in this position' (i.e., any value 1–255).

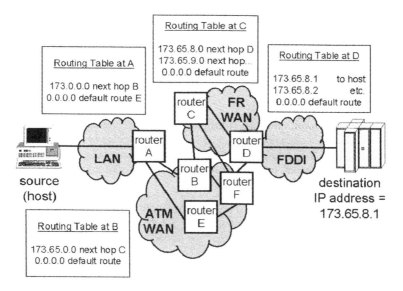

Figure 5.2 Network and associated routing tables.

- at router C: the longest match is '173.65.8.0' and the chosen next hop is via router D
- at router D the full destination address is recognised as a directly-connected destination.

Routing table updating and maintenance

Routing tables in routers are kept up-to-date as follows:

- Entries in a router's routing table are either based on *static routing* (i.e., only occasionally updated manually by humans) or on *dynamic routing* (in which they are kept up-to-date automatically by means of routing table updates sent between different routers by means of *routing protocols*).

- In the case of dynamic routing, network configuration changes (e.g., new links, removed links or failed links) and the addition or removal of *reachable addresses* (e.g., an end-device or subnetwork has been added to or removed from the local network) are monitored automatically by means of a *routing protocol*.

- Routers must normally support at least one *interior gateway protocol (IGP)* for advertising routing changes to other routers within the immediate network (called an *administrative system* or *autonomous system—AS*). Examples of commonly used IGPs are *RIP (routing information protocol), OSPF (open shortest path first),* Cisco's *IGRP, interior gateway routing protocol* and ISO's *IS-IS* and *ES-IS* protocols.

- Routers (acting as *border nodes*) at the boundaries of different networks (*autonomous systems*) exchange routing information with neighbouring networks by means of an *exterior gateway protocol (EGP)*. The most commonly used protocol is *BGP4 (border gateway protocol version 4)*.

- A router supporting dynamic routing uses routing *metrics, routing algorithms* and *routing policies* to sift through and analyse the information it receives by means of the routing protocols. By doing so, it can determine the best current next-hop route to each *reachable* IP address or range of addresses and thus keep its routing table up-to-date.

Internet protocol (IP) and other protocol processing

At a minimum, all routers must support the Internet protocol (IP) and the related protocol *ICMP (Internet control message protocol)* which is an integral part of it. IP allows for the processing and forwarding of packets. ICMP allows for monitoring the correct functioning of IP:

- Each received IP-packet must be verified. The header *checksum* (i.e., *frame check sequence — FCS*[7] needs to be checked, as do all the other fields in the packet header. *Time-expired* packets and those with invalid destination addresses (e.g., the *loopback address* 0.0.0.0) should be discarded, returning a notification to the source (if appropriate) by means of ICMP.

- Source routers or source hosts should perform *discovery* of the *path maximum transmission unit (PMTU)*. By doing so, fragmentation of packets along the path can be avoided by only sending packets of a size less than or equal to the PMTU. This maximises the network efficiency.

- In the case of *source routing*, the *destination IP address* appearing in the IP header may not be the *ultimate destination* of the packet. Nonetheless the *next hop* is determined by the router based upon the indicated destination address rather than upon the ultimate destination. This allows the source router to place a string of IP addresses in the *source routing field* of the IP header to control the exact path taken by the packet across the internetwork.

- In general, *transport layer protocols* need not be supported by intermediate routers except as required for internal network management purposes (e.g., for carriage of management information by means of *SNMP, simple network management protocol*).

- *Multicast* and *anycast* routers must support *IGMP (Internet group management protocol)* for the appropriate updating of multicast address lists.

Network congestion and admission control

In some cases it is appropriate for routers to perform *network admission controls*. By limiting the amount of data *traffic* admitted to the network, we can minimise the possibility of network congestion:

- Some routers perform congestion control by *prioritisation* of forwarded packets according to the *type of service (TOS), IP precedence* or *differentiated service (DiffServ)* information contained in the IP packet header.

- Some modern routers which support the optional *RSVP, Resource ReSerVation Protocol* perform *admission control* at the time when a *host* initially requests a *connection* to be set up or a given quantity of bandwidth to be reserved (e.g., for a real-time *stream* application). Only if sufficient bandwidth is adjudged to be available is the new connection permitted, otherwise the connection is rejected (rather like returning the busy tone on a telephone call to a busy line).

---

[7] See Chapter 2.

Network security and access control

Since some of the packets which arrive at routers connected to the router are unwelcome — having originated from maliciously-minded intruders, most modern routers provide various means of *access control*:[8]

- Most routers offer access control based on *filtering* and *access control lists* (*ACLs*, also called simply *access lists*). A *filter* or *access list* checks the *source IP address*, only forwarding packets from selected (i.e., permitted) source addresses to given allowed destinations. This provides a measure of security — allowing only certain outsiders to get access to certain data and computer files.

- Non-transparent routers may perform *network address translation (NAT)*. Such routers translate the IP addresses in the IP packet headers from public Internet addresses into private Internet addresses as used within the 'local' network. The use of NAT has two benefits — first, the number of IP addresses available for use within the 'local' network is almost unlimited (public IP addresses, on the other hand) are difficult to come by; second, only those local addresses converted by NAT can be reached by outside parties (e.g. intruders).

- Some sophisticated types of routers are also able to act as *proxies*. A *proxy* acts as a gateway between an outside host and a *client* or *server* within the local network. The outside host wishes to communicate directly with the client or server within the local network, but it is not allowed to do so directly. Instead the *proxy client* or *proxy server* (at the network boundary) acts on behalf of the real client or server. The proxy interprets the higher layer protocol requests of the outside host and decides which requests will be dealt with and which will not. The proxy (as a *trusted party*) then communicates with or forwards request to the real client or server as it sees fit. The use of a proxy client or proxy server concentrates the network security measures in a single device, thereby avoiding the need for specific security measures in each internal client and server. The disadvantage is that it requires sophisticated router hardware and software.

- Some types of routers include *firewall* security functions. These usually include NAT and proxy server functions as described above. We shall discuss firewalls in more detail in Chapter 13.

Network configuration, monitoring and administration

Routers must support certain network and self-administration functions:

- Routers are expected to monitor network status and any communications errors.

- Network status is normally reported to a remote network management control station (e.g., a network management centre) by means of standard managed objects, *MIBs* and the *simple network management protocol (SNMP)*. We shall discuss these in Chapter 9.

- Routers typically *boot* (i.e., set their basic settings) automatically using *BOOTP (bootstrap protocol) or TFTP (trivial file transfer protocol)*. These protocols are used to load appropriate network settings and other configuration files.[9]

---

[8] Some access routers (*access routers* are those which typically connect a local network such as a LAN to the public Internet) are *transparent routers*. By transparent we mean they use the public Internet addressing scheme even within their local network and forward packets freely in both directions. The LANs connected by means of such transparent routers are relatively prone to intruder attacks from the Internet.

[9] See Chapter 6.

- Routers are often used as *address servers*: allocating or advising IP addresses to host devices when they are switched on using protocols like BOOTP and DHCP. We shall talk more about this in Chapter 6.

- Hosts may also have to act as routers, or at least have simple *embedded router* functionality. At a minimum, a *static gateway* or a *default gateway* needs to be configured to direct all outgoing IP traffic to the LANs main router.

5.3 Unicast, broadcast, multicast and anycast forwarding

There are four types of communication supported by the Internet protocol (Figure 5.3):

- *unicast* communication;

- *broadcast* communication;

- *multicast* communication; and

- *anycast* communication.

Unicasting (Figure 5.3a) is the simplest form of Internet packet (*datagram*) carriage. A datagram requiring to be unicast is analogous to a letter being sent to a single addressee as its destination.

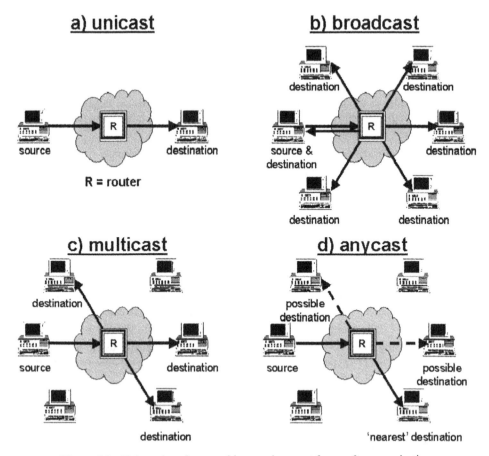

Figure 5.3   Unicast, broadcast, multicast and anycast forms of communication.

When an IP-packet (datagram) is to be broadcast (Figure 5.3b), the *source* of the message will only submit one copy of the datagram to the network, but wishes the datagram to be copied and sent to *all systems* (i.e., all addresses) connected to the network. There are a number of standard IP addresses for various types broadcast:

- the address 224.0.0.1 is a broadcast to *all systems on this subnet*;

- the address 255.255.255.255 is a general *all systems broadcast*.

Sometimes it is useful to be able to deliver the same message to a number (i.e., a *group*) of destinations simultaneously. Such a delivery could be performed using a *broadcast*, if we made the assumption that host stations for which the message was not intended simply ignored the contents. Such a broadcast, however, is not usually undertaken, since it leads to three problems:

- It confuses the host stations which had not expected the message and don't know how to react to it.

- There is a possible security risk associated with the fact that unintended recipients may be able to read the message.

- The broadcasting of the message unnecessarily congests the network, particularly (as in the case of the Internet) when a very large number of host stations are attached.

Messages required to be sent to groups of destinations within the Internet are usually multicast (Figure 5.3c). A specific address for a multicast group has to be requested from the *Internet Assigned Numbers Authority (IANA)*, and messages required to be multicast to all the members of the group are sent only once — addressed to the multicast *group address*. The routers use *IGMP (Internet group management protocol)* to keep track of the members of the various multicast groups and use this knowledge to perform the multicast forwarding.

Multicasting is typically used between routers to distribute network configuration changes and other *routing information*. There are a number of standard multicast addresses, which we shall list more fully later, but here are a couple of examples:

- 224.0.0.2 is the multicast address to address *all routers on this subnet*;

- 224.0.0.5 is the multicast address to address *all routers using OSPF* (open shortest path first) protocol;

- 224.0.0.9 is the multicast address to address *all routers using RIP2* (routing information protocol version 2).

Anycasting (Figure 5.3d) is a variation on multicasting. An anycast message is addressed to all the members of a particular (multicast) group of recipients (e.g., routers), but will only be delivered to one member of the group — the one adjudged by the router to be the 'nearest' destination (according to the router's calculation of proximity).

Broadcast, multicast and anycast addresses may not be used as source IP addresses.

5.4   Routing table format — static and dynamic routing

The principal task of a router is to forward Internet protocol (IP) packets according to its *routing table*. The routing table itself can be thought of as a two-column table (see Table 5.1). In the first column of the table, individual Internet addresses or address ranges are listed in

Table 5.1   Detailed routing table for router C in figure 5.4

Table Ref.	Address or group of addresses	Next hop of path to destination
i	0.0.0.0 (default route)	To Router B
ii	173.0.0.0 (except subnetworks 173.65.0.0 & 173.66.0.)	To Router B
iii	173.65.0.0 (except 173.65.8.0 & 173.65.9.0)	Destination within LAN connected to C
iv	173.65.8.0	To Router D
v	173.65.9.0	To Router F
vi	173.66	To Router D
vii	195.52	To Router B

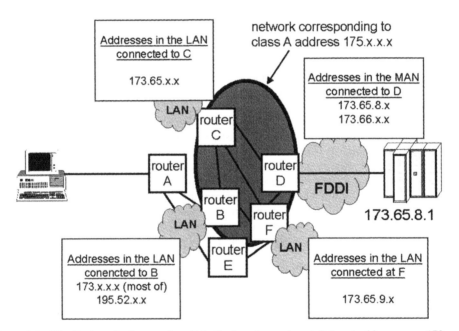

Figure 5.4   Distribution of subnetworks within the Imaginary class A Internet address range 173.x.x.x.

numerical order. In the second column is listed the appropriate next hop of the path to the corresponding address.

Figure 5.4 is a more detailed illustration of the network of Figure 5.2, showing how the LANs and MANs connected to the routers B, C, D and F together provide the network for the Internet *class A* address space corresponding to 173.x.x.x. The figure shows how the various sub-ranges of addresses (i.e., the *subnetworks* of 173.x.x.x) are distributed across the individual routers. Table 5.1 is the detailed routing table for router C.

Each of the routers in Figure 5.4 has been configured (probably manually) to recognise the range of directly-connected IP-addresses (i.e., the 'addresses in the LAN connected to' as it appears in Figure 5.4). All the addresses in each configured address range are considered to be *reachable addresses*[10] and will thus appear in the routing table of all the routers in the network.

Critical for the proper operation of router networks is that the routing tables are kept up-to-date. Any incorrect or out-of-date entries lead to mis-routing of packets, with the consequence of delivery to the wrong destination, or to the inability of the network to deliver the packets at all.

The maintenance of the routing table in a router (or in the *embedded router* of a host device) may be conducted either manually by a human, or automatically by the router itself. In the case of manual updating, the router and the routing table are said to undertake *static routing*. In other words, the routes remain static (unchanged) except for human intervention. In the case of automatic updating of the routing table (in response to network configuration and routing information received from other routers by means of routing protocols), the router is said to undertake *dynamic routing*.

Methodology and advantages of dynamic routing

The advantage of dynamic routing (automatic updating of routing tables) is that the network adjusts its routing to all reachable destination addresses on an up-to-the-minute basis. Dynamic routing automatically accommodates for:

- addition of new devices or addresses to the network;

- removal of devices or addresses from the network;

- moving of devices or address from one location in the network to another; and even for

- re-adjustment of routes when links between routers fail.

In a complex network using dynamic routing, the network can be subjected to continuous re-configuration, even by many different un-coordinated parties, and yet the routing tables always remain up-to-date, thus minimising the problems of mis-routing. This is perhaps the greatest quality of the Internet — the ability to allow many different uncoordinated networks to be internetworked with one another, while always maintaining an up-to-date routing scheme. Understandably, *dynamic routing* is highly complex.

Routers undertake dynamic routing by sifting through lots of *routing information* received from other routers by means of routing protocols. The analysis of the routing information is carried out according to a *routing policy* (programmed into the router by a human) and using a *routing algorithm* (pre-defined calculation method). The development of such *routing policies* and routing algorithms is a highly complex mathematical challenge which we shall discuss in Chapter 6.

For our current discussion on the Internet protocol (IP) and how routers forward packets, it is sufficient for us to know that a routing table exists in each router. The routing table provides an important resource for the Internet protocol in determining the next hop which packets should be forwarded, but is not always 100% reliable. For this reason, the design of the Internet protocol (IP) and associated routing protocols includes a number of measures intended to:

- minimise *misrouting* arising from out-of-date information held in routing tables;

- discard packets which get lost or turn out to be *undeliverable*;

---

[10] As far as *routing protocol*s are concerned, an address is considered to be reachable if a router recognises it as belonging to one of its directly connected IP address-ranges. But this does not means that the address has actually been configured for use by an end-user device, and does not exclude the possibility that a destination device is *unavailable*, for example: switched off!

- optimise routes to the shortest possible path, but adjust these routes (and the handling of packets as necessary) to take account of prevailing network traffic demand; and

- cater for the different urgency of packet delivery required by different types of traffic.

To prevent against *misrouting* arising from out-of-date routing information, the routing table is *aged* — routing table entries are simply deleted if they are not used within a given period of time. On deletion of the 'special route' to a particular destination, routing to that destination reverts to the *default route* valid for all destinations. Effectively, this turns the network into a *hierarchical* network — concentrating routes via a few centralised and very powerful *core* routers. But just in case any packets do get misrouted and get lost on an endless trek through the network, or take too long trying the reach their destination, the Internet protocol uses a simple mechanism to eradicate them. Packets which exceed their allowed *time-to-live* (*TTL* — a field in the IP version 4 header) or exceed their maximum permitted *hop count* (a field in the IP version 6 header: a *hop* is a link between two routers) are simply discarded. This simple mechanism prevents the possibility of a gradual accumulation of undeliverable packets causing network congestion.

Optimisation of routes to the shortest possible paths and adjusting routes according to prevailing network traffic demand are highly complex challenges. Delay sensitive communications must be given priority. The IP-header packet fields *type of service (TOS), IP precedence and DiffServ (differential services)* are used for this purpose.

For real-time communications flows (e.g., live video), it is important that all the packets making up the communication *stream* take nearly the same time to propagate across the network. A variation in the delay incurred by different packets would otherwise lead to *jitter* of the signal. In this case, the connection as a whole is usually subjected to *path-oriented routing*, and the individual packets are given a *flow label* (in IP version 6) which ensures that all take the same route through the network.

Applications of static routing

While dynamic routing may be the only practicable means of keeping routing tables up-to-date in fast-changing networks, the use of static routing may sometimes be more appropriate. The main advantage of static routing is that network routes do not change quickly. This leads to very predictable performance and easier troubleshooting of network performance problems.

Static routing is often employed by the embedded routers in end-systems (i.e., hosts) and in access routers (e.g., a router connecting a single small LAN to the public Internet by means of a single connection). The routing table of a static router may contain only two entries: one entry accepts traffic from the Internet to the IP address range of the local network and forwards it to the LAN. Meanwhile the *default route* simply forwards IP packets originated by devices in the LAN to all other 'unknown' addresses via a standard *default* route — to the public Internet.

5.5   Routing table conventions

It is usual to denote the default route entry in a routing table with the address entry 0.0.0.0. The default route is the route which will be used for all IP-addresses which do not otherwise specifically appear in the routing table. The '0' values of the IP address ranges in the routing table are used to denote 'any valid numerical value' (the valid values are 1–255 in the case of IPv4 addresses). Thus the entry '173.65.8.0' (Table 5.1) of router C of Figure 5.4 represents all the addresses in the IP address range between values 173.65.8.1 and 173.65.8.255. All of these addresses are reachable via router D.

When looking up a particular address in a routing table, the principle of finding the *longest match* is applied. Let us 'look up' four addresses in Table 5.1: first, the address 173.65.8.2; then, the address 195.52.3.5; third, the address 184.5.45.233; finally, the address 195.51.46.27.

The address 173.65.8.2 matches four of the entries in the routing table of Table 5.1 (entries i, ii, iii and iv). The longest match is with entry iv, so the route chosen will be 'next hop via D'.

The address 195.52.3.5 matches two of the entries in Table 5.1 (entries i and vii). The longest match is entry vii, so that the chosen route in this case will be 'next hop via B'.

The address 184.5.45.233 does not appear in the routing table of Table 5.1 other than as represented by the default address 0.0.0.0 (entry i), so in this case the default route to router B will be selected. Similarly, the address 195.51.46.27 also only matches the default route entry. (Note that entry vii of Table 5.1 only applies to the 195.52.x.x subnetwork portion of the 195.x.x.x-address space.)

5.6   Simple IP routing control mechanisms: time-to-live (ttl) and hop limit fields

Extreme care must be used when setting up routing tables (no matter whether they are set up manually using static routing or automatically using dynamic routing). In particular it is important to avoid the possibility of circular routing arising from routing loops. Figure 5.5 illustrates a simple example in which the default routes of routers E and F create an endless loop. Imagine sending in a packet to router A of Figure 5.5 with the address 184.5.45.233. This address is known to all the routers, but only by means of the default route address (0.0.0.0). The packet is therefore forwarded from router A to router E to router F, and from then on, in a continuous loop F-to-E-to-F-to-E-to-F … etc.

In the relatively straightforward example of Figure 5.5 it might be relatively easy to eliminate the routing loop, but more complex networks, especially those using dynamic routing are much harder to check manually or administer with very carefully selected routing policies and

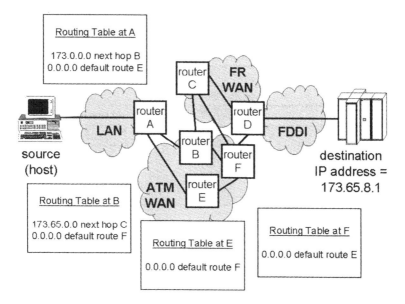

Figure 5.5   Circular routing arising from inappropriate default routing.

routing algorithms. For this reason, the Internet protocol provides a field called time-to-live (TTL) in IP version 4 and an equivalent field in IPv6 called *hop limit*. If the TTL or hop limit is exceeded, then the packet should be discarded by any router which subsequently receives it. This at least prevents the packet from trekking endlessly through the network.

Enough of the background! Let's now get down to describing the Internet protocol (IP) itself.

5.7   Internet protocol version 4 (IPv4)

Internet protocol version 4 (IPv4) is a *network layer* protocol for packet-switched datagram communication. Figure 5.6 illustrates the standardised format of an Internet protocol datagram (*IP datagram* or *IP packet*) conforming to IPv4, as defined in RFC 791.[11]

Internet protocol version 4 (IPv4) is quite a straightforward packet-switching protocol. In some ways it is similar to the ITU-T packet-switching protocol, recommendation X.25, which we met in Chapter 3. But the fact that it is a *connectionless*, protocol rather than a *connection-oriented* protocol makes it different. We shall explain IP in detail by considering the contents, meaning and use of the various IP-header fields (Figure 5.6). But if you need any extra assistance on exact details of the protocol operation or precise field codings, RFCs 791 (Internet protocol version 4) and 1812 (Requirements for IPv4 routers) provide the definitive documentation.

IP Version 4 (IPv4) header format

RFC 791, in line with most other IP-suite protocol documentation, specifies the protocol headers and packet formats in terms of *words* of data rather than in terms of single *octets*[12].

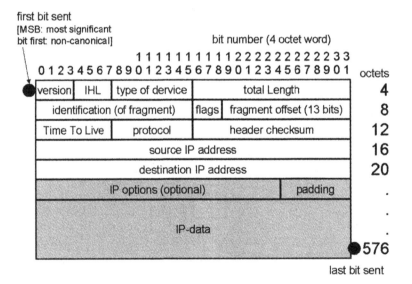

Figure 5.6   IP datagram structure showing IP-header fields and IP data.

---

[11] Though IP does not correspond exactly to OSI layer 3, it nonetheless can be considered approximately equivalent.
[12] An *octet* is the same as a *byte* — i.e., 8 bits. The name octet rather than byte is commonly used in protocol specifications.

In the case of the IPv4 itself, RFC 791 specifies the packet or datagram format in terms of a number of 4 *octet* (i.e., 32-bit) *words*.

The IP header is always transmitted in the *non-canonical* format. In other words, the header fields are always sent *most significant bit (MSB)* first. Thus transmission generally commences at the left-hand corner of the illustrated packet format (as shown in Figure 5.6) and continues line-by-line (i.e., word-by-word) until reaching the bottom right-hand corner of the packet header. But the transmission order of bits in the *IP data* (i.e., *user data*) field (whether transmitted in *canonical* or *non-canonical* format) is not specified.

IP Version field

The *version field* of the IPv4-header indicates the format of the Internet header. The value is set to binary '0100' [decimal value 4] for IP version 4. The only other allowed values are 5 (IP version 5 was an experimental *stream protocol* (also called *ST2* and *ST2+* which was not widely deployed) and 6 (IP version 6, which is currently commencing large-scale deployment).

IHL (Internet header length) field

The *Internet header length (IHL)* field indicates the length of the packet header in 32 bit words. The minimum value of this field is 5 words (5 × 4 octets), but with additional *options*, the header could be much longer, as is apparent from Figure 5.6. If necessary, *padding* can be used to fill out any remaining empty final portion of the final 4-octet word of the *options field*.

Type of service (TOS) field

The *type of service (TOS)* field (Figure 5.7a) is intended to indicate the relative priority to be afforded to the *forwarding* of the particular packet. The intention is that intermediate routers along the packet's path handle packets not on a simple *first-in-first-out (FIFO)* basis but instead

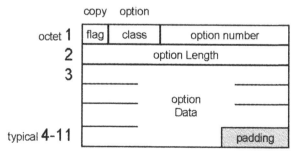

Figure 5.7   Type of Service (TOS) and options fields of the IPv4 header.

Table 5.2   IP Precedence field determines relative priority for packet processing and forwarding

IP Precedence Value decimal (binary in brackets) TOS bits 0, 1 and 2	Meaning
7 (111)	Network Control
6 (110)	Internetwork Control
5 (101)	CRITIC/ECP (critical/exceptional)
4 (100)	Flash Override
3 (011)	Flash
2 (010)	Immediate
1 (001)	Priority
0 (000)	Routine

Table 5.3   Delay, throughput and reliability — bits (D-, T- and R-bits) of the type of service (TOS) field

TOS bit	Meaning	Binary value '0'	Binary value '1'
D-bit (TOS bit 3)	Delay	Normal Delay	Low Delay
T-bit (TOS bit 4)	Throughput	Normal Throughput	High Throughput
R-bit (TOS bit 5)	Reliability	Normal Reliability	High Reliability
TOS bit 6	Reserved for future use	Always set to '0'	Not used
TOS bit 7	Reserved for future use	Always set to '0'	Not used

will process them in an order more appropriate to their relative priorities. The first three bits of the TOS-field (bits 0, 1 and 2) contain the *IP precedence* indication (Table 5.2). This sets a relative priority for the packet's processing and forwarding through the network. In addition, the TOS-field includes the *D (delay), T (throughput)* and *R (reliability)* bits (Table 5.3) which define additional specific needs of the communication. At most, two of the D, T, and/or R-bits in the TOS-field should be set. An alternative use of the TOS-field is to support *differentiated services (DiffServ)* — as we shall discuss later in the chapter.

It is not mandatory that the type of service (TOS) field should be processed by all routers. In the case that a router does not interpret or process this field, all packets must be processed normally (i.e., with equal precedence).

Total length of IPv4 datagram and the relationship to the path maximum transmission unit (PMTU)

The *total length* field indicates the total length of an IP datagram (i.e., IP packet) including the IP header. The value is coded in binary and represents the total number of octets. The maximum value is 65 535. All hosts and routers which support IPv4 must be able to accept IP datagrams of at least 576 octets length without requiring fragmentation. The value 576 corresponds nominally to 512 octets of data plus a 64-octet header.

By first discovering the *maximum transmission unit (MTU)* length (i.e., the maximum IP-data field length — see Table 5.4) that can be carried by all the links of a given end-to-end *path*, a host or source router can ensure that fragmentation will not be necessary. This process is called *path MTU (PMTU) discovery*. Fragmentation is avoided by sending packets corresponding to a packet length equal to or shorter than the PMTU. A standard method for PMTU discovery for IPv4 is defined in RFC 1191 (and for IPv6 in RFC 1981). A more pragmatic approach assumes that the PMTU is the lower of the first hop MTU and 576 octets (the recommended minimum MTU of an IPv4 router).

Table 5.4 Path maximum transmission unit (PMTU) size for common types of networks

Network type	Maximum message transmission unit	Specification
Minimum IP MTU	68 octets (= 68 bytes)	RFC 791
Standard IP MTU	576 octets	RFC 791
X.25	576 octets	RFC 877
IEEE 802.3/802.2 Ethernet LAN	1492 octets	RFC 1042
X.25	1500 octets	RFC 1356
PPP (Point-to-Point Protocol)	1500 octets	RFC 1661
FDDI (Fibre Distributed Data Interface)	4352 octets	RFC 1390
4 Mbit/s Token Ring LAN (IEEE 802.5)	4464 octets	RFC 1042
ATM (Asynchronous Transfer Mode)	9180 octets	RFC 1626 & RFC 2225
16 Mbit/s Token Ring	17914 octets	RFC 1191
Maximum IP MTU	65535 octets	RFC 791

Table 5.5 Fragmentation flag field of IPv4 header

Flag bits	0	1	2
Bit meaning	Reserved	DF — Do not Fragment	MF — More Fragment
Binary value '0'	Always set to '0'	May be fragmented	No more fragments follow
Binary value '1'	Not used	May not be fragmented	More fragments follow

Fragmentation-related fields

The second 4-octet word of the IPv4 header (octets 5–8) contains information regarding the fragmentation of the communication, i.e., whether the packet is entire (unfragmented) or only a single fragment of a larger message. The fragmentation *flags* (Figure 5.6 and Table 5.5) indicate whether the packet is or may be *fragmented*. Typically, the use of path MTU discovery is employed to try to avoid fragmentation, and the flags are then set to '010'.

In the case that fragmentation is undertaken, the fragment *identification field* is used to mark all the fragments of an original packet with a common 'packet number'. In addition, flag bit 2 (MF-bit) indicates whether more fragments will follow, or whether the current fragment is the last. The *fragment offset* is used during reassembly. It indicates where the fragment belongs in the original packet. The first fragment has offset value 0. Subsequent ones indicate the position of the fragment in multiples of 8 octets (64 bits) from the start of the original packet.

Time-to-live (TTL) field

The *time-to-live (TTL)* field contains an integer binary value corresponding to the duration of time in seconds for which the packet is still allowed to 'live'. Each time the packet header is processed by a router, the value in the TTL field must be checked, and be reduced by at least 1 second. As soon as the TTL value reduces 0, the packet <u>must</u> be destroyed. In practice, although the TTL field is nominally the lifetime in seconds, in reality it reflects the maximum number of hops allowed to be traversed before a packet is considered to be 'lost' and *undeliverable* (and therefore ripe to be destroyed). A common initial default value setting for the TTL field is 64 (possible values are 0–255 seconds). Such a value would allow for the safe traversal of a path comprising around 20 routers (with a 'safety margin' to spare).

Table 5.6 Protocol field values (IANA protocol numbers) and their meanings

Protocol field value (possible range 0–255)	Protocol used at next higher layer (i.e., content type of IP data field)
1	ICMP — Internet Control Message Protocol
2	IGMP — Internet Group Management Protocol
4	IP in IP encapsulation (RFC 2003)
6	TCP — Transmission Control Protocol
7	CBT — Core Based Trees
8	EGP — Exterior Gateway Protocol
9	IGP — Interior Gateway Protocol (e.g., Cisco's IGRP Interior Gateway Routing Protocol)
10	UDP — User Datagram Protocol
27	RDP — Reliable Data Protocol
28	IRTP — Internet Reliable Transaction Protocol
30	NETBLT — Bulk Transfer Data Protocol
35	IDPR — Inter-Domain Policy Routing Protocol
42	SDRP — Source Demand Routing Protocol
45	IDRP — Inter-Domain Routing Protocol
46	RSVP — Resource ReSerVation Protocol
47	GRE — Generic Routing Encapsulation
48	MHRP — Mobile Host Routing Protocol
54	NARP — NBMA Address Resolution Protocol
80	ISO-IP — ISO Internet Protocol
81	VMTP — Versatile Message Transaction Protocol
88	EIGRP — Extended Interior Gateway Routing Protocol (Cisco)
89	OSPF — Open Shortest Path First
92	MTP — Multicast Transport Protocol
95	MICP — Mobile Interworking Control Protocol
97	ETHERIP — Ethernet-within-IP encapsulation
98	ENCAP — Encapsulation header / private encryption
103	PIM — Protocol Independent Multicast
108	IP Comp — IP Payload Compression Protocol
111	Novell IPX-in-IP
112	VRRP — Virtual Router Redundancy Protocol
115	L2TP — Layer 2 Tunnelling Protocol

Note: For an up-to-date and complete listing refer to www.iana.org/assignments/protocol-numbers.

Protocol field

The *protocol field* indicates the format of the data held in the packet's IP data field (see Figure 5.6). This is the protocol used at the next higher layer of the communication. During the data-transfer phase between two hosts communicating end-to-end across an IP network, the next higher layer protocol will typically be either *TCP (transmission control protocol)* (protocol value = 6) or *UDP (user datagram protocol)*(protocol value = 10). But a much wider range of protocols may be contained in the packet as is apparent from Table 5.6. Most of these protocols transfer internal network control, routing information, security, management and monitoring information between different network devices. Official protocol field values are assigned following application to the Internet assigned numbers authority (IANA) and a current list of all allocated protocol numbers can be viewed on their website at www.iana.org.

Header checksum

The *header checksum* has an equivalent role to the *frame check sequence (FCS)* we discussed in Chapter 3, but uses a different (simpler) error detection algorithm. As for most IP-suite protocols (including IP itself), the frame check algorithm is the *16-bit one's complement of the one's complement sum of all 16-bit words in the header*. For purposes of calculating the checksum, the checksum field is assumed to be all 0s.

Source and destination addresses

The *source* and *destination IP address* fields usually indicate the <u>original</u> source and <u>ultimate</u> destination of the IP packet. The format (for IP versions up to IPv4) is a 32-bit value, usually written in the form d.d.d.d where each d is a decimal value between 0 and 255 representing one of the four 8-bit octets of the number. We explain the numbering structure in more detail later in the chapter.

The source address may not contain an address assigned as a multicast or broadcast group address. The destination address may not contain the *unspecified address* (0.0.0.0) and should only contain the *loopback address* (0.0.0.1) when used for test purposes.

The destination IP address field does not always contain the address of the *ultimate destination*. In the case of packets which follow paths controlled by *source routing*, the destination IP address field will contain the IP address of the next intermediate point along the way.

IP Packet options

The IP packet header may (but need not) include the *IP options* fields. When included, IP options are indicated by a corresponding increase in the value of the *IHL (Internet header length)* field. The format of the IP options fields conforms to that illustrated in Figure 5.7b. In cases where the option field does not exactly fill exactly the last 4-octet word of the header, padding is used to fill the remaining space.

The *copied flag* in the options field header indicates that the option is copied into all fragments (if the packet has been fragmented).

The possible options are *security; source routing; timestamping; route recording* and *router alerting*, and the coding of the various options fields according to these options can be seen in Table 5.7.

The *security option* field indicates the security level (unclassified, confidential, restricted, secret, top secret, etc.) of the IP packet contents, the compartmentation, handling restrictions and closed user group or *TCC (transmission control code)* information. The terminology reflects the initial use of the Internet protocol for government and military communication. Most recently, the security option field is used to carry *authentication* and *encryption* information to ensure that only the intended recipient can decode the packet contents. IPv4 routers are expected to support the security options.

Strict source and route records (SSRR) or *loose source and route records (LSRR)* are composed of a series of Internet addresses. These occupy the IP options data field (if the option number field is set to one of the values 3, 7 or 9: see Table 5.7). The list of consecutive IP addresses which appear in a source route record specify the precise route the packet is to take through the network. As each destination in turn of a *source-routed* path is reached, the destination IP address in the packet header is amended to the next indicated destination in the list. Similarly, a route record (option number = 7) records the actual route taken by a packet. This option is valuable to network administrators who are trying the trace the cause

Table 5.7   IP Packet options and coding of the IP packet header options field

Option number value (0–31)	Option	Option length	Option class value (values 1 & 3 reserved)	Copy flag
0	End of Option List	Option length field omitted, total option fields only 1 octet	0 = control	Not used
1	No operation	Option length field omitted, total option fields only 1 octet	0 = control	Not used
2	Security	11	0 = control	0 = 1 =
3	Loose Source Routing	Variable	0 = control	0 = 1 =
4	Internet Timestamp	Variable	2 = debug & measurement	Not used
7	Route Record (traces route taken)	Variable	0 = control	Not used
8	Stream ID	4	0 = control	0 = 1 =
9	Strict Source Routing	Variable	0 = control	0 = 1 =
20	Router Alert	4	0 = control	0 = 1 =

of network performance problems in IP networks or the Internet. It is important to note that source routing and route recording options may not be supported by all routers along the path.

The *stream ID* was used in certain experimental networks. The value is carried transparently by networks which do not support the stream concept.

The *timestamp* is a right-justified 32-bit timestamp in milliseconds since midnight UT (universal time). Routers may (but need not) support this option.

5.8   ICMP (Internet control message protocol)

The Internet protocol (IP) is an *unreliable* protocol since delivery of packets and end-to-end user-information is *not guaranteed*. The problem lies in the fact that IP packets may be submitted to a router network, without checking whether the destination is ready and able to receive them and without even checking whether the destination address exists. IP is a connectionless protocol.[13] *ICMP (Internet control message protocol)* is designed to report such delivery problems. But while ICMP reports most of the errors and delivery problems which can be encountered by IP datagrams (packets) or fragments, and while it is an 'integral part' of the Internet Protocol, it nonetheless does not make IP *reliable* and delivery of packets remains *unguaranteed*. The problem is that the delivery of the ICMP messages themselves (which are carried by IP) cannot be guaranteed. In the case of network congestion, for example, the time-to-live value could expire, whereupon an ICMP message would be destroyed.

An ICMP message is generated when a packet cannot reach its destination; when no buffering is available at a gateway or destination host; or when the destination host could

---

[13] See Chapter 3.

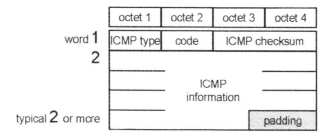

Figure 5.8   Internet Control Message Protocol version 4 (ICMPv4): protocol format.

be reached via a shorter route. The ICMP message is generated in the format illustrated in Figure 5.8, this field being packed into the IP data field of an IP packet (Figure 5.6). The destination and source addresses of the original IP packet (the one which encountered the problem) are reversed, so that the ICMP message is returned to the original source.

So that the source host can identify which packet encountered the problem, the ICMP information field (Figure 5.8) is usually filled with the IP-header of the original packet plus at least 64 bits (8 octets) of the original IP data (i.e., user data).

The coding of the various fields in the ICMP protocol is documented in Table 5.8. As you will see from Table 5.8, ICMP has a number of uses other than merely reporting *destination unreachable, time exceeded* or other IP *parameter errors*. It can also be used to acquire status information about the network.

Table 5.8   Internet control message protocol version 4 (ICMPv4): messages and coding

ICMP Type value (0–255)	ICMP Type meaning	ICMP code value (0–255)	ICMP Code meaning and ICMP information field content
0	Echo message (e.g., PING request)	0	ICMP-info field gives identifier and sequence number
3	Destination unreachable	0	Network unreachable
		1	Host unreachable
		2	Protocol unreachable
		3	Port unreachable
		4	Fragmentation needed and DF set
		5	Source route failed
		6	Destination network unknown
		7	Destination host unknown
		8	Source host isolated
		9	Communication to network prohibited
		10	Communication to host prohibited
		11	Network unreachable for Type of Service
		12	Host unreachable for Type of Service
		13	Communication administratively prohibited
		14	Host Precedence violation

(continued overleaf)

Table 5.8 (*continued*)

ICMP Type value (0–255)	ICMP Type meaning	ICMP code value (0–255)	ICMP Code meaning and ICMP information field content
		15	Precedence cut-off in effect (packets with lower precedence are being discarded)
4	Source Quench message	0	N/A
5	Redirect message (via gateway IP address given in ICMP-info field)	0	Redirect datagrams for the network
		1	Redirect datagrams for host
		2	Redirect for TOS and network
		3	Redirect for TOS and host
8	Echo reply message (e.g., PING response)	0	ICMP-info field gives identifier and sequence number
11	Time exceeded	0	Time to live exceeded
		1	Fragment reassembly time exceeded
12	Parameter problem	0	Pointer in first octet of ICMP information field indicates error
		1	Fragment reassembly time exceeded
13	Timestamp message	0	ICMP files includes 2 octet identifier, 2 octet sequence number, 4 octet originate timestamp, 4 octet receive timestamp and 4 octet transmit timestamp
14	Timestamp reply message	0	ICMP files includes 2 octet identifier, 2 octet sequence number, 4 octet originate timestamp, 4 octet receive timestamp and 4 octet transmit timestamp
15	Information request	0	ICMP info is 2 octet identifier and 2 octet sequence number
16	Information reply message	0	ICMP info is 2 octet identifier and 2 octet sequence number
17	Address mask request	0	
18	Address mask reply	0	

One of the most commonly used features of ICMP used by network administrators is the *PING (Packet INternet Groper)* service. When a device is *PINGed* by a network administrator, he/she sends a short message (an *echo request*) from a remote location across the Internet (or other IP network) to the device. The device is merely required to reply (with an *echo reply*). If the echo reply returns, the network administrator is able to confirm that the remote device is *reachable* from the location from which he/she sent the original *echo request*. Most people refer to a PING request and reply (rather than an *ICMP echo request/reply*: ICMP types 0 and 8). In some forms of PING, a timestamp is requested (ICMP types 13 and 14). In addition, the IP route record option (previously discussed) can be set, in order to gain information about the path taken to the destination. The PING service is thus a valuable but simple tool for network troubleshooting.

5.9   Internet addressing (IPv4)

In order that the Internet protocol (IP) can deliver IP packets to their intended destinations, each device connected to an IP network must be uniquely identified by means of an IP address.

Internet addresses for Internet protocol (IP) versions prior to IPv6 have a fixed length of 32-bits. These addresses (IPv4 addresses) are used in the IP-packet header to identify both the source and destination of the packet and are critical for data communication using the Internet protocol.

IP addresses can be classified into two types: <u>public IP addresses</u> and <u>private IP addresses</u>. Public IP addresses are unique. Both versions have a common format, usually written as a series of four decimal integers (each of value between 0 and 255) separated by 'dots', e.g.,

$$173.65.8.1$$

or more generally:

$$d.d.d.d$$

The Internet addressing scheme used for Internet protocol version 4 is defined in a number of different RFCs as follows:

- the original *classful addressing* scheme is documented in RFC 1812;

- the management of the address space and the top level allocations are covered in RFC 1466; and

- *classless inter-domain routing (CIDR)* is documented in RFCs 1517–1519.

We describe *classful addresses* and the *classless* address scheme in turn.

Classful IP addressing scheme

The original *classful addressing scheme* foresaw dividing the fixed length number (of 32 bits) into two separate portions, the first part of the address to be known as the *network address* and the second (remaining) portion as the *host address*.

The *classes* defined in the classful addressing scheme correspond to different lengths of network- and host-address portions, as illustrated in Figure 5.9 and detailed in Table 5.9. A *class A address-range* thus has a network address of 1 byte length (8-bits) and a host address portion of 24 bits (corresponding to the last three bytes of the address). Each individual address within a class A address range is typically written in the form A.d.d.d, where A is a decimal value between 1 and 127 (identifying the particular operator's network) and the d values are decimal values in the range 1–255 representing the precise host address value. Routers outside of the network corresponding to the class A address range need only have routing table entries 'A.0.0.0' if the access to the class A network is by means of a single gateway router. The

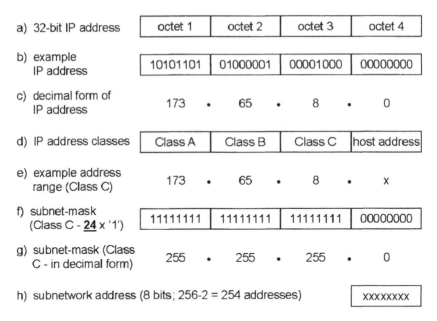

Figure 5.9   Classful IP addressing scheme.

Table 5.9   IPv4 classful addressing scheme

	First bits of address	Number of bits of network address	Number of bits of host address	Number of network address ranges available	Number of hosts per network address
Class A	0	7	24	126	16.8 million
Class B	10	14	16	16 383	65 536
Class C	110	21	8	2 097 151	256
Class D (Multicast)	1110	28	Non-aggregatable Multicast address	268 million	0
Class E	1111	Experimental use		268 million	0

gateway address for this address range is thus 'A.0.0.0'. Each full class A address range has scope for nearly 16.8 million host addresses.

A *class B address* has the number format B1.B2.d.d where B1 and B2 are the permanently assigned numbers corresponding to the network address and d.d forms the host address assigned to the end-user device by the class B address range owner. A class B address thus has a 16-bit sub-address corresponding to the last two bytes of the public IP address. This allows a class B network to comprise up to 65 536 hosts. Routers outside the network can make do with a single routing table entry B1.B2.0.0 if the access to the class B network is by means of a single gateway router. The gateway address is B1.B2.0.0.

A *class C address* has the form C1.C2.C3.d. The host address comprises only the last 8 bits of the address, allowing a maximum of 256 hosts to be connected. The routing table entry for routers external to the network will be C1.C2.C3.0 (gateway address).

Class A and class B address ranges can be subdivided by their owners into *subnetworks* corresponding to class B and class C address ranges respectively. By using such subnetwork address ranges, the network operator helps to simplify the task of routing and administering his or her network by keeping groups of similar destination address in defined geographical locations (e.g., attached to a particular router, or within a particular LAN or campus network).

Class D addresses are *multicast addresses*. These have the form D.d.d.d where D is a decimal value between 224 and 239.

Class E addresses are reserved for experimental use. These have the form E.d.d.d, where E is a decimal value between 240 and 255.

Classless inter-domain routing (CIDR) and subnet masks

By 1992 it became clear that the IPv4 address space would be exhausted, and that the *classful* method of allocation of addresses could not be sustained. This led to a new method of public IP address range allocation based on *classless inter-domain routing (CIDR)*. The CIDR address scheme (introduced by RFCs 1517-9 in 1993) is a more flexible address allocation scheme which allows for host or subnet (subnetwork) address ranges of any length (rather than the strict 8-bit, 16-bit or 24-bit host and subnetwork address lengths of the classful address scheme). By introducing CIDR, the remaining unallocated class B and class C addresses could be shared between more business enterprises than would otherwise have been possible.

The main feature of classless inter-domain routing (CIDR) is the separation of the network address and the subnet address (i.e., host address) by means of a *subnet-mask*. It is the subnet-mask which reveals where the first bit of the subnetwork address begins.

It is useful to commence with an example from the classful address scheme to explain the idea of the subnet-mask, and we return to Figure 5.9. The example class C address range shown in Figure 5.9b and 5.10c is the address range including all the addresses between 173.65.8.0 and 173.65.8.255. The host address (or subnet address) corresponds to the last 8-bits of the address (i.e., the value between 0 and 255 at the end). The subnet mask is a series of 24 bits of binary value '1', followed by 8 bits of binary value '0'. By comparing the subnet mask with the full IP address, the subnet address can be obtained by ignoring all the bits of the IP address which align with a value '1' in the subnet-mask.

Whereas in the classful addressing scheme a whole address range could be identified by means of the gateway address (of the form A.d.d.d [class A addresses] or B1.B2.d.d [class B addresses] or C1.C2.C3.d [class C addresses]), it is necessary in the case of classless addresses to specify the gateway address and the subnet mask. The subnet mask can be denoted in one of two ways: either as a number similar to an IP-address comprising four decimal numbers equivalent to the 'value' of the subnet mask (the value of the subnet mask in the example of Figure 5.9g is 255.255.255.0) or as a 'slash character' (/) and a decimal number equal to the number of '1's in the mask: thus in our example /24.

The routing table entry for the class C address 173.65.8.0 can thus also be denoted in a CIDR format as follows, either as:

- 173.65.8.0 / 24, or as:

- 173.65.8.0 subnet mask: 255.255.255.0

Figures 5.10 and 5.11 provide further illustrations of subnet address ranges conforming to the CIDR addressing scheme. These clearly illustrate how classless inter-domain routing (CIDR) brought the possibility of *variable length subnet masks (VLSMs)*. No longer are the mask lengths fixed at 8 bits (Class A), 16 bits (Class B) and 24 bits (Class C). It is worth studying them to ensure you have understood the principle of the gateway address and the subnet mask and can derive from these two the full address range represented by them.

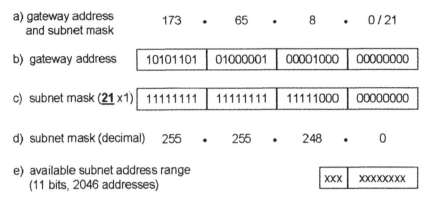

Figure 5.10 CIDR example: gateway address 173.65.8.0–2048 subnet addresses available 173.65.8.0–173.65.15.255.

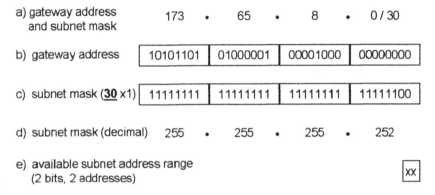

Figure 5.11 CIDR example: gateway address 173.65.8.0–4 subnet addresses available 173.65.8.0–173.65.8.3.

Physical and logical addresses

Addresses used in telecommunications are sometimes *physical addresses*[14] and sometimes they are *logical addresses*. The 48-bit unique IEEE identifiers commonly used as MAC addresses in ethernet and other LAN networks are examples of *physical addresses*.

Sometimes it is convenient to use unique physical addresses. But on occasions where a given address (e.g., an *IP-address*) is used to direct a number of remote computers to a given software application running on a computer, the use of a logical address is better.

A logical address provides a <u>permanent</u> means for remote hosts to address a particular software application, without the software application having to reside for all time at the same physical location (i.e., on the same computer). The benefit of using a logical address is that the destination address used by the remote hosts to access the application need not be changed even if the software application is moved from one computer to another (for example,

---

[14] the term *physical address* may appear somewhat confusing here, since the MAC address is an address used at the layer 2 or datalink layer (rather than being a physical layer or layer 1 address). Nonetheless, the MAC address is permanently associated with a particular *physical* port on the LAN. Messages sent to this physical address will always find their way to the same physical network end-point (unless, of course, the particular ethernet interface card is transferred from one computer to another).

because of a 'hardware upgrade'). But in order for the message to be delivered to the physical device which runs the application corresponding to the logical address, the address must first be *resolved* into a physical address. As we will discover in Chapter 6, a protocol like *ARP (address resolution protocol)* is used to carry out the *resolution*.

IP addresses can be allocated either as physical or as logical addresses. In other words, an IP address could be assigned to physical hardware (e.g., a server) or to a software application (which can be moved from one server to another). Most common is to use the IP address as the logical address of a server application and the ethernet MAC address as the physical address of a host or server.

Internet address assignment — IANA and the regional registries

Public IP addresses are allocated on application to the *Internet Assigned Numbers Authority (IANA* — www.iana.org) or one of its *regional registries*:

- *APNIC (Asia-Pacific Network Information Centre* — www.apnic.net);

- *ARIN (American Registry for Internet Numbers* — www.arin.net); or

- *RIPE (Réseaux IP Européens* — European IP Network Coordination Centre — www.ripe .net).

Public and private IP addresses

Local networks which use *transparent routers* for their connection to the public Internet need to use public IP addresses, in order that all the numbers are unique: see Figure 5.12a. Meanwhile,

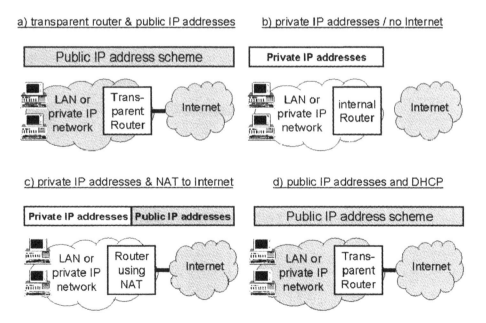

Figure 5.12   Alternative public/private IP addressing schemes and related local network/Internet configuration.

private IP networks which are either not connected to the Internet (Figure 5.12b) or are connected by means of routers which perform NAT (network address translation — Figure 5.12c) may use their own privately conceived network addressing scheme. IETF (in RFC 1918) and IANA recommend the use of the following assigned private Internet ranges for private IP networks:

- 10.0.0.0   to   10.255.255.255   (10.0.0.0/8)   [65 535 host addresses]
- 172.16.0.0   to   172.31.255.255   (172.16.0.0/12)   [4 095 host addresses]
- 192.168.0.0   to   192.168.0.255   (192.168.0.0/16) [255 host addresses]

A business enterprise or private individual that decides to use private IP addresses from the above ranges may do so without coordination with either IANA or any of the regional registries, but these addresses can only be used within the private IP network (e.g., as in Figure 5.12b or Figure 5.12c), as they are not unique addresses.

Why has the use of private IP address space proliferated? Primarily because the public address space was rapidly exhausted during the mid-1990s as the 'Internet boom' took off. As a result, it became difficult for businesses and private individuals to get as many public IP addresses assigned as they required. The solution was to use a 'private IP-network' in one of the configurations of Figure 5.12b, 5.13c or 5.13d. Figures 5.13c and 5.13d illustrate alternative methods of connecting private IP networks to the public Internet.

Network address translation (NAT)

Figure 5.12c shows a private IP network using a private IP addressing scheme, connected to the Internet by means of a router employing *NAT (network address translation)*. Each time a packet traverses the NAT router (no matter whether leaving the private network to the Internet or entering it from the Internet), the private IP address in the IP packet header is translated to or from (i.e., swapped for) a public IP address. Thus a device in the private IP network is 'known' in the private IP network under its private IP address and in the public Internet under a different public IP address. What is the benefit of using NAT then, you might ask? The answer is that only those devices within the private network which need to communicate with the public Internet need have a public IP address. So, while the configuration of Figure 5.12c

a) differentiated services field (DS-field)

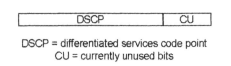

DSCP = differentiated services code point
CU = currently unused bits

b) functions of a differentiated services router

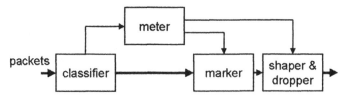

Figure 5.13   Differentiated services (DiffServ): DS field format and functions of DS-router.

allows for data communication between any internal company private IP network devices, this might be achieved with only a very small number of public IP addresses. A secondary benefit of the configuration of Figure 5.12c is that the private IP network is much more secure than the network of Figure 5.12a. This is because all the private network devices which do not require public Internet access do not have public IP addresses and therefore cannot be accessed by malicious 'hackers' in the outside public Internet community.

Dynamic address allocation using DHCP (dynamic host configuration protocol)

Figure 5.12d shows the use of a router employing *DHCP (dynamic host configuration protocol)* as a gateway between a private IP network and the public Internet. In this configuration, the IP address is only temporarily assigned to the communicating host within the private IP network. By such means it is possible to share the public IP address space as shown in Figure 5.12d, given that the number of hosts requiring addresses at any one time does not exceed the total number of public addresses available to the DHCP for *dynamic address assignment*.

It is becoming increasingly common to employ *DHCP servers* for the allocation of IP addresses in private LANs and private IP networks since this makes for much easier administration of the numbering range. Hosts and software applications which are configured to accommodate DHCP are prepared to accept temporary IP address assignments (e.g., for a 24-hour period or for the course of a particular communication). The fact that all addresses are only temporarily assigned makes for much easier management of the numbering scheme, allowing individual hosts or addresses easily to be moved from one part of the network (subnetwork) to another.

Another common configuration is to use a router at the gateway which combines both NAT (Figure 5.12c) and DHCP. The router acts as both a *DHCP server* (Figure 5.12d) and as a *DHCP client*. As a DHCP server, the router allocates private IP addresses to hosts within the private IP network. As a DHCP client, the router may receive public IP address allocation on a temporary basis from an Internet service provider (ISP) to which it is connected for public Internet access. The public IP address allows a host in the private IP network temporarily to access the public Internet. NAT is needed (as in Figure 5.12c) to translate between the two temporarily assigned private and public IP addresses. This configuration is very efficient in its use of public IP addresses as well as being quite secure, for now there is not a fixed public IP address always used to access a given private host from the public Internet.

5.10   Differentiated services (Diffserv and DS field)

The Internet and IP-based networks have become the technology of choice for data communications but there has been ever-increasing demand to use these networks for carriage of communications *flows* — multimedia services, including real-time *voice-over-IP (VOIP)* and *video streaming*. Such *multimedia* information flows or *streams* make much heavier demands upon the network and have led to the development of a new *integrated services model* for the Internet. The model is presented in RFC 1633.

The integrated services model is a further development of the *type-of-service (TOS)* and *IP precedence* concepts of IP. The model seeks to give higher priority for the use of network capacity to critical services at times of network congestion. Done away with is the, best effort' principle of data delivery and there is an attempt to provide a greater guarantee of reliable and predictable delivery for high priority traffic *streams*. Instead of queuing packets for forwarding across the network on a *FIFO (first-in-first-out)* basis, packets which can afford to wait will be held in the queue (i.e., output buffer) until after more critical packets have already been sent.

The integrated services architecture for *differentiated services (DiffServ)* is defined in RFCs 2474-5 and RFC 3260. It defines a new *flow-specific state* intended to give differential priority for stream services and delay-critical signals like real-time video, and aims to mirror the differential *quality of service (QOS)* needs for different communications types established by IEEE 802.1p and RFC 2386.

Differentiated services codepoint (DSCP) and per-hop behaviour (PHB)

When *differentiated services (DiffServ)* are enabled and in use, a *DS field (differentiated services field* — as depicted in Figure 5.13a) is included in all relevant IP packet headers. (In IPv4 the *DS field* replaces the *TOS-field*: in IPv6 the *traffic class* octet is used — see Figure 5.14). A 6-bit *codepoint* value appears in the DS field (the *differential services codepoint, DSCP*). The DSCP identifies the priority of the traffic stream to which the packet belongs. In the jargon the traffic stream is a *flow-specific state*, and the particular characteristic needs of a given type of flow are called a *behaviour aggregate (BA)*.

Analysis of the DSCP causes a *differentiated services router* receiving the packet to apply the corresponding *per-hop behaviour (PHB) group* to it. The PHB defines the means and method by which network resources will be assigned to different traffic flows (i.e., *behaviour aggregates*). In particular the PHB affects the relative delays encountered by packets and the relative packet *discard priority* (i.e., which packets should be thrown away first should network congestion cause packet discarding to become unavoidable).

As an example: a simple PHB could reserve X% of a given link's for a given *behaviour aggregate* (i.e., traffic stream). Following the assignment, the given traffic stream can 'rely' upon a minimum bit rate equivalent to X% of the bandwidth. Only packets bearing the correct *DS codepoint* will gain access to this reserved bandwidth. The bandwidth reservation can be undertaken as a permanent assignment (for example, giving the more important customers of

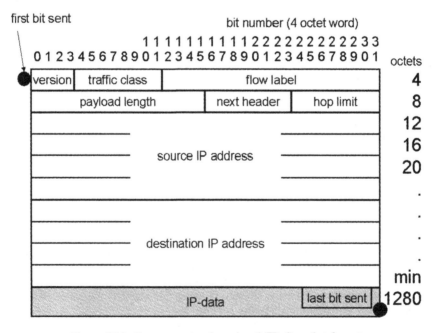

Figure 5.14   Internet protocol version 6 (IPv6) packet format.

Table 5.10 Differential services: codepoints (DSCP) and PHB (per-hop behaviour groups)

PHB type	First three bits of DSCP			Second three bits of DSCP		Actions of PHB
Default PHB	000			000		RFC 1812
Assured forwarding	001	Class 1	A l t	010	LDP (Low Drop Priority)	RFC 2597
(AF) PHB	010	Class 2	e r n	100	MDP (Medium Drop Priority	
	011	Class 3	a t i	110	HDP (High Drop Priority)	
	100	Class 4	v e s			
Expedited forwarding (EF) PHB	101			110		RFC 2598
Network control traffic	11x			xxx		Highest priority

an Internet service provider better service than the rest) on a temporary basis (e.g., at the time of connection set-up) using a protocol like *RSVP (Resource ReSerVation Protocol*[15]).

Table 5.10 lists the DSCPs (differentiated services codepoints) which have been defined so far. These correspond to four different PHB types:

- *default PHB* (DSCP = 000 000). The default PHB is defined by RFC 1812.

- *assured forwarding (AF) PHB* (DSCP = ccc ddd, where ccc is the class and ddd is the drop probability). The assured forwarding PHB is defined by RFC 2597.

- *expedited forwarding (EF) PHB* (DSCP = 101 110). The expedited forwarding PHB is defined by RFC 3246.

- network control traffic.

Default PHB

The routing and forwarding actions to be undertaken by the *default PHB* are the same as the actions required to be undertaken by a normal IPv4 router which does not support differential services. In other words, 'best effort' forwarding and FIFO (first-in-first-out queuing) as defined in RFC 1812.

Assured forwarding (AF) PHB

The *assured forwarding (AF) PHB* (RFC 2597) is specifically designed for the purpose of enabling *ISPs (Internet service providers)* to divide their customers into different quality *classes*. For each different class of customer, specific resources (e.g., bandwidth) within the network can be reserved. A given level of resources (e.g., bandwidth) is, guaranteed' to be

---

[15] See Chapter 7.

available. Over and above the 'guaranteed' bandwidth, customers may send further data (*excess data*), but only on the understanding that there is no 'guarantee' of delivery.

The different classes allow the ISP to give different priority to different customer groups dependent upon the charges the customers are prepared to pay for service. In addition, within the individual classes, different traffic types can be divided into separate streams, depending upon whether these streams are to be treated with *low drop priority (LDP), medium drop priority (MDP)* or *high drop priority (HDP)*. (see Table 5.10). At a time of congestion a packet with low drop priority will only be discarded after all high drop priority packets have already been discarded. Thus a low drop priority packet is least likely to discarded.

Expedited forwarding (EF) PHB

The *expedited forwarding (EF) PHB* (RFC 2598) is specifically intended to cater for *real-time* traffic (i.e., signals or information streams like high-quality video) which require packets to be carried with:

- very low loss (i.e., very few packets to be misrouted or discarded due to congestion);

- low *latency* (i.e., minimal propagation delay);

- low *jitter* (i.e., there should be very little difference in the delay experienced by different packets);

- assured bandwidth.

The expedited forward (EF) PHB is based upon the premise that the loss, latency and jitter are minimised, provided that the packets are subject to no buffering delays (or at least only minimal buffering delays) along the path of the communication. Provided the packets are forwarded immediately after they are received, they will encounter no such buffering delays (so the theory goes, at least).

Since the development of differentiated services is still at a comparatively early stage, we can expect further refinement of PHBs. Perhaps a technique which will be 'imported' from *ATM (asynchronous transfer mode)* will be the limitation of MTU (maximum transmission unit) size during times of network congestion. Consider trying to expedite a packet from a high priority stream, but finding that a lower priority packet was currently in course of being transmitted to the required outgoing path. Our high priority packet might get uppermost priority once the buffer is empty again but we may have to wait until the entire low priority packet has been sent unless we are willing to simply abort it in the middle! Waiting for the buffer to empty can take quite a long time, as Table 5.11 shows. For example, a message

Table 5.11   Time in milliseconds to transmit a single packet to line based on packet size and line bit rate

Bit rate of line	Message transmission unit size			
	576 bytes (normal IPv4 packet)	1280 bytes (normal IPv6 packet)	1500 bytes (ethernet packet)	65535 bytes (maximum size)
64 kbit/s	72 ms	160 ms	188 ms	8192 ms
1.544 Mbit/s (T1)	3 ms	7 ms	8 ms	340 ms
2.048 Mbit/s (E1)	2 ms	5 ms	6 ms	256 ms
34 Mbit/s (E3)	0.1 ms	0.3 ms	0.3 ms	15 ms
45 Mbit/s (T3)	0.1 ms	0.2 ms	0.3 ms	12 ms
155 Mbit/s (STM-1)	0.03 ms	0.1 ms	0.1 ms	3 ms

transmission unit of 65 535 bytes length, when sent on a 64 kbit/s line will take 8 seconds to transmit. This is certainly not low latency!!

For high speed lines (T3, E3 or above), the maximum latency might be kept around 15 ms without any particular further precautions, so this could suffice in a high-speed differential services *backbone* network. On the other hand, a backbone network is more likely to experience high volumes of equally important other high priority traffic!

Operation of a differential services router

A router which supports differential services undertakes three basic functions (Figure 5.13b):

- *classification*;
- *metering*; and
- *conditioning* (comprising *marking, shaping* and *dropping*).

Arriving packets are *classified* according to the *differential services codepoint (DSCP)* value held in their DS-field and sorted accordingly into the appropriate *behaviour aggregate (BA)*. The *meter* function analyses the *profile* of the relevant packet stream, checking whether the packet arrival rate and demanded bit rate is *in-profile* or *out-of-profile* according to the relevant *traffic conditioning agreement (TCA)*. Should the flow condition be out-of-profile, then the router is permitted to *condition* the traffic to bring it back into profile. The *conditioning* can include any or all of the actions: marking, shaping and dropping.

Marking changes the value of the *DS codepoint* in the IP packet header. This will typically reduce the priority given to a specific packet (judged to be out-of-profile) by subsequent routers. By marking the out-of-profile-packet in this way, the router can ensure that the remaining packets (judged to be *in-profile*) all receive the appropriate high-priority handling along the rest of the path. If a single packet were not singled out for subsequent low-priority handling by later routers in the path, it could turn out that different packets were adjudged to be the out-of-profile offenders at the different routers. This could mean that many more of the individual packets are degraded by *conditioning* along the way.

If the outgoing route from the router is in congestion or its buffer is overflowing, it might be necessary to undertake more draconian *conditioning* upon the offending out-of-profile packets. These measures are *shaping* and/or *dropping*. Shaping means delaying the sending of packets. Thus an imaginary packet stream of 2 packets per second which is contravening its *traffic conditioning agreement (TCA)* for a guaranteed bit rate of 64 kbit/s by submitting 128 kbit/s can be shaped back to the traffic agreement by slowing the packet forwarding rate to 1 packet per second. Alternatively, when things get really congested, out-of-profile packets can be simply discarded (i.e., *dropped*). Naturally the lowest priority out-of-profile packets will be dropped first.

Differentiated service (DiffServ) is an asymmetric service. The differential (i.e., priority) service afforded to the packets corresponding to one direction of communication (e.g., A-to-B) is not necessarily also afforded to the packets sent in the reverse direction (i.e., B-to-A). For a 'duplex' differentiated service, both directions of communications *flow* need individually to be coded for differentiated service.

Differentiated services domains and regions

A *differential services domain (DS domain)* may comprise one or more networks (usually all under the same administration). The differential services philosophy of one administration (e.g., a given Internet service provider, ISP) may be to afford given customer groups better than average service when higher subscription fees are paid. Meanwhile, another administration might conduct differential services purely based upon the traffic or *stream flow* type (e.g., video, VOIP, games, data, etc.).

A *differentiated services region (DS region)* comprises one or more contiguous DS domains. *Service level agreements (SLAs)* between the different administrations govern the handling of differential service flows which traverse the boundaries from one domain to another within the DS region.

We shall return to the subject of how to use differentiated services (DiffServ) as part of an overall network strategy to ensure end-to-end *quality-of-service (QOS)* in Chapter 14.

5.11   Internet protocol version 6 (IPv6)

Internet protocol version 6 (IPv6) is defined in RFC 2460. It does not represent a fundamental change in the architecture of IP-based networks from the architecture of IPv4. The retention of the same basic architecture and functioning of the protocol was a conscious decision of the IETF (Internet Engineering Task Force) in order that the migration of router networks from IPv4 to IPv6 could be more gradual, and in order that the two different IP protocol versions could co-exist. One of the main reasons for developing IPv6 was simply the need to extend the available addressing range.

The deployment of IPv6 routers has already commenced. As expected, IPv6 is being used in the *core routers* of high performance IP networks, while IPv4 continues to be used at the periphery of the network — providing access to the established large base of IPv4 hosts. Between the IPv4 and IPv6 parts of the network IPv6 routers act as IPv4-to-IPv6 gateways.

The new capabilities of IPv6 (over and above IPv4) are:

- an expanded address-range and address-length (128-bit address instead of 32-bit address);

- a simplified IP packet header — one in which the largely unused 'options' of IPv4 have been removed;

- improved extensions and options;

- *flow labelling* capability — specifically designed to cater for the growing need for carriage of *real-time* communications streams and flows, and for *differential services*; and

- improved *authentication* and privacy.

IPv6 packet format

The IPv6 packet format is illustrated in Figure 5.14. The basic structure of the packet is very similar to that of IPv4 (Figure 5.6), the most notable differences being that:

- the header has a standard length of 40 octets (and therefore no longer needs a header length indicator field);

- the *traffic class* field has replaced the IPv4 type-of-service (TOS) field;

- a *flow label* has been added specifically for the purpose of carrying communications *streams* and *flows*;

- the *payload length* field has replaced the *total length* field of IPv4;

- the *next header* field has replaced the *protocol* field of IPv4;

- the *hop limit* field has replaced the *time-to-live (TTL)* field of IPv4;

- the *fragmentation* fields have become an optional header extension field (fragmentation is generally avoided nowadays by using path MTU discovery, as we discussed earlier);

- the *header checksum* has been removed (the 'overhead' of the checksum is not considered worthwhile, given the high quality of modern high-speed fibre and digital networks);

- the *address fields* take up much more space — thereby considerably extending the header length.

Another important change is that the <u>minimum</u> *message transmission unit (MTU)* length to be supported by IPv6 routers is 1280 octets (the recommended value is 1500 octets. Compare this with the recommended MTU of 576 octets for IPv4 routers). With an MTU of 1500 octets even complete ethernet LLC frames can be carried without requiring fragmentation.

The following sections explain the coding of the individual fields in detail.

IP Version field

The *version field* of the IPv6-header indicates the format of the Internet header. The value is set to binary '0110' (decimal value 6) for version 6. The only other allowed values are 4 (IP version 4) and 5 (IP version 5 was an experimental *stream protocol* also called *ST2* and *ST2+* which was not widely deployed).

Traffic class field

The *traffic class* field of IPv6 is an 8-bit field conceived for uses similar to those of the *type-of-service (TOS)* field of IPv4. The most likely uses of this field are for differentiated services (DiffServ — Figure 5.13a) *or* IP precedence, as we discussed earlier.

Flow label field

The *flow label field* is a 20-bit field used expressly for the purpose of allowing *fast-forwarding* of packets making up a given traffic *flow* or communication *stream*. Faster forwarding is made possible by *label switching*.[16]

During the set-up of a *connection* between the two network end-points which wish to communicate *real-time* information, bandwidth along the entire path is reserved (typically by network management action or on-demand using a protocol such as *RSVP, Resource ReSerVation Protocol*). Once the bandwidth is reserved, a *flow label* is allocated to identify the flow at each of the interfaces (i.e., router to router hops) along the path. Fast forwarding of packets making up the flow can then be undertaken simply by inspecting the (relatively short) flow label. This can be undertaken by a specialised label switching hardware within the router. The IP address routing table look-ups, and the associated delays which 'normal' IP packets have to endure, can be avoided.

The forwarding function inspects the flow label, changes the label to the relevant value for the next interface along the path and forwards the packet (Figure 5.15). The 20-bit field allows

---

[16] See Chapters 3 and 7.

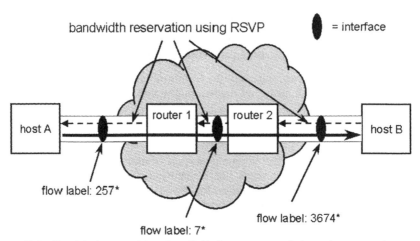

Note: *flow labels on each interface (link) do not necessarily have the same value.

Figure 5.15   Flow labelling.

up to 1 million flows to be simultaneously established across any given network interface (i.e., router-to-router or router-to-host link).

Payload length

The *payload length* field replaces the *total length* field of IPv4. The payload length field of IPv6 records the length (in octets) of the packet content <u>following</u> the header in octets. Since the field is a 16-bit field, we can conclude that the maximum payload length (of the IP data field of Figure 5.14) is 65 535 octets. This corresponds with the maximum transmission unit (MTU) of IPv4. With the IPv6 header, the total length is 65575 octets. In contrast the total length field of the IPv4 header includes the header. The change from total length to payload length was necessitated by the increased length of the IPv6 header: otherwise the maximum payload length would have had to be reduced.

Any extension headers (for fragmentation or IPv6 options — we shall come to these shortly) are considered to be part of the *payload* (i.e., the IP data in Figure 5.14).

Next header field and extension headers

The *next header* field is equivalent to the *protocol* field of IPv4. Where the payload of the packet (i.e., the IP data field of Figure 5.14) contains only protocol information of a higher layer protocol, the next header field is coded using the same values as used in IPv4 (these are listed in Table 5.6). But since some of the optional functions of the IPv6 header have also been made part of the payload, these also may be indicated in the next header field. The additional next header values made necessary by IPv6 are listed in Table 5.12.

The optional functions of IPv6 are appended to the header by means of *extension headers* which are included in the first part of the payload (the IP data field of Figure 5.14). An IPv6 packet may carry none, one or multiple extension headers. The format of the extension headers takes one of the forms shown in Figures 5.16 (*IPv6 options header* and *general routing header*), 5.17 (*routing type 0 header*) and 5.18 (*fragmentation header*).

Table 5.12 Next header field values in ipv6 (all the protocol values of table 5.6 may also be used)

IPv6 next header field value (possible range 0–255) [IPv4 protocol values for upper layer protocols may also be used: see Table 5.6]	Protocol used at next higher layer (i.e., content type of IP-data field)
0	HOPOPT — IPv6 Hop-by-Hop Option
43	IPv6-Route — Routing Header for IPv6
44	Ipv6-Frag - IPv6 Fragment header follows
50	ESP — Encapsulation Security Payload for IPv6
51	AH — Authentication Header for IPv6
58	IPv6-ICMP — ICMP for IPv6
59	IPv6-NoNxt — No next header — payload should be ignored
60	Ipv6-Opts — Destination Options for IPv6
98	ENCAP — Encapsulation header / private encryption

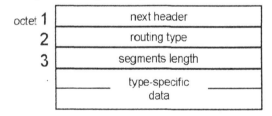

Figure 5.16 IPv6 general options and general routing header formats (optional extension headers).

When more than one extension header is used, the headers should be 'cascaded' in the following order:

- *hop-by-hop options* header [first];

- *destination options* (placed in this position when the options are required to be processed along the route);

- *routing options* (Type 0);

- *fragmentation header*;

- *authentication header (AH)*;

- *encapsulating security payload* (ESP);

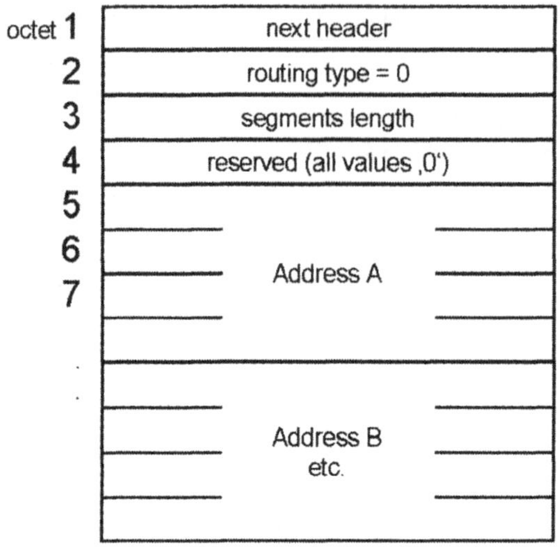

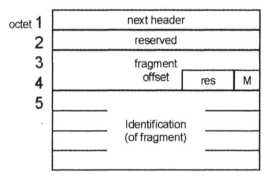

Figure 5.17   IPv6 routing type 0 header (optional extension header).

Figure 5.18   IPv6 fragmentation header (optional extension header).

- *destination options* (should be placed here when they are to be undertaken at final destination);

- upper-layer *protocol* header (last).

In other words, if all the headers are in use, the next header field in the IPv6 header bears the value 0 (see Table 5.12) for *hop-by-hop options*. The next header field of the hop-by-hop options header will bear the value 60 for destination options, and so on, until the last extension header, when the next header should be set to the protocol value of the upper layer protocol according to Table 5.6.

IPv6 optional extension header fields

The *hop-by-hop options (HOPOPT)* and *destination options* fields of IPv6 follow the simple *type-length-value (TLV)* extension header format of Figure 5.16a. The options indicate to intermediate routers how the traffic packet should be processed. Options are merely listed in the options field (not many are yet defined).

The optional *routing header* allows different variants (Figures 5.16b and 5.17). The *type 0 routing header* (Figure 5.17) is a version of *source routing*. As in the source routing options of IPv4, the type-specific data field of a type 0 routing header is merely a list of IP addresses, which identify the path to be taken by the packet as it traverses the IP network. As each intermediate destination is reached in turn, the destination header in the main IPv6 packet header is adjusted to the next IP in the source route listing of the type 0 routing header data field.

Correspondingly, this address is deleted from the source route listing in the routing header field, so that the field is reduced progressively — one address at a time until the routing header can finally be removed.

IPv6 fragmentation

The optional IPv6 header extension for fragmentation (Figure 5.18) contains similar fields to the fragmentation fields of the IPv4 header (Figure 5.6). The method of fragmentation and reassembly is also largely the same as for IPv4, except for the adjustments necessary in the IPv6 next header and payload length fields.

The *M-bit* (last bit of the fourth octet) of the IP fragmentation header is the *more fragments bit*. It is equivalent to bit 2 of the IPv4 fragment *flags* field. When M = 1, *more fragments* will follow, otherwise when M = 0 the current packet is the last fragment.

Packets which have to be fragmented are broken up into their *unfragmentable* and *fragmentable* parts. The unfragmentable part is the IPv6 header and any extension headers. These unfragmentable parts are copied into the header and first part of the payload (if necessary) of each of the fragment packets. During fragmentation, the *payload length* value in the original IPv6 packet is changed to reflect the length of the individual fragment. Otherwise all other header options fields remain the same as in the original packet, apart from the next header field of the very last header of the unfragmentable part, which is changed to value 44 (to mean 'fragment header follows' — see Table 5.10).

On reassembly all headers up to but not including the fragment header are re-instated. The next header is reset to that in the fragment header *next header* field and the payload length is corrected appropriately.

Hop limit field of IPv6

The IPv6 *hop limit* field is equivalent to the time-to-live (TTL) field of IPv4. It was renamed because the TTL field is reduced by each router (i.e., after each hop) by the minimum decrement of 1 second, even though the time elapsed during the hop is in practice much less than 1 second. When the hop limit is reduced to 0, an IPv6 packet <u>may</u> but <u>need not</u> be destroyed. This is a change from IPv4, in which routers are <u>expected</u> to destroy packets with a TTL value which has been reduced to 0.

IPv6-to-IPv4 compatibility issues

An IPv6 packet of 1280 octets length (the minimum recommended to be supported by IPv6 routers) sent to an IPv4 destination is likely to receive an ICMP (Internet control message protocol) message 'packet too big'. This is because while IPv6 routers are required to support packet lengths of at least 1280 octets, IPv4 routers need only support 576 octet packet-lengths as a minimum. In this case, the IPv6 router need not immediately fragment the packet, but must include a fragment header, in order that IPv6-to-IPv4 translating router can use a suitable fragment *identification* value in IPv4 fragments.

Upper layer checksums of upper layer protocols may also require adjustment. Any transport layer protocol originally conceived for IPv4, that includes the addresses in its checksum computation (e.g., TCP or UDP) must be modified for use over IPv6. The checksum must include the full 128-bit IPv6 address instead of the 32-bit IPv4 address.[17]

5.12   ICMP for IPv6

The *Internet control message protocol for IP version 6 (ICMPv6)* is an integral part of the IPv6 protocol, in the same way that ICMPv4 is an integral part of IPv4. ICMPv6 is specified in RFC 2463 and has a similar role to that of ICMPv4 in reporting network errors and delivery problems. Since the message format (Figure 5.19) and functioning of ICMPv6 are largely the same as ICMPv4 and self-explanatory, we shall restrict our coverage of it here largely to a presentation of ICMP messages in tabular format (Table 5.13) and a few notes.

In the case of an ICMPv6 message *destination unreachable*, the first 4 octets of the ICMP message (Figure 5.19) carry the message type (value = 1) and code (one of values 0–4), as well as the ICMP checksum. The second 4 octets of the message are unused. Thereafter (in the remainder of the message), as much of the original message as will fit into the IPv6 maximum transmission unit (typically 1280 octets) is returned to the sender. The returned part of the message is intended to help the sender identify more easily which message was not delivered.

The *packet-too-big* and *time exceeded* messages have a similar format to that of the destination unreachable message (although the code and type values will be different). Again the second four octets of the ICMP message (Figure 5.19) are left blank, and as much of the original message as possible is appended thereafter.

In the *parameter problem* message, the second 4 octets of the ICMP message are used as a *pointer*. The pointer is a value which indicates the position number of the octet within the packet header where the parameter problem occurred. Again a copy of the original packet is returned for reference of the sender.

Echo request and *echo reply* messages have a more complicated format than other ICMP messages. The first four octets are again the ICMP type, code and checksum. The next two octets are an echo request identifier, and the next two are a sequence number. The remainder is data, which may include further requests relating to the IPv6 *PING (Packet INternet Groper)* service.

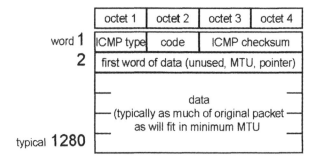

Figure 5.19   Internet Control Message Protocol version 6 (ICMPv6) format for IPv6 data field.

---

[17] See Chapter 7.

Table 5.13   Internet control message protocol version 6 (ICMPv6)

Message type	Message type value	Meaning	Message code value and meaning
Error	1	Destination Unreachable	0 — no route to destination
			1 — communication administratively prohibited
			2 — not assigned
			3 — address unreachable
			4 — port unreachable
	2	Packet too Big	Always set to 0
	3	Time Exceeded	0 — hop limit exceeded in transit
			1 — fragment reassembly time exceeded
	4	Parameter Problem	0 — erroneous header field encountered
			1 — unrecognised next header type
			2 — unrecognised IPv6 option encountered
Informational	128	Echo request (PING)	Always set to 0
	129	Echo reply (PING reply)	Always set to 0

5.13   IPv6 addressing

The IPv6 addressing scheme and the basic high level allocations are defined in RFC 2373. The scheme defines separate address ranges for unicast, multicast and anycast addresses (Table 5.14).

The main difference between IPv6 and IPv4 addresses is the length of the address. IPv6 addresses are 128 bits in length. IPv4 addresses are only 32 bits in length. The much longer format of IPv6 addresses gives more flexibility for re-introducing a structured format of numbering (somewhat like the original class-structure of addresses intended originally for IPv4 addresses). The length also demands a new shorthand method of writing IPv6 addresses. Otherwise the same principles of *variable length subnet masks (VLSMs) and classless inter-domain routing (CIDR)* are applied to IPv6 addresses in just the same way as they are to IPv4 addresses.

IPv6 address format and standard notation

The 128-bits of an IPv6 address of unicast (i.e., 'normal') IPv6 addresses are split up into six different fields, as shown in Figure 5.20a.

The *format prefix (FP)* is a three bit code. So far IPv6 number ranges have only been allocated corresponding to the three codes '000' (reserved for special uses),, 001 'for *unicast* addresses and, 111' for *multicast* addresses and *local-site* (i.e., private) IPv6 addresses (see Table 5.14).

The *TLA (top level aggregation identifier), NLA (next level aggregation identifier) and SLA (site level aggregation identifier)* of IPv6 *unicast-addresses* are equivalent to the 'area code' of a telephone number. The TLA is expected to identify the *Internet service provider (ISP)* who has assigned the number to a customer and the other two *aggregation* codes allow a further subdivision of the total number range into smaller number ranges (i.e., geographic zones or other groups of numbers).

Since it is rather cumbersome (and very prone to mistakes) to have to write out 128-bit address in the full binary form, it is usual instead to consider the number to comprise 16 octets

Table 5.14 IPv6 addresses: allocated ranges and special use addresses

Allocated usage	Prefix or address		Fraction of address space
	(Binary)	(Hexadecimal)	
Reserved (IANA/IETF)	0000 0000	00xx: ...	1 / 256
Unassigned	0000 0001	01xx: ...	1 / 256
Reserved for NSAP allocation	0000 001	02xx: ...	1 / 128
		03xx: ...	
Reserved for IPX allocation	0000 010	04xx: ...	1 / 128
		05xx: ...	
Aggregatable unicast addresses	001	2xxx: ...	1 / 8
		3xxx: ...	
Link-local unicast addresses	1111 1110 10	FE8x: ...	1 / 1024
		FE9x: ...	
		FEAx: ...	
		FEBx: ...	
Site-local unicast addresses	1111 1110 11	FECx: ...	1 / 1024
		FEDx: ...	
		FEEx: ...	
		FEFx: ...	
Multicast addresses	1111 1111	FFxx: ...	1 / 256
All nodes multicast		FF01::1	2 addresses
		FF02::1	
All routers multicast		FF01::2	3 addresses
		FF02::2	
		FF05::2	
Unspecified address (used by a source node enquiring about ist address)	0:0:0:0:0:0:0:0	0::0 or ::	1 address
Loopback address (returns a packet to ist source for test purposes)	0:0:0:0:0:0:0:01	::1	1 address
Embedded IPv4 address (IPv4 node supports IPv6)	0:0:0:0:0:0:d.d.d.d	::d.d.d.d	IP v4 address range
Embedded IPv4 address (IPv4 node does not support IPv6)	0::FFFF:d.d.d.d	::FFFF:d.d.d.d	IP v4 address range

and to represent each of the octets by two hexadecimal characters (as we first encountered in chapter 2). This reduces the address to 32 hexadecimal characters. But since even this format is still a cumbersome length, the number is conventionally formatted into eight blocks of four characters each, each block separated by a colon ':'. Thus an IPv6 address will normally have the format:

xxxx : xxxx : xxxx : xxxx : xxxx : xxxx : xxxx : xxxx

where each x is a hexadecimal character with one of the values: 0, 1, 2, 3, 4, 5, 6, 7, 8, 9, A, B, C, D, E or F.[18]

In the example of Figure 5.20, the IPv6 address with the binary value:

0001 0000 1000 0000 0000 0000 0000 0001 0000 0000 0000 0000 0000 0000 0000 0000

[18] IPv4 addresses, in contrast, are usually expressed as a series of four decimal number values (0–255) separated by dots thus: d.d.d.d.

a) 128-bit IPv6 address

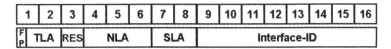

b) hexadecimal value of IPv6 address

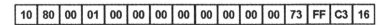

c) normal representation of IPv6 address

10 80: 00 01: 00 00: 00 00: 00 00: 00 00: 73 FF: C3 16

d) alternative shortened form of IPv6 address (strings of 0s suppressed)

1080 : 0001 : : 73FF : C316

Figure 5.20   Format and notation of IPv6 unicast addresses.

0000 0000 0000 0000 0000 0000 0000 0111 0011 1111 1111 1100 0011 0001 0110 0000

is represented in its hexadecimal format in Figure 5.21b and then in the standard IPv6 address notation in Figure 5.21c.[19]

Finally, Figure 5.20d illustrates an allowed *shortened form* of the IPv6 address, in which strings of 0s need not be written. Instead of writing . . .0000 : 0000 : 0000 : 0000 : 0000. . . it is

a) 128-bit IPv6 gateway address and subnet-mask

1080 : 0001 : : 73FF : C200 / 111

b) full text form of IPv6 address

1080 : 0001 : 0000 : 0000 : 0000 : 0000 : 73FF : C200 / 111

c) hexadecimal/octet form of IPv6 address

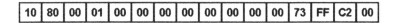

d) subnet mask

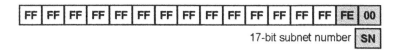

17-bit subnet number   SN

Figure 5.21   Variable length subnet mask (VLSM) and classless inter-domain routing (CIDR) with IPv6 addresses.

---

[19] The conversion between binary and hexadecimal is easy (take four binary bits at a time and convert to hexadecimal or vice versa.

acceptable to shorten the string to two immediately following colons thus::. In addition, *leading zeros* (i.e., left hand zeros) may be removed from each of the blocks of four hexadecimal characters. *Trailing zeros* (i.e., right-hand zeros), however, must never be removed. Thus the simplest version of the address used in the example of Figure 5.20 is:

$$1080 : 1 : : 73FF : C316$$

Understandably, various IPv6 address prefixes have been conceived to allow IPv4 addresses to be *embedded* into a standard IPv6 format. These are listed in Table 5.14. There is also a notational convention which combines the IPv6 address notation (eight hexadecimal character [x] blocks with colons) and the older IPv4 address notation (four decimal number values [d] with dots). The standard notation of embedded IPv4 addresses allows the last two four character hexadecimal blocks to be replaced with the IPv4 address notation, thus:

$$xxxx : xxxx : xxxx : xxxx : xxxx : xxxx : d . d . d . d$$

Allocated IPv6 address ranges and special-use addresses

Table 5.14 lists the prefix allocations of the IPv6 address ranges and also some of the special-use addresses (e.g., common *global multicast addresses, unspecified address* and *loop-back address*).

Variable length subnet masks (VLSM) and classless inter-domain routing (CIDR) in IPv6

Variable length subnet masks (VLSMs) and classless inter-domain routing (CIDR) are undertaken with IPv6 in the same manner as in IPv4, as we discussed earlier in the chapter — Figures 5.10 and 5.11. Figure 5.21 illustrates some example subnet masks in the IPv6 address format.

Allocation of IPv6 addresses to network interfaces

Any device or software application which requires to communicate across an IPv6 network must have its own IPv6 address — either a permanently allocated address or an address temporarily allocated (e.g., using a protocol like *DHCP — dynamic host configuration protocol*). Even the routers and other nodes making up an IPv6 network must have certain addresses (for routing protocol and network management purposes, among others). But the devices themselves do not have addresses, individual hardware interfaces or software ports have addresses.

As we discovered in the case of ethernet interface cards (Chapter 4), it is often helpful to assign unique *physical addresses* to hardware interfaces at their time of manufacture. IPv6 defines a specific format, as shown in Figure 5.22, as a means of incorporating IEEE unique identifier addresses. Specifically, IEEE.

EUI-64 (64-bit *extended unique identifiers*) can be incorporated into IPv6 addresses — thereby avoiding the need for a separate IPv6 address allocation. The IEEE EUI-64 (*extended unique identifier*) includes the same IEEE 48-bit MAC address we learned about in Chapter 4, and uses the same *organisational unique identifiers (OUI)*, including G/I and U/L bits. As is the case with 48-bit IEEE MAC addresses, the G-bit, when set to '1' represents a *group* rather than an *individual* address. But in IPv6, the value of the U/L bit is inverted. When set $U = 0$, the address is *universally* recognised (i.e., an IEEE RAC — registration authority committee — assigned

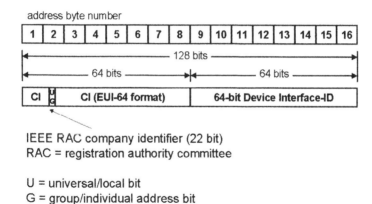

IEEE RAC company identifier (22 bit)
RAC = registration authority committee

U = universal/local bit
G = group/individual address bit

Figure 5.22   IPv6 address based upon the IEEE RAC EUI-64 address (extended unique identifier).

identifier). U = 1 is a *locally* defined number. You might be wondering why the U- and G-bits appear in the middle of the *company identifier (CI)* portion of the address. This is because these values are defined to be the least significant bits. In the *canonical transmission format* used by ethernet LANs (as we discussed in Chapter 4), these bits conveniently come at the front of the transmission, but here (Figure 5.22) where the CI field has been extended, they end up in the middle.

5.14   Multicasting

The forwarding of IP packets to *multicast address* (i.e., groups of destinations) is basically the same as forwarding of packets by routers to a *unicast address* destination... the router looks-up the destination address in its routing table and selects the *next hop* to which the packet will be forwarded. The difference with the handling of a *multicast address* is that the routing table of a multicast router contains not a single *next hop* entry (as in the case of a unicast address) but instead a whole set of different distribution paths (if you like 'parallel next hops') according to a *multicast distribution tree*.

   To be able to take part in a multicast, a router must be a *multicast router*. There are at least two phases in the delivery process of a multicast message. First, the message (i.e., packet to be multicast) must be multicast by the Internet protocol (IP) so that it reaches all the routers of destination local networks (i.e., routers in LANs to which one or more of the destination hosts is attached). Thereafter, each of these routers must broadcast or multicast the message within the local network using normal LAN layer 2 protocols (e.g. ethernet).

   The scope of a multicast is usually limited. It is rare to multicast messages to destinations spanning the entire Internet. Instead, messages are typically limited to the network of a single network administration (called an *AS — administrative* system or autonomous *system*). Alternatively, multicast message may be limited to a maximum hop count (this is done by setting the IPv4 time-to-live (TTL) field or IPv6 hop limit field), thereby limiting the 'range' of the message.

   There are two basic types of multicasting:

- *sparse mode (SM)* multicasting, and;
- *dense mode (DM)* multicasting.

In both forms of multicasting, the multicast packet can be thought of as being distributed by the multicast *source* throughout the network and transferred to all relevant destination networks.

The actual paths used for the distribution have a tree-like form, continually branching into multiple paths — so that multiple copies of the original packet have to be created. This is part of the forwarding process of a multicast router. Apart from copying packets to multiple branches, the process of multicast forwarding is otherwise much the same as unicast forwarding — and carried out by the Internet protocol (IP).

The *multicast distribution tree* is a 'map' of the routes which the multicast packet will follow on the way to its multiple destinations (Figure 5.23). In order to create the, map', the end points of the branches (i.e., the destinations making up the multicast *group*) need to be known and the optimum branching points (where packets are to be copied by *multicast routers*) need to be calculated. These functions are undertaken automatically by multicast routers, using the *Internet group management protocol (IGMP)*, a *multicast routing protocol* and a *multicast routing algorithm*.

IGMP (Internet group management protocol) is used between multicast routers and from multicast hosts to multicast routers to identify the end destinations associated with a particular multicast address. In essence, IGMP is used to identify which destination routers wish to receive particular multicast messages and where exactly these routers are located. A multicast routing protocol is then used between the multicast routers to share network topology information with the other multicast routers in the network. Together with the use of a multicast routing algorithm, one or more of the multicast routers can then calculate the entire multicast distribution tree for reaching all required destination routers. When only one router undertakes the calculation, this single centralised router propagates the necessary information for creating the routing table entries to all other multicast routers. Once the routing tables are complete, so is the distribution tree.

In *dense mode (DM)* multicasting, the assumption is that the group of individual destinations which together share the multicast address are all densely situated in the network. In this case, the multicast packet is literally *flooded* through the network. Multicast routing protocols and algorithms designed for *dense mode* multicasting concentrate on ensuring that all routers within the network are reached, but that there are no *reverse broadcasts* (Figure 5.23a). An example

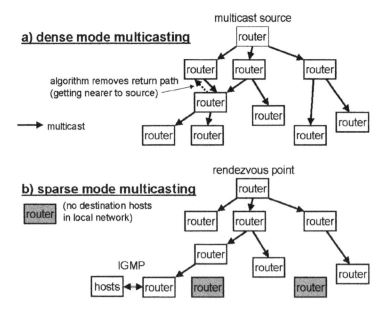

Figure 5.23   Dense (DM) and sparse modes (SM) of multicasting.

of such a multicasting routing algorithm is the *truncated reverse path broadcasting* algorithm (*TRPB*). The best-known dense-mode multicast routing protocols are:

- host extensions for multicasting (RFC 988 and RFC 1054);

- *DVMRP (distance vector multicast routing protocol*—RFC 1075);

- *MOSPF (multicast extensions for OSPF*—RFC 1584); and

- *PIM—DM (protocol independent multicast—dense mode)*.

In *sparse mode (SM)* multicasting, the assumption is that the individual destinations making up the multicast address group are spread thinly (i.e., sparsely) through the network, so that a widespread broadcast would be an inefficient use of network resources. In this case, multicast packets are only forwarded by routers to neighbouring routers which have previously expressed an interest in receiving messages to the specified multicast address. This is called *reverse path forwarding*. In other words, 'if you don't advise me (in the reverse path) that you wish to receive a given multicast address message, I won't forward it to you'). IGMP (Internet group message protocol) is used to *report* the desire of a host to receive given multicast group messages. IGMP is also used when given hosts wish to *leave* a group.

In *sparse mode (SM) multicasting* (Figure 5.23b) everything is coordinated by a centralised multicast router called the *rendezvous point (RP)*: there must be at least one RP in each *autonomous system (AS)* network making up the multicast domain. The *rendezvous point* develops the central knowledge about the valid multicast addresses and the groups of individual destinations which wish to receive the packets sent to these addresses. An individual destination which wishes to start receiving messages to a given multicast address must register its interest with the rendezvous point (RP) router by means of *IGMP (Internet group management protocol*—Figure 5.24). The rendezvous point develops a multicast distribution tree based upon the number and location of the individual destinations and the best calculated route to each destination.

The best-known routing protocols for sparse mode multicasting are:

- *core-based trees (CBT*—RFC 2201); and

- *PIM—SM (protocol independent multicast—sparse mode*—RFC 2362).

Older multicast routing protocols (e.g., DVMRP) are carried within IGMP messages, so that IGMP messages are used between multicast routers as well as between the end hosts and the routers. However, more modern multicast schemes (e.g., PIM-SM) use IGMP only for host-to-router reporting and querying. Inter-router communication for establishment of the multicast broadcast tree is entirely the domain of the multicast routing protocol.

It is important to note that multicast addresses may not be used as source addresses. In addition, no ICMP (Internet control message protocol) messages are returned for multicast packets (e.g., to report the unreachability of one of the individual destinations).

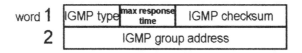

Figure 5.24   IGMP (Internet group management protocol) message format.

Table 5.15 IGMP (Internet group management protocol)

Type code	Type	IP-Destination address	IGMP Group address
0x11	General Membership Query	All Systems Multicast 224.0.0.1	0.0.0.0
	Group-Specific Query	Relevant Group Address	Relevant Group Address
0x12	V1 Membership Report	Relevant Group Address	Relevant Group Address
0x16	V2 Membership Report	Relevant Group Address	Relevant Group Address
0x17	Leave Group Message	All Routers Multicast 224.0.0.2	Relevant Group Address

IGMP (Internet group management protocol)

The *Internet group management protocol (IGMP)* is used to support the multicast function of IP (Internet protocol)-based router networks. It is specified in RFC 1112 (version 1) and RFC 2236 (version 2). IGMP, like ICMP is an integral part of the Internet protocol (IP), but need only be implemented if the router is to be a *multicast router*.

Multicast routers use IGMP to learn which groups have members on each of their physically attached networks. They maintain a list and a timer for each membership.

The format of IGMP messages is shown in Figure 5.24 and the coding of the messages is listed in Table 5.15. This field is inserted into the IP data field of the IP packet (Figure 5.6).

A *general membership query* is *multicast* to all hosts to find out who's out there. It is the only type of IGMP message for which the *maximum response time* field has any meaning. In response to the general query, hosts or routers which desire to receive specific multicast address messages must return a *V2 membership report*, listing the relevant multicast address(es). In order that not all the response reports are sent at once, each host responds after a random amount of time (but limited by the maximum response time).

When a host no longer wishes to receive multicast messages for a given address, it issues a *leave group message*. Following a leave message, the multicast router may issue a *group-specific enquiry* to the affected multicast group to check whether any members remain in the local branch network, or whether the branch may be *pruned* (i.e. removed).

Well-known multicast addresses

Table 5.16 lists some of the best-known IP multicast addresses. Note that *static local scope* multicast addresses may not be routed or forwarded outside the routing domain in which they originate. These are typically used for *interior gateway protocol (IGP)* routing protocol updates, as we shall encounter in Chapter 6.

Source-specific multicasting (SSM) and IGMPv3

Before we leave the subject of multicasting, we should also mention that a new form, *source-specific multicasting (SSM)* is under development by IETF. As the name suggests, when using *PIM-SSM (protocol independent multicast — source-specific multicasting)*, the delivery points of the source-specific multicast message depends not only on the multicast *group address* but also upon the source of the message. The combination of the multicast *group address* and the *IP source address* defines a *multicast channel*. Thus even though the destination group addresses used to direct messages originating from two different sources simultaneously may be the same, the *multicast channel* will not be, so that the points of actual delivery of the two different messages will also differ.

Table 5.16   Well-known multicast and other special addresses

	IP Address	Multicast group
Static Global Scope	224.0.1.22	Service location protocol (SLP) general multicast (RFC 2165)
	224.0.1.35	SLP directory agent discovery (RFC 2165)
	224.0.1.24	Microsoft WINS server autodiscovery
	224.0.1.39	Cisco PIM rendezvous point (RP) announce
	224.0.1.40	Cisco PIM rendezvous point (RP) discovery
	224.0.1.41	H.323 (ITU-T rec.) gatckeepei discovery
	224.0.1.75	All SIP (session initiation protocol)
	224.2.127.254	SAP (session announcement protocol) announcements
Static Local Scope	224.0.0.1	All systems on this subnet
	224.0.0.2	All routers on this subnet
	224.0.0.4	DVMRP routers (distance vector multicast routing protocol)
	224.0.0.5	OSPF all routers (open shortest path first)
	224.0.0.6	OSPF all designated routers
	224.0.0.9	RIP2 routers (routing information protocol)
	224.0.0.10	IGRP (interior gateway routing protocol)
	224.0.0.13	All PIM routers (protocol independent multicast protocol)
Special addresses	255.255.255.255	Limited broadcast
	192.168.10.x	NET.ALL1 Network broadcast address
	127.x.x.x	Loopback addresses

The advantage of *source-specific multicasting (SSM)* is the greater security afforded to multicast services (it is not so easy to pretend to be the source of the message). In addition, it allows for the re-use of addresses. The use of a given destination group address does not have to be coordinated between all possible multicast message sources.

IPv4 addresses in the range 232.0.0.0–232.255.255.255 (232/8) are allocated as source-specific multicast (SSM) addresses. A similar IPv6-address range is also planned.

As in the earlier example of *PIM-SM (protocol independent multicast—sparse mode)*, which, incidentally, is an example of *ASM—any source multicasting*, destinations may *subscribe* to or *unsubscribe* from individual multicast group address messages. This is done in the case of SSM using IGMPv3.

6

Routing Tables and Protocols

We have seen how routers, armed with the Internet protocol (IP), are able to provide for end-to-end packet communication and internetworking of complex data networks. We have also learned of the critical role in the IP forwarding process played by the routing table held within the router. Forwarding of IP packets (or datagrams) takes place by 'looking up' the IP-destination address in the routing table to determine the next hop on which the packet should be sent. But up until now, we have not discussed how the routing table is created in the first place, and how it is kept up-to-date. This is the realm of routing protocols, metrics, routing algorithms, the calculation of routing distance (or cost) and routing policies. These topics are the focus of this chapter. We shall discuss all the common routing protocols and go on to set out in detail the function and use of the most popular ones: RIP (routing information protocol), OSPF (open shortest path first) and BGP4 (border gateway protocol 4). We also discuss the related topics of address resolution, IP-address assignment and IP-parameter configuration of hosts, since these are also an important means by which routers collecting the information necessary to complete the routing table.

6.1 Routing tables: static and dynamic routing — a recap

Modern data communications take place by *packet switching*. *Packets* of data (up to 65 535 bytes long[1]) are prepared for transmission across a data network by the addition of a *packet header* (the IP packet header). The packet header contains important information which is required to control the network nodes as they *forward* the packet on a *hop-by-hop* basis to its destination. Each node in the internetwork (called a *gateway* or *router*) acts like a postal sorting office. It looks at the destination address in the packet header and, from this and its *routing table*, determines the best outgoing route (the *next hop*) towards the destination and forwards the packet accordingly.

The routing table contains the information for *routing* packets to their *next hop*. In its most basic form, a routing table comprises a list of all possible destinations (identified by their IP addresses) in one column and the next hop for each destination in a second column (Figure 6.1). In addition, the routing table usually includes a *default route* entry corresponding to: 'send all packets with addresses unknown to me to another router which is more likely to be familiar with this destination'.

---

[1] The maximum IP packet size is set by the packet length field (16 bits) of the IPv4 header. The minimum packet size, meanwhile, is typically set by the minimum frame size of the Ethernet SNAP-frame format (see chapter 4) — 28 bytes (20 byte IP header plus 8 byte LLC/SNAP header.

Data Networks, IP and the Internet: Protocols, Design and Operation   Martin P. Clark
© 2003 John Wiley & Sons, Ltd   ISBN: 0-470-84856-1

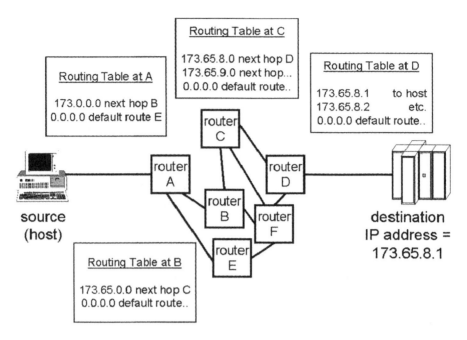

Figure 6.1 Next hop determination by routers using their routing table.

The creation and maintenance of the routing table are critical to the operation of a router. The creation and updating of the routing table may be undertaken manually (by human network administrators — in which case the router is said to perform *static routing*), but this is impractical in most networks. Alternatively, routing table creation and maintenance may be undertaken automatically by the routers (in which case the routers perform *dynamic routing*). Most routers even allow for a mix of the two methods — allowing some routing table entries to be maintained statically (e.g. the *default route* and local route *preferences*) while leaving the rest to be dynamically maintained. In this chapter we are concerned nearly exclusively with dynamic routing.

6.2 Choosing the best route by comparing the routing distance or *cost* of the alternatives

In order to perform dynamic updating of its routing table, a router has to gather *routing information* about the ever-changing topology of the network. It then calculates the shortest routing *distance* (or route *cost*) to each *reachable* destination. *Routing information* is exchanged between the routers in a network by means of a *routing protocol*.

Which information is used for the routing table calculation, how it is 'weighted', and the calculation method used are determined by a set of rules (laid out in the *routing policy* and/or routing protocol). Which policies and protocols are to be used are decided by the human network administrator during configuration of each router.

The policy might include particular preferences when choosing between alternative routes to a given destination (e.g. 'use the cheapest route rather than the one with the least hops', 'use the route with the most bandwidth', 'ignore routing information received from party X').

The determination of the best route to a particular destination is normally based upon the shortest routing *distance* (or path *cost*) from the router in question. First, the individual link *distances* (*link costs*) for all possible paths to the destination are added together. These are the alternative *path distances* (or *path costs*). The path with the lowest overall path distance (or path cost) is chosen to be the *shortest path*. The first hop of this path is inserted into the router's routing table as the *next hop* for purpose of forwarding packets.

Distances and *costs* are usually integer values. Paths with the lowest distance or cost are most preferable. The distance or cost associated with each link in the network (i.e. between each pair of routers) must be either assigned manually or calculated automatically by the router. When assigned manually by the human router during the configuration of the router, the value is known as the *administrative distance* or *administrative cost*. When calculated automatically by the router, the distance or cost value can be changed over time to reflect the link bit rate, current transmission delay, link loading, link reliability or some other *metric*. Typical *default costs* are either the value '1' for each link or a value equal to 1000 million divided by the link bit rate. A standard link cost of '1' will result in selection of the path with the least number of hops. A link cost inversely proportional to the link bit rate will instead tend to favour higher bit rate paths over lower bit rate paths.

Each router will either be configured to 'know', or will calculate, the distances or costs of the links to which it is directly connected. Thus taken together, the routers have enough information to calculate the shortest paths to all destinations. But before any single router can determine its routing table, the routers must 'pool' their information — sharing what they know with all the other routers in order that they can all create a 'map' of the network.

The communication of routing information from one router to another is the task of the *routing protocol*. It allows information about link distances and the network topology to be collated and used to create a routing table. Figure 6.2 illustrates a network 'map' in which the entire routing information has been collated. Table 6.1 shows the associated shortest path calculations and next hop determination as they will appear in the routing table.

You may be surprised to discover in the example of Figure 6.2 that not all the links have the same distance associated with them. Some of the links shown also have a different distance associated with them, dependent upon the direction in which they are traversed (e.g- routes A-B; B-C and D-F). Such differences may also occur in practice (dependent upon the manner in which the link distance or cost is calculated). But you may be most surprised to discover which routes have been selected as having the shortest distance. The chosen routes for C-to-B and C-to-A transit the routers D and F rather than taking what might seem to be the more

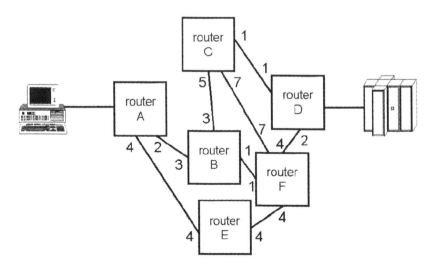

Figure 6.2 Example of network topology and link *routing distances* (integer values).

Table 6.1   Example routing table calculated from distances and topology of figure 6.2

Source router	Destination router	Shortest distance	(Best path)	Next hop
A	B	2	A–B	B
A	C	5	A–B–C	B
A	D	6	A–B–C–D	B
A	E	4	A–E	E
A	F	3	A–B–F	B
B	A	3	B–A	A
B	C	3	B–C	C
B	D	4	B–C–D	C
B	E	5	B–F–E	F
B	F	1	B–F	F
C	A	7	C–D–F–B–A	D
C	B	4	C–D–F–B	D
Etc.	Etc.	Etc.	Etc.	Etc.

obvious direct route C-to-B!! You may think this is a dumb decision! But consider a road analogy, in which the route C-D-F-B is a motorway on which you can travel at high speed while the 'direct road' C-B is a winding back lane with hold-ups. Which route would you rather take then?

6.3   Storage, updating and recalculation of the routing table and routing database

In order that the routing table of a router remains current, it must be continuously re-calculated, using up-to-date routing information. This is an onerous task, requiring continuous *routing protocol* exchange with other routers in the network and repeated re-calculation of routes. Because of the heavy demands upon router processing and storage power for routing table maintenance, it is usual for routers to be sub-divided functionally into two separate parts: a *forwarding engine* and a *routing engine* — with dedicated hardware, processors and software for each (Figure 6.3).

The forwarding engine is designed to be optimised for the processing and forwarding of IP (Internet protocol) packets in the manner we discussed in Chapter 5. At all points in time (after its initial calculation), the 'current version' of the routing table is stored in the forwarding engine and is in permanent use. The routing engine, meanwhile, is fully occupied with preparing an updated version of the routing table — as soon as it becomes necessary.

The routing engine stores the network topology database (or its otherwise-named equivalent). This is a collection of information about the locations of destinations, the available links and the routing distances in the network. The database may take a form similar to Table 6.1 or may include additional fields, but at all times will keep an 'image' of the network topology data on which the current routing table is based. The routing table itself is usually an extract of this database. (The columns of most importance for the forwarding engine's routing table are those which are not shaded grey in Table 6.1. These columns link the destination with the appropriate *next hop*.)

The routing engine keeps track of network topology changes by means of *routing information updates* received from other routers in the network by means of a routing protocol. The routing protocol is used as the communication directly between routing engines. In effect, it provides for an internal 'network control channel'.

The routing protocol conveys the routing updates from one router's routing engine to another by copying either the entirety, or extracts, of the network topology database. Thus

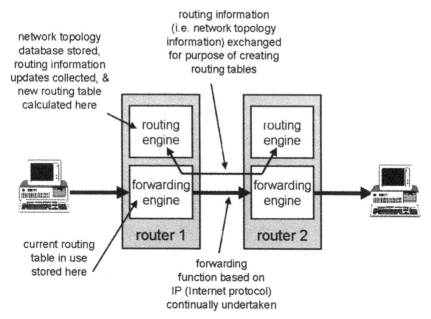

network topology database stored, routing information updates collected, & new routing table calculated here

routing information (i.e. network topology information) exchanged for purpose of creating routing tables

routing engine

routing engine

forwarding engine

forwarding engine

router 1

router 2

current routing table in use stored here

forwarding function based on IP (Internet protocol) continually undertaken

Note: The communication 'channel' for the routing protocol is actually an IP 'connection' between the routers passing via the normal forwarding engine part of the router (Figure 6.3). You may like to compare it with ITU-T's c-plane and m-plane communications model—Figure 3.31.

Figure 6.3   Typical router architecture — subdivided into forwarding and routing engines.

the routing engine of router A in Figure 6.2 would *advertise* the first five rows of Table 6.1 (these are the rows relevant to router A). The other routers would meanwhile independently advertise the other rows. This enables each routing engine to maintain a complete and up-to-date topology database (by listening to all relevant *advertisements*). As each update is received, the router compares the new information with its existing database to determine whether any topology changes have taken place. If so, a new routing table will be calculated, loaded into the forwarding engine part of the router and *activated*.

By storing a complete image of the data used for the previous routing table calculation, the re-calculation process can be simplified, and so speeded up. Furthermore, by storing all the previously known data, the routing protocol usually needs only to send information about real 'changes'. This reduces the workload of the routing protocol and offloads traffic from the IP forwarding network (with obvious cost and performance benefits).

There have been a number of different routing protocols developed over the years, and each has its advantages and disadvantages. Each protocol is optimised for the particular mathematical method by which the shortest distance routes will be calculated. The format of the protocols, as we shall see later in the chapter, usually directly reflect the data records which are held in the related network topology or routing database.

6.4   The accuracy and stability of routing tables

Router networks work most efficiently when their routing tables are both accurate and stable. Inaccurate routing tables lead to *misrouting* of packets — delivering them to the wrong address, losing them or sending them in endless circular routes around the network.

Ageing routing table entries to ensure they are still current

One of the easiest ways in which routing tables can become inaccurate is as a result of the withdrawal or failure of a previously operational route. Imagine driving along your favourite country road to a particular destination. You have always driven that way. But today it's different — the road has been closed and barred, because it is to be dissected by a brand new motorway and the roadworks have already started. How do you get through? You back up to the previous junction and check the roadsigns. But these have not been updated, and there are no 'diversion' signs — so you are lost, and you watch helplessly as other drivers follow the signs and head down the old road to the same fate as you!

In router networks, the 'severed country road problem' (our analogy above) regularly occurs — either as devices are taken out of service, moved from one location to another or because they simply 'fail' as a result of an equipment fault. We cannot rely on being informed about such link severences because typically the device which would have informed us about the change in link status can no longer 'talk' to us (the communications path has been severed!).

In the example of Figure 6.4, router D has failed because of an equipment fault — and naturally router D is no longer able to inform the other routers of its fate! So how do the other routers learn about the 'change in topology' of the network? The answer is they don't — directly. But if router D does not continue to *advertise* routing updates by means of the routing protocol, then eventually the other routers delete the routing table entries which have derived from router D's previous advertisements. This is called *ageing*. (We met *ageing* before — when talking about bridge *source address tables (SATs)* in Chapter 4.)

The *ageing* process demands that all routes are reconfirmed on a regular basis. Routes which are not reconfirmed as still existing are assumed to have fallen out of use, and are deleted from the routing table. Reconfirmation must occur at least once within the *ageing time* period (typically 180–1800 seconds). Thus at least once every 3–30 minutes all the routes must be advertised using routing protocols. Between these updates, the routing protocols send only more urgent 'change' updates regarding network topology alterations.

The process of maintaining routing tables is a very onerous one! And the extra traffic load on the network created by the routing protocols is considerable!

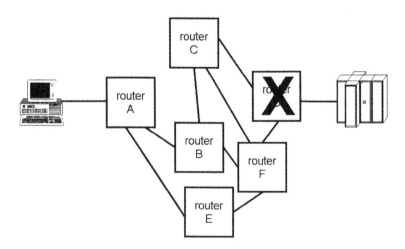

Figure 6.4   Reacting to unadvertised route withdrawals by means of routing table entry *ageing*.

Route flapping

Routers are repeatedly recalculating their routing tables and advertising the amendments they make to them. The amendments, in turn, lead to routing table recalculation in other routers and so to more advertisements of amendments and so to a new recalculation in the first router and so on.

It is conceivable that a situation could arise in which the route choice of a given router to a given destination oscillates between one path and another because of the repeated recalculation. This is an effect called *route flapping*. Route flapping can be very destructive — leading rapidly to poor network performance. Counter-measures to prevent route flapping are called *route damping*.

Figure 6.5 illustrates an imaginary network in which route flapping can occur for one of two different reasons. We are concerned with the choice of route from router A to router C.

At low link loadings, both links BC and DC will register a link distance of 1. So that both routes ABC and ADC have equal routing distances of 2. Let us imagine router A randomly chooses route ABC. Sometime later, router D will re-advertise its route to C as having a distance of '1'. Prompted by the re-advertisement, router A recalculates the route via D and discovers it has a total distance of 2 — i.e. equal to its currently established route via B. If router A were to respond by replacing the route ABC with the route ADC, route flapping would commence, because the next route advertisement by router B of its route to C would lead to reversion to the original route and so on. So the first lesson in route damping is that only routes with <u>lower</u> distances may replace existing routes.

Now let's consider what happens when the total traffic between A and C reaches 50% of the link loading. The initial route taken is ABC. The load on link BC exceeds 50% so router B advertises the route BC with a distance of 2 while router D is still advertising its route DC with a distance of value 1. Naturally A changes its preferred route to C to be the route ADC. Whereupon the load on link DC immediately exceeds 50% and the load on BC drops back below 50%. This, of course, leads to new route distance advertisements: BC distance = 1 and DC distance = 2. Router A dutifully recalculates the best route.. and reverts to the route ABC.. and so on ad infinitum. Another case of route flapping! And a much harder one to spot and to resolve!

A pragmatic approach to route damping is to restrict the frequency with which new routes to a given destination are accepted (this is called *holddown*). But in doing so, we must be careful to ensure that the network acts fast enough to real topology changes. When a link fails, there must be quick action to divert around the problem.

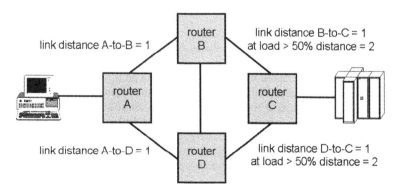

Figure 6.5   The problem of route flapping.

Routing table convergence

The *convergence time* is the time taken by a router to develop its routing table after first being introduced to the network or during its recovery after a network failure. The most efficient modern routing protocols achieve *fast convergence*. Fast convergence is beneficial not only when a router first comes into service, but also for re-routing (adapting routing tables) during a period of network failure. Some older routing protocols (e.g. *routing information protocol, RIP*) are *slow converging* because several cycles of advertising of routing information and re-calculation of routing tables are necessary before all the routes finally stabilise. Newer, *fast converging* routing protocols, meanwhile (e.g. *open shortest path first, OSPF*) allow the entire information necessary for the calculation of a new router's routing table to be requested immediately by means of a *hello procedure*.

6.5   Representation of destinations in a routing table

In the discussion of this chapter so far, we have illustrated how routing protocols are used to aid the process of routing table calculation — determining the best route from one router (the source router) to another (the destination router). Ensuring the most efficient routing of IP packets between routers is indeed the main purpose of a routing protocol, and hopefully the discussion has given you a good understanding of the basic methodology of routing table production. But what we have not yet discussed is how the destinations themselves are represented in the routing table. After all, the ultimate destination is not the destination router (as we implied in the example of Figure 6.2 and Table 6.1), but an IP address — typically one of a range of IP addresses within a given LAN connected to the destination router. The information held in Table 6.1 is actually insufficient to create any routing tables, because it does not relate the ranges of IP addresses which pertain to each of the destination routers!

Routing protocols differ in how they convey the IP address and other destination information. In the earliest routing protocols (based on *classful IP addressing* — see Chapter 5), the *class address range* was carried by the routing protocol to identify the range of addresses connected to the destination router. Thus for example 173.65.8.0 meant a *class C address-range*, equivalent to the addresses in the range 173.65.8.1 to 173.65.8.255. (There were also *class B* and *class A* address-ranges). With the advent of *classless interdomain routing (CIDR)* and *variable length subnet masks (VLSMs)*, it became necessary to adapt routing protocols to identify address ranges by carrying both the *gateway address* of the destination network and the associated *subnet mask*.[2]

The destination network *gateway address* is the IP address of the router port connecting to the destination LAN (e.g. 173.65.8.0). The *subnet mask* is conveyed simultaneously by the routing protocol in the IP-address-like-format (for a 2-bit subnet address the mask is 255.255.255.252). The gateway address 173.65.8.0 and the 30-bit subnet mask 255.255.255.252 thus represent a *reachable* address range at the destination LAN and router of: 173.65.8.0–173.65.8.3.

Each IP address range is treated by the routing protocol as if it were a different destination, even if multiple different address-ranges are all used in the same destination LAN and all are connected to the same destination router. In other words, there may be multiple routing table entries for the different ranges of reachable addresses attached to the same destination router.

Some routing protocols (and their associated routing table calculation methods) concern themselves only with calculating the 'shortest' route which will reach a given address range, and repeat the same calculation for all the address ranges on the same destination router. Other

---

[2] See Chapter 5.

protocols, meanwhile, associate all the destination IP address ranges with a given *router-ID (router identification)* and perform the calculation only once — how best to <u>reach the router</u>.

6.6 Routing protocols and their associated algorithms and metrics

Routing protocols, as we have seen, provide a basis for routers to share *routing information* and to calculate routing tables. Each routing protocol defines the data elements of a *network topology database* (routing database, or equivalent otherwise-named database) which can be communicated by means of the routing protocol and will provide the basis of the *routing table* calculation. The routing protocol specification also defines the *routing algorithm* (i.e. calculation methodology) which is to be used to calculate the 'shortest' route to any given destination.

The earliest routing protocols (e.g. *routing information protocol, RIP*) calculated the 'shortest' route as that requiring to traverse the least number of IP (Internet protocol) forwarding *hops*. In the case of the network shown in Figure 6.2, for example, the route with the smallest number of hops from router C to a destination connected to router B has two hops[3] (the first is the router-to-router link CB; the second is the hop from router B to the destination in the locally-connected LAN).

As time has progressed, routing protocols have become more sophisticated. Network administrators have demanded that dynamic routing schemes adapt not just in the case of routing topology changes (i.e. routing according to hop count) but also according to prevailing network operational conditions. Some modern routing protocols and modern routers thus allow the distance values associated with outgoing links (sometimes also called the link *cost*) to be made dependent on one or more of a number of other *metrics*:

- bandwidth (i.e. bit rate);
- delay;
- least cost;
- load;
- path MTU;
- reliability.

In their default configuration for the OSPF (open shortest path first) routing protocol, Cisco routers set the link cost (link distance value) to a value equal to 1000 million divided by the link bandwidth (i.e. bit rate). Thus a link between two routers based on Gigabit ethernet has a default link cost of 1. Meanwhile, a 64 kbit/s line has a link cost of 15625. By so biasing the route distance (or total cost) calculation, the network administrator can force a preference for the use of high bit rate routes. We saw such an example of route preference in Figure 6.2 and Table 6.1. The 'shortest' route (measured according to the defined distance parameters) from C-to-B in Figure 6.2 and Table 6.1 was the route CDFB and not the single hop route CB.

6.7 Distributing routing information around an internetwork

For the distribution of routing information around a network, a complete topology of routing protocol 'connections' between *neighbouring* routers in the network has to be established. The topology usually, <u>but does not always</u>, reflect the topology of the actual physical links between the routers.

---

[3] Directly connected hosts in the LAN local to the router are one hop away.

To simplify the task of distributing information and the complexity of calculating routing tables, it is customary to divide large complex networks into smaller, more manageable *areas, subnetworks* or *domains* and to create a hierarchy among the different routers. Specialised routing protocols can then be used in the different levels of the network hierarchy to address the different types of problems which arise. Figure 6.6 illustrates the basic hierarchy of routing protocols and node types. We discuss these in turn in the following sections.

Autonomous systems (AS) and administrative domains (AD)

An *autonomous system (AS)* (sometimes also called an *administrative domain (AD)*) is a network of routers all under the same operational administration (and typically all owned by the same network operator). The underlying assumption of routing protocols designed for *interior* use within a single autonomous system (AS) is that no routing information need be *hidden* or kept secret from other *internal routers (IRs)*, and that the 'shortest' route through the network will always be preferred.

Interior gateway protocol (IGP)

Interior gateway protocols (IGPs) are used for the distribution of routing information between the routers within a single autonomous system (Figure 6.6). IGP is not a particular protocol in itself, but a generic name used to denote any of a number of alternative protocols. The best-known IGPs are *RIP (routing information protocol)*, OSPF (open shortest path first), *IS-IS (intermediate system-intermediate system)* and the Cisco proprietary IGPs: *IGRP (interior gateway routing protocol)* and *EIGRP (enhanced interior gateway routing protocol)*. The first ARPANET IGP was the *gateway-to-gateway protocol (GGP)*. This is now obsolete.

Border nodes

Border nodes (BN) (previously also known as *exterior gateway nodes*) assume the responsibility of exchanging routing information between different autonomous systems (Figure 6.6).

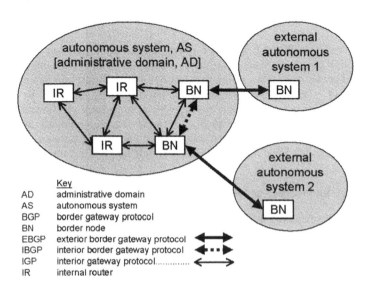

Figure 6.6   Basic routing hierachy and routing protocol types.

Border nodes always communicate on a *peer-to-peer* basis with their partner in the other autonomous system, but are not necessarily directly physically connected to one another. [But a *TCP, transmission control protocol* connection must exist between the *BGP-speaking peer* partners].

Border gateway protocol (BGP) or exterior gateway protocol (EGP)

The routing protocol used to communicate inter-AS (autonomous system) routing information directly between border nodes is called the *border gateway protocol (BGP)*. The modern version is version 4 (BGP4) and this is the default version in which all BGP sessions are commenced (a negotiation for the use of an earlier version of BGP can be conducted if necessary). A now-defunct protocol — the *exterior gateway protocol (EGP)* was replaced by BGP, although the term EGP continues to be used sometimes to describe 'generic' protocols of this nature. In this sense, BGP is a particular example of an EGP.

The border gateway protocol (BGP) introduces the concepts of *reachability* and routing *policy*. Reachability is the term used to describe whether a route to a particular range of exterior IP addresses is known by the border node or not. If the address range is not known, then devices within the source AS (autonomous system) will be unable to *reach* the destination.

In attempting to reach a remote destination, a *border node* will try to discover the best possible path, no matter how many different autonomous systems (AS) need to be transmitted along the way. But maybe the owner of one of the autonomous systems along the way does not want his network to be transmitted? After all, why should he or she have to carry third-party traffic? To cater for such a situation, the *border gateway protocol (BGP)* introduced the idea of *routing policy*. An *import policy* governs which incoming *routing information updates* will be considered when calculating routes, and which ones will be ignored. The routes suggested by an 'untrusted' partner, for example, could be *filtered* out. Similarly, the *export* or *advertising policy* will govern which reachable destinations are made known to other parties. By *hiding* the reachability of some destinations from other autonomous systems, unwanted transit traffic can be avoided.

Within any given autonomous system (AS) network, it is normal to be running at least one IGP (interior gateway protocol) internally, and also to be running BGP to neighbouring autonomous systems (Figure 6.6). Some of the routing information gleaned from BGP needs to be input to the IGP process (so that exterior address ranges are also reachable from *internal routers (IRs)*. Conversely, address ranges reachable within the AS need to be *advertised* by means of BGP to external hosts which may want to communicate with them. The sharing of routing information between BGP and the IGP (which is not straightforward, because of the different mode of operation of the different protocols) is undertaken directly by the border node and is called *route redistribution*.

In a case where there is more than one border node within an AS (as in the case of Figure 6.6), some of the information derived from one of the BGP relationships may need to be advertised via the other border node. In this way the 'external autonomous system 1' in Figure 6.6 can learn about 'external autonomous system 2'. The BGP routing information can be transferred from one border node to another either by means of the IGP or, much better, by using BGP internally within the AS. The latter approach avoids the need for a double-conversion (back-to-back *route redistribution*) between different routing protocol formats. Used internally between border nodes in this way, BGP is sometimes referred to as *interior border gateway protocol (IBGP)*. This distinguishes from the use of BGP externally: sometimes called *exterior border gateway protocol (EBGP)*. There are not two separate protocols!

When an autonomous system (AS) network is broken down into separate routing *areas* (Figure 6.7), the routing tables for each area are worked out separately. Routing information updates pertaining to internal routing within the *area* are not advertised to other areas, but

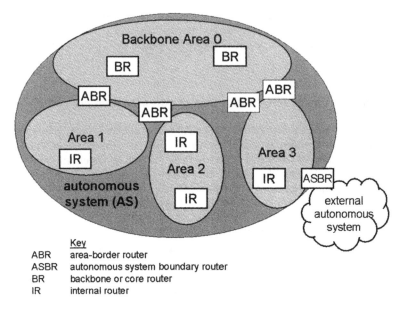

Figure 6.7   Breaking an autonomous system into routing areas.

instead are *hidden*. As a result, the routing protocol traffic between areas can be drastically reduced, and the tasks of storing the network topology database and calculating the routing table are greatly simplified.

By breaking an autonomous system into routing *areas* we impose a rigid hierarchy upon the network. Some people see this as a drawback, since now not only the routing protocol traffic, but all the IP forwarding traffic as well, can only pass from one area to another via the *backbone* (or *core*) network (area 0 in Figure 6.7). But the rigid hierarchy can be a benefit too, since only the backbone (or core network) routers need to store an extensive routing table with a knowledge of all reachable IP addresses! Not only this, but the traffic flows and paths in a hierarchical network are more predictable, thus simplifying the job of troubleshooting network performance problems and tracing bottlenecks.

Route redistribution

In any network in which more than one *routing protocol* is in use, *routing information* obtained by means of one of the protocols may need to be used by the other protocol in calculating route distances or costs (in practice nearly all networks use BGP in addition to at least one IGP). This raises the question of the relative value we should place on routing information, possible paths and path costs obtained from one routing protocol in comparison to similar information obtained from the other. In particular, routing information obtained from third-party ('untrusted') network operators may be considered less reliable than that obtained internally. In another case, we may prefer using certain third-party *exterior* routes more than others (for cost, performance or other reasons).

Route redistribution allows for the transfer of routing information from one protocol type to another and for the simultaneous 'weighting' of the information (i.e. purposely scaling-up or scaling down the value of the relative distance or cost of the route). This allows the network administrator to influence the preferred routes.

6.8 Distance vector and link state protocol routing methodologies

Routing protocols can be split into two categories — according to whether they are *distance vector protocols (DVP)* or *link state protocols (LSP)*. The difference between the two types is the way in which they store and advertise the routing information (i.e. network topology information), and in the way they calculate routes.

Distance vector protocols (DVPs)

As the name suggests, a distance vector protocol works by calculating the distance and the direction (i.e. the *vector* — the *destination*, or the *next hop*) from each possible source router to each possible destination. Once the distance and vector have been determined, they can be *advertised* as *routing information* to other neighbouring routers within the network, which in turn can then calculate their routing tables. We shall use the example of Figure 6.8 to explain the basics of high a DVP works.

The example of Figure 6.8 comprises a network of 6 routers in which the source router is router A and the destination we are concerned with is connected to router F. A simple hop count is being used as the *routing algorithm*. We shall assume that router A's *neighbours* — routers B, C and D have already calculated their routing tables for the shortest path to reach the destination and each is advertising this information to router A (as well as to all their other directly-connected neighbours) by means of the routing protocol. For the calculation of the best route to the destination, router A only needs to:

- know that the destination is not directly *reachable* (i.e. is not in the range of addresses configured into router A as being in the locally connected network); and then can

- deduce from the lowest advertised hop count of one of its neighbours, which of these neighbours represents the best next hop for reaching the destination.

In our example, router A will conclude that its next hop to the destination should be via router C, and that the hop count from router A to the destination will be $2 + 1 = 3$ hops. A simple process of deduction reaches this conclusion: the route advertised to the destination by router

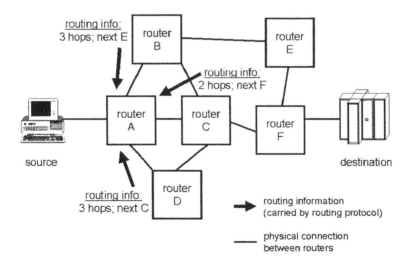

Figure 6.8 Distance vector protocol — routing information.

C has two hops. Therefore a route from router A via router C will have three hops (since an additional hop is required from A-to-C). Meanwhile, the minimum hop count via either router B or router D (using similar logic) is $3 + 1 = 4$ hops.

The chosen 'shortest route' via router C will now be adopted into router A's routing table and advertised using the routing protocol. In addition, the hosts directly connected to router A will be advertised as being *reachable* within 1 hop. Other routers can then deduce that router A is the *destination router* for these IP-address-ranges.

The main advantages of distance vector protocols are:

- the great simplicity of the algorithms required to calculate routing tables from the routing information received from other routers;

- the low amount of processing effort and data storage involved in creating routing information in preparation for advertising. (The router simply advertises the contents of its routing table).

Put simply: distance vector protocols are not demanding in their use of router processor power or data storage capacity. But they are rather crude in their derivation of routes and each router has to keep on advertising its entire routing table (at least once per *ageing time period* — typically 30–300 seconds for a DVP). This can represent a considerable additional traffic load for the network.

Most of the early routing protocols were distance vector protocols (DVPs). Examples of DVPs are: *RIP (routing information protocol* — both *RIP-1* and *RIP-2*), Cisco's *IGRP (interior gateway routing protocol)* and *BGP (border gateway protocol)*. We shall discuss RIP and BGP in more detail later.

Link-state protocols (LSPs)

In contrast to distance vector routing protocols, *link state routing protocols* operate by distributing routing information about the state of the individual links of the network. By listening to the routing protocol advertisements of other routers (which are broadcast or *flooded to all routers throughout the network*), a router is able to build a complete 'map' of the topology of the network. Using its own 'map' (i.e. network topology database), each router can work out the shortest path to each individual destination and so build an optimum routing table.

The main advantages of link state protocols over distance vector protocols are:

- the quality and efficiency of routes selected across a network;

- the ability to provide for reliable routing even in very large, complex networks (without circular routes or similar routing instabilities); and

- the greatly reduced amount of routing information which has to be advertised by each router (each router only advertises information about its own links — not the entire contents of its routing table). The advertisements are called *link state advertisements (LSAs)*.

The disadvantage of using a *link state protocol* is that routers must be equipped with large amounts of data storage capacity to store the entire network topology database. They also require high power processing capability to undertake the complex mathematics associated with deriving the shortest path to each possible destination.

Examples of link state protocols are *IS-IS (intermediate system-intermediate system)* and OSPF (open shortest path first). We shall discuss OSPF in detail later in the chapter.

6.9   Initiating router protocols: neighbour discovery and the hello procedure

All routing protocols — no matter whether they *are distance vector protocols or link state protocols* — rely on routers knowing who their neighbours are, or at least having been configured in such a way that they know how to *discover* them for themselves. As long as all routers know who their neighbours are, a complete topology 'map' of the network can be worked out!

In the case of simple distance vector protocols (e.g. RIP), routers must be configured by human network administrators to advertise routing information to their neighbours. This involves *activating* the protocol (e.g. RIP) on relevant (point-to-point) links at all relevant routers and configuring the neighbour's IP address, so that it may be used as the destination address for routing protocol messages. In the case of router A in Figure 6.8, the relevant *links* are AB, AC and AD. Once the links are *activated* for RIP and the neighbour IP addresses are known, router A will automatically advertise its routing table to the neighbouring routers: B, C and D. And by knowing from which link (i.e. which neighbour) a given remote destination was advertised as being reachable, the router can determine its own best next hop for this destination.

In contrast to distance vector protocols, which exchange routing information only between *neighbouring* routers, the *link state advertisements (LSAs)* of *link state routing protocols* are flooded to all the *internal routers* within the network or network *area*. Nonetheless, the details of the neighbour need to be known, so that these details can be advertised as *link data* in the *link state advertisements*. Rather than have to configure the information about neighbouring routers, many modern routers conduct *neighbour discovery* for themselves. For the purpose, a *hello procedure* is defined as part of many modern routing protocols. It usually works something like this.

When first introduced to the network, a router broadcasts a *hello packet* message across all the interfaces which have been configured for the particular routing protocol (e.g. OSPF). The interface might be either a point-to-point link or some kind of *shared medium* (e.g. a LAN). The message says 'hello ... my router identification is xxxxx3 ... I currently do not know my neighbours'.[4] All the routers neighbours (i.e. those which are reached by means of a single IP *hop* — only *neighbours* will have received the *hello packet*) reply with their own hello packets. The responses go something like 'hello, my router identification is xxxxx1; my IP-address is n.n.n.n; my neighbours are xxxxx2 and xxxxx3'. On receipt of this message, the new router (identification xxxxx3) knows that one of its neighbours is router xxxxx1! The exchange of hello packets to enable *neighbour discovery* is called the *hello procedure*.

6.10   Routing protocols and their relationship with the Internet protocol (IP)

Routing protocols play an important role in enabling the correct forwarding of Internet protocol (IP) packets. But unlike the layered protocols of the OSI model as we discussed in earlier chapters, routing protocols do not directly take part in the transfer of end-user data across the network. Instead, routing protocols are *network control* protocols, used between the different nodes of a network for steering the packet forwarding process. Nonetheless, they share the routers' IP *forwarding engines* and lower layer *networking* protocols for basic carriage of packets (as we illustrated in Figure 6.3).

---

[4] The first *hello packet* is directed to an appropriate 'all neighbouring routers' IP *multicast address*. If necessary, IEEE MAC-addresses necessary for the *datalink* (i.e. layer 2) transmission may be determined by means of the address resolution protocol, ARP, as we shall discuss at the end of the chapter. Alternatively, a LAN-broadcast address can be used.

Different routing protocols are variously carried either directly by IP, or by one of the transport (layer 4) protocols — *TCP (transmission control protocol)* or *UDP (user datagram protocol)*. When carried directly by IP, routing protocols are assigned a protocol value (see Chapter 5 — Table 5.6). Otherwise, when carried by a *transport protocol* such as TCP or UDP, a *port* value must be assigned (we shall learn more about these in Chapter 7). But in all cases, the *IP precedence* value in the IP packet header (Figure 5.6) is set to *Internetwork control* (i.e. high priority), for it is more important for routing tables to be updated than for user messages to be forwarded (possibly incorrectly)!

The forwarding of packets in an IP-based network is generally carried out in a *hop-by-hop* manner. Under hop-by-hop forwarding each router concerns itself only with deciding upon the next hop to which each received packet should be forwarded and develops its routing table accordingly. Most routing protocols are expressly designed with hop-by-hop forwarding in mind. But for certain types of traffic, it is important to dictate certain transit points of the path. In this case, a *source route* listing needs to be included in the *options field* of the IP packet header (as we discussed in Chapter 5).

The use of source routing for steering IP packet paths can be a useful way, for example, of ensuring that packets pass through a given format *conversion* or *gatekeeper* function on their way to a final destination. But even when source routing is used, the forwarding process is still carried out on a hop-by-hop basis. (The IP packet is forwarded to each consecutive destination in the source route listing by means of hop-by-hop forwarding.[5])

The main dangers associated with hop-by-hop forwarding are *circular routing* and *routing instability*. Both problems can arise because the routing tables held in individual routers are calculated independently of one another and at different points in time. Circular routing arises when an IP packet is forwarded endlessly around a loop of interconnected routers without ever reaching its destination. Routing instability, meanwhile, results in unpredictable and wildly fluctuating traffic flows around a network. Both circular routing and routing instability are highly undesirable, because of the extremely damaging effects on the IP forwarding performance of the network. We shall see in subsequent sections of this chapter how different routing protocols go about trying to avoid them.

6.11   The different internetwork routing protocols — when to use them

During more than 30 years of the life of the Internet, many different routing protocols have been developed and used. Each was designed with a particular purpose in mind, and many of these protocols remain in use today. Later in the chapter we shall discuss the most popular protocols — *RIP (routing information protocol)*, *OSPF (open shortest path first)* and *BGP (border gateway protocol)*. But first of all, let us get an overview of all the other common protocols, and their relative strengths and weaknesses. If you ever have cause to connect an old router network to a modern one, you may need to be familiar with all of them! Table 6.2 provides an overview.

RIP (routing information protocol) is the mother of all routing protocols. It derived from the *routed* (pronounced Route-Dee) *daemon* of the UNIX *Berkeley System Distribution* (we spoke about BSD in Chapter 1). RIP is a simple *distance vector routing protocol* which is suitable as an *interior gateway protocol (IGP)* for small networks. Its *routing* algorithm is based on a single parameter — the smallest *hopcount* to a given destination. The maximum hopcount of a path allowed under RIP is 15 hops. Destinations with paths longer than

---

[5] Although theoretically possible using a link state routing protocol, it is not normal to calculate entire IP routing paths from source-to-destination and include all the intermediate router IP addresses in a source route listing. Source routing of this nature is largely limited to bridge networks (chapter 4).

Table 6.2 Comparison of commonly used routing protocols

Routing protocol type	Interior gateway protocols (IGPs)							EGP
Routing /protocol name capability	Static routing	RIP-1	RIP-2	IGRP	EIGRP	IS-IS	OSPF	BGP4
Suitable for large and complex networks	Yes	No	No	No	Yes	Yes	Yes	Yes
Easy to implement	No	Yes	Yes	Yes	Yes	No	No	No
Routing protocol/algorithm type	—	DVP	DVP	DVP	DUAL	LSP	LSP	DVP
Supports classful addressing	Yes	Yes	Yes	Yes	Yes	Yes	Yes	Yes
Supports CIDR and VLSM	Yes	No	Yes	No	Yes	No	Yes	Yes
Supports load sharing	No	No	No	Yes	Yes	Yes	Yes	Yes
Supports authentication	No	No	Yes	No	Yes	Yes	Yes	Yes
Allows weighted cost or distance metric	No	No	No	Yes	Yes	Yes	Yes	No
Fast converging	No	No	No	Yes	Yes	Yes	Yes	Yes
Hello procedure for neighbour discovery	No	No	No	No	Yes	Yes	Yes	No
Uses multicasting for routing information updates	No	Yes*	Yes*	Yes #	Yes	Yes	Yes	No

Notes: DUAL = diffusing update algorithm; DVP = distance vector routing protocol;
LSP = link state routing protocol; EGP = exterior gateway protocol

* RIP-1 *broadcasts* routing updates to hosts, RIP-2 multicasts to hosts. # IGRP *broadcasts* updates.

15 hops are deemed *unreachable*. The beauty of RIP is the ease with which it may be implemented — there is little configuration work which the human operator has to undertake. For this reason alone, it is likely to remain in use in small-scale networks for years to come. Another reason for its likely continued use is its wide availability on different manufacturers' router equipment. The main drawback of RIP is the heavy load of routing protocol traffic which is inflicted on the network: each router must repeatedly advertise its complete routing table to all its neighbours.

The original version of RIP (RIP-1) was specified in RFC 1058. It supported only *classful IP addressing* and routing protocol messages were not *authenticated*. With the advent of *CIDR (classless inter-domain routing)* and *variable length subnet masks (VLSMs)*, a revised version of RIP was devised in 1994 (RIP-2), which is defined in RFC 1723. RIP-2 supports not only CIDR and VLSM but also *authentication* of RIP messages. But while these new capabilities will extend the life of RIP, the same basic routing methodology and constraints on network size (maximum 15 hops) remain.

IGRP (interior gateway routing protocol) is a Cisco-proprietary adaptation of RIP which can be used in networks comprising only Cisco routers. The two main differences between IGRP and RIP are an increased maximum hop count (255 instead of only 15 for RIP) and the incorporation of other *metrics* when calculating link *distances*. In particular, IGRP allows link distances to be calculated on a dynamic basis — based on link *bandwidth, delay, load* and/or *reliability*. Alternatively, link distances may be set to fixed values by the human network operator (so-called *administrative distances*).

While the maximum hopcount of 255 allows IGRP networks to be somewhat larger than the maximum size of networks possible with RIP, the nature of IGRP, in particular the need to keep advertising the complete routing table of each of the routers means it is still unsuited to very large networks. Like RIP, IGRP suffers the disadvantage that it is incapable of supporting CIDR (classless inter-domain routing) and variable length subnet masks (VLSM).

EIGRP (enhanced interior gateway routing protocol) is, as its name suggests, an enhanced version of IGRP. It is also a Cisco-proprietary routing protocol intended for use in Cisco-only networks as an *interior gateway protocol (IGP)*. EIGRP is a routing protocol based on the *DUAL (Diffusing Update ALgorithm)* algorithm which supports both CIDR and VLSM. DUAL is a hybrid routing algorithm, comprising features of both a *distance vector protocol* and a *link state protocol*. In particular, it is designed to be fast converging and to 'guarantee' avoidance of circular routing.

The *IS-IS (intermediate system-intermediate system)* routing protocol was developed by *ISO (International Organization for Standardization)* for use in conjunction with the *OSI (open systems interconnection) connectionless network service (CLNS)*. The original version was defined in ISO 10589. It developed from the *link state* routing protocol developed initially by the Digital Equipment Corporation (DEC) for DECnet. IS-IS is a hierarchical *link state* routing protocol similar to OSPF (which succeeded it).

In the original version of *IS-IS* (ISO 10589), the routing protocol could be configured to support either OSI *CLNS (open systems interconnection — connectionless network service)* or IP (Internet protocol) but not both in the same *routing domain* at the same time. Subsequently, an enhanced version, called *integrated IS-IS* (or *dual IS/IS*) allowed for dual operation of both OSI CLNS and IP protocols in the same routing domain. With the advent of OSPF, IS-IS has largely fallen out of use.

OSPF (open shortest path first) is a *link state routing protocol* based on the *shortest path first (SPF)* or *Dijkstra algorithm*. It supports both CIDR (classless inter-domain routing) and VLSM (variable length subnet masks), as well as authentication. Route calculation is undertaken according to a link *cost*-parameter, which in turn can be made dependent on other link *metrics* such as *bandwidth* (i.e. bit rate), *delay, load, reliability*, etc. OSPF has become the protocol of choice as an interior gateway protocol (IGP). This is because it is an *open* protocol (supported by many different manufacturers). It is robust and capable of calculating reliable and efficient routes through large and complex networks. It is less demanding of network bandwidth than RIP, but requires huge router storage and processing capacity for dealing with the network topology database and for calculating routing tables.

Following the demise of the original *exterior gateway protocol (EGP* — RFC 0904) the border gateway protocol (BGP) has become the routing protocol of choice for sharing routing information across the boundaries of different autonomous systems (i.e. different networks owned and operated by different administrations). It is a *distance vector routing protocol* which finds the shortest path in terms of the least number of autonomous system (AS) networks which have to be transited in order to reach the destination. The current version (and the first version considered to work reliably) is BGP version 4 (BGP4).

6.12   RIP (routing information protocol)

Routing information protocol (RIP) is one of the oldest and simplest interior gateway (IGP) routing protocols. It developed from the *routed* (pronounced route-dee) *daemon* of the Berkeley System Distribution (BSD) of UNIX. RIP works by determining the shortest path distance from router to destination as measured in terms of the hop count. It is thus a *distance vector routing algorithm*. The particular hop count algorithm used is also known as the *Bellman-Ford algorithm*.

There are two versions of RIP and both are still in use. *RIP version 1 (RIP-1)* is documented in RFC 1058. *RIP version 2 (RIP-2)* was developed in 1994 as an extension to RIP-1 in order to cope with classless inter-domain routing (CIDR) and variable length subnet masks (VLSMs). RIP-2 is specified in RFC 1723. In addition, RIP-2 includes the option for authentication of routing protocol messages. This prevents 'untrusted' third parties from maliciously manipulating network routing.

Basic functioning of the RIP-routing algorithm

The basic functioning of the routing algorithm, the calculation of routing tables and the advertising of routing tables in *RIP (routing information protocol)* take place exactly as we described in the example of Figure 6.8. All the neighbours of a given router *advertise* their entire routing table to the router (router A of Figure 6.8), saying which destinations are *reachable* and how many hops need to be transited in order to reach them. The router then chooses the shortest route of those advertised to it as its own preferred route to the given destination and sets its own *next hop* choice accordingly. This next hop route is in turn then advertised (along with the rest of the routing table) to this router's neighbours.

Making each router share its entire routing table with its neighbours is sufficient to enable all routers' routing tables to be developed. And provided there are no network problems, the optimum *shortest hopcount* path will always be chosen. But to guard against possible *circular routing* and network *instability* at the time of network failures and topology changes, two additional basic 'rules' are included in the *Bellman-Ford algorithm*:

- there is a maximum allowed hopcount of 15. This prevents the possibility of circular routing; and

- only routes with <u>lower</u> hopcounts are allowed to replace previously established routing table entries.

The example of Figure 6.9 illustrates how circular routing can arise (when using RIP) following a network failure or topology change. Prior to the failure of router C, the destination D was *reachable* via router C, which was advertising a *reachability* with hopcount 1. As a result, both routers A and B had reachability to the destination, both with a hopcount of 2. This route was being mutually advertised: A advertised its route with hopcount 2 to B and B its route with hopcount 2 to A. After the failure of the route via C, there is no longer a possible route to D via router C, so router A selects instead the route last advertised by router B! It starts sending traffic via B and even *advertises* the new route! B, meanwhile, has also stopped receiving the advertised route to D via C. It also reverts to the next best alternative: the route being advertised by router A! So a circular route emerges. Packets intended for the destination D

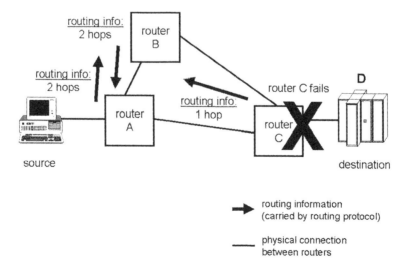

Figure 6.9 Routing calculation problems arising in RIP from network failures.

are caught in a continuous loop between routers A and B. Thankfully, the maximum hopcount of 15 allowed by RIP-1 eliminates the *circular route*.

After the failure of router C, router A in Figure 6.9 reverts to the route of 2 hops to destination D being advertised by router B and starts forwarding packets to D via B. But meanwhile its own hopcount to D has increased to 3. When router A's routing table is advertised by the routing protocol again, the revised hopcount will be 3. Nonetheless, router B knows of no better route than the 3 hop route to D via A and so chooses this route, updating its own routing table and subsequent advertisements to reflect the new hopcount of 4. For a while, the cycle of circular routing goes on, but meanwhile the hopcount is increasing with each new route *advertisement* of the two routers A and B. Soon, the hopcount increases to 16. This value (the *infinity value* — actually binary value '0') is defined to equate to an *unreachable* destination. As a result, both routers A and B remove their routing table entries corresponding to destination D. This is the prime manner in which RIP eliminates circular routing. It is known as the *count-to-infinity* algorithm.

More sophisticated routing protocols based on hop counting (including RIP-2) advertise not only each reachable destination and the hopcount to the destination, but also the next hop taken to reach the destination. But while knowledge of the next hop would help to prevent a 2-node circular route as illustrated in Figure 6.9, it is not effective in eliminating possible circular routes which emerge over 3-node, 4-node or longer paths.

Other techniques designed for eliminating circular routing in networks using RIP are called *split horizon* and *poisoned reverse*. The split horizon technique prohibits a router from advertising reachability to a given destination to the neighbour from which a path was learned. Relevant entries in the router's routing table are deleted from the copy advertised to that particular neighbour. Thus router A of Figure 6.9 would not be allowed to advertise routes to router B where the next hop is via B. *Route poisoning* goes one step further than the split horizon method — it advertises the route to the neighbour but sets the hopcount at 16 (i.e. *unreachable*). But while the split horizon and poisoned reverse techniques prevent routing loops between router pairs, three-cornered triangles and larger routing loops are still possible.

Route flapping (i.e. oscillation of routes between two or more alternative paths) is prevented in RIP by only accepting new routes with a lower hopcount than the route currently installed in the routing table.

Updating, ageing and convergence of RIP routing table entries

To ensure that withdrawn routes and withdrawn destinations are removed from routing tables, RIP employs ageing, as we discussed in detail earlier in the chapter (in conjunction with the 'severed country road' example and Figure 6.4). Even if there have been no changes in their routing tables, routers using RIP must nonetheless advertise their entire routing table to all their neighbours at least once every 30 seconds to prevent correct routing table entries from being deleted! Should no update for a route currently appearing in the routing table be received within a period of 180 seconds, then routing table entries are marked invalid. If still no update has been received after a further 60 seconds (240 seconds in total), then the route is deleted. Once a route is marked invalid a new route will be selected (if available) to the same destination. Of course, the new route will most likely have a higher hopcount.

For the purpose of ageing, each entry in the routing table is associated with two timers — the *expiration timer* and the *flush timer* (or *garbage collection timer*). The *expiration timer* of a given routing table entry is set to 180 seconds when the entry is first created and reset to this value each time a routing *update* is received to confirm that the router which is the next hop in the path is still active. The timer slowly reduces over time. Should the value reach zero (because no new update is received in the meanwhile), then the routing table entry is marked invalid and an alternative path to the destination is selected. The table entry is finally deleted once the flush timer reduces to zero. The initial value of the flush timer is 240 seconds. Like the expiration timer it is reset to its initial value (240) each time the route is confirmed as being still active.

Should the topology of the network around the router or its selected routes change in the 30-second period between normal routing table advertisements, then a router is expected to send a complete update immediately. This is known as a *triggered update* (an update is *advertised* because it is *triggered* by a routing change at the router sending it.) Another method of *triggering* an update is for a router to send a *routing information request* to its neighbour. (It can request either the whole routing table or a particular part of it to be sent). Triggering the sending of an update request can be a useful way of speeding up the calculation of the first routing table in a router newly introduced to a network. It also ensures that the *count-to-infinity algorithm* works properly. But while triggered updates help to speed up the rate at which the entire network's routing tables are adjusted to reflect changes in network topology (this process is known as *convergence*), they do not necessarily lead to *fast convergence*. The problem is that the routing tables may take a little while to stabilise as different routers re-calculate and re-advertise new routes after a network change. All the *count-to-infinity* and other iterative route calculation processes have to run their course. This can take a little while!

RIP protocol format

Figure 6.10 illustrates the basic protocol format of a RIP-2 packet. The format is basically the same as the format used by RIP-1, except that a number of additional fields have been added. The *authentication* fields (shaded grey), for example, cannot be added to RIP-1 packets. Instead the 5th octet (i.e. second word) commences directly with the *route records*.

20-byte *route records* as shown in Figure 6.10 form the 'core' of the routing informa- tion updates transmitted by both RIP-1 and RIP-2 for communicating the entire contents of

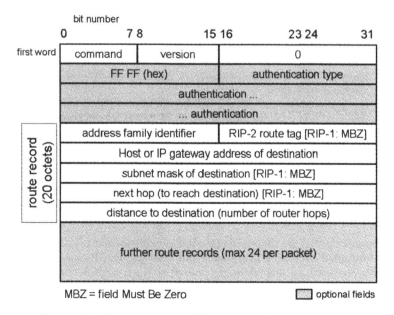

Figure 6.10 Protocol format of RIP (routing information protocol).

a router's routing table to its neighbours. The *address family identifier* (16 bits — four hex-adecimal characters) identifies the network *protocol* to which the routing updates pertain. In practice, only two values are ever used: 0002 (Internet protocol routing information update) and FFFF (RIP-2 authentication in use).

The *route tag* (used only in RIP-2: in RIP-1 this field must be set to all '0's) is intended to identify the type of route (e.g. *internal* route, *external* route). This reveals a little about how the router learned about the route and could be useful in helping to select routes, but is not always used.

The IP address in the route record identifies the destination group of IP addresses to which the routing table entry applies. In RIP-1 this field must be filled with a classful-type address of the form: A.0.0.0 (class A address range); B1.B2.0.0 (class B address range); C1.C2.C3.0 (class C address range) or D1.D2.D3.D4 (destination host IP address) and the subnet mask field in RIP-1 must be set to all '0's. In RIP-2, the *gateway address* and the *subnet mask* (taken together) allow for the use of *classless inter-domain routing (CIDR)* type addresses as well as *classful*-address ranges. In both RIP-1 and RIP-2 there is no formal distinction indicated in route records which pertain to *networks* or *subnetworks* and route records which pertain to single destination *hosts*.

The next hop field (in the case of RIP-2 only) contains the IP address of the next hop of the relevant path from the router advertising the route. When set to the value 0.0.0.0, this field indicates that the destination is directly reachable (i.e. within 1 hop) from the advertising router. The destination hop count field indicates how many hops need to be traversed to reach the destination. When set to value hopcount $= 1$, then the destination is directly reachable from (i.e. directly connected to) the advertising router. When set to hopcount $= 16$ (binary value '0'), the destination is unreachable.

Unlike other more recent routing protocols, both RIP-1 and RIP-2 <u>always calculate</u> the shortest distance calculation based upon the <u>smallest number of hops</u>. There is no possibility to 'bias' link costs towards other route *metric* qualities such as higher bandwidth, lower delay, higher reliability, although some realisations do allow the hop distances to be manipulated manually between values of 1 and 15. When used, such values are called *administrative distances*.

The *command* field of the RIP header has two allowed values. When set to value 1 (*request*), the message is requesting the router which is the target of the message to return a partial or full routing table update. To request a full routing update (e.g. when a new router is added to the network) the *address family identifier* needs to be set to all '0's and the distance set to 16 (infinity). When the command field is set to 2 (*response*), it indicates that the RIP packet contains a *routing information update* of the router sending it. Most RIP update messages are *response* messages.

The valid *version* number values of the RIP-header are 1 (RIP-1) and 2 (RIP-2).

RIP-messages have a maximum allowed length of 512 octets and are carried by the *user datagram protocol (UDP)* in the UDP-data field.[6] The UDP port value is set to 520. In order to ensure that RIP messages reach all the router's neighbours, the destination IP address of RIP messages (indicated in the IP packet header of the corresponding message) is set as follows:

- RIP-1 response updates are addressed by *unicasting* to the known (i.e. pre-configured) neighbour IP addresses or sent to the broadcast address (255.255.255.255) with the *IP time-to-live (TTL)* field set to allow a maximum of only one hop;

- RIP-2 response updates are addressed to the RIP-2 multicast address 224.0.0.9. This address is never forwarded over more than one IP-link (because the IP *time-to-live* field is set to allow only one hop), so that messages only ever reach neighbouring routers and hosts connected either to direct links or local LANs.

---

[6] See Chapter 7.

Occasionally, RIP-1 updates may be sent to the IP *local broadcast* address (255.255.255.255) or in RIP-2 to the RIP-2 local multicast address (224.0.0.9) to allow *hosts* in locally connected LANs *silently* to acquire routing information for the purpose of updating internal routing tables without actually taking active part in the main RIP process [For this broadcast, the LAN MAC address is also set to the broadcast value].

Usage, benefits and limitations of RIP (routing information protocol)

The main benefit of RIP is the simplicity with which it can be realised. It is both easy to design into the basic functionality of a simple router and easy for human network administrators to configure. The human network administrator need only *enable* the protocol on each relevant network interface of the router which is to use it.

From a design point-of-view, the routing table created by RIP is quite simple, containing only a small number of relatively simple fields (i.e. IP address of destination, subnet mask, next hop address, distance hopcount, route timers). There is very little processing which the router has to undertake in preparing router update messages, since the routing table is sent largely in its 'raw' form. Similarly, little processing effort needs to go into calculating new outgoing routes from the routing information updates which a router receives.

The disadvantage of RIP is the maximum hopcount of 15 allowed in reaching a destination. This limits its use to small and medium-sized networks only. Even if a hopcount of more than 15 were possible, its suitability for use in large networks would be limited by the need to keep broadcasting full routing table updates from each router every 30 seconds. This can add considerably to the overall traffic load.

When configuring routers for the use of RIP, human network administrators should bear in mind that the normal default version is RIP-1, and that RIP-1 does not support classless inter-domain routing (CIDR) and variable length subnet masks (VLSMs). During configuration it is common practice to set a default network/default route to identify a 'gateway of last resort'. This is the next hop to which packets should be sent if no other routing table entry identifies the destination in question or if these routing table entries have *aged* and been deleted.

It is prudent (if possible) to configure RIP routers to ignore routing updates from certain 'untrusted' or 'unreliable' sources. This may significantly improve the stability of network routing.

6.13   OSPF (open shortest path first)

Open shortest path first (OSPF) is a routing protocol designed to be an interior gateway protocol (IGP). It is nowadays the 'IGP of choice' for modern networks. OSPF is a *link state routing protocol*, which evolved from the ARPA's 'experimental new routing algorithm' of 1980 and the *subsequent IS-IS (intermediate system-intermediate system)* routing protocol. The current version of OSPF is version 2. The first version was defined in 1989 in RFC 1131. OSPF2 is specified in RFC 2328.

The name open shortest path first derives from the fact that the protocol is based upon the *shortest path first (SPF)* routing algorithm (also known as the *Dijkstra algorithm*). The additional word *open* reflects the fact that the standard is an 'open standard' rather than a 'proprietary standard'. Much of the initial work in the use of *link state* protocols and the *SPF algorithm* was undertaken by the Digital Equipment Corporation as part of its *proprietary* DECnet computer networking architecture.

The principal benefits of the OSPF routing protocol over RIP (routing information protocol) are:

- OSPF uses a link *cost* parameter as the basis for *shortest path* calculations. The parameter may be weighted to accommodate other *metrics* including link bandwidth, delay, load, reliability, etc.

- by using a link state routing protocol and by limiting routing information messages to 'real' update information, the network traffic load created by OSPF routing messages is much lower than that of RIP;

- the subdivision of the OSPF routing *domain* into separate routing *areas* (as in Figure 6.7) provides for hierarchical routing and a further reduction in routing protocol traffic. It also allows OSPF to cope with large, very complex networks;

- OSPF allows for the use of equal cost multipaths to the same destination by *load sharing* traffic between alternative paths if desired;

- OSPF is a *fast converging* protocol requiring minimum routing message traffic: but, as a result, is much more demanding of router processing power and memory capacity; and

- OSPF traffic is always authenticated — so only trusted routers may take part in the routing process.

How does a link state routing protocol work?

Routers which employ *link state routing protocols* maintain a *link state database*. This is a database containing information about the network as a whole and the *state* (i.e. link cost) of each of the links. Using the database, each router separately calculates a *shortest path tree* with itself as the *root*. This enables the router to determine its routing table by working out the shortest path route (i.e. the lowest *cost* route) to every reachable destination on the branches of the tree. We shall use the example network of Figure 6.11 to explain in detail how the method works.

The network of Figure 6.11 comprises 7 routers (R1 to R7), 8 networks (N1 to N8: these might be ethernet LANs, ATM or Frame Relay networks, etc.) and one point-to-point (PP) link (from router R3 to router R6).

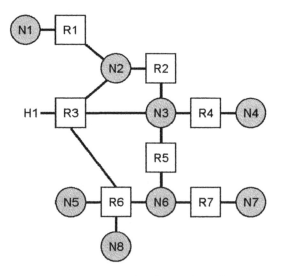

Figure 6.11   Example of network topology.

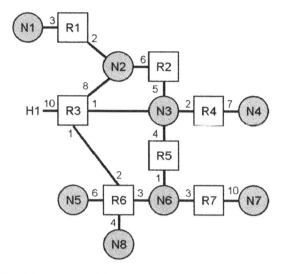

Figure 6.12   Link states (i.e. link costs) for the network example of Figure 6.11.

There are also a number of different hosts connected to the different networks and routers. For our example we only illustrate one of these (H1), which is connected directly to router R3.

The first stage in the creation of the link state database which will represent this network is the assignment of states (i.e. link costs) to each of the links in the network. This step is illustrated in Figure 6.12. The diagram uses the normal convention, that a hop or link from one router to another has a given link cost. The cost for the link traversed in one direction may be different from the cost of the same link when traversed in the opposite direction. The cost of a link is assigned by the associated router for each relevant <u>outgoing</u> interface. Thus, if we study the costs to the routers (R1, R2 and R3) of using their respective <u>outgoing</u> interfaces to network N2 these are values 2, 6 and 8. The cost value is a dimensionless parameter (i.e. a simple integer value) as far as the routing protocol is concerned. The lower the value, the 'shorter' (and thus more preferable) the route.

The values assigned as link costs to the relevant router outgoing interfaces are assigned by the individual routers themselves. The assignment may take place automatically (e.g. as the result of a calculation based on other bandwidth, delay, load or reliability metrics) or alternatively the value may be set as an *administrative distance* — assigned manually by a human operator when configuring the router. Automatically calculated values may vary over time (as the link delay, load and reliability change). Typically the *default cost* value of 1000 million divided by the link bit rate is assigned. (Thus a Gigabit ethernet network would have a link cost of 1, while a 64 kbit/s point-to-point line would have a value of 15 625.) It is important to note (as in Figure 6.12) that there is no cost for exiting a network to enter a router. Thus the total cost of the route from router R1 to router R2 is 2.

Using Figure 6.12 we are now in a position to create a 'database' to represent the network and the link states illustrated. This appears in Table 6.3. The table illustrates a matrix of the individual links in the network which interconnect all the various routers, networks and hosts. The numerical values which appear within the table are the link costs according to the direction of packets across the link, and all the network components (routers, networks hosts) are represented.

As an example, Table 6.3 shows the link from router R1 to router R2 broken down into two separate sub-links: FROM router R1 TO network N2 (with a cost of 2) and a second sub-link

Table 6.3 Typical OSPF link state database corresponding to the example of Figure 6.12

	F	R	O	M												
	R1	R2	R3	R4	R5	R6	R7	N1	N2	N3	N4	N5	N6	N7	N8	H1
T R1								0	0							
O R2									0	0						
R3						2			0	0						
R4										0	0					
R5										0			0			
R6			1									0	0		0	
R7													0	0		
N1	3															
N2	2	6	8													
N3		5	1	2	4											
N4				7												
N5						6										
N6					1	3	3									
N7							10									
N8						4										
H1			10													

FROM network N2 TO router R2 (with a cost of 0). The empty positions in Table 6.3 indicate non-existent links.

In its completed form, Table 6.3 contains as much information as Figure 6.12, and we can use it alone to calculate all available routes through the network. It is this link state database which serves as the basis for each router to calculate its routing table.[7]

Having received all relevant routing information updates (by collecting the routing protocol broadcasts of all other routers in the network), each router is able to collate a complete *link state database* like Table 6.3. Using it, each router applies an OSPF *process* using the *Dijkstra algorithm* to calculate the *shortest path tree* with itself at the root. Two examples of shortest path trees, respectively with routers R1 and R2 as their roots are shown in Figures 6.13 and 6.14.

It is perhaps surprising to discover that the shortest path routes from the two routers R1 and R2 appear to follow totally different 'spinal' paths. Router R1 tends to reach destinations in the lower portion via routers R3 and R6 (Figure 6.13), while router R2 instead accesses the same destinations by way of network N3 and router R5 (Figure 6.14). The fact that the shortest routes for the two routers should follow such different paths may not have been obvious at first (i.e. from Figure 6.12)! This illustrates the major strength of *link state algorithms* — they always find optimal routes through the network — no matter how complex the network is. There is no chance of selecting a circular route, and routes of equal shortest path cost will both be used — by employing traffic *load sharing* across all relevant paths.

Once the shortest path tree has been calculated from the link state database (using the shortest path first, SPF, algorithm), the router can create its own routing table, which is stored separately.

The need to store the routing table as well as the entire link state database demands that routers using OSPF have much more data storage capacity than is needed for distance

---

[7] You might like to use Table 6.3 to work out the possible alternative routes (and their costs) when traversing the network from router R1 to network N7. Having done so, you will appreciate that the 'database' of Table 6.3 is not as easy to use as the 'map' of Figure 6.12. But the database form is all that the routers have to work with. Being in a numerical form, it is easy to transfer the information in the database from one router to another by means of a routing protocol, but the calculation of the possible routes requires a special mathematical routing algorithm. The algorithm used is the *Dijkstra* or *shortest path first (SPF)* algorithm.

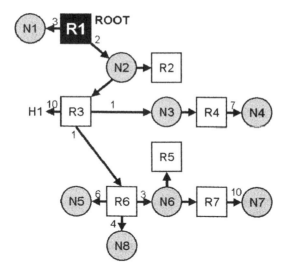

Figure 6.13 Shortest path tree from router R1 as the root.

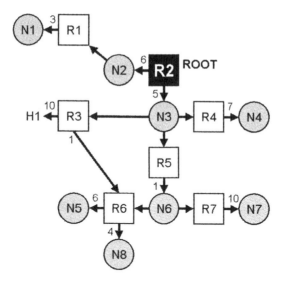

Figure 6.14 Shortest path tree from router R2 as the root.

vector routing protocols such as RIP. Furthermore, the complexity of the calculations necessary to determine the shortest path tree demands not only that a powerful router processing capability be available but also that the routing table calculation should not be repeated too often.

The big benefit of link state protocols is that the level of network topology detail held in the link state database means that only 'real' topology changes need normally be notified by the routing protocol — rather than having to keep on re-broadcasting the entire routing table (as undertaken by RIP). This means much larger and more complex networks can be handled efficiently by OSPF than by RIP.

Information content of routing updates (link state advertisements — LSAs)

The routing updates of *link state* routing protocols take the form of *link state advertisements (LSAs)*. Each router restricts the origination of routing information which it floods to all other OSPF routers to its *router link state advertisement (router-LSA)*. This contains only information about <u>outgoing</u> access possibilities from the router (e.g. directly connected routers, networks and hosts). Thus the information which the router R6 of Figure 6.12 would advertise in its *router-LSA* is as listed in Table 6.4.

Using the same logic as we used to create the router link state advertisement (router-LSA) of Table 6.4 we could create a *network link state advertisement (network-LSA)* table for each of the networks of Figure 6.12. Table 6.5, for example, shows the network link state advertisement (network-LSA) of the 'multipoint' network N3. The only problem is that the network is unable to create and advertise the *network-LSA*, since the network does not take an active part in the *OSPF protocol* or *OSPF process*. To get around this problem, we designate one of the routers to undertake the task on behalf of the network. In the case of network N3 of Figure 6.12 we could designate any of the routers R2, R3, R4 or R5 to perform the task. The router chosen for the task is termed the *designated router (DR)* and its backup (which ensures continued operation of OSPF in the case of failure of the DR) is called the *back-up designated router (BDR)*. The designated router creates and advertises the *network-LSA* for the network (e.g. Table 6.5 for network 3 of Figure 6.12).

Collect all the router link state advertisements (router-LSAs) and network-link state advertisements (network-LSAs) together, and you are in a position to create the complete link state database of Table 6.3! (All the routers in an OSPF collect this <u>identical</u> link state database). In principle, this is exactly the type of information contained in routing updates and exactly the manner in which the OSPF protocol works. But before we go on to explain the OSPF protocol in detail, we should explain *network-LSAs* more thoroughly.

Network link state advertisements (network-LSA)

To ensure the creation of network-LSAs, both a *designated router (DR)* and a *back-up designated router (BDR)* are *elected* from the routers connected to a given *broadcast* (i.e.

Table 6.4 Router R6's router-link state advertisement (LSA)

		FROM
		R6
T	R3	2
O	N5	6
	N6	3
	N8	4

Table 6.5 Network N3's network-link state advertisement (network-LSA)

		FROM
		N3
T	R2	0
O	R3	0
	R4	0
	R5	0

shared medium network such as an ethernet LAN) or *non-broadcast multiple access (NBMA)* network.[8]

Rather than making all the routers separately issue the (same) network-LSA, the task is left to the *designated router*, which also takes on the task of coordinating incoming and outgoing link state advertisements. Thus, for example, neighbouring routers within the broadcast- or NBMA-network need not exchange LSAs with each other and all possible other routers on a bilateral basis. Instead, each router within the local network simply updates the *designated router (DR)* which then broadcasts the consolidated updates to all the other local and remote routers on their behalf.

Both a designated router (DR) and a *back-up designated router (BDR)* are *elected* for each *broadcast-* and *NBMA-network*. This ensures smooth continued operation of the OSPF network in the case of failure of the DR. But smooth operation does not extend to immediate adoption by the BDR of the DR's duties. First the BDR will have to create a new version of the DR's router-LSAs and network-LSAs as well as re-collecting the details of neighbouring routers. A hot standby copy (i.e. *mirror* copy) of the various DR databases is not maintained by the BDR, as this is considered an undue effort during normal operation of the protocol. As soon as the BDR takes over the role of the DR, a new BDR is elected.

Both the DR and BDR *elections* are undertaken as part of the *hello procedure* (which we shall discuss later). The router with the highest *router priority* is elected to become the DR. The router with the second highest router priority becomes the BDR. Where more than one router has the same router priority (which is set during configuration of the router or left at the factory default value), the router with the higher *router identifier (RID)* is elected. The router identifier is a 32-bit value used to identify uniquely the source of link state advertisements and router-relevant entries in the link state database.

OSPF operation — the detailed protocol functions and their chronology

We have seen how the OSPF protocol is designed to work by advertising link state advertisements (LSAs) which describe the link topology of the network surrounding a particular router. *LSAs* are *flooded* by all OSPF routers to all other routers within the relevant routing domain or area (Figure 6.7), and enable each router to create and maintain a full copy of the link state database (network topology database) from which the shortest path tree and ultimately, the routing table are calculated. But this is only a small part of the story. There is a lot more complex detail involved in initialising the protocol, conducting the *hello procedure, discovering neighbours* and *synchronising data*, as we shall discover next.

Figure 6.15 illustrates the functions which must be undertaken and the chronological order of the steps, as a new OSPF router is introduced into, and subsequently maintained in, operation. As with other Internet devices, OSPF routers are designed to find out as much as they can about the network for themselves — thereby minimising as far as possible the need for human operators to configure them. This minimises initial human installation effort, but also maximises the capability of the OSPF *routing process* to cope with network failures. As we shall see, this means there are a number of complicated functions all taking place at the same time.

Before a router can be introduced to a network using the OSPF routing protocol, it must first be configured by the human operator for OSPF. This involves:

---

[8] An NBMA (non-broadcast multiple access) network is a data network based on a technology like *X*.25, ATM (asynchronous transfer mode) or *frame relay* which has multipoint access but does not have a simple, single address method for multicasting. The designated router in an NBMA network provides functionality which makes the network act like a broadcast or multicast network.

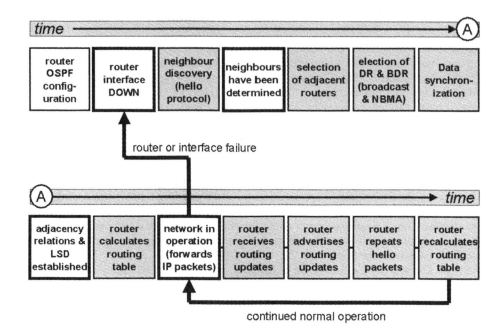

Figure 6.15   Functions and chronology of the OSPF protocol [LSD = link state database].

- defining the *OSPF process* in the router and identifying it with a *process ID* (one process is required for each network area of Figure 6.7 in which the router is to take part in the *routing process*);

- defining the network area (*area-ID*) to which each *OSPF process* is to be assigned;

- defining the *router priority* and *router-ID*;

- defining the network interfaces which are to be subject to the OSPF routing process: associating the IP address-ranges and setting up the *metrics* which are to be used to define the link costs associated with OSPF-controlled interface (otherwise the default is 1000 million divided by the interface bit rate); and

- defining the *default route* (gateway of last resort for the address 0.0.0.0) if required.

When first introduced to the network, the router will be in the *down* (i.e. non-active) state. The first function it must perform is to *discover* who its neighbours are. Neighbours are routers which can be reached by means of a single IP forwarding hop. In other words, neighbours are all those other routers 'directly connected' to the router, by means of a point-to-point link or single network (e.g. LAN) connection. The process of neighbour *discovery* of other OSPF routers is carried out in the OSPF protocol by means of the *hello procedure*.

The OSPF Hello Procedure (hello protocol)

Straight after a router is first switched on, configured for OSPF and introduced to a network, it is said to be in a *down* state. It is down because it is not yet participating fully in either the OSPF *routing process* or in the forwarding of IP traffic. To come 'on line', the first thing an OSPF router does, is to discover its neighbours. It does so by *multicasting* a

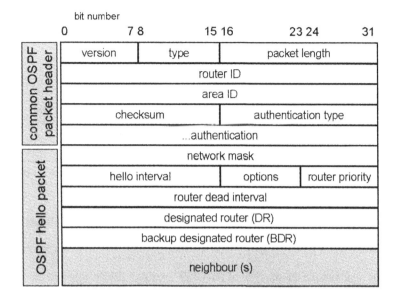

Figure 6.16 OSPF common packet header and hello packet format.

hello packet (Figure 6.16) on all its outgoing interfaces which have been configured to be controlled by OSPF.

The first *hello packet* sent by a new router will let the neighbouring routers know its details — its IP address will be in the IP packet header, its *router-ID* and routing *area-ID* will be identified in the corresponding hello packet fields (Figure 6.15), as will other parameters locally configured within it (authentication type, hello options, router priority, router dead interval, etc.). Although the hello packet is addressed to the *all OSPF routers multicast address* (224.0.0.5), this first *hello packet* will only reach directly connected neighbour routers by virtue of the fact that the number of allowed IP forwarding hops is limited to one by setting the IP packet header field *time-to-live (TTL)* to one.

All the routers in an OSPF network are obliged to continue sending hello packets to their neighbours at regular intervals. By so doing, the routers continually re-confirm that they are still available and 'alive' — that their routes and links are still valid and should not be aged and deleted. So ... sometime after the receipt of the hello packet from the new router, it will be the turn of each of the neighbouring routers to broadcast their own hello packets back. In these hello packets, the 'new router' will be identified as being one of the neighbouring routers (since a hello packet contains the information: 'my neighbours are...'). Following the receipt of all the hello packets of neighbouring routers, the 'new router' will thus be able to determine who all the neighbours are, to which link (i.e. interface) they are connected and other IP address details.

Next, the router has to decide with which of the neighbours an *adjacency relationship* will be established. We shall discuss this shortly, but before we do, we describe the information fields in the common *OSPF packet header* and OSPF hello packet fields, as used for the OSPF process described so far.

Common OSPF packet header

All OSPF routing protocol messages are preceded by the common OSPF packet header of 20 octets (Figure 6.16). The first four octets of the header identify the OSPF *version* number

Table 6.6 OSPF common packet types

OSPF packet type value	Meaning
1	Hello packet
2	Database Description (DD)
3	Link State Request (LSR)
4	Link State Update (LSU)
5	Link State Acknowledgement (LSAck)

Table 6.7 OSPF authentication types

OSPF packet authentication type value	Authentication method in use
0	Null authentication
1	Simple password
2	Cryptographic authentication
3- 65 535	Reserved values

(either version 1 or version 2), the *type* of the OSPF message (Table 6.6) and the *packet length* (the total length of the packet including the common header fields).

The 32-bit *router-ID* uniquely identifies the router which generated the OSPF message. This value is manually configured into the router. The *area-ID* is also a 32-bit value, identifying an area within the overall OSPF routing domain (as shown in Figure 6.7).

The *checksum* applies to the entire OSPF-packet (excluding the 64-bit authentication field).

The *authentication* fields reveal the type of authentication in use (Table 6.7) and the authentication codeword and algorithm.

The entire OSPF packet is carried directly by the Internet protocol, with the IP packet header *protocol* field value set to protocol number 89 (OSPF).

Hello packet format

The hello packet format is as illustrated in Figure 6.16. It carries the *common OSPF packet header*, with the OSPF packet type field set to value '1'.

The *network mask* field is the subnet mask of the link interface across which the hello packet is being sent. The *hello interval* indicates the time between transmission of successive hello packets in seconds. The *router dead interval* is the period of time for which a neighbouring router will still be considered to be 'alive' even if it does not send a hello packet. After expiry of the router dead interval without receipt of a hello packet, the router is assumed to be 'dead' and relevant database entries are amended accordingly. The *options* field we discuss later.

Designated routers (DR) and *back-up designated routers (BDR)* are identified by their router-ID if the interface on which the hello packet is being sent is either a *broadcast-* or a *NBMA (non-broadcast multiple access)*-network.

The *neighbours* field comprises a list of the various 32-bit *router IDs* of all *neighbouring routers*, who have sent a hello packet within a period of time equal to the router dead interval.

Turning a neighbour relationship into an adjacency relationship

A *neighbour relationship* has been created with a given *neighbour router* as soon as a hello packet has been received from that neighbour with the first router listed as one of its neighbours. This is the first step towards full OSPF operation. The next step (as Figure 6.15 shows) is the selection of *neighbouring routers* with which an *adjacency relationship* will be created. Not all *neighbour relationships* are turned into adjacency relationships, but only adjacency relationships take part in the main *OSPF process*. An OSPF router forms adjacency relationships with the following routers:

- all neighbour routers at the opposite end of point-to-point links;

- each of the neighbour routers at the opposite end of point-to-multipoint network interfaces (each neighbour in this case is treated as if it were a separate point-to-point neighbour);

- all *designated routers (DRs)* and *back-up designated routers (BDR)* of *broadcast* or *non-broadcast multiple access (NBMA)* networks to which the router is directly attached; and

- to any *backbone* (i.e. *area 0*) *neighbour router* to which the router may be connected by a real or *virtual link* (more about virtual links later).

Routers only exchange routing information in adjacency relationships (i.e. only with particular neighbours). Any information which needs to be propagated to all other routers in the OSPF network is cascaded through the network across one adjacency relationship to the next in a process known as *flooding*. Thus OSPF routing information received on one interface will be *flooded* by *advertising* it to all other adjacent OSPF interfaces of the router.

An adjacency relationship first exists when adjacent routers have *synchronised* their link state databases (i.e. network topology databases). But before the *synchronisation* can commence, it is necessary to *elect* the *designated router (DR)* and *back-up designated router (BDR)* for any *broadcast* or *non-broadcast multiple access (NBMA)* networks to which the 'new router' is connected. The designated router (DR) saves the routers in broadcast and NBMA networks from having to create adjacency relationships with all other routers in the network. Instead each router has an adjacency relationship with the designated router (DR), which acts to broadcast all the relevant link state advertisements (LSAs) to all routers. In the case of an NBMA, there is no actual broadcast capability for messages provided by the normal network technology, but the designated router is configured in such a manner to simulate broadcasting or multicasting. This might involve, for example, being pre-configured with the IP unicast addresses of all the routers in the network and progressively sending the same LSAs to each one in turn.

The process of automatic neighbour discovery is critical to the operation of OSPF, if there is not to be a heavy load placed upon the human network operator to configure OSPF routers by hand (and keep them up-to-date) so that they always 'know' their neighbours. While neighbour discovery in most 'pure IP' networks is covered by the provisions of the hello protocol, there are occasions when this alone is not sufficient. For example, when a number of routers are all connected to *a* non-broadcast multiple access (NBMA) network such as *frame relay*, they may not be capable of undertaking either neighbour discovery or the election of a designated router (DR) without further assistance (either manual configuration or help from a further protocol). For this reason, the development of protocols for neighbour discovery continues. RFC 2461 defines a specific procedure for *neighbour discovery in IPv6* while the *inverse address resolution protocol* (*inARP* — RFC 2390) defines a process for neighbour discovery in NBMA networks (specifically designed for frame relay).

Data synchronisation

Once all the neighbouring routers which are to enter adjacency relationships have been selected, the process of *data synchronization* can commence. This is triggered by a request made in a hello packet, and involves comparing the entries in the link state databases (i.e. network

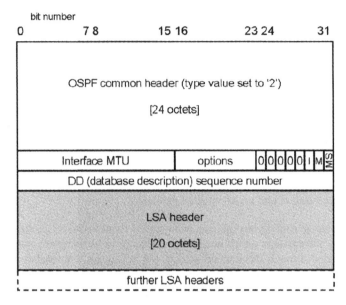

Figure 6.17   Format of the OSPF data description (DD) packet.

topology databases) of the two adjacent routers and making sure that the two are identical and as up-to-date as possible. The synchronisation process takes place by means of OSPF *database description (DD)* packets. A *DD* packet contains a list of all the *link state advertisements (LSAs)* (identified by their *LSA-headers*) making up the link state database (Figure 6.17).

One or more DD packets may necessary to list all the *LSA-headers* making up the link state database of a given router. When more than one packet is necessary, this is indicated by the M-bit (More DD-packets to follow).

OSPF *database description (DD)* packets are transmitted using the unicast IP-address (learned during neighbour discovery) of the selected neighbour router. The *data synchronisation* process (also called the *data exchange process*) is controlled by one of the routers as the *master* (the router which acts as master first is the one with the highest *router priority*). The master is the router allowed to send DD packets. It sends DD packets repeatedly until they are acknowledged. Meanwhile, the other 'soon-to-be-adjacent' router acts as a *slave*: it acknowledges the receipt of each DD packet (by returning an empty DD packet with the same sequence number) and checks the list of the LSAs contained in the master's link state database against the list of LSAs in its own.

Should the database description (DD) packet indicate to the slave router that the master has a link state advertisement (LSA) not previously known to the slave or more up-to-date than the LSA-version held by the slave, then the slave is obliged to generate a *link state request (LSR)*. The link state request (LSR) demands that the master provide the full LSA (link state advertisement). This may, in turn lead the slave to *flood* the same LSA to other routers.

Of course, there is bound to be some new topology information for newly connected adjacent routers to share between themselves. The link state databases cannot possibly have been the same beforehand! Consider the example of Figure 6.18. The network comprising routers A and B is being newly-connected to the existing network consisting of routers C, D and E. Let us assume that router B acts as *master* first during the process of data synchronisation — then router E is the slave — and receives database description (DD) packets. While router E acts as the slave it will generate LSRs for the LSAs pertaining to the network comprising routers A and B. The LSAs (link state advertisements) describe each of the links in the A/B part of

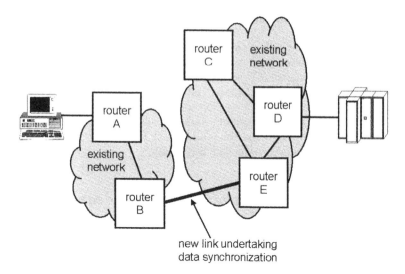

new link undertaking
data synchronization

Figure 6.18   Routers forming a new adjacency relationship having been newly interconnected.

the network. Because this information is found to be new to router E, it will also flood the LSAs to all its outgoing interfaces. In this way, routers C and D will also keep their link state databases up-to-date. Once router B has sent DDs describing its entire link state database, it ceases to be the master. Router E now is free to become master and correspondingly router B takes on the task of being slave. Now it is the turn of router B to *synchronise* its database and ensure (by means of link state requests, LSRs) that router A is also flooded with the LSAs describing the topology of the C-D-E part of the network.

Once both routers in Figure 6.18 have acted both as master and slave during data synchronisation process and all the resulting link state requests (LSRs) have been fulfilled, then the two routers are said to have formed an adjacency relationship. The adjacency relationship exists once both routers of the new adjacency relationship have both sent and received DD packets with the M-bit (more packets follow) set to '0' and had these packets acknowledged.

Database description (DD) packet format

The format and fields of the DD (database description) packet are illustrated in Figure 6.17. Briefly, the fields of the DD packet indicate:

- the Interface MTU (maximum transmission unit) — the value in this field indicates the maximum size of IP (Internet protocol) packet which can be sent over the interface without requiring *fragmentation*;

- the options field describes which *OSPF options* are supported by the router. We describe the options and the format of this field in more detail later;

- the *I* (*initialise*)-bit is set to value '1' to indicate the first of a series of DD packets;

- the *M* (*more packets* follow) bit is set to value '1' when more DD packets will be sent. In this case, the DD-sequence number is also used to identify the individual DD packets; and

- the *MS* (*master*)-bit, when set to value '1' indicates the packet was generated by the master. The slave sets this bit to value '0'.

The link state database and the calculation of the routing table

The link state database is the collection of the full set of link state advertisements (LSAs). Taken together, the LSAs describe the entire topology of the network.

Having undertaken the process of synchronisation with each of its adjacent routers, a router has a complete copy of the link state database and is in a position to calculate its routing table to each reachable destination. The calculation uses the shortest path first (SPF) algorithm (otherwise known as the Dijkstra algorithm). First the shortest path tree (SPT) is calculated (like Figures 6.13 and 6.14). From this, the routing table is easily derived. Packet forwarding and 'normal' operation of the router can then commence.

Normal network operation: maintaining the neighbour database, link state database and routing table

In normal operation, an OSPF router has established and will maintain three separate databases:

- the *neighbour database* — with details of the identities, router priorities, IP-addresses, designated routers (DRs) and back-up designated routers (BDRs);

- the link state database — containing all the link state advertisements (LSAs): this represents a complete 'map' of the current network topology; and

- the *routing table* — calculated using the shortest path first (SPF) algorithm and the link state database.

For correct normal operation of the router, each of the above databases must be kept up-to-date. This involves four continuing duties:

- receiving new routing updates (in the form of new *link state advertisements*) from other routers;

- *flooding* link state advertisements (LSAs) to all outgoing OSPF interfaces (i.e. via *adjacent routers* to all other routers in the routing domain or area) to inform other routers of routing updates;

- regularly repeating *hello packets* to ensure that other routers are aware that the router is still in operation; and

- recalculating the routing table as made necessary by routing updates.

When a router receives a routing update (correctly called a *link state update, LSU* — see Figure 6.19), it inserts any new or updated LSAs (Figure 6.20) into its link state database. It will only receive such updates from its adjacent routers. Should the update prove, according to a set of pre-determined criteria to represent a topology 'change' (this is called a route *discovery*) then the router will flood the update to all its OSPF neighbours (in some cases, even sending it back across the interface on which it learned the update). In this way, the new information 'cascades' to all OSPF routers in the routing *domain* or *area* by means of the flooding process.

Hello packets must be sent by every active OSPF router to its adjacent routers at least once per hello interval. This forms part of the ageing process. The hello packet confirms that the router is still active — and by inference, that all its links are still active and the link state database as last updated is still current. The repeated hello packets also, of course, help routers newly introduced to the network to discover their neighbours. If necessary, they also

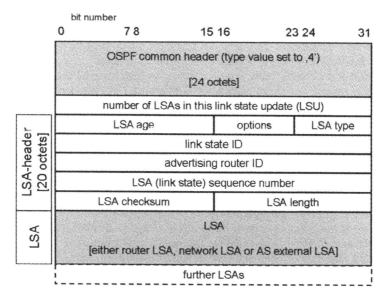

Figure 6.19 Format of OSPF link state update (LSU) packet and the link state advertisement header (LSA-header).

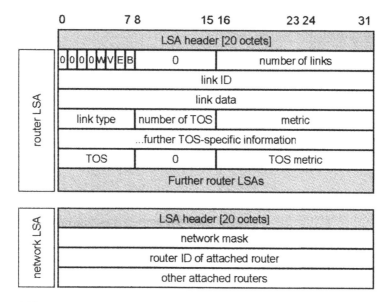

Figure 6.20 OSPF link state advertisement (LSA) formats: router LSAs and network LSAs.

allow the re-election of designated routers (DRs) or back-up designated routers (BDRs) should these fail.

Should a hello packet not be received from a neighbouring router within the hello interval then all LSAs relevant to this router will be immediately aged (in effect marked invalid and no longer used for route calculation). Should the router dead interval also expire without any further hello packet, then the database entries will be completely deleted.

A further important component of the *ageing* process in OSPF is that all LSAs are associated an *age* timer. When the age timer expires (typically after 30 minutes), the LSA is no longer used for route table calculation. This demands that the LSA-originating router repeat all LSAs (i.e. re-flood them to the network) at least once every 30 minutes.

Recalculation of the routing table is undertaken after each material change in the link state database. The complexity of the re-calculation is minimised by considering only the changes to the shortest path tree and routing table which need to be undertaken.

The flooding process

A *link state algorithm* such as OSPF relies on each router maintaining an up-to-date copy of the link state database, and thus upon knowing at all times the full topology of the routing *domain* or routing *area*. Any changes in domain or area topology are required to be *updated* to all other routers in the domain (or area) by means of *link state updates (LSUs)* containing link state advertisements (LSAs—Figure 6.19). The notification takes place by a process known as *flooding*. The end result is similar to an IP (Internet protocol) multicast or broadcast, but despite the fact that the process makes use of an IP multicast address, it nonetheless operates somewhat differently.

The critical difference between *flooding* and normal IP multicasting, is that OSPF packets are processed by each successive OSPF router before being *flooded* (on all outgoing OSPF interfaces) to all adjacent routers. In this way, *link state updates (LSUs)* can be thought of as 'cascading' through the network. The use of flooding (as opposed to normal multicasting) regulates the updating process and ensures, in particular, that each link state advertisement (LSA) in each update is *acknowledged* by each router as having been received. It is thus a more secure way of keeping all the link state databases synchronised than straightforward multicasting would be.

The format of link state updates (LSUs) and link state advertisements (LSAs)

Link state updates (LSUs) are generated by routers or designated routers (DRs) and flooded via adjacent routers to all other routers in the same routing area or domain. They are sent by a router, either:

- Because the router has detected a network topology change; or

- To 'refresh' previously sent link state advertisements (LSAs), thereby preventing those LSAs from being aged and deleted by other routers; or

- In response to a link state request (LSR)—which is part of the initial data synchronisation process of a router new to a network.

A link state update (LSU) contains a number of link state advertisements (LSAs) as shown in Figure 6.19. Each LSA (the different types of which are listed in Table 6.8) must be separately acknowledged with an *LSack (link state acknowledgement)* message. The format of *router-LSA* and *network-LSA* packets is illustrated in Figure 6.20. The *LSack* message has a similar format to the LSU message (Figure 6.19) except that the 4 octets 'number of LSAs in this message' is omitted, as are the main part of the LSAs (only the LSA-header is sent). In an LSack message, the OSPF type in the OSPF common packet header is set to value '5' (Table 6.6).

The meanings and coding of the various fields in LSU- and LSA-headers (Figure 6.19) are as follows:

- *LSA age* — this is the elapsed time since the origination of the link state advertisement. The time increases until the maximum age, whereupon the LSA is aged and no longer used for routing table calculation;

- *LSA type* — this field is coded according to the type of link state advertisement as listed in Table 6.8;

- *link state ID* — this field identifies the link described by the LSA. The exact format of the link-state-ID depends upon the *LSA-type* as listed in Table 6.8;

- *advertising router-ID* — this is the router-ID of the router which generated the LSA;

- *LSA sequence number* — this distinguishes an LSA from a previous version of the same LSA;

- *LSA checksum* — this applies across the entire LSA (including the header but not the LSA age field);

- *LSA length* — the length of the LSA (including header).

The meanings and coding of the various fields in the router-LSA and network-LSA messages (Figure 6.20) are as follows:

- *W-bit (wildcard bit)* — when set to value '1' this represents a *wildcard multicast receiver* (as defined for multicast OSPF in RFC 1584);

- *V-bit (virtual link bit)* — when set to value '1' the router issuing the LSA is connected by means of a *virtual link* (Figure 6.22). At least one of the routers in area 5 of Figure 6.21 will be connected by means of a virtual link to a neighbour router in area 0, since no direct physical link between the areas is available (and each area must have a 'direct' link of some sort to area 0 — as we shall see shortly);

- *E-bit (external bit)* — when set to value '1' the router issuing the LSA is an *autonomous system boundary router* (*ASBR* — Figure 6.21);

- *B-bit (border bit)* — when set to value '1' the router issuing the LSA is an *area border router* (*ABR* — Figure 6.21);

- *link type* — this describes the type of router interface, coded according to Table 6.9. The *link-ID* and *link-data* fields are coded accordingly (Table 6.9);

- *metric* field — this relays the link cost (as assigned by the router to its associated link). As we discussed earlier in the chapter, the value is usually related to a combination of the link bit rate, delay, load and reliability. A typical default value is 1000 million divided by the link bit rate;

- (if included) the various *type-of-service (TOS)* fields include information about the suitability of the link for different types of service as we discussed under *differential services (DiffServ)* in Chapter 5.

When flooded through a network, link state updates (LSUs) are usually addressed to the following IP addresses:

- designated routers (DRs) and back-up designated routers (BDRs) are addressed by means of the multicast address 224.0.0.6 (*all OSPF designated routers* multicast);

- all other routers reachable by means of a multicast network are addressed by means of the multicast address 224.0.0.5 (*all OSPF routers* multicast);

Table 6.8 OSPF link state advertisement (LSA) types

LSA type value	LSA type	LSA Purpose	LSA generated by	Link state ID	LSA flooded to
1	Router LSA (Intra-Area)	Describes router links	All OSPF routers	Router ID of LSA-generating router	Immediate routing area
2	Network LSA (Intra-Area)	Describes broadcast or NBMA network links	Designated routers (DR)	Interface IP-address of designated router	Immediate routing area
3	Summary-LSA (Inter-Area network)	Describes one OSPF routing area to another	Area border routers (ABR)	IP-address of target network	Areas throughout the entire AS
4	Summary-LSA (ASBR) (Inter-Area)	Describes routes to external ASs (i.e. to the relevant ASBR)	Autonomous system boundary routers (ASBR)	Router ID of the ASBR router	Areas throughout the entire AS
5	AS external LSA (External Type)	Describes a route within an external AS	Autonomous system boundary routers (ASBR)	IP address of the external AS	Immediate routing area
6	MOSPF (multicast OSPF) group membership LSA	Describes multicast group membership	Multicast routers	IP address of the multicast group	Immediate routing area
7	NSSA (not so stubby area) external LSA	Used as alternative to LSA-type 5 in NSSA areas	Autonomous system boundary routers (ASBR)	IP address of the external AS	Only in NSSA areas
9	Opaque LSA	Opaque LSAs are intended for future extension of OSPF	—	8-bit *opaque* type and 24-bit *opaque*-ID	Only in the local subnetwork
10	Opaque LSA	Opaque LSAs are intended for future extension of OSPF	—	8-bit *opaque* type and 24-bit *opaque*-ID	Immediate routing area
11	Opaque LSA	Opaque LSAs are intended for future extension of OSPF	—	8-bit *opaque* type and 24-bit *opaque*-ID	Throughout the entire AS

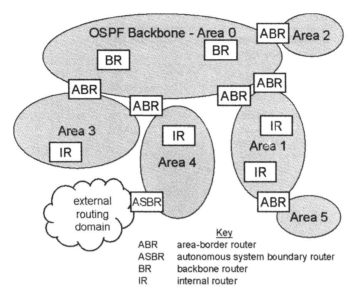

Figure 6.21   OSPF routing areas and router types.

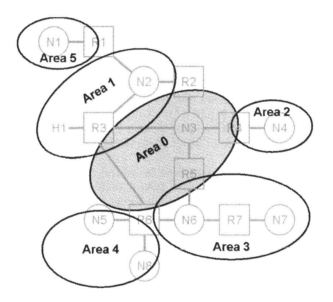

Figure 6.22   Example segregation of an OSPF domain into separate areas (example of Figure 6.11).

- other routers which cannot be addressed by means of a multicast are separately addressed by means of a unicast message. Replies to link state requests (LSRs) are also unicast in this manner, as are unacknowledged *retransmissions*.

Hierarchical network and routing structure of OSPF networks

OSPF has become the *interior gateway protocol (IGP)* 'of choice' because it is much more efficient and effective in dealing with routing around large complex networks than *RIP (routing*

Table 6.9   OSPF router-LSA: link type, link-ID and link-data field coding

Link type value	Link (i.e. Router interface) type	Link ID	Link data
Type 1	Point-to-point connection	Router-ID of neighbour router	IP-address of the interface (or MIB-2 interface index value)
Type 2	Connection to a broadcast or NBMA network with multiple remote routers	IP address of designated router	IP address of the interface
Type 3	Connection to a broadcast or NBMA network with only one remote router	IP network address	IP network mask
Type 4	Virtual link	Router-ID of neighbour router	IP address of the interface

information protocol) and other routing protocols which pre-dated it. One of the reasons why OSPF is suited to being used in large networks is the efficiency of the protocol and routing algorithm. But another important capability which comes with OSPF is the ability to segregate a very large routing domain into a number of smaller routing areas (Figure 6.21).

By splitting an OSPF routing domain into multiple *areas*, a network administrator is able to reduce the load on both the network and the OSPF protocol, since link state advertisements (LSAs) are only flooded within the local area and not across the domain as a whole. The domain takes on a hierarchical nature, with a *backbone area (area 0)* and all other areas attached to this *backbone* (either directly or by means of a *virtual link*). The task of determining routes through the network now reduces to maintaining 'full mesh' routing within each individual area (so-called *intra-area routing)*, and establishing *inter-area routes* via pre-selected *area border routers (ABRs)* as shown in Figure 6.21. *ASBRs (autonomous system boundary routers)* meanwhile are the routers within the routing domain which are directly connected to other *autonomous systems (AS —* i.e. other routing domains). Different *OSPF processes* deal with these various tasks.

OSPF areas are defined to fall into four basic categories: the backbone area (area 0 — there can be only one backbone area), *stub areas* (which have no connection to *external* routing domains), *totally stubby areas* (which do not accept routing updates from any *external* areas — OSPF or otherwise) and *not so stubby areas (NSSAs —* which accept some types of LSAs from *external* autonomous systems). The different types of areas have slightly different routing update needs, which are reflected in the different types of LSAs (Table 6.8). The *LSA-type* field is used to control the LSA-flooding process through the area and across the different ABR and ASBR routers as appropriate.

It is a requirement that each OSPF area has at least one *ABR (area border router)*, and that at least one of its ABRs is either directly connected to the backbone area (area 0) or connected by means of a *virtual link*. In the example of Figure 6.21, the *area border router (ABR)* of area 5 requires to be connected by means of a virtual link transmitting area 1. The ABR will treat the virtual link as if it were a point-to-point connection to the *backbone router* (in our example one in area 0 and bordering area 1).

The structuring of the areas in an OSPF routing domain is a planning task requiring careful thought. The more areas which are created, the more hierarchical the network will be. Remember, since all the routes go only via the pre-determined ABRs and ASBRs, so does all the network traffic (i.e. IP forwarding traffic). It is not just the OSPF routing network which is made hierarchical, it is the IP-forwarding network as well! Thought out carefully, the network hierarchy can be a major benefit — helping to 'stream' the network traffic flows across particular chosen routes and routers. Figure 6.22 illustrates a possible OSPF area scheme for the previous example of Figure 6.11 (although it is unlikely that the segregation into separate areas is really necessary for so few routers).

Route redistribution in OSPF

Real IP-networks nearly always comprise a number of different routing *domains*, using different *interior gateway protocols (IGPs)* as well as the *border gateway protocol (BGP)* between the domains. This leads to the need for *route redistribution* — feeding routing and link *cost* information learned from one routing protocol into the routing table calculations and information exchanges of a different routing protocol. Naturally, route redistribution of routing information inserted into or extracted from OSPF by other routing protocols is possible as is a 'weighting' or 'scaling' of the link cost value. But in general, the philosophy of OSPF is to keep internally learned *link state* data separate from that learned from exterior sources, e.g. that derived from BGP or another RIP. The separation is maintained by means of the different LSA-types (Table 6.8).

OSPF options

You will recall that OSPF hello packets, database description (DD) packets and link state advertisements (LSAs) contain an options field, which we have so far not explained. This field indicates whether the router generating the message is capable of supporting various extensions to the OSPF protocol as listed in Figure 6.23. The options, as other parameters in the OSPF hello packet, may be negotiated between neighbouring routers when establishing the adjacency relationship.

OSPF summary

By virtue of being a *link state routing protocol*, OSPF (open shortest path first) is an efficient and robust interior gateway protocol (IGP), suitable for the control of routing in large complex networks. It uses a single link cost metric and a shortest path tree for calculating the shortest path to a destination, but the link cost can be made dependent upon a number of other link attributes (e.g. bit rate, delay, load and reliability etc.).

0	O-bit	DC-bit	EA-bit	N/P-bit	MC-bit	E-bit	T-bit

KEY

DC-bit	OSPF extension for demand circuits (RFC 1793)
E-bit	stub (external) area flag
EA-bit	OSPF external attribute LSA
MC-bit	Multicast extensions for OSPF (RFC 1584)
N/P-bit	Not so stubby area LSA (RFC 1587)
O-bit	OSPF opaque LSA option (RFC 2370)
T-bit	Type & quality of service OSPF extensions (RFC 2676)

Figure 6.23 The OSPF options field (contained in hello, DD and LSA messages).

Each router in an OSPF network stores information about the complete topology of the network in its link state database, and uses this as the basis of the route calculation. Because this database is directly available to each router, the calculation of routes is very efficient. As a result, the possibility of *circular routing* and other *routing instabilities* is much reduced and there is the possibility of traffic *load sharing* across equal cost paths. Furthermore, the routing protocol itself need only be used to update routers of changes in the overall topology. It is not (as in RIP) necessary for each router to keep advertising its entire routing table every 180 seconds. OSPF link state advertisements (LSAs) typically need only be re-flooded once every 30 minutes.

Although the link state advertisements (LSA) of OSPF have to be advertised to all routers within an OSPF routing domain or area (the process is known as flooding), each individual router only 'talks' with the direct neighbours with which it has an adjacency relationship. For this purpose, OSPF routers must maintain a close understanding of their neighbours, which are automatically discovered and kept in constant close touch by means of the hello protocol. The hello protocol caters for the initial neighbour discovery when a new router is added to a network and for the negotiation of network operation parameters and routing service capabilities. In addition, by noticing the lack of the neighbours' hello packets (which should be regularly repeated) a router can spot the failure of a neighbour and trigger commensurate routing changes.

The flooding process of distributing information about the network topology changes in link state advertisements (LSAs) works by 'cascading' messages progressively from one *adjacent neighbour* to the next. By cascading in this way, the different types of link state updates (LSUs) can be sorted and forwarded appropriately, and each router individually can be made to acknowledge receipt of the update. In this way each router can be assured of having a complete and up-to-date link state database.

OSPF, like other link state protocols, is a fast converging protocol. In other words, routing tables throughout the network reach their stable and 'fully calculated' state quickly after either the introduction of a new router to the network or after the recovery of a router from a fault. Only three predictable actions need to take place to reach *convergence*:

• the link state database of the 'new' router must be established. This is done by copying the database from one of its neighbours in a process called data synchronisation which takes place as part of the hello protocol and during the forming of the adjacency relationship;

• the change in the network topology (resulting from the introduction of the 'new' router) is recorded in a link state advertisement (LSA) which is flooded to all other routers in the routing domain or area by means of the routing protocol. As a result of this advertisement all routers' link state databases are brought up to date about the 'new' router;

• all routers recalculate their routing tables based on the updated link state database.

The convergence process is achieved with only one re-calculation of each router's routing table. This is in stark contrast to the slow convergence which can be encountered with routing protocols *like RIP (routing information protocol)*. *The* problem with the calculation of routing tables in RIP is that each neighbour needs to recalculate its routing table each time it receives a routing update from its neighbour. The recalculation can lead in turn to a fresh advertisement back to the neighbour, which in turn leads to a recalculation of the routing table, and so on. In complex networks, the convergence process of RIP is slow and of unpredictable duration!

Like most other IP-suite protocols, the OSPF routing protocol is equally well suited to use in IPv6 networks as well as in IPv4 networks, although it will require adaptation to cater for the much longer address fields demanded by IPv6 addresses.

6.14   BGP4 (border gateway protocol version 4)

The *border gateway protocol (BGP)* is designed to provide for loop-free *inter-domain routing* between autonomous systems (ASs). In other words, it is an *exterior gateway protocol (EGP)*. The current version of BGP is version 4 (BGP4). It is defined in RFC 1771. An autonomous system (AS), otherwise known as an administrative domain, is a homogeneous router network typically owned and administrated by a single network operator.

In order for the routing between two different *autonomous systems (ASs)* to be controlled by BGP, at least one router in each system must be configured to *speak* BGP. BGP *speakers* exchange *reachability* information by means of the border gateway protocol (BGP) — in the case of speakers in <u>different</u> autonomous systems, the BGP is termed *exterior border gateway protocol (EBGP)*. All *reachable* destinations are identified by means of their AS number, the IP address-ranges associated with the destination AS and the complete path to the destination AS from the BGP router advertising the route. The path is described as a list of networks (i.e. autonomous systems) which must be transited along the way. In addition, a list of path attributes is also provided. This identifies the types of services which can be carried and any limitations on the use of the path.

The route calculation and path decision process undertaken by BGP will select the single shortest path to the destination, the shortest path being determined as that 'requiring to transit the fewest intermediate autonomous systems (ASs)'. BGP is thus a *distance vector routing protocol (DVP)* — but one much more sophisticated than mere *hop count*-based DVPs. By virtue of the fact that the entire transit path is identified in the routing *update* and used in the path calculation, BGP suffers neither from circular routing problems nor from slow convergence in the way that RIP (routing information protocol) and other mere hop-count-based DVPs do.

Unlike routers using interior gateway protocols (IGPs), *BGP-speaking* routers are not said to have neighbours but instead have *BGP peers*. This is because BGP routers need not be directly connected to one another. In the case of BGP peer routers interconnected across the boundary between two (exterior) autonomous systems it is normal for the BGP peer routers to be directly interconnected. However, provided the EBGP router can reach all relevant addresses without the help of a second routing protocol (e.g. by using static routing), it is also possible for EBGP peers not to be directly interconnected.

Apart from using BGP for *speaking* to BGP peers exterior to the 'home' autonomous system to *learn* about *reachable* destinations outside the AS (this usage is termed *exterior border gateway protocol, EBGP*), it is also common to use BGP between different boundary routers within the same AS (autonomous system) in cases where the AS is to be used as a transit system between other ASs. (The example of Figure 6.6, for instance, might allow transit from 'external AS 1' to 'external AS 2'). In this usage (*called interior border gateway protocol, IBGP*), BGP provides an efficient manner of transferring *exterior* routing information across the AS in line with the local administrative preferences laid out in the *BGP policy*. IBGP peer routers are rarely directly connected to one another.

Instead of using IBGP between the two EBGP routers in the same AS, it would also be possible to use an interior gateway protocol (IGP) and *redistribute* (i.e. convert) the exterior routing information from BGP into the IGP and then back again. But using IBGP provides a more powerful and flexible solution — avoiding the loss of BGP-learned path information which would result if 'back-to-back' route redistribution were to be undertaken.

To provide the scope for reliable interconnection of BGP peers without the constraint of direct interconnection, the BGP protocol employs the *transmission control protocol (TCP)* and TCP port 179.[9]

---

[9] See Chapter 7.

a) BGP common message header

marker	16 octets
length	2 octets
type	1 octet

b) BGP open message

BGP common header	19 octets
version	1 octet
my autonomous system (AS)	2 octets
hold time	2 octets
BGP identifier	4 octets
optional parameter length	1 octet
parameter type	1 octet
parameter length	1 octet
parameter value	variable
further parameters	

Figure 6.24   BGP common message header and open message format.

a) BGP update message

BGP common header (type=2)	19 octets
unfeasible routes length	2 octets
withdrawn routes	variable
total path attribute length	2 octets
path attributes	variable
NLRI (network layer reachability information)	variable

b) BGP path attributes

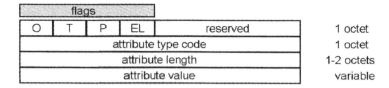

	flags				
O	T	P	EL	reserved	1 octet
attribute type code					1 octet
attribute length					1-2 octets
attribute value					variable

Figure 6.25   BGP update message format and path attributes.

There are three types of messages defined by BGP: *open, update* and *notification/keepalive* messages. The various formats of these message types are illustrated in Figures 6.24, 6.25 and 6.26. Their functions are described next.

BGP common message header

All BGP message types are preceded by the *BGP common message header* as illustrated in Figure 6.24a. The fields are coded as follows:

BGP notification message

BGP common header (type=3)	19 octets
error code	1 octet
error subcode	1 octet
data	variable

Figure 6.26 BGP notification message format.

Table 6.10 BGP message types

BGP message *type* value	BGP Message type
1	Open message
2	Update message
3	Notification message
4	Keepalive message

- the *marker* field is either set with all bits of value '1' or alternatively contains authentication information;

- the *length* field indicates the length of the BGP-packet in octets (including header). Allowed values are 19–4096; and

- the *type* field value indicates the type of BGP message and therefore the format of the remainder of the message following the BGP common header. The permitted type values are listed in Table 6.10.

BGP open message

A *BGP open* message is used to negotiate the mutual configuration of BGP peer routers after establishment of the TCP connection. On acceptance of the BGP open message (opening the connection and establishing the BGP peer relationship), the router sends a *keepalive* message.

The meaning and coding of the various fields of the BGP open message (Figure 6.24b) are as follows:

- the *version* field identifies the version number of BGP in use (the current version is version 4);

- *my autonomous system* is the AS number of the network of which the transmitting BGP speaker is a part. (AS numbers are allocated by IANA (Internet assigned numbers authority) and may be looked up on www.iana.org);

- the *hold time* is the time period for which a BGP peer connection will be held (i.e. considered active) even if no BGP message is received within this period. Should the hold time expire, the remote BGP speaker will be considered to have failed or otherwise been removed from service. This is the method used for ageing BGP learnt routing information (we discussed ageing earlier in the chapter);

- the *BGP identifier* is a unique identifier for the BGP-speaking router. The same value is used in all BGP peer relationships of a given router; and

- the *parameter* fields provide the potential for optional services to be added to the BGP *open message*. Parameter type 1 corresponds to the use of authentication information.

BGP update message

Once a *BGP peer* relationship has been *opened*, the BGP peers exchange their complete routing tables by means of *BGP update* messages. Subsequent BGP update messages are then sent to advise of changes in the topology or an increase in the number of reachable destinations. Like other routing protocols, regular repeats of each update are also sent to prevent routing information from ageing and being deleted.

The format of a BGP update message is shown in Figure 6.25a. Each individual update describes the path to reach a given remote autonomous system (AS), the nature and attributes of the path (i.e. which services it is capable of carrying, limitations, etc.) and a list of IP address ranges which can be reached in the target AS. The meaning and codings of the various fields in a BGP update message are as follows:

- the *withdrawn routes* field comprises a list of IP address ranges which may no longer be reached in the AS;

- the *unfeasible routes length* field merely indicates the number of octets making up the withdrawn routes field;

- the *network layer reachability information (NLRI)* is a list of the IP address ranges which can be reached via the BGP router transmitting the update by means of the path described in the *path attributes* field. Taken together with the path attributes field, the NLRI represents one or more entries in the routers routing table; and

- the *path attributes* field describes the route to a particular destination (or group of desti-nations as defined in the NLRI field) as it appears in the routers routing table. Normally the entire route (a list of transit autonomous systems, ending with the destination AS) will be listed, but it could be that only the *next hop* is listed (depending upon the coding of the path attributes fields — Figure 6.25b).

The BGP path attributes field is formatted as illustrated in Figure 6.25b and coded according to Tables 6.11 and 6.12. Attribute *flags* (O, T, P and EL) are set according to Table 6.11. *Well-known attributes* (flag bit O = '0') must be recognised by every *BGP-speaker*. Certain *well-known attributes* are mandatory and must always be included. These include the ORIGIN, NEXT_HOP and AS_PATH fields as set out int Table 6.12. Optional attributes (bit O = 1) need not be recognised by the BGP speaker, though the related information should be forwarded to other BGP peers if appropriate. Optional attributes are further divided into *transitive* and *non-transitive* types (flag bit T). Information received about transitive type attributes is forwarded to other peers, but non-transitive type attributes should not be forwarded.

Table 6.11 BGP attribute flags in path attributes field

Flag abbreviation	Flag name	Meaning	
		Bit value set to '0'	Bit value set to '1'!
O	Optional bit	Well-known attribute	Optional attribute
T	Transitive bit	Non-transitive attribute	Transitive attribute
P	Partial bit	Complete	Partial
EL	Extended length bit	1 octet attribute length field (Figure 6.25b)	2 octet attribute length field (Figure 6.25b)

Table 6.12 BGP path attributes

Path attribute	Attribute type code value	Attribute length field value	Possible Attribute values
ORIGIN	1	1	0 = IGP (*interior* destination is in local AS) 1 = EGP (*exterior* path learned by means of EGP 2 = path or destination learned otherwise (e.g. by route redistribution)
AS_PATH	2	Variable	Path segment type value as follows: 1 = AS_SET: unordered list of transit ASs 2 = AS_SEQUENCE: ordered list of transit ASs 3 = AS_CONFED_SET: unordered list of transit ASs within the local BGP confederation 4 = AS_CONFED_SEQUENCE: ordered list of transit ASs within the local BGP confederation
NEXT_HOP	3	4	IP address of the border router which forms the next hop of the path
MULTI_EXIT_DISC	4	4	This value serves to distinguish between multiple possible boundary router connections to a neighbouring AS. The route with the lowest MULTI_EXIT_DISC will be preferred
LOCAL_PREF	5	4	This field is used within an network (i.e. in IBGP) to inform another external BGP router in the same AS of the administratively preferred route to a given destination.
ATOMIC_AGGREGATE	6	–	This field indicates to other BGP speakers that several overlapping routes have been combined into a single summary route.
AGGREGATOR	7	6	The AS number and IP address of a BGP speaker which conducted aggregation of this route.
COMMUNITIES	8	4	The BGP COMMUNITIES attribute allows the speakers of multiple routes to be combined into a single community. This simplifies the creation and administration of BGP routing policy. Destination communities may then be classified into types like NO_EXPORT (route may not be made known to external ASs) or NO_ADVERTISE (route may not be advertised to other BGP peers). The BGP communities attributes is defined by RFC 1997.
ORIGINATOR_ID	9	4	The ORIGINATOR_ID attribute is related to BGP route reflection (defined by RFC 1966)
CLUSTER_LIST	10	Variable	The CLUSTER_LIST attribute is related to BGP route reflection (defined by RFC 1966)

The *partial (P) bit* indicates whether the list of path attributes is complete or not, or whether further attributes follow. The *extended length (EL) bit* indicates whether the attribute length field (Figure 6.25b) is 1 octet or 2 octets long.

Table 6.12 lists the path attributes used by BGP to describe routes. The ORIGIN, NEXT_HOP and AS_PATH fields are mandatory fields. This information must always be provided. Together with the *NLRI (network layer reachability information)* field in the BGP update message, it enables complete routing table information to be advertised to BGP peers for their calculation of the best route. The best route calculation is then undertaken according to the local *BGP policy* and the selection of the shortest available route (least number of AS transit hops) to the destination. BGP does not support load balancing across shortest routes of equal 'length'.

BGP notification message

BGP notification messages are used to notify protocol errors which may occur during a BGP connection. Should such an error occur, then the connection is cleared immediately after sending the notification message. The format of BGP notification messages is as illustrated in Figure 6.26. Message meanings and field coding are according to Table 6.13.

BGP keepalive message

BGP *keepalive* messages have a similar function to the *hello packets* of OSPF. The *keepalive* message informs the peer BGP router, that despite not having sent a routing update message, the router is still 'alive' and fully operational — and that all previously advertised reachable destinations are still available. This prevents routing information from being aged and therefore deleted and removed from routing table calculations. The keepalive message consists merely of the BGP common header (Figure 6.24a, with the type field set to value '4').

BGP route decision-making and BGP policy

Route calculation and decision-making under BGP are according to the shortest route. The shortest route is that with the smallest number of autonomous system (AS) transit hops. No load

Table 6.13 BGP notification messages: meaning and coding of error code and error subcode

Error message category	Error code	Error message	Error subcode
Message header error	1	Connection not synchronised	1
		Bad message length	2
		Bad message type	3
Open message error	2	Unsupported version number	1
		Bad peer AS	2
		Bad BGP identifier	3
		Unsupported optional parameter	4
		Authentication failure	5
		Unacceptable hold time	6
Update message error	3	Malformed attribute list	1
		Unrecognised well-known attribute	2
		Missing well-known attribute	3
		Attribute flags error	4
		Attribute length error	5
		Invalid ORIGIN attribute	6
		AS routing loop	7
		Invalid NEXT_HOP attribute	8
		Optional attribute error	9
		Invalid network field	10
		Malformed AS_PATH	11
Hold timer expired	4		
Finite state machine error	5		
Cease	6		

sharing is possible across alternative paths of a similar shortest length. In the case of multiple paths being found to have the same 'length', the path with the lowest MULTI_EXIT_DISC value (see Table 6.12) will be chosen.

The network administrator can influence the choice of routes by BGP by setting up a *BGP policy*. The BGP *policy* affects which routing information received from BGP peer partners is considered in the routing table calculation and also affects which routes are *advertised*.

By ignoring routing information from certain sources (this is done by incoming *filtering*), the BGP router can be forced to use alternative routes. And by not advertising certain routes, the use of these routes by other parties can be restricted. Two forms of non-advertisement may be defined:

- NO_EXPORT will prevent routing updates from being advertised to external BGP peers (outside the BGP confederation);

- NO_ADVERTISE will prevent routing updates from being advertised to any BGP peers (this information will only be shared with interior gateway protocols (IGPs)).

BGP communities

A BGP *community* allows a single BGP policy to be applied to a whole group of different destinations. This makes for easier creation and maintenance of the policy. Each BGP community is classified according to whether routing updates may be advertised or not (i.e. as subject to NO_EXPORT, NO_ADVERTISE or Local AS only policies as explained above). An individual destination may belong to multiple BGP communities.

Management of BGP: route maps, route redistribution and route aggregation

A *route map* sets the conditions for *route redistribution* between different routing protocols. The route map consists of *set* and *match* commands which create a *defined condition*, according to which, routing information learned by one routing protocol (e.g. BGP or an IGP) will be transferred to the other. *Unmatched* routing information is not transferred.

Obviously some level of *route redistribution* is required in order that externally *reachable destinations* can be advertised to relevant internal routers and internal reachable destinations similarly notified to exterior routers. For this, the link cost or distance metrics of different *interior* and *exterior gateway protocols* need to be converted. To avoid the problems created by this conversion, BGP route redistribution is discouraged as far as possible. In particular it is recommended that BGP information relevant to a transit AS (autonomous system) not be redistributed to an IGP, but instead be transferred from one BGP router to another by means of *IBGP (interior border gateway protocol)*.

Route aggregation is the term used to describe grouping a number of IP address ranges together to share a single outgoing route. This is typically achieved by the use of a *static route*. The fact that such manual intervention might still be considered by network administrators reflects the desire to apply administrative 'preferences' to external routes rather than rely entirely upon destination reachability information automatically received using BGP. On its own, automatic route calculation by BGP is unable to take into account factors in route choice such as the financial cost or reliability of a given third-party AS (autonomous system) network.

BGP confederation

A *BGP confederation* presents a group of autonomous systems (AS) externally as if they were a single AS. This has the benefit of reducing the mesh of BGP peer relationships which might otherwise have to exist and enables a tighter control of routing policies.

Route reflection

When *BGP route reflection* is used, an autonomous system (AS) is divided into multiple areas called *clusters*, each cluster being assigned a CLUSTER_ID (Table 6.12). The route reflector *clients* (i.e. routers within the cluster) are only permitted to establish IBGP sessions with the *route reflector (RR)* of the cluster. In this way, the requirement for full-meshing between IBGP routers within an AS is reduced to only requiring a full mesh between the route reflectors (RR). Like BGP confederation, this simplifies the number of BGP connections which must be established, so leading to more easily manageable networks.

Route flap dampening

Route flapping is the term applied to unstable routes, in particular when a given destination or path to that destination is seen to oscillate between an 'available' and a 'down' state. Route flapping is undesirable because it can lead to instability of the network as a whole, and so BGP attempts to eliminate it using a technique called *route flap dampening*. In simple terms, BGP 'penalises' flapping routes by making them appear artificially 'longer' during the routing table calculation.

6.15 Problems associated with routing in source and destination local networks

So far in this chapter, we have learned how routers employ routing protocols between themselves to exchange routing information and so build routing tables for the forwarding of IP packets. Such mechanisms ensure that an IP packet can be delivered from a router in the source network to the router in the appropriate destination network. The destination router knows which networks are connected to it, using which type of physical interfaces (e.g. ethernet, serial line etc.). In addition, the router is aware which range of IP addresses are associated with each directly connected interface (because this information is configured manually into the router when the interface is set up). But our routing table is not quite complete! What the router may not yet know, is the layer 2 address (sometimes called *hardware address*) which is associated with a given IP address. Thus it could receive an IP packet destined for one of the IP-ranges which it recognises as being 'directly connected' but be unable to deliver the packet to the correct end-device because it does not know the LAN address (MAC address, IEEE address or hardware address) which corresponds with the destination IP address. To overcome the problem, the *address resolution protocol (ARP)* was devised.

Resolving a destination IP address for the associated hardware address

When the router in a given destination LAN receives an IP packet for an IP address which it can identify as being 'directly connected' by means of a given *shared medium* interface (typically a LAN) but does not already know the *hardware address* (i.e. layer 2 address, MAC address or IEEE-address) of the destination device, then it broadcasts an ARP request to all stations in the LAN (using the LAN or other layer 2 datalink protocol). The *ARP request* indicates the IP address for which a hardware address is being sought (the *target IP address*). The station which recognises its IP address (which in this case will have been manually configured in its 'network settings') replies to the *ARP request* with an *ARP response*. The ARP response packet replies with the *target IP address* and the related hardware address (Figure 6.27). This

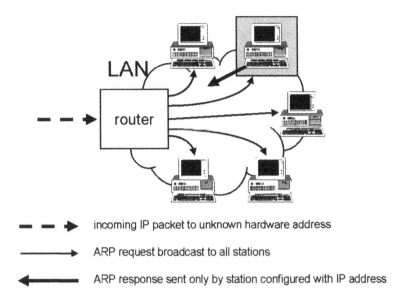

— — ➤ incoming IP packet to unknown hardware address

————➤ ARP request broadcast to all stations

◀———— ARP response sent only by station configured with IP address

Figure 6.27   Identifying the hardware address associated with a *target* destination IP address using ARP.

enables the router to build yet another entry in its routing table — the hardware address of a *directly reachable* destination IP address.

How a device wishing to communicate gets a source IP address and discovers the nearest router

Using routing protocols and *ARP (address resolution protocol)*, routers are able automatically to determine the best route to any reachable IP address. Nonetheless, communication using the IP protocol cannot commence until the sending device also knows its IP address and the *default gateway address* of the first router. The source IP address must be included in all sent packets, which in the first instance need to be forwarded by the router to the first router in the connection. This presents two problems: determining the IP address of a source host, and knowing which is the first router (i.e. *gateway*) to which this host should forward IP packets.

There are two basic ways in which to assign IP addresses to hosts or other end-user devices: either by manual configuration or by automatic assignment. Manual configuration is normally used in the case of permanently assigned addresses, while temporarily assigned addresses are normally automatically configured. There are various different protocols available for the automatic assignment of IP addresses by a router to a source host. The best-known of these are:

- *DHCP (dynamic host configuration protocol)*; and

- *BOOTP (bootstrap protocol)*; and

- *RARP (reverse address resolution protocol)*; and

- *ES-IS (end system-intermediate system)*.

The job of 'finding' the first router can similarly either be manually performed (by configuring a so-called *default gateway* IP address) or automatically (when the process is called *router discovery*).

Nowadays, DHCP (as defined in RFC 2131) has become the normal and most widely used method of dynamically and automatically assigning IP addresses to source stations and hosts for temporary periods and of router discovery. DHCP itself is really just a further development and refinement of the *bootstrap protocol (BOOTP)*. Before DHCP, BOOTP was originally intended to assign an IP address to a host being newly booted (i.e. switched on) and to deliver a *boot file* from the *BOOTP server* (typically a UNIX server in the past). The boot file includes the *default gateway* address and other network settings. One of the disadvantages of BOOTP is the continuing need for manual updating of the boot file. DHCP extends the capabilities of BOOTP by arranging for automatic generation of configuration settings and the control of a much wider range of IP-suite protocol parameters.

Once the IP address of the gateway (i.e. first router) is known, *ARP (address resolution protocol)* may need to be used to derive the hardware address (i.e. the MAC address) of the router. Following this, IP packets generated by the host can be forwarded to the first (*default gateway*) router, from where they can be successfully forwarded to their destination.

Each host (e.g. PC in a LAN) is typically configured nowadays as a *DHCP client*. The 'network settings' of the PC are configured to receive an 'automatic IP address assignment' and to discover the 'gateway IP address'. Meanwhile, the access router in the LAN acts as the *DHCP server*. Dynamic allocation of IP addresses in this way makes for much easier administration of IP addresses. It is much easier to move or sub-divide subnetwork ranges of IP addresses to different parts of a network — without having manually to reconfigure the IP addresses in each individual host. In addition, the temporary assignment of IP addresses alleviates to some extent a worldwide shortage of IPv4 addresses.

Reverse ARP (RARP) is an alternative to BOOTP or DHCP. As the name suggests, it is an adaptation of *ARP (address resolution protocol)* to allow the process to work 'in reverse'. Rather than wishing to resolve a *target IP address* for the corresponding hardware address (as ARP does), hosts using RARP already know their own hardware address but need to request the assignment of an IP address.

Router discovery

The *IRDP* protocol (*ICMP router discovery protocol*) in IPv4 or *neighbour discovery* in IPv6 (RFC 2461) provides a means for hosts automatically to discover routers within their local network. By making use of such automatic discovery, hosts need no longer be manually configured with a static *default gateway* routing table entry. But while IRDP allows for the automatic discovery of neighbouring routers, it does not allow the host to determine which of possible multiple routers represents the best route to a given destination. For this purpose there are two alternatives available to the network administrator:

- implement a static default gateway route after all; or

- configure the host *silently* to *listen* to routing information broadcasts of the neighbouring routers using *interior gateway protocols (IGPs)* such as *RIP (routing information protocol)* (We discussed earlier in the chapter how RIP-2 specifically includes a multicast address for this purpose).

ES-IS (end system-intermediate system) protocol

ES-IS is another protocol designed for router detection and address resolution. It was developed principally as the protocol to be used in conjunction with the OSI (open systems interconnection) *connectionless network service (CLNS)*.

ARP (address resolution protocol), RARP (reverse ARP) and inARP (inverse ARP) message format

Much of the functioning of the *address resolution protocol (ARP)* can be understood merely by inspecting the message format (Figure 6.28) and understanding the field codings:

- the *hardware type* is the type of network interface for which a hardware address is sought (Table 6.14);

- the *protocol type* has the value 08-06 for the Internet protocol (IP) in the case of an ethernet LAN;

- the *hardware length* field indicates the length of the hardware address;

- the *protocol length* field indicates the length of the protocol type field;

- the *operation* field indicates the ARP message type (Table 6.15); and

- the remaining fields are IP address and hardware fields of the sender and *target* (i.e. destination).

ARP (address resolution protocol) is used to discover the hardware address (i.e. layer 2 address, MAC-address or IEEE address) associated with a given *target IP address*. It is defined by RFC 0826. Messages are broadcast in the format of Figure 6.28 within the IEEE 802.2 frame (LAN or other layer 2 datalink frame).

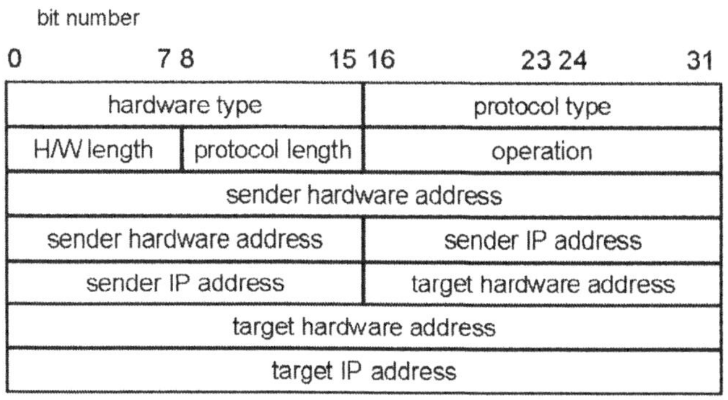

bit number

0	7 8	15 16	23 24	31
hardware type		protocol type		
H/W length	protocol length	operation		
sender hardware address				
sender hardware address		sender IP address		
sender IP address		target hardware address		
target hardware address				
target IP address				

Figure 6.28 ARP (address resolution protocol) message format.

Table 6.14 ARP (address resolution protocol), RARP (reverse ARP), inverse ARP (inARP), BOOTP (bootstrap protocol) and DHCP (dynamic host configuration protocol): hardware type field coding

Network hardware type	Network hardware type code value
Ethernet (pre-IEEE 802)	1
IEEE 802-networks	6
Frame Relay	15
ATM (asynchronous transfer mode)	19

Table 6.15   ARP (address resolution protocol), RARP
(reverse ARP) and inverse ARP (inARP):
operation field coding

Message type	Operation field code value
ARP request	1
ARP response	2
RARP request	3
RARP response	4
Inverse ARP (inARP) request	8
Inverse ARP (inARP) response	9

RARP (reverse address resolution protocol) is an adaptation of ARP used for triggering the assignment of an IP address to a given source host which already knows its own hardware address. The message format is the same as ARP (Figure 6.28 and Tables 6.14 and 6.15). Like ARP request messages, RARP request messages are also broadcast to all stations on the LAN, but using the *LLC (logical link control) protocol type* 80-35. RARP is defined in RFC 0903.

A further variation of ARP is the *inverse address resolution protocol (inARP)*. Somewhat like RARP, inARP is used to request a protocol address corresponding to a given hardware address. Specifically, inARP is designed to be used in conjunction with frame relay stations, to discover the IP addresses of remote stations for which only the frame relay connection address (the DLCI — data link connection identifier) is known by the local router. This is important for the discovery of neighbouring routers and in the election of designated routers and back-up designated routers as we discussed earlier in the chapter. inARP is defined in RFC 2390.

BOOTP (bootstrap protocol) and DHCP (dynamic host configuration protocol)

The *bootstrap protocol (BOOTP)* is a protocol by which a client (i.e. a host in an ethernet or other LAN) may obtain an IP address, the IP address of a server and the name of a bootfile. It typically ran on a UNIX server. *DHCP (dynamic host configuration protocol)* is an extension of *BOOTP* to allow dynamic configuration of the entire IP-suite protocol software of a host (DHCP client) by a DHCP server. Typically DHCP servers are routers. BOOTP is defined in RFC 0951. DHCP is defined in RFC 2131. The interoperation of BOOTP and DHCP is defined in RFC 1534. BOOTP and DHCP share the same packet format (Figure 6.29).

The fields of the BOOTP/DHCP message are coded as follows:

- the *message type* is either a BOOTP request (value '1'), a BOOTP reply (value '2');

- the *hardware type* is usually 'IEEE 802.2'-based network (value '6');

- the *hops* field records the number of hops across which the packet has been forwarded in an attempt to find a BOOTP or DHCP server. A maximum of 16 hops are allowed before the message must be discarded;

- the *transaction ID (XID)* contains any random value to link BOOTP/DHCP requests and replies by a common identifier. This enables recognition of the reply;

- the *seconds* field records the elapsed time in seconds since the client sent the request;

- the first bit (ie. most significant bit) of the *flags* field is the *broadcast flag*. When set, this indicates that the reply should be sent to the network broadcast address rather than the client unicast address. All other bits in the flag field should be set to value '0';

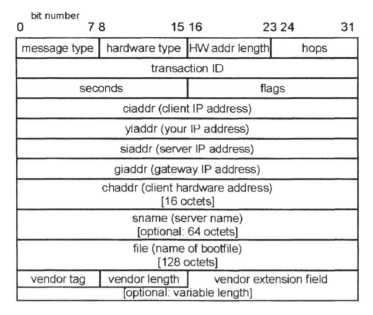

Figure 6.29 BOOTP (bootstrap protocol) and DHCP (dynamic host configuration protocol) packet format.

- the *client IP address (ciaddr)* is a particular IP address which the client would like to have formally assigned to its use. If the client is unconcerned about the address value, then it sets the field value in its request to 0.0.0.0;

- *your IP address (yiaddr)* is the address assigned to a client as your address by the server;

- the *server IP address (siaddr)* is the IP address of the BOOTP or DHCP server;

- the *gateway IP address (giaddr)* is the IP address of the first router (or other relay agent) which must be transited when the host (i.e. BOOTP/DHCP client) wishes to communicate with the BOOTP/DHCP server;

- the *client hardware address (chaddr)* is the hardware address (i.e. layer 2 address, MAC address, IEEE address or equivalent of the client host);

- the *sname* field is an optional field which may contain the name of the BOOTP/DHCP server;

- the *file* field contains the name of a bootfile which the client should load; and

- the *vendor* field allows for the inclusion and configuration of further parameters. This comprises a one octet vendor extension *tag,* a one-octet extension length field and the extension itself. The various different extensions are listed without further comment in Table 6.16. All DHCP messages are *vendor field extensions.*

DHCP may be operated in one of three modes, providing for the configuration of host (i.e. DHCP client) IP addresses:

- *automatic allocation* — in which the DHCP client is allocated a permanent IP address;

- *dynamic allocation* — in which the DHCP client is allocated an IP address which is only *leased* on a temporary basis. After expiry of the *lease*, the address must be re-applied for; or

Table 6.16   BOOTP and DHCP vendor extensions field

Type	Vendor tag value	Vendor extension length	BOOTP/DHCP extension or option
Vendor extensions	0	1	PAD
	255	1	This extension is the last in a string of extensions
	1	4	IPv4 subnet mask
	2	4	Time offset
	3	N	Router (list of routers in the subnetwork)
	4	N	Time server
	5	N	Name server
	6	N	Domain name server (DNS)
	7	N	Log server
	8	N	Cookie server
	9	N	LPR server
	10	N	Impress server
	11	N	Resource location server
	12	N	Host name (client name)
	13	2	Boot file size
	14	N	Merit dump file
	15	N	Domain name (indicates the domain name the client should use)
	16	N	Swap server
	17	N	Root path
	18	N	Extensions path
Host IP configuration	19	1	IP forwarding enable/disable
	20	1	Non-local source routing enable/disable
	21	N	Policy filter
	22	2	Maximum datagram reassembly
	23	1	Default time to live
	24	4	Path MTU ageing timeout
	25	N	Path MTU plateau table
Interface IP Configuration	26	2	Interface MTU (maximum transmission unit)
	27	1	All subnets are local
	28	4	Broadcast address
	29	1	Perform mask discovery
	30	1	Mask supplier
	31	1	Perform router discovery (IDRP — RFC 1256)
	32	4	Router solicitation address
	33	N	Static route
Interface Link Layer Parameters	34	1	Trailer encapsulation
	35	4	ARP cache timeout
	36	1	Ethernet encapsulation (ethernet version 2 or IEEE 802.3)
Transmission Control Protocol (TCP) Parameters	37	1	TCP default TTL
	38	4	TCP keepalive
	39	1	TCP keepalive garbage
Network Information Services	40	N	Network information service (NIS) domain
	41	4	Network information servers
	42	N	NTP (network time protocol) server
	43	N	Vendor-specific information
	64	N	Network information service + (NIS+) domain
	65	N	Network information service + servers

Table 6.16   (*continued*)

Type	Vendor tag value	Vendor extension length	BOOTP/DHCP extension or option
NetBIOS Parameters	44	N	NetBIOS over TCP/IP name server (NBNS)
	45	N	NetBIOS over TCP/IP datagram distribution server
	46	1	NetBIOS over TCP/IP node type
	47	N	NetBIOS over TCP/IP scope
X-Windows	48	N	X-Windows system font server
	49	N	X-Windows system display manager
DHCP Options	50	4	Requested IP address
	51	4	IP address lease time
	52	1	Option Overload
	53	1	DHCP message type
	54	4	Server identifier
	55	N	Parameter request list
	56	N	Server error message
	57	2	Maximum DHCP message size
	58	4	Renewal (T1) time value
	59	4	Rebinding (T2) time value
	60	N	Vendor class identifier
	61	N	Client identifier
	66	N	TFTP (trivial file transfer protocol) server name
	67	N	Bootfile name
Mobile, Email and Web Services	68	N	Mobile IP home agent
	69	N	SMTP (simple mail transfer protocol) server
	70	N	POP3 (post office protocol) server
	71	N	Network news transport protocol (NNTP) server
	72	N	Default Worldwide Web (www) server
	73	N	Default finger server
	74	N	Default internet relay chat (IRC) server
	75	N	StreetTalk server
	76	N	StreetTalk directory assistance (STDA) server

- *manual allocation* — in which the role of DHCP is reduced to informing the DHCP client of an IP address manually assigned (within the DHCP server) to a particular DHCP client.

In addition, DHCP can be used for the automatic configuration of a large number of protocol and communications, as well as service parameters, as will be clear from Table 6.16. This removes the effort and the potential for errors associated with manual configuration of each host station software.

When a DHCP client (i.e. a host) requires its IP configuration in order to connect to the network it transmits a *DHCP discover* message. This is answered by the DHCP server with a *DHCP offer* message (see Table 6.17). The DHCP message is sent by means of a UDP (user datagram protocol) broadcast message [e.g. to IP address 192.168.10.x] on port 67 (to the BOOTP server) or on port 68 (to the BOOTP client). Alternatively, instead of configuring the IP-suite protocols, DHCP can be used to communicate with and configure other services too, for example the domain name service (DNS) (on UDP port 53), one of the NetBIOS services (on UDP ports 137–139), TACACS (terminal access controller access control system) (on UDP port 49) or TFTP (trivial file transfer protocol) (on UDP port 69). More about these protocols in later chapters.

Table 6.17   DHCP messages (used with BOOTP/DHCP vendor extension option 53)

DHCP message type	DHCP message type code	Message function	Message generated by
DHCP discover	1	Broadcast message to discover a DHCP server	DHCP client
DHCP offer	2	Server reply with TCP/IP configuration parameters	DHCP server
DHCP request	3	Request for specific parameters or renewal	DHCP client
DHCP decline	4	Requested IP address is already in use	DHCP server
DHCP ack	5	Acknowledges client configuration parameters	DHCP server
DHCP nak	6	Notification that client IP address is incorrect	DHCP server
DHCP release	7	Release of IP address by client after use	DHCP client
DHCP inform	8	Client already has IP address, but needs other configuration parameters	DHCP client

6.16   Routing management issues

Routing protocols enable routers to share routing information about the reachability and exact whereabouts of IP address ranges, and create the database for a router's calculation of its routing table. *Address resolution protocols (ARPs)* provide for the location of individual IP addresses in target networks, while *DHCP (dynamic host configuration protocol)* or *BOOTP (bootstrap protocol)* provides for the automatic allocation of IP-addresses and the configuration of IP-suite protocol parameters in IP source hosts. Taken together, these protocols and mechanisms provide a complete framework for managing the allocation of IP addresses and ensuring that routers can create comprehensive routing tables to forward them appropriately. The whole series of different steps is illustrated in Figure 6.30.

   Much of the routing control process runs automatically but there are still a number of complex tasks for the human network administrator to undertake, as listed below:

- allocating IP address ranges to individual router interfaces (and thus to the associated *subnetwork* connected by means of the interface);

- designing the network routing hierarchy (e.g. defining OSPF areas, BGP communities, BGP confederations, etc.) and configuring this information into relevant routers;

- configuring routers to be capable of neighbour discovery (either programming the IP and hardware addresses of all neighbours or ensuring that adequate protocols are available to enable automatic discovery);

- determining the metrics to be associated with link cost or distance parameters (e.g. administrative cost, bandwidth, delay, load, reliability) and configuring relevant information into all routers;

- determining the routing policy to be applied in each relevant router (i.e. which routing information learned from external networks will be *filtered* and ignored, and which routes will be administratively *preferred*); and

- managing the process of route redistribution (i.e. determining which routing information will be transferred from one routing protocol to another, and which weightings will be applied to relative link costs learned by the different protocols).

How well the various tasks above are performed will determine the efficiency of the network routing scheme which results.

Before we finally leave the subject of routing, we should point out that the principles used for the exchange of routing information and calculation of routing tables will remain largely the same in IPv6 as in IPv4. Nonetheless, each of the protocols we have spoke of will require adaptation — at a minimum to enable the carriage of the longer (128-bit) addresses of IPv6.

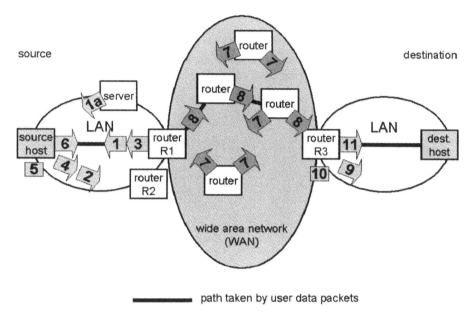

———— path taken by user data packets

Figure 6.30 Routing processes contributing to the carriage of user data packets from source to destination.

Key: the stages of routing and packet-forwarding:

1. DHCP server (typically the access router of the LAN) delivers a source IP address to the source host.

1a. (Alternatively) an IP address is provided by BOOTP or RARP (typically delivered by a server).

2. Router discovery undertaken — IRDP (ICMP router discovery protocol) — this is typically part of the DHCP message delivered to the host, but the host may have to initiate the process. (Alternatively, this step is avoided by programming a static route into the host for the gateway router.)

3. By listening to the routing protocol, the host may determine which is the best router (in our example router R1) to reach a particular destination. (Alternatively, this step is avoided by programming a static route into the host for the gateway router.)

4. ARP used by host to *resolve* the hardware address (i.e. MAC-address) of the *gateway* router.

5. The hardware address of the gateway router is cached by the source host for use on subsequent occasions — thereby avoiding the need to repeat using ARP each time.

6. User data packets can now be forwarded by the source host to the gateway router across the source LAN (using the LAN protocol).

7. The routing protocol used in the *wide area network (WAN)* keeps all the routers in the network constantly advised of the WAN topology and shortest routes to all *reachable* destinations.

8. Using the routing tables of the routers (created using the routing protocol), the user data packets are forwarded to the destination router (in our case router R3) in a *hop-by-hop* manner.

9. The hardware address (i.e. MAC address) of the destination host is *resolved* using ARP.

10. The hardware address of the destination host is cached by router R3 for use on subsequent occasions — thereby avoiding the need to repeat using ARP each time.

11. User data packets can now be forwarded by router R3 to the destination host across the destination LAN (using the LAN protocol).

7

Transport Services and Protocols

The Internet protocol (IP) provides for end-to-end carriage of data through the routers and across the hops of an internetwork, but it is the role of a transport protocol to control and manage the end-to-end communication (between the hosts). The transport protocol provides a transport service to the software application running in the host — ensuring that data is delivered to the right application in the destination device, that individual packets are in the right order and that none have gone missing en route. In this chapter we shall consider three different types of transport services: connectionless transport service (CLTS); connection-oriented transport service (COTS) and a special form of connection-oriented service called flow- or stream-oriented transport service. We then go on to describe in detail the IP-suite protocols which provide for these services: UDP (user datagram protocol — provides for connectionless service); TCP (transmission control protocol — provides for connection-oriented service), MPLS (multiprotocol label switching — provides for flow-based communication) and RSVP (resource reservation protocol — provides for bandwidth reservation, as particularly needed in associated with flow-based communication). The real-time application transport protocol (RTP) appears in Chapter 10.

7.1 Transport services and end-to-end communication between hosts

The function of a *transport (layer 4) protocol* is to manage the end-to-end communication between software applications in the two end devices (usually called *hosts*) which are interconnected by means of a data network. The *network (layer 3) protocol* provides for the actual carriage of the data across the network, but the *transport protocol* is necessary to ensure that all the individual packets of information making up the application's message to its peer arrive at the destination and (as appropriate) are presented in the right order.

While at the *network layer* there is only one communication path between the source network interface and the destination network interface (the interfaces being uniquely identified by means of their IP addresses), the transport protocol allows multiple communication *sessions* to be in progress at the same time — all using different commands and message structures (i.e., different protocols). The transport protocol thus *multiplexes* multiple application protocols (*multiprotocols*) across a single path between two network interface points. As Figure 7.1 illustrates, different software modules and data structures in the two communicating end-devices (hosts) might be conducting separate parallel communication sessions using different protocols for *email*, network management, *network file system (NFS), Worldwide Web (www)* or other applications.

Data Networks, IP and the Internet: Protocols, Design and Operation   Martin P. Clark
© 2003 John Wiley & Sons, Ltd   ISBN: 0-470-84856-1

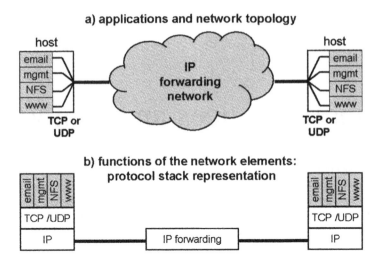

Figure 7.1   Multiplexing function of TCP and UDP transport protocols.

The two main transport protocols used in conjunction with IP are TCP (transmission control protocol) and UDP (user datagram protocol). They provide respectively for a connection-oriented transport service (COTS) and a connectionless transport service (CLTS). By connection-oriented, we mean that a connection must be established before transport of user data can commence. When setting up the connection, no data may be sent until there is confirmation that the destination device exists, is switched on, and is ready to receive it. Subsequently, during the data-transfer phase of a transport connection, the blocks of user data sent (at layer 4, called segments) must be acknowledged — thereby guaranteeing reliable delivery. In contrast, a connectionless transport protocol carries each message (segment of data) independently and without a connection set-up phase. A connectionless transport protocol is akin to posting a letter — there is no guarantee the message will arrive: you may have spelt the address wrongly; it might get lost or corrupted. But while *connectionless transport* is *unreliable*, it has the advantage of being simpler, less demanding of network capacity and quicker when conveying short (one segment) messages. If all you want to say is 'Help!' it is quicker to just say 'Help' rather than first have to set-up a connection — something like: 'Are you listening?'...'Yes'...'May I start?'...'Yes, I'm ready'...'**Help**!!'. Just because connectionless transport is *unreliable* doesn't mean it is a bad thing!!

Protocol layers and their configuration parameters: critical for communication between hosts

A pair of hosts wishing to communicate data with one another across a data network must be equipped with suitable *networking* and *transport* protocols, and have the parameters of these protocols configured in a compatible manner. The same transport protocol must be used at each end of the communication, since the two hosts are *peer partners* at the transport layer. But the network interfaces and protocols used at the two ends may be different, as Figure 7.2 illustrates. The diagram will help us to understand the benefits of *layered protocols* and to consider the prerequisites for successful end-to-end communication.

Figure 7.2 illustrates the same example network as Figure 7.1, except that it also shows the lower layer communications protocols. In Figure 7.2 it is apparent that the path from one host

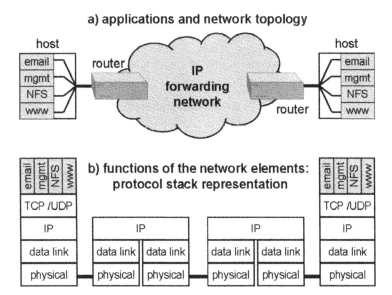

Figure 7.2 Detailed network topology and communications protocol functions involved in the example of Figure 7.1.

to the other traverses two routers along the way. We shall assume that each host is connected to its corresponding router by means of a LAN interface, and that the connection between the two interfaces is by means of a point-to-point leaseline. As a result, the routers perform not only IP layer forwarding but also *physical* interface and *datalink* layer adaptation. The datalink layer protocol used in the LAN between the hosts and their corresponding routers will most likely be IEEE 802.2 (*LLC—logical link control* protocol), while the datalink layer protocol used on the point-to-point link between the two routers will probably be either *PPP (point-to-point protocol)* or *HDLC (higher level data link control)*. The two physical LAN interfaces will most likely conform to ethernet (e.g., IEEE 802.3, 10baseT, 100baseT, etc.), while the point-to-point connection will have a leaseline interface format, e.g., X.21, V.35 etc. as we discussed in Chapter 3.

The communication of one of the application protocols in the host on the left-hand side of Figure 7.2 with the corresponding peer application protocol in the host on the right-hand side (e.g., email-to-email) has to pass through 18 protocol processing steps along its complete path!! First, the transport layer (TCP/UDP) creates *segments* of data containing the email message. Each segment of data is then packed into a *packet* (at layer 3 — the network layer) and then into a *frame* (at layer 2 — the data link layer) before being coded into the appropriate physical format for transmission. But after each hop of the way, each frame is unpacked for the IP packet *forwarding* process, before being re-packed into another frame and recoded for the next hop. At the final destination, the frames and packets are all unpacked and the individual segments are *reassembled* in the correct order for presentation of the message to the destination email application.

Each of the 18 protocol functions of Figure 7.2 has to be configured correctly, or the communication will fail!

The configuration of the parameters of each of the various different protocol functions (18 in all) of Figure 7.2, as in any real network, has to be undertaken carefully. Typically the different functions will comprise individual software and/or hardware modules, each of which will have been added to the network at a different time — as a new router or host has been connected to the network, or as a new interface or link has been added. Most of the modules are likely to pre-configured in a *default configuration*, so that much of the network can be built in a *plug-and-play* fashion, but there are still configuration duties for the human administrator to undertake and check (if the plug-and-play approach does not bring immediate success):

- *physical layer interface*: cable lengths and types, connector pin-outs and DTE/DCE functional assignments must all be correct;

- *datalink* layer: the MAC address is usually a factory-configured globally unique identifier in the case of ethernet or other LANs, but HDLC or PPP will require address assignments. On a PC, configuration changes are usually to be found in a window labelled something like 'control_panel/system/ equipment_manager/network interface adapters';

- IP layer: *TTL (time-to-live)* and other parameters must align, IP addresses must be assigned manually or a protocol such as DHCP or BOOTP must be organised to undertake dynamic allocation. Routing protocols must be set up and configured in the routers. IP parameters in host devices (e.g., PCs or workstations) are typically to be found in a 'control_panel/networking' configuration window; and

- TCP or UDP transport layer capabilities and parameters must align with the peer host.

Things may seem to be getting frightfully complex. There are lots of different protocols to configure in all the different devices — and much scope for error — but mainly because of the large number of devices and software functions involved in a data network. Anyone who has tried to configure a network, or simply tried to configure the network settings in his own PC to enable him to connect to the Internet will know about the potential for mistakes. Nonetheless, all experienced data network engineers and software developers will tell you that the use of *layered protocols* makes things much easier to administer than they might otherwise have been — at least the configuration parameters are not embedded in the application software! It would be a nightmare to have to keep recompiling application software each time a network parameter or topology change was undertaken!!

But maybe you are wondering how a software application can initiate data communication with a distant partner? The answer is by invoking a *transport service*. This is done at an internal software interface called a *port* or a *socket*. (This is the *service access point (SAP)* of the *transport layer*.) An *application protocol* (such as one of those listed in the right hand column of Table 7.1) uses the port or socket.

Transport layer multiplexing — ports and sockets

Table 7.1 lists the most commonly used port (or socket) numbers made available by *TCP (transmission control protocol)* and *UDP (user datagram protocol)*. The use of different port numbers allows a single point-to-point communications path between two end-devices to be used simultaneously for a number of different applications listed in the table.

Transport of communication flows

Today, *TCP (transmission control protocol)* and *UDP (user datagram protocol)* are the most widely used data transport protocols. Despite their age (the current versions date back to 1981 and 1980 respectively) they have proved to be very durable for all manner of data communications applications. But to meet the needs of modern *multimedia* communications, there has

Table 7.1   TCP (transmission control protocol) and UDP (user datagram protocol) port numbers

UDP/TCP port number	TCP or UDP as carriage protocol	Application protocol or service
21	TCP	FTP (file transfer protocol)
22	TCP	SSH (secure shell) remote login
23	TCP	Telnet
25	TCP	SMTP (simple mail transfer protocol)
53	TCP/UDP	DNS (domain names service)
65	TCP/UDP	TACACS (terminal access controller access control system) database service
67	UDP	BOOTP (bootstrap protocol) / DHCP (dynamic host configuration protocol) server
68	UDP	BOOTP (bootstrap protocol) / DHCP (dynamic host configuration protocol) client
69	UDP	TFTP (trivial file transfer protocol)
80	TCP	World wide web HTTP (hypertext transfer protocol)
111	UDP	Sun remote procedure call (RPC)
119	TCP	NNTP (network news transfer protocol)
123	UDP	NTP (network time protocol)
137	UDP	NetBIOS name service
138	UDP	NetBIOS datagram service
139	UDP	NetBIOS session service
161	UDP	SNMP (simple network management protocol)
162	UDP	SNMP trap
179	TCP	BGP (border gateway protocol)
194	TCP	IRC (Internet relay chat)
213	UDP	Novell IPX (internetwork packet exchange)
443	TCP	HTTPS (secure hypertext transfer protocol)
512	TCP	rsh (BSD — Berkeley software distribution) remote shell
513	TCP	RLOGIN (remote login)
514	TCP	cmd (UNIX R commands)
520	UDP	RIP (routing information protocol)
540	TCP	UUCP (UNIX-to-UNIX copy program)
646	TCP/UDP	LDP (label distribution protocol): LDP hello uses UDP, LDP sessions use TCP
1080	TCP	SOCKS (OSI session layer security)
1645	UDP	RADIUS (remote authentication dial-in user service) authentication server
1646	UDP	RADIUS accounting server (Radacct)
1701	UDP	L2F (layer 2 forwarding)
2049	TCP/UDP	NFS (network file system)
2065	TCP	DLSw (data link switching) read port
2066	TCP	DLSw (data link switching) write port
5060	UDP	SIP (session initiation protocol)
6000-4	TCP	X-windows system display
9875	UDP	SAP (session announcement protocol)

Note: For a complete and up-to-date listing see also www.iana.org/assignments/port-numbers

been recent intensive effort to further develop both the transport and network layer protocols further — in order to support the much more stringent demands made by the transport of *real-time* voice and video signals. Communication of real-time signals requires the carriage of a continuous signal (a communication *flow* or *stream*), with predictable and very high quality of service — with low and almost constant propagation delay.

Communication flows or streams require very high grade *connections* and are carried in IP-based networks by means of a new type of *connection-oriented* network layer protocol based on *label switching* (formally called *tag switching*). We encountered the basic principles of label switching in Chapter 3. Label switching takes place by assigning a *flow label* to IP packets making up a given communications flow. By doing so, the normal IP-header processing for determining the routing of packets at each router can be dispensed with, and replaced by a much faster *label switching* function. Such a label switching function is defined by *MPLS (multiprotocol label switching)* and is capable of direct incorporation into the *flow label* field of the IPv6 header (as we saw in Chapter 5). In effect, *MPLS* provides an optional *connection-oriented* extension to the IP network layer protocol by means of an additional *shim* layer between the network layer (layer 3) and the data link layer (layer 2). We shall discuss it in greater detail later in the chapter.

As a *connection-oriented network service (CONS)*, MPLS is used extensively to provide for *VPN (virtual private network)* services — in which a public *Internet service providers (ISPs)* offer 'leaseline-like' point-to-point links as router-to-router connections for private IP router networks. In addition, MPLS is often combined with special transport layer protocols used to reserve bandwidth reservation and *quality of service (QOS)* assurance along the entire path of the label-switched connection. The most important of these transport layer protocols is the *resource reservation protocol (RSVP)*. RSVP is a control channel protocol, used before or during the set-up phase of a connection to reserve bandwidth and other network resources for the handling and forwarding of the packets making up the communication *flow* or *stream*. Unlike the TCP and UDP transport protocols, RSVP is not used directly for the end-to-end carriage of user-information, and it may often be used in addition to TCP. We return to discuss RSVP later in the chapter.

7.2 User datagram protocol (UDP)

The *user datagram protocol (UDP)* is the IP-suite transport protocol intended for connection-less data transport. It is defined in RFC 0768. The *segment* format of UDP is illustrated in Figure 7.3: there are only four header fields: source and destination *port* fields, a *length* field and a *checksum*.

The source and destination port (also called *socket*) fields are coded according to Table 7.1. These provide for a multiplexing function to be undertaken by UDP — allowing segments intended for different application software protocols to share the same network path but still be kept apart from one another.

The *length* field indicates the length of the UDP segment (as shown in Figure 7.3), including the four header fields as well as the UDP data field.

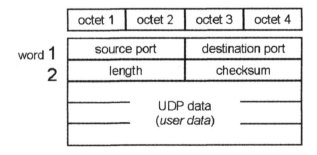

Figure 7.3 Segment format of the user datagram protocol (UDP).

The UDP checksum (which is optional) is applied not only across the *UDP segment* but also includes the *pseudo-UDP-header* fields taken from the IP header (see Figure 7.9). The *pseudo-header* reflects the manner in which some application software sets up a communications path using UDP and simultaneously issues commands to the IP forwarding software. The application software packs the data and includes a 'label' (the UDP header and parts of the IP header) to indicate the coding of the data (IP *protocol* and UDP *port* number) as well as the address (IP-address) to which it is to be delivered.[1]

The user datagram protocol (UDP) is little more than 'IP with a port number added'. For some purposes, a similar multiplexing effect to the UDP port number can be obtained using only the Internet protocol (IP) and relying upon the *protocol* field for *multiplexing*. This is the approach used, for example, for the carriage of OSPF (open shortest path first) routing protocol.

7.3   Transmission control protocol (TCP)

Transmission control protocol (TCP) is the IP-suite *transport* protocol which provides for reliable *connection-oriented* transport of data with guaranteed delivery. It is defined in RFC 0793. The protocol has come to be the most commonly used of the two available IP-suite transport protocols (TCP and UDP) and for many people come to be synonymous with IP — hence the frequently used terminology: TCP/IP. TCP only need be supported by the two hosts at either end of the connection, and the requirements of these hosts are detailed in RFC 1122 (which is essentially an implementation guide for TCP/IP).

Basic operation of the transmission control protocol (TCP)

As shown in Figure 7.4, the transmission control protocol (TCP) provides for a three-stage process of data transport. In the first phase, a connection is set up between the two hosts which are to communicate. This stage is called *synchronisation*. Following successful synchronisation of the two hosts (shown as the 'A-end' and the 'B-end' in Figure 7.4), there is a phase of data transfer. In order that the data transfer is *reliable* (i.e., that the delivery of *segments* can be guaranteed), TCP undertakes acknowledgement of segments, as well as *flow control* and *congestion control* during the data transfer phase. After the completion of data transfer, either of the two ends may *close* the connection. If at any time there is a serious and unresolvable network or connection error, the connection is *reset* (i.e., *closed* immediately).

The acknowledgement of segments confirms to the sender of the data, that the receiver has received them successfully — without errors being detected by the checksum. Any segments lost during transmission are *retransmitted* (either automatically by the sender having not received an acknowledgement or on request of the receiver).

The transmitting end takes the prime responsibility in ensuring that data is reliably delivered. . .using a technique known as *retransmission*. The transmitter sets a *retransmission timer* for each TCP segment sent. Should the timer exceed a pre-determined period of time, known as the *retransmission timeout (RTO)*, without having received an *acknowledgement (ACK)* for the segment, then the segment is automatically *retransmitted* (i.e., re-sent). A typical value for the retransmission timeout (RTO) is 3 seconds. Retransmission of the same segment continues until either an acknowledgement (ACK) is received or until the *connection*

---

[1] Because the IP address fields are included in the UDP checksum, the protocol has to be adapted when used in conjunction with IP version 6 (IPv6) because of the longer length of IPv6 addresses (128 bits) — IPv4 addresses are only 32 bits long.

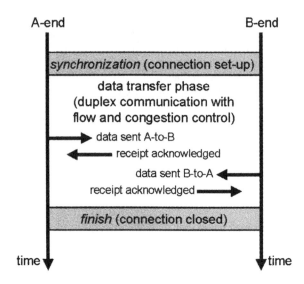

Figure 7.4   Basic operational phases of a TCP connection.

timeout expires — in which case the connection is *reset*. The reset may be necessary to recover from a remote host or network failure.

To prevent inappropriate retransmission of packets on long delay paths (which might otherwise only lead to network congestion without improving the *reliability* of delivery), the retransmission timeout (RTO) is corrected according to the *round trip time (RTT)*. In theory, the round trip time (RTT) is the minimum time taken for a message to be sent in one direction and immediately acknowledged in the other direction. In practice, it is determined by the transmitter by means of the *Jacobsen algorithm* (described in RFC 1323 and RFC 2988), using a *sample* of measured RTTs of previous segments. The actual calculation continually adapts two parameters called the *smoothed RTT (SRTT)* and the *RTT variance (RTTVAR)*. The RTO is then accordingly set (or adjusted) to correspond to a slightly longer time period than the RTT. In other words, RTO = RTT + small margin (Figure 7.5).

The aim of constantly adjusting the RTO is to avoid the network performance degrading effects of excessive retransmission, without compromising the reliability of guaranteed segment delivery.

The use of the TCP *timestamp option (TSOPT* — described in RFC 1323) is helpful in calculating the RTT more exactly. The *timestamp* (i.e., the time the original data segment was sent) is returned in the acknowledgement (ACK) of the data segment. This avoids a possible problem (following retransmission of a given segment) of not knowing for which segment an acknowledgement has been received (the first transmission or the retransmission). When not using a timestamp, no new calculation of RTT is undertaken for retransmitted segments. In other words: the RTT of a retransmitted segment is not a valid *sample* for recalculation of the RTT. (This is called *Karn's algorithm*.)

It is important to note that not every TCP segment is individually acknowledged (ACK). Instead, *cumulative acknowledgement* is undertaken. The acknowledgement does not indicate directly which segments have been received, but instead which segment is the next expected! By inference, all previous segments have been received. The use of such a cumulative acknowledgement procedure has a number of advantages. First, not so many acknowledgement (ACK) messages need be sent, thereby reducing the possibility of network congestion. Second, the return path need not be kept permanently free for sending acknowledgements. In this way it is possible to use the connection in a *duplex* mode — for sending data in both directions

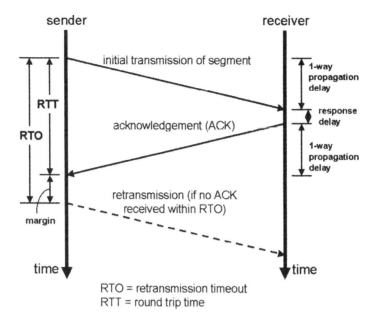

Figure 7.5 TCP acknowledgement of data receipt: the round trip time (RTT) and retransmission timeout (RTO).

simultaneously. A data segment sent along the 'return' path can simultaneously be used to acknowledge all previous received data in the 'forward' path. Finally, since the ACK message indicates the 'next expected segment' it provides a useful 'trick' for re-ordering of segments (i.e., requesting their immediate retransmission). The re-ordering is signalled by duplicating ACKs — four identical ACKS (i.e., with the same value) in a row are interpreted by the transmitter as a request for a *fast retransmit.*

Rather than acknowledging each received segment, the cumulative acknowledgement is undertaken regularly according to a timer (an acknowledgement must be sent at least every 0.5 seconds). In addition, at least every second segment of *maximum segment size (MSS)* must be acknowledged. Having too many outstanding (unacknowledged) packets is not a good thing, since if a retransmission is necessary, the repeated transmission of large packets adds significantly to the network load — and thus to the chance of congestion.

At the transmitting end of the connection, segments are sent either when the *maximum segment size (MSS)* is reached, when the software application forces the transmission (for example, with a *push, PSH,* command) or after a given timeout expires. Push commands are used in terminal-to-host communications protocols (e.g., *Telnet*). They allow terminal commands to a mainframe computer (something typed and *entered* with a *carriage return* key) to be immediately sent to the host. If a push command is not used, TCP segments are forwarded regularly during the connection, according to a timer. If necessary (when there is no 'real data' to be sent), *keepAlive* segments may be sent to advise the remote application that the connection is still live.

For data *flow control* and *congestion control,* the *transmission control protocol (TCP)* maintains a transmit *window.* The size of the window determines how many unacknowledged octets of data the transmitter is allowed to send before it must cease transmission and wait for an acknowledgement. By limiting the window, TCP ensures that the receiver input buffer is not overwhelmed with data — so preventing a receive buffer overflow. Should overflow occur, segments would not be acknowledged, and would be retransmitted, so the reliability of data delivery is not at stake—but network congestion is. Avoiding retransmission of data (when possible) is a prudent manner of preventing network congestion.

The size of the window can be adjusted by the receiver during the course of the TCP connection, as will become clearer later—when we review the TCP segment format and discuss TCP flow and congestion control.

TCP connection set-up: the three-way handshake

The setting-up of a TCP connection takes place by means of a *three-way handshake*, as illustrated in Figure 7.6. The process uses a TCP *synchronisation message*. Unlike a telephone call, the connection does not comprise a dedicated physical circuit, and unlike X.25 (which we discussed in Chapter 3), there is not really a *virtual circuit* either. Instead the *connection set-up* establishes simply that a transmission path exists between the 'caller' and the 'destination' and that both are ready to 'talk' to one another. The three messages which form the *three way handshake* effectively have the following functions:

- first synchronisation message ('can we talk on port x? If so, I will number messages in order, starting with message number y');

- synchronisation acknowledgement message ('yes, I understand the protocol used on this port number and am ready to talk now. My messages to you will be numbered too, starting with message number z'); and

- handshake acknowledgement ('I confirm that I have understood your messages'). The third and final message of the *handshake* confirms to both parties that both directions of transmission are working correctly—that duplex communication is possible. Without duplex communication, the protocol will not work.

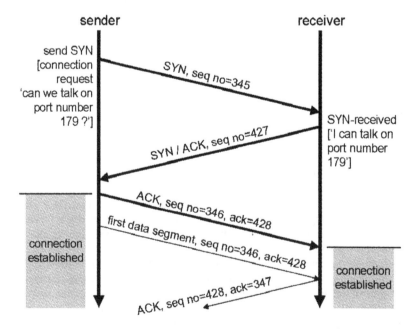

Figure 7.6   TCP connection set-up (synchronisation)—by means of a 3-way handshake.

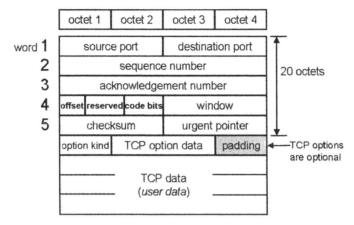

Figure 7.7   Transmission control protocol (TCP) segment format.

At the time of connection set-up, the initial connection parameter values are set: the retransmission timer (RTO) typically set to 3 secs; the *round-trip time (RTT)* is set to 0 secs; and the *initial window (IW)* size is set to a value in octets equivalent to one segment. In addition, the *initial sequence numbers (ISNs)* are set. The *sequence numbers (sequence* and *acknowledgement numbers)* are the segment message numbers. Segments are numbered consecutively according to how many octets they contain. The ISNs are random numbers chosen by the two transmitters.

Transmission control protocol (TCP) segment format

Figure 7.7 illustrates the TCP segment format. The standard segment[2] header is 20 octets long.

Source and destination port fields

The *source port* field (possible values 0–65 535) identifies the application software in the transmitting host. The *destination port* field (possible values 0–65 535) identifies the intended destination application software in the receiving host. Both fields are normally coded with the same value. The most commonly used *port* values (also called *socket numbers*) are listed in Table 7.1.

Sequence number and acknowledgement number fields

The *sequence number* commences at the *initial sequence number (ISN)* number, which is chosen randomly by the transmitter, and each subsequent data message (i.e., segment) has a sequence number accordingly greater than the previous segment. The increment in the sequence number depends upon the number of octets in the previous segment. The sequence number

---

[2] At protocol layer 4 (transport layer) a *block* of data is called a *segment*, whereas at protocol layer 3 (network layer) we refer to *packets* and at protocol layer 2 (datalink layer) we refer to *frames* — the data starts life as a *segment*, gets packed into a *packet* which in turn gets packed into a *frame* — as we saw in chapter 3 — Figure 3.28.

thus counts *octets*, but its value also uniquely identifies a particular segment, thereby enabling the receiver to:

- check that it has received all octets (and therefore all segments) of the user data;

- unambiguously acknowledge the receipt of all segments back to the transmitter;

- order the re-transmission of any missing segments, and

- re-sequence segments into the correct order, should they be received out-of-order.

The *acknowledgement number* field is used to return the sequence number of the next awaited octet (by inference indicating the cumulative octets successfully received so far).

Both the sequence number and the acknowledgement number can be incremented up to the value $2^{32} - 1$ (which equals around 4 billion octets — around 4 Gbytes), before the value is *wrapped* (i.e., reset to zero) and counting starts from '0' again. Sometime or other, the total number of octets sent may exceed the 4 Gbytes, whereupon a previously used sequence number will be re-used (see Table 7.2).

The re-use of sequence numbers can cause a problem, if there is any chance that two segments might be in existence at the same time with the same segment number. Bear in mind here, that the TCP specification defines a *maximum segment lifetime (MSL)* of 120 seconds — thus limiting the data transfer rate to 4 Gbytes in 120 seconds. But working at its maximum speed of data transfer, a Gigabit ethernet interface will be able to transfer 4 Gbytes of data in around 34 seconds. In other words, high speed interfaces present a real risk of multiple different segments bearing the same sequence number. A solution to the problem is defined in RFCs 1185 and 1323. The solution is called *PAWS (protect against wrapped sequences)*.

PAWS (protect against wrapped sequences) is intended to be used on very high-speed TCP connections. Instead of simply increasing the size of the TCP sequence number (and thereby reducing the frequency of sequence number wrapping), the PAWS solution to wrapped sequences is to include a *timestamp* with each transmitted segment. The timestamp creates clear distinction between segments with duplicate *wrapped* sequence numbers without having to increase the sequence number field length.[3]

Table 7.2 Time elapsed per TCP sequence number wrap cycle at full speed of transmission

Network or interface type	Bit rate	Time to sequence number wrapping
ARPANET	56 kbit/s	7.1 days
Digital leaseline	64 kbit/s	6.2 days
T1-line	1.544 Mbit/s	6.2 hours
E1-line	2.048 Mbit/s	4.7 hours
10baseT ethernet	10 Mbit/s	57.3 mins
E3-line	34 Mbit/s	16.8 mins
T3-line	45 Mbit/s	12.7 mins
Fast ethernet (100baseT)	100 Mbit/s	5.7 mins
STM-1 (OC-3) line	155 Mbit/s	3.7 mins
Gigabit ethernet (1000baseX)	1 Gbit/s	34 secs

---

[3] Increasing the length of the sequence number field would have meant that devices with this 'new version of TCP' would be incompatible with existing devices using 'previous version TCP'. By instead using a protocol option field — the timestamp field — protocol compatibility can be maintained. Of course there is a chance that older devices may not correctly interpret and use the option — but there again, these are unlikely to be devices with high speed interfaces!

Offset field (length of the TCP header)

The *offset* field indicates the length (in 32-bit *words*) of the TCP header (in effect, whether and how many *TCP options* are in use). The value also indicates the start of the TCP data field, and thus allows the receiving software application to assign memory buffer space for the header and the data fields accordingly.

Reserved bits field

The reserved bits are reserved for an as-yet undefined purpose.

Code bits field

The *code bits* indicate the nature of the TCP segment (e.g., whether it is a *connection control message* such as a *synchronisation (SYN)* message or connection *finish (FIN)* message). The *code bits* also indicate which fields within the TCP segment header are in use. The order of the bits is illustrated in Figure 7.8. Their meanings are explained in Table 7.3.

Window size field

By means of the *window size* field, each receiver is able to inform the corresponding TCP transmitter how many octets of TCP-data may be sent before an acknowledgement (ACK) message must be awaited from the receiver. We discussed earlier in the chapter, how, by limiting the window size, the receiver is able to ensure that the transmitter does not send more data than the receiver is currently able to accept into its receive buffer. The *window size* may be changed during the course of the TCP connection, thereby adjusting to the instantaneous capability of the receiver to accept data. Dynamic adjustment of the window size is a critical part of the TCP flow control, congestion control and segment loss recovery procedures, as we shall discover later.

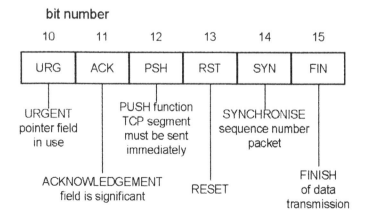

Figure 7.8   TCP code bits field.

Table 7.3 TCP code bit interpretation

Code bit name	Meaning/full name	Interpretation of bit value when set to '1'	Interpretation of bit value when set to '0'
ACK	Acknowledgement	Acknowledgement number field is in use	Acknowledgement number field not in use
FIN	Finish	This segment is a request to close the connection	This segment is not a finish message
PSH	Push	This segment should be sent immediately	Wait for a timeout before transmitting this segment
RST	Reset	Reset connection (i.e., close immediately following an error)	Maintain connection
SYN	Synchronise	This segment is a connection set-up request message	This segment is either a data segment or a finish message
URG	Urgent	The first part of the TCP data field contains an urgent command (e.g., an 'escape' or 'break' command), and the user pointer field is in use. The pointer points to the octet position of the first 'real' user data in the TCP data field (following the urgent data).	The user pointer field is not in use.

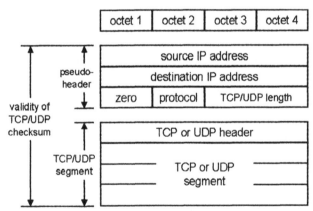

Note: As in UDP, when used in conjunction with IPv6 rather than IPv4, the TCP checksum calculation has to be adapted to cater for the longer length of IPv6 addresses.

Figure 7.9 TCP/UDP pseudo-header and validity range of the checksum calculation.

TCP checksum field

The checksum provides a means of detecting errors in the TCP segment. The value (as always in the IP-suite) is *the one's complement of the one's complement sum of the message* and is applied across the entire TCP segment including the *pseudo-header* (Figure 7.9: i.e., including the IP source address, IP destination address and *protocol* fields of the IP header). As in the case of UDP, the pseudo-header (Figure 7.9) provides a means for a software application to instruct the IP protocol layer where the segment is to be forwarded to.

Urgent pointer field

When the *urgent bit* (in the *code bits* field) is set, then the *urgent pointer* field is in use. The value contained in this field indicates the octet number of the first octet of 'real' TCP user data, following any *urgent data* in the TCP-data field. Urgent data typically corresponds to a computer control command — a command like ⟨*abort*⟩, ⟨*break*⟩, ⟨*escape*⟩, etc.

TCP options field

If included, the *TCP options* field is indicated by means of the *offset field* — to show that the TCP header is longer than the standard 20 octets. The TCP options field comprises one or more options, listed one after the other. As necessary, padding is suffixed to the TCP options to make the total length of the TCP header an integral number of 32-bit *words*.

Each *TCP option* comprises an *option kind* field and some option data. The option kind field is coded according to Table 7.4, which also lists the main TCP options.

The *maximum segment size (MSS)* option allows the maximum segment size to be changed from the default value of 536 octets set out in RFC 0879 and RFC 1122. The default value corresponds to the IP minimum transmission unit size of 576 octets: 576 minus 20 octets for the IP header and 20 octets for the standard TCP header.

The *window scale option (WSOPT)* allows the window parameter size to be extended from 16 (65 536 octets — 65 kbytes) to 32 bits (429 million octets — 429 Mbytes). The potential for an increased window size was added to ensure that efficient and high performance could be obtained from high speed network interfaces. Without such an extended window, the transmitter would be forced to keep waiting for acknowledgements, when the line (and probably the receiver as well) could handle almost permanent transmission.

The *selective acknowledgement (SACK)* option (option '5'), which is negotiated during TCP connection set-up (using option '4'), allows the receiver to acknowledge the receipt of individual (i.e., selected) segments rather than using cumulative acknowledgement. By means of such selective acknowledgement, a receiver can acknowledge a segment, even though an intervening one is missing. Only the missing segment (rather than all segments) need be retransmitted.

The *timestamp option (TSOPT)* is also negotiated during connection set-up. As we discussed previously it has two main functions: assisting in the accurate calculation of the *round-trip time*

Table 7.4   TCP options

Option kind value	Option length in octets	Option	Defined in
0	–	End of option list	RFC 0793
1	–	No operation	RFC 0793
2	4	MSS (maximum segment size)	RFC 0793
3	3	WSOPT (window scale option)	RFC 1323
4	2	SACK (selective acknowledgement) permitted	RFC 2018
5	Variable	SACK	RFC 2018
8	10	TSOPT (timestamp option)	RFC 1323
9	2	Partial order connection permitted	RFC 1693
10	3	Partial order service profile	RFC 1693
14	3	TCP alternate checksum request	RFC 1146
15	Variable	TCP alternate checksum data	RFC 1146
19	8	MD5 signature (encryption) option	RFC 2385

(RTT) and supporting PAWS (protect against wrapped sequences). The timestamp is provided by the transmitter and returned by the receiver in the acknowledgement segment.

The *signature option* ('19') allows for encryption of the TCP segment data contents.

TCP segment header processing and the TCP control block

In order to process data and communicate with a remote host by means of the *transmission control protocol (TCP)*, a host must maintain a *TCP control block*. A TCP control block is required for each live TCP connection: it comprises stored information about the connection state, the local software process associated with sending and/or receiving data, communication feedback parameters and such like. It also includes *pointer* values which indicate the beginning of send and receive buffers within the computer's memory, as well as pointers to transmission and retransmission queues etc. Once a TCP segment has been loaded into the transmission queue part of the memory, the IP forwarding software takes over. Meanwhile, data is transferred from the retransmission queue to the transmission queue if the relevant *retransmission timeout (RTO)* expires.

TCP flow control and congestion control

The acknowledgement process and the use of a transmission window to limit the number of octets which may be sent before having to wait for an acknowledgement, form the critical elements of *flow control* and *congestion control* in the *transmission control protocol (TCP)*.

RFC 2581 defines four separate algorithms for TCP flow and congestion control. A further method is defined in RFC 3168. All five methods are listed below:

- *slow start*;

- *congestion avoidance*;

- *fast retransmit*;

- *fast recovery*; and

- *explicit congestion notification (ECN)* [RFC 3168].

The first of these five schemes work by allowing the receiving host to adjust the *window size* during the duration of the TCP connection, depending upon the amount of memory space it has available for use as a receive buffer and upon the recent performance of the communication.

First, let us take time to consider the basic flow control mechanism using acknowledgement messages and the function of the window. Both are illustrated in Figure 7.10.

Let us assume that the window size in Figure 7.10 has been set at 1608 octets. This is equivalent to three segments of the default *maximum segment size (MSS)* of 536 octets. Now let us assume that the sender is sending segments of 536 octets in length. Then the sender is permitted to send three segments (a total of 1608 unacknowledged octets), but must then wait for an acknowledgement (ACK) before sending any more (this is the effect of the window size of 1608 octets).

We shall further assume that acknowledgements are returned by the receiver for each segment received, and this is also illustrated in Figure 7.10. (This will not generally be the case, since cumulative acknowledgement is used, but would be necessary if the line bit rate were only 9600 bit/s. At 9600 bit/s, the requirement to send an acknowledgement at least every 0.5 seconds, requires an acknowledgement of each 4800 bits — 600 octets.)

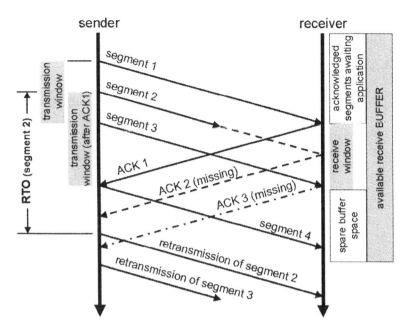

Figure 7.10   TCP flow control by means of acknowledgement messages: the effect of the transmission window.

At the start of our example in Figure 7.10, everything is working as planned, the first segment sent ('segment 1') is acknowledged by the receiver (with ACK 1). On receipt of the acknowledgement, the sender slides the transmission window and works out that a further 536 octets may now be sent (the other 1072 octets still in the window correspond to the as-yet unacknowledged segments '2' and '3'). The newly recalculated position of the transmission window after the receipt of the ACK is as shown on the left-hand side of Figure 7.10. So, using the free space in the window, the sender sends a further 536 octets as 'segment 4'. But since 'segment 2' got lost in transmission and did not reach the receiver (such a *packet loss* could happen, for example, due to a corruption of the destination IP address), the receiver stops sending acknowledgements. Thus the 'dotted' arrow representing the incomplete path of 'segment 2' and the missing ACK2 message are shown dotted in Figure 7.10. Not even 'segment 3' (which does turn up successfully at the receiver) can be acknowledged, since to send an acknowledgement for it would signal a *cumulative acknowledgement* of the successful receipt of both 'segment 2' and 'segment 3'.

Eventually the *retransmission timer* for the data corresponding to 'segment 2' expires (i.e., exceeds the retransmission timeout, RTO), and the sender recovers the situation by commencing retransmission of segments '2', '3' and '4'.

The window size is continuously adjusted during the period of the TCP connection (see Table 7.5). The window size advised by the receiver to the transmitter reflects its current capability to receive and store data. At the start, the *initial window (IW)* size is usually set to be the equivalent of one segment of the maximum segment size (MSS). But if communication proceeds smoothly, the window size is usually increased quickly to the maximum window size the receiver is able to handle. This ensures the quickest and most efficient transport of data. The maximum window size the receiver is likely to permit will depend upon the total memory space available for the receiver's buffer and the current level of its usage. The total memory

Table 7.5   TCP congestion control and window parameters

Congestion control, flow control or window parameter abbreviation	Full parameter name	Parameter relevance	Calculation of value
Awnd	Allowed window	The window size which the sender is currently allowed to use (this is the value set by the receiver).	The minimum value of Cwnd and Rwnd
Cwnd	Congestion window	The current window theoretically allowed window size as determined by calculation based on network performance.	Depends on previous receipt of segments and thus on network performance
Flight size	Flight size	The number of octets sent but not yet acknowledged	This value is tracked by the sender
IW	Initial window	The initial window size established during connection set-up	IW = 1 segment of length equal to MSS
LW	Loss window	The value to which the window size is reduced after a packet loss	Depends upon the loss recovery method
MSS	Maximum segment size	The maximum allowed size of the TCP-data field in octets	Negotiated at connection set-up
Rwnd	Receiver window	The maximum window size currently permitted by the receiver	Depends upon current state of receiver
SSTHRESH	Slow start threshold	The maximum window size allowed under the slow start procedure. Above this window size, the window control must revert to congestion avoidance	Pre-configured

space available for the receiver's buffer can be considered to comprise three subdivisions (as shown in Figure 7.10):

- a part storing received TCP data, which has already been acknowledged to the sender, but has not yet been shifted into another part of the computer's memory by the relevant receiving application;

- the window part of the receive buffer which has been made available to the sender for receiving segments which are yet to be acknowledged; and

- some spare space.

The receiver cannot afford to make its entire memory space available as a *receive window*. A certain proportion is taken up by data waiting to be taken up by an *application protocol*. The slower the rate of take-up, the smaller the space available for the receive window. In addition, it is prudent always to retain some spare space.

Next, we describe in detail how different TCP *congestion control algorithms* allow the receiver to adjust the window size to optimise the speed of TCP communication, given a maximum window size the receiver is prepared to make available (the *receiver window — Rcwd*).

Slow start procedure

Under the *slow start procedure*, the *initial window (IW)* is set quite small (equivalent to 1 segment of maximum segment size, MSS) in order to prevent the sender instantly swamping the receiver with data. But provided the transmission goes well and acknowledgements are received as expected, the window size is increased exponentially by the sender (as shown in the first part of Figure 7.11). This is called the slow start procedure. A slow start procedure is also often combined with *congestion avoidance*, once the window size exceeds a threshold termed *SSTHRESH* (*slow start threshold* value), as we shall see next.

Congestion avoidance

Once the window size has been increased by the slow start procedure to a value exceeding SSTHRESH (the slow start threshold value), the *congestion avoidance* procedure for controlling the window size takes over. Under *congestion avoidance*, the window size continues to be increased by the sender — but <u>linearly</u> instead of exponentially. In other words, the rate at which the window size is increased is not as great as in the slow start procedure. This is a prudent precaution if such a rapid increase in window size might lead to either or both of receiver and network congestion. The maximum size to which the window is increased corresponds to the maximum buffer space the receiver is prepared to make available. This is called the *receiver window (Rcwd)*. Congestion *avoidance* is illustrated in Figure 7.11.

Adjustment of the TCP window size after a packet loss

Should a packet be lost, the window size is immediately reduced to a pre-determined value called the *loss window (LW)*. This prevents the continuing transmission of a large amount of data across the network towards the receiver at a time when it may be unlikely to arrive successfully and might simply worsen network congestion. If the loss window value is less

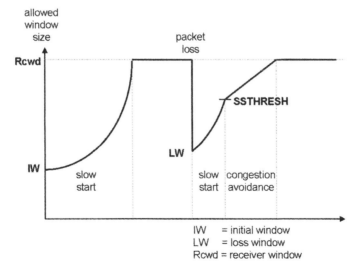

Figure 7.11   Slow start and congestion avoidance procedures.

than the SSTHRESH value, then the window is once again increased according to the slow start and congestion avoidance procedures (Figure 7.11).

The detection of a packet loss takes place either by expiry of the retransmission timeout (RTO) in the sender or as the result of a *fast retransmit request (FRR)*, which we discuss next.

Fast retransmit

The *fast retransmit* procedure allows a receiver to request the retransmission of segments by duplicating acknowledgement (ACK) messages. Thus in the example of Figure 7.10, once the receiver had detected that 'segment 2' had gone missing (because 'segment 3' had turned up before it), it could request a *fast retransmission* of 'segment 2' by repeating ACK 1 three further times. (This is called a *fast retransmit request, FRR*). (Remember that the ACK 1 message is an indication that the first octet of 'segment 2' is the next octet expected to arrive at the receiver).

By generating a fast retransmit request (FRR) a receiver may be able to reduce the time which will elapse before the sender retransmits the missing segment, particularly in the case that the *retransmission timeout (RTO)* is set to a relatively long duration.

Fast recovery

Fast recovery is a method of quickly recovering (retransmitting) lost segments before adjusting the transmission into the congestion avoidance mode. Following a packet loss, the window size is immediately reduced by the sender without requiring advice from the receiver, and the sender continues to calculate a theoretical maximum allowed *congestion window (Cwnd)*. The procedure is illustrated in Figure 7.12.

The calculation by the sender of the theoretical congestion window value (Cwnd) during fast recovery follows a three-stage procedure (Figure 7.12).

- The *loss window (LW)* is calculated as $Cwnd = SSTHRESH + 3 \times MSS$
  Where: $SSTHRESH = $ max of $(1/2 \times $ flight size [4] and $(2 \times MSS)$

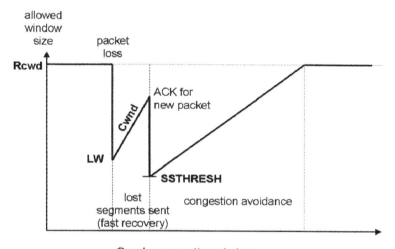

Cwnd = congestion window

Figure 7.12 TCP fast recovery procedure.

---

[4] See Table 7.5.

- The *fast recovery* phase window size is: $Cwnd = Cwnd$ (previous) $+$ MSS
  Whereby: the increment takes place for each subsequent ACK

- As soon as all the packets have been recovered, (i.e., the ACK received indicates that the receiver is expecting a new segment), the congestion window is *deflated* to the value SSTHRESH and the window size control procedure is returned to the congestion avoidance mode.

Two slightly different versions of TCP fast recovery are defined. The first (defined in RFC 2581) is the standard method of TCP fast recovery. It makes use of the *selective acknowledge (SACK)* message to acknowledge receipt of all those TCP segments which have actually arrived, thereby restricting the number of segments which need to be retransmitted during the fast recovery phase to those which were actually lost. RFC 2582 defines an alternative fast recovery technique without using the SACK message. This method is better suited to packet losses in which more than one packet/segment goes missing.

Selective acknowledgement (SACK)

The selective acknowledgement option for TCP is defined in RFC 2018. It was developed for use in high speed networks, where large window sizes are necessary to ensure efficient utilisation of the line bit rate. (A small window size would mean that most of the line bit rate would be wasted — while waiting for acknowledgements.) The problem with a large window size, when using cumulative acknowledgement, is that a large number of segments may need to be retransmitted after each packet loss, even though some of the original packets arrived successfully at the receiver. The retransmission, in turn, will unnecessarily congest the network and unnecessarily load the receiver. Selective acknowledgement (in contrast to cumulative acknowledgement) allows the individual segments which were received successfully to be acknowledged, thereby reducing the number of segments requiring to be retransmitted to those actually lost.

If it is to be used, the selective acknowledgement (SACK) option (TCP option '5' — see Table 7.4) must be *permitted* (TCP option '4') by negotiation during the connection set-up phase.

Silly window syndrome (SWS)

The *silly window syndrome (SWS)* and its resolution are described in RFC 813. Silly window syndrome is the inefficient use by a sender of much smaller segment sizes than allowed by the *MSS (maximum segment size)* and *window size* limits. SWS can arise when a sender is trying to send a large quantity of data without much interruption. In such a case, the sender usually has transmitted as much information as allowed by the transmission window, and is waiting continually for new acknowledgements, so that further segments may be sent.

Let us imagine an example in which the window size is 1000 octets and the MSS is 200 octets. The sender has a lot of data to send, so it naturally starts sending segments all of the MSS of 200 octets. It sends five segments immediately and then has to wait for an acknowledgement. Following each acknowledgement, it sends another segment. But maybe there is a momentary drop in the amount of data to be sent, or a segment is *pushed* by the application software (as explained in Table 7.3 a pushed segment has to be sent immediately rather than waiting for further data). A short segment results (let us assume of 40 octets in length). After the momentary lapse, transmission returns to the full rate, so the sender makes use of the full capacity of the window and sends a further segment of 160 octets, and continues sending segments immediately in response to each acknowledgement....but now every sixth acknowledgement only allows a segment of a maximum of 40 octets to be sent!! This effect is called the silly window syndrome (SWS). Without preventive action, the use of very small windows (silly windows) will continue until the volume of data required to be transported by the connection subsides.

The preventive action against silly window syndrome (SWS) is provided by means of the *Nagle-algorithm*. The transmission of data by the sender is delayed slightly during times of heavy load, until sufficient previous segments have been acknowledged to allow a segment of length equal to the MSS (maximum segment size) to be sent.

Increased initial window (IW) size

As we have discussed, the normal default for the *initial window (IW)* size is the equivalent of one segment of the maximum segment size (MSS). But RFC 2414 defines an increased initial window size for TCP. The increased initial window size has two specific benefits:

- the receiver is less likely to have to wait for a timeout before *acknowledging* the receipt of the first segment (if, for example, the IW size is more than twice the MSS, then an acknowledgement <u>must</u> be sent straight away). This increases the efficiency of the line utilisation — particularly in the case of very high speed lines; and

- a larger proportion of data messages can be sent as single segment messages. This greatly increases the speed of communication. Single segment messages are relatively common when 'surfing' the Worldwide Web, when individual *webpages* are sporadically downloaded from different *websites*.

Congestion window validation

The need for *congestion window validation* (RFC 2861) arises after long periods of no activity on the TCP connection. During such idle periods, the previously recorded value of the congestion window no longer reflects either the current state of the network or the receiver. So to prevent possible problems which might result from a sudden large transmission of data, the congestion window is slowly *decayed* (i.e., reduced) over time.

Explicit congestion notification (ECN)

The *explicit congestion notification (ECN)* procedure for controlling congestion is defined in RFC 3168. It is an inheritance from *frame relay*.[5] Any time when a packet encounters congestion (the exceeding of a given waiting time or a given queue length) on its way to its destination, then the packet is tagged with a notification message called the *forward explicit congestion notification (FECN)*. This notifies the receiving device and router of the congestion, which may be communicated back to the source by means of a *backward explicit congestion notification (BECN)* message. The sending device may then respond by reducing its rate of data output, or may be forced by the network to do so.

Closing a TCP connection

Just like its setting-up (i.e., synchronisation), the *closing* of a TCP connection is undertaken (after completion of data transfer) by means of a three-way handshake, which may be initiated by either party.

---

[5] See Appendix 9.

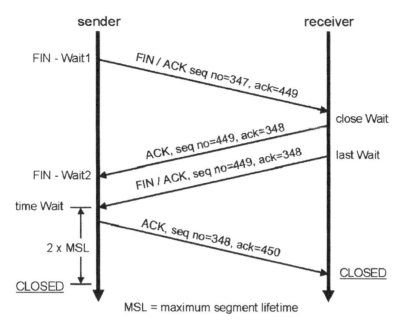

Figure 7.13 Closing TCP connections using a three-way handshake and a waiting period.

As Figure 7.13 illustrates, the request to close the connection is requested by means of a *FIN-message (finish-message)*. This is immediately acknowledged, but the critical FIN/ACK message is not sent until a subsequent waiting period expires. The connection is finally considered closed after a period equal to 2 × MSL (maximum segment lifetime) after receipt of the FIN/ACK message. By waiting more than the *maximum segment lifetime (MSL)*, the communicating TCP peer partners aim to avoid the possibility that a segment 'belonging to' the connection might subsequently turn up (and be ignored)—having been delayed by a multiple hop IP forwarding path.

TCP (and UDP) use with IPv6 jumbograms

Finally, (for completeness) we should note that RFC 2147 presents a method for using TCP or UDP in conjunction with IPv6 jumbograms (IP packets longer than 65535 octets). The normal UDP length field cannot cope with such jumbograms, because the UDP length field is constrained to 16 bits. And while TCP does not use a length field, both the *MSS (maximum segment size)* and *urgent* field are also constrained to 16 bits. When sent via jumbograms, the UDP length field is set to value '0' ('ignore this value') while the TCP MSS value is set to 65 535 (to be interpreted as 'infinity').

7.4   Resource reservation protocol (RSVP)

The resource *reservation protocol (RSVP)* defined by RFC 2205 is a control signalling protocol which allows two communicating hosts to arrange with the network for reservation of bandwidth and other resources (e.g., buffering and forwarding resources) as necessary to ensure a minimum *quality of service (QOS)* for the data transported between the hosts. In this sense,

RSVP is a protocol used to assure high quality of data transport, and in particular for the transport of *flow* or *stream data* (e.g., *real-time* voice, video or multimedia signals). But unlike other *transport protocols*, RSVP is not itself used to carry the actual user data. The user data will typically be carried by *TCP (transmission control protocol), MPLS (multiprotocol label switching)* or even a combination of the two.

The basic process of resource reservation using RSVP

Having established a connection by normal means (i.e., with TCP and IP), RSVP is used in a 'second stage' of connection set-up to reserve the bandwidth necessary to meet the *quality of service (QOS)* as defined by the *flow specification (flowspec)*. The reservation of bandwidth (or other resources) by RSVP is carried out for each direction of communication separately. As far as RSVP is concerned there is a *sending host* (at the *upstream* end of a one-directional connection) and a *receiving host* (at the *downstream* end of the connection). And if reservation of capacity for both directions of communication is necessary, then RSVP needs to be used twice — both ends being established as both upstream and downstream ends of two separate one-way connections!

The reservation process starts with an RSVP *path* message being issued by the *sending host*. The *Path* message describes the nature of the data or *data flow* which will be sent in the form of *path state* information. In particular, the RSVP *Path* message includes the *sender_template* and the *sender_tspec*. The sender_template provides information by which the packets of the relevant data flow (for which capacity is to be reserved) can be unambiguously distinguished from other data which might be sharing the same path. The *sender_tspec* describes the *traffic specification* of the data which is to be sent (the bit rate, quality of service required, etc.).

The *Path* message must follow the actual path that the subsequent data will be transported along. It cascades along the connection between RSVP routers, collecting information about the path in an *RSVP_hop* list of IP addresses and in the *adspec (advertising specification)*. Finally, the *Path message* arrives at the *downstream host* (or in the case of a *multicast* message, at a number of *downstream hosts)*. Each downstream host which requires bandwidth reservation must originate its own reservation request based on the path information received.[6] This is done by means of an RSVP *Resv* (reservation request) message.

The *Resv* message is sent upstream from the *receiving host* (using the *RSVP_hop* list to guide it). The *Resv* message makes the *reservation request* by defining its needs in terms of the *style, flowspec* and *filter_spec*. The *style* describes the nature of the reservation required. The flowspec describes the quality of service (QOS) being requested and the filter_spec provides the means by which the data packets will be able to be unambiguously identified (it includes not only information drawn from the *sender_template* but also information about the receiver).

The *reservation request* message (*Resv*) is subjected by each router to an *admission control* procedure. This determines whether the reservation as requested is *administratively permitted* and whether sufficient resources are currently available to be able to meet the request. *Admission control* also tries to ensure that bandwidth is not excessively reserved and subsequently not used. If the reservation is successful, the packets making up the data flow will be subjected to *flow admission control (FAC)* during the main period of data transfer. This ensures that the sender does not exceed his reservation. *Admission control, flow admission control (FAC)* and the enforcement of *policy* are critical to the good functioning of networks offering RSVP. An important part of the *policy* needs to be the creation of a 'back-pressure' on users to dissuade them from excessive reservation. Charging for the reservation is probably the best method of achieving this.

---

[6] An RSVP reservation request (*Resv*) may also be made even if a *Path* message is not received. In this case, the details of the path must already be known to the receiver.

If requested at the time of the reservation request (*Resv*), a *reservation confirmation (Resv-Conf)* will be generated by the *sending host* (once the *Resv* message successfully reaches it) and sent downstream to the receiver.

If there are any message or procedural errors during the RSVP *signalling* process (or a rejection of the reservation), then a *PathErr (path message error)* or *ResvErr (reservation request message error)* will be generated as appropriate. Finally, the reservation will be *torn down* after the communication is finished. The reservation may be torn down either by the sender (by issuing a *PathTear* message *downstream*), by the receiver (by issuing a *ResvTear* message upstream), or by any of the intermediate RSVP routers (which must send both a *PathTear* message downstream and a *ResvTear* message upstream).

What sort of resources and for which types of services can I make an RSVP reservation?

RSVP (resource reservation protocol) allows for the reservation of network resources as are necessary to fulfil, either a guaranteed *quality of service (QOS)* specification (as defined in RFC 2212) or the *controlled-load network element service (CLNES* or *CLS)* (as defined in RFC 2211). Both QOS and CLNES schemes 'assure' bandwidth and other network resources by creating a priority scheme for the packet header processing and forwarding of network layer packets. Higher priority packets get to use the resources first. The principles of how quality of service (QOS) is 'guaranteed' we discussed under *differentiated services (DiffServ)* in Chapter 5. Thus RSVP is the 'transport protocol' which allows for reservation of bandwidth to ensure that the prioritisation of packets using DiffServ will actually achieve the quality of service (QOS) specification required for the data *flow*. The additional *control signalling* and resource reservation organised by RSVP eliminates the need for *hop-by-hop* packet *filtering* and re-*prioritisation* of data flow packets which are necessary with straightforward DiffServ (differentiated services). The *flow admission control (FAC)* and policing only need to be undertaken once (at the first router in the connection), after which packets are simply forwarded along the reserved path. This makes for faster forwarding with less effort.

Resource reservations can be undertaken using RSVP either for a specific user (i.e., *receiving host*), or for a specific service (i.e., specific application protocol, port or socket) being used by that user (host). The user will typically be using one of the following combinations of protocols:

- Internet Protocol version 4 (IPv4) TCP/UDP and differentiated service (DiffServ). In this case, the reservation is based upon the IPv4 address and/or the DiffServ *behaviour aggregate*;

- Internet Protocol version 6 (IPv6) TCP/UDP and differentiated service (DiffServ). In this case, the reservation is based upon the IPv6 address and/or the DiffServ behaviour aggregate;

- a label-switched network connection (e.g., an MPLS (multiprotocol label switching) connection — as we shall discuss shortly). In this case the reservation is associated with the connection *label*; or

- an IPv6 label-switched network connection. In this case the reservation is associated with the IPv6 *flow label* (which we saw in chapter 5 — Figure 5.14). (In effect the IPv6 flow label is the same as an MPLS flow label, as we shall see later).

It is important to note that RSVP controls the reservation of resources which will affect the network layer forwarding of packets making up a given data flow. The reservation is necessary in order that the *transport layer* (i.e., directly between the communicating hosts) meets the required *quality of service* of the *application layer*. In this sense, RSVP is a 'transport protocol' (but not one used for the carriage of the actual user data). (In reality, RSVP is an *Internet control protocol*). Transmission control protocol (TCP) will most likely also be in use.

Reservation style

The reservation *style* reflects both the type of reservation (called the *reservation option*) and the type of sender (called the *sender selection*) which are required. The reservation option may be either a *distinct* reservation (i.e., specifically intended by the requesting receiver to apply to a distinct sender) or be *shared* reservation (in which case the reserved resources can be used to carry data from a range of different senders). The *sender selection* (i.e sender identification) may be either by means of an explicit IP address (identifying a single sender) or by means of a *wildcard* (identifying all senders within a specific IP subnet address range). The three different styles (combinations or reservation option and sender selection) defined so far are shown in Table 7.6.

A *fixed filter (FF) style* reservation is a *distinct* reservation intended for a single point-to-point data flow, with no sharing. This is the type of reservation used typically for real-time video and high quality multimedia data flow reservations.

A *shared explicit (SE) style* reservation makes a single reservation of bandwidth, but allows the bandwidth to be used for any of a number of explicitly identified senders. Such a reservation will typically reserve resources sufficient for the most demanding pair of sender/receivers. It could, for example, allow a receiver to 'listen' to any number of different source senders (assuming he or she does not want to 'listen' to them all at the same time). In a shared explicit (SE) style reservation request, a filter_spec must be provided for each sender.

A *wildcard filter (WF) style* reservation provides for a 'shared pipe' whose size is the largest of the resource requests from all receivers. The resources reserved by means of the reservation may be used by any sender within the IP-address range identified by the *wildcard*.[7] A wildcard filter (WF) reservation is automatically extended to new senders as they appear. No filter_spec is required in a request for a wildcard filter (WF) style reservation.

RSVP message format

The RSVP message format (as defined by RFC 2205) is illustrated in Figure 7.14a. Messages are carried directly by the Internet protocol (IP) with *protocol number* 46. They are therefore not delivered *reliably*.

Table 7.6 RSVP reservation attributes and styles

Sender selection	Reservation option	
	Distinct	**Shared**
Explicit	Fixed Filter (FF) style	Shared Explicit (SE) style
Wildcard	(none defined)	Wildcard Filter (WF) style

---

[7] A wildcard is in effect a 'reverse subnet-mask' — in a wildcard all the '1's of the subnet mask are changed to '0' and all the '0's are changed to '1' — so that the '1's range of the wildcard 'blacks out' the least significant bits of the IP address. As long as all the remaining more significant bits of the sender's IP address match the pre-defined value, all the senders with IP addresses in this range will be treated the same.

a) **RSVP message format**

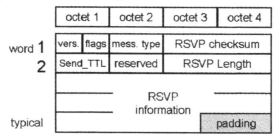

b) **RSVP object format**

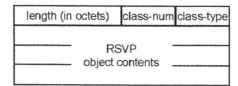

Figure 7.14 RSVP message format and object format.

Table 7.7 RSVP message types

Message type value	Message type	Message purpose
1	Path	*Path* indication and data flow specification message generated by upstream sending host and sent downstream
2	Resv	Quality of Service (QOS) resource reservation req*uest* generated by each receiving host and propagated upstream to the sending host
3	PathErr	Notification of path message error sent upstream
4	ResvErr	Notification of reservation request message error sent downstream
5	PathTear	Teardown message (resource release message) sent downstream
6	ResvTear	Reservation teardown message propagated upstream
7	ResvConf	Confirmation message

The *version number* is a 4-bit field describing the RSVP version number. The current version is version 1. No *flag bits* have yet been defined. The *message type* is one of the seven types listed in Table 7.7. The two fundamental message types, as we discussed earlier at the *Path* and *Resv* message types.

The checksum (as standard for IP-suite protocols) is the 16-bit 'one's complement of the one's complement sum of the message' (the checksum field being replaced by all '0's for the purposes of the calculation). The *send_TTL* field contains the IP TTL value with which the message was sent. The *length* field contains a value reflecting the total length of the RSVP message in octets.

The main information content of an RSVP message is a series of RSVP *objects*, constructed according to the *Backus-Naur Form (BNF)*. The objects are presented as one or more 32-bit words in the *RSVP information field* of Figure 7.14a, according to the format

Table 7.8   RSVP object classes and related class types

Class-num value	Object class type	Meaning	Carried by	Class types
0	NULL	Null class type — to be ignored	Any RSVP messages	
1	SESSION	Contains destination IP address and IP protocol-ID for the session to which the RSVP message applies	All RSVP messages	1 = IPv4/UDP 2 = IPv6/UDP
3	RSVP_HOP	Contains IP address of RSVP node that sent the message. Might be either a previous hop (PHOP, of a downstream message) or the next hop (NHOP, for an upstream message)	Path or Resv messages	1 = IPv4 2 = IPv6
4	INTEGRITY	Cryptographic data	Any RSVP message	—
5	TIME_VALUES	The refresh period used by the creator of the message	All Path and Resv messages	Always value '1'
6	ERROR_SPEC	Specifies and error in a PathErr or ResvErr message or a confirmation for a ResvConf message	PathConf, PathErr or ResvErr messages	1 = IPv4 2 = IPv6
7	SCOPE	A list of sender hosts to which the information is to be forwarded	May appear in any Resv, ResvErr or ResvTear message	1 = IPv4 2 = IPv6
8	STYLE	Defines the reservation style	All Resv messages	Always value '1'
9	FLOWSPEC	Defines the QOS being requested	Resv messages	1 = reserved 2 = integrated services
10	FILTER_SPEC	Defines the subset of session data packets which should receive the desired QOS	Resv messages	1 = IPv4 2 = IPv6
11	SENDER_TEMPLATE	The sender IP address and other demultiplexing information (e.g., port number) to identify the sender in a path message.	All Path messages	1 = IPv4 sender 2 = IPv6 sender 3 = IPv6 flow label identifies sender
12	SENDER_TSPEC	Defines the traffic characteristics of the sender's data flow (carried in a path message)	All Path messages	2 = integrated services
13	ADSPEC	Carries OPWA (one-pass with advertising) data	Path messages	2 = integrated services

Table 7.8 *(continued)*

Class-num value	Object class type	Meaning	Carried by	Class types
14	POLICY_DATA	Carries information to determine whether the reservation is permitted by the administration	May appear in Path, PathErr, Resv or ResvErr messages	Always value '1'
15	RESV_CONFIRM	The IP address of the receiver which requested a confirmation	May appear in a Resv or ResvConf message	1 = IPv4 2 = IPv6

shown in Figure 7.14b. Table 7.8 lists the different object classes (identified by numbers — the *class-num*) and the *class-types* associated with each class. We discussed previously the use of the most important of these object classes.

An *RSVP session* (the *object* for which the reservation is made) is defined by the IP destination address, the *ProtocolID* and the destination *port* (TCP or UDP port), but the *port* may optionally be omitted.

In the case of reservations made for multicast flows, *Resv flowspecs* are *merged* — as the reservation requests from the multicast destinations converge on the source, each RSVP node selects only the largest of the requests received to be forwarded as a request to the next previous hop.

RSVP aggregation is the combining of individual RSVP reserved *sessions* into a common differentiated services (DiffServ) class. This can be useful for defining a reservation *aggregation region* in the case that the reservation is to transit different networks or *domains* as defined in RFC 3175.

7.5 MPLS (multiprotocol label switching)

Multiprotocol label switching (MPLS) is a recent addition to the *network layer* protocol (i.e., IP layer) to enable much faster 'switching' of packets making up a data flow (or stream). The main MPLS processing and sorting of packets takes place just once — at the beginning of a connection. A *label* is added at this stage to the packet header, and the label alone will control the switching process along the entire length of the connection (called the *label-switched path, LSP*) — steering it along the previously determined path without requiring further unpacking or processing of the normal IP-header. By pre-determining the path and using the label, the intermediate routers are thereby spared the tasks not only of IP-header processing, but also the onerous task of *hop-by-hop forwarding* according to their individual routing tables. The result is much faster 'switching' of the data stream and much lower processing load for the intermediate routers.

MPLS connections are well suited to the fast-forwarding (also called switching) of any type of *network layer* protocol (not just IP), hence the word *multiprotocol* in the name. It will be widely used for two main types of application:

- switching of connections for real-time data streams (such as video, multimedia or voice-over-IP [VOIP]). In this application, MPLS is likely to be combined with *RSVP (resource reservation protocol)* to reserve bandwidth along the *label-switched path (LSP)*, thereby assuring a given *quality of service (QOS)* for transport of data; and

- creating *virtual private networks (VPNs)*. Because the labels used to identify and switch all the packets making up a given MPLS communication are assigned by the network routers (as opposed to IP-addresses in normal IP packets, which are provided by end-users), it is not possible for 'hackers' to 'spoof' the label and thereby maliciously 'muscle in' on the communication. As a result, MPLS provides a reliable way of creating secure point-to-point connections across a public router network (i.e., the Internet). Such *label-switched path (LSP)* connections (marketed as either *VPN services* or *tunnels*) are now offered by all the major Internet backbone network service providers for use in building economic enterprise or other private router networks.

The principles of *label switching*[8] owe much to the *X.25*, *frame relay and ATM (asynchronous transfer mode)* protocols which pre-dated it. The specific technique and network architecture used in MPLS (which, during early development was also called *tag-switching*) is defined in RFC 3031. Packets are sorted into so-called *forwarding equivalence classes (FEC)* only once at the beginning of the *LSP (label-switched path)* and *labelled*. (Packets in the same *FEC* belong to the same *connection*, data *flow* or data *stream*.) The packets then follow a pre-determined path, along the course of which the label is known — and used by the routers for fast forwarding of packets. The MPLS connection is set-up using a *label distribution protocol (LDP)*, which both determines the path which will be used and advises all the affected routers of the label or labels which will be used to identify the *forwarding equivalence class (FEC)*.
    There are three main ways of realising an MPLS network:

- *MPLS-LDP* (MPLS using *label distribution protocol*). This is the 'purest' form of MPLS, and may be implemented either using IPv6 (by employing the *flow label* field of the IPv6 header — see Chapter 5, Figure 5.14) or as a network layer protocol *encapsulation* scheme (called *MPLS-SHIM*). *MPLS-LDP* is defined by RFCs 3031, 3032 and 3036.

- *MPLS-FR* (MPLS over *frame relay*). This is a version of MPLS based upon a *frame relay* transport network. It is defined by RFC 3034.

- *MPLS-ATM* (MPLS over *ATM*). This is a version of MPLS base upon an *ATM (asynchronous transfer mode)* transport network. It is defined by RFC 3035.

And there are two main applications:

- *MPLS-BGP* (MPLS using the *border gateway protocol* for VPN path management). It is defined in RFC 2547).

- *MPLS-VPN* (*virtual private networks* by means of MPLS, as defined in RFC 2917).

The 20-bit flow label and its capacity constraints

The *flow label* used in MPLS-LDP is a 20-bit label (corresponding to approximately 1 million different labels). The relatively short label (compared with a 32-bit IPv4 address or 128-bit IPv6 address) adds to the speed with which packets can be processed and switched by intermediate routers. But you may wonder how only one million labels could possibly be sufficient for all the different possible connections which might be necessary. The answer is that the labels are either only *interface-significant-labels* or *domain-significant-labels*.
    An interface-significant-label is recognised only at a single interface — by the two routers at each end of that interface. Because interface-significant-labels are being used in the example of Figure 7.15, the MPLS path (router 1-router 2-router 3-router 4) uses three separate labels to

---

[8] See Chapter 3.

MPLS domain

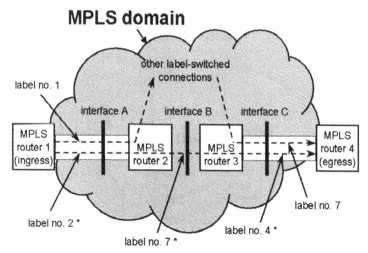

Note: *The label numbers used at each interface making up the path do not share the same value (except when using domain-signicant labels).

Figure 7.15 MPLS interface-significant labels.

traverse the three interfaces (label number '2' at interface A; label number '7' at interface B; and label number '4' at interface C). The limitation of the labelling space range under interface-significant-labelling is one million MPLS connections simultaneously in use per MPLS interface.

In contrast to interface-significant-labelling is MPLS domain-significant-labelling. Under domain-significant-labelling, the same label will be used for the same FEC by all routers within the *MPLS domain*. The advantage of domain-significant labelling (over interface-significant labelling) is that the label does not have to be *swapped* by each of the routers along the path of the connection during the 'switching' process. This saves time and effort. The disadvantage is that much fewer simultaneous MPLS connections are possible with domain-significant labelling (only one million per domain).

MPLS network architecture and basic functioning

MPLS domains typically comprise a number of MPLS-capable routers within the *backbone* part of a given administration's network. The MPLS domain is thus typically surrounded by an IP-routing domain, used at its periphery to connect more remote devices. There are four different types of nodes which together make up an MPLS domain (see Figure 7.16):

- *MPLS nodes* are routers capable of supporting MPLS services;

- *MPLS edge nodes* are nodes at the edge of an MPLS domain. Edge nodes perform the conversion of other network layer protocols into the MPLS format or provide for gateway functions between different MPLS domains;

- *MPLS ingress nodes* are MPLS edge nodes at the point where MPLS traffic is origi-nated — usually by entering from a non-MPLS routing domain. (MPLS is only concerned with one-directional paths — two separate one-directional paths are created to form a 'duplex' path);

- *MPLS egress nodes* are MPLS edge nodes at the point where the data flow leaves the MPLS domain for delivery via a non-MPLS domain.

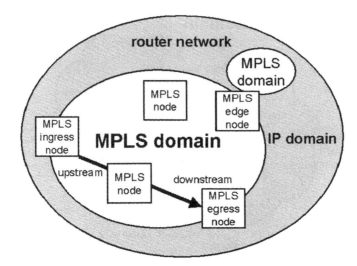

Figure 7.16   MPLS network architecture and node types.

Upstream and downstream MPLS nodes

With relation to a particular MPLS connection, the MPLS nodes which are transmitted are either upstream or downstream of one-another. Upstream nodes are those nodes nearer the source of the traffic (i.e., nearer the *ingress node*) while downstream nodes are those nearer the destination (i.e., the *egress* node).

It is the downstream MPLS node (like the *receiving host* or *downstream host* in RSVP — resource reservation protocol) which controls the setting up of an MPLS connection. The process involves assigning connection labels and *distributing* them (i.e., notifying the next upstream node which label should be used to identify the connection). After the connection is no longer needed, the MPLS connection and its corresponding label are *withdrawn* by the downstream node.

An upstream MPLS node may *request* the assignment of a label by the downstream node, but otherwise merely responds to the controls of the downstream node while setting up and withdrawing the connection. Messages issued by the upstream node (to the downstream node) include: *NotAvailable* (label or connection cannot currently be used); *release* and *labelUse* notifications.

The selection of a route for the MPLS connection and the distribution (i.e., notification) of the labels to be used at each interface (or within each MPLS label-domain) is undertaken using the *label distribution protocol (LDP)*. LDP (for MPLS-LDP) is defined in RFC 3036. The route selection itself (which is carried out before the user data is sent) can either be determined in a *hop-by-hop* (this is the normal method) or using *explicit routing*. The route decision will be based on the IP address (or other network layer address) and can therefore use the routing information obtained using standard routing protocols. The route of a *LSP (label-switched path)*, however, will typically differ from the normal IP-packet path — by virtue of service or administrative routing constraints. When using *explicit routing* the intermediate points of the connection are all dictated by the downstream node (this is equivalent to source based routing — except that it is carried out at the destination).

The label stack and label format

In order to facilitate the fast swapping of labels which must be undertaken in MPLS domains using *interface-significant labels* (Figure 7.15), MPLS allows for the creation of a *label stack*.

The label at the top of the *stack* will be used to identify the *forwarding equivalence class (FEC)* to which the associated packet belongs and perform the necessary *label-switching*. This label is then removed from the label stack, so that a new label identifies the FEC of the packet to the next router in the *LSP (label-switched path)*.

In the 'purest' form of MPLS (the IP-suite version), the labels are 20-bit numbers, which are indicated either in the *flow label* field of the IPv6 protocol header (Figure 5.14) or in the *MPLS-SHIM* header defined in RFC 3032.[9]

When labels are stored in the label stack, each is stored as a label stack entry of 4 octets (32 bits — Figure 7.18a). The last entry in the stack is the *NULL entry* — indicating the bottom of the stack. When this label is reached, the packet has progressed all the way along the *label-switched path (LSP)* to the MPLS egress node. The label must now be *popped* (i.e., removed) and the normal network layer protocol must be passed to the appropriate *protocol engine* for processing.

But not all implementations of MPLS allow the creation of a label stack or the use of the standard 20-bit labels defined in RFC 3032. ATM (asynchronous transfer mode) and *frame relay* implementations instead use their normal *VPI/VCI (virtual path identifier/virtual channel identifier)* and *data link channel identifier (DLCI)* respectively. The ATM *VPI/VCI* 'label' is a 28-bit identifier.[10] The frame relay *DLCI* is either a 10-bit or 23-bit field.[11] The MPLS-LDP (label distribution protocol) as defined in RFC 3036 is not used in the case of either frame relay (FR) or ATM. Instead, RFCs 3034 (MPLS-FR) and 3035 (MPLS-ATM) apply. No label stack is used. Instead, both FR and ATM employ label swapping at each node (since both employ *interface-significant-labelling*).

Use of MPLS in conjunction with RSVP

For *on-demand* MPLS connections, it is common to use *RSVP (resource reservation protocol)* in addition to the *label distribution protocol (LDP)* at the time of connection set-up. Used in conjunction with MPLS, RSVP policing is made much simpler, since the filter specification (RSVP filter_spec) need only be applied once at the beginning of the connection and not at each hop along the way.

Example MPLS connections, 'nested' connections and their usage

MPLS connections (sometimes also called *tunnels*) are a useful way of connecting devices for fast, reliable and secure (i.e., private) communication. Although the MPLS connection actually traverses a router network, it appears to the devices at its two ends as if it were a 'transparent' direct physical connection of the two devices.

As far as the *level 1* MPLS nodes of Figure 7.17 are concerned, nodes R2 and R3 are directly connected with one another. In reality, there is a *level 2* connection between them. During carriage of the MPLS packets across the level 2 connection, the MPLS level 1 label has itself been *encapsulated* by an MPLS level 2 label. Meanwhile routers A and B at either end of the MPLS connection are unaware that any MPLS is taking place at all. These two routers think they are a single network layer *hop* (e.g., one IP-hop) apart. The *level 1* MPLS connection traverses routers R1-R2-R3-R4; the level 2 MPLS connection traverses routers R2-R21-R22-R3.

---

[9] The MPLS-SHIM header is an extra network layer protocol encapsulation header inserted between the normal network layer packet header (e.g., IP header) and the datalink frame header.
[10] See Appendix 10.
[11] See Appendix 9.

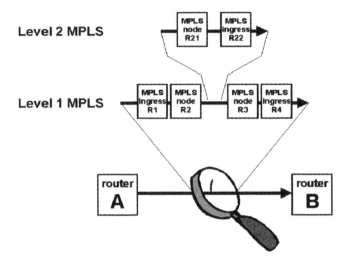

Figure 7.17   'Nested' MPLS connections of different levels.

Planned and used carefully, MPLS connections have the capability to simplify the topology and management of IP-layer networks, as well as providing for *VPN (virtual private network)*[12] connections or tunnels. On the other hand, the MPLS network itself could become exceedingly complex and unwieldy!!

MPLS will also play an important role in adapting the powerful switching capabilities of frame relay and ATM networks for use in IP-based backbone networks for the switching of multimedia and real-time multimedia signals.

MPLS-SHIM label format

Figure 7.18a illustrates the standard encoding of *label stack entries* under the MPLS-SHIM encapsulation scheme defined by RFC 3032. This is the 'normal' MPLS label scheme. The same basic label format is also used in the IPv6 header (Figure 5.14). Each *label stack entry* comprises 4 octets (32 bits), including 20-bits for the label itself, a *bottom-of-stack* bit (S) and a *TTL (time-to-live)* field.

When using the *SHIM* version of MPLS, the label stack is inserted between the *datalink frame header* and the *network layer packet header*.

The label values are coded according to Table 7.9. A null label is used at the bottom of the label stack, to indicate that the label should be popped and the network layer packet within it passed to the relevant network layer protocol 'engine' for normal network layer header processing and forwarding.

MPLS Label distribution protocol (LDP)

The *MPLS-LDP (label distribution protocol* defined by RFC 3036) provides the means by which *LSRs (label switching routers)* establish *LSPs (label-switched paths)* and notify one another of the labels to be used to identify packets belonging to a particular *forwarding equivalence class (FEC)*.

---

[12] See Chapter 13.

a) label stack entry

b) TLV (type length value) information

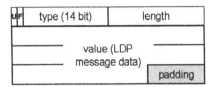

EXP= bits reserved for experimental use
F = forward unknown TLV type message
S = bottom of stack indicator bit
U = unrecognised TLV type

Figure 7.18 MPLS-SHIM label stack entry format and LDP (label distribution protocol) TLV information format.

Table 7.9 MPLS label values

Label value (20-bits: 0–1 048 576)	Meaning
0	IPv4 explicit NULL label (bottom of label stack — pass to IPv4 header processing)
1	Router alert label (forwarding determined by next lower label in stack)
2	IPv6 explicit NULL label (bottom of label stack — pass to IPv6 header processing)
3	Implicit NULL label (used only in label distribution protocol but never appears in label stack)
4-15	Reserved
16-	Normal label values

The labels used by MPLS and communicated by means of LDP are stored in a *label information base (LIB)* by each relevant *MPLS router*. Exchanges of LDP information between these routers take place between *LDP peers*.

The information transmitted by an LDP message has an informational *object* called a *TLV (type length value)*, as illustrated in Figure 7.18b. The different types of *TLV object* enable the *LDP peers* to exchange all relevant information about the *forwarding equivalence class (FEC* — i.e., how to identify the packets making up the MPLS connection), the label to be used to identify the FEC (whether a *generic label*, an *ATM label* or a *frame relay label*), about the session which will be conducted, the path which will be taken and the current status of the network. The various TLV object types (and thus the codings of the various fields of Figure 7.18b) are described in Table 7.10.

Table 7.11 lists the various types of MPLS-LDP messages. The four most important message types are:

Table 7.10 MPLS LDP (label distribution protocol) TLV (type length value) information field coding

TLV information type	U-bit value	F-Bit value	Type value	Length value	Value field content and meaning
FEC TLV	0	0	0x0100	Multiples of 4 octets	FEC elements (Wildcard, prefix of host address)
Generic label TLV	0	0	0x0200	8 octets	20-bit label and padding
ATM label TLV	0	0	0x0201	8 octets	2 reserved bits; 2 V-bits (indicated whether VPI and/or VCI are significant) 12-bit VPI and 16-bit VCI
Frame Relay label TLV	0	0	0x0202	8 octets	7 reserved bits, then 2-bit length field (0 = 10 bit DLCI, 2 = 23 bit DLCI) then 23-bit DLCI field
Address list TLV	0	0	0x0101	Variable	2-octet address family (address family numbers as RFC 1700) then addresses
Hop count TLV	0	0	0x0103	5 octets	1-octet hopcount (HC) value
Path vector TLV	0	0	0x0104	Multiples of 4 octets	List of 4-octet LSR IDs
Status TLV	U	F	0x0300	15 octets	4-octet status code, 4-octet message ID, 2-octet message type
Extended status TLV	U	F	0x0301	Variable	Included in notification message
Returned PDU TLV	U	F	0x0302	Variable	Included in notification message
Returned message TLV	U	F	0x0303	Variable	Included in notification message
Common hello parameters TLV	0	0	0x0400	8 octets	2-octet hold time, T-bit (targeted hello), R-bit (request send targeted hellos), 14-bits reserved
IPv4 transport address TLV	0	0	0x0401	8 octets	4-octet IPv4 address in hello message
Configuration sequence number TLV	0	0	0x0402		Included in hello message
IPv6 transport address TLV	0	0	0x0403	20 octets	16-octet IPv6 address in hello message
Common session parameters TLV	0	0	0x0500	18 octets	2-octet protocol version, 2-octet keepAlive time, A-bit (label advertisement discipline), D-bit (loop detection), 6-bits reserved, 1-octet PVLim (path vector limit), 2-octet max PDU length, 6-octet LDP identifier)
ATM session parameters TLV	0	0	0x0501		Included in initialisation message

Table 7.10 (*continued*)

TLV information type	U-bit value	F-Bit value	Type value	Length value	Value field content and meaning
Frame Relay session parameters TLV	0	0	0x0502		Included in initialisation message
Vendor-private TLV	U	F	0x3F00-0x3EFF	Variable	4-octet vendor ID field followed by proprietary format of data fields.
Experimental TLV	U	F	0x3F00-0x3FFF	Variable	Experimental extension formats

Table 7.11   MPLS-LDP message types

Message type	U-bit value	Type value	Length	Mandatory parameter field	Optional parameter field
Notification message	0	0x0001		Status TLV	0x0301 extended status 0x0302 returned PDU 0x0303 returned message
Hello message	0	0x0100		Common hello parameters TLV	0x0401 IPv4 transport address 0x0402 configuration sequence number 0x0403 IPv6 transport address)
Initialisation message	0	0x0200		Common session parameters TLV,	0x0501 ATM session parameters 0x0502 FR session parameters
KeepAlive message	0	0x0201		Field not used and omitted	None defined
Address message	0	0x0300		Address list TLV	None defined
Address withdraw message	0	0x0301		Address list TLV	None defined
Label mapping message	0	0x0400	Variable	FEC TLV followed by Label TLV	Label request message ID TLV Hop Count TLV Path vector TLV
Label request message	0	0x0401	Variable	FEC TLV	Hop Count TLV Path vector TLV
Label abort request message	0	0x0404		FEC TLV followed by Label request message ID TLV	None defined
Label withdraw message	0	0x0402	Variable	FEC TLV followed optionally by Label TLV	Label TLV
Label release message	0	0x0403		FEC TLV followed optionally by Label TLV	Label TLV

a) LDP message header

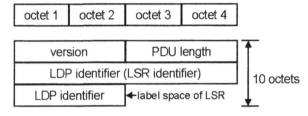

b) LDP general message format

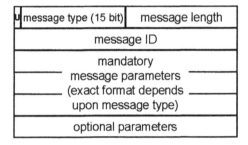

Figure 7.19   MPLS-LDP message format.

Table 7.12   MPLS-LDP Status TLV: status codes

Status data coding (4 octets — here shown in hex)	Status
00 00 00 00	Success
00 00 00 01	Bad LDP identifier
00 00 00 02	Bad protocol version
00 00 00 03	Bad PDU length
00 00 00 04	Unknown message type
00 00 00 05	Bad message length
00 00 00 06	Unknown TLV
00 00 00 07	Bad TLV length
00 00 00 08	Malformed TLV value
00 00 00 09	Hold timer expired
00 00 00 0A	Shutdown
00 00 00 0B	Loop detected
00 00 00 0C	Unknown FEC (forwarding equivalence class)
00 00 00 0D	No route
00 00 00 0E	No label resources
00 00 00 0F	Label resources / available
00 00 00 10	Session rejected / no hello
00 00 00 11	Session rejected / parameters advertisement mode
00 00 00 12	Session rejected / parameters max PDU length
00 00 00 13	Session rejected / parameters label range
00 00 00 14	KeepAlive timer expired
00 00 00 15	Label request aborted
00 00 00 16	Missing message parameters
00 00 00 17	Unsupported address family
00 00 00 18	Session rejected / bad KeepAlive time
00 00 00 19	Internal error

- LDP *advertisement messages* (label request and withdrawal) — used to create, change and delete mappings for FEC (forwarding equivalence classes) as needed to establish and withdraw MPLS connections;

- LDP *notification messages* — used to provide advisory information and signal errors;

- LDP *session messages* — used to establish, maintain and terminate sessions between LDP peers for the purpose of exchanging LDP messages between MPLS LSRs (label-switched routers); and

- LDP *discovery messages* — used to announce and maintain presence of an *LSR (label switched router)* in the network, to operate a *hello protocol* and maintain *adjacencies* (rather like the OSPF routing protocol — LDP is after all, itself the *routing protocol* used for MPLS networks).

All LDP messages (Table 7.11) have the same message format, as illustrated in Figure 7.19. They comprise the LDP message header (Figure 7.19a) and a series of TLV objects, each coded in the same general message format (Figure 7.19b). Each mandatory or optional message parameter of Figure 7.19b is an object TLV (Figure 7.18b Table 7.10).

LDP discovery messages are carried on UDP (user datagram protocol) on port 646. *LDP session messages* (including advertisement and notification messages) are carried on TCP (transmission control protocol) port 646.

Finally, Table 7.12 lists the status codes returned in notification messages.

8

IP Networks in Practice:
Components, Backbone
and Access

This chapter is concerned with building real IP networks; with the structure and components of such networks. We shall start by considering the architecture of a typical IP-based data network. We discuss in detail the different types of routers, the WAN technologies available for interconnecting them and the considerations which should go into backbone network topology design. We then move on to discuss the access network, the technologies available to connect end-users to the network and the relative strengths of each: leaselines, dial-in, xDSL and wireless. And during the discussion we shall introduce two important further protocols used in the access arena: PPP (point-to-point protocol) and PPPoE (point-to-point protocol over Ethernet). These protocols are important for discovering and configuring access network connections.

8.1   The components and hierarchy of an IP-based
data network

An IP-based data network comprises a number of interconnected routers of two broad types: *backbone routers* and *access routers*. As their name suggests, access routers are specifically designed to provide the connection of groups of end-users to the network. The end-user devices themselves are typically connected to access routers by means of *local area networks (LANs)*. Backbone routers, meanwhile, are the routers used in the main wi *de area* part of the network (Figure 8.1). Both types of router come in all different shapes and sizes, attuned to the different traffic and service needs of a particular usage.

No matter whether the network is a public *Internet service provider (ISP)* network, an *Internet backbone* network, an *enterprise*-wide IP network or a campus network it is likely to have a hierarchical structure. The nearer a router is to the *core* of the network, so the greater its traffic-carrying capacity has to be and the more sophisticated its routing table has to be. The *core routers* need to be able to cope with the most amount of packet forwarding and require routing tables which are able to recognise all the *reachable* destinations.

In a small campus or even an enterprise-wide IP network, the number of routers may be relatively small, so that a high degree of 'meshing' is possible. By directly interconnecting many of the routers, a very robust network can be built which is relatively insensitive to individual

Data Networks, IP and the Internet: Protocols, Design and Operation   Martin P. Clark
© 2003 John Wiley & Sons, Ltd   ISBN: 0-470-84856-1

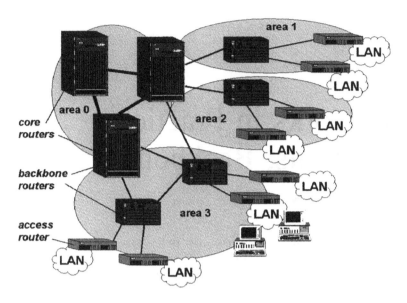

Figure 8.1   Typical hierarchy of an IP-based router network.

router or link failures. Provided the network is relatively small, it may also be possible to operate it as a single *routing protocol domain*,[1] so that the optimum routes through the network will be chosen and maintained automatically by an *interior gateway protocol (IGP)* such as *OSPF (open shortest path first)*. This keeps life pretty straightforward, as does standardising the network and transport protocols used. But as networks grow larger, things get much more complex.

In a 'public' IP network — such as that operated by an Internet service provider (ISP) or an Internet backbone operator, the scope to standardise and limit the number of different protocols in use is much more limited, the traffic is much greater, as are the distances between routers. The network of some ISPs and backbone operators even comprises multiple different networks (called correctly: *autonomous systems — AS*).

A well thought-out network hierarchy is critical to reliable and efficient operation of all networks. Particularly in large networks, it is necessary to divide the network into different hierarchical areas, simply so that the *routing protocol* is able to function correctly.

Figure 8.1 illustrates the typical hierarchical structure of an IP-based network. Note how the largest routers (the *core routers*) are fully interconnected (fully *meshed*) with one another. Routers in the next layer typically have at least two connections to other network routers (to ensure network redundancy), while the access routers may have only one connection (corresponding to the *default gateway* route). Back-up connections for access routers are typically either by means of dial-up connections or other temporary alternative connections.

Different types of routers

To suit the different requirements placed on routers requiring to perform different network and service functions, there are naturally all manner of different types of routers, e.g.:

- *access* or *edge* routers
- *desktop* routers

---

[1] See Chapter 6.

- *workgroup routers*

- dial-access routers

- *firewall*/secure access routers

- *xDSL* access routers

- network routers (both IPv4 and IPv6 routers)

- *VPN (virtual private network)* routers

- *MPLS (multiprotocol label switching)* routers

- high performance routers

- *enterprise switches* and *routers*

- *backbone* or *core* routers

- *border node* or *peering* routers

- *multicast* routers

Each of the different routers is designed for a different speciality — depending upon whether the router is intended for a public IP backbone network, an *enterprise network* or a small private network. Different types of router also reflect different combinations of functions required by a given type of user or application (e.g. desktop user, workgroup user, private residential user, enterprise backbone network, etc.). Thus routers differ greatly in terms of which *routing protocols* they support (e.g. RIP, OSPF, BGP) and whether or not they support certain 'optional services' such as *multicasting, VPN (virtual private network)*, *MPLS (multiprotocol label switching)*, high performance, *border node* functionality, accounting, and so on. Routers differ in size and price based upon the number and speed of network interfaces which can be connected, the level of performance (maximum packet throughput) and upon their service functionality. As a complement to Figure 8.1 — which shows a typical public IP-network structure, Figure 8.2 illustrates the typical *desktop/workgroup/enterprise* switches and routers to be found in private or enterprise IP networks.

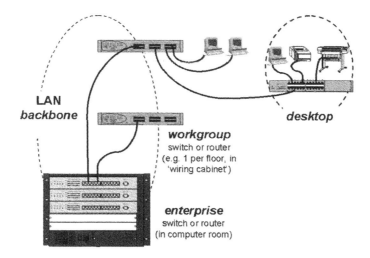

Figure 8.2   Different types and sizes of router for the enterprise environment.

8.2   The Internet, intranets, extranets and VPN

While the Internet protocol (IP) has become the *de facto* standard for modern data networking, the *Internet* itself is considered by many enterprises to be an inadequately secure medium for the transport of confidential company data. In consequence, many companies have built their own internal company-wide IP-based networks, using routers of the same technology as used in the Internet itself. Such networks have come to be known as *intranets*.

Intranets typically use private IP addressing schemes (using the 192.168.x.x and 10.x.x.x number ranges allocated by IANA for the purpose) and are only interconnected with the public Internet at strictly regulated gateway locations, protected by *network address translation (NAT), firewalls, proxy agents* and *demilitarized zones* as we shall discuss in Chapter 13.

In cases where it might otherwise be uneconomic to connect directly certain remote locations to an enterprise *intranet*, the *intranet* is sometimes extended by means of an *extranet*. An *extranet* is a connection carried by a public Internet from a remote location to an *intranet*. *Extranet* connections typically employ secure networking technology — either *tunnels, label-switching* (e.g. *MPLS*) or some other *VPN (virtual private network)* topology.

Such technology is intended to make it impossible for normal 'public users' of the Internet to enter a company's intranet without permission (Figure 8.3). In effect, an *extranet* location is a remote connection to the intranet.

VPN and tunnel connections are also sometimes used as a cost-effective means of realising normal or back-up trunk link connections between routers within an *intranet*.

Internetworking: network interconnection and peering

To provide the widespread *interconnectivity* which end-users expect, many modern computers are connected to IP-networks and the Internet. But have you ever stopped to think what exactly the Internet is? There is certainly no single network administration which owns and operates a router network connected to everyone in the world and called the

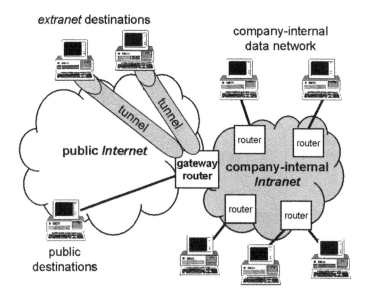

Figure 8.3   Explanation of terms: Internet, Intranet and Extranet.

Internet. Instead there are a collection of 'public' IP networks, which by virtue of their interconnection provide for global interconnection of all possible users. It's a 'network of networks'. As an end-user you gain access to this *internetwork* (the Internet) by means of an *Internet service provider (ISP)*. Well-known ISPs include AOL (America Online), British Telecom's BTInternet, Compuserve, Deutsche Telekom's T-Online, France Telecom's Wanadoo.

The ISP guarantees that all public IP addresses worldwide are *reachable* via his network. But few ISPs have router networks which extend beyond national or even regional boundaries. And some ISPs networks do not extend even beyond a given metropolitan area. So how does the ISP provide for global *reachability*? The answer is by means of *peering* agreements. The ISP sets up *peering* connections to other ISPs. The *peering* may be either by means of direct connections between the ISPs (called *direct peering*) or at special exchange called a *peering point* or an *Internet exchange (IX or INX)*. Both types of peering are illustrated in Figure 8.4.

Direct peering will typically be arranged between larger ISPs who share a lot of common traffic (i.e. the customers of one of the ISPs communicate a lot with the customers of the other ISP). Peering points (Internet exchanges), meanwhile, will generally be used to gain access to a wide range of other ISPs by means of a single network connection. Most ISPs are connected to at least one Internet exchange or peering point.

Using a combination of Internet exchanges (IXs) and direct peering connections we can achieve extensive global interconnectivity. But one problem still remains: two ISPs (we shall assume in different countries) might both be interconnected to *Internet exchange peering points* — but to different ones (as in the case of ISPs 1 and 4 in Figure 8.4). So how can these ISP networks be interconnected? The answer is by means of an *Internet backbone* or *transit* network between the two different peering points (the 'Internet backbone provider' in Figure 8.4).

The best known Internet backbone networks are those of Cable & Wireless (formerly MCI), Sprint and UUNET (formerly ANS). Unsurprisingly, all of these companies started their

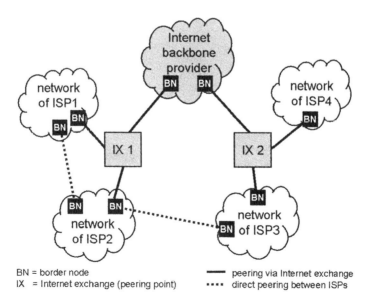

BN = border node
IX = Internet exchange (peering point)

—— peering via Internet exchange
···· direct peering between ISPs

Figure 8.4 Internet peering: direct peering, peering points (Internet exchanges) and Internet backbone providers.

Internet activities in the USA — the birthplace of the Internet. Indeed ANS grew out of the NSFnet (National Science Foundation network) — in effect the first 'public' Internet network. Lately, other national carriers have also come to offer Internet backbone services — AT&T, British Telecom, France Telecom's equant, Deutsche Telekom, Telia, etc. — most of them drawn by the lucrative market for providing enterprise and multinational company IP-data networks and *virtual private networks (VPNs)*.

Internet exchanges (IX or INX) — managed ethernets (MAE)

The first *Internet exchange (IX)* points were set up under the cooperation of the first Internet backbone providers — ANS, MCI and Sprint. It was viewed as a necessity and for the good of the general public (a spirit which determined much of the early development of the Internet) that the major US IP-backbone networks be combined into a single Internet. Each network operator, it was assumed, would receive as much traffic as he/she himself/herself 'handed off' to other networks, so the exchange points were therefore organised on a 'pay-your-own-costs' basis. Each operator would pay his or her own line costs of connecting to the Internet exchange (IX) site, where traffic was exchanged on a 'quid-pro-quo' (no money changes hands) basis. The IX sites themselves were run on a 'non-profit' or shared-cost basis.

An Internet exchange (IX) typically comprises a *MAnaged Ethernet (MAE)* LAN — indeed the first Internet exchanges in the US were called *MAE-east* and *MAE-west* (Figure 8.5). The IX organisation (of which there are now many — e.g. *London Internet eXchange [LINX]*, *AMSterdam Internet eXchange [AMSIX]* etc.) assumes responsibility for providing, operating and managing the *managed ethernet* LAN. *Each ISP* (Internet service provider) or Internet backbone provider who wishes to connect to the IX installs a *border node* router in the IX site, and connects it to the managed ethernet. The border node *speaks* BGP with the other ISPs' routers in the same managed ethernet, and provides an interconnection point to all the other ISP networks (Figure 8.6).

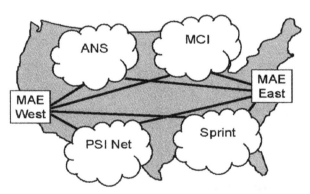

Note: ANS (Advanced Network & Services) was a company set up by IBM, MCI and MERIT (Michigan Research and Network Group) to administer the IP backbone of the NSFNET (National Science Foundation Network). It was subsequently subsumed into MCI Worldcom and became part of UUNET (nowadays simply called Worldcom). The MCI Internet backbone network is today owned by Cable & Wireless. PSINet (Performance Systems International) is still used as a brand name, but it is nowadays owned by Cogent Communications.

Figure 8.5   MAE-east and MAE-west provided for interconnection of the first Internet backbone providers.

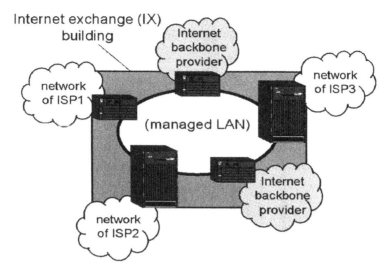

Figure 8.6   The managed ethernet (MAE) network topology of an Internet exchange.

8.3   Network technologies typically used in IP-backbone networks

In considering the challenge facing a network designer planning a new router network, we first consider the range of available network technologies (i.e. interfaces) which can be used to realise the connections between neighbouring routers in a network. There are a wide-range of different technologies to choose from:

- *point-to-point* transmission lines or *leaselines*;

- *point-to-multipoint* technologies such as *frame relay* or *ATM (asynchronous transfer mode)*;

- *metropolitan area network (MAN)* technologies such as FDDI and SMDS/DQDB; and

- ethernet (particularly fast ethernet and Gigabit ethernet) LAN technology.

Naturally, the different technologies have different advantages, disadvantages and economics, as we shall discuss next.

Point-to-point transmission line interfaces up to 155 Mbit/s (STM-1 or OC-3)

Point-to-point lines are the standard means of connecting routers in a *wide area network* (i.e. routers more than a few hundred metres apart — typically not on the same campus). Such point-to-point transmission lines are the basis of the networks operated by public telecommunications carriers and are freely available to third parties as *leaseline* connections.

A point-to-point line serves as a 'reserved and private' connection between two neigh-bouring routers. A number of different interfaces and bit rates are available (as illustrated in Figure 8.7). Each interface is typically used with *HDLC (high level datalink control)* or *PPP (point-to-point protocol)* as the layer 2 (*datalink*) and IP (Internet protocol) as the layer 3 (*network*) protocol. Which particular interface of Figure 8.7 is the best choice for any given case depends upon the bit rate of the traffic to be carried, the relative costs of different leaseline

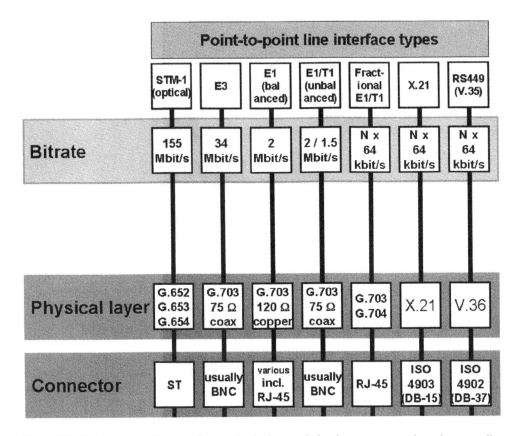

Figure 8.7 Point-to-point line interface technologies used for inter-router trunk and access line connections.

types and the cost of a new interface card for the router (assuming that a spare port is not already available).

One word of caution — network planners and operators should not forget the special challenge presented by very high speed lines. As we learned in Chapter 2, a large number of bits are always 'in transit' on the line — so that the efficient usage of a very high speed line requires that the protocols and application software used accommodate specially adapted data *flow control* techniques. In particular, in Chapter 7 we discussed how the TCP window size (in the end-user hosts) may require appropriate adjustment.

Many routers intended to be high performance *core network* routers nowadays offer a packet-over-SONET (or packet-over-SDH) interface option. This is an interface designed to allow direct interconnection of routers with OC-3 (SONET) or STM-1 (SDH) lines. Cisco calls this interface POSIP. We discussed SONET and SDH in a little more detail previously in Chapter 3.

The line connection of a packet-over-SONET (or packet-over SDH) interface may be either an optical (i.e. fibre) interface or the alternative electrical interface. The optical interface allows two routers to be directly interconnected by fibre cables (so-called *dark fibre*). The electrical interface, meanwhile, may be the cheaper alternative, if the STM-1 or OC-3 connection between the routers is to be multiplexed with other SONET or SDH connections by collocated SONET or SDH multiplexors (Figure 8.8). The *datalink layer* protocol is HDLC and the network protocol is IP.

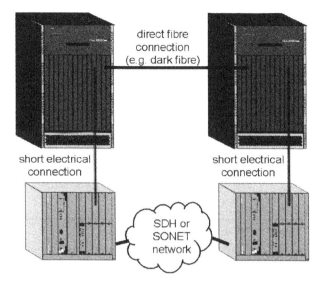

Figure 8.8 Packet-over-SONET: use of optical and electrical interface variants of OC-3/STM-1.

Point-to-point transmission at bit rates above 155 Mbit/s

Modern transmission technology developed for traditional carriers' telecommunications networks offers rates above 155 Mbit/s (called OC-3 [SONET hierarchy] or STM-1 [SDH hierarchy]) in power-of-4 multiples of 155 Mbit/s, thus:

- STM-1 (OC-3) bit rate: 155 Mbit/s

- STM-4 (OC-12) bit rate: 622 Mbit/s

- STM-16 (OC-48) bit rate: 2.5 Gbit/s

- STM-64 (OC-192) bit rate: 10 Gbit/s

Some router manufacturers are beginning to introduce line interface cards for interfaces up to STM-64 (OC-192) and even claiming to have developments underway for even higher speed interfaces. On the one hand, user demand will drive the major Internet backbone operators to use these very high speed interfaces — simply to be able to handle the total volume of traffic (and even if they are very expensive!). But on the other hand, the basic IP protocol and router design will probably need upgrading and overhauling to handle such high speeds efficiently. Alone the packet header processing requirements are a phenomenal challenge. If you expect to run your network at these rates you need to consider carefully the overall packet-per-second processing capabilities of the routers to be used. Even if the maximum packet size of 65 535 octets is used, the packet-throughput of a 1 Gbit/s line (15 259 packets/second) is still four to five times greater than that of an older T1- or E1- interface carrying the 'standard' IPv4 packet length of 576 octets (Table 8.1).

What is the maximum achievable line bit rate for a router-router connection?

In advance of another technology breakthrough, the practical linespeed limitation of a single interface is realistically around 1 Gbit/s to 2.5 Gbit/s (10 Gbit/s at a stretch). Try to go to a

Table 8.1   Packets-per-second demands of different router interfaces

Router interface speed	Packets received per second (assuming 576 octet packet size)	Packets received per second (assuming 65 535 octet packet size)
T1 (1.544 Mbit/s)	2 681	24
E1 (2.048 Mbit/s)	3 556	31
E3 (34 Mbit/s)	59 028	519
T3 (45 Mbit/s)	78 125	687
100baseT (100 Mbit/s)	173 611	1 526
OC-3 or STM-1 (155 Mbit/s)	269 097	2 365
OC-12 or STM-4 (622 Mbit/s)	1 079 861	9 491
1000baseX (1 Gbit/s)	1 736 111	15 259
OC-48 or STM-16 (2.5 Gbit/s)	4 340 278	38 148
OC-192 or STM-64 (10 Gbit/s)	17 361 111	152 590

much higher line bit rate than this at the moment and you are reaching the limits of current day optical communication. Greater aggregate bit rates can be achieved on a single pair of fibre optic cables — but only by using *WDM (wave division multiplexing)* or *DWDM (dense wave division multiplexing)*.

Both WDM (wave division multiplexing) and DWDM (dense wave division multiplexing) work to increase the aggregate bit rate which can be carried on a single fibre pair, by using multiple lasers of different *colours* or wavelengths (so-called *lambdas*, after the Greek letter λ which is used in scientific notation to represent signal *wavelength*). Each laser is typically used to carry a single STM-16 (OC-48) signal of 2.5 Gbit/s. Forty or even more different colours or lambdas may be available, giving a total aggregate bit rate for a single fibre pair equipped with DWDM of more than 100 Gbit/s. A device called a *transponder* is used to convert the *baseband* optical STM-16 into the relevant lambda bandwidth.

DWDM greatly increases the aggregate bit rates possible to be carried a single fibre pair. As a result, it is already in widespread use in long-distance transmission networks of public telecommunications carriers. And no doubt: DWDM is already transporting some of the few STM-16 inter-router trunks which are already in service. But DWDM is not an obvious interface to include directly on the routers themselves, since it affords an increase in bit rate above the 2.5 Gbit/s of an STM-16 router port only by (in effect) using multiple STM-16 ports.

Various optical networking manufacturers have set about developing very high speed optical switching (as opposed to electronic switching), optical cross-connects and even optical packet-forwarding engines. There are even plans for an updated version of MPLS based on lambda optical switching — provisionally called MPλS. But it will take a little while before these developments are 'stable' and ripe for wide-scale deployment. In the longer term, higher bit rate interfaces are the likely solution to the fast and efficient transport of ever increasing data traffic volumes, but in the meantime, we will need to 'make do' with 2.5 Gbit/s STM-16 (OC-48) or 1 Gbit/s Gigabit ethernet connections or multiples thereof as the highest practical interface bit rate.

Metropolitan trunks: Gigabit ethernet over 'dark fibre'

For very high speed network connections in metropolitan areas, Gigabit ethernet is becoming the interface of choice (Figure 8.9). As a full duplex connection, the Gigabit ethernet interface, as we discovered in Chapter 4, has a range of 3 km. Added to this, Gigabit ethernet cards are cheaper than other interface cards offering a similar bit rate. And perhaps most important of all, ethernet is a favoured interface in the data-communications community. By using a Gigabit

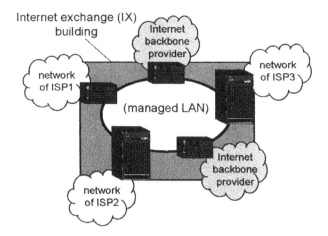

Figure 8.9   Gigabit ethernet switch used as a metropolitan area IP network: ethernet-in-the-first-mile (EFM).

switch like a 'local exchange' a high speed public metropolitan data networking service can be achieved without a router and you have *ethernet-in-the-first-mile (EFM)*. A router is used to interconnect the metropolitan network with the rest of the Internet. Alternatively, new switch/router technologies are appearing (e.g. from Extreme Networks and Foundry Systems).

'Fully meshed' router backbone networks — by means of frame relay or ATM

Although for some IP zealots, it is anathema to include either *frame relay*[2] or *ATM (asynchronous transfer mode)*[3] into an IP-based data network, it can make a lot of sense — as the network practice of the large Internet backbone providers in the late 1990s testifies. Before we explain why, let us briefly introduce the technologies themselves.

Introduction to frame relay and ATM (asynchronous transfer mode)

Frame relay and *ATM (asynchronous transfer mode)* are both *connection-oriented* data-transport technologies developed by *ITU-T (International Telecommunications Union — standardization sector)*. Frame relay connections were widely offered by public telecommunications carriers during the 1990s as cheaper alternatives to point-to-point leaseline services. The frame relay *service* provides for the transport (*relaying*) of data *frames* (i.e. datalink frames — layer 2 protocol frames) across *virtual circuits* between two *UNI (user-network interface)* endpoints (Figure 8.10). Frame relay developed from the earlier X.25 packet-switching standard as a technique better suited to higher bit rates. (X.25 was best suited for end-user connection speeds up to about 256 kbit/s. Frame relay, meanwhile, was suited to speeds between 64 kbit/s and 34 Mbit/s.)

   ATM (asynchronous transfer mode) was a further development of *frame relay*, undertaken by the public telephone companies under the auspices of ITU-T and intended to provide both for even higher data connection speeds (2 Mbit/s up to 34 Mbit/s) but also optimised for efficient

---

[2] See Appendix 9.
[3] See Appendix 10.

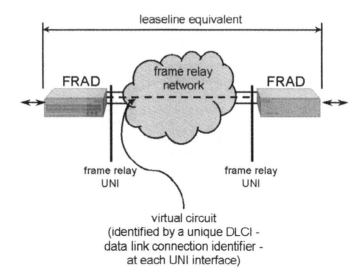

Figure 8.10   The frame relay UNI (user-network interface).

and simultaneous carriage of both voice and data signals. ATM also operates connection-oriented switching of virtual channels. Both frame relay and ATM were revolutionary in their time.

Frame relay

The *virtual circuits* of frame relay[4] are identified by a label (see Chapter 3 and compare with the label-ID of MPLS) called the *DLCI (data link connection identifier)*. With frame relay developed the first principles of *quality-of-service (QOS)* guarantees for data connections. In particular, frame relay provided for a guaranteed minimum transport bit rate of the connection, called the *committed information rate (CIR)*, but also allowed for temporary periods of higher-speed data transfer called *excess bursts* (which are carried at the *excess information rate, EIR*) provided that the network capacity required for such an excess burst is not required by other users.

The frame relay interface was, and still is, widely available as an interface card for routers and other computer end-devices, but if necessary a special device called a *FRAD (frame relay access device)* can be used at the two ends of the connection to convert the *frame relay UNI (user-network interface)* into the equivalent of a 'transparent' leaseline interface (appearing to create a direct connection of the devices — Figure 8.10).

ATM (asynchronous transfer mode)

Like frame relay, ATM[5] is a connection-oriented frame-relaying technique using virtual circuits. In ATM, the connection label is a combination of the *VCI (virtual channel identifier)* and *VPI (virtual path identifier)*.

---

[4] See Appendix 9.
[5] See Appendix 10.

ATM was revolutionary in its use of a fixed length packet (called a *cell*) for the carriage of the data. By fixing the length of the packets at 53 octets (48 octets of *user data* plus a 5 octet ATM *cell header*), the packet forwarding or relaying process could be optimised for the predictable transport of real-time signals such as voice and video. In addition, the fixed packet-length lends itself well to very high speed and predictable electronic switching (the fixed packet length can be 'built into' a high speed hardware switching design). As a result, many ATM switching chipsets have appeared.

ATM also furthered the development of mechanisms to control connection *quality of-service (QOS)*. The limited ATM cell length of only 53 octets created the scope to intersperse cells from different connections, giving greater priority to certain classes of connections. In this way, the cells of the highest priority could be 'guaranteed' a maximum 'waiting time' delay at intermediate ATM switches equal to the time it might take to complete the sending of another ATM cell which was already in progress. This period of time, as Table 8.2 shows, is quite short.

Given that the total end-to-end propagation delay of a packet will largely depend upon the accumulated buffering delays accrued as the result of waiting for the outgoing line to become free, only the short 53 octet cell size of ATM is likely to achieve end-to-end propagation results in the low millisecond or sub-millisecond range. Such a quality is critical to carrying delay-sensitive signals such as real-time voice and video. In addition, the fact that the variation in the propagation delay (called *jitter*) is very small gives humans a good impression of the transmitted picture and audio quality.

Multimedia services to be carried by ATM were subdivided into different types according to whether they required *CBR (constant bit rate), variable bit rate (VBR), unspecified bit rate (UBR)* or could make do with the *available bit rate (ABR)*. The priorities of cell forwarding were accordingly set. And to make sure the network as a whole was not subjected to more traffic than it could handle while maintaining its *quality-of-service (QOS)* guarantees, a *connection admission control (CAC)* process was developed. *Connection admission control* requires that users declare their bit rate requirements during the time of connection set-up. At times of subsequent network congestion, an ATM network will hold the user to this contract: *policing it* and *discarding* any excess cells, in order to maintain the QOS of other users' connections.

Though ATM was revolutionary in its handling of multimedia signals and its guarantees of quality-of-service (QOS), many IP zealots criticised it because of the necessity to *segment* larger IP packets (up to 65 535 octets in length) into multiple ATM cells. The main accusation was the 'inefficient' 10% cell overhead created by the 5-octet cell header on the 48-octet cell payload. But while the 'religious war' (IP v. ATM) saw to it that ATM got a tarnished name in the data-switching world, the IP community further developed the ideas into new protocols of its own — MPLS (multiprotocol label switching) and RSVP (resource reservation protocol) — as we described in Chapter 7.

Table 8.2 Maximum waiting time to next opportunity to send high priority packet

Line bit rate	Maximum waiting time for high priority packet or cell (i.e. maximum time required for full packet transmission)		
	53 octet ATM cell	**576 octet standard IPv4 packet**	**65 535 octet IP-packet of maximum transmission unit (MTU) size**
2 Mbit/s (E1)	207 μs	2 ms	256 ms
34 Mbit/s (E3)	12 μs	136 μs	15 ms
45 Mbit/s (T3)	9 μs	102 μs	12 ms
155 Mbit/s (STM-1 or OC-3)	3 μs	30 μs	3 ms

How to achieve an efficient full-mesh router backbone

By using either frame relay or *ATM* (or nowadays also *MPLS — multiprotocol label switching*) in the main core of the network, the effect of a 'full mesh' topology of routers can be achieved, even if each router in an IP network is only connected to the *frame relay* (or ATM or MPLS backbone) switch by a single physical connection. As Figure 8.11 illustrates,[6] the full meshing of the routers is achieved by the creation of *virtual circuit* connections between each individual pair of routers. But why bother, you might say? Why not let the routers sort it all out automatically using their dynamic routing protocols? The answer could be one of two reasons:

- by directly interconnecting each pair of routers, the routing table look-up and the IP-forwarding process in Figure 8.11 have been limited to a maximum of two look-ups. This would still be the case even if we added many more routers to the network; or

- each router only requires one (high-speed) physical connection to the network, so that overall less equipment is required from the router manufacturer. If the frame relay, ATM or MPLS technology is cheaper, this has obvious economic benefits.

As the development and deployment of MPLS (multiprotocol label switching) testify, even the IP zealots have accepted that frame-relaying and cell-switching make sense!

Satellite links and other links with long propagation delays

Before we leave the subject of lines used in the backbone part of IP-networks, we should remember that links with long propagation delays can lead to protocol and application difficulties. We explained in Chapter 2 how a poorly designed application may run very slowly and inefficiently in the presence of long end-to-end propagation delays. Such delays will be encountered on very long terrestrial and satellite connections.

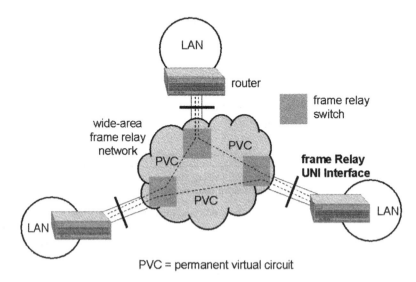

Figure 8.11   Creating a full router mesh using frame relay (or ATM or MPLS) in the IP-network backbone.

---

[6] See Chapter 14, Figure 14.15.

The distance from the earth to a *geostationary* satellite in orbit and back again, when traversed at the speed of light, takes around a quarter-of-a-second to cover — the loop-delay (there-and-back-again) is around half a second). This is a huge delay in data communications terms, and can cause protocol problems. You may recall from Chapter 7 that indeed TCP (transmission control protocol) includes a measurement of the *round-trip time (RTT)* to deal with such long delay paths. And don't get the idea that the problem is limited to satellite communications! Given the electronic delays of repeaters, regenerators, etc., the effective speed of transmission on a terrestrial leaseline is around a third to a half of the speed of light. At half the speed of light, the one-way delay on a fibre connection from London to Sydney (Australia) [17 000 km] is 113 ms!

Summary of backbone network interfaces used between routers

Complementing Figure 8.7, Figure 8.12 illustrates the network, datalink and physical layer protocols commonly used in router backbone networks. The diagram is intended to consolidate

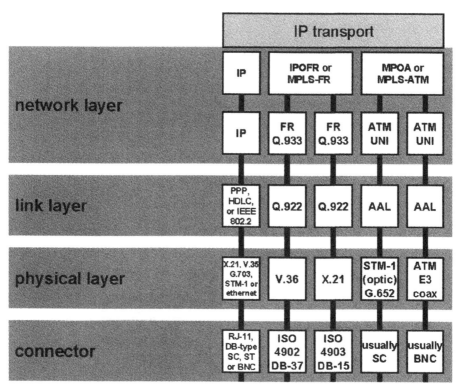

Key: AAL = ATM adaptation layer; ATM = asynchronous transfer mode; BNC = bayonet connector; FR = frame relay; HDLC = higherlevel datalink control; IEEE 802.2 = LLC = logical link control; IP = Internet protocol; IPOFR = Internet protocol over frame relay; MPOA = multiprotocol over ATM; MPLS = multiprotocol label switching; PPP = point-to-point protocol; STM-1 = synchronous transport module-1; UNI = user network interface.

The following are ITU-T recommendations for interfaces and protocols: G.652, G.703, Q.922, Q.933, X.21, V.35, V.36.

The following refer to standardised connector types: DB-15, DB-37, RJ-11, SC, ST.

Figure 8.12   Common backbone network interfaces used in router networks.

and summarise the previous discussion. The diagram shows the relationship of the various interface and protocol specifications, and succinctly illustrates the fact we explained in more detail in Chapter 3 — that a network interface is not fully defined when simply referred to as 'an X.21-interface', an 'RJ-11-interface' or 'ATM UNI'. Such a reference to only one of the protocol layers may be adequate to convey the right meaning in a given circumstance, but to be certain it is better to describe the entire protocol stack, e.g.: IP/100baseT ethernet/RJ-11.

8.4 Access network technologies

That part of an IP network which is intended to provide for the connection of end-user devices to the nearest *backbone* router node is commonly called the *access network*. Various common access network configurations are illustrated in Figure 8.13:

- dial-in access;

- dedicated access; and

- xDSL or cable modem access.

We shall discuss each in turn.

Dial-in access

Figure 8.13a illustrates a dial-in access connection. This is the type of access connection used by most residential Internet users. The user's PC is equipped with a *modem* or *ISDN (integrated*

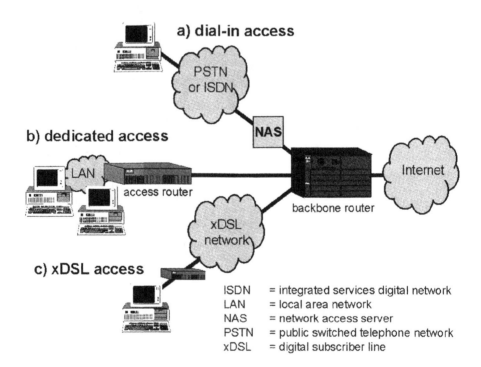

ISDN	= integrated services digital network
LAN	= local area network
NAS	= network access server
PSTN	= public switched telephone network
xDSL	= digital subscriber line

Figure 8.13 Common access network configurations used for IP network or Internet access.

services digital network)-card, which is used to establish a dial-in connection across the public switched telephone network (PSTN) or the digital telephone network (called the ISDN) to a *network access server (NAS* — also called a *remote access server, RAS* or simply a *modem pool)* which is connected to the nearest Internet backbone router. The bit rates possible with dial-in access connections are typically 56 kbit/s (analogue); 64 kbit/s or 128 kbit/s (ISDN).

The *de facto* standard for modems used in *dial-in* (also called *dial-up)* access across analogue telephone lines is nowadays the ITU-T V.90 modem. This provides for an auto-negotiated bit rate (depending upon the quality of the line) of up to 56 kbit/s. Such modems are also backwards-compatible with earlier slower modems (e.g. V.32 or V.34), 'falling back' to the highest speed possible. V.92 modems allow for even higher speeds (up to 64 kbit/s downstream and 48 kbit/s upstream — as opposed to 56 kbit/s downstream and 33.6 kbit/s upstream for a V.90 modem). (See Figure 8.14 and Table 2.4 in Chapter 2 which lists other related modem and data compression standards in the ITU-T V-series recommendations).

For higher dial-in access speeds, it is normal to use an *ISDN (integrated services digital network)* type telephone line rather than an analogue line. The advantage of ISDN is not only the higher speed of the line (usually 64 kbit/s (duplex) or 128 kbit/s (duplex) — by aggregating two separate dialled-up connections to act like a single line) but also the much lower *bit error ratio (BER)*. Less errors means higher net bit rate, as we discovered in Chapter 2.

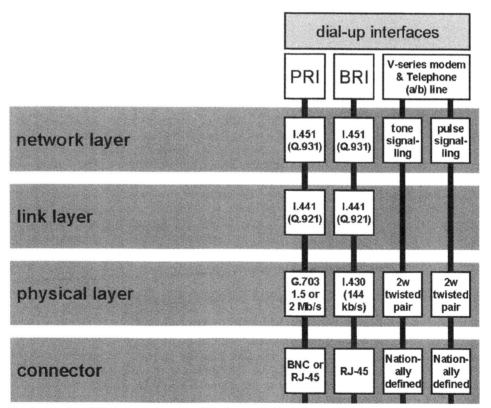

Key: 2w = 2-wire; BNC = bayonet connector; BRI = basic rate ISDN; ISDN = integrated services digital network; PRI = primary rate ISDN. The following are ITU-T recommendations for interfaces and protocols: G.703, I.430, I.441, I.451, Q.921, Q.931. RJ-45 is an 8-lead connector type.

Figure 8.14   ISDN and analogue telephone line interfaces used as dial-in access connections to IP backbone networks.

a) physical network configuration

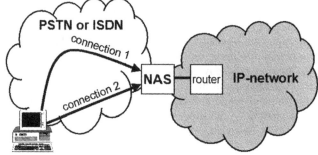

b) reverse multiplexor function

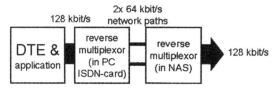

Figure 8.15   Datalink aggregation: the need for reverse multiplexing to overcome different propagation delays.

Data link aggregation and reverse multiplexing

Data *link aggregation* (Figure 8.15) is a useful way of providing high bit rate connections. Thus it is common for *basic rate ISDN (BRI)* cards to be able to aggregate both of the 64 kbit/s (B-channels) to create the effect of a single connection of 128 kbit/s duplex. But it is important to note that the aggregation of links is not a trivial task. As shown in Figure 8.15a, the two separate connections (created by two separate dial-set-up procedures) typically demonstrate different propagation characteristics — in particular different connection delays. A special procedure called *reverse multiplexing* (Figure 8.15b) has to be used to ensure that the bits carried by the two separate paths are *reassembled* at the receiving end and presented to the application in the correct order. Without the reverse multiplexing functionality the individual bits could get 'shuffled' out of order by the different propagation delays encountered on the two separate paths.

Public telephone network configuration for dial-in Internet access

Figure 8.16 illustrates typical configurations of a public telephone company's network for dial-in Internet access. Figure 8.16a shows the 'standard' configuration using a *network access server (NAS* — a 'modem pool') as the 'interface' between the public telephone network part of the connection and the Internet backbone. User B of Figure 8.16a would thus be told to dial a telephone number corresponding to the nearest NAS in the network. Sometimes the number of the nearest NAS has to be determined by the human user and programmed into the computer (labelled 'user B' in Figure 8.16a). In other cases, a single number is published by the telephone company for Internet access. Calls to the number are simply forwarded by the network to the nearest NAS.

'User A' in Figure 8.16a also requires to be connected to the nearest NAS of the network operator offering the Internet access service (labelled 'inter-exchange carrier network'). But

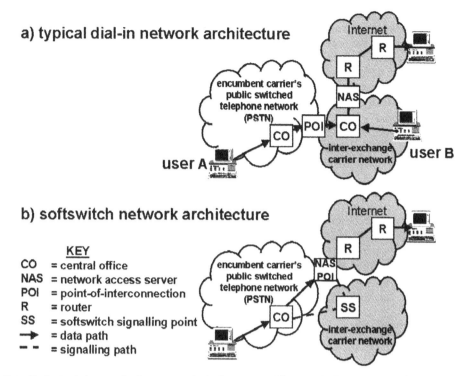

a) typical dial-in network architecture

b) softswitch network architecture

KEY
CO = central office
NAS = network access server
POI = point-of-interconnection
R   = router
SS   = softswitch signalling point
→   = data path
- - = signalling path

Note: Both shaded networks (Internet and telephone network) are typically operated by the same public telecommunications carrier.

Figure 8.16   Public telephone network configuration for dial-in Internet access.

user A's telephone (or ISDN) connection is itself connected to a different local telephone company's network (typically the old 'incumbent's' telephone network). This is a very common occurrence. To reach the alternative telephone company's network (the inter-exchange carrier network of Figure 8.16a), the call first has to cross a *point-of-interconnection (POI)* between the two different carriers' networks.

In Figure 8.16a the *point-of-interconnection (POI)* used is shared with other 'normal' telephone traffic crossing between the two carriers' networks and is thus connected directly between two public telephone switches (labelled *CO* for *central office*) in the diagram. This is a standard means of interconnection of the two carrier networks, but it can be inefficient when large volumes of dial-in Internet access traffic are involved. In the case of large volume Internet dial-in access, the *softswitch* configuration of Figure 8.16b may be used instead.

By using a softswitch configuration, the first central office in the connection (Figure 8.16b) can be directly connected to the NAS, thereby bypassing the second central office (CO) — thus saving the cost of the transit connection and the two ports (one 'in' and one 'out') on the second CO which would otherwise be necessary to carry it.

The softswitch configuration can only be used as a means of *interconnection* between different carriers' networks where both carriers are prepared to accept different POIs (points of interconnection) for voice and data traffic. The 'incumbent' carrier of Figure 8.16 has to be prepared to 'sort out' the Internet dial-in data traffic destined for the 'inter-exchange carrier' from the voice traffic. This is done by analysis of the dialled number. The 'cost' of splitting the original POI into two — 'data POI' and 'voice POI' (which still directly connects central offices in the two different carrier networks) is the larger number of POI circuits required. Thus where one 2 Mbit/s POI connection may have sufficed in the past, two separate connections may be necessary for the softswitch configuration.

Finally, we should note that softswitch configurations are not limited to use for network interconnection of different carriers' networks. Increasingly, hardware manufacturers (including both central office manufacturers and router manufacturers) are offering solutions which allow Internet dial-in access traffic to bypass even the customer's 'home' *local telephone exchange* (i.e. central office). By bypassing even the *local exchange*, exchange ports and costs can be saved and the exchange is spared possible traffic congestion. One means of achieving this is using the xDSL access architecture which we explain later.

Use of dial-up lines for 'back-up' service

Before we finally leave the subject of dial-up lines, we should note that they are commonly used as a means of *back-up* — a fallback connection set-up on demand should a dedicated access line (or even an inter-router trunk circuit) fail. Furthermore, by aggregating (Figure 8.15) different numbers of dial-up lines at different times of day, connections of 'variable' bit rate can be achieved to carry data traffic volumes which might fluctuate greatly during a 24-hour cycle.

Dedicated access

Figure 8.13b illustrates the typical *dedicated access* configuration used to connect most business premises to the Internet (or to an enterprise-wide IP-based router backbone network). In this configuration, a number of end-user devices at the customer premises site share the same high speed connection to the backbone network. These devices are usually connected by means of a LAN (local area network) to an access router on the customer's premises. The access router performs one or more of the following functions:

- forwarding of outgoing packets from the LAN to the *default gateway* (in this case, the Internet service provider's first backbone router);

- *filtering* of packets allowed to pass into and out of the LAN;

- *Network address translation (NAT)* as necessary to convert local IP network addresses to public IP-addresses which can be recognised by the public Internet;

- selection of connection and bit rate to be used when connecting to external networks; and

- keeping track of *reachable* destination IP-addresses either by 'listening to' or 'participating' in a routing protocol.

The access router typically provides for a default gateway function — it forwards packets to remote IP addresses to the backbone network, but leaves packets destined for local addresses within the LAN. In addition, the access router may also be used to *filter* traffic and to provide for a *firewall*.

In order to filter traffic, a router may be equipped with a set of *access lists (ACLs)*. An access list records the combination of source-IP and destination-IP addresses which are to be forwarded by the router. Packets with combinations of source- and destination-IP address which are not held in the access list will simply be discarded by the access router. Thus an access list (ACL) can be used to filter out employees attempts to access pornographic or sports websites on the public Internet. Similarly, an access list used to filter incoming traffic might allow only pre-determined 'public' Internet users to access only a limited number of devices on the company LAN. A router equipped with such access list (ACL) filter functionality is a simple firewall.

A firewall, as we shall discover in more detail in chapter 13, is a device intended to protect a local network from malicious intrusion by outsiders. The simplest type of firewall is an access router equipped with access list (ACL) or filter functionality. More sophisticated firewalls also check the content of files passed between the *protected* and *unprotected* (or *untrusted*) parts of the network.

Company premises on which multiple computer end-users need simultaneous access to remote data processing resources (via an enterprise *intranet, extranet* or via the *Internet*) are typically connected by means of an access router using a *dedicated connection* as shown in Figure 8.13b. (The line is *dedicated* in the sense that the capacity of the line is not shared with other companies or private users — but the line is shared between the various end-users within the company.) Each of the end-users connected via the dedicated line is able to send and receive files quickly at the full speed of the line — typically 1.5 Mbit/s, 2 Mbit/s or more (this is a lot better service than the employees would experience if they each had a dial-in connection). A dedicated access line is usually complemented by a range of dedicated IP addresses, allocated to the customer-company by the *ISP (Internet service provider)*.

xDSL and cable modem access

Figure 8.13c illustrates a new type of connection, which is effectively a 'hybrid' dial-in/dedicated type of connection. The connection itself is usually realised using either an *xDSL (digital subscriber line)* or a *cable mode*m technology. It typically has a speed of between 128 kbit/s and 2 Mbit/s. Some of these technologies are said to be *symmetric* (i.e. offering the same bit rate in both upstream (from the end-user to the network) and downstream from the network to the end-user directions). But other technologies trade off some of the upstream bit rate to enable a much higher downstream bit rate to be available. Such *asymmetric* technologies are considered to be well suited to individual end-users of the Internet, who tend to 'download' web pages, videos and other files from the Internet but send comparatively little.

The 'generic' name xDSL is generally used to apply to modem devices intended to be used on standard public telephone network grade cabling and providing for a high-speed digital subscriber line (DSL). Otherwise the term cable modem describes modem devices intended to achieve a similar high-speed network connection line, but in this case by means of a coaxial cable-TV distribution network.

The basic techniques of modulation and line coding used in both *xDSL* and *cable modems* are similar to those of other modems as described in chapter 2 (though the actual procedures are more sophisticated in order to 'squeeze' a higher bit rate from the line). There are various different techniques offered by different manufacturers: offering different bit rate combinations upstream/downstream; suited to different types of cable; and requiring different numbers of cable pairs (some techniques used 1-pair [2-wires], whereas others require 4- wires or more). We shall not describe the transmission techniques in detail here, but, with the help of Figure 8.17 will discuss the typical network configuration of an xDSL access connection line.

Figure 8.17 illustrates a typical configuration of an xDSL network access line. Telephone cabling (provided by the local telephone company) directly connects an xDSL *line splitter* installed on the customer's premises with an xDSL *head end* equipment at the next local telephone exchange site. Typically the existing 2-wire (1-pair) copper cable connection used for the household telephone line is taken for the purpose.

Both the xDSL line splitter and the head-end device perform a multiplexing function, enabling an ISDN telephone connection and a highspeed broadband data connection to share the same line. At the line splitter end (i.e. customer end) of the connection, the customer

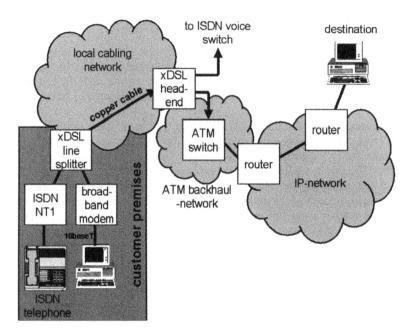

Figure 8.17 Typical network configuration of an xDSL network access connection.

derives a basic rate ISDN (BRI) line for telephone use and a broadband data connection (terminated using an *NTBBA — network terminator broadband access*). The data terminal equipment (typically a PC) is usually connected using 10baseT ethernet, as shown in Figure 8.17. The ISDN telephone, meanwhile, is connected to the line splitter via an ISDN *NT1 (network termination 1)*.

At the network end of the access line, the xDSL head-end device separates the ISDN and high-speed data connections. The ISDN access line is connected directly to the collocated public ISDN local exchange. Meanwhile, the data connection is typically *backhauled* by means of an ATM network to the nearest *Internet* backbone router.

Different manufacturers and service providers use different marketing names for their versions of xDSL. The following are examples of a few of the names in common usage:

- *ADSL — asymmetric digital subscriber line* — this is the generic term for devices which offer a higher downstream bit rate than upstream bit rate;

- *HDSL — high-speed digital subscriber line* — this is a generic term initially used for devices offering symmetric 2 Mbit/s data carriage;

- *SDSL* — a proprietary ADSL technique offered by Siemens;

- *T-DSL* — the marketing name used for a 768 kbit/s downstream and 128 kbit/s upstream ADSL service offered by Deutsche Telekom.

In the alternative case of using cable modems (rather than xDSL), the *DOCSIS (data-over-cable service interfaces specifications)* apply.

8.5   Link establishment and control

A problem presented particularly by dial-up lines is the need to establish the datalink connection, determine bit rates and protocols to be used and provide for layer 2 (datalink layer) transport of data between the customer terminal or modem and the NAS (network access server) or equivalent. On dedicated lines and leaselines, *HDLC (higher level datalink control)* is typically used as the layer 2 protocol. But on dial-up lines in particular, it is more normal to use *PPP (point-to-point protocol)*. In former times, the SLIP (serial line *Internet protocol*) was also used prior to the development of PPP.

Point-to-point protocol (PPP)

The point-to-point protocol (PPP) is defined in RFC 1661. The version for use in conjunction with IP version 6 (IPv6) is defined in RFC 2472 (see Table 8.3). PPP is intended to be easy to configure and provides for:

- datalink (i.e. layer 2 protocol) establishment and control (using the *link control protocol, LCP*), once a physical path (e.g. dial-up connection) has been established;

- network control (the assignment of IP addresses for communication and the management of the connection using the *network control protocol, NCP*);

Table 8.3   PPP LCP (and IPv6CP) configuration options

Configuration option type value	Configuration option and other options
0	Vendor-specific (RFC 2153)
1	Maximum receive unit (MRU) size
2	Asynchronous control character map
3	Authentication protocol to be used (PAP or CHAP)
4	Link quality monitoring protocol to be used (link quality report RFC 1989)
5	Magic number (otherwise the value ' 0' will be inserted in relevant fields)
6	Deprecated (quality protocol)
7	Protocol field compression (default value is 2 octet protocol field in PPP packets)
8	Address and control field compression — type of compression is negotiated
9	FCS (frame check sequence) alternatives (RFC 1570)
10	Self-describing PAD (RFC 1570)
11	Numbered mode (RFC 1663)
12	Deprecated (multi-link procedure)
13	Callback (RFC 1570)
14	Deprecated (connect-time)
15	Deprecated (compound frames)
16	Deprecated (nominal data encapsulation)
17	Multilink MRU (RFC 1717)
18	Multilink short sequence number header (RFC 1717)
19	Multilink endpoint discriminator (RFC 1717)
20	Proprietary
21	DCE (data circuit terminating equipment) identifier
22	Multilink plus procedure
23	Link discriminator for bandwidth allocation control protocol (BACP — RFC 2125)
24	LCP (link control protocol) authentication option
25	Consistent overhead byte stuffing (COBS)
28	Internationalisation (RFC 2484)
29	Simple data link on SONET/SDH (packet over SONET — RFC 2823)
IPv6CP value 1	Interface identifier (64-bit unique identifier for IPv6 address autoconfiguration — RFC 2472)
IPv6CP value 2	IPv6-compression-protocol (RFC 2472)

- simultaneous full-duplex transport of data on a point-to-point basis by means of *encapsulation*; and

- a layer 2 (datalink) protocol for transfer of data frames, flow control and error detection.

PPP has nowadays become the 'standard' link protocol used for dial-in or 'remote access' from remote devices to an Internet or other IP-based data network. It is thus in widespread use, for example, by millions of private Internet users worldwide for accessing their local ISP (Internet service provider). Unlike HDLC (higher level datalink control — which we described in Chapter 3), PPP does not undertake acknowledgement of data frames or and is thus said to be an *unreliable* protocol. The benefit is that a lower protocol *overhead*. TCP (as we described in Chapter 7) can be used to provide acknowledgement and reliable data transport, should this be needed.

Link establishment and control under PPP

PPP is conceived to establish and control links which are not permanently configured. It therefore defines a number of connection phases, as illustrated in the PPP phase diagram of Figure 8.18. The connection passes through up to five distinct phases:

- link dead;

- link establishment;

- authentication;

- network layer protocol phase; and

- link termination.

Link establishment is triggered by an event such as a carrier detect (as notified by the physical layer, i.e. the modem). In other words, PPP tries to set-up a datalink (layer 2) protocol layer, having been triggered by the creation of a physical connection. During link establishment, the *link control protocol (LCP)* is used to configure and test the datalink. Various configuration options (Table 8.3) are negotiated by means of the LCP and the negotiation *automoton* (Table 8.5).

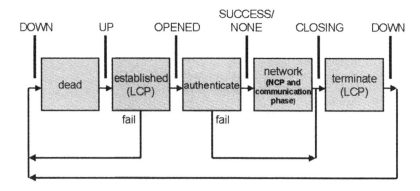

LCP = link control protocol
NCP = network control protocol

Figure 8.18   Point-to-point protocol (PPP) phase diagram.

After the link has been *opened* by LCP, the *peer* may be *authenticated* (this is not manda-tory but typical). Following authentication, the *network control protocol (NCP)* configures each network protocol (e.g. IP) individually. This typically involves obtaining a temporary IP address from a DHCP server for identifying the remote user and managing the IP 'con-nection'. But various different network protocols (listed in Table 8.4) and their corresponding NCPs (also listed in Table 8.4) may be in use at the same time. Each NCP may be *opened* and *closed* at any time provided the datalink is live.

After the end of the main data transfer phase of the communication, the link control protocol (LCP) is once again used to *close* the connection. After the exchange of LCP *terminate* packets, a signal is passed to the *physical layer* to enforce termination of link. The forced termination of the link is particularly important in the case of an authentication failure.

PPP frame format

As a layer 2 protocol, PPP is structured as a series of frames. A very simple encapsula-tion frame format and HDLC-like framing is usually used (RFC 1662 — as illustrated in Figure 8.19[7]) in which the network layer protocol *packets* are simply prefixed by an 8-bit or 16-bit header (called the PPP encapsulation header). Whether 1-octet (8-bit) or 2-octet (16-bit) encapsulation is used is optional. The standard method is for 2-octet encapsula-tion but the 1-octet encapsulation option may be negotiated during link establishment, and is used in particular to reduce the protocol *overhead* when only slow speed data links are in use.

The address and control fields have a similar function to the equivalent fields of HDLC (higher level datalink control) described in Chapter 3. The address-field indicates whether the frame is a command or a response. The control field provides for flow control at the datalink layer.

The *PPP encapsulation header* (Figure 8.19) is coded with an 8- or 16-bit value (as listed in Table 8.4) to represent the network protocol being used in the *PPP-information* field. It is important to note that not all the 65536 possible values may be used, since the least significant bit is required always to be of binary value '1', and the last bit of the first octet is always required to be of value '0'. This reduces the possible range of different protocols to 16 384 (Table 8.4).

When the *PPP protocol field compression* (configuration option '7' of Table 8.3) is in use, it is the first octet (shaded grey in Figure 8.19) which is omitted. The use of protocol field compression will be apparent even during the data transfer phase because of the appearance of a binary value '1' in the eighth bit position of the PPP frame. Only certain protocol field values are allowed to be compressed (those with two leading '0s' in their hexadecimal format).

The *PPP-information* field (see Figure 8.19) may contain zero or more octets of data, coded according to the protocol format indicated in the *PPP encapsulation header*. In addi-tion, padding data may of one or more octets may be present — up to the *maximum PPP receive unit (MRU)*. The default MRU size is 1500 octets but this can also be changed by negotiation during the link establishment using configuration option ' 1' (see Table 8.3). In cases where padding is being used, it is up to the *protocol* (Table 8.4) to determine where the information finishes and the padding starts. Finally, the *frame check sequence (FCS)* provides for error detection.

---

[7] Alternatively, other PPP frame formats based on frame relay, ATM, ISDN, SONET/SDH or X.25 may be used. These formats are defined in the RFCs as listed in the abbreviations appendix.

Table 8.4   PPP Protocol value field — common values

Protocol type	Protocol value range	Allocated value	Protocol
Network layer protocols	0xxx–3xxx	0001	Padding protocol
		0003	Robust Header Compression (ROHC small-CID, context identifier: RFC 3095)
		0005	Robust Header Compression (ROHC large-CID, context identifier: RFC 3095)
		0007 - to - 001F	Reserved (transparency efficient)
		0021	Internet Protocol v4
		0023	OSI network layer
		002B	Novell IPX
		002D	Van Jacobsen compressed TCP/IP (RFC 2508)
		002F	Van Jacobsen uncompressed TCP/IP (RFC 2508)
		0031	Bridging PDU
		003D	PPP Multilink protocol (MP — RFC 1717)
		003F	NetBIOS framing
		0041	Cisco systems
		0049	Serial data transport protocol (SDTP — RFC 1963)
		004B	IBM SNA over IEEE 802.2
		004D	IBM SNA
		0057	Internet Protocol v6
		0059	PPP muxing (RFC 3153)
		0061–0069	RTP (Real-Time Transport Protocol) Internet Protocol Header Compression (IPHC — RFC 2509)
		007D	Reserved (control escape — RFC 1661)
		007F	Reserved (compression inefficient — RFC 1662)
		00CF	Reserved (PPP NLPID)
		00FB	Single link compression in multilink (RFC 1962)
		00FD	Compressed datagram (RFC 1962)
		00FF	Reserved (compression inefficient)
	02xx -1Exx (compression inefficient)	0201	IEEE 802.1p hello protocol
		0203	IBM source routing BPDU
		0207	Cisco discovery protocol
		0281	MPLS unicast
		0283	MPLS multicast
		2063–2069	Real-time Transport Protocol (RTP) Internet Protocol Header Compression (IPHC — RFC 2509)

Table 8.4 (*continued*)

Protocol type	Protocol value range	Allocated value	Protocol
Low volume traffic protocols with no NCP	4xxx–7xxx	See www.iana.org	
Network control protocols (NCPs)	8xxx–Bxxx	8001–801F	Unused
		8021	IPv4 control protocol (RFC 1332)
		8023	OSI network layer control protocol (RFC 1377)
		802B	Novell IPX control protocol (RFC 1552)
		802D	Reserved
		802F	Reserved
		8031	Bridging control protocol (BCP — RFC 2878)
		803D	Multilink control protocol (RFC 1717)
Network control protocols (NCPs)	8xxx — Bxxx	803F	NetBIOS framing control protocol (RFC 2097)
		8041	Cisco systems control protocol
		8049	Serial data control protocol (SDCP)
		804B	IBM SNA over IEEE 802.2 control protocol
		804D	IBM SNA control protocol (RFC 2043)
		8057	IPv6 control protocol
		8059	PPP muxing control protocol (RFC 3153)
		807D	Unused (RFC 1661)
		80CF	Unused (RFC 1661)
		80FB	Single link compression in multilink control (RFC 1962)
		80FD	Compression control protocol (CCP — RFC 1962)
		80FF	Unused (RFC 1661)
Link control protocols (LCPs)	Cxxx — Fxxx	C021	Link control protocol (LCP)
		C023	Password authentication protocol (PAP — RFC 1334)
		C025	Link quality report
		C029	CallBack control protocol (CBCP)
		C02B	Bandwidth allocation control protocol (BACP — RFC 2125)
		C02D	Bandwidth allocation protocol (BAP — RFC 2125)
		C223	Challenge handshake authentication protocol (CHAP — RFC 1994)
		C227	Extensible authentication protocol (RFC 2284)

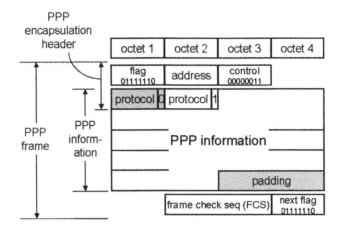

Figure 8.19   Point-to-point protocol (PPP) frame format (in HDLC-like framing (RFC 1662)).

a) LCP message format

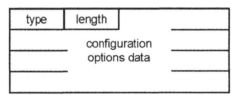

b) LCP data field format - configuration options

Figure 8.20   PPP link control protocol (LCP) message format (LCP messages are encapsulated in the PPP information field of Figure 8.19).

PPP link control protocol (LCP)

The PPP link control protocol (LCP) is also defined by RFC 1661. The general frame format of an LCP message is as shown in Figure 8.20 and the different *events* reported by, and *actions* requested by the LCP automoton are listed in Table 8.5.

For simplicity, there is no 'version number' in an LCP packet. This was considered necessary to keep the protocol 'overhead' to an absolute minimum (as is critical on low speed lines). The *code* field in the LCP packet header indicates what type of link control message the packet contains. There are three classes of link control protocol (LCP) packets:

Table 8.5 PPP LCP (link control protocol) automoton

Events		Actions	
Events related to link configuration packets			
Up	Lower layer is up	Tlu	This layer up
Down	Lower layer is down	Tld	This layer down
Open	Administrative open	Tls	This layer started
Close	Administrative close	Tlf	This layer finished
TO+	Timeout with counter >0	Irc	Initialise restart count
TO−	Timeout counter expired	Zrc	Zero restart count
RCR+	Receive configure request (good)	Scr	Send configure request
RCR−	Receive configure request (bad)		
RCA	Receive configure ACK	Sca	Send configure ACK
RCN	Receive configure NAK/REJ	Scn	Send configure NAK/REJ
Events related to link termination packets			
RTR	Receive terminate request	Str	Send terminate request
RTA	Receive terminate ACK	Sta	Send terminate ACK
Events related to link maintenance packets			
RUC	Receive unknown code	Scj	Send code reject
RXJ+	Receive code (permitted) or protocol reject		
RXJ−	Receive code (catastrophic) or protocol reject		
RXR	Receive echo request, echo reply or discard request	Ser	Send echo reply

- *link configuration packets* — used to establish and configure links;

- *link termination packet* — used to terminate connections after the data transfer (or after and authentication failure); and

- *link maintenance packets.*

The individual message code field values are listed in Table 8.6. The *identifier* field is used to match requests with their corresponding replies. An arbitrary value is used to indicate the first message. The value is incremented for each subsequent message. The *length* field indicates the length of the LCP-packet including the header (code, identifier and length fields). The length may not exceed MRU value as negotiated at link establishment (or the default value of 1500). Specific parameter values (e.g. authentication passwords, encryption algorithm keys, etc.) may be included if necessary in the LCP-data field. A PPP-information field (Figure 8.19) may contain only one LCP.

PPP network control protocol (NCP)

The *network control protocol (NCP)* arranges for allocation of relevant network layer addresses and for general management of the PPP connection as necessary to support a given network layer protocol. Various different optional NCPs (network control protocols) have been defined, as listed in Table 8.7. As an example of an NCP, the *IP control protocol (IPCP)* uses the same basic frame format as LCP (Figure 8.20) except that only codes 1 to 7 (Table 8.6) are allowed. IPCP establishes an IP-network layer on top of the datalink layer established by LCP.

PPP configuration options; PPP security and authentication using PAP and CHAP

A number of configuration options are available for PPP connections. These are negotiated using the link control protocol (LCP) during connection establishment. The various options

Table 8.6 LCP message types

LCP code value	LCP message type	Message use and purpose
1	Configure-request	Mechanism for opening an LCP connection
2	Configure-ACK (acknowledgement)	
3	Configure-NAK (non-acknowledgement)	Requested options incorrect/unclear
4	Configure-reject	Requested options not allowed
5	Terminate request	Mechanism for closing an LCP connection
6	Terminate-ACK (acknowledgement)	
7	Code reject	Indicates peer is operating with different version
8	Protocol reject	Identifies protocol rejected and reason (this code is not used in PPP/IPv6)
9	Echo request	Loopback mechanism for testing both directions of communication (uses a magic number) (these codes are not used in PPP/IPv6)
10	Echo reply	
11	Discard request	Packet sent to exercise the link for debugging, performance testing, etc. (this code is not used in PPP/IPv6)

Table 8.7 PPP network control protocols (NCPs)

Protocol abbreviation	Full protocol name	PPP Protocol number	Defined in
ATCP	PPP Appletalk control protocol	8029	RFC 1378
BCP	PPP bridging control protocol	8031	RFC 2878
CCP	PPP compression control protocol	80FD	RFC 1962
DNCP	PPP DECnet phase IV control protocol	8027	RFC 1762
IPCP	PPP Internet protocol control protocol	8021	RFC 1332
IPXCP	PPP Novell internetworking packet exchange control protocol	802B	RFC 1552
OSINLCP	PPP OSI network layer control protocol	8023	RFC 1377

Note: See also www.iana.org/assignments/ppp-numbers

are summarised (LCP data field of configuration messages) in Table 8.3. Apart from the basic configuration options which we have already explained, we shall concern ourselves here only with three further types of option:

- *authentication* options (in particular PAP — password authentication protocol and CHAP — challenge handshake authentication protocol);

- *magic numbers*; and

- *PPP multilink protocol.*

PPP's *authentication* options are used at the time of connection establishment to check that the remote user is permitted to log on to the IP-network. The user (or his computer at least) is required to identify himself with a password (in the cast of PAP) or an *authenticator value*

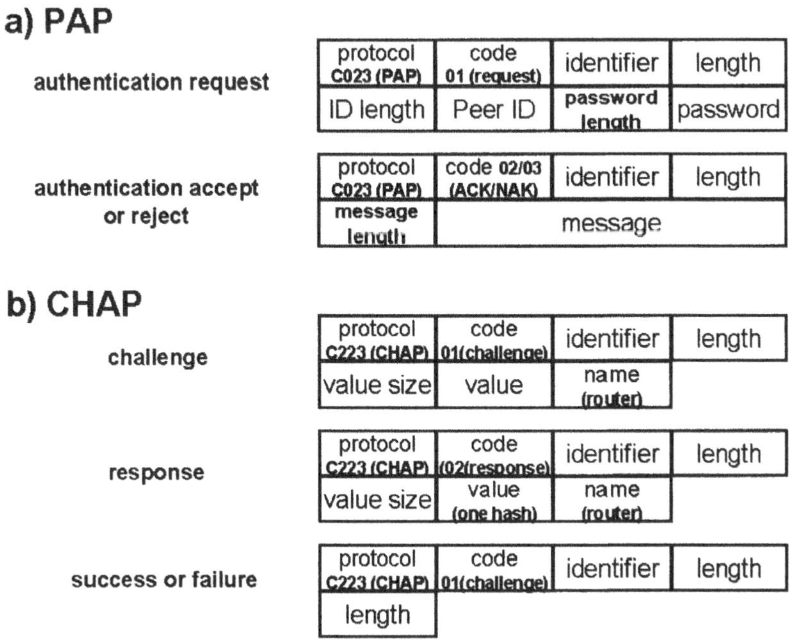

a) PAP

authentication request

protocol C023 (PAP)	code 01 (request)	identifier	length
ID length	Peer ID	password length	password

authentication accept
or reject

protocol C023 (PAP)	code 02/03 (ACK/NAK)	identifier	length
message length	message		

b) CHAP

challenge

protocol C223 (CHAP)	code 01(challenge)	identifier	length
value size	value	name (router)	

response

protocol C223 (CHAP)	code (02(response)	identifier	length
value size	value (one hash)	name (router)	

success or failure

protocol C223 (CHAP)	code 01(challenge)	identifier	length
length			

Figure 8.21   LCP authentication messages: PAP (password authentication protocol) and CHAP (challenge handshake authentication protocol).

(in the case of CHAP). The format of LCP-PAP and LCP-CHAP authentication messages are illustrated in Figure 8.21.

PAP (password authentication protocol) is nowadays considered to be a relatively insecure method of providing access control of remote users to an IP network, since it employs a simple 2-way handshake (consisting of an *authentication request* and an *authentication accept* or an *authentication reject* — Figure 8.21). The acceptance or rejection is dependent only upon the use of a valid password.

CHAP (challenge handshake authentication protocol) is considered more secure than PAP since it uses a 3-way handshake procedure (challenge/response/succeed) and communicates the password in an encrypted form (using the MD5 encryption algorithm). The use of encryption makes the task of overhearing or interception of the password (for a later log-on by a hacker) much harder.

A magic number (contained in an LCP message) may be used to perform special tests on a PPP connection. In particular, a magic number may be used to detect looped-back links and other data layer anomalies.

PPP multilink protocol (as defined in RFC 1990) allows multiple physical links to be *bundled* into a single logical link (somewhat like *link aggregation* and *reverse multiplexing* as we saw in Figure 8.15).

Point-to-point protocol (PPP) over ethernet (PPPoE)

With the increasing popularity of high speed xDSL connections for Internet access, it has become increasingly common to share the line between multiple end-user devices which are connected to the xDSL modem by means of a *local area network (LAN)*. Now, since the xDSL connection is generally treated as if it were a dial-up connection (requiring connection

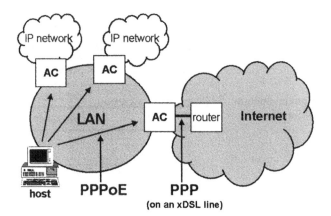

AC = access concentrator

Figure 8.22   Network connection to different IP-based networks across a LAN using PPPoE.

establishment each time it is set up and the allocation of an IP address for the duration of the communication session), it is also common to employ the point-to-point protocol (PPP) on the xDSL connection. So now the question arises, how can the end-user devices 'talk' PPP across the LAN, in order to access the xDSL line? The answer is — by means of *PPPoE (PPP over ethernet)* as defined in RFC 2516.

Figure 8.22 illustrates a typical LAN in which the PPPoE protocol is in use. An end-user device (labelled 'host') is connected to a LAN, to which a number of *access concentrators (ACs)* are also connected. In reality, the AC functionality is typically just part of the functionality of a device sold as an *xDSL access router*. The PPPoE protocol operates in the ethernet LAN to which both the host and the AC (access concentrator) are connected. PPPoE allows the host to discover the access concentrator (AC) for accessing a given service (e.g. a particular ISP's — Internet service provider's — network) and to transfer *session packets* thereafter.

The *PPPoE discovery* stage comprises four steps (Table 8.8). First, the host broadcasts *a PPPoE active discovery initiation (PADI)* message to all devices on the LAN. This in effect says, 'are there any access concentrators out there?' (i.e. router devices in the LAN with access to external networks). All access concentrators in the LAN reply with a *PPPoE active discovery offer (PADO)* message which informs the host which external network services they serve (i.e. which Internet service providers they are connected to). The host decides which access concentrator it wishes to use and sends a *PPPoE active discovery request (PADR)* message to the relevant access concentrator (to set up a PPP connection). If the selected access concentrator is able to accept a new connection at that moment then the concentrator returns a *PPPoE active discovery session confirmation (PADS)* message. Now the discovery process is complete and both PPPoE *peers* know the PPPoE session_ID. Having received the confirmation (PADS) packet, the communication may proceed to the *PPP session* stage, during which the PPPoE payload contains *PPP session packets* conforming to the standard PPP information message format (as we saw in Figure 8.19).

PPPoE frame format

The PPPoE version 1 frame format (as defined in RFC 2516) is as illustrated in Figure 8.23. The MAC address fields identify the host to the access concentrator and vice versa. The

Table 8.8   PPPoE Discovery and PPP sessions: PPPoE message field codings

Connection state	Message name	Full message name	Sent by	Sent to	Ethernet type field setting	PPPoE Code setting	PPPoE Session_ID
PPPoE Discovery step 1	PADI	PPPoE active discovery initiation packet	Host	Broadcast address	8863	09	0000
PPPoE Discovery step 2	PADO	PPPoE active discovery offer packet	All access concentrators in host's network	Host	8863	07	0000
PPPoE Discovery step 3	PADR	PPPoE active discovery request packet	Host	Selected access concentrator	8863	19	0000
PPPoE Discovery step 4	PADS	PPPoE active discovery session-confirmation	Selected access concentrator	Host	8863	65	Unique value chosen by access concentrator
PPP session stage	PPP Establishment phase	Conducted by LCP (see Figure 8.11)	Concentrator or host	PPPoE Peer partner	8864	00	Value indicated in PADS packet
	PPP Authenticate phase	Conducted by PPP authentication protocol (see Figure 8.11)	Concentrator or host	PPPoE Peer partner	8864	00	Value indicated in PADS packet
	PPP Network phase	Conducted by NCP and network layer protocol (see Figure 8.11)	Concentrator or host	PPPoE Peer partner	8864	00	Value indicated in PADS packet
	PPP Terminate phase	Conducted by LCP (see Figure 8.11)	Concentrator or host	PPPoE Peer partner	8864	00	Value indicated in PADS packet
PPPoE Discovery termination (after session)	PADT	PPPoE active discovery terminate packet	Host or selected access concentrator	PPPoE peer partner	8863	A7	0000

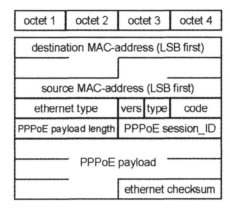

octet 1	octet 2	octet 3	octet 4

destination MAC-address (LSB first)

source MAC-address (LSB first)

| ethernet type | vers | type | code |
| PPPoE payload length | PPPoE session_ID |

PPPoE payload

ethernet checksum

Figure 8.23   PPPoE (point-to-point protocol over ethernet): frame format.

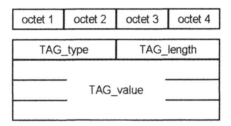

octet 1	octet 2	octet 3	octet 4

| TAG_type | TAG_length |

TAG_value

Figure 8.24   PPP discovery packet payload: TAG format.

ethernet type field is set at value 8863H during the *discovery stage* and at value 8864H during the *PPP session* stage. The *version* field is set to value '1' (for the version of PPPoE defined in RFC 2516). The *type* field is not used in version 1 and must always be set at value '1'. The PPPoE *code* and PPPoE *session_ID* fields are coded as detailed in Table 8.8. The *length* field contains a value corresponding to the length of the PPPoE payload field in octets (not including either the PPPoE header or the ethernet header).

PPPoE discovery packets contain a payload of one or more *TAGs* (see Figure 8.24), each of which contains *TLV (type length value)* objects as defined in Table 8.9. These objects allow the host and access concentrator (AC) to exchange information about the external network services which are available via the access concentrator and how to configure the PPPoE session.

The emergence of PPPoE is bound to further widen the usage of gigabit ethernet as an access technology for high speed Internet and IP network access in metro networks (as illustrated in Figure 8.9).

8.6   Wireless technologies for Internet access

Wireless technologies are becoming increasingly popular as a means of connecting end-user devices (particularly laptop PCs) to data networks. Wireless offers greater mobility of the end user and avoids the need for installing cables or moving cables around. Figure 8.25 illustrates a number of the wireless technologies used nowadays for data network access:

- Data communications via special mobile handsets is possible using *modern GPRS (general packet radio service), UMTS (universal mobile telephone service)* or *WAP (wireless*

Table 8.9 PPPoE TAG-Messages

Tag_type	Tag name	Purpose of TAG	Tag_length	Tag_value
0000	End-of-list	Indicates there are no further TAGs in the list	0 octets	—
0101	Service-name	Typically contains ISP name or class or quality of service name	Unspecified	Service name in UTF-8 format
0102	AC-Name	Used to uniquely identify access concentrator from all others in the same ethernet network	Unspecified	Name, usually a combination of trademark, model and serial number ID or the MAC address in UTF-8 format
0103	Host-uniq	Used by a host to uniquely associate an access concentrator response (PADO) or PADS to a host request (PADI or PADR)	Unspecified	Number of any value
0104	AC-cookie	Used by the access concentrator in protecting against DOS (denial of service attacks)	Unspecified	Binary number of any value
0105	Vendor-specific	Used to pass proprietary vendor messages and information	<4 octets	First four octets contain vendor_ID
0110	Relay-session-ID	May be added to discovery packet by an intermediate relaying agent. It will be included in any PPPoE response message	12 octets	Number of any value
0201	Service-name-error	Indicates that the requested service name cannot be honoured	Typically 0 octets	—(if non zero length a printable UTF-8 string)
0202	AC-system-error	Indicates access concentrator encountered error performing host request	Typically 0 octets	—(if non zero length a printable UTF-8 string)
0203	Generic error	Indicates an error and can be included in any PADO, PADR or PADS packet	Typically 0 octets	—(if non zero length a printable UTF-8 string)

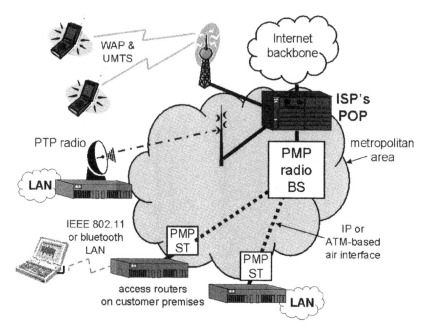

Figure 8.25 Wireless data networking technologies used for IP data network access.

application protocol) mobile telephone handsets. Generally such devices allow full mobility of the end user but are limited to relatively slow speeds, although UMTS is designed to allow short bursts of transmission at up to 2 Mbit/s;

- *Point-to-point (PTP)* microwave radio is offered as a means of fixed wireless access by some public telecommunications 'fixed-network' providers. Using relatively small and compact technology (including parabolic antennas of 30 cm diameter) a highly reliable radio transmission service at rates of up to 155 Mbit/s and over ranges of up to about 5 km can be achieved at a capital outlay under $ 30 000 per link;

- *Point-to-multipoint (PMP)* microwave radio is a technology reserved for *wireless local loop (WLL)* application by public telecommunications network operators. Such technology is also known as *LMDS (local multipoint distribution service)* and *LMCS (local multipoint communications system)* in North America. Although initial interest in PMP and LMDS was high in the late 1990s, it has not experienced the uptake that manufacturers had hoped for.

- Wireless LAN technologies (particularly IEEE 802.11-based systems which are now available as PCMCIA slide-in cards for laptop PCs and the bluetooth LAN standard (see www.bluetooth.com). These have been among the most popular of radio technologies for IP communication so far.

There are clear advantages for wireless, but a major disadvantage is its fundamental unsuitability for data communications. Wireless communication is particularly prone to noisy connections, *interference* and *fading* (i.e. loss of signal), all of which lead to high *bit error ratios (BER)* when carrying data traffic.

A particular challenge for wireless data network designers is the problem of *burst errors* caused by fading (periods of poor reception caused by heavy signal attenuation during propagation). Unlike the usual isolated single *bit errors* normally encountered on wireline communications media, wireless is prone to long streams of contiguous errors — called burst errors. Burst errors can generally not be effectively dealt with by standard error detection and correction methods using frame check sequences (FCS) and retransmission of errored frames. Instead special, more efficient techniques optimised for wireless media have had to be developed. These usually employ *forward error correction (FEC)* — powerful error detection and correction codes based on *Reed-Solomon* and *Viterbi convolutional* error correction codes and *byte-interleaving*. (The techniques are too mathematically complex to consider addressing here — refer to a specialist book on wireless if you need more detail.)

The *range, reliability* and *link-availability* of all radio systems are inter-dependent and rely upon the BER (bit error ratio) target, the propagation characteristics and any interference of other radio users in the same area. The propagation, in turn, depends upon:

- the radio *frequency band* of operation (the higher the frequency the lower the range);

- the weather or other atmospheric conditions (the more water around, the more attenuation that will be suffered); and

- the nature of the radio path (surfaces which might cause reflections may cause severe degradation due to so-called *multipath* interference).

You should be aware that the *range* of many older telecommunications radio systems (including PTP and PMP microwave radio) is typically quoted for a target BER of only 10^{-6} and an annual *availability* of only 99.99% (i.e. an out-of-service time per year of 52 minutes a year.[8] For the most critical data network applications these targets may not allow sufficiently reliable service. In practical terms, this means that your maximum allowed operational radio system range may be lower than that quoted in the radio manufacturer's datasheet. Take advice if you are not sure..or try it out and see how it works!

Wireless application protocol (WAP)

The *wireless application protocol (WAP)* was conceived as an Internet access technique for handheld *PDAs* (*personal data assistants* — so-called 'palmtop' PCs), mobile phones and similar devices with less powerful CPUs (central processor units), less memory, small displays, restricted keyboard and input potential, as well as limited power availability. It is a layered protocol (somewhat similar to IP) which has been developed by the WAP forum (see www.wapforum.org) to cater for the needs of data communication under the constraints of restricted bandwidth connections, limited link *availability* and poor transmission reliability.

The specifications for WAP version 2 (WAP 2.0) were issued in July 2001. It is an extensive set of specifications and we do not have the capability to deal with it in detail here, although the basic protocol layer model is presented in Figure 8.26 and Table 8.10. It is mainly concerned with access to the Internet from mobile telephone and other handheld devices — a subject we shall return to in Chapter 11.

---

[8] While 'only' 52 minutes downtime per year may appear attractive, in reality the 52 minutes is comprised of a large number of short interruptions rather than a single interruption of 52 minutes duration. In practice, the large number of disturbances are very disruptive for data communications — particularly if applications or computer systems have to be rebooted each time!

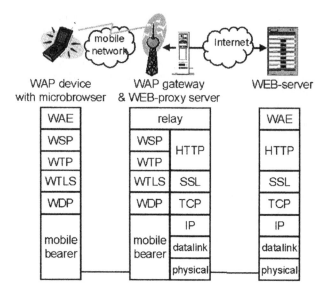

Figure 8.26   The wireless application protocol (WAP) suite.

Table 8.10   Components of the wireless application protocol (WAP) suite

	Abbreviation	**Full protocol name**
WAP	WAE	Wireless application environment
Wireless	WDP	Wireless datagram protocol
Application	WSP	Wireless session protocol
Protocol	WTLS	Wireless transport layer security protocol
Suite	WTP	Wireless transaction protocol

8.7   Host functionality and software for communication via IP

To ensure that a data network functions correctly is to make sure that the communicating
hosts are correctly equipped and configured. Indeed, as we shall see later in the chapter on
performance optimisation (Chapter 14), the incorrect configuration of hosts can be a major
source of problems.

RFC 1122 defines the requirements of hosts intended to communicate across IPv4 networks.
In brief, you need to make sure that the host is equipped with a network interface card (NIC)
of the right type, that the cables are all correct (just because the cable fits in the socket does
not imply that it is the right cable, as we learned in Chapter 3!). The software configuration of
the various communications protocols (typically altered by means of a 'networking' window)
needs to have been configured correctly, and relevant transport protocols (e.g. TCP — transport
control protocol) and application protocols need to have been configured.

Finally, as one of the most important communications application software types used in
modern PC communications, we end the chapter with a short review of *TAPI (telephone application programming interface)*, a Microsoft Windows application program used to program and
configure telecommunications line devices to communicate on behalf of Windows application
software programs.

TAPI (telephone application programming interface)

TAPI provides for control of telecommunications and telephone network termination devices, including modems and IP telephony in a device-independent manner by applications programs written for Microsoft Windows. TAPI allows for the setting-up of data network connections and dial-up networking in a standard manner, assuring control of the quality of service (QOS) of the communication and using a common application interface (despite the different types of connection and signals used on different types of networks). TAPI is integrated into Windows95, Windows98, WindowsNT, Windows2000 and WindowsXP.

To date, three versions of TAPI have been developed (TAPI, TAPI 2.0 and TAPI 3.0). TAPI 3.0 was announced in September 1997. As well as for basic dial-up and IP communications functions, it provides *multimedia* streaming control functionality integrated with telephony, support of the ITU-T H.323 standard for *VOIP* (*voice-over-IP*: as we shall discuss in Chapter 10), for conferencing via *multipoint control units (MCUs)*, as well as supporting *multicasting*, the use of *directory access protocols* (such as *LDAP — lightweight directory access protocol*) and packet scheduling. TAPI includes functionality both for initiating a communications session (i.e. a call) with a remote host or for answering an incoming call.

9

Managing the Network

A large part of the job of updating network topology and configuration data in router networks is undertaken automatically — by means of routing protocols. This capability, together with the Internet Protocol's capability to divert traffic around failed nodes and links in a network, have allowed the Internet to become a remarkably resilient and reliable worldwide data network, even though there is no centralised network management centre and operations staff to oversee the whole network. Meanwhile, the development of management tools and capabilities for human 'network managers' to oversee and manage large IP networks has been relatively slow. Initially, the simple network management protocol (SNMP) was conceived to monitor and control individual items of network equipment. Remote MONitoiring (RMON) followed. Nowadays, the development of management tools is gaining pace — driven by the network operators' need to meet the ever-increasing expectations of Internet users for predictable network performance and secure and guaranteed quality of service (QOS), while efficiently dealing with fault management and fairly accounting for the services used. In this chapter we review the current state-of-the-art means available for IP network management.

9.1 Managing and configuring via the console port

Most IP-networking equipment can be configured by means of its *console port*. This is a serial-interface port. It usually appears on the back of the network equipment and is equipped with a female DB-9 or DB-25 connector. By means of a standard serial cable (DTE-to-DCE as described in Chapter 3) a *terminal* (i.e. *console*), laptop or other computer may be connected to the console port (Figure 9.1). The terminal or *terminal emulation* software (e.g. *Hyperterminal*) is then used to issue textual commands in ASCII format from the PC to the network equipment. A password is normally required as part of a simple log-on procedure, after which the technician usually receives either a *command line prompt (command line interface — CLI)* or a simple text-based *command menu* from the equipment, for example:

```
PASSWORD?>       XXXX
Select option:
1      Configure system
2      System and Alarm status
3      Operating system update
4      Quit
>
```

Data Networks, IP and the Internet: Protocols, Design and Operation   Martin P. Clark
© 2003 John Wiley & Sons, Ltd   ISBN: 0-470-84856-1

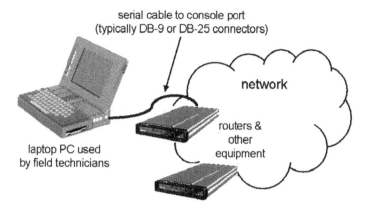

serial cable to console port
(typically DB-9 or DB-25 connectors)

network

routers &
other
equipment

laptop PC used
by field technicians

Figure 9.1   Use of a laptop computer and the console port for network equipment configuration.

In order to change the equipment configuration, text commands are issued by the technician via the terminal or PC to the equipment. The meanings and purpose of the commands are usually specific to the equipment and difficult for a newcomer to decipher (e.g. 'atur', 'debug', 'load 4.x.5', 'resetxr' etc.). Commands are typically confirmed either by the response 'OK' or by a *prompt* for another command. The prompt in the example above is the character '>'.

Since routers rely largely on routing protocols to learn automatically about the surrounding network topology, there is no need for a great deal of manual configuration when they are first installed. By allowing the manual configuration to be undertaken from a 'simple' data terminal (or a PC with terminal emulation software) via the console port, both the network equipment and the tools required by the technician can be kept relatively simple, and thus relatively cheap. But there are three main drawbacks of using the console port for configuring or monitoring/diagnosing equipment. The first is that the console (i.e. technician's terminal) has to be connected locally to the equipment. In other words, to reconfigure an entire network, the technician will have to visit each location and reconfigure each network component individually. The second drawback is that the text commands used at a console interface are usually quite complex and not intuitive — a skilled and trained specialist is required for the task of configuration. The final problem is that any errors in spelling or other command errors may lead to serious mis-configuration. (There is usually no helpful *graphical user interface (GUI)* software to check the commands issued, sort the commands into the right order, prompt for forgotten information and keep an 'audit trail' of the changes made so far.)

Using the console port for equipment configuration and fault diagnosis requires only cheap equipment, but is rather cumbersome and requires quite a high level of training! As a result, alternative methods have emerged which have been specifically designed to provide for:

- remote control of equipment — for purpose of configuration, monitoring equipment status or diagnosing faults;

- specialised software-based tools for network management. Such tools present network status information in a graphical format and convert comprehensible commands given by the human network operator into the 'gibberish' which the network components expect as control commands.

Nowadays, most network operators prefer to use a remote and centralised network management for configuration, monitoring and fault diagnosis of network components. Such remote network management relies heavily upon the *simple network management protocol (SNMP)*, and the related *managed objects* and *management information bases (MIBs)*...as we shall discover in the next sections of this chapter. But despite the availability of remote management methods, the console port (or an equivalent) will always be retained for at least two critical functions:

- configuring the name and IP address of network components prior to first installation in a network. A router, for example, has to know its own IP address and the subnet range of addresses which are directly connected to it before it can be added to a network. It also has to have been configured to know which routing protocols will be used on which interfaces and to know the relevant *routing area* to which it belongs (refer back to Chapter 6 if necessary);

- local connection of a PC for diagnosing a fault, re-loading software or similar measures: should the network component become isolated from the rest of the network as the result of a fault.

9.2   Basic network management: alarms, commands, polling, events and traps

While the use of the console port may be a simple, convenient, cheap and effective means of configuring routers and other IP network nodes and components at their time of first installation, it would be impractical to use all the console ports for diagnosing faults in a large network.

In a small network (maybe you have a small LAN at home) it may be adequate to configure the LAN switch, Internet access router and firewall using their respective console ports. As long as the network works, you may think it unnecessary to track the daily network statistics, alarm status and performance reports. When a fault occurs, you simply check each of the components in turn — using the console port as necessary. Or maybe you simply 'reset' them all by quickly switching them off and back on again. But as the network administrator of a large data network in a major bank, you could not consider visiting each of the branches any time a fault or network performance problem arose.

As the early Internet developed and the number of interconnected routers rapidly increased, the need for remote network monitoring, fault diagnosis and equipment configuration became rapidly apparent. The solution was the development of an *architecture* for network management based on a standardised communications interface between a centralised network *manager* and a remote *network element* (i.e. network node or other network component). The standardised communications interface, as illustrated in Figure 9.2, is a protocol called *SNMP (simple network management protocol)*. We shall describe it in detail later in this chapter. The centralised network manager is an *SNMP manager* and the *network element* (which 'speaks' to the SNMP manager by means of SNMP) is termed an *SNMP agent*.

The SNMP manager, which receives network *status* information and *alarms* from SNMP agents and can issue commands for network configuration, is usually realised in software on centralised server hardware. Various manufacturer-proprietary systems are available as SNMP managers. Perhaps the best known is the Hewlett Packard company's *OpenView* system. We shall return to the subject of so-called 'umbrella management systems' later in the chapter.

SNMP (simple network management protocol) is a communications protocol by which the network element (the SNMP agent) can exchange status, configuration and command information with the *SNMP manager*. It is a standardised format in which information may be sought by the *manager* from the *agent* using a process called *polling*. Alternatively, the agent may send *solicited* or *unsolicited* reports to the manager.

As the name suggests, the SNMP manager is in control of the communication. This is important, in order that the manager is not swamped by continual status updates from the

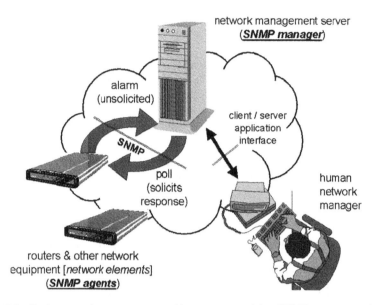

network management server
(**SNMP manager**)

alarm
(unsolicited)

client / server
application
interface

SNMP

poll
(solicits
response)

human
network
manager

routers & other network
equipment [*network elements*]
(**SNMP agents**)

Figure 9.2   Basic network management architecture, comprising SNMP manager and agent.

agents which it may be uninterested in, and anyway it may be unable to store or process at the time when they are received.

Under normal circumstances, the SNMP manager software is designed to track the basic operation of the network and to respond to the requests and commands of the human network managers using it. In order to fulfil this function, it *polls* the network elements (regularly or sporadically — as it sees fit), *soliciting* information about their current operations and performance status. The information received from the different network elements as a result of *polling* is correlated and usually presented to human network operators as a series of different graphs, network topology diagrams, network load and performance charts as well as other graphics.

Should a fault occur in any of the network elements, then the corresponding SNMP agent generates an *alarm* or *event* message which it sends to the SNMP manager. This message (called an *SNMP trap*) is said to be *unsolicited*. Unsolicited messages (SNMP traps) are usually of an urgent nature and are treated with priority by the SNMP manager. Indeed, SNMP trap messages use a different UDP (user datagram protocol) *port number* to differentiate them from 'ordinary' SNMP messages, and thus help their priority treatment.

Depending upon the seriousness of the alarm, SNMP trap messages are typically presented immediately by the SNMP manager to the human network managers as either 'yellow' or 'red status' alarms. The affected network element is typically illustrated in yellow or red colour on the network topology graphic, and the human network manager may additionally be notified by means of an audible alarm, an immediate screen 'pop-up message' or by an urgent email.

As well as being used for network status monitoring, the SNMP manager can also be used for network configuration. The three basic communications processes undertaken between SNMP managers and *network elements* (i.e. SNMP agents) are thus:

- solicited 'informational/status' responses of the SNMP agents caused by polling (get-messages) of the SNMP manager;

- *unsolicited* messages (SNMP traps) sent by SNMP agents to advise the SNMP manager of an alarm status, and

- *command* messages (so-called *set*-messages) issued by the SNMP manager to instruct the SNMP agent to change the configuration of the associated network element and the related confirmational responses.

The five basic types of SNMP message (SNMPv1) are thus:

- *get-request* (by which the SNMP manager polls information from the SNMP agent);

- *get-next-request* (this is a particular type of get request message, used by the SNMP manager to solicit the next data value from a 'table' or 'matrix' of values held within the SNMP agent);

- *get-response* (by which the SNMP agent sends solicited information to the SNMP manager);

- *set-request* (by which the SNMP manager issues commands for changing the configuration of the network element associated with the SNMP agent), and;

- *trap*-messages (by which an SNMP agent advises the SNMP manager of alarms and/or events and provides associated pre-defined additional status information).

The protocol specification of SNMP defines the format of each of the message types above and how both the SNMP manager and SNMP agent are to conduct themselves during the exchange of information. But the real informational meaning of the requests and the content of the responses is not defined within SNMP. This is the domain of a *management information base (MIB)* and the *managed objects* which it defines. This is our next subject for consideration, but we shall return to a detailed discussion of SNMP later.

9.3   Management information base (MIB) and managed objects (MOs)

On its own, SNMP (simple network management protocol) is not able to convey any useful information from SNMP agent to the SNMP manager or vice versa. SNMP is simply a set of 'rules of orderly conversation'. In the same way that the project manager of a major building project can make sure that instructions are issued to the electrician, the bricklayer and the plumber and that each of the craftsmen understands his instructions and completes them on time, so SNMP is able to 'project manage' the task of managing a network. The actual content of the network management 'work instructions' depends upon the particular network component being managed, and is coded according to the appropriate management information base (MIB).

The management information base (MIB) for a given type of *network element* defines the functions which the network element is capable of, the configuration options which are possible and the information it can provide in terms of a set of managed objects (MOs). A managed object is a standardised definition of a particular feature or capability of the network component, and the normalised states in which it can exist.

Imagine a traveller's water bottle to comprise two standard managed objects. The first managed object — the bottle itself — might be 'a vessel for storing one litre of liquid' (exactly what shape it is, is unimportant). The second object — the 'stopper' — might be a screw top, a cap-top or a cork, but as a managed object all three are simply 'watertight stoppers' which may be in any of the states 'secured', 'opened' or 'being opened'. Standardised states of the bottle, meanwhile, might be 'full', 'empty', 'half full', etc. Standard management commands relevant to the bottle object might be 'fill up', 'pour out' while 'secure' or 'undo' commands might be sufficient for the 'stopper' object. The definitions of both the 'bottle' and 'stopper' objects, together with the related commands form the management information base (MIB) of the 'travellers water bottle'. Provided all manufacturers produced traveller's water bottles and stoppers which are able to respond to the management commands, you could buy whichever bottle and stopper combination was the most aesthetically pleasing and be confident that a standard command procedure will get the stopper out!!

Management information base (MIB): definition and object hierarchy

Management information bases (MIBs) and the managed objects which they define are nowadays standardised for most types of new computer and telecommunications products. They may be defined either on an industry-standard basis and a large number have been defined by IETF (Internet Engineering Task Force) and similar standards bodies as a standardised means of managing:

- IPv4 routers;
- IPv6 routers;
- IP (Internet protocol), TCP (transmission control protocol) and other networking protocols;
- OSPF (open shortest path first) and other routing protocols;
- ethernet, token ring and other LAN (local area network) devices;
- ISDN (integrated services digital network), ATM (asynchronous transfer mode), FR (frame relay), SDH (synchronous digital hierarchy), SONET (synchronous optical network) and a range of other types of network interfaces.

Having well-defined, standard MIBs for each of the network components is a critical prerequisite for centralised (remote) management of an entire network. In the case where industry standards do not already exist for a given network component, protocol or interface, most manufacturers nowadays define their own MIB (management information base) but publish it as an intended 'standard'.

The structure and syntax of management information bases (MIBs) and managed objects (MOs)

The definition of managed objects (MO) and managed information bases (MIBs) is carried out using the *abstract syntax notation 1 (ASN.1)*. The procedure of definition, including advice on how to group and subdivide MOs is defined by ISO (International Organization for Standardization) and ITU (International Telecommunications Union) and is documented in ITU-T recommendations X.722 (ISO 8824, 8825) and X.739 (ISO 10164-11). These are the *guidelines for the definition of managed objects (GDMO)*. They include strict rules about the naming and spelling conventions of objects and their hierarchical structure. The adaptations and specific use of these guidelines for the definition of Internet-related managed objects for use with SNMP are defined as the *structure of management information (SMI)*. SMIs are defined in various RFCs:

- the *structure of management information version 1 (SMIv1)* is defined in RFC 1155;
- a methodology for defining concise MIB (management information base) modules for Internet usage is defined in RFC 1212;
- the core set of managed objects for IP-suite protocols (called mib-2 or MIB-II) is defined by RFC 1156 and various updates are defined in RFC 1213; and
- the *structure of management information version 2 (SMIv2)* is defined in RFCs 2578-80. SMIv2 was introduced concurrently with SNMPv3. In particular, SMIv2 requires the

definition of security-related parameters about each managed object and provides for more powerful network management. Its use is obligatory with SNMPv3.

Managed objects and their relation to the 'root' object

The management information base (MIB) for a particular network component, protocol or other function usually comprises a number of managed objects, structured in a hierarchical manner and defined using the abstract syntax notation (ASN.1). Each managed object has a name (the *object identifier*), a *syntax* and an *encoding*.

Depending upon the *object type*, the data used to represent an object may be formatted as an integer, an octet string, an address value, a network mask, a *counter*, a *timer* or some other special format — as best describes the object. This is defined by the syntax assigned to the object type. The *encoding* defines how specific *instances* of the object type are represented using the syntax. The syntax and encoding define how the information about the object will be represented while being carried by SNMP (simple network management protocol). Thus, in the case of our traveller's water bottle the object with the name 'bottleVolume' might be declared to be a single 'floating point' number within a given range, the value being defined to be equal to the volume of the bottle in cubic centimetres (cm^3). On the other hand, an IPv4-type network address is best defined by the syntax of four decimal integer values (each between 0 and 255), separated by dots.

Objects are defined in a hierarchical manner, all possible objects being derived from a standard *root* object (which is administered by ISO — International Organization for Standardization). Figure 9.3 illustrates the top level of the *managed object inheritance hierarchy*,

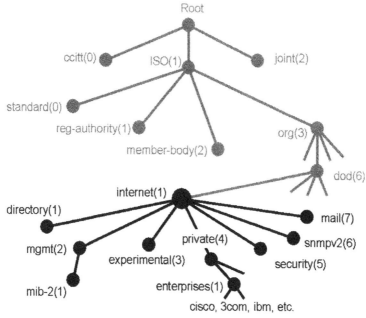

Key: 'ccitt' stands for *international telephone and telegraph consultative committee* (the old name for ITU-T—International Telecommunications Union standardization sector), 'joint' stands for 'shared ISO/CCITT objects', 'org' stands for 'independent organisation' and 'dod' stands for the United States 'Department of Defense'.

Figure 9.3   The object inheritance hierarchy of the Internet group of objects from the ISO root object.

showing how objects relating to the Internet are derived from the *root* object. As can be seen in Figure 9.3, the Internet management information base (the current version is mib-2) is one of the main object *groups* or *modules* within the Internet object group.

Usually objects are referred to by their name (e.g. mib-2), but it is also important to understand their position within the object inheritance hierarchy. This can be denoted in one of two ways:

- either by showing its direct parent thus   mib-2 ::={mgmt 1}

- or by showing its absolute position in the inheritance hierarchy thus   1.3.6.1.2.1 [mib-2]

In the first of the above notations, the object mib-2 is defined to be the first object (i.e. denoted '1') within the 'mgmt' group of objects. In the second notation the complete inheritance from the root object is recorded (compare with the values with the 'branches' of the inheritance 'tree' in Figure 9.3!).

We shall illustrate some the actual managed objects and groups within the Internet mib-2 (management information base) later in this chapter, but we shall not cover all the managed objects which have been defined for Internet-related component and protocol management. For those readers who need this, there are a number of commercial software products available which provide an 'encyclopaedia' of the objects and their inheritance hierarchy (these are called *GDMO browsers*).

9.4   Structure of management information (SMIv1 and SMIv2)

As we learned above, network management of the Internet and IP-based internetworks is intended to be carried out using the simple network management protocol (SNMP) and a series of managed objects defined within the Internet management information base (the current version is mib-2). The managed objects are defined according to the abstract syntax notation 1 (ASN.1), but the full capabilities of ASN.1 are not supported by SNMP. Instead a 'simple and workable' subset of ASN.1 is defined and called the *structure of management information (SMI)*. In the same way that SNMP was defined by IETF (Internet Engineering Task Force) to be a 'simple and workable' alternative to the ISO/ITU-T management protocol *CMIP (common management information protocol)*, so SMI is the 'simple and workable' subset of ASN.1. Together, SNMP and SMI form the *architecture and management framework* for the network management of IP-based internetworks.

The first version of SMI (structure of management information) — SMIv1 — was introduced with SNMP version 1 and is defined in RFC 1155 (SNMPv1 is defined in RFC 1157). SMIv1 sets out the format in which MIB and managed object definitions should be presented and the various information which is required. A typical managed object definition (in ASN.1/SMI format) of an object-type called 'tcpMaxConn' is shown below:

```
tcpMaxConn          OBJECT-TYPE
        SYNTAX      INTEGER
        ACCESS      read-only
        STATUS      mandatory
        DESCRIPTION "The limit on the total number of TCP
                    connections the entity can support. In
                    entities where the maximum number of
                    connections is dynamic, this object should
                    contain the value -1"
        ::= {tcp 4}
```

The first line of the SMI definition defines the name of the object (called the *object identifier* — in this case tcpMaxConn) and identifies it as an object (with the words OBJECT-TYPE). The syntax explains that this object is a data element of type INTEGER. The *access* line explains that the object value held within the SNMP agent may only be read (and thus not altered). The status line explains that all network elements with the tcp-MIB must 'mandatorily' be able to respond to a get-request for the value of tcpMaxConn. Finally, the description explains the meaning of the object and how it is to be encoded.

The second (and current) version of SMI — SMIv2 — is an extension of SMIv1 to support more powerful capabilities of network management made possible by SNMPv3. SMIv2, which is defined in RFCs 2578-80, requires more information to be defined for each managed object. Among other things, SMIv2 provides more access mode types for each object: *read-only, read-write, notify,* etc. — in line with the extra security built into SNMPv3.

9.5   Management information base-2 (mib-2 or MIB-II)

Figure 9.4 illustrates the basic modules which make up the sub-hierarchy of the Internet management mib (mib-2 or MIB-II). The mib (mib-2) is defined in RFC 1156 and RFC 1213. Note how the mib-2 *modules* already reflect both the network and interface components of an IP-based network, as well as the various protocols used for data transport. Take a look at the sub-modules in the lower levels of the mib-2 hierarchy (Figure 9.5) and you will start to find objects suitable for *setting* or reading (i.e. *getting*) protocol parameter settings or for controlling the actions of the protocols or other network devices. Thus, for example:

● The object ipDefaultTTL [::= {ip 2}] might allow the default time-to-live parameter value in the Internet Protocol (IP) to be checked or changed; similarly

● The object tcpMaxConn [::= {tcp 4}] allows the maximum number of connections to be established simultaneously by a given device using transmission control protocol (TCP) to be checked. This is the object for which we illustrated the full object definition in SMIv1-format earlier.

As a quick reference to the main modules and higher layer managed objects in mib-2 we present Figure 9.5 and Table 9.1, but we shall not discuss these in detail.

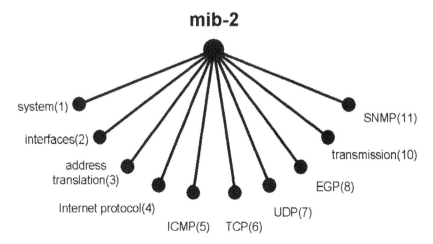

mib-2

system(1)

interfaces(2)

address
translation(3)

Internet protocol(4)

ICMP(5)   TCP(6)

UDP(7)

EGP(8)

transmission(10)

SNMP(11)

Figure 9.4   Internet management information base version 2 (mib-2).

9.6   Remote network monitoring (RMON)

The RMON-MIB was specifically established for the purpose of *Remote network MONitoring (RMON)*. It is designed to be used by monitors, *probes*[1] and other 'RMON devices'. The RMON objects set out in the RMON-MIB provide for a 'common language' by which an

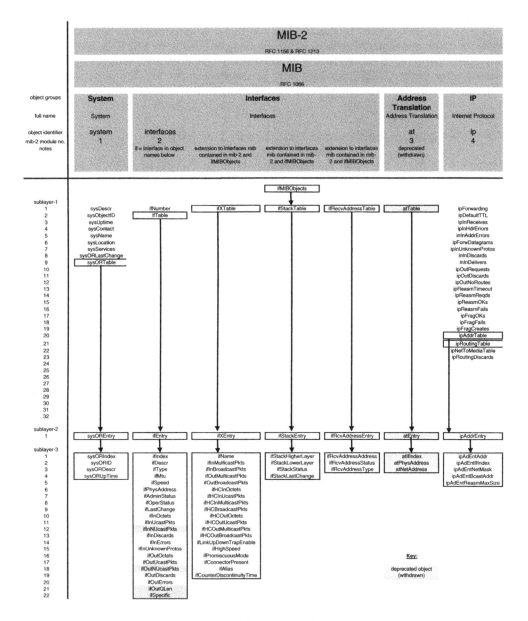

Figure 9.5   mib-2 structure and modules.

---

[1] A *probe* is a special device typically used for monitoring traffic on a LAN or internetwork. Network administrators use such probes to investigate poor network performance.

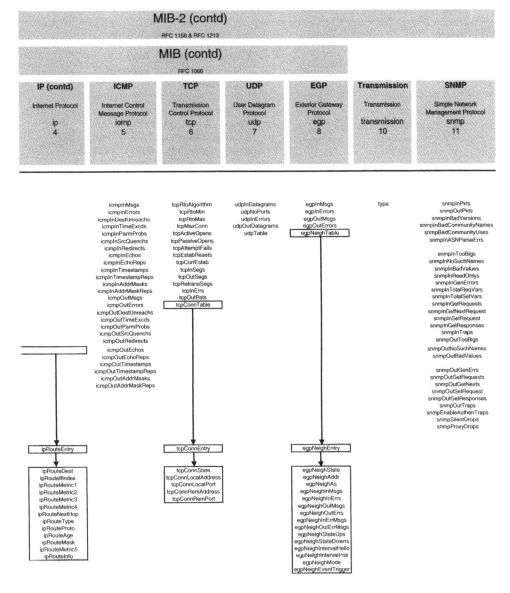

Figure 9.5 (*continued*)

RMON management application (such as a *probe*) can communicate using SNMP with an *RMON agent* (typically a network component). The current version of RMON is set out in SMIv2 (structure of management information version 2) format in RFC 2819.[2]

Many of the objects defined by RMON are intended to be suitable for different types of IP-based networks, but they may be too general to give full control of certain types of network or protocol. Initially, specific objects were developed for monitoring ethernet networks, and

---

[2] RFC 1757, now superseded by RFC 2819, is a semantically identical version of RMON to RFC 2819 (using identically named objects). The difference is that RFC 1757 is defined in SMIv1 rather than SMIv2-format.

Table 9.1   Module extensions of the basic internet management information base (mib-2)

mib-2 module	Module name	MIB name
[9	CMOT	CMIS/CMIP over TCP/IP]
12	GenericIF	Generic Interface Extensions (RFC 1229, 1239)
14	Ospf	Open Shortest Path First (RFC 1253)
15	Bgp	Border Gateway Protocol (RFC 1657)
16	Rmon	Remote Network Monitoring
17	Bridge	Bridge Objects (RFC 1286)
23	rip-2	Routing Information Protocol (RFC 1389)
24	Ident	Identification Protocol (RFC 1414)
25	Host	Host resources (RFC 1514)
26	SnmpDot3MauMgt	IEEE802.3 medium attachment units (RFC 2668)
27	Application	Network services monitoring (RFC 2248)
28	Mta	Mail monitoring (RFC 2249)
29	Das	X.500 Directory Monitoring (RFC 1567)
30	IANAifType	Interface types (RFC 1573)
31	IfMIB	Interface types (RFC 1573)
32	Dns	Domain Name System (RFC 1611)
33	UpsMIB	Uninterruptible power supplies (RFC 1628)
35	EtherMIB	Ethernet-like generic objects (RFC 2665)
37	AtmMIB	Asynchronous Transfer Mode objects (RFC 1694)
38	MdmMIB	Dial-up modem objects (RFC 1696)
39	RdmsMIB	Relational database objects (RFC 1697)
40	FlowMIB	Traffic flow objects (RFC 2064)
43	PrintMIB	Printer (RFC 1759)
44	MipMIB	Mobile IP MIB (RFC 2006)
45	dot12	IEEE 802.12 MIB (RFC 2020)
46	DlswMIB	Data link switching MIB (RFC 2024)
47	EntityMIB	Entity MIB (RFC 2037)
48	IpMIB	Internet Protocol MIB module (RFC 2011)
49	TcpMIB	Transmission Control Protocol MIB module (RFC 2012)
50	UdpMIB	User Datagram Protocol MIB module (RFC 2013)
51	Rsvp	ReSerVation Protocol MIB (RFC 2206)
52	IntSrv	Integrated Services MIB (RFC 2213)
53	VgRptrMIB	IEEE 802.12 Repeater MIB (RFC 2266)
54	SysSpplMIB	System application MIB (RFC 2287)
55	Ipv6MIB	Internet Protocol version 6 MIB (RFC 2465)
56	Ipv6IcmpMIB	Internet Control Message Protocol v6 MIB (RFC 2466)
57	MarsMIB	Multicast address resolution MIB (RFC 2417)
58	PerfHistTCMIB	Performance History TC-MIB (RFC 2493)
59	AtmAccountingInformationMIB	ATM Accounting MIB (RFC 2512)
60	AccountingControlMIB	Accounting control MIB (RFC 2513)
61	IANATn3270eTCMIB	3270 emulation TC-MIB (RFC 2561)
62	ApplicationMIB	Application management MIB (RFC 2564)
63	SchedMIB	Schedule MIB (RFC 2591)
64	ScriptMIB	Script MIB (RFC 3165)
65	WwwMIB	Worldwide Web service MIB (RFC 2594)
66	DsMIB	Directory Server monitoring MIB (RFC 2605)
67	RadiusMIB	Remoter Authentication Dial-In User Service MIB (RFC 2618)
68	VrrpMIB	Virtual Router Redundancy Protocol MIB
72	IanaAddressFamily	IANA Address Family Numbers (RFC 2677)

Table 9.1   (*continued*)

mib-2 module	Module name	MIB name
73	Ianalanguagemib	IANA Language MIB
80	PingMIB	Packet Internet groper MIB (RFC 2925)
81	TraceRouteMIB	Traceroute MIB (RFC 2925)
82	LookupMIB	Look-up MIB (RFC 2925)
83	IpMRouteStdMIB	IP multicast route standard MIB (RFC 2932)
84	IanaioRouteProtocolMIB	IANA Route Protocol MIB (RFC 2932)
85	IgmpStdMIB	Internet Group Management Protocol MIB (RFC 2933)
87	RtpMIB	Real-time application Transport Protocol MIB (RFC 2959)
92	NotificationLogMIB	Notification log MIB (RFC 3014)
93	PintMIB	PSTN/Inter NeTworking MIB (RFC 3201)
94	CircuitIfMIB	Circuit Interface MIB (RFC 3202)

the intention was that equipment designers could define similar objects for network types other than ethernet. Subsequently, many further modules have been added to RMON, and it is nowadays possible to remotely monitor nearly all the components and protocols of an IP-based network using RMON and SNMP. In particular, RMON is designed to provide for:

- offline operation of remote devices;

- multiple different managers operating simultaneously;

- proactive monitoring of remote devices;

- problem detection and reporting; and

- 'value-added data' — a range of statistics helpful to human network managers in operating a network.

RMON objects are organised in the groups and according to the hierarchy shown in Figure 9.6 which is presented as a quick reference without detailed discussion. But before we leave the subject of RMON, it is important to understand the references made to *RMON-1* and *RMON-2*. RMON-1 and RMON-2 are not versions one and two of RMON, but instead are mutually exclusive subsets of a single MIB called *rmon*. RMON-1 is used to denote the basic RMON modules (the current definition of which is contained in RFC 2819). RMON-2 is merely an optional extension MIB defined in RFC 2021, which allows RMON additionally to be used to monitor the application layer and related functions of remote hosts. The relation of RMON-1 to RMON-2 should be clear from Figure 9.6. Finally, SMON MIB (also illustrated in Figure 9.6) extends RMON for Switched network MONitoring. SMON is defined in RFC 2613.

9.7   MIB for Internet protocol version 6 (ipv6MIB)

Finally, as a quick reference, and as a comparison with the Internet Protocol version 4 MIB (mib-2 modules 4 and 5 — as shown in Figure 9.5), Figure 9.7 presents the MIB (management information base) for the v6 versions of both the Internet Protocol (IPv6) and the Internet control message protocol (ICMPv6). The definition of a new MIB for IPv6 reflects the different structure and operation of the new protocol (as compared with its predecessor IPv4).

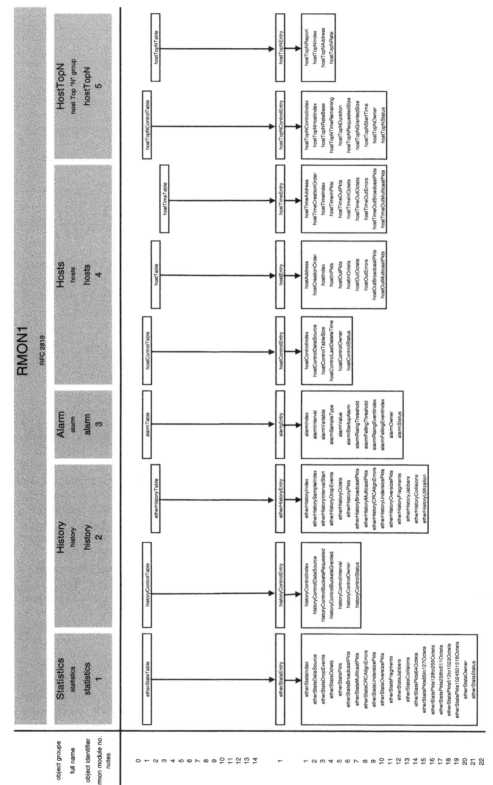

Figure 9.6   Remote MONitoring management information base (RMON MIB): RMON-1, RMON-2 and SMON.

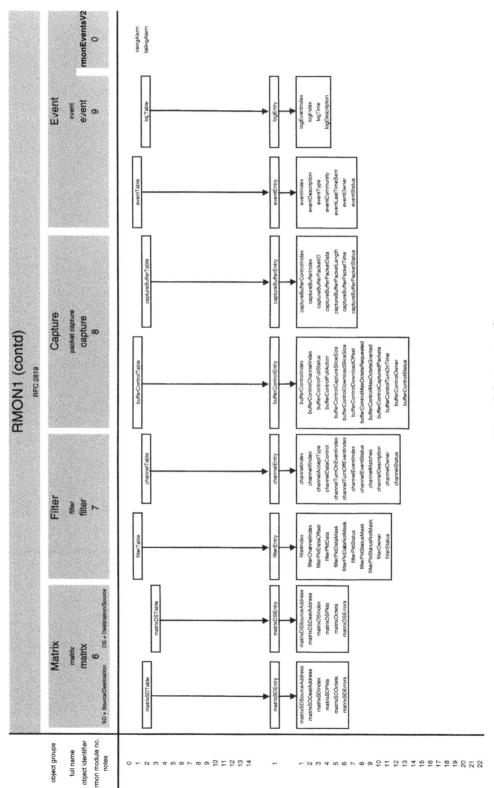

Figure 9.6 *(continued)*

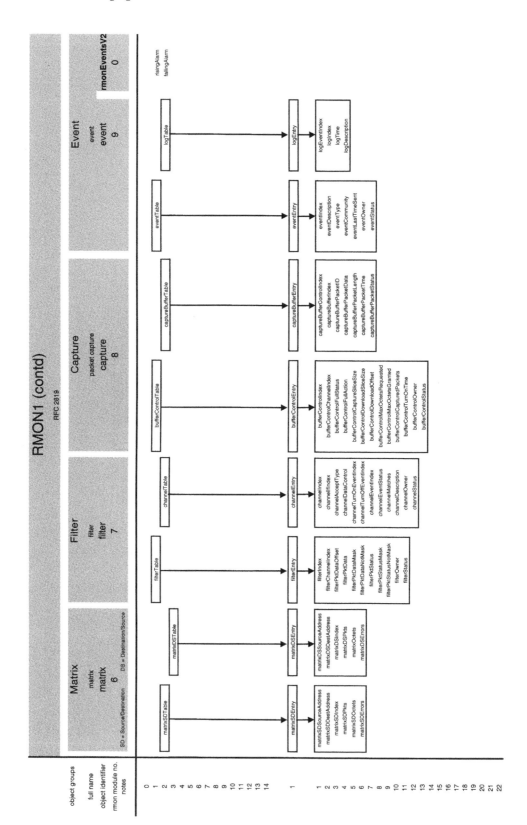

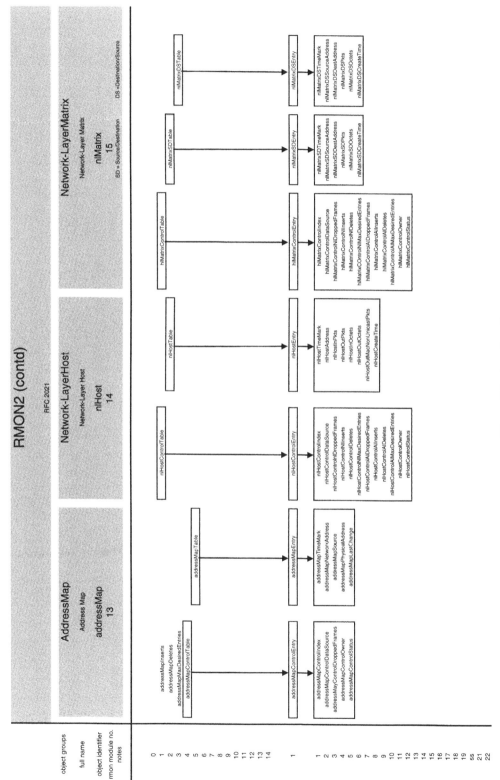

Figure 9.6 (*continued*)

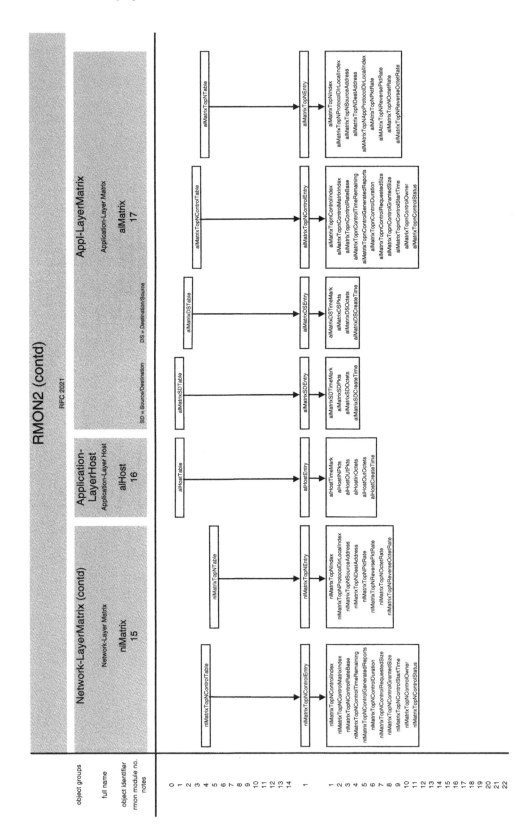

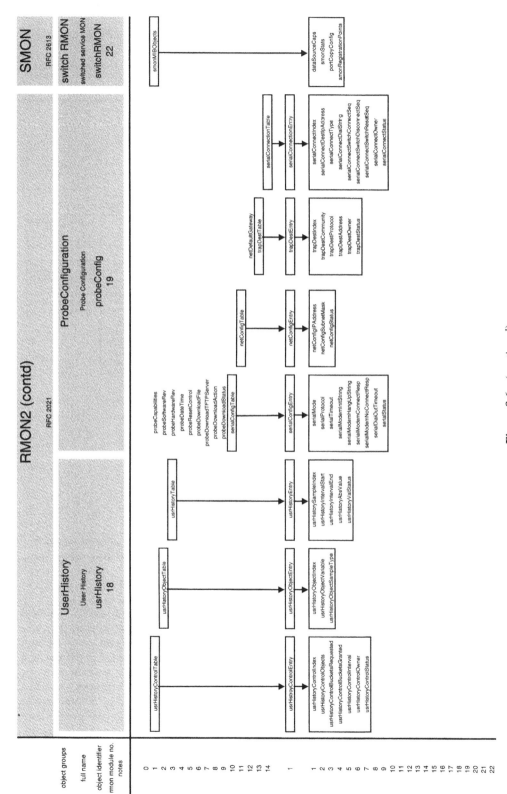

Figure 9.6 (*continued*)

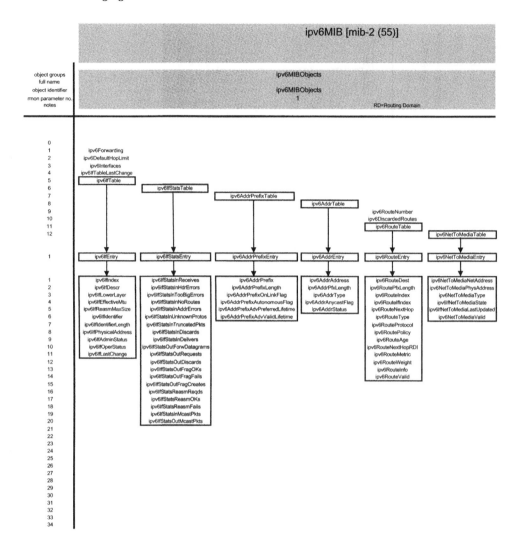

Figure 9.7   MIBs for Internet Protocol version 6 and Internet control message protocol version 6.

9.8   Simple network management protocol (SNMP)

As its name suggests, the *simple network management protocol (SNMP)* is a 'simple' protocol designed to enable remote network management (monitoring and control) of Internet networking devices and protocols. It is an *application layer* (i.e. layer 7) protocol carried by the user datagram protocol (UDP) on port 161 (SNMP traps on port 162) and formatted in the standard manner for modern application layer protocols—in *ASN.1 (abstract syntax notation 1)*.

SNMP was designed to provide a standard means for the carriage of information between a network manager and a remote networking device. The information carried by SNMP may be either a command to *set* a given parameter in the remote device, a request to *get* current status information (for purpose of remote network monitoring) or a response to one of these requests. Alternatively, certain alarm and other trap messages are sent unsolicited (i.e. without a previous request) to notify the network manager of faults or other *events* (changes in network status).

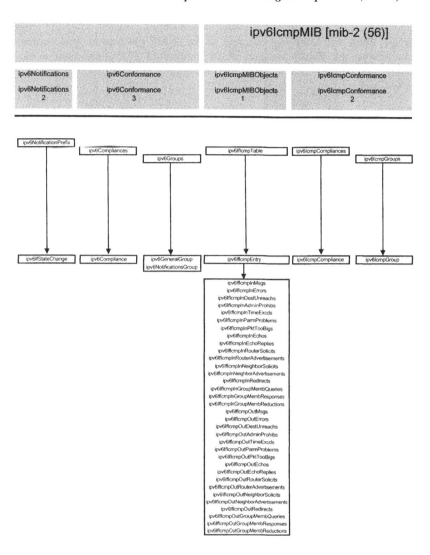

The fact that SNMP provides a standardised interface allows network management software and hardware provided by one manufacturer to be used to manage remote network devices (so-called *network elements*) provided by a different manufacturer.

SNMP itself only sets out the basic rules of the communication between the manager and the network element, allowing a small set of basic functions to be performed — *setting* configuration parameter values, *getting* (i.e. monitoring) parameter values, etc. The parameters which are subjected to the set and get commands are the standard managed objects which are defined in the network element's associated management information bases (MIBs). Examples of MIBs were presented in Figures 9.5, 9.6 and 9.7. An example of a managed object which we have previously discussed is tcpMaxConn — this object represents the maximum number of TCP (transmission control protocol) connections which a remote device (in this case a *host*) can support.

Without the MIBs (management information base) and managed objects which are related to a particular network element, SNMP is useless. And alone the MIBs are also useless. Together they create a request or a response in a format which object-oriented computer programmers will recognise. Thus, the following example of an SNMP *get-request* will request that the remote host advises the SNMP manager of the number of TCP connections it can support simultaneously. The request makes use of the managed object typMaxConn, the definition of which we saw earlier:

```
get-request RequestID, 0, 0, tcpMaxConn
```

SNMP protocol definition in ASN.1 format

Like the managed objects and MIBs (management information bases), the commands of application layer protocols (layer 7) are defined using the *ASN.1 (abstract syntax notation 1)* language. The following part of the SNMPv1 definition (RFC 1157 — in ASN.1 format) defines there to be five (and only five) different SNMP message types:

```
-- protocol data units
PDUs ::=
                        CHOICE{
                get-request
                        GetRequest-PDU,
                get-next-request
                        GetNextRequest-PDU,
                get-response
                        GetResponse-PDU,
                set-request
                        SetRequest-PDU,
                trap
                        Trap-PDU,
                }
```

The five messages (called *protocol data units* or *PDUs* — we met these in earlier chapters too) are called:

- get-request

- get-next-request

- get-response

- set-request

- trap

These different messages are also *objects* in an ASN.1-sense, and thus have small letters at the start of their names. The five objects listed above are actually 'PDU type identifiers'. The PDUs which go with them (called GetRequest-PDU, GetNextRequest-PDU, GetResponse-PDU, SetRequest-PDU and Trap-PDU) are the names of pre-defined 'data structures' or 'data formats'. In this case, these structures correspond to the 'protocol fields' which provide the main details and 'content' of the SNMP message. In line with ASN.1 practice, these data structure names (PDU names) have a name commencing with a capital letter.

The structure of each PDU is also defined using ASN.1. Thus, for example, the form of the GetRequest-PDU (SNMPv1 and SNMPv2 — as defined in RFCs 1157 and 1905) is:

```
GetRequest-PDU ::= [0]
        IMPLICIT SEQUENCE {
```

```
request-id
      RequestID,
error-status                  --- always 0
      ErrorStatus,
error-index                   --- always 0
      ErrorIndex,
variable-bindings
      VarBindList
}
```

In turn, the GetRequest-PDU definition (above) refers to four more object types (*request-id, error-status, error-index* and *variable-bindings*) and their related data structures (*RequestID, ErrorStatus, ErrorIndex, VarBindList*). To format the complete PDU and be able to program an SNMP message you will need to refer to all the object and command definitions which crop up along the way. We shall not cover this here, but hope nonetheless that our brief explanation of ASN.1 and the few examples above will help when referring to other reference sources. (For your assistance, Figure 9.8 is presented in the 'classical manner' for illustrating protocol formats to aid the interpretation of the ASN.1 manner of presenting the same information).

A complete SNMP message in SNMPv1 (and SNMPv2) is defined to be structured as follows:

```
-- top-level message
      Message ::=
            SEQUENCE {
                  version
                        INTEGER {
                              Version-1(0)      ---value 0 for SNMPv1
                                                ---value 1 for SNMPv2c
                        }
                  community
                        OCTET STRING,
                  data                          --- e.g. PDUs is trivial
                        ANY                     --- authentication is
                                          --- being used
            }
```

a) SNMPv1 and SNMPv2 PDU-format (RFCs 1157 & 1905)

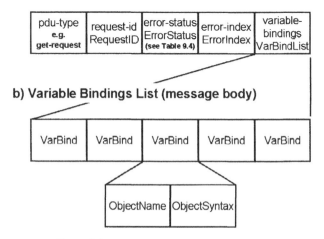

b) Variable Bindings List (message body)

Figure 9.8   Format of SNMPv1 and SNMPv2 PDUs.

A simple SNMP get-request message in its full form might thus be something like below:

version   community   get-request   RequestID, 0, 0, tcpMaxConn

The actual coding of the message will depend upon the values of each of the objects and data formats. Thus, for example, the version number will be an INTEGER value in the range (0..2147483617 — a 32-bit value). The value '0' represents SNMP version 1. The community is an alphanumerical value (i.e. an OCTET-STRING) identifying the *community name* to which the SNMP message belongs. We shall discuss *SNMP communities* later. 'get-request' identifies the *PDU-type* contained in the message, and the four data values are the pre-defined structure of the standard GetRequest-PDU-format. The RequestID is a unique integer value identifying the get-request message. The get-response will be identified by the same identifier. This helps the associated *get* and *set* messages to be easily related to one another. The two '0' values are the *error-status* and *error-index* as defined in the GetRequest-PDU definition. (The allowed values of ErrorStatus appear later in Table 9.4.) Finally, the object name tcpMaxConn identifies the requested parameter value.

Using the example SNMP get-request message presented above, an SNMP manager could request the SNMP agent of the addressed network element to respond with the current value of the tcpMaxConn parameter. The response will come as a get-response message with the same RequestID and including the value. In this response, the value of the object (in our example 'tcpMaxConn') will be formatted in the manner defined in the MIB for that object (e.g. integer value, octet-string, pre-defined special data-format etc.). As we illustrated earlier in the chapter, 'tcpMaxConn' is defined as an object with INTEGER value. Thus the get-response message which is returned to the manager will contain an INTEGER value corresponding to the current maximum number of TCP connections which the remote *network element* (in this case a host) can support.

A side observation about the use of ASN.1 for describing application layer objects and protocols

It is common practice to define *application layer* protocols using the abstract syntax notation 1 (ASN.1). This gives the protocol formats a very different appearance from the 'box-like' formats of octets, bits and bytes which we encountered in earlier chapters (e.g. Internet Protocol format as explained in Chapter 5). The protocol format appears much more like a 'computer program command' than like the 'bits/bytes/octets' Internet Protocol message format (e.g. Figure 5.14 of Chapter 5). This is appropriate, given the need to 'integrate' layer 7 datacommunications protocols into 'ordinary' computer software, but it would be possible (if you had the time) to work out the exact string of octets and even bits which go to make up the message. In assisting you to try to understand the manner of protocol definition using ASN.1, Figures 9.8 and 9.12 have been added in the 'classical style' of protocol format presentation. These diagrams can be compared with the corresponding ASN.1-format protocol format definitions which appear in the text.

It is normal practice to code alphanumeric (i.e. text-type and octet-string) fields of relevant ASN.1 objects according to ISO/IEC alphabet ISO 10646-1, otherwise known as the universal transformation format-8 or UTF-8. This is also equivalent to an 8-bit 'padded-out' version of the original 7-bit ASCII code.

Goals of SNMP as a 'simple' network management protocol

Critical for the correct operation of SNMP network management is that:

- both network manager (*SNMP manager*) and network element (*SNMP agent*) are using the same version of SNMP, and

- the relevant MIBs (and thus managed objects) are known to both parties.

Just in case a given MIB (management information base) or managed object is unknown to one of the parties, the simple network management protocol (SNMP), of course, is able to return suitable error messages.

Newcomers to the subject of network management may not find any of the foregoing explanation about the workings of SNMP very <u>simple</u>, so why the name <u>simple</u> network management protocol, you might ask? The answer is that it was developed by IETF for the Internet as a simpler alternative to the common management information protocol (CMIP). CMIP is a very robust, but complex protocol intended for remote network management strictly conforming to the *open systems interconnection (OSI)* model. CMIP was developed by CCITT (International Telephone and Telegraph Consultative Committee — now called ITU-T: International Telecommunications Union Standardization sector) and ISO (Organization for International Standardization).

In contrast to CMIP, SNMP was designed to eliminate the 'elaborate access control policies' of CMIP and to use only a restricted (i.e. simplified) subset of ASN.1 (abstract syntax notation.1). The goals of SNMP were to:

- make the development of management agent software for network elements as simple as possible, and thereby to

- maximise the management functionality which can be provided,

- minimise the restrictions on the controls and monitoring functions which can be supported, and

- minimise the costs of management agent software development.

SNMP managers, SNMP agents and basic network management architecture

Simple network management protocol (SNMP) is used as a protocol for communication only between an *SNMP manager* and an *SNMP agent*. As Figure 9.9 illustrates, the *SNMP manager* is a software program, usually running on powerful *workstation* or *server* computer hardware. Typically this hardware, and the complete set of software running on it is called a *network management system (NMS)*. The *SNMP manager* software will be only one of many software applications running on this system. The *SNMP agent*, meanwhile, is a software module running on a remote *network element* (e.g. a router, a host or some other network device). A network management *data communications network (DCN)* provides for the communication path between manager and agent. Typically the same IP router network as that being managed will be used for this purpose (although sometimes dedicated management networks are used in order to minimise the risk of unwanted network intervention by unauthorised third parties).

The human network managers (operators) are usually equipped with PCs or workstations and are located in a *network operations centre (NOC)* or *network management centre (NMC)*. The NMS hardware may or may not be located within the NOC or NMC, but immaterial of the location, a standard client/server or equivalent protocol interface is used to connect the operator workstations to the NMS.

The NMC hardware typically maintains an extensive database of information about the network topology, status and performance. This database is kept up-to-date by means of the SNMP messages received from the various SNMP agents (network elements). The messages might be generated by the SNMP agents because of a change of status — such as an event

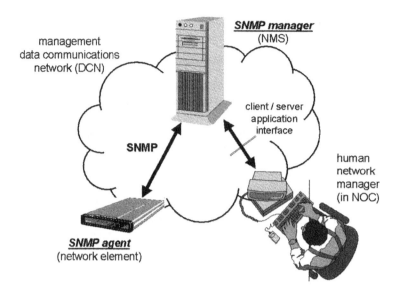

management
data communications
network (DCN)

SNMP manager
(NMS)

client / server
application
interface

SNMP

human
network
manager
(in NOC)

SNMP agent
(network element)

Figure 9.9   Basic network management architecture showing SNMP manager and SNMP agent.

or alarm (these are sent as SNMP trap messages) — or might be polled on a regular basis by the SNMP manager (these are *response* messages). Polling might be used, for example, every few minutes to keep track of the current traffic loading on a trunk, or perhaps on a daily basis to transfer a complete statistical performance record.

An enquiry from a human network operator about current network status or historical performance will generally be dealt with by the NMS software by making an analysis of the database and presenting the results in a suitable graphical format. A command from the human operator for a network configuration change, on the other hand, will be converted to an SNMP message by the SNMP manager software and sent to the SNMP agent in the relevant network element(s). Exactly which information may be monitored or which configuration parameters may be changed using SNMP is defined by the SNMP access policy — which we will explain in the following section.

Components, terminology and basic functioning of SNMP

SNMP communication is initiated and controlled by *SNMP application entities*. While there are application entities in both the SNMP manager and in the SNMP agent, we shall be most concerned with those in the SNMP manager. The manager's SNMP application entity is the application software which is the main 'brains' of the process. A number of different SNMP application entities may exist in the SNMP manager — each designed for a particular network management function (Figure 9.10). The application entities communicate with *peer entities* in the SNMP agent, and are designed to monitor and control the different functions, capabilities and status of the network element as represented by its various MIBs and managed objects (Figure 9.10). In particular, it might be that one application entity is concerned with 'network configuration' while another is concerned with 'network monitoring', a third with 'alarm reporting', a fourth with 'historical network performance', etc. Thus, depending upon the responsibilities of individual groups of operations staff, different employees may be allowed to use some application entities but not others. For example, customer service staff may be able to 'monitor' the network, while only 'network administration' staff may be permitted to undertake network configuration.

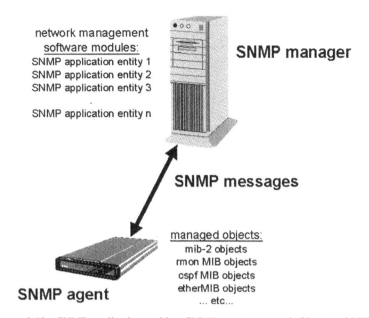

Figure 9.10   SNMP application entities, SNMP agent, managed objects and MIBs.

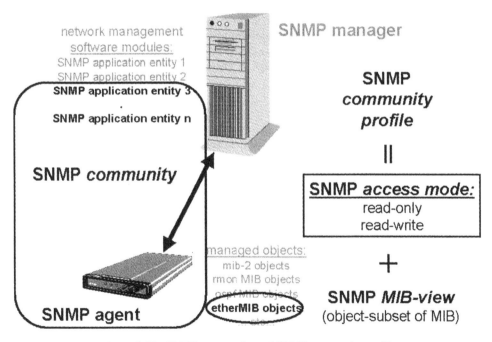

Figure 9.11   SNMP community and SNMP community profile.

A collection of closely related SNMP application entities and the particular *network elements* (i.e. SNMP agents) with which they are allowed to interact is called an *SNMP community* (Figure 9.11). The idea is that the different operations staff units correspond to different SNMP communities — and thus have different capabilities of network management.

Table 9.2   The effect of the SNMP access policy on the SNMP messages allowed

MIB access-view	SNMP community access view	SNMP messages allowed
None	Does not matter	None
Read-write	Read-write	Get, Set and Trap
Read-write	Read only	Get and Trap only
Read-only	Read-write or read only	Get and Trap only
Write-only	Read-write	Get, Set and Trap

The particular subset of a MIB within the SNMP agent which the SNMP application entities are allowed to deal with (and thus which the particular SNMP community is allowed to manage) is called the *SNMP MIB-view* (Figure 9.11). The *SNMP access-mode* defines which management actions are permitted, for example objects may only be monitored (*read-only*), monitored and/or changed (*read-write*) or may be alerted should a certain event occur (*notify* mode). Both the SNMP community and the individual managed objects within the SNMP MIB-view have a defined SNMP access-mode. In the case of the managed objects within the SNMP MIB-view the access-view is defined in the MIB definition. We saw in our example earlier in the chapter, how the SNMP access-mode for the object tcpMaxConn is read-only. Together the access modes of the SNMP MIB-view are called the *SNMP community profile* (Figure 9.11 and Table 9.2).

The SNMP community and its corresponding SNMP community profile are together termed the *SNMP access policy*. The establishment of such an SNMP access policy is a useful means by which SNMP-based network management can be controlled — allowing for a simple authentication and checking of messages by the application entities. This might ensure, for example, that unintended network configuration changes are not undertaken by unauthorised employees (i.e. those that do not belong to an appropriate operations 'community'). The SNMP community name is included in each SNMP message. It ensures that only appropriate SNMP agents will respond to each message.

Different versions of SNMP

The original version of simple network management protocol — SNMPv1 — is defined in RFC 1157 (issued in 1990). SNMPv1 defines only five *different PDU (protocol data unit) types* of SNMP message to allow the *managed objects* of an SNMP agent's MIB to be inspected or altered. The five types of PDU defined by SNMPv1 (and the 'basic functions' of SNMP) are:

- **GetRequest-PDU** (this is sent from SNMP manager to SNMP agent to request status information for the purpose of monitoring the network);

- **GetNextRequest-PDU** (this is similar to a GetRequest message, but is specifically designed to allow the SNMP manager to scan through a 'table' or 'matrix' of data in the SNMP agent to locate a particular value);

- **GetResponse-PDU** (this is the response message sent by an SNMP agent to an SNMP manager in response to either a GetRequest-PDU or GetNextRequest-PDU);

- **SetRequest-PDU** (this is sent by the SNMP manager to the SNMP agent to initiate a change in the configuration of the associated network element);

- **Trap-PDU** (this is sent by the SNMP agent to the SNMP manager unsolicited — i.e. without a request — and notifies of an event, i.e. change of status, or alarm condition).

The second version of SNMP, SNMPv2 was first proposed in RFCs 1441-50 (issued in 1993) but it was not until 1996 that a stable version was agreed (RFCs 1901-10). SNMPv2 added extra PDU-types, as we shall discuss in detail later in the chapter, and is significantly more powerful, but more complex than SNMPv1. Unfortunately, during the process of trying to agree on the SNMPv2 definition, a number of slightly different versions appeared, which held-up its widescale deployment:

- SNMPv2c (community-based SNMP) (RFC 1901) — this version was a popular 'experimental' version of SNMPv2 which tried to maintain close compatibility with SNMPv1, and in particular, retained the use of SNMP communities as the main means of ensuring security of the network and its management information;

- SNMPv2u (RFC 1909–1910) — this version included more security and administration features than SNMPv2c, but lacked the endorsement of IETF;

- SNMPv2* also lacked the endorsement of IETF.

As a result of the confusion over SNMPv2, development of the protocol has continued — and SNMPv3 has appeared. In its basic operation, SNMP3 is the same as SNMPv2 (as defined in RFC 1905), and the PDUs used are the same. The main difference introduced by SNMPv3 are specific security and administration procedures:

- a *user-based security model (USM)* — this sets out the elements of procedure for SNMP message level security (RFC 2574) — allowing a network administration to define different management information access rights (for monitoring or changing network parameters) for each individual employee; and

- a *view-based access control model (VACM)* (RFC 2575) designed to give 'generic' access rights to particular groups of employees (i.e. organisational units).

The structure of management information (SMI — the object data definition language) to be used with SNMPv3 must conform to SMIv2 (RFC 2578-80). Coexistence of both SMIv1 and SMIv2 types would have required conversion from SMIv1 to SMIv2 and is not supported. SNMPv3 is defined by RFC 2570-5 (issued in 1999).

SNMPv3 shares the same basic structure and components as its predecessors, SNMPv1 and SNMPv2, but further develops it. In particular, SNMPv3 introduces a new message format as well as introducing the idea of PDU-classes. Seven different PDU classes are defined, and the idea is that over time, different types of each of the new PDU classes will appear. Individual PDUs are classified into one of the classes 1–5 as follows:

1   Read class

2   Write class

3   Response class

4   Notification class (SNMP traps and event notifications)

5   Internal class (these PDUs are exchanged internally between SNMP devices)

In addition, each PDU is additionally classified according to whether or not a response is expected:

6   Confirmed class

7   Unconfirmed class

In the first instance, the SNMPv2 PDUs (RFC 1905) are retained and classified according to the new structure and no new PDU-types are added. The basic operation of the PDUs also

Table 9.3   SNMP protocol data unit (PDU) types

PDU name and format	Function	SNMPv1	SNMPv2	SNMPv3
get-request GetRequest-PDU	SNMP manager requests information from the SNMP agent (polls the agent)	Yes	Yes	Yes (READ class and CONFIRMED class)
get-next-request GetNextRequest-PDU	SNMP manager requests that the agent send the next value in a 'table' or 'matrix' of values.	Yes	Yes	Yes (READ class and CONFIRMED class)
get-bulk-request GetBulkRequest-PDU	SNMP manager sends a single request to generate a response from the SNMP agent containing potentially a very large amount of data.	No	Yes	Yes (READ class and CONFIRMED class)
get-response GetResponse-PDU	SNMP agent response to a GetRequest or GetNextRequest PDU	Yes	No	No
Response Response-PDU	SNMP agent response to a 'get' type message, a confirmation of a 'set' message or a response to an 'InformRequest-PDU'.	No	Yes	Yes (RESPONSE class and UNCONFIRMED class)
set-request SetRequest-PDU	SNMP manager sends a command to the SNMP agent for reconfiguration of the associated network element.	Yes	Yes	Yes (WRITE class and CONFIRMED class)
Inform-request InformRequest-PDU	Request by the SNMP manager to be informed by the SNMP agent should a given event occur.	No	Yes	Yes (NOTIFICATION class and CONFIRMED class)
Trap Trap-PDU	Unsolicited message sent by the SNMP agent (SNMPv1) to advise of the occurrence of a given alarm or other pre-determined event.	Yes	No	No
Snmpv2-trap SNMPv2-Trap-PDU	Unsolicited message sent by the SNMP agent (SNMPv1) to advise of the occurrence of a given alarm or other pre-determined event.	No	Yes	Yes (NOTIFICATION class and UNCONFIRMED class)
Report Report-PDU	An SNMP message containing message in the form of a report.	No	Yes	Yes (RESPONSE class and UNCONFIRMED class)

remains unaffected by SNMPv2. The classification of SNMPv2 PDU-types into the SNMPv3 PDU-classes appears in Table 9.3.

Other new terminology introduced by SNMPv3

Although the basic network management framework employed by SNMPv3 is shared with its predecessors, SNMPv1 and SNMPv2, the protocol has become considerably more complex — requiring a new structure of structured management information (SMIv2 — RFC 2578-80) and a new message structure. These changes have largely been necessitated by the introduction of the more robust security and information access controls.

SNMPv3 is considerably more complex than its predecessors. To keep it manageable, it was decided to break it down into a number of smaller functional modules, and these have lead to new terminology. In particular, the following new terms have been introduced...

A device employing SNMP comprises an *SNMP entity* and an *SNMP engine*. The SNMP entity is an application which generates or receives SNMP messages. There are two main types of SNMP entity (corresponding to what was previously called an SNMP manager and an SNMP agent):

- One type of SNMP-entity comprises a *command generator* and a *notification receiver* (formerly this combination was called an SNMP manager).

- The second SNMP-entity comprises a *command responder* and a *notification originator* (formerly this combination was called an SNMP agent).

An SNMP engine serves to process, dispatch and receive the SNMP messages on behalf of the SNMP entities (in effect, this subdivides the SNMP protocol into a number of sub-layers). The 'layered' functions are:

- a dispatcher;

- a message processing subsystem;

- a security subsystem; and

- an access control subsystem.

The idea is that, in principle, the same SNMP engine hardware and software can be used in both 'manager' and 'agent'. In addition, by using a modular structure, individual components (e.g. access control or security) can be more easily modified without affecting the rest of the SNMP system. But by introducing sub-layer protocols, there are many more protocols and *primitives* (standardised signals and commands used between the different sub-layers) which have to be defined. These are covered in RFC 2571, but we do not have space to cover them in detail here.

SNMPv3 dispenses with the idea of SNMP communities for access control in favour of a user-based security model (USM — RFC 2474) and a view-based access control model (VACM — RFC 2475). In SNMPv3 the term *context* is used instead to refer to management information which is accessible by an SNMP entity.

As you will have gathered, the word 'simple' does not appropriately reflect the complexity of SNMPv3!

SNMP message format

The SNMP protocol and the management information which it is used to carry are coded according to the abstract syntax notation 1 (ASN.1 — ISO 8824/5). The actual messages have the appearance of object-oriented computer programs (applications), and it is this syntax which we shall use to illustrate the actual structure of messages in this section. The actual coding of data within the messages is covered by the definition of the objects or commands, but typically either an integer value (in binary coding) or an octet value (in UTF-8 — universal transformation format) is used to code the octets of the actual message.[3]

Format of SNMPv1 and SNMPv2c messages

SNMPv1 and SNMPv2 messages comprise a version number (the version of SNMP), an SNMP *community name* and the message itself. The message is made up of one of a number of pre-defined PDUs (protocol data units). The most important PDUs, as discussed earlier in the chapter, are those related to the get, set and trap messages. Get messages are used for monitoring purposes. Set messages are used for changing the configuration of the network element, and trap messages report alarms, specific events or other network status changes.

Defined in ASN.1 notation, the message format and structure of an SNMPv1 message are as follows (RFC 1157):

```
Message ::=        SEQUENCE {
                   version        INTEGER {version-1(0)},
                   community      OCTET STRING    --- community name
                   data           ANY             --- PDUs
                   }
```

Format of SNMPv3 messages

SNMPv3 introduced a new format for SNMP messages. As you will see from the following definition (in ASN.1-format), it introduces a number of new fields into the message 'header' (also called the message 'wrapper'):

```
SNMPv3Message   ::=        SEQUENCE {
                msgVersion               INTEGER
                msgGlobal Data           Header
                msgSecurityParameters    OCTET STRING
                msgData                  ScopedPduData
                }
```

The value of *msgVersion* refers to the version of SNMP in use. This is presented first, to retain compatibility with earlier versions of SNMP. Value '0' represents SNMPv1. Value '1' represents SNMPv2c. Value '2' represents SNMPv2u. Value '3' represents SNMPv3. The *msgGlobalData* object includes a number of 'header' fields in a special data format called 'Header'. This includes the message identifier (msgID), the maximum allowed message size (msgMaxSize), various header flags (msgFlags) and an identification of which security model (for information protection/access) is in use (msgSecurityModel). The main 'substance' of the SNMP message is included in the msgData field, which is structured in the standard-format (as defined in ASN.1 and called ScopedPduData). The actual PDUs used in SNMPv3 are the same as those of SNMPv2 (RFC 1905). These are listed in Table 9.3.

---

[3] When 'padded out' to 8 bits, standard 7-bit ASCII is equivalent to UTF-8.

Table 9.4   SNMP error messages

ErrorStatus Meaning	Error status (INTEGER value)	Error message meaning
NoError	(0)	The associated PDU does not contain an error.
TooBig	(1)	The associated PDU is too big to be received.
NoSuchName	(2)	There is no known object with this name.
BadValue	(3)	The value indicated is not understood.
ReadOnly	(4)	The parametei requested may not be changed.
GenErr	(5)	Generic error message (unspecified cause).
NoAccess	(6)	The user or community is not allowed access to this field.
WrongType	(7)	Data is of the wrong type.
WrongLength	(8)	Data is of the wrong length.
WrongEncoding	(9)	Data is wrongly encoded.
WrongValue	(10)	Wrong data value.
NoCreation	(11)	There is no facility to create further objects.
InconsistentValue	(12)	Value is inconsistent.
ResourceUnavailable	(13)	This object or other resource is unrecognised or not available.
CommitFailed	(14)	The attempted configuration change commit failed.
UndoFailed	(15)	The attempted 'undo' command failed.
AuthorizationError	(16)	The SNMP message failed due to an authorisation error.
NotWritable	(17)	The object is not currently writable.
InconsistentName	(18)	The name provided is inconsistent.

SNMP PDU (protocol data unit) types

Table 9.3 lists all the the different PDU types used in SNMPv1, SNMPv2 and SNMPv3 messages.

SNMP message errors

Just like everything else to do with SNMP, the error message formats are specified in ASN.1. Thus the data format *ErrorStatus* is defined to be an INTEGER with the possible values as appear in Table 9.4. We encountered the data type ErrorStatus earlier in the chapter when we presented the ASN.1-format definition of the SNMPv1 GetRequest-PDU. The error message is sent simply by returning the PDU with the ErrorStatus value reset from '0' (noError) to one of the values in Table 9.4.

SNMP traps

SNMP traps (Figure 9.12) are messages sent unsolicited (i.e. not in answer to a direct get-request of the SNMP manager). They report alarm conditions or other pre-defined changes in network status (often called events). In order that they can be dealt with by the SNMP manager as a matter of priority, they use a dedicated *UDP (user datagram protocol)* port number. SNMP traps are carried on UDP port 162. ('Ordinary' SNMP messages are carried on UDP port 161). As Table 9.5 lists, there are a number of standard definitions for 'generic' traps, but in addition, equipment manufacturers will usually arrange that their equipment send

a) Trap-PDU [SNMPv1] (RFC 1157)

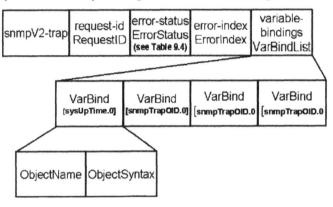

b) SNMPv2-Trap-PDU [SNMPv2 & SNMPv3] (RFC 1905)

Figure 9.12   Trap-PDU (SNMPv1) and SNMPv2-Trap-PDU (SNMPv2 and SNMPv3) formats.

Table 9.5   Generic SNMP traps

Generic traps		Meaning
ColdStart	(0)	Sending protocol entity is re-initialising itself. This may result in an alteration of the configuration.
WarmStart	(1)	Sending protocol entity is re-initialising itself. This will not result in an alteration of the configuration.
LinkDown	(2)	SNMP manager has detected a failure in the communication link represented in the agent's configuration.
LinkUp	(3)	SNMP manager has detected that the failed communication link represented in the agent's configuration has come up again.
AuthenticationFailure	(4)	The addressee of the message cannot be authenticated properly.
EgpNeighborLoss	(5)	The EGP neighbour has been marked down and the peer relationship no longer exists.
EnterpriseSpecific	(6)	A pre-defined but enterprise-specific event has occurred.

alarm messages and other notification messages (event messages) to advise network operators of current network fault conditions or to warn them of growing congestion or other likely imminent problems.

In most cases, there will be a default set of SNMP traps defined and pre-configured within a particular network element. Thus if a communication link fails, the appropriate SNMP trap is sent. Similarly, if a trunk goes into 'congestion' this might be notified as a 'warning' — a 'yellow' alarm, as it were, instead of a 'red' one. In many cases the traps themselves are configurable. Thus the exact percentage of trunk usage which is deemed to correspond to 'congestion' might be able to be adjusted by the network operator.

In order to ensure that the SNMP traps reach the appropriate SNMP manager, it is necessary to configure the SNMP traps in the network element when it is first installed. This involves configuring the SNMP agent to 'know' the IP address of the SNMP manager for each relevant type of SNMP trap. It may be that all SNMP traps are to be sent to the same SNMP manager, or certain types of messages may be required to be sent to different, or even to multiple SNMP managers.

The SNMP manager also needs to be configured to receive SNMP traps. Typically SNMP managers allow alarms and notifications of a nature considered by the human network management to be of no or little importance to be 'filtered out' (alarm filtering). Remaining alarms, meanwhile, are presented immediately to the human network manager — typically by turning the 'icon' of the affected network element on a network topology map from a green ('OK status') to a yellow (warning) or red (critical alarm) status.

In the case that the SNMP manager is equipped with management software specifically designed for the management of a particular type of network element, the meaning of the various standard SNMP traps will already be known, and the job of configuring the SNMP manager to receive them will be easy. But in a case where previously unknown SNMP traps are being sent (for example, SNMP traps configured by the network operator for his own special use), the SNMP manager will have to be specially configured or programmed to deal with them. This can be a very onerous task.

SNMP proxy agents

The fact that most IP networking equipment is designed to be capable of being managed using SNMP (simple network management protocol) has led to the emergence of a wide range of 'umbrella manager' solutions for the network management of entire data networks. The idea of the 'umbrella manager' is that a single network management solution collects and correlates network status and configuration information from all the different devices within the network, irrespective of their type and whether they were manufactured by different manufacturers (Figure 9.13).

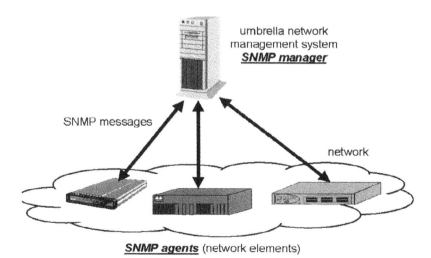

Figure 9.13   Use of an 'umbrella network management system' using SNMP to manage a wide range of different equipment making up an internetwork.

Unfortunately, not all network elements are capable of supporting SNMP (simple network management protocol). In other words, not all network elements have a resident SNMP agent. Without an SNMP agent entity (the relevant software for acting as an SNMP agent), a network element is unable to exchange SNMP messages with the SNMP manager. In this case, the network element cannot be managed or even monitored by the 'umbrella network management system' (Figure 9.13). This, of course, rather defeats the object of an 'umbrella network manager'. The pragmatic solution is often to use a SNMP *proxy agent* (Figure 9.14).

An SNMP proxy agent is usually a 'proprietary' network management system, provided by the manufacturer of the network element which is unable to support SNMP. Manufacturers who provide such 'proprietary' network management systems typically argue that by their use of a 'proprietary' management protocol between the management system and the network element, a greater range of network management functionality is made possible and the 'over-head' of network management traffic on the network can be minimised. The SNMP *proxy agent* (the 'proprietary' network management system) merely converts the standard SNMP messages exchanged between itself and the SNMP manager into the 'proprietary' message format required to 'talk' to the network component (Figure 9.14).

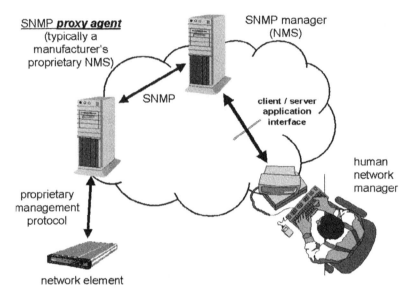

Figure 9.14   An SNMP proxy agent.

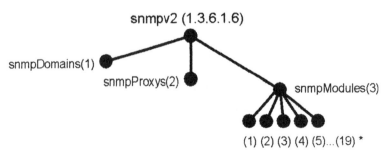

Figure 9.15   MIB for Simple network management protocol versions 2 and 3.

Table 9.6 SNMP modules included in SNMP MIB

	Name	Description	Specified by:
1	snmpMIB	MIB for SNMPv2	RFC 1907
2	snmpM2M	SNMPv2 M2M MIB	RFC 1451
3	partyMIB	SNMPv2 Party MIB	RFC 1447
6	usecMIB	User-based security for SNMPv2	RFC 1910
10	snmpFrameworkMIB	SNMP management architecture	RFC 2571
11	snmpMPDMIB	SNMP message processing	RFC 2572
12	snmpTargetMIB	SNMP target	RFC 2573
13	snmpNotificationMIB	SNMP notification	RFC 2573
14	snmpProxyMIB	SNMP proxy	RFC 2573
15	snmpUsmMIB	SNMP user-based security for SNMPv3	RFC 2574
16	snmpVacmMIB	SNMP view-based access control	RFC 2575
18	communityMIB	SNMPv2 community MIB	
19	snmpv2tmMIB	SNMPv2-TM MIB	

Network management of SNMP itself

Finally, before we leave the subject of SNMP, we should note that the devices and software functions which enable the network management of network elements using SNMP can themselves be network managed. SNMP has its own MIB (management information base) in order that it can itself be remotely managed. Figure 9.15 and Table 9.6 present the highest level of the SNMPv2 MIB — as used to manage SNMPv2 and SNMPv3.

9.9 The ISO management model: FCAPS, TMN, Q3 and CMIP/CMISE

The ISO (International Organisation for Standardization) model for classifying the functional processes which must be undertaken in the management of computer and telecommunications networks is the model which underlies all network management protocols and frameworks. All management tasks are classified into one of the following five categories:

- fault management;

- configuration management;

- accounting management;

- performance management, and;

- security management.

This model is referred to widely by ITU-T's recommendations on *telecommunications management network (TMN)*, and is reproduced in the ITU-T X.700-series recommendations (Figure 9.16). It is sometimes called the *FCAPS*-model — this name having been derived from the first letter of each management category.

FCAPS

Fault management is the logging of reported problems, the diagnosis of both short- and long-term problems, 'blackspot' analysis and the correction of faults.

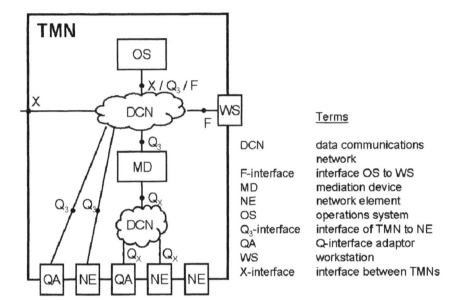

Figure 9.16   ITU-T telecommunications management network (TMN) model for network management architecture (ITU-T/M.3010) [reproduced courtesy of ITU].

Configuration management is the maintenance of network topology information, routing tables, numbering, addressing and other network documentation and the coordination of network configuration changes.

Accounting management is the collection and processing of network usage information as necessary for the correct billing and invoicing of customers and the settlement with other operators for the use of their interconnected networks to deliver packets.

Performance management is the job of managing the traffic flows in a live network — monitoring the traffic flows on individual links against critical threshold values and extending the capacity of links or changing the topology by adding new links. It is the task of ensuring performance meets design objective (e.g. *service level agreement*).

Security management functions include:

- identification, validation and authorisation of network users and network management operators;

- security of data;

- confirmation of requested network management actions, particularly commands creating or likely to create major network or service upheaval;

- logging of network management actions (for recovery purposes when necessary, and also for fault auditing); and

- maintenance of data consistency using strict *change management*.

TMN management model

The *telecommunications management network (TMN)* is the management infrastructure developed by ITU-T for full-scale network management of complicated networks — and in particular, of carrier networks. While it is not directly relevant to IP-based networks, many network operators may in practice encounter its terminology and interfaces. For this reason, we present it here. Figure 9.16 illustrates the architecture of management system components in a TMN. The most important components of the architecture are as follows:

- the *operations system (OS)*—this is a combination of hardware and software—what in common language is called a 'network management system' or 'network management server'.

- the *data communications network (DCN)*—this is generally conceived as a dedicated data-communications network for collecting and distributing network management information. (This function could also be provided by the data network being managed.)

- the *network element (NE)* being managed.

- the Q_3-*interface*—this is the interface over which the *CMIP (common management information protocol)* is intended to be used. CMIP is a protocol similar to SNMP, but more powerful and more complex;

- the *mediation device (MD)*—this is the equivalent of an SNMP proxy agent. It converts management messages from the standard Q3-format to non-standard 'proprietary' formats.

- the *workstation (WS)*—this is the computing device (e.g. PC or workstation) used by human network managers to access the network management system (i.e. the OS).

Figure 9.17 illustrates the typical split of TMN functions between a *network manager* (typically server hardware and software) and the *agent* (network management functions residing in the network element being managed). It also shows the interfaces (F, Q_3, Q_x, X) intended to be standardised as part of TMN.

Figure 9.18 shows the five layers of functionality defined by the OSI (open systems inter-connection) management model for a TMN. The layers are intended to help in the clear and rational definition of TMN *operating system* boundaries, thus simplifying the definition, design and realisation of software applications and management systems, by simplifying their data and communication relationships to one another.

At the lowest functional layer of Figure 9.18 (*network element layer, NEL*) are the *network elements* themselves. These are the active components making up the networks. Above them, in the second layer of the hierarchy, is the *element management layer (EML)* containing

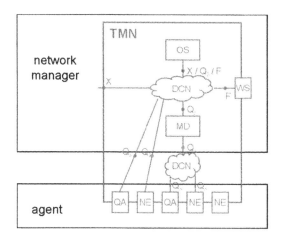

Figure 9.17 Typical division of functionality between a TMN network manager and its agents.

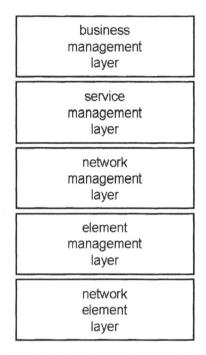

Figure 9.18   The functional layers of TMN.

element managers. Element managers are devices which provide for network management and control of single network components or subnetworks (e.g. a local management terminal or a proprietary network management system).

At the third layer, the *network management layer (NML)* are managers which control all the subnetworks of a given network type (e.g. ISDN or ATM).

The *service management layer (SML)* contains *service managers* which monitor and control all aspects of a given service, on a network-independent basis. Thus, for example, it is possible in theory to conceive a *frame relay service* provided over three different network types — *packet switched*-type network, *ISDN* and *ATM*. The service manager ensures installation of a given customer's connection line needs on the appropriate network(s) and monitors the service delivered against a contracted frame relay *service level agreement*.

The *business management layer (BML)* contains functionality necessary for the management of a network operator's company or organisation as a whole. Thus purchasing, bookkeeping and executive reporting and information systems are all resident in this layer.

The Q3-interface and the common management information protocol (CMIP)

Crucial for the successful communication between manager and agent over the Q_3-interface of the TMN (telecommunications management network) architecture is:

- the definition of a standard protocol (set of rules) for communication (this is the common management information protocol, CMIP), and;

- the definition of standardised network status information and control messages for particular standardised types of network components (the managed objects and management information base — MIB).

CMIP (common management information protocol) delivers the *common management information service (CMIS)* to the operating system of Figure 9.16. CMIS is the service allowing a *CMISE (common management information service entity* — i.e. a software function) in the manager (the 'OS' of Figure 9.16) to communicate status information and network control commands with a CMISE in each of the various agents. CMIP itself is an OSI layer 7 protocol (rather like SNMP), which sets out the rules by which the information and commands (CMISE services) may be conveyed from manager to agent or vice versa. These basic CMISE services are restricted in number. They are listed in Table 9.6. Note the similarity with the basic protocol messages (PDUs — protocol data units) of SNMP (simple network management protocol).

9.10   Tools for network management

Over the years, a range of different software-based tools has emerged for network-managing large and complex data networks. The market demand for good tools is strong, and a number of the tools have become 'household names' in the telecommunications industry, but there is still no single tool which is up to the 'whole job'. Instead *de facto* standard solutions for specific tasks have appeared (i.e. specialised solutions for fault management, order-processing and configuration, network service performance management, etc.).

Tools are beginning to mature, but nonetheless continue to be developed. Specialist network management software manufacturers are typically trying to expand their areas of 'specialism' into neighbouring areas — so that they may take market share from competing products. Meanwhile, many manufacturers of network components (so-called *network elements*) do not invest much effort in their 'proprietary' *element manager (EM)* network management software. Their belief (rightly or wrongly) is often that an expensive network management solution may count against the choice of their networking products.

Some network management tool manufacturers would like you to believe that they can offer a 'complete' or 'umbrella' solution for everything — order processing, provisioning,

Table 9.7   The basic CMISE (common management information service entity) services [CMIP protocol commands]

Service name	Function
M-ACTION	This service requests an action to be performed on a managed object, defining any conditional statements which must be fulfilled before committing to action.
M-CANCEL-GET	This service is used to request cancellation of an outstanding previously requested M-GET command.
M-CREATE	This service is used to create a new instance of a managed object (e.g. a new port now being installed)
M-DELETE	This service is used to delete an instance of a managed object (e.g. a port being taken out of service)
M-EVENT-REPORT	This service reports an event, including managed object type, event type, event time, event information, event reply and any errors.
M-GET	This service requests status information (attribute values) of a given managed object. Certain filters may be set to limit the scope of the reply.
M-SET	This service requests the modification of a managed object attribute value (i.e. is a command for parameter reconfiguration).

configuration, performance management, billing, fault management and *SLA (service level agreement)* management, but even the most comprehensive solutions are rarely little more than a 'patchwork' of different 'specialised function' systems (e.g. for fault management, configuration management, etc.) which have been loosely tied together.

Perhaps the best-known and most widely used system is Hewlett Packard's OpenView. OpenView started life as a popular SNMP manager for collecting SNMP alarms and monitoring the current network performance status. It is still one of the most popular network management products for this function. However, over time, Hewlett Packard have started to offer a range of complementary software for other network management functions under the same marketing name — 'OpenView'. Some of these products were originally developed by partner companies or start-up companies which got taken over by Hewlett Packard and have been adapted for 'integration' into the original OpenView software.

A product with similar functionality to Hewlett Packard's OpenView is the IBM company's *NetView*. NetView is popular among large enterprise organisations with established large networks of IBM mainframe computers, since it allows for an effective management of both the network and the computer software applications running across it — allowing human system operators to diagnose quickly the root cause of any problems which might arise.

Configuration management

Historically, the systems available for end-to-end *configuration management* of complex networks have not been sophisticated enough to cope with the complexity of the task. Configuration management requires that a single consistent database be maintained about the complete topology and configuration of the network. For many years, it was impossible to even create such a database, let alone maintain it, because of the lack of standardised MIB (management information base) definitions of all the possible managed objects which go to make up a network.

While nowadays many MIBs and managed objects have been defined, there are endless possibilities for how different objects may be related to one another. Thus, for example, an end-to-end data connection between two inter-communicating computers might traverse a large number of different types of connections represented by different types of managed objects (Figure 9.19). To understand the entire end-to-end connection, each of these objects needs to have been appropriately related to one another. But how do you easily relate objects of a different nature to one another? Let us consider Figure 9.19 in detail.

The 'connection' between the Internet user's PC and the server in Figure 9.19 traverses a number of different networks and configuration types. There is a dial-up connection across the public switched telephone network (PSTN), followed by an IP 'connection' across a router network and an ethernet LAN 'connection' at the far end. But to complicate things, the two routers forming the IP part of the network are actually interconnected by means of a frame relay network. Next let us consider the effect of a failure in the frame relay network at the centre of this connection. Ideally, as the network manager responsible for the connection from the PC to the server I would like to be immediately informed of the cause of the failure. But how can I relate the frame relay network line failure to the knock-on service problems it will cause? The answer is — with great difficulty.

A pragmatic approach to working out the list of customer connections affected by a particular frame relay network connection failure in Figure 9.19 might appear to be to make a list of customer names or network addresses and associate this list with the frame relay line 'object' in the network configuration database. In a few cases, such an approach (coupled with a large amount of effort to maintain the database) might work. But in practice, such an approach is impractical in most cases. Each time the PC user in Figure 9.19 dials in to the network he/she will arrive at a different port on the router and be allocated a different IP address by DHCP

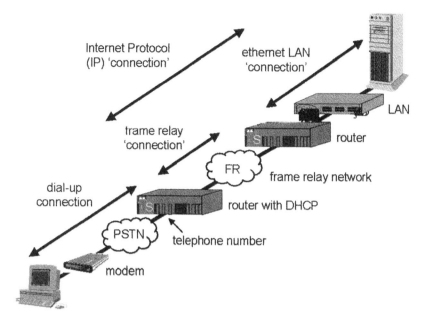

Figure 9.19   Example of an end-to-end 'connection' traversing different types of networks.

(dynamic host configuration protocol). In addition, the IP path to the destination router may change according to traffic conditions or due to changes in topology of the router network. In short, the connection may take a different path through the frame relay network each time. So the knock-on effect of any individual frame relay network trunk will be difficult to predict.

The network manager of a network like that illustrated in Figure 9.19 faces three main problems when either trying to configure new 'connections' ordered by customers or when trying to trace the cause of faults. These are:

- the different components of the network (corresponding to the different managed objects in the MIB of the relevant network element) can be combined together in different combinations to create an overall end-to-end connection (there is not a set of simple 'hard-and-fast' rules about how to combine the different components); and

- the relationship of different components to one another may be on a 1 : 1 basis, or alternatively on a 1:n, n:1 or n:n basis (This makes it very difficult to conceive a simple network configuration database structure which is capable of recording the correct one of many possible different network topology permutations). Worst of all:

- the network topology is changing all the time. It is affected by the provision of new customer lines and trunks, new nodes, current traffic loading and current network failures.

The complexity of creating a configuration management system is indeed awesome. But the potential reward of significantly higher levels of network service and quality have spurred many network operators and network management software developers to attempt a solution. Much money has been spent and some interesting approaches have been developed but much work still has to be done.

Perhaps one of the most effective tools for end-to-end management of 'connections' across a network is provided by the Syndesis company tool *NetProvision*. This aims to provide a solution for end-to-end network connection 'service creation and activation'.

There is a wide range of 'provisioning support' and 'service activation' software tools available, but many of these are oriented to the realisation of a 'simple' customer network access line rather than for the provision of the end-to-end connection across an entire network. Typically these systems are designed to:

- check that a network port is available at the relevant first network node to connect the customer premises (order and schedule new equipment if necessary);

- check that a line is available within the local cabling network to connect the customer premises to the port;

- allocate an available network address (as appropriate);

- schedule the installation manpower to undertake the task; and

- confirm the installation date to the customer.

Some 'provisioning' systems are linked to 'customer service' databases, and are intended to track the quality achieved on customer lines while in operation. In this case, the quality of the line is assumed to be adequate except during times when the 'fault' or *'trouble ticket'* database system has received a 'live' fault report for the connection. Such an approach assumes that the customer will complain when his service is not working and thus that the customer is the main means of quality monitoring.

The number of tools available for assisting the planning and configuration of networks increases daily, but the task remains heavily dependent on skilled network engineers and technicians; their knowledge of how to combine different types of components into reliable network services and their experience of diagnosing and tracing faults encountered with such networks. The ultimate 'umbrella network manager' (the 'glue' between the different network management tools) is still a human engineer!

Below, we review some of the common tools used for configuration of Internet network components:

- Cisco *IOS (Internetworking Operating System)* is a largely text-based control language used for configuring network services and networked applications in a standard manner on Cisco routers and other Cisco devices. It is intended to provide a unified and homogeneous manner of configuring devices and of controlling and unifying complex and distributed network information;

- Cisco works is a tool designed for optimising network traffic and managing router access lists;

- Cisco ConfigMaker is a software tool intended for the configuration of small router networks;

- Juniper's JunOS (Juniper Operating System) is the Juniper equivalent of Cisco's IOS and is used to configure Juniper's routers.

An approach used in some SNMP-based 'umbrella' network manager systems (and possible with HP OpenView, among other systems) is to enable the configuration of all the different network device types from a 'single workstation'. For some network operators, such a 'single workstation' approach has been important, because of the impracticability and costs of multiple video screens and keyboards for each network management operator. Nowadays it is becoming increasingly common to find that such a 'single workstation' approach is based on the standard use of SNMP to monitor and configure all the network components directly. In the past, however, it was not uncommon for the different configuration softwares of the different network components simply to be 'hidden' behind a shared *graphical user interface*. Thus the 'click' to configure one component of the network (shown on a common topology diagram) would activate a different configuration software than the 'button' for configuring a different kind of device.

Fault management tools

Faults in an IP-based data network are usually discovered as the result of either:

- the receipt of an SNMP alarm or event message (an SNMP trap); or

- the reported complaint of a customer or end-user to the help desk.

It is usual that both types of 'fault report' be recorded by the issue of a trouble ticket by a fault management system. Probably the best-known fault management and trouble ticket system is the *Remedy* system (nowadays marketed by Peregrine systems).

Faults reported by humans are entered into the trouble ticket system by hand either by helpdesk or customer service representatives. Sometimes this software is also integrated into call centre system software (in order that customer details can be automatically filled out by derivation from the calling telephone number). A trouble ticket number is issued and the ticket remains 'open' until a technician has diagnosed the cause of the fault, rectified the problem and 'handed over' the use of the network back to the customer or end-user. At this point, the trouble ticket is closed with an explanatory report classifying the problem and its resolution.

The trouble ticket system provides valuable statistics for analysing the quality of network service achieved, and can thus be used to manage *service level agreements (SLAs)* made with end-users.

By integrating an SNMP-based 'umbrella' network management system (such as Hewlett Packard's OpenView) into a trouble ticket system, certain 'critical' SNMP trap messages (network alarms) can be made to generate a trouble ticket automatically. Such automatic generation of trouble tickets is standard practice in large scale networks. It can be a handy way of 'calling out' a technician using the 'dispatch' and 'scheduling' functions of the trouble ticket system. In addition, the automatic generation of trouble tickets leads to a much more precise measurement of the achieved standard of service quality and availability.

For the monitoring and diagnosis of network faults, SNMP-based 'umbrella' network management systems are generally used. These typically filter and *correlate* the plethora of information received by means of SNMP (SNMP traps, alarms, events as well as other information) in an attempt to locate the 'root cause' of a problem. One of the greatest problems is sifting through the deluge of information which a single network failure can lead to. Thus, for example, a frame relay connection failure in the network of Figure 9.19 will lead to a whole range of different alarm, event and other SNMP messages being reported. The two routers at either end of the frame relay connection will report the loss of the trunk on a given port. The PC meanwhile will lose its end-to-end connection with the server. It might notice immediately, or may have to conclude as the result of a timeout (the server having failed to respond) that this connection has been severed. The PC will report that the 'connection to the server has been lost'. By filtering and alarm correlation, the 'umbrella' network management system is programmed to conclude that the fact that the 'connection to the server has been lost' because of the 'link failure on router port X'. In consequence it prioritises the link failure fault for the immediate attention of the human network manager.

The correlation and filtering of alarms usually rely on human experience and judgement, though some 'expert' software in fault diagnosis systems is able over time to 'learn from experience' (by working out which was the most frequently determined root cause and resolution determined by human technicians on previous occasions). A good dose of human input is required in the job of tracing, diagnosing and correcting network faults!

Two of the most popular network management software tools for monitoring and managing network faults are Hewlett Packard's OpenView and Micromuse's *Netcool* products. As a simple tool for small networks, CiscoWorks is also widely used for troubleshooting and network optimisation.

Localisation of network faults

The localisation of faults within a network is often carried out by checking sections of the end-to-end path in turn, using *loopbacks*. Working from one end (say, the PC in Figure 9.20), the technician attempts to locate the point in the end-to-end connection where continuity has been lost. First he or she checks that the PC is getting a response from the modem and can 'talk' in both directions with the modem. This is done by setting a loopback condition at the modem and then sending a test signal. Provided the test signal is returned by the modem in the loopback condition, then the technician concludes that the PC, the modem and the line between them are all working 'OK'. Next, the *loopback* at the modem is removed and replaced with a *loopback* at the first router. If the line between the modem and the router is faulty, then the new test signal from the PC will not be returned. If, on the other hand, the signal is returned, then the line from the PC as far as the router is assumed to be 'OK'. Steadily the technician checks each progressive link in the connection until the faulty link is found. More detailed checks can then go into determining the precise cause of the fault and the most appropriate remedy.

There are a number of different ways in which loopbacks (or equivalent tests) can be conducted. Since I have myself discovered technicians struggling to interpret the results of different types of loopback tests, I think it may be worth explaining how some of the different types work. Figure 9.20 illustrates a number of different loopback-type tests.

The PING (packet Internet groper) test is specific to IP networks. The other tests shown in Figure 9.20 are commonly used standard loopback tests used in telecommunication line transmission testing. The tests function in very different manners. Understanding how they work is critical to understanding the results of the test!

Figure 9.20a shows the configuration of a connection in 'normal operation'. The connection commences at node A and traverses node B. Since the communication is *duplex* there is a separate path used for *transmit* and *receive* directions of transmission.

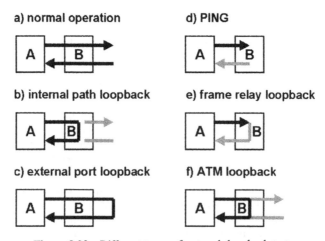

Figure 9.20   Different types of network loopback tests.

Figures 9.20b and 9.20c both show physical path loopbacks. Telecommunications line transmission equipment in particular (including line terminating devices such as CSUs, DSUs, NTEs and NTs — see Chapter 3) typically allow such loopbacks to be applied by means of remote network management commands to the device (device B in both Figures). Such loopbacks return the physical layer (layer 1) protocol signal unchanged to the sender (device A). A tone-generating device or a BERT (bit error ratio tester) are the correct types of test device to send signals to such a loopback. An IP (Internet protocol) packet, on the other hand, is an inappropriate test signal in the case of a physical loopback.... The problem is that the device A has to send a packet with the same source as destination address (its own IP address). IP devices such as routers usually discard such 'unallowed' packets. You therefore might conclude (incorrectly) that the path from node A to node B was broken, when in fact the packet only did not return because it was (rightly) discarded. Use care when using this type of loopback! During the period of the loopback, the transmission path from node A to the remote destination is cut.

Figures 9.20d, 9.20e and 9.20f are all special types of loopback tests developed for testing the transmission continuity of different types of data network and *layered protocols*. Figure 9.20d illustrates the widely used PING (packet Internet groper) test used to check the continuity of IP (Internet protocol) paths between routers, hosts and other IP-network components. An IP packet containing an ICMP (Internet control message protocol) message is sent to the IP address of a particular node in the network (in the case of Figure 9.24 the addressed node is node B). When the packet is received at node B, node B returns a confirmation message as a reply, including the time when it received the initial PING *request* in its PING *response*. Such PING messages can be a useful way not only of checking continuity, but also of determining network delays along the route (calculated from the *timestamp* included in the response). The response message can also be programmed to record all the intermediate nodes traversed along the route back from node B to node A. So in addition, the PING message can be used to determine the exact path of IP 'connections' through a router network. PING is a simple, but very valuable means of localising problems in IP-based networks!

Figure 9.20e illustrates a type of loopback available in some types of frame relay network. The loopback comprises not only a *physical layer* loopback to cause received frames to be returned, but in addition, reverses the source and destination addresses in the frame. This allows the node A to receive the packets it itself sent. Without this reversal of source and destination addresses, frames would be discarded as invalid. (As we discussed above, node A will usually discard as 'invalid' frames it receives in which it appears to be the source.)

Figure 9.20f illustrates a type of loopback available in ATM (asynchronous transfer mode) networks. In this case, the node B is able to return (i.e. loopback) special PL-OAM (ATM physical layer operations and maintenance) cells while simultaneously allowing the 'normal' operation of the end-to-end connection to continue undisturbed. In many ways the operation of the PL-OAM cell is analogous to the PING procedure used in IP networks (Figure 9.24d).

Performance management tools

For everyday performance monitoring of large networks, the same network management systems as used for receiving SNMP traps come into question — Hewlett Packard's OpenView, Micromuse Netcool, Cisco Works, Cisco Netsys, etc.

Major network operators typically expect the following abilities from performance management tools: the ability to:

- view network traffic demand history and trends;
- analyse network traffic statistics in a variety of ways;

- determine quickly areas of the network in congestion (where new nodes, new or upgraded trunk circuits need to be added); and

- report and manage the network quality achieved and compare this with the contracted service level agreements (SLAs) regarding network performance and availability with individual end users and customers

Specialised analysis tools are used widely to analyse network performance problems. Typically such problems are reported by users rather vaguely as 'slow application response times'. They do not necessarily result from specific network failures and SNMP alarms, and tend to exhibit symptoms of 'malaise' rather than of identifiable 'illness'. They can be hard (but important) to trace, and in consequence a number of different manufacturers offer tools, variously called probes, *sniffers* and such like. Such devices typically aim to help the network operator identify and measure:

- overall statistics of usage;

- network usage of *Top Talkers* (i.e. the main sources of network traffic);

- average transaction delay;

- link utilisation, including peak packet rates and packet sizes; and

- software application activity.

We shall return to the subject of network performance optimisation in detail in Chapter 14.

Accounting tools

In comparison to the range of accounting and billing tools available for charging of telephone network usage, the range of tools available for the accounting and billing of IP data network usage is rather sparse, and somewhat primitive. This mostly reflects the Internet 'culture' of the network being a 'good thing' — something for which users should pool and share their resources without charging one another: 'you can use my network if I can use yours.'

Internet service providers (ISPs) still apply charges largely based on the telephone network usage — a monthly subscription service (to cover the Internet network access) and 'per-minute' charges (to cover the telephone network costs used for dial-in) — rather than based on the volume or the value of the data transmitted. But the situation is bound to change, as the major Internet service providers and backbone network providers look for more ways of generating revenue from their networks. We can expect to see the introduction of new network tariffing models as new types of services (for example, voice-over-IP, VOIP) and different grades of service are offered by the providers.

An IP network operator has a number of questions to consider in deciding how to tariff his services and how to collect and process the necessary accounting records for billing (if billing is to be network usage-based):

- Which usage should be billed? — e.g. connected minutes, number of transported bytes or Megabytes, the number of simultaneous connection established, the bit rate at which data is carried;

- Can I handle the volumes of accounting data records which my chosen method of usage charging will generate? (e.g. If you tried to count individual bits, the counter would rapidly increment, and there might be a danger of 'over-running' the counter. Alternatively, the network might collapse in congestion at the extra network load resulting from collecting the accounting records to a central billing database server.)

- How and when shall I collect the accounting data records? Are the network components capable of the reliable delivery of accurate records?

- Do the accounting records need to be *mediated*? (i.e. converted to a standardised format prior to *rating* — the actual tariffing process).

Network security management

Chapter 13 deals in detail with the design of IP-based networks for secure transport of user-data across them, and we shall therefore not cover the subject of network security management here, other than to stress the huge importance of the secure protection of access to the network management system and network analysis tools. As I heard someone once say: 'the most dangerous potential hackers are employed in the network management organisation' — disgruntled employees or former employees can be the greatest risk of all! The network manager, or an ex-employee with a password which has not been cancelled can cause a huge amount of havoc!

Valuable and important security precautions for the design and installation of network management systems are:

- individual passwords and other authentication for all individual employees with access to network management systems or tools;

- use of one-time passwords (see Chapter 13) as far as possible, in order to ensure that the same password is never used twice;

- restricted authorisation of individual employees for performing critical network configuration changes; and

- segregation of the data transport network used for carrying network management messages (as far as possible). The use of VPN, IPsec and or other techniques as described in Chapter 13 can be effective ways of ensuring that outside 'hackers' cannot tamper with, misuse or reconfigure the network maliciously.

The main challenge of network management

Despite the progress made with the standardisation of MIBs (management information bases) for many different types of network component, one of the biggest problems faced by designers of network management software and operators of network management systems is the collation of configuration and status data for all relevant network entities in a consistent database type and format (e.g. relational database, hierarchical file system, UNIX free text files, etc.).

10

Data Networking and Internet Applications

In nine chapters so far, we have discussed in detail the protocols and transmission techniques used to convey data between inter-communicating computers across a modern Internet protocol IP-based network. We have addressed the physical and electrical interfaces used to transfer the bits which make up a binary-encoded data signal. We have learned about the various layered protocols which combine to ensure the reliable, efficient and error-free transmission of data from one end of a network connection to another. And we have also covered the routing and network management protocols and techniques needed for administration. But surprisingly perhaps, even all the foregoing techniques put together are still an inadequate basis for sending data from one computer software program (application) to another on a different computer. The final missing pieces of the data communications 'jigsaw' are the session, presentation and application layer protocols. In this chapter we introduce the most important application layer protocols used with IP (Internet protocol)-networks and explain how these provide the main foundation of modern 'networked computing'. In particular, we shall explain in detail: telnet, FTP (file transfer protocol), TFTP (trivial file transfer protocol), SSH (secure shell) and RTP (real-time application transport protocol). We shall refer to a number of 'proprietary' networking protocol suites and programs and explain their uses (e.g., Microsoft Networking, Appletalk, Sun Microsystems' NFS (network file system), samba, Novell's IPX and Netware). We shall also introduce the DNS (domain name system) and SMTP (simple mail transfer protocol) protocols. The detailed protocol functions and formats of DNS and SMTP are covered in following chapters.

10.1 Computer applications and data networks: application layer protocols

It is first at the *application layer* that computer users and software programmers start to recognise the commands and syntax associated with communications protocols. Thus standardised application layer protocols (OSI layer 7 protocols) undertake a range of valuable functions of use to computer users and programmers, including:

- transmitting keyboard commands from a computer terminal to a remote computer or printing device;

Data Networks, IP and the Internet: Protocols, Design and Operation   Martin P. Clark
© 2003 John Wiley & Sons, Ltd   ISBN: 0-470-84856-1

- allowing users or computer programs to search through the file system of a distant computer and transfer files to or from this 'file server';

- allowing users to send text messages and attached files as *electronic mail (email)* messages; or

- allowing users to send voice, video or other data on a 'real-time' basis from one computer to another.

The first two standard application layer protocols — *telnet* and *FTP (file transfer protocol)* developed in the early 1980s as part of the original development of ARPANET (ARPANET laid the foundation of the modern *Internet*). At the time the objectives were to enable the 'sharing of files' among the US scientific and computing community and 'encourage the use of remote computers'. Telnet and FTP laid the foundation for modern computer networking by establishing a standardised system for the representation, filing and transfer of data between computer hardware and operating systems of differing design and manufacture.

In the early 1980s, computer applications were largely confined to numerical 'number-crunching' tasks with text-oriented input and output. Simple text messaging was also possible, but mainly used for communication between computer system users and operators about system status and 'job scheduling'. A large proportion of computer program output was simple printed onto paper. In consequence, you will find that the basic terminology of the telnet and FTP protocols reflects the simple computer devices of the time — the terminal (or *teletype, TTY*), the printer and the file directory system. The number of functions which could be undertaken with such simple devices and protocols was somewhat limited.

During the intervening 20 years since the early 1980s, the capabilities of computers (the functions or *applications* they are able to perform) have developed enormously. The number of computers, particularly the number of *personal computers (PCs)* has grown exponentially — as have the expectations of the many millions of users. In consequence, the range of computer applications software on offer has increased dramatically, as have the number and capabilities of different applications layer communications protocols. Indeed, the range of modern applications layer protocols is so large that we cannot even name them all here — never mind explaining how they work in detail. We shall thus restrict our coverage to only the best-known protocols and to explaining the standard principles used to define modern applications protocols. In addition, we shall explain the use of the *abstract syntax notation 1 (ASN.1)* as the standard means of defining a modern applications layer protocol.

Standard application layer protocols

Figure 10.1 illustrates a typical computer data network of the early 1980s and the main demands placed upon them. The emergence of data networks, the capability to perform a remote login (rlogin) and to transfer files from one location to another sealed the end of the era in which 'jobs' were sent away to the computer centre for processing (e.g., on punched cards or magnetic tape) and received back as a large computer printout. The earliest requirements of application layer protocols were thus the support of terminal-to-mainframe and terminal-to-printer communications, as well as computer-to-workstation *file transfers*. The first robust and stable protocols designed for these functions *(Telnet* and *FTP — file transfer protocol)* remain in use today.

As Figure 10.2 illustrates, the telnet and *file transfer protocols (FTP)* are designed to use the *reliable* data transmission *service* provided by the combination of the *transmission control protocol* and the *Internet protocol (TCP/IP)*. Telnet uses TCP port 23, while FTP uses TCP ports 20 and 21. At the same time in the early 1980s as the telnet and FTP protocols were being developed, a number of other important application layer protocols emerged, as also shown

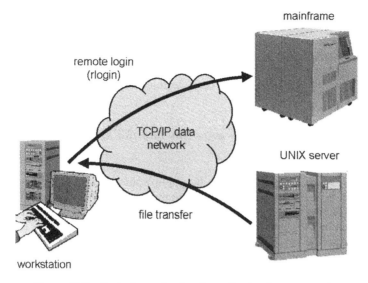

Figure 10.1   Basic demands of early application layer protocols.

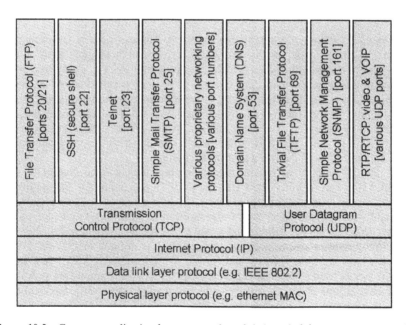

Figure 10.2   Common application layer protocols and their underlying transport protocols.

in Figure 10.2. Together these form the foundation of modern IP-based computer networking. *TFTP* (*trivial file transfer protocol*—RFC 783) appeared in a stable version in June 1981 (TFTP version 2), followed by *SMTP* (*simple mail transfer protocol*—RFC 788) in November 1981. The telnet protocol was developing at the same time, but continued to be amended until a stable version appeared in May 1983 (RFC 854). The stable version of *FTP* (*file transfer protocol*) did not appear until October 1985 (RFC 959)—even though its low TCP port number belies its earlier conception. FTP was adapted several times to accommodate the needs of the UNIX operating system, the 'minicomputers', 'mid-range' computers, 'servers' and 'workstations' which were appearing at the same time. SSH (TCP port 22), as we shall discuss in detail later in the chapter, is a 'modern and secure version of telnet' — a protocol used for secure remote login to a distant server.

The first *domain name system (DNS)* was defined in 1984 — concurrently with a new version of SMTP. This laid the foundation of modern electronic mail (email).

Object-oriented application layer protocols and the abstract syntax notation 1 (ASN.1)

Up to 1981, most computing had remained largely text-based — using teletypes and terminals to access remote 'timeshared' mainframe computers. But in 1981 came a revolution — the IBM personal computer. The computing world changed 'overnight'. Suddenly nearly every desk had a PC on it, and many more people were writing their own 'application' programs. There was a clamour for 'shared data', 'shared files', graphical presentation of data and rapid file transfer — but the programming styles and data formats used by individual programmers made this nearly impossible. Meanwhile, computer system management staff were struggling to manage the local area network devices, routers and workstations which had appeared — they needed a means of remote monitoring and management. There was only one way forward — a standardised method for defining computer information, file formats and other such computing *objects*.

Object-oriented computer programming emerged. Object-oriented programming allows computer programmers to define the data characteristics of a tangible item, a function, a file or a process in a standard manner which other programmers can easily interpret and use. Once defined in this way, the item, function, file or process is called an object. Each type of different object uses a standard data format to describe it (e.g., a circle might be defined by means of its radius, its location coordinates and its colour. Meanwhile a 'bank balance' object might be defined by means of its account number, owner, address.) The parameter definitions of objects, as well as the actual data values which can be easily passed from one computer program to another, make for much easier data 'sharing'.

In 1988, for the purpose of remote monitoring and management control of widespread network and computing devices, ISO (International Organization for Standardization) and ITU-T (International Telecommunications Union telecommunications sector — formerly called CCITT) established a hierarchical set of standardised objects and a standard procedure by which further objects could be defined and added to the set. The language for object definition is called *ASN.1 (abstract syntax notation one)*. (ASN.1 is defined in ISO 8824-5 and ITU-T recommendations X.208-9, X.680-3 and X.690). We encountered ASN.1 in Chapter 9. We illustrated the syntax used to describe different types of network components as ASN.1 *objects* and learned how groups of objects corresponding to a single network component or protocol are called a *management information base (MIB)*. We also learned in Chapter 9 how quite a simple protocol — *SNMP (simple network management protocol)* could be used in conjunction with the standardised MIBs to provide for very powerful monitoring and remote management of telecommunications networks. SNMP appeared in parallel with ASN.1 in 1988.

The appearance of object-oriented computer programming and of ASN.1 has allowed much faster development of a wide range of different applications software and complementary application layer communications protocols. Today there are more than 5000 TCP and UDP port allocations for different application protocols, most of which have been defined using ASN.1 and we cannot even list them all here. But with an understanding of ASN.1, most modern application protocols can be deciphered (refer to Chapter 9 for coverage of ASN.1 or other specialist books if necessary).

Other modern application layer protocols

The most recent demands made upon IP data networks are for the transport of *multimedia* signals. Multimedia, and in particular 'real-time' video and *voice-over-IP (VOIP)* signals, place very stringent demands upon the telecommunications networks which carry them. As a result, it has been necessary to develop a new range of application layer protocols. Later in the chapter, we shall review the techniques used to carry video and voice-over-IP (VOIP) signals, and in particular will describe the real-time application transport protocol (RTP) and the associated real-time application transport control protocol (RTCP) which are used for this purpose.

Review of application layer protocols

Because of the importance of the application layer protocols illustrated in Figure 10.2, we shall explain the uses and protocol operation of each in detail. In this chapter we shall cover the following protocols in turn:

- *telnet*;
- *file transfer protocol (FTP)*;
- *trivial file transfer protocol (TFTP)*;
- *secure shell (SSH)*; and
- *real-time application transport protocol (RTP), real-time application transport control protocol (RTCP)* and their use for carriage of video and *voice-over-IP (VOIP)*.

Our detailed coverage of the *domain name system (DNS)* appears in Chapter 11 and the *simple mail transfer protocol (SMTP)* is discussed in Chapter 12. Coverage of the *simple network management protocol (SNMP)*, meanwhile, appeared in Chapter 9.

10.2 Telnet

The current version of the *telnet* protocol is defined in RFC 854, issued in May 1983, though a number of adaptations and extensions of the protocol have been issued subsequently (as listed in the abbreviations appendix of this book).

Telnet was designed to provide for 8-bit (byte)-oriented bidirectional communication between a computer terminal device and an associated remote computer 'process'. Telnet uses TCP (transmission control protocol) port 23.

The designers of the telnet protocol were faced with the need to define a protocol which allowed the connection of a remote terminal provided by one manufacturer to a computer process (typically called a computer control 'shell') running on a mainframe computer provided by a different manufacturer (Figure 10.3). The main problem lie in the definition of a common format in which data could be exchanged. In consequence, Telnet comprises a standardised combination of *session layer, presentation layer* and *application layer* protocol functions.

A *telnet session* provides for a communications connection (on TCP port 23) across a TCP/IP network between a *virtual terminal (VT)* and a remote device such as a mainframe computer (Figure 10.3). The telnet protocol itself assumes that there is a terminal and a printer at both ends of the connection. This configuration is called *network virtual terminal (NVT)*. Figure 10.4 illustrates the imaginary NVT connection between a virtual terminal and its *virtual printer* in one of the directions of communication. A similar NVT connection to Figure 10.4 is imagined to exist in both other directions of communication. Thus the real

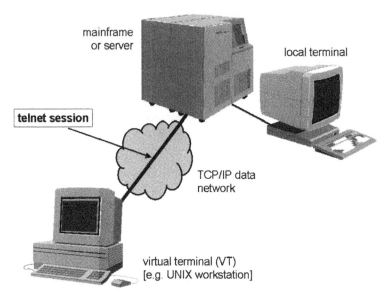

Figure 10.3   The main objective and usage of the telnet protocol.

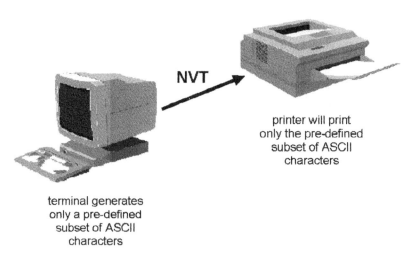

Figure 10.4   The imaginary network-virtual terminal (NVT) communication relationship using the telnet protocol.

terminal of Figure 10.3 acts as both virtual terminal (when sending data) and virtual printer (when receiving data).

The data actually sent on a telnet connection reflects the imaginary NVT (network virtual terminal) relationship of the two devices. In particular, the data is sent with the relevant printer control commands to reproduce the information as if it is required to be formatted and printed on a sheet of paper. The telnet NVT data file format was used by early computer programming methods, and in addition is ideally suited to program input or output of text-type data.

The data sent over a telnet NVT (network virtual terminal) connection is a standardised character set called *NVT-ASCII (network virtual terminal — American standard code for information interchange*). NVT-ASCII is based upon the original 7-bit ASCII code, but each ASCII-character is sent as an 8-bit (1 byte) sequence. The most significant bit (MSB) is set at '0' and followed by the normal 7-bit ASCII code value for individual alphanumeric characters. In this way, NVT-ASCII set the foundation for the modern UTF-8 (universal transformation format-8) and ISO 10646 character sets. However, unlike ASCII and other character sets, only a limited number of printer control characters are valid (as listed in Table 10.1).

The transport of data by a telnet connection is *byte-oriented* (i.e., framed as a series of bytes representing individual characters). The characters themselves are first stored by the terminal until a complete line of data is ready for transmission. Typically the end of the line of data is signalled by means of *the* ⟨*CRLF*⟩ *(carriage return line feed)* key nowadays usually called the *enter* key. Thus, for example, the terminal user in Figure 10.3 might type a command to the mainframe computer 'e.g., delete file.com' and then hit the enter key. At this point, the entire command will be forwarded via the telnet connection using a single TCP segment to the mainframe computer. Waiting in this way until an entire line of data is ready to be sent has two important benefits:

- the most efficient use is made of the data network, since only a single TCP segment/single IP packet is necessary for the transmission; and

- the mainframe computer processor need only be *interrupted* to handle the incoming data once — at the time when the complete command has arrived.

Individual bytes (representing individual NVT-ASCII characters or commands) are sent in the *canonical format* (i.e., least significant bit first).

As well as introducing the concept of a standard presentation format (NVT), telnet also laid the groundwork for *peer* protocols by developing a communication protocol based on a symmetric view of terminals and the processes with which they communicated. Another concept introduced in telnet and now commonly used in other communications protocols is the ability to negotiate options. The negotiated options of telnet (listed in Table 10.1) allow additional services over and above those of the basic NVT (network virtual terminal) character set to be requested at one end of the connection and either accepted or rejected at the other. Such options allow, for example, computers from the same manufacturer to intercommunicate via a telnet connection using additional control characters which may not be supported by all types of computer. The telnet commands used for the negotiation of options are called *WILL, WON'T, DO* and *DON'T* (character values 251–254). Their use is explained in Table 10.1.

An interesting feature of the telnet protocol is the SYNCH signal. As detailed in Table 10.1, the SYNCH signal is sent by coding the data mark (DM) (value 242) character into the outgoing telnet connection, while simultaneously setting the *TCP (transmission control protocol) urgent marker*.[1] The SYNCH signal is designed to clear the data path from a terminal to a distant mainframe computer. To understand why we might need it, consider the following example.

Table 10.1 The telnet network virtual terminal (NVT) character and command set (NVT-ASCII)

Telnet control functions and signals	Code (decimal)	Code (hexa-decimal)	Function
NULL (NUL)	0	00	No operation.
Bell (BEL)	7	07	(Option) Produces an audible or visible signal without moving the print head.

(continued overleaf)

---

[1] See Chapter 7.

Table 10.1   (*continued*)

Telnet control functions and signals	Code (decimal)	Code (hexa-decimal)	Function
Back space (BS)	8	08	(Option) Moves the printer one space towards the left margin, remaining on the current line.
Horizontal tab (HT)	9	09	(Option) Moves the printer to next horizontal tab stop (undefined is where exactly this is).
Line feed (LF)	10	0A	Moves the printer to the next line, but retaining the same horizontal position.
Vertical tab (VT)	11	0B	(Option) Moves the printer to next vertical tab stop (undefined is exactly where this is).
Form feed (FF)	12	0C	(Option) Moves the printer to the top of the next page, retaining the same horizontal position.
Carriage return (CR)	13	0D	Moves the printer to the left margin of the current line.
Alphanumeric characters and punctuation	32–126	20–7E	Alphanumeric text (ASCII) characters making up the main portion of the telnet data.
Subnegotiation end (SE)	240	F0	This signal indicates the end of subnegotiation of option parameters.
No operation (NOP)	241	F1	This signals that no operation is possible.
Data mark (DM)	242	F2	The data stream part of the SYNCH mechanism—for clearing a congested data path to the other party.
Break (BRK)	243	F3	The break command is an additional command outside the normal ASCII character set required by some types of computers to create an 'interrupt'. It is not intended to be an alternative to the IP command, but instead used only by those computer systems which require it.
Interrupt process (IP)	244	F4	This command suspends, interrupts, aborts or terminates a user process currently in operation on the remote host computer. (This is the equivalent of the 'break', 'attention' or 'escape' key.)
Abort output (AO)	245	F5	This command causes the remote host computer to jump to the end of an output process without outputting further data.
Are you there? (AYT)	246	F6	This command causes the remote host computer to reply with a printable message that it is still 'alive'.

Table 10.1   *(continued)*

Telnet control functions and signals	Code (decimal)	Code (hexa-decimal)	Function
Erase character (EC)	247	F7	This command deletes the last character sent in the data stream currently being transmitted.
Erase line (EL)	248	F8	This command deletes all the data in the current 'line' of input.
Go ahead (GA)	249	F9	Continue.
Subnegotiation (SB)	250	FA	This signal indicates that the following codes represent the subnegotiation of telnet features.
WILL (option code)	251	FB	This signal is a request to start performing a given (indicated) option.
WON'T (option code)	252	FC	A rejection of the requested option.
DO (option code)	253	FD	An acceptance of the requested option.
DON'T (option code)	254	FE	An instruction to the remote party not to use, or to stop using the indicated option.
IAC	255	FF	Data byte 255.
SYNCH	—	—	The SYNCH signal is a combination of a TCP 'urgent notification' coupled with a 'data mark' character in the data stream. SYNCH clears the data path to a remote 'timeshared' host computer. The 'urgent' notification bypasses the normal TCP flow control mechanism which is otherwise applied to telnet connection data.

Imagine that the mainframe computer of Figure 10.3 is *timeshared* between a number of remote terminal users (this was a common practice in the 1980s). As users, we are therefore sharing the capacity of the mainframe's *central processing unit (CPU)* with a large number of other users. Now let us assume, that things are busy, so that the CPU is heavily loaded and not responding very quickly. To make things worse, all the commands we are sending to the computer are simply being 'stacked up' to wait their turn for processing. Unless we have a way of 'jumping the queue' we will have to wait for all our previous commands to be executed before changing our minds to do something else. How do we achieve this? By using the SYNCH command to *purge* the data path — deleting as-yet-unexecuted commands sent in the outgoing path.

The telnet protocol was rapidly adopted and integrated directly into computer operating system software. Thus, for example, the UNIX operating system incorporated new commands (telnet and rlogin) to enable UNIX workstation users to conduct remote logins to distant servers using a single *command line* input. UNIX took off as the favoured operating system for networked computing and *virtual terminal (VT)* and *terminal emulation* flourished for UNIX workstations and PCs alike.

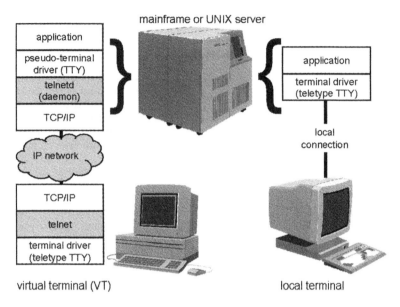

Figure 10.5   UNIX realisation of telnet protocol: terminal and telnet daemon functions.

Figure 10.5 illustrates the protocol functions which must be undertaken by both the *virtual terminal (VT)* and the *server* of a UNIX-based telnet connection. The server runs a telnet software program called *telnetd (telnet daemon)* to respond to incoming users who wish to establish telnet connections to the server. Each virtual terminal first establishes a TCP (transmission control protocol) connection to telnetd. Thereafter, telnetd takes over control and gives the terminal user the impression of being a locally connected terminal (VT) to the distant UNIX server, host or mainframe. Within the 'telnet window' which appears on the virtual terminal PC or workstation, the user sees commands to and responses of the distant server. The actual content and syntax of these operating system commands and responses (e.g., 'delete', 'edit', 'execute', etc.) are called the *shell*.

In the years since 1983 and the issue of the telnet protocol specification in RFC 783 there have been many optional extensions made to telnet. Noteworthy are the versions specifically designed for IBM terminal-to-IBM mainframe communications (*TN3270*) and for *VT100* terminal emulation. (VT100 was originally defined as part of the DECnet suite of protocols, but became perhaps the most widely-used terminal character set.)

10.3 FTP (file transfer protocol)

The file transfer protocol (FTP) is defined in RFC 959 (issued in October 1985). The protocol was intended to provide a reliable and efficient means of data transfer to encourage:

- the 'sharing of files' and

- the 'use of remote computers'.

FTP allows a remote user (typically a UNIX workstation user) to search through the file system of a remote file *server* (typically a UNIX server) and to insert, remove or copy files to the directory. It provides a standardised system for such file transfer and for simple file and directory manipulation of remote servers (e.g., adding and removing file directories).

By means of the standard FTP-defined file system, the two communicating devices at either end of the FTP connection need not be concerned with any 'proprietary' file directory structure or file naming conventions used by their correspondent. Unlike the equivalent 'proprietary' protocols which went before it, FTP was specifically designed to be suitable for all types of computers and manufacturers. It is implemented on mainframes, minicomputers, mid-range computers, workstations and personal computers alike. In common with telnet, FTP commands are embedded into the UNIX operating system.

Principles of FTP (file transfer protocol)

The file transfer protocol (FTP) may be used to set up access for a *user (FTP user)* to a remote file server *(FTP server)* system (Figure 10.6). *Access controls* define the FTP user access privileges, i.e., which systems and files are allowed to be accessed.

The FTP connection actually comprises two separate connections:

- an FTP *control connection* (on TCP port 21); and

- an FTP *data connection* (on TCP port 20).

The FTP control connection connects the *user-PI (protocol interpreter)* with the *server-PI (protocol interpreter)* for the purpose of exchange of FTP *commands* and *replies*. Meanwhile, the data connection is used for the transfer of the user's actual data (i.e., the file to be transferred). A *data transfer process (DTP)* in each of the ends (user and server) coordinates the actual communication across the data connection. At any point in time, the user and/or server DTPs are either *active* (currently transferring user data) or *passive* (not transferring data).

As shown in Figure 10.7, the *protocol interpreter (PI)* and *data transfer process (DTP)* are rather like different 'layers' of the *OSI*-model *(open systems interconnection* model). Indeed,

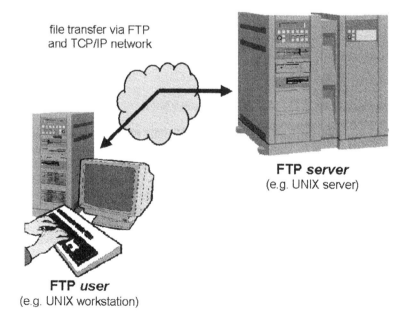

file transfer via FTP
and TCP/IP network

FTP *server*
(e.g. UNIX server)

FTP *user*
(e.g. UNIX workstation)

Figure 10.6   Use and terminology of the file transfer protocol (FTP).

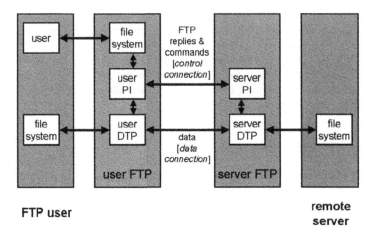

Figure 10.7   Sublayers of FTP (file transfer protocol): PI (protocol interpreter) and DTP (data transfer process).

the experience gained in developing protocols in the early 1980s led to the definition of the OSI model. But FTP pre-dates the OSI model and includes a mixture of *session layer* (layer 5), *presentation layer* (layer 6) and *application layer* (layer 7) functionality.

The protocol interpreter (PI) is the software function within FTP (file transfer protocol) which sends FTP *commands* and receives FTP *replies* by means of the *control connection*. By means of the protocol interpreter, the FTP user (Figure 10.7) is able to indicate to the distant server which file he or she wishes to transfer to the server or receive from the server. A standardised file directory system (called *NVFS—network virtual file system*) is used by the protocol interpreter to locate the storage position of the file on the remote server. Once everything has been prepared, the user's data file is transferred by the FTP *data transfer process (DTP)* by means of the *data connection*.

The network virtual file system (NVFS) has much in common with the UNIX *network file system (NFS)*. This is hardly surprising, given the hand-in-hand development of UNIX and the TCP/IP protocol suite!

Protocol operation of FTP (file transfer protocol)

As its presentation layer (for user data transfer across the data connection), the FTP specification recommends that both FTP commands and user files are transferred across the network in the standard *NVT* (telnet *network virtual terminal*) format (Table 10.1). The two computers communicating by means of FTP are assumed to convert between their internal data formats and the 8-bit NVT format called the data *transfer size* as necessary. (At the time, the different computer manufacturers used very different character set formats and data storage formats: e.g., five 7-bit characters stored in a 36-bit data *word* (with one bit 'empty'); 8-bit characters stored as an 8-bit *byte* (the IBM approach—using the *EBCDIC* alphabet) or four 9-bit characters in a 36-bit word).

The *logical byte size* is the term given to describe the length of data words used to store the data in one of the two computers at either end of the FTP connection. The conversion between the logical data size (computer-internal data storage format) and the data transfer size (always 8-bit NVT-ASCII) is the responsibility of the relevant PI or DTP function—before and/or after data transfer.

FTP commands and the FTP control connection

FTP (file transfer protocol) commands have the style of an old computer programming command! Indeed, UNIX experts are likely to be familiar with the commands, some of which are also built directly into the UNIX operating system. FTP commands (which are sent via the FTP *control connection* — TCP port 21) allow a remote user to *access* and *navigate* around the file directories of a remote file server and then arrange either to send or receive a file.

FTP commands (Table 10.2) are sent on the *control connection* using the Telnet (NVT-ASCII) character set (Table 10.1) for coding the individual characters which make up the command. Thus to send the control command, the sequence ACCT⟨CRLF⟩ is sent (where ⟨CRLF⟩ is the telnet EOL character). If parameter values are required by the commands, these follow the command and a 'space' character.

FTP replies (Table 10.3) ensure that requests and actions are coordinated and that recovery is possible (after a problem or error). Every command must generate at least one reply. The replies consist of a three-digit number (coded as standard 8-bit NVT-ASCII characters) followed by the space character, one line of printable text and the telnet end-of-line character ⟨CRLF⟩. FTP commands and replies are intended to alternate.

Table 10.2 FTP commands

FTP command code	Full command name	Operation performed
Access Control Commands		
ACCT	Account	Telnet string identifying user's account, not necessarily the same or related to USER
CDUP	Change to parent directory	A special command within CWD. This command simplifies the change of working directory in a case that the two different communicating computer operating systems have different syntax for naming the parent directory
CWD	Change working directory	Allows user to work with a different directory without altering login or accounting information
LANG	Language	New parameter introduced by RFC 2640 to identify in which language to present server greetings and text of server responses
PASS	Password	Telnet string with user's password
QUIT	Logout	Command terminates a user and closes control connection if file transfer is not still in progress
REIN	Reinitialize	Command terminates a user, flushing all input, output and account information
SMNT	Structure mount	Allows user to mount a different system data structure
USER	User name	Telnet string identifying the user — the name required to access the server file system
Transfer parameter commands		
MODE	Transfer mode	This parameter indicates the mode to be used for data transfer: B — block-mode C — compressed-mode S — stream-mode

(continued overleaf)

Table 10.2   (*continued*)

FTP command code	Full command name	Operation performed
PORT EPRT	Data port	This indicates the host port to be used for the data connection PORT replaced by EPRT in IPv6 (extended address — RFC 2428)
PASV EPSV	Passive	Command requests server to 'listen' on a data port (other than its default) and to wait for a connection rather than initiate one PASV replaced by EPSV in IPv6 (extended address — RFC 2428)
STRU	File structure	This parameter indicates the data structure of the file being transferred: F — file-structure P — page-structure R — record-structure
TYPE	Representation type	This parameter indicates the data type being transferred: A — ASCII C — carriage control E — EBCDIC I — image L — local byte size N — non-print T — telnet format effectors
	FTP service commands	
ABOR	Abort	Command tells server to abort the previous FTP service command and any associated data transfer
ALLO	Allocate	Command may be needed by some servers to reserve storage prior to data transfer
APPE	Append (with create)	Command causes the server to accept the data transferred and to store them, appending them as necessary to any existing file with the same name.
DELE	Delete	Command causes the indicated file to be deleted at the server site
HELP	Help	Server will return helpful information regarding its status and how to initiate next action
LIST	List	Command causes a list of all files and their details in the specified current directory or current information on the specified file to be returned
MKD	Make directory	Command causes the specified directory to be created
NLST	Name list	Command causes a list of all the names (only) of files in the current directory or a specified directory to be returned
NOOP	Noop	No action or operation need be performed other than to send an OK reply
PWD	Print working directory	Command causes current directory name to be returned in reply
REST	Restart	This parameter indicates the server marker at which the file transfer is to be restarted
RETR	Retrieve	Command causes the server to transfer a copy of the file to the other end of the connection.

Table 10.2 (*continued*)

FTP command code	Full command name	Operation performed
RMD	Remove directory	Command causes the specified directory to be removed
RNFR	Rename from	Command indicates the old pathname of the file which is to be renamed.
RNTO	Rename to	Command indicates the new pathname of the file which is to be renamed
SITE	Site parameters	Command used by server to provide system-specific services essential to file transfer
STAT	Status	The reply to this command, sent over the control connection, indicates the status of the operation in progress
STOR	Store	Command causes the server to accept the data transferred and to store them.
STOU	Store unique	Command has similar effect to STOR — except that the file is to be stored in the current directory with a unique name
SYST	System	The reply to this command indicates the operating system in user on the server

Table 10.3 FTP replies

Reply types and numbers	Meaning
x0z	Syntax error; unrecognised or superfluous command or simply OK confirmation
x1z	Request for further information
x2z	Reply refers to control and data connection
x3z	Authentication and accounting replies during login
x4z	Unspecified
x5z	Status of file system
Positive preliminary reply (1yz)	The requested action is being initiated — wait for a further reply before issuing a new command
110	Restart marker reply
120	Service ready in nnn minutes
125	Data connection open, transfer commencing
150	File status OK
Positive completion reply (2yz)	The requested action had been completed successfully
200	Command OK
202	Command not implemented: superfluous
211	System status or system help ready
212	Directory status
213	File status
214	Help message
215	NAME system type
220	Service ready for new user
221	Service closing control connection
225	Data connection open, no transfer in progress

(*continued overleaf*)

Table 10.3   *(continued)*

Reply types and numbers	Meaning
226	Closing data connection — file action successful
227	Entering passive mode
230	User logged in
250	Requested file action OK and completed
257	PATHNAME created
Positive intermediate reply (3yz)	The command has been accepted but the requested action has not been commenced, while awaiting further information — which should be sent by the user
331	User name OK, need password
332	Need account for login
350	Requested file action awaiting further information
Transient negative completion reply (4yz)	The command was not accepted and the requested action not undertaken, but the action may be requested again — if so by returning to the beginning of the command sequence
421	Service not available, closing control connection
425	Data connection cannot be opened
426	Connection closed, transfer aborted
450	Requested file action not undertaken, file not available
451	Requested action aborted due to local processing error
452	Requested action not possible due to insufficient storage
Permanent negative completion reply (5yz)	The command was not accepted and the requested action not undertaken. The same request command sequence should not be repeated
500	Syntax error, command unrecognised
501	Syntax error, parameter or argument error
502	Command not implemented
503	Bad sequence of commands
504	Command not implemented for parameter specified
530	Not logged in
532	Account required for storing files
550	Requested file action not undertaken, file not available
551	Requested action aborted, page type unknown
552	Requested file action aborted, exceeds storage allocation
553	Requested action not undertaken, file name not allowed

FTP data transfer

Data transfer using the *file transfer protocol (FTP)* takes place by means of the *data transfer process (DTP)* and the FTP *data connection* (TCP port 20). The *user-DTP* runs at the user (human user or terminal) end of the connection, while the *server-DTP* runs on a file server (e.g., UNIX server) at the other end (Figure 10.7). Alternatively, a server-DTP may exist at both ends of the connection.

It is recommended that data sent over the FTP *data connection* be coded as NVT-ASCII *(network virtual terminal-ASCII* — Table 10.1). This is effectively the 'presentation layer' protocol of FTP. The computer sending a file may need to convert data stored internally in another format into the NVT-ASCII format, and likewise the receiving computer will convert from NVT-ASCII to its own internal data storage format if necessary. But NVT-ASCII is not the only allowed data type which may be used. The following data types are also allowed:

- *ASCII* (the default character set is NVT-ASCII — network virtual terminal-American standard code for information interchange);

- *EBCDIC* (extended binary coded decimal interchange code — this is an 8-bit character code originally defined by the IBM company);

- *IMAGE-type* (this is data coded as a simple stream of contiguous bits); and

- *LOCAL-type* (this data type allows identical computers at either end of the FTP connection to exchange data in a 'proprietary' format, e.g., of a non-standard logical byte size, without the need to convert it to the NVT-ASCII format).

The exact format of ASCII and EBCDIC types may be adapted by means of a second parameter which indicates whether the file includes any *vertical format control* characters (i.e., whether the file includes the page or print format). Files with vertical format control indicate *non-print characters* (i.e., characters which do not appear in a printed form of the file). These characters are also called *format controls*. The standard set of vertical format control characters are:

- ⟨CR⟩   [carriage return];
- ⟨LF⟩   [line feed];
- ⟨NL⟩   [new line];
- ⟨VT⟩   [vertical tab];
- ⟨FF⟩   [form feed];
- ⟨CRLF⟩   [carriage return line feed — this is the normal EOL (end-of-line) character]

No matter what the logical byte size of the data is, the *transfer byte* size used by FTP is always an 8-bit byte.[2]
    The files to be transmitted by means of FTP may have one of three different file structures:

- *file structure*;

- *record structure*; or

- *page structure*.

Both *user-DTP* and *server-DTP* have a default *data port* on which they *listen* (when otherwise they are *passive* — i.e., not *actively* transferring data) across a connection. This is called the *passive data transfer process*. The *user data port* is the *control connection*. In other words, the user-DTP listens on the control connection for activity on the part of the server. (Alternatively, the user-DTP can be replaced by a second server-DTP.) Meanwhile, the server DTP listens on the data connection. Prior to sending a *transfer request*, both user-DTP and server-DTP must confirm by listening on their respective data ports to confirm that there is no current activity. The direction in which the first FTP *request command* is sent determines the direction of the data transfer which will follow. Upon receipt of a request, the FTP file server *initiates* the *data connection* (swapping the use of the port connections appropriately).
    Files (in any of the three formats) may be transferred according to one of the three available *transfer modes*:

- *stream mode*;

- *block mode*; or

- *compressed mode*.

---

[2] In the early days of computing, different manufacturers used different character sets — in particular different numbers of bits (the logical byte size) to represent alphanumeric characters, as we saw earlier.

The transmission of files sent in all three transfer modes is always terminated by the indication of an *EOF (end-of-file)*. This can be done either explicitly (by the inclusion of an *EOF* character or sequence of characters) or *implied* by *closing* the data connection.

When files are transferred in the FTP *stream mode*:

- *file-structure EOF (end-of-file)* is indicated by closing the data connection;

- *record-structure EOF* and *EOR (end of record)* are both indicated by a two-byte control character code. The first byte is set as all '1's (hexadecimal FF — the escape character). The second byte is coded 01(hex) for *EOR (end-of-record)*; 02 (hex) for *EOF (end-of-file)*; or 03 (hex) for EOR/EOF on last byte.

- *page-structure EOF* is indicated either by an explicit EOF or by closing the data connection.

When transferred in the FTP *block mode*, files are split up into *blocks*, each of which is preceded by a *block header*. The FTP block header (as illustrated in Figure 10.8a) includes a descriptor field of 1 byte (or octet) length, a *byte count* field of 16 bits (value 0–65535) and the *data field*. The descriptor defines the coding used to represent EOR (end-of-record), EOF (end-of-file) and the *restart marker* (The restart marker is used for error recovery, as we shall explain shortly). The coding of the descriptor field is as shown in Table 10.4.

When sent in the FTP compressed mode, files are sent by means of three kinds of data:

- data and text — is sent as a string of bytes (the standard format uses the NVT-ASCII character set — Table 10.1);

- compressed data — allows repetitions of the same character in a string (e.g., lots of 'tabs' or 'spaces') to be sent more efficiently. In this format, up to 63 consecutive occurrences

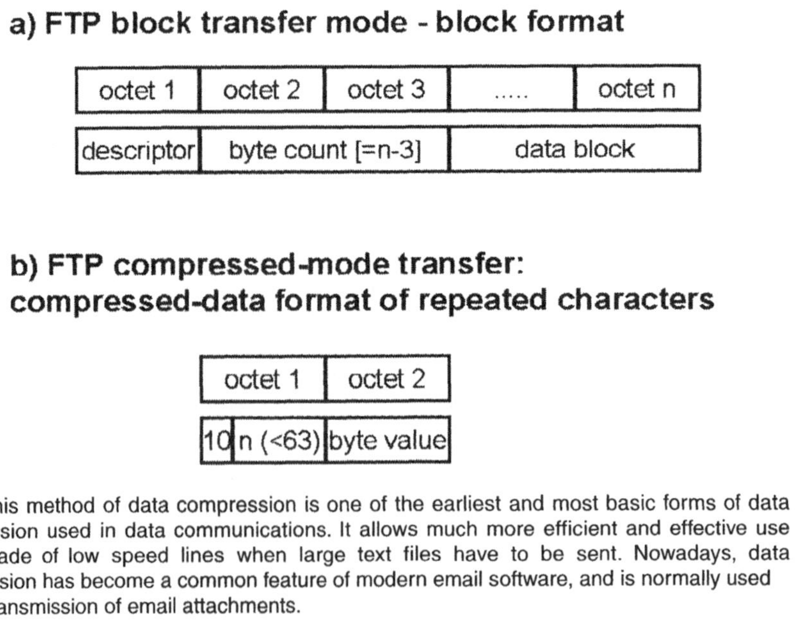

a) FTP block transfer mode - block format

octet 1	octet 2	octet 3	.....	octet n

descriptor	byte count [=n-3]	data block

b) FTP compressed-mode transfer:
compressed-data format of repeated characters

octet 1	octet 2

10	n (<63)	byte value

Note: This method of data compression is one of the earliest and most basic forms of data compression used in data communications. It allows much more efficient and effective use to be made of low speed lines when large text files have to be sent. Nowadays, data compression has become a common feature of modern email software, and is normally used for the transmission of email attachments.

Figure 10.8   FTP block transfer mode and compressed-mode transfer formats.

Table 10.4 FTP block mode transfer: block header — descriptor field meanings

Descriptor code value (decimal — hex)	Meaning
16 (10 hex)	This data block is a *restart marker* (i.e., the value in the data field is used to indicate a particular block in the data file, in case a recovery should become necessary from this point)
32 (20 hex)	There are *suspected errors* in a data block
64 (40 hex)	The end of this data block is the *end-of-file (EOF)*
128 (80 hex)	The end of this data block is the *end-of-record (EOR)*

of the same character in the text (e.g., 'spaces' or 'tabs'), can be transmitted using only 2 bytes of code — as shown in Figure 10.8b.

- control information — is sent in two-byte *escape-code* sequences. The first byte is always the escape character (00). The second byte is coded in the same way as the descriptor codes of the FTP block mode (see Table 10.4).

FTP includes a simple mechanism for 'error' *recovery* and *restart*. The procedure is intended to protect against major system failures — indeed it is better to think of the recovery as being a 'failure' recovery, rather than a recovery from a bit *error* or a small transmission hiccup. The detection of bits lost or scrambled during data transfer is assumed to be handled by *TCP (transmission control protocol*[3]*)*. In the case that the two communicating file systems become 'out of synchronism' during the transmission (as the result of a system failure or other problem), the FTP block-transfer-mode and compressed transfer mode allow transmission to be recommenced starting at the position of the indicated restart marker.

In the FTP *block-mode*, the restart marker is an extra block of data, periodically inserted into the data stream. It has a unique number value which is indicated in the *data field* (see Figure 10.8a) of the restart marker block (a block in which the descriptor value is set at decimal value '16' (hexadecimal value '10')). The range of the number value can be set in accordance with the chosen byte count length set for the restart marker clock.

In the FTP compressed mode the restart marker is indicated by a two-byte *escape-code* sequence.

10.4 TFTP (trivial file transfer protocol)

TFTP (trivial file transfer protocol) is a very simple but *unreliable* file transfer protocol. It is not intended for 'normal' file transfer, but instead is typically used to transfer *boot* files or keyboard *font* files to terminals, diskless PCs or diskless workstations (Figure 10.9). TFTP is defined in RFC 783 and operates on *UDP (user datagram protocol)* port 69.

Three modes of transfer are defined by TFTP:

- *netascii-mode* (this mode of transfer indicates that the file is coded in 8-bit NVT-ASCII — Table 10.1);

- *octet-mode* (this mode of transfer, previously called *binary mode* is the equivalent of FTP stream mode — a string of binary bits transmitted with an 8-bit *transfer size*);

- mail-mode (in this mode *netascii* characters are sent to a *user* rather than to a *file*. This allows, for example, a message to appear on a terminal screen.)

---

[3] See Chapter 7.

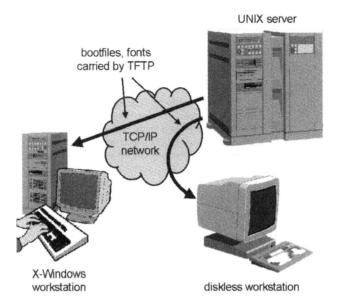

Figure 10.9 Typical uses of TFTP (trivial file transfer protocol).

a) RRQ and WRQ packet format

octet 1	octet 2	octet 3	octet 4

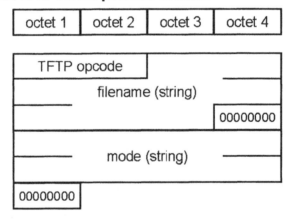

b) ACK packet format

octet 1	octet 2	octet 3	octet 4

TFTP opcode (0004)	block number

Figure 10.10 TFTP (trivial file transfer protocol): RRQ, WRQ and ACK message packet formats.

The TFTP file transfer process is established by first sending a request message—either a *WRQ (write request)* or a *RRQ (read request)*. Provided a positive reply is received, the file transfer may begin. An *ACK (acknowledgement)* packet represents a 'permission' to write (i.e., to send a packet to the remote end). The positive reply to a read request (RRQ) is a response containing the first *data packet* of the file requested.

The header of all TFTP packets (Figure 10.10) consists of a 2-byte *opcode* field (Table 10.5) which indicates the TFTP packet-type.

Table 10.5 TFTP packet types and opcode values

TFTP opcode	Operation	Full name
1	RRQ	Read request
2	WRQ	Write request
3	DATA	Data packet
4	ACK	Acknowledgement
5	ERROR	Error

RRQ (read request) and WRQ (write request) packets, as shown in Figure 10.10a, include the TFTP opcode (value '1' or '2'), the name of the file to be transmitted (in a text *string* format), a *NULL octet* and then the mode in which it is to be transmitted (again indicated as a character string in NVT-ASCII character format). A further NULL octet completes the RRQ or WRQ packet.

The ACK (acknowledgement) packet (Figure 10.10b) is initially returned with the *block number* set at 1. The acknowledgement indicates the number of the next expected data block. Should the response to an RRQ or WRQ message be an error packet (opcode = '5'), then the request has been denied.

Each data packet includes the TFTP block number, consecutively numbered starting at 1. This is used for data *flow control*, as we discussed in detail in Chapter 3. Following the opcode (set at value '3') and the block number (which is incremented for each successive packet), comes the data itself (Figure 10.11a). A *TID (terminal identification)* is chosen randomly by

a) DATA packet format

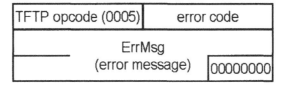

b) ERROR packet format

Figure 10.11 TFTP (trivial file transfer protocol): data and error message packet formats.

Table 10.6   TFTP (trivial file transfer protocol) error message code
values

Error code value	Meaning
0	Not defined, see textual error message (if any)
1	File not found
2	Access violation
3	Disk full or allocation exceeded
4	Illegal TFTP operation
5	Unknown transfer ID
6	File already exists
7	No such user

both ends and these values are used for duration of connection as the values of the UDP
source and destination ports. The initial request (RRQ or WRQ), on the other hand, is sent to
the UDP port corresponding to TID = 69 (decimal).

Error messages have the format shown in Figure 10.11b, where the *error code* value is
set according to Table 10.6. Error messages are typically extended by means of a short text
message which is intended to appear on the user's terminal screen.

10.5   Secure shell program and protocol (SSH or SECSH)

Because the traditional telnet protocols (RFC 854) and related UNIX remote login commands
(e.g., *rsh, rlogin* (RFC 1282), *rcp*, etc.) are vulnerable to different kinds of network *attacks*,
there has been an increasing focus on improving the security of protocols (as we shall discuss
in more detail in Chapter 13). A third party who has network access to hosts or servers
connected to the Internet, or simply with physical access to communications lines, can gain
unauthorised access to systems, steal passwords and wreak havoc in a variety of ways. For
this reason, the *ssh (SSH or secure shell)* program and protocol was designed as a secure
replacement for telnet, rlogin and similar data network services.

SSH (secure shell) is a protocol and program for secure remote login and other secure
network services over an insecure network. It is used in particular as 'secure version of telnet'
or as a secure *shell* — a program which allows commands to be executed in a remote machine,
for the transfer of files between machines and other secure network services. It was invented
by SSH Communications Security Oy, and ssh is a trademark of this company. It is currently
an Internet draft (being prepared as an RFC by IETF — Internet Engineering Task Force),
where it is also known as *secsh (secure shell)*.

SSH is designed to protect against:

- interception of passwords and other data by intermediate hosts;

- manipulation of data by people in control of intermediate hosts or network transit points;

- IP *spoofing* (in which a targetted machine is 'conned' by a remote host which sends IP
packets which pretend to come from another, *trusted* host);

- *DNS spoofing* (in which an attacker forges domain name system (DNS)[4] server *resource
records*.

---

[4] See Chapter 14.

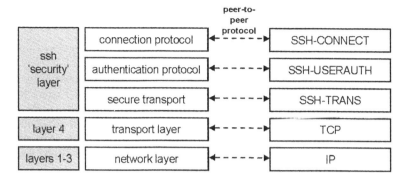

Figure 10.12 SSH protocol architecture (SSH-ARCH).

SSH considers the network to be a hostile and evil environment — not to be trusted — and protects data to be carried by the network accordingly. Even though a hostile third party who manages to wrest control of the network can force SSH connections to disconnect, they will not be able to decipher or play back messages, and will not be able to take over (i.e., hijack) connections. SSH achieves this by using encryption. The basic principles of SSH are similar to those of IPsec.[5] In essence, SSH adds a new 'data security (authentication and encryption) layer' between the transport layer (layer 4) and the application data and protocols (see Figure 10.12).

SSH protocol architecture

SSH has a three-layer protocol architecture comprising the following components:

- the *SSH transport layer protocol* (*SSH-TRANS*) provides for a secure, low-level transport connection — a confidential connection (called a *session*). It provides for encryption, message authentication and compression too (if required). It is usual for the SSH transport layer protocol to be run over a TCP (transmission control protocol)[6] connection [port 22], but any other reliable data stream protocol may be used instead;

- the *SSH user authentication protocol* (*SSH-USERAUTH*) runs over the transport layer protocol and is used by the client to authenticate itself to the server. If authorisation is successful, then an SSH *tunnel* results;

- the *SSH connection protocol* (*SSH-CONNECT*) provides for the multiplexing of several logical *channels* into the single tunnel provided by the transport and authentication protocols. It runs over the user authentication protocol.

The layered protocol architecture of SSH (called SSH-ARCH and illustrated in Figure 10.12) is intended to allow for easy adaption of the protocol suite and incorporation of other protocols.

SSH transport layer protocol

Once a TCP/IP connection has been established on port 22 for the SSH protocol, both the SSH client and SSH server will begin to listen on this port. The first action is for the two

---

[5] See Chapter 13.
[6] See Chapter 7.

ends to check their SSH version compatibility. This is done by each end sending a plain text message as a single TCP segment/IP packet in the format:

```
SSH-protoversion-softwareversion-comments followed by <CR-NL>
```

Note how the SSL specification documents refer to ⟨CR-NL⟩ (carriage return, new line) rather than to ⟨CRLF⟩ (carriage return line feed). Nonetheless the same ASCII characters (13 and 10) are used.

Once compatibility has been established (the current version of SSH incidentally is 2.0), SSH packets are composed in the SSH transport protocol *binary packet protocol* format as shown in Figure 10.13. The individual fields have the following meanings and uses:

- *packet_length* — this is the length of the SSH packet (in number of *bytes* or *octets*), but not including MAC field or the packet_length field itself.

- *padding_length* — this is the number of bytes (octets) of padding which has been added to the *payload* field to make the total length of the SSH packet (minus MAC field) equal to a multiple of 8 bytes or a multiple of the encryption code *cipher block* size. There must be at least 4, but no more than 255 bytes.

- *payload* — the *payload* is the useful contents of the packet. Normally, this will contain both *compressed* and *encrypted* information. The use of encryption on the payload is what makes SSH a secure protocol. But, at the start, while the SSH transport layer protocol is still establishing the secure connection — a process which involves exchange of security encryption keys and negotiation of compression algorithms, plain text (unencrypted and uncompressed text) may be used in this field. The plain text takes the format of an SSH message type (a one byte field, coded with the message numbers according to Table 10.7 and followed by relevant message parameters). The character set used to code such *plain text* messages must be explicitly specified. In most places, ISO 10646 with UTF-8 encoding is used (RFC-2279). When applicable, a field is also provided for the language tag (RFC-1766) — to indicate the human language in which text messages appear.

- *random_padding* — this is a random pattern of data used to fill out the payload to the standard length required by the encryption algorithm.

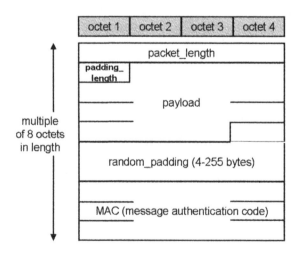

Figure 10.13   SSH transport protocol binary packet protocol format.

Table 10.7 SSH message numbers

Used by SSH protocol:	Message name	Message number	Message parameters
SSH transport protocol	SSH_MSG_DISCONNECT	1	Reason code
			Description (RFC 2279)
			Language tag (RFC 1766)
			Reasons:
			1 host not allowed to connect
			2 protocol error
			3 key exchange failed
			4 reserved
			5 MAC (message authentication) error
			6 compression error
			7 service not available
			8 protocol version not supported
			9 host key not verifiable
			10 connection lost
			11 disconnect by application
			12 too many connections
			13 authorisation cancelled by user
			14 no more auth. methods available
			15 illegal user name
	SSH_MSG_IGNORE	2	Message-specific data
	SSH_MSG_UNIMPLEMENTED	3	Packet sequence number of reject message
	SSH_MSG_DEBUG	4	Always display (Boolean value)
			Message (RFC 2279)
			Language tag (RFC 1766)
	SSH_MSG_SERVICE_REQUEST	5	Service name
			Service names:
			Ssh-userauth
			Ssh-connection

(continued overleaf)

Table 10.7 (*continued*)

Used by SSH protocol:	Message name	Message number	Message parameters
	SSH_MSG_SERVICE_ACCEPT	6	Service name
	SSH_MSG_KEXINIT	20	Service names: Ssh-userauth Ssh-connection (key exchange using following parameters:) cookie kex algorithms server_host_key_algorithms encryption_algorithms_client_to_server encryption_algorithms_server_to_client mac_algorithms_client_to_server mac_algorithms_server_to_client compression_algorithm_client_to_server compression_algorithm_server_to_client languages_client_to_server languages_server_to_client first kex packet follows (Boolean value)
	SSH_MSG_NEWKEYS	21	(confirmation of end of key exchange — no parameters)
	SSH_MSG_KEXDH_INIT	30	Value e of encryption algorithm
	SSH_MSG_KEXDH_REPLY	31	Value f of encryption algorithm
	Reserved for other kex (key exchange) packets	32–49	(message specific)
SSH authentication	SSH_MSG_USERAUTH_REQUEST	50	Username Service name Method name
(general authentication)	SSH_MSG_USERAUTH_	51	Authentications_that_can_continue Partial success (Boolean value)

FAILURE SSH_MSG_USERAUTH_SUCCESS	52	(no parameters)
SSH_MSG_USERAUTH_BANNER	53	Text message Language tag
(method specific) These messages are only sent by the server (client sends only SSH_MSG_USERAUTH_REQUEST messages). Different authentication methods reuse the same message numbers.	60–79	(message specific)
SSH connection protocol SSH_MSG_GLOBAL_REQUEST	80	Request name Want reply (Boolean value) Request-specific data
SSH_MSG_REQUEST_SUCCESS	81	Response-specific data (usually none)
SSH_MSG_REQUEST_FAILURE	82	(no parameters)
Reserved for other generic connection protocol messages	83–89	—
SSH_MSG_CHANNEL_OPEN	90	Channel type (e.g., 'session') Sender channel number Initial window size Maximum packet size
SSH_MSG_CHANNEL_OPEN_CONFIRMATION	91	Recipient channel number Sender channel number Initial window size Maximum packet size
SSH_MSG_CHANNEL_OPEN_FAILURE	92	Recipient channel number Reason code Additional text Language code

(continued overleaf)

Table 10.7 *(continued)*

Used by SSH protocol:	Message name	Message number	Message parameters
		94	Reason codes: 1 administratively prohibited 2 connect failed 3 unknown channel type 4 resource shortage Recipient channel number Data
	SSH_MSG_CHANNEL_DATA	95	Recipient channel number Data type code Data
	SSH_MSG_CHANNEL_EXTENDED_DATA	96	Recipient channel number
	SSH_MSG_CHANNEL_EOF	97	Recipient channel number
	SSH_MSG_CHANNEL_CLOSE	98	Recipient channel number Channel-type required Want reply (Boolean value) Channel-type-specific data
	SSH_MSG_CHANNEL_REQUEST	99	Recipient channel number
	SSH_MSG_CHANNEL_SUCCESS	100	Recipient channel number
	SSH_MSG_CHANNEL_FAILURE	101–127	—
	Reserved for other connection protocol channel-related messages	128–191	—
	Reserved for client protocols	192–255	—
	Reserved for local extensions		

- *MAC (message authentication code)* — this field contains the *message authentication code* bytes (of a length appropriate to the message authentication algorithm). The *MAC* is also called a *fingerprint* or *digital signature*. It serves the same function as a hand-written signature at the bottom of a letter — it confirms the authenticity of the message and the sender. Initially the MAC value will be set as *none*.

Once the TCP connection has been set up and once both SSH client and server have agreed to use the same SSH version, the SSH transport protocol can start its work in earnest. The first step is to undertake *key exchange* to establish the *encryption, MAC (message authentication code)* and compression algorithms which will be used to code the packet payload contents. The SSH client starts the process by sending an SSH_MSG_KEXINIT to *initiate* the *key exchange (kex)*. The message informs the SSH server of all the encryption, MAC and compression algorithms which it supports. The server replies with a similar SSH_MSG_KEXINIT message.

The encryption, MAC algorithm and compression initialisation process takes place by exchange of SSH_MSG_KEXDH_INIT and SSH_MSG_KEXDH_REPLY messages (see Figure 10.14). The client selects its chosen algorithms, calculates certain encryption code values (we shall discuss encryption keys and MAC digital signatures in more detail in Chapter 13) and communicates relevant *initialisation vector (IV)*[7] values for the encryption algorithm chosen. Since the encryption in the reverse direction (server-to-client) may be different, the server also chooses encryption, compression and MAC algorithms and similarly informs the client of initialisation vector values. (The algorithms used in the two directions are recommended to be run independently of one another if not by different algorithms — this gives greater security.)

The different encryption algorithms (ciphers) which may be used in association with SSH are listed in Table 10.8. The different types of *keys* which may be used in association with these algorithms may conform with any of the standards listed in Table 10.9.

The *message authentication code (MAC, integrity check value, fingerprint* or *digital signature)* associated with the message may conform to any of the standards listed in Table 10.10.

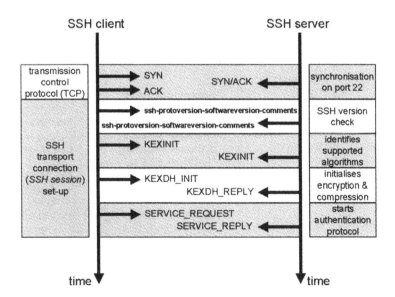

Figure 10.14 SSH transport protocol: keyexchange and establishment of a secure transport connection.

---

[7] See Chapter 13.

Table 10.8   SSH Transport layer protocol: alternative encryption algorithms and ciphers

Cipher name	Use in SSH protocol	Cipher code name	Specified by:
3des-cbc	REQUIRED	Triple-DES (Defense encryption standard — 3DES) [CBC — cipher block chaining — mode]	RFC 1851
aes128-cbc	RECOMMENDED	AES (advanced encryption standard) [CBC mode; 128-bit key]	FIPS-197 (NIST Federal Information Processing Standard)
aes192-cbc	OPTIONAL	AES (advanced encryption standard) [CBC mode; 192-bit key]	FIPS-197
aes256-cbc	OPTIONAL	AES (advanced encryption standard or *Rijndael code*) [CBC mode; 256-bit key]	FIPS-197
arcfour	OPTIONAL	ARCFOUR stream cipher	(Internet draft, December 1999)
blowfish-cbc	RECOMMENDED	Blowfish (CBC mode)	Bruce Schneier: *Fast Software Encryption* [Cambridge Security Workshop Proceedings, Dec 1993: Springer-Verlag 1994]
cast128-cbc	OPTIONAL	CAST-128 (CBC mode)	RFC 2144
idea-cbc	OPTIONAL	IDEA (International Data Encryptions Algorithm CBC mode) [ASCOM AG company-patented encryption mechanism]	RFC 3058
none	OPTIONAL (NOT RECOMMENDED)	No encryption	—
serpent128-cbc	OPTIONAL	Serpent (CBC mode: 128-bit key)	www.cl.cam.ac.uk/ ~rja14/serpent.html
serpent192-cbc	OPTIONAL	Serpent (CBC mode: 192-bit key)	(see above)
serpent256-cbc	OPTIONAL	Serpent (CBC mode: 256-bit key)	(see above)
twofish-cbc or twofish256-cbc	OPTIONAL	Twofish (CBC mode; 256-bit key)	*The Twofish Encryption Algorithm* (Schneier, Kelsey, Whiting, Wagner, Hall, Ferguson) John Wiley & Sons
twofish128-cbc	RECOMMENDED	Twofish (CBC mode; 128-bit key)	(see above)
twofish192-cbc	OPTIONAL	Twofish (CBC mode; 192-bit key)	(see above)

Note: CBC = cipher block chaining mode. An excellent reference work on encryption is *Applied Cryptography* (Bruce Schneier), published by John Wiley & Sons.

Table 10.9 SSH public key and certificate formats

Public key name	Use in SSH protocol	Certificate and key name	Specified by:
ssh-dss	REQUIRED	Simple DSS	www.datasecuritysolutions.com
ssh-rsa	RECOMMENDED	Simple RSA	www.rsasecurity.com
x509v3-sign-rsa	OPTIONAL	X.509 certificates (RSA key)	RFC 2459
x509v3-sign-dss	OPTIONAL	X.509 certificates (DSS key)	RFC 2459
spki-sign-rsa	OPTIONAL	SPKI certificates (RSA key)	RFC 2692-3
spki-sign-dss	OPTIONAL	SPKI certificates (DSS key)	RFC 2692-3
pgp-sign-rsa	OPTIONAL	OpenPGP certificates (RSA key)	RFC 1991
pgp-sign-dss	OPTIONAL	OpenPGP certificates (DSS key)	RFC 1991

Table 10.10 SSH message authentication code (MAC) algorithms

MAC name	Use in SSH protocol	MAC (message authentication code) algorithm name	Specified by:
Hmac-md5	OPTIONAL	HMAC-MD5 (digest length = key length = 16)	RFC 2104, RFC 1321
Hmac-md5-96	OPTIONAL	HMAC-MD5 truncated to 96 bits (digest length = 12, key length = 16)	RFC 2104, RFC 1321
Hmac-sha1	REQUIRED	HMAC-SHA1 (digest length = key length = 20)	RFC 2104, RFC 3174
Hmac-sha1-96	RECOMMENDED	HMAC-SHA1 truncated to 96 bits (digest length = 12, key length = 20)	RFC 2104, RFC 3174
None	OPTIONAL (NOT RECOMMENDED)	No MAC	—

As we shall discuss in more detail in Chapter 13, the *message authentication code* is a value calculated using the message itself, an encryption-like algorithm (called a *message digest algorithm*) and a shared secret code (the key). The resulting code (MAC) is unique and it is almost impossible for a third-party to work out how it was derived. Only a recipient who knows the 'unlocking' key is able to tell whether the received *integrity check value* (i.e., MAC value) corresponds to the message. Thus if either the message or the MAC are tampered with during transport through the network, the recipient will be able to detect the tampering. If tampering has occurred, the SSH connection will be disconnected and re-established.

The *compression* algorithms supported by the SSH transport layer protocol are either *none* or *ZLIB* (*LZ77 algorithmn*—as defined in RFCs 1950 and 1951).

Once a secure SSH transport layer connection (an *SSH session*) exists between SSH client and SSH server, the authentication process can commence. The SSH client requests the authentication service *(ssh-userauth)* by means of a SSH_MSG_SERVICE_REQUEST message

(Figure 10.14). The SSH server responds with the SSH_MSG_SERVICE_REPLY if successful. Alternatively, the server may reply with a failure or rejection message, or simply disconnect.

SSH authentication protocol

The SSH authentication protocol runs on top of the SSH transport layer protocol and provides a single authenticated tunnel for the SSH connection protocol. The service name for the authentication protocol is ssh-userauth. *Authentication* is the means by which the SSH server confirms that the identity of the SSH client is genuine — and is authorised to use the relevant service.

There are three alternative methods for user authentication:

- public key;
- password; or
- host-based.

The most secure method is the public key method. When using the public key method of authentication, the client sends a message and generates an appropriate *message authentication code (MAC)* for that message using its *private key* for encryption.[8] The server uses the client's *public key*[9] to decode and verify the correctness of the MAC received. If correct, this serves to confirm the client's authenticity.

Should the client not have a pair of public and private keys,[10] then a simple *password* authentication and authorisation method may be used. But while the password method is not as secure as the public key method, at least the encrypted SSH transport layer connection sees to it that the password is not carried across the network in 'plain text'. This makes interception of the password much harder.

The authentication process is conducted by the client sending an SSH_MSG_USERAUTH_REQUEST message (Figure 10.15 and Table 10.7). The message identifies the service which the client subsequently wishes to access and contains sufficient information to identify and authenticate the client to the SSH server. The SSH server will respond either with an SSH_MSG_USERAUTH_SUCCESS message or an SSH_MSG_USERAUTH_FAILURE message. A success message allows the client to generate an SSH_MSG_SERVICE_REQUEST message for the SSH connection service (*ssh-connection*), while a failure message advises the client of the authentication methods available to the client. (This may be bogus information if desired.)

Authentications that can continue (the main parameter of the message SSH_MSG_USERAUTH_FAILURE [message number 51]) is a comma-separated list of authentication method names that may be used to continue attempting the authentication process. This is 'helpful information' sent by the server to assist the client in authenticating itself.

Only once the server is satisfied about the authenticity of the client's identity, will the client be allowed to progress to the connection service. Should the client take too long during the authentication process (typically a timeout is set at 10 minutes), then the SSH server will *disconnect* the SSH transport layer connection. The server also limits the number of failed authentication attempts (recommended: 20) which a client may perform in a single session.

There is also an option for host-based authentication (i.e., authentication of the server by the client) but this is not normally recommended.

---

[8] See Chapter 13.
[9] See Chapter 13.
[10] See Chapter 13.

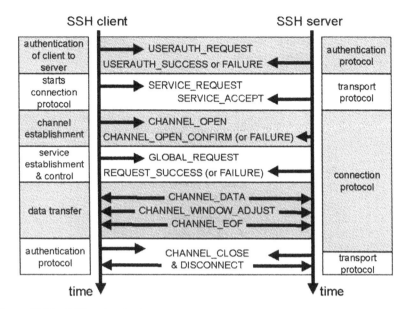

Figure 10.15   SSH authentication (ssh-userauth) and connection (ssh-connection) services.

If the authentication is successful, then the service name in the SSH_MSG_USERAUTH message specifies the service which will now be started by the SSH server and an SSH *tunnel* is said to exist between the SSH client and the relevant service on the SSH server. Alternatively, the client may send an SSH transport layer protocol message — SSH_MSG_SERVICE_REQUEST to request the SSH connection service (ssh- connection). If the requested service is not available, the server may disconnect immediately or at any later time. If the requested user does not exist, the server may disconnect or send a bogus list of acceptable authentication methods. This makes it possible for the server to avoid disclosing information useful to an intruder.

SSH connection protocol

Once the SSH transport layer protocol has established an SSH session and the authentication protocol has converted this into an SSH tunnel, then the SSH connection protocol is activated to provide the *ssh-connection* service, which in turn may be used to set up SSII *channels* (Figure 10.15).

The connection protocol, which is designed to run on top of both the SSH transport layer and authentication protocols, provides channels which can be used for a wide range of purposes. Standard methods provide for the setting up of secure *pseudo-terminal (pty)* session or interactive shell sessions or for the secure forwarding (i.e., *tunnelling*) of an arbitrary TCP/IP port or an X-Windows (e.g., X11) connection. These allow for interactive login sessions of clients to remote servers, remote execution of commands, forwarded TCP/IP connections, and forwarded X11 connections. A number of channels (as required by the SSH client and server) can be multiplexed into a single encrypted SSH tunnel.

The SSH *channel types* which may be *opened* are listed in Table 10.11. Either the SSH client or the SSH server may open a channel and multiple channels may be *multiplexed* into a single *tunnel* connection. The different channels are identified by numbers at each end and the numbers used for the same channel will usually differ between the sender and receiver end of the tunnel (this adds to security).

Table 10.11 SSH channel types and related channel requests

Channel type	Relevant channel requests	Purpose
direct-tcpip		Channel opened for transfer of a given TCP port connection
Forwarded-tcpip		Channel opened to forward a given TCP port to the other side of the SSH connection (following request for the relevant TCP port forwarding using the relevant SSH_MSG_GLOBAL_REQUEST message
Session	Env	Setting environment variables
	Exec	Executing a given command
	exit-signal	Causes a remote command or process to terminate violently. Possible signals are
		• ABRT abort
		• ALRM alarm
		• FPE floating point exception
		• HUP hangup
		• ILL illegal instruction
		• INT interrupt
		• KILL kill process
		• PIPE write to pipe if no process to read
		• QUIT quit process
		• SEGV segmentation violation
		• TERM software terminal signal
		• USR1 user-defined signal 1
		• USR2 user-defined signal 2
	exit-status	Returns the exit status of a given requested command. After this message, the channel is closed.
	pty-req	Pseudo-terminal mode required
	Shell	Request to start a shell program
	Signal	A means of sending a signal to a remote process or service
	Subsystem	A means of starting a remote subsystem program
	window-change	Notifies remote end of a change in the space available for the window
	xon-xoff	A request to use XON/XOFF data flow control.
x11	x11-req	A request for the x11 version of the X-Windows protocol

A request to open a channel (using the SSH_MSG_CHANNEL_OPEN message — see Figure 10.15 and Table 10.7) includes the channel number the sender will use. The SSH_MSG_CHANNEL_OPEN_CONFIRM message sent as a reply indicates the channel number which will be used by the receiver. Any subsequent messages from sender to receiver regarding the channel will refer to the recipient's channel number.

The data carried by an SSH channel is subject to flow-control using a windowing mechanism similar to that of TCP (transmission control protocol). The *initial window (IW)* size (defined in the SSH_MSG_CHANNEL_OPEN message) specifies how many bytes of channel data may be sent to the sender of the message before the channel window must be *adjusted* (the data is not *acknowledged* as in TCP with an ACK message — instead the window size must be adjusted). The window size is frequently adjusted (as appropriate) by means of the SSH_MSG_CHANNEL_WINDOW_ADJUST message. No data may be sent across a channel until a message is received to indicate that the appropriate window space is available.

The maximum packet size indicated in the SSH_MSG_CHANNEL_OPEN message specifies the maximum size of an individual SSH data packet which may be sent to the sender of the

OPEN message (smaller packets may be chosen for interactive connections to give a faster response on slow links).

Following the SSH_MSG_CHANNEL_OPEN request, the remote SSH device decides whether it can open the channel, and responds appropriately — either with an SSH_MSG_CHANNEL_OPEN_CONFIRMATION or a failure message (SSH_MSG_CHAN-NEL_OPEN_FAILURE). The failure message is usually accompanied by a short text message (and relevant language tag) to provide the human user with a readable message explaining the situation. If the recipient does not support the requested channel type, it may (if desired and appropriate) initiate a disconnection of the SSH transport layer connection (as a security precaution).

Once the SSH channel has been opened, relevant SSH_MSG_CHANNEL_REQUEST messages may be generated (appropriate to the channel type) to configure the channel or request given service options or subsets (Table 10.11).

In the case that the SSH tunnel is to be used for TCP forwarding of a particular TCP port connection (should the connection be required), this can be initiated by means of an SSH_MSG_GLOBAL_REQUEST for the service: *tcpip-forward*. The port number and other TCP connection details (window size, etc.) must be provided with the request. The request may subsequently be cancelled by means of another GLOBAL_REQUEST — this time for *cancel-tcpip-forward*.

Data transfer

Data is transferred in conjunction with the SSH_MSG_CHANNEL_DATA message label. The window size dictates how many bytes may be sent before the sender must wait for the window to be adjusted (by means of the SSH_MSG_CHANNEL_WINDOW_ADJUST message). Having received the adjustment message, the recipient may send the specified additional number bytes window size (over and above the previously authorised limit).

When there is no more data to be transferred, the SSH_MSG_CHANNEL_EOF message is sent. Finally an SSH_MSG_CHANNEL_CLOSE message is sent in both directions to close the channel, and if appropriate the SSH connection should be *disconnected.*

Interactive sessions

Interactive sessions are used for the remote execution of a program or for remote access to a server. The program may be a shell, an application, a system command, an X-windows session or some type of built-in software subsystem. Alternatively, SSH can be used like a 'secure alternative to telnet (RFC854) or UNIX rlogin (RFC 1282)'. In this case, the client *opens* a *session* channel-type and *requests* a *pseudo-terminal(pty-req = pseudo-terminal request).* The request has the following syntax:

```
Byte                    SSH_MSG_CHANNEL_REQUEST
32-bit value            recipient_channel
String                  "pty-req"
Boolean value           want_reply
String                  TERM environment variable value (e.g., vt100)
32-bit value            terminal width, characters (e.g., 80)
32-bit value            terminal height, rows (e.g., 24)
32-bit value            terminal width, pixels (e.g., 640)
32-bit value            terminal height, pixels (e.g., 480)
String                  encoded terminal modes
```

In *terminal-mode*, data is transferred as a stream of bytes (i.e., characters) and consists of pairs of opcodes and their respective *arguments* (so-called *opcode-argument pairs*). The stream is terminated by the opcode TTY_OP_END (value 0). The standard *POSIX (portable operating system interface)*-like tty interface which is used is reproduced in Table 10.12.

Table 10.12 SSH terminal mode opcodes (mostly the same as the equivalent POSIX terminal mode flags)

Opcode	Code name	Purpose
0	TTY_OP_END	Indicates end of options.
1	VINTR	Interrupt character; 255 if none.
2	VQUIT	The quit character (sends SIGQUIT signal on POSIX systems).
3	VERASE	Erase the character to left of the cursor.
4	VKILL	Kill the current input line.
5	VEOF	End-of-file character (sends EOF from the terminal).
6	VEOL	End-of-line character in addition to ⟨carriage return line feed⟩.
7	VEOL2	Additional end-of-line character.
8	VSTART	Continues paused output (normally control-Q).
9	VSTOP	Pauses output (normally control-S).
10	VSUSP	Suspends the current program.
11	VDSUSP	Another suspend character.
12	VREPRINT	Reprints the current input line.
13	VWERASE	Erases a word left of cursor.
14	VLNEXT	Enter the next character typed, even if it is a special character.
15	VFLUSH	Character to flush output.
16	VSWTCH	Switch to a different shell layer.
17	VSTATUS	Prints system status line (load, command, pid etc.).
18	VDISCARD	Toggles the flushing of terminal output.
30	IGNPAR	The ignore parity flag. The parameter is 0 (FALSE) or 1 (TRUE).
31	PARMRK	Mark parity and framing errors.
32	INPCK	Enable checking of parity errors.
33	ISTRIP	Strip 8th bit off characters.
34	INLCR	Map NL (new line) into CR (carriage return) on input.
35	IGNCR	Ignore CR (carriage return) on input.
36	ICRNL	Map CR (carriage return) to NL (new line) on input.
37	IUCLC	Translate uppercase characters to lowercase.
38	IXON	Enable output flow control.
39	IXANY	Any character will cause a restart after stop.
40	IXOFF	Enable input flow control.
41	IMAXBEL	Ring bell on input queue full.
50	ISIG	Enable signals INTR, QUIT, [D]SUSP.
51	ICANON	Canonicalize input lines (i.e., convert to least significant bit first).
52	XCASE	Enable input and output of uppercase characters by preceding their lowercase equivalents with '\'.
53	ECHO	Enable echoing.
54	ECHOE	Visually erase characters.
55	ECHOK	Kill character discards current line.
56	ECHONL	Echo NL (new line) even if ECHO is off.
57	NOFLSH	Do not flush after interrupt.
58	TOSTOP	Stop background jobs from output.
59	IEXTEN	Enable extensions.
60	ECHOCTL	Echo control characters as ⟨Char).
61	ECHOKE	Visual erase for line kill.
62	PENDIN	Retype pending input.
70	OPOST	Enable output processing.

Table 10.12   (*continued*)

Opcode	Code name	Purpose
71	OLCUC	Convert lowercase to uppercase.
72	ONLCR	Map NL (new line) to CR-NL (carriage return new line).
73	OCRNL	Translate carriage return to newline (output).
74	ONOCR	Translate newline to carriage return-newline (output).
75	ONLRET	New line (NL) performs a carriage return (output).
90	CS7	7-bit character set mode.
91	CS8	8-bit character set mode.
92	PARENB	Parity enable.
93	PARODD	Odd parity (otherwise even parity).
128	TTY_OP_ISPEED	Specifies the terminal input baud rate in bits per second.
129	TTY_OP_OSPEED	Specifies the terminal output baud rate in bits per second.

Global requests

Several kinds of requests affect the state of the remote end of the SSH connection globally — i.e., independently of specific channels. An example is a request to start TCP/IP forwarding for a specific TCP port number. Such *global requests* use the SSH_MSG_GLOBAL_REQUEST message format.

Using the SSH (secure shell) protocol

The SSH protocol was designed to be simple and flexible protocol for providing a secure means of remote computer operation. It allows parameter negotiation, but still minimises the number of request-reply 'round trips' required before 'real data transfer' can commence. Nonetheless, the 'price' of security is heavier traffic loading for the network and longer response times for the human user and computer application. The addition of the encryption process quite simply adds to the volume of data (i.e., the overhead) which must be carried. A typical remote shell or telnet command is increased from 33 bytes (using plain text TCP/IP) to 51 bytes (using SSH). Where the communication is carried across a slow speed line (e.g., one using PPP — point-to-point protocol[11]), this could cause up to 54% reduction in speed! On the other hand, in an ethernet environment (where the minimum frame size is anyway 46 bytes), the degradation in performance is only around 10%.

And one final thought — how easy is it to 'break' the *secure shell*? Well, encryption is generally based on the assumption that messages are long and complicated. The encrypted form of the message appears to be 'gibberish' and cannot be distinguished from other encrypted messages. Encrypted messages are therefore difficult to decode. On the other hand, when messages are relatively simple and short, and where there is a relatively restricted set of different messages in use (as maybe is the case with simple terminal operations), cracking an encryption code is somewhat easier. Just because a message is encrypted doesn't mean it is safe!

Imagine you know that there are only two messages ever sent — 'YES' and 'NO' — and that you know that 'YES' is more commonly used than 'NO'. Listening to the relevant conversation you may note that only two different messages are ever sent 'GIBBERISH1' and 'GIBBERISH2' and that 'GIBBERISH2' is most common. Any guesses with which message 'GIBBERISH2' is related? ('YES' of course. Now you can understand the conversation — and

---

[11] See Chapter 8.

even join in if you want to!). This is the main reason why *random padding* is added to short messages before encryption is applied. It is important that the padding is *random* if the encrypted messages are not to be duplicated!

10.6   RTP/RTPC: real time signal carriage over IP networks

In recent years it has become popular to use data networks, and IP (Internet protocol) data networks in particular for the transport of *multimedia* — including not only 'classical' data files but also real-time video and voice signals. Thus live videoconferencing or telephony can be carried by the Internet. There is excitement about the potential offered by voice-over-IP (VOIP).

For the end-to-end delivery of data with real-time characteristics (Figure 10.16), the *real-time application transport protocol (RTP)* was developed. It first appeared in 1996 and is defined in RFC 1889.

Figure 10.16 illustrates two particular real-time applications which might run on an IP network. The first is a *video-streaming* application in which a PC-user is viewing a video or newsclip being downloaded from a video server. This might be a real 'live' application (such as a *webcam* — Worldwide Web camera) or it might involve viewing a film previously stored on the server. Alternatively, if the server is also a *multipoint control unit (MCU)*, maybe the video user is taking part in a videoconference! The second application allows the PC user to make a 'telephone call' via his PC and the Internet. The telephone call is carried as *VOIP (voice-over-IP)* across the IP-network (or the Internet), via a *gatekeeper* to the public telephone network.

A whole set of new protocols and signal encoding techniques have started to emerge for the carriage of real-time signals, the most important of which are those in ITU-T's H.323 protocol suite. H.323 establishes a framework for carriage of video, voice (VOIP) and conferencing (audio and videoconferencing) via non-guaranteed links (such as that offered by the *user datagram (UDP)* and IP-based networks). The framework (Figure 10.17) includes call *signalling*

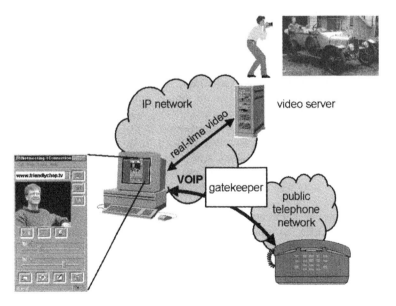

Figure 10.16   Real-time communications applications running via a data network.

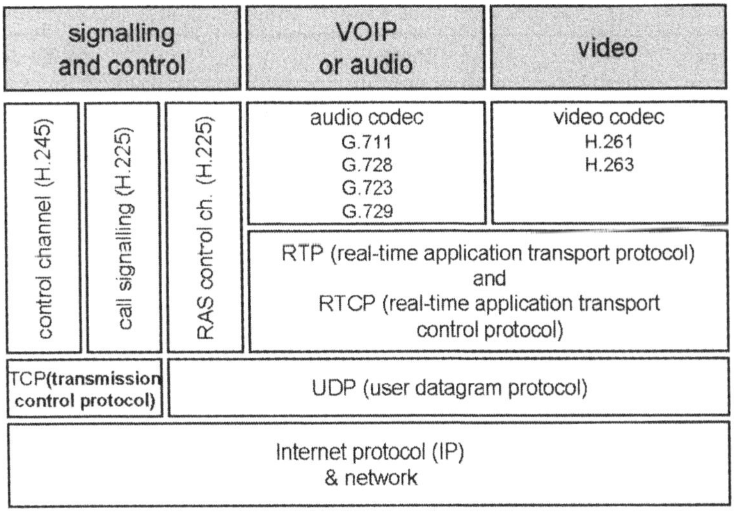

signalling and control			VOIP or audio	video
control channel (H.245)	call signalling (H.225)	RAS control ch. (H.225)	audio codec G.711 G.728 G.723 G.729	video codec H.261 H.263
			RTP (real-time application transport protocol) and RTCP (real-time application transport control protocol)	
TCP(transmission control protocol)		UDP (user datagram protocol)		
Internet protocol (IP) & network				

Figure 10.17   ITU-T recommendation H.323: protocols for carriage of real-time voice and video via IP-networks.

and *control channel* mechanisms for establishing *connections* and efficient coding techniques for low bit rate audio and video signals. RTP (real-time application protocol) provides the *transport protocol* for the real-time data stream.

RTP (real-time application transport protocol) and RTCP: profiles and payload types

The *real-time application transport protocol (RTP)* and the *associated real-time application transport control protocol (RTCP)* together create a *transport layer* for carriage of real-time signals across an IP-based network. The two protocols are both defined in RFC 1889. RTP is used for the carriage of the actual user data (equivalent to the *data connection* of FTP). RTP relies on the lower layer protocols, including the user datagram protocol (UDP) and IPv4 *DiffServ* or IPv6 for timely delivery of packets, *QOS (quality of service)* guarantees and the like. RTP makes use of the UDP multiplexing and checksum functions. RTCP, meanwhile, continuously monitors the actual QOS (quality of service) of the received signal and can be used to modify the transmission (for example, by reducing the bit rate used for a video signal).

Various different payload formats (*profiles*) may be carried by RTP. Each individual payload format is separately defined in individual RFCs or ITU-T recommendations (Table 10.13). We shall not cover these in detail here, but instead shall cover the basic protocol format and protocol operations of RTP/RTCP.

RTP (real-time application transport protocol) and RTCP: packet formats and protocol operations

RTP (real-time application transport protocol) packets comprise the *fixed RTP header* (Figure 10.18) and the *RTP payload*. RTCP (real-time application transport control protocol) packets comprise a fixed header part (Figure 10.18) followed by other *structured elements* dependent upon RTCP packet type.

Table 10.13   RTP (real-time application transport protocol): profiles and payload types

Payload type	Profile	Profile defined by
	RTP profile for audio and video conferences with minimal control	RFC 1890
0	Audio digital telephony (North American μ-law PCM coding)	ITU-T Rec. G.711
2	Digital audio	ITU-T Rec. G.721
3	Mobile telephony (GSM — global system for mobilecommunication)	GSM 6.10 audio
4	Low bit rate telephone speech encoding	ITU-T Rec. G.723.1
8	Audio digital telephony (European A-law PCM encoding)	ITU-T Rec. G.711
9	16 kHz audio	ITU-T Rec. G.722
12	Low speed digital telephony (CELP — code excited linear prediction)	ITU-T Rec. G.723
15	Low delay CELP audio — code excited linear prediction at 16 kbit/s	ITU-T Rec. G.728
18	CS-CELP audio (conjugate structure algebraic CELP) at 8 kbit/s	ITU-T Rec. G.729
26	JPEG (joint photographic experts group) video	RFC 2435
31	Video (signal encoding according to ITU-T Rec. H.261)	RFC 2032 /H.261
34	Video (signal encoding according to ITU-T Rec. H.263)	RFC 2190 /H.263
	MPEG1/MPEG2 video (motion picture experts group)	RFC 2250

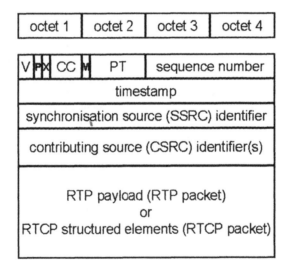

Figure 10.18   RTP and RTCP: fixed header field format.

The *synchronisation source (SSRC)* is the source of the stream of RTP packets. Packets must be *synchronised* for playback. This is done by means of the *timestamp*, which is usually coded using *wallclock* time (i.e., absolute time — represented by means of the timestamp format of the *network time protocol (NTP)* (a value representing the number of expired seconds since 1 January 1900).

A *contributing source (CSRC)* is a second (i.e., not the primary source) of a real-time signal. Contributing sources take part in conferencing sessions between multiple users. In this case, the synchronisation source is the main conference controller unit. Contributing sources are added by means of an *RTP mixer*.

An *RTP monitor* receives RTCP packets and is used to inform the source of the real-time data stream of the received signal quality. The RTCP packets are in the form of *reports* which indicate the received packet count, the number of lost packets, the signal *jitter, delay* and other quality measurement parameter values. Should the received signal *quality of service (QOS)* be too low, the signal source may choose to adapt the transmission (for example, by reducing the bit rate and/or switching to a *profile* with a lower bit rate signal encoding scheme).

RTP/RTCP packet format

The individual protocol fields within the RTP/RTCP fixed header (Figure 10.18) are coded as follows:

- V is a 2-bit *version* number version number. The current version (RFC 1889) is version 2.

- P is the *padding bit* field. If the P bit is set (to value '1') then the packet includes one or more padding octets at end of last 32-bit word of the payload. The padding octets are not part of the payload and should be removed by the RTP receiver before being passed to higher layer protocols.

- X is the *extension bit*. If set, the header is followed by one header extension.

- CC is the *CSRC count (contributing source count)*. This value indicates the number of CSRC identifiers included in the header (and thus the number of additional participants in addition to the synchronisation source (SSRC) which are taking part in a conference session).

- M is the *marker bit*. The M-bit is intended to be used to mark frame boundaries in packet stream and can be used like the restart marker in FTP (file transfer protocol).

- The *payload type (PT)* identifies the RTP profile being used (see Table 10.13).

- The *sequence number* is used to detect packet loss and restore packet sequence. The initial value is set to a randomly chosen number. This helps prevent 'computer hacker' attacks on encryption methodologies.

- The *timestamp* reflects the sampling time instand of the first octet in the payload. This is used to reassemble and synchronise the real-time signal at the receiving end.

Functions of RTCP (real-time application transport control protocol)

RTCP (real-time application transport control protocol) is used to control an RTP session. RTCP is used to establish the *session* and the addition or removal of members to or from a conference. There are five different RTCP packet types:

- SR (sender report);

- RR (receiver report);

- SDES (source description items);

- BYE (end of participation of a conference participant); and

- APP (application specific functions).

An RTCP report includes fields to indicate the total packet count (sent or received), the number of lost packets or measures of received signal quality — such as the jitter, delay, etc.

An SDES packet always contains the CNAME field. This is a *binding* name for an RTP session which is independent of the source network address and is used to unambiguously define the session. The CNAME can take a number of forms to describe the source — for example, a web address name, an email address, a telephone number or a geographical location.

10.7   Applications, protocols and real networks

The ability to send and receive files between remote computers across a data network heralded the era of 'networked' computing. Sharing of data and messaging began to get popular and *electronic data interchange (EDI)* was born.

The first computer operating system to fully embrace the new era of networked computing was UNIX. The servers used in the early TCP/IP networks were nearly always UNIX servers. By means of a TCP/IP data network, users with UNIX workstations could now access different databases, file servers and applications spread across different UNIX servers (Figure 10.19). The potential of computing grew, as did the range of software applications and the demand for them.

The UNIX operating system incorporated the TCP/IP protocol suite, and for application layer protocols such as FTP (file transfer protocol) even incorporated special command sets. Thus the familiar UNIX FTP-commands listed in Table 10.14 were born. Below we show an example of the UNIX prompts and commands that a UNIX computer user may be familiar with when setting up (i.e., *opening*) an FTP connection to a remote server called 'aserver1'. The user logs in under the name 'books 2' using a password and then requests a listing of the server's file directory. Maybe he/she *gets* a file from the server or sends (i.e., *puts*) a file onto it. After the file transfer, the user *quits* the session. The responses from the server are a mixture of UNIX messages and standard FTP replies (with their 3-digit reply numbers as we saw in Table 10.2).

```
$ ftp -d
ftp> open aserver1
Connected to aserver1.
220 aserver1 FTP server (Version 4.4 Tue Dec 20 1988) ready.
```

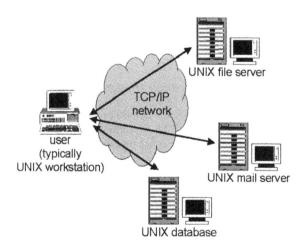

Figure 10.19   Networking UNIX workstations and servers in the early days of TCP/IP networks.

Table 10.14 UNIX operating system: FTP commands

Unix FTP command	Action undertaken
binary	Changes data transfer to the binary (octet) mode
cd	Change directory (on the remote server)
del	Deletes the named file on the server
dir	Lists the server file directory
get	Requests and receives a file from the remote server
hash	Confirms activity by displaying a '#' character for each block of data transferred
lcd	Changes directory at the local (client) end of the connection
open	Opens an FTP session to a named remote server
put	Sends a file from the local client to the server for storage
pwd	Print the working directory of the remote server
quit	Close the FTP session

```
Name (aserver1:books2) :books2
331 Password required by books2.
Password: *****
230 User books2 logged in.
ftp> dir
.
(...etc..etc...)
.
ftp> quit
221 Goodbye
```

The UNIX commands associated with FTP (file transfer protocol) at the user end are included in the ftp-command set. A similar command set for the server is included in ftpd (ftp *daemon*). The prompts are accordingly labelled 'ftp)'. The telnet and SMTP (simple mail transfer protocols) are included as similar command sets in UNIX. The telnet protocol is used by the UNIX telnet command set — which uses the 'telnet)' prompt. SMTP, meanwhile, is used by a range of UNIX programmes, including /usr/lib/sendmail.

We have now reached a level of understanding of data communications which spans the basic physical connections and electrical representation of signals right through to the integration of network protocols and controls in computer applications and operating systems! And since our objective is not to explain the computing operating systems and applications as well, we shall venture no further. There are plenty of books about UNIX!

By the mid-1990s, the growing importance of data networking and the increasing reliance of companies upon electronic data interchange (EDI), along with the growing numbers of personal computers (PCs) led to the adoption of the TCP/IP protocol suite, including the telnet, FTP (file transfer protocol), SMTP (simple mail transfer protocol) and SNMP (simple network management protocol) by nearly all computer operating systems, though nowadays most users will be unfamiliar with the protocol and operating systems commands directly, as they tend to be 'hidden' by a software program with a graphical user interface. Nonetheless even today, PC users are still able to commence a telnet session directly by entering telnet on the command line (e.g., MS-DOS command line) or by calling a program such as 'hyperterminal' (e.g., held by Windows under 'programs/accessories/hyperterminal').

The ability to 'network' computers led to the appearance of much more powerful computer programs, drawing on files and data input from a wide range of sources and almost in 'real time'. Transactions previously undertaken with paper forms transmitted by post or fax and then input to computers by hand became fully automated. Networked computers were suddenly able accurately to track stock and resources through the entire supply and production chain — from receipt of raw materials to customer delivery, confirmation and payment. Individual computer systems which had previously each served separate functions — bookkeeping, personnel records, order-taking and stock control, etc. — could be coordinated to create a uniform consistent business management system. Networked applications were born (Figure 10.19). Peer-to-peer computing (e.g., IBM's AS400-series of minicomputers, workstations, servers and client/server software architectures all appeared. But the biggest explosion in demand for networked computing — the Internet boom itself, was prompted by the development of the *Worldwide Web (www)* and Internet *electronic mail (email)*. Given the importance of the protocols associated with the Worldwide Web (www) and email, we have dedicated a separate chapter to each of them — chapters 11 and 12.

10.8   Other network/application protocols of note

There are many different application protocols, and we do not have the space to cover them all in this book. Many of the 'proprietary' manufacturer-specific networking protocol suites (e.g., *Appletalk* and Novell's *IPX — internetworking packet exchange*) are being supplanted by the standard IP and web-based protocols. But this has not stopped computer and software manufacturers from attempting to develop and adapt their 'proprietary' application layer protocols to run over standard IP networks while simultaneously giving their computer products capabilities which other manufacturers are unable to match. In the long term such efforts are unlikely to succeed, but in the meanwhile, some software applications continue to require the use of 'proprietary' higher layer networking protocols such as those defined by:

- *Appletalk* (the Apple computer company's networking protocol suite), or

- *IPX* (*internetworking packet exchange* — the Novell company's networking protocol suite — part of the Novell *NetWare* product range), or

- *Microsoft networking* (an integral part of the Windows operating system since Windows95 and WindowsNT), or

- NFS (network file system) and NIS (network information service) — networking protocols developed by Sun Microsystems, or

- *SNA* (*systems network architecture* — the IBM company's networking protocol suite).

Microsoft Networking: Windows remote console server, SMB (server message block) and CIFS (common Internet file system)

Windows *remote console server* is specially designed to allow terminal-driven applications on Windows NT machines to be operated from remote terminals in a LAN environment. If you like, it is the 'Microsoft version of telnet'. It allows console-based applications to be executed and controlled on remote Windows NT or Windows 2000 host machines and supports multiply independent users sessions simultaneously.

SMB (server message block) is a *Microsoft networking* protocol (initially designed by Microsoft, IBM and Intel) which allows PC-related machines to share files and printers and related information such as lists of available files and printers. Operating systems which support SMB in their 'native versions' include Windows NT, IBM OS/2 (operating system /2 for PCs), LINUX, Apple Macintosh, web browsers, etc.

The latest and enhanced version of SMB is called the *common Internet file system* (CIFS). This was first published by Microsoft in 1996. SMB or CIFS are typically used for the following purposes:

- integrating user's Microsoft Windows or IBM OS/2-style desktop PCs as *clients* into enterprise computing environments comprising UNIX or other servers;

- integrating Microsoft NT and Windows2000 servers into enterprise networks also comprising UNIX or VMS servers; or for

- replacing 'proprietary LAN operating system' protocols like NFS (network file system), DECNET, Novell Netware, Banyan Vines, etc.

Since interworking with Microsoft-based machines is unavoidable in the modern computing world, add-on packages which perform SMB functions are available for most computer operating systems. (UNIX, DOS etc.).

Alternatives to SMB include 'proprietary' LAN network operating software such as Novell Netware, Sun Microsystems' NFS (network file system), Appletalk, Banyan Vines, DECNET. Each of the different alternatives has its strengths and weaknesses, but none are both public specifications and widely available in desktop machines by default. A further alternative is *samba*.

Sun's protocols for UNIX networking

The networking suites for Sun Microsystems computers include the important *NIS (network information service)* and *NFS (network file system)*. NIS is a method of centralising user configuration files in a distributed computing environment. NFS, meanwhile, is a distributed file system protocol (for file search, *binding*[12] and *locking*) which includes the well-known procedures:

- *RFS (remote file system)*;

- *RPC (remote procedure call)*;

- *XDR (external data representation)*; and

- YP (yellow pages).

SAMBA

Samba is an initiative (www.samba.org) and open source/free software suite that provides seamless file and print services to SMB/CIFS clients. The initiative was started by Andrew Tridgell with the intention to 'open Windows to a wider world'. The samba software has been developed as a cooperative effort is freely available under a general public licence. It is claimed to be 'a complete replacement for Windows NT, Warp, NFS or Netware servers.' It provides a LAN-operating system capable of:

- shared file and printing services to SMB clients (e.g., Windows users);

- NetBIOS nameserver service (as defined by RFCs 1001 and 1002);

- ftp-like SMB client services — enabling PC resources (files, disks and printers) to be accessed from UNIX, or subnetworks using 'proprietary' LAN operating systems such as Novell Netware.

---

[12] See Chapter 13.

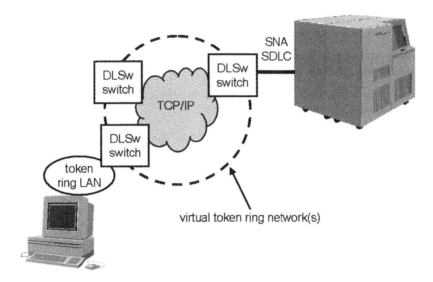

Figure 10.20   Data link switching (DLS or DLSw).

FTAM

FTAM stands for *file transfer, access and management*. The FTAM protocol is roughly equivalent to FTP (file transfer protocol) and is used for accessing remote file servers; managing the file directory; creating, deleting and renaming files; as well as sending and receiving them. It was developed after FTP as a protocol fully compliant with the open systems interconnection (OSI) model and thus intended to be used with the standardardised OSI networking and transport layer protocols. Many users considered the FTAM and other OSI protocols to be rather cumbersome and in consequence FTP remains in widespread use in modern IP-based networks.

Data link switching (DLS or DLSw)

Finally, Figure 10.20 illustrates *data link switching* (*DLS* or *DLSw*). DLS is a method by which IBM computers may be networked using a TCP/IP-based router network while retaining the 'proprietary' IBM network protocols (SNA and token ring) for connection of the IBM computer devices. In effect, DLS is an *encapsulation* protocol. It takes the 'proprietary' SNA (systems network architecture) packets or token ring frames of IBM computer devices, packs these in an 'IP envelope' for carriage across an IP-based data network, before unpacking them at the remote end. DLS makes the TCP/IP network appear to the end computing devices to be a virtual' token ring network — and thus part of a 'proprietary' IBM network architecture. DLS is defined in RFC 1795 and uses TCP ports 2065 and 2067. The use of DLS is only likely to be considered by enterprises with a large installed base of IBM computing equipment. DLS follows an IBM tradition of 'SNA protocol encapsulation' protocols. NPSI (network control point packet switching interface), for example, was the IBM protocol for SNA encapsulation over X.25 networks.

11

The Worldwide Web (www)

The Worldwide Web (www) is a huge 'shared library' of information, stored on many millions of different computers around the world and readily accessible from anywhere else in the world via the Internet. It came about through the initiative of the academic community, and their desire to develop the Internet as a means of 'sharing information'. But while basic 'sharing of information' was possible by means of file transfer across the Internet as early as 1980, the Worldwide Web (www) did not appear until the early 1990s. And not until the Worldwide Web appear did the demand for the Internet explode. So what, exactly, is special about the Worldwide Web? The answer is: a combination of four different technologies which enable easy searching for and browsing of information on remote computers. The four technologies which emerged by 1990 to create the Worldwide Web are: the domain name system (DNS), the hypertext transfer protocol (http), the hypertext markup language (html) and the web browser. The domain name system (DNS) allows the use of 'human-friendly' Worldwide Web addresses in the form www.company.com rather than having to remember long numerical Internet addresses. The hypertext transfer protocol (http) arranges for the rapid transfer of files from remote computers by means of hyperlinks. The hypertext markup language (html), meanwhile, allows these hyperlinks to be written into familiar text-like documents, so that documents, images, calculation routines and other files can be easily linked with one another. The web browser allows the human user to view web 'documents' (actually these documents may be collections of different files from different sources). In this chapter we describe in detail each of the four technologies in turn. Afterwards we also illustrate how the use of web technology has revolutionised the design of modern 'distributed computing' applications.

11.1 The emergence of the Worldwide Web (www)

The *Worldwide Web (www)* emerged in the early 1990s and has rapidly become the world's most powerful resource for sharing information. It has revolutionised business — making possible things that were previously inconceivable — *online shopping, online banking, online television, online auctions*, the ability to 'search' for a given product, supplier or *keyword*, the ability to look up detailed product specifications and handbooks *online* etc.

Over and above the basic data communications transport capabilities of the Internet, four technologies emerged by the early 1990s to create the Worldwide Web. These are:

Data Networks, IP and the Internet: Protocols, Design and Operation   Martin P. Clark
© 2003 John Wiley & Sons, Ltd   ISBN: 0-470-84856-1

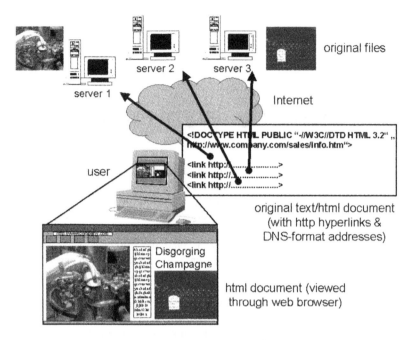

original files

server 2   server 3

server 1

Internet

user

<!DOCTYPE HTML PUBLIC "-//W3C//DTD HTML 3.2" „ http://www.company.com/sales/info.htm">

<link http:/..............>
<link http://.............>
<link http://................>

original text/html document
(with http hyperlinks &
DNS-format addresses)

Disgorging
Champagne

html document (viewed
through web browser)

Figure 11.1 The Worldwide Web (www): a combination of DNS, http, html and web-browser.

- the *domain name system (DNS)* — which allows Worldwide Web (www) addresses to be presented in the 'human-friendly' form www.company.com rather than as the string of numbers and 'dots' which make up an Internet address (e.g., 37.168.153.1);

- the *hypertext transfer protocol (http)* — which allows computer documents and files on different servers to be linked with one another by means of *hyperlinks*;

- the *hypertext markup language (html)* — which allows hyperlinks to be 'written into' document-like computer text files intended to be viewed by humans; and

- the *web browser* — which allows complex html-based web 'documents' to be viewed as *web pages*, even though the individual blocks of text and images which appear on the *web page* have been drawn by means of *hyperlinks* from different computers spread across the world.

Figure 11.1 illustrates crudely how the four different components combine to give the web user's familiar view of a *web page*. The document itself is actually written in a text/html format file. Hyperlinks (using http — hypertext transfer protocol and DNS-format web addresses) provide links to the remote computers where the individual pictures and text files are stored. Only when viewed through the web browser (such as Netscape *Navigator* or Microsoft *Internet Explorer*) does the document appear in the *web page* format which humans are familiar with (which appears in the lower left-hand corner of Figure 11.1).

We discuss each of the four technologies in turn.

11.2 Domain name system (DNS)

The first 'foundation stone' for the *Worldwide Web (www)* was laid in the early 1980s, by the development and specification of the *domain name system (DNS)* (RFC 819: August 1982 and

RFCs 882 and 3: November 1983). The current version is based upon RFCs 1034, 1035 and 1591 (1994).

Prior to the *domain name system*, it was already common under the UNIX operating system to create *name bindings*. In effect, a *binding* is a line in a computer operating system configuration file which relates a computer resource (known to the human computer user by a name) to the network location of that resource (a numerical address which computer programmes use to access that resource). Thus a binding might reveal that the UNIX server known to human users as 'server 1' to be at the Internet address 37.168.153.1. Such binding information is held in a UNIX operating configuration file (specifically that file usually called /etc/hosts). In effect a binding says 'if I use this name, I mean the computer resource found at this network location/number'. Bindings are similar in function to *logical addresses* in networking. Bindings allow software to be written using a 'human-friendly' naming convention (which makes for easy remembering and reduced likelihood of mis-typing). In addition, the computer hardware accessed for a given function can be changed simply by altering the binding in the configuration file (rather than having to change all the different programmes which use the resource).

Since UNIX was the prevalent operating system among the ARPANET community which developed the Internet, it was natural to extend the idea of the UNIX *hosts* table to enable any UNIX server in the ARPANET to be able to locate all the other servers (or hosts). Thus was born the NIC/DOD (network information centre/Department of Defense) *host table* (called HOSTS.TXT).[1] By *caching* the NIC/DOD host table (i.e., copying it using the file transfer protocol) into the local hosts table, each UNIX host in the early Internet/ARPANET could be sure of locating any other host connected to the network by means of a 'lookup' in its local *hosts table*.

Initially, all the hosts in the ARPANET were administered from the network information centre (NIC) and the host table update procedure ran relatively smoothly. However, as the size of the network grew and the number of hosts connected to the network began to multiply, the NIC hosts table became impractical — both to administer and to copy to all the hosts. As a replacement for the NIC hosts table, the *domain name system (DNS)*, a decentralised directory service was developed.

The domain name system (DNS) is primarily used to map a server's *hostname* onto the IP (Internet protocol) network address required to *locate* the server and communicate with it. Thus, by means of an enquiry to a relevant domain name server, it might be possible to *resolve* the web address www.company.com to the Internet address (e.g., 37.168.153.1) of the related server. Using the IP address, the server can be contacted using any of the normal IP-suite protocols. Initially, the main protocols employed for communication following a DNS query were telnet, FTP (file transfer protocol) or SMTP (simple mail transfer protocol). As a result, the popularity of Internet email grew rapidly during the 1980s and 1990s, so that most people are nowadays familiar with email addresses of the form:

```
martin.clark@company.com
```

The domain name system (DNS) is critical to *resolving* the latter half of such an email address (the part after the @-sign — in this example: 'company.com') into the Internet address of the relevant destination mail server.

While the primary use of the domain name system (DNS) is to resolve server and email names to the Internet addresses of website and email servers, this is not its only use. In fact, DNS provides a powerful *directory service*, linking the server name not only to the

---

[1] Actually the first DNS implementation was incorporated in an operating system called TOPS10. TOPS10 was the first operating system to have the hosts.txt file.

Internet address, but also to a wide range of other possible *resource records* which reveal other characteristics of the server and related services.[2]

The root domain and the hierarchical relationship between all domains

The domain name system (DNS) stores the address database associated with the Worldwide Web (www) and the Internet electronic mail service. The *namespace* administered by means of the domain name system (DNS) is segregated into a hierarchical structure of domains. A domain is a network of computing devices administered, owned and/or run by a single organisation. Each domain is characterized by its own domain-name, and may be sub-divided into further sub-domains. As an example, Figure 11.2 illustrates a computer network which has been sub-divided into four different domains.

Because there are now many millions of networks and hosts (i.e., domains) connected to the Internet, there is also a huge number of domain names which have been allocated to identify each of them. It would not be feasible to consider a single directory server to store all the domain name information in one place. The domain name system (DNS) is thus designed to hold the database in a distributed manner.

The domain name system (DNS) assumes that all devices connected to the Internet form a single domain — called the *root domain*. As shown in Figure 11.3, the *root domain* is sub-divided into a number of different *top-level domains (TLDs)*. Perhaps the best known of the *top-level domains* are the .com and .org domains. Other well-known *top-level domains* are the *country code top-level domains (ccTLDs)*, which are based on the ISO 3166-1 two-letter country codes.[3]

Each domain is administered by a single organisation (the domain *authority*). Thus the root domain is administered by IANA (Internet Assigned Numbers Authority — www.iana.org). To be allocated a name within a given domain namespace requires an application to the domain authority. Table 11.1 lists the already allocated top-level domains which together comprise the root domain, listing the usage allowed for each and the name allocation authority for each. Thus, for the allocation of a .com (dot-com) or .org (dot-org) domain name, you

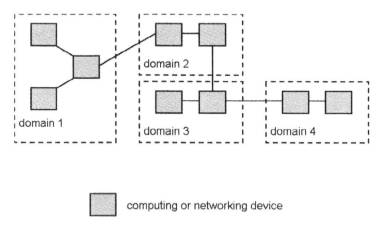

computing or networking device

Figure 11.2   The sub-division of a large network into domains.

---

[2] The NICNAME/WHOIS service (defined in RFC 812) is a powerful directory service based on looking-up information in the extensive database of the domain name system (DNS). NICNAME/WHOIS allows a wide-range of different queries to be made.

[3] See Appendix 3.

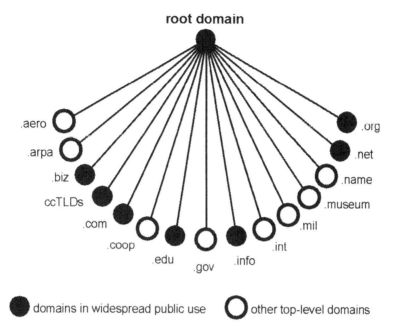

Figure 11.3 The root domain and the top-level domains (TLDs).

Table 11.1 Internet top-level domains (TLDs)

Domain	Domain usage	Domain operator/authority
.aero	aeronautical and air-transport industry	Société Internationale de Télécommunications Aéronautiques (SITA)
.arpa	address and routing parameter area	IANA / Internet Architecture Board
.biz	restricted to businesses	NeuLevel, Inc
.com	commercial organisations	VeriSign Global Registry Services
.coop	reserved for cooperative associations	Dot Cooperation LLC
.edu	reserved for higher educational institutions	Educause
.gov	reserved for government use (the top-level domain is the US government)	United States General Services Administration
.info	information domains	Afilias Limited
.int	used only for international organisations established by international government treaties	IANA .int Domain Registry
.mil	reserved exclusively for the US military	United States Department of Defense Network Information Center
.museum	reserved for museums	Museum Domain Management Association
.name	reserved for individuals	Global name registry
.net	network organisations	VeriSign Global Registry Services
.org	Organisations	VeriSign Global Registry Services

need to apply to VeriSign Global Registry Services or one of its authorised partners. The form of the domain name allocated to you will be correspondingly in the form of either 〈company.com〉 or 〈company.org〉. You can then subdivide this (your) domain by adding further sub-domains in a hierarchical fashion (by prefixing further sub-domain-names and 'dots'; e.g.: 〈sales.company.com〉, 〈marketing.company.com〉 and 〈ops.company.com〉).

The country code top-level domains (ccTLDs) are generally administered and operated by national Internet domain registry authorities. Some of these country level domains are sub-divided along similar lines to the .com/.org/.edu/.gov structure of the root domain. Thus the .au (Australia) domain is subdivided into separate domains: 〈.com.au〉, 〈.edu.au〉, 〈.gov.au〉, 〈.org.au〉 etc. Meanwhile, other country domain operators [including those responsible for the .il (Israel), .jp (japan) and .uk (United Kingdom) domains] have elected for two-letter sub-domain names thus:

- co (commercial), e.g., 〈company.co.uk〉

- ac (academic). e.g., 〈ox.ac.uk〉

There are no hard and fast rules other than the naming convention for sub-domain names: which are created by prefixing the parent domain name with the sub-domain name and a further 'dot'.

Resolution of names using the domain name system (DNS)

The domain name system (DNS) specifications (RFCs 1034, 1035 and 1591) define the hierarchical structure of the *namespace* as well as the *query* protocol (*DNS protocol*) which can be used to *resolve* (i.e., look up) unknown names. The name space defined by DNS is used by a number of different application protocols within the IP-protocol family. Its best known use is for Internet email and www (Worldwide Web) — but, as we encountered in Chapter 10, for example, it also provides the basis for *name bindings* in the *real-time application transport protocol (RTP)*.

The assumption which underpins the hierarchical and distributed structure in which data making up the DNS directory is stored, is that the data changes only very slowly. This means that the database is relatively stable over a long period of time and is not over-worked with update routines. In addition, copies of parts of the database remain valid for relatively long periods of time. The stability of the database is important to the correct functioning of the application protocols which rely on DNS.

Users (either human users or software programs) who need to 'look up an address' in the DNS (domain name system) do so by making a *query* to a domain *name server* (in effect the query poses the question 'what is the IP network address for the server with the domain name 〈sales.company.com〉? (see Figure 11.4)). In response to the query, the server returns a copy of the relevant *resource records* in the directory database using a file transfer protocol. The response allows the user subsequently to set up an IP (Internet protocol) communications path to the relevant (and now known) IP network address associated with the domain name. From this point on, communication continues between the two end-points using other standard application protocols (e.g., telnet, FTP, SMTP, RTP, www, etc.) without requiring further use of DNS. Using the DNS protocol is like calling the human telephone network operator for *directory assistance service* prior to making your call (Figure 11.4)!

DNS queries work basically in the manner illustrated in Figure 11.4, though in reality things are a little more complex. First of all, there is not one DNS *name server*, but instead a large number of name servers — at least one for each domain. Thankfully, all the name servers are linked according to a tree structure, and queries about unknown domain names can be channelled from the top (or *root*) of the tree downwards as appropriate. Let us consider

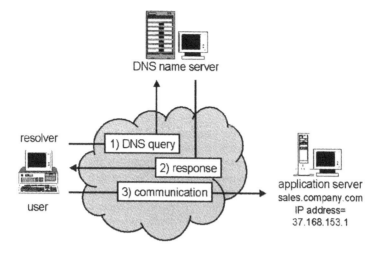

DNS name server

resolver

1) DNS query

2) response

3) communication

user

application server
sales.company.com
IP address=
37.168.153.1

steps (1) and (2) constitute a directory look-up

Figure 11.4 The domain name system (DNS) provides a directory look-up service.

a query to *resolve* the domain name ⟨sales.company.com⟩. A first query could be made to
the *root name server* to locate the ⟨.com⟩ name server. A subsequent query to the ⟨.com⟩
name server will help us to locate the ⟨company.com⟩ name server. A third DNS query to the
⟨company.com⟩ name server may provide us with the *resource record* (address information)
we require about the ⟨sales.company.com⟩ domain. If not, we might have to make a further
enquiry to a specific ⟨sales.company.com⟩ name server.

By storing the DNS *resource records (DNS_RR)* provided by the DNS name server, the
user PC of Figure 11.4 is able to avoid the need for subsequent DNS queries regarding the
server ⟨sales.company.com⟩. Such storage is called *caching*. There are two main benefits of
caching:

- the user PC is able to set up communication to the destination more quickly on subsequent
  occasions, since it does not first have to undertake a DNS query. In addition;

- the processing load on the DNS name server and the traffic load on the network are both
  kept to a minimum.

But cached information cannot be assumed to remain valid for ever. Occasionally the cached
information needs to be refreshed by means of a repeat DNS query.

The basic components of the domain name system (DNS)

There are three basic components of the domain name system (DNS). These are:

- the domain *name space* as recorded in DNS *resource records (DNS_RR)*;

- DNS *name servers*; and

- DNS *resolvers*.

The domain name space defines a hierarchical (tree-structured) naming scheme for all hosts and subnetworks within a given domain. *Nodes* and *leaves* of the *domain space tree* correspond to the information (called *resource records*) pertaining to a given host or subnetwork. Queries to *DNS name servers* (which store the *route records*) indicate the domain name of interest and the type of *resource* information which is required. The most common usage of the DNS is to identify hosts and servers; queries for *address resources* return Internet host addresses.

Name servers (domain name servers — DNS) are server programs and data bases which store information about the domain tree structure. The name server is the *authority* for the given part of a name space, which may be subdivided into *zones*. The name server stores copies of the *resource record* files for the zones for which it is the *authority*.

Resolvers are programs that run on user machines. They extract, use and cache information from name servers in response to DNS client requests (called *queries*). A resolver is typically a *system routine* within the client operating system or software which is directly accessible to other client user application programs. Web browser software usually includes the resolver functionality, for example.

Computer application software being used by the human computer user accesses the DNS (domain name system) through an operating system *call* to the local resolver. To the resolver, the complete DNS appears to be a large and unknown number of name servers, each of which contains only part of the 'DNS directory database'. As we discussed in conjunction with Figure 11.4, the *resolver* may need to make a number of *queries* to different DNS name servers and receive various *referrals* in order to resolve a particular address. Subsequently it will cache (i.e., store) the information which it learns.

The DNS specification defines:

- a standard format for domain name space data;

- a standard method for querying the domain name server database; and

- standard methods for refreshing local data from foreign name servers.

Human system administrators are responsible for:

- defining domain boundaries;

- maintaining and updating master data files relating to the relevant domain; and

- defining and administering the refresh policies relevant to data cached from the domain name server.

The domain name space tree and DNS resource records (DNS RR)

The top level of the domain name space tree is illustrated in Figure 11.3. All nodes (i.e., 'branches') of the domain name space tree have a label (i.e., a name of length 0–63 octets). The *null label* is reserved for the *root domain*. Names are coded in case-insensitive ASCII. But, when cached or stored, names should store the case of letters in the name (this is intended to allow the introduction of case-sensitive spellings in later developments of the DNS).

Labels are written in order and separated by 'dots'. Thus the example domain name of Figure 11.4 ⟨sales.company.com⟩ is correctly referred to as 'sales-dot-company-dot-com'. The total number of octets that may be used to represent a domain name is limited to 255.

Internet mail addresses can be converted into domain names (for the purpose of a DNS query) by removing the @ symbol and replacing with a 'dot'.

Table 11.2 Domain name system resource record (DNS RR) parameters

Resource record feature	Description
Class	A 16-bit value which identifies the protocol family for which the resource record is relevant:
	• IN *Internet* system protocols
	• CH the *Chaos* system
Owner	This is the domain name corresponding to the resource record (this field is often omitted, in which case the owner name is said to be *implicit* — i.e., the same as the domain name server name.
RDATA	The main data comprising the resource record. This depends upon the type of the resource record, as explained under 'type' below.
TTL	A 32-bit field representing the remaining lifetime in seconds (the time-to-live) of the resource record. TTL is primarily used by resolvers which cache resource records.
Type	A 16-bit value that specifies the type of the resource. Main types are:
	• A an IPv4 host address (RDATA field contains a 32-bit IPv4 address)
	• CNAME canonical name of an alias (RDATA field contains a domain name)
	• HINFO host information — the CPU and OS used by the host
	• MX mail exchange server used for the domain (RDATA field contains a 16-bit preference value (the lower the better) and a host name of a mail exchange server for the domain
	• NS the authoritative name server for the domain (RDATA field contains a host name)
	• PTR a pointer to another part of the domain name space (RDATA field contains a host name)
	• SOA identifies the start of a zone of authority

Note: A full-listing of up-to-date DNS parameters may be found at www.iana.org/assignments/dns-parameters

A domain name identifies a node of the domain name space tree. In practice, each node is a computer host or server in the data network (e.g., ⟨sales.company.com⟩). Each node is associated with a set of resource information, which is stored on the relevant domain name server (although this information may be 'empty').

When present, the resource information (e.g., an Internet host address, etc.) is composed as a series of resource records. The *resource records* are formatted according to a standard format according to Table 11.2. The order of the parameters within the individual records and the order of the records themselves is not significant.

The *canonical name (CNAME)* is the primary name of a given domain or device, but in addition, the device may have a number of *aliases* (i.e., duplicate domain names). In other words it might respond to a number of different domain names (aliases).

DNS queries and responses

DNS queries and responses are carried out using a standard message format (*DNS protocol*) as illustrated in Figure 11.5 and detailed in Table 11.3. The protocol is carried on TCP/UDP port 53.

DNS queries and responses comprise four sections: *question, answer, authority* and *additional* information. The content of the different types of messages varies (according to the header *opcode*) but the basic message format (Figure 11.5) is always the same.

The question field comprises the three sub-fields QNAME, QCLASS and QTYPE. The QNAME identifies the domain name (e.g., ⟨sales.company.com⟩) of the device which is the

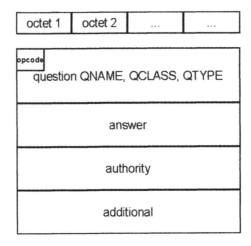

Figure 11.5 Format of DNS queries and responses.

Table 11.3 Domain name system: protocol header opcode (4-bit field)

Opcode value (4 bits)	Meaning	Purpose of message	Sub-fields	Sub-field meaning
0	Standard DNS query QUERY	Requests resource records relevant to the requested domain name (as defined in RFC 1035)	QNAME QTYPE QCLASS	Target domain name Query type Query class
1	Inverse query (optional) IQUERY	Maps a particular resource to domain names that have that resource (as defined in RFC 1035)	—	—
2	Status query (optional) STATUS	Delivers information about the status of a domain name or resource record (as defined in RFC 1035)	—	—
(3)	Completion (obsolete)	Completion services (now obsolete) as defined in RFCs 882 and 883	—	—
3	(reserved)	—	—	—
4	Notify NOTIFY	Notification services as defined in RFC 1996	—	—
5	Update UPDATE	Update services as defined in RFC 2136	—	—
6-15	Unassigned values		—	—

focus of the query. The QTYPE field allows *resource records (RRs)* of only a given type to be requested, e.g., MAILB requests that only mailbox-related RRs be delivered (for example, the name and address of the Internet mail *postmaster* serving a given mail domain), '*' meanwhile requests the delivery of <u>all</u> available RRs. QCLASS requests the delivery of only those RRs which are relevant to the RR class requested.

The principal activity of *name servers* is to answer standard *queries*. The manner in which they respond to queries depends upon whether they have been programmed to operate in the *recursive* or in the *non-recursive mode*. A server operating in the non-recursive mode which is unable to answer a particular query from its own 'local' database will answer with a *referral* to a server *'closer'* to the answer. A *recursive server*, on the other hand, will itself refer to other databases if necessary and will never reply with a *referral*. If the query cannot be answered, an error message is returned.

Serving the DNS client: the operation of the DNS resolver

The DNS resolver acts on behalf of the *DNS client* to resolve a given domain name into an address or other resource record (RR) information (as detailed under the 'type' field of Table 11.2 — e.g., MX: mailbox exchange server, etc.). The DNS client is typically a computer program serving a human computer user. The DNS client might thus be a web-browser software attempting to look up a 'web-address' (e.g., www.company.com) or an electronic mail software looking up the address of a destination mail server).

The DNS resolver is also a computer software usually resident on the same computer as the DNS client. Indeed both DNS client and DNS resolver software modules are included in a standard *web browser* software. The DNS resolver conducts the DNS query process and caches the resource records which it receives as responses from the queried DNS name servers. Table 11.4 lists the data structures maintained by the DNS resolver during its processing of the enquiry on behalf of the DNS client. The values held in these data structures represent the 'current status' of the enquiry.

The following steps are normally undertaken by the DNS resolver in answering the DNS client's query:

1) First, the resolver checks its 'local data' (in the case that the DNS resolver is also the local DNS name server — the 'local data' is the sum of the cached data received as a result of previous queries and any *authoritative* zone data).

2) Using its 'local data', the resolver determines the best DNS name server to query.

3) The resolver submits *DNS queries* to one or more servers until a response is received.

4) The resolver analyses the response:

   (a) in the case of an answer or name error, the response is cached and also forwarded to the DNS client;

Table 11.4   State of the DNS resolver

Resolver	Description
SNAME	The domain name about which information is sought
STYPE	The QTYPE of the search request
SCLASS	The QCLASS of the search request
SLIST	A list of name servers and the zone which the resolver is currently querying. The 'best guess' of the name server holding the relevant information based upon information already collected
SBELT	'Safety belt' in the same form as SLIST which is initialised by a configuration file
CACHE	A store of information collected from previous responses and not yet expired as a result of TTL (time-to-live) deletion

(b)   in the case of a referral, the response is cached (in SLIST) and the resolver returns to step (3) above, submitting a new request to the 'referred' name server;

(c)   in the case that the response provides a CNAME which is not the answer, the DNS resolver must change the SNAME to the CNAME and return to step (1). In this case, the original domain name sought may have been an alias (i.e., duplicate, but not primary name of the target resource);

(d)   in the case of a server failure or incomprehensible reply, the DNS resolver deletes the server from the SLIST and returns to point (3), querying other servers (if available).

The final answer supplied by the DNS resolver to the DNS client will be a copy of the relevant DNS resource records. In the example below, the response indicates that the mail exchange server for ⟨sales.company.com⟩ is the server ⟨mail.company.com⟩ and provides the IPv4- Internet address of this server:

```
sales.company.com    MX    10    mail.company.com
mail.company.com     A     37.168.153.3
```

11.3   Internet cache protocol (ICP)

Before we leave the subject of locating resources, files and other objects from the Internet, we should also cover the *Internet cache protocol (ICP)* as defined in RFCs 2186 and 2187.

We have seen how, by *caching* the IP address associated with a given resource or object, we are able to avoid the need for DNS queries and responses when a subsequent request is made for the same resource. But not only the address of the resource can be cached: a local copy of the resource itself could also be cached. This might significantly reduce the load on the wide area network, particularly if a large number of users in a particular location (e.g., a company headquarters) make use of the same resource.

We need not restrict ourselves only to using our own cache, we could also make use of our neighbour's cache too. In this way we might be able to reduce further the traffic on the long distance part of the network. For this purpose the Internet cache protocol (ICP) was developed.

ICP (Internet cache protocol) allows the use of multiple caches, by establishing a hierarchical relationship structure with *neighbour* caches. ICP is a lightweight message protocol used between neighbour *peer* caches to determine the nearest source or copy of a given resource of object within a nearby cache.

ICP queries and replies are exchanged in a lightweight and fast manner to gather information which will help determine which is the best location (e.g., source location or cache) from which to retrieve a given file (or other object). Queries which find a given object in a local cache are said to *hit*, while a *miss* indicates that the source object is not cached.

11.4   WINS; Windows2000 ADS; Novell NDS

WINS (Windows Internet name service) is a name resolution service somewhat similar to DNS which *resolves* Windows computer names to IP addresses. The WINS server is typically a *WindowsNT* server within *the local area network (LAN)*. The *Windows2000 active directory service (ADS)* extends WINS to provide a comprehensive directory of other LAN users and of files and other resources available within the Windows 2000 network domain.

The *Novell directory service (NDS)* is intended to be used within Novell *Netware*-based corporate local area networks to provide access to common files and other resources. Meanwhile *NIS (network information service)* is the SUN Microsystems version used in many UNIX networks.

Though proprietary directory service mechanisms (such as WINS/ADS, NDS or NIS) may be used in some types of 'private' networks (based on 'Microsoft networking', Novell Netware or SUN/UNIX), these directory services cannot be used as 'substitutes' for the DNS (domain name system), when the domain is connected to the Internet.

11.5   Hypertext transfer protocol (http)

The *HyperText Transfer Protocol (http)* is an *application layer* protocol allowing 'collaborative information systems' to be built based upon distributed storage of information. In simple terms, the hypertext transfer protocol allows 'documents' to be created from multiple different text and image files, where each of these files may be stored on a different computer. A *hyperlink* is a kind of 'pointer' used to mark the position in the 'document' where a given text, image or other file should appear and to 'point at' the location where the relevant file is stored). Often the hyperlink appears in text as a domain name address, prefixed by http://www, thus:

`http://www.sales.company.com`

HTTP is nowadays so widely used that word processing software like Microsoft *Word* automatically assumes you intend to insert a hyperlink into a document whenever you type a character string in this familiar format. Microsoft Word automatically underlines the hyperlink, shades the text blue and underlines it. Subsequently, if you click anywhere on the hyperlink, your computer immediately tries to access the corresponding *website*.

HTTP is the protocol which retrieves the file indicated by the hyperlink. The current version of http is version 1.1 (HTTP/1.1). It is defined in RFC 2616 (issued in June 1999).

The hypertext transfer protocol (http) allows for:

- raw data (i.e., text) transfer based on a 'pointer' (and thus for data retrieval, search capabilities or annotation of scientific papers with bibliographical references to other documents or reports);

- hyperlinking data in a *MIME (multipurpose Internet mail extension)*-like message format (the MIME format is used by Internet mail to code the attachments to mail messages[4]).

The hypertext transfer protocol (http) is used between an http client (also called the *user agent* — *UA*) and the http *origin server (0)* (Figure 11.6). The hyperlink exists to 'connect' a 'pointer' resident in the client with the actual data file held at the origin server. The hyperlinked file is retrieved using a series of http *requests* and *responses*.

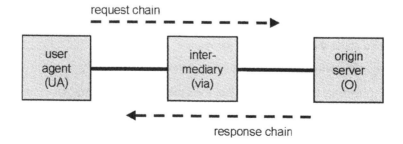

Figure 11.6   Hypertext transfer protocol (http): request and response chains.

---

[4] See Chapter 13.

HTTP user agent, origin server and http intermediaries

HTTP *requests* are generated by the client (user agent) and passed up the *request chain* (see Figure 11.6) to the *origin server (O)*. The http *response* is returned by means of the *response chain*. Both requests and responses are normally carried by means of TCP *port* 80.

The http intermediaries illustrated in Figure 11.6 may be http *proxies* or caches. Such devices are not always present, but are sometimes used either to increase the security of access to a given http server (use of a proxy) or to improve the speed of response to an http request (use of a cache).

When present (e.g., in a firewall[5]), an http proxy is usually located near the origin server. The proxy vets the incoming http requests to the server and decides which ones will be allowed. Only those allowed are forwarded to the origin server for a response. Unallowed requests are answered with an error by the proxy server. An http proxy is a type of http *forwarding agent*, which might rewrite part or all of the http message.

A *cache server* is usually provided near, and for the benefit of, the user agent. The cache server stores all the http responses received in the response chain, thereby enabling it to respond to subsequent requests for the same file without requiring a repeat request to the origin server. Caching removes the need to send some requests across the network, thereby reducing network load and improving the response time preceived by the customer. Because cached data must be kept 'fresh' enough to be reliable, a *time to live (TTL)* parameter is used, after the expiry of which the cached copy is deleted and a new copy retrieved from the server on the occasion of the next client request.

Other http intermediaries include *gateways* and *tunnels*. A gateway may be used when the origin server does not directly support http. In this case, the gateway performs an application protocol conversion from http to the native protocol of the origin server. A tunnel is a relay mechanism intended to improve the security of http transport. We shall encountering tunnelling in more detail in Chapter 13.

HTTP requests and responses

HTTP requests (generated by the http client or *user agent — UA*) include the following information:

- a request *method* (i.e., a command like *PUT, GET, DELETE*, etc.);

- a *universal resource identifier (URI)* (also called a *universal document identifier*) A URI is equivalent to the combination of a *universal resource locator (URL)* and a *universal resource name (URN)*. This is the file locator or 'pointer' which indicates where http can locate the requested file.

- the http protocol version;

- information about the client making the request; and

- additional information forming part of the request (if required).

The first three elements listed above together form the http *Request-Line*. An example Request-Line might be:

```
GET http://www.company.com/sales/orders.html HTTP/1.1
```

The http server (the origin server) responds to an http request with an http response including:

---

[5] See Chapter 13.

- a response *Status-Line* (which comprises the http protocol version number and a success or error code);

- a datafile formatted in one of the standard MIME-formats containing information provided by the server in response to the request; and/or

- other response information.

HTTP protocol coding

The commands of the hypertext transfer protocol (http) have the appearance of a 'classical computer programming language', typically comprising a command or keyword, followed by colon and a list of parameter values (called arguments), and concluded at the end-of-line (EOL) by the ⟨CRLF⟩ (carriage return, line feed sequence ASCII 13, 10).

The http generic message format is defined by RFC 822 and RFC 2616. Request and response messages consist of:

- a start line (comprising either a Request-Line (http request) or a Status-Line (http response),

- zero or more header fields;

- an empty line (i.e., two consecutive ⟨CRLF⟩ (carriage return line feed) sequences) to indicate the end of the header); and

- a message or *entity-body* (if appropriate). This is typically an attached file coded in one of the MIME (multipurpose Internet mail extension)[6] formats conceived for Internet mail attachments. The type of file attached is indicated in the entity header.

Header fields include the *general header*, together with the *request header*, the *response header* and/or the *entity header* (Figure 11.7). Header fields comprise a *field-name* (as detailed in Table 11.5) followed by a colon ':', then one or more 'spaces' followed by the *field-value* and any other relevant *field-content*. If necessary, header lines can be extended over multiple horizontal lines by preceding the line with a space or a HT (horizontal tab) character.

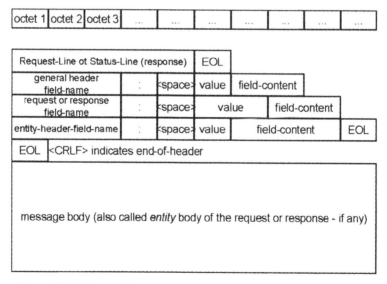

Figure 11.7   General format of hypertext transfer protocol (http) messages.

---

[6] See Chapter 12, Table 12.2.

Table 11.5 Hypertext transfer protocol (http): header field names listed in order of appearance in http messages

Header field type	Field-type or field name	Purpose
Request-Line	Method [field type]	OPTIONS [field name] (requests information about the options available on the request/response chain). GET [field name] (retrieves information in the form of an entity). HEAD [field name] (identical to GET except the server must not return a message body in the response). POST [field name] (requests the server to make the request entity a new subordinate of the request-URI). PUT [field name] (requests that the server store the request entity under the request-URI). DELETE [field name] (requests the server to delete the resource identified in the resource-URI). TRACE [field name] (this request causes an application layer loopback of the request message for purpose of testing). CONNECT [field name] (this request instructs a proxy to switch to become a tunnel (e.g., SSL — secure sockets layer tunnelling). Extension-method
	Request-URI	Absolute URI (universal resource identifier expressed relative to the root domain). abs_path (absolute path) Authority
	HTTP-version	e.g., HTTP/1.1
	⟨CRLF⟩	This indicates the end of the Request-Line.
Status-Line (i.e., response)	HTTP-Version	e.g., HTTP/1.1
	Status-Code	1xx: informational — the request was received — process is continuing. 2xx: success — the request was received and accepted. 3xx: redirection — the request needs to be referred to a different server. 4xx: client error — the message syntax is wrong or cannot be undertaken. 5xx: server error — the server could not complete the valid request.
	Reason-Phrase	An additional text message for the user to explain the Status-Code (if provided in the response).
	⟨CRLF⟩	This indicates the end of the Status-Line.
General-header	Cache-Control	These are explicit commands issued to a cache server (e.g., 'no-cache' 'no-store' 'no-transform' 'max-age' allowed, etc.)
	Connection	Allows the sender to specify options for the connection. These should not be forwarded by proxies.

Table 11.5 (*continued*)

Header field type	Field-type or field name	Purpose
	Date	This field indicates the date and time at which the message originated.
	Pragma	This field includes directives for controlling the action of recipients in the response chain.
	Trailer	This field indicates that the given set of header fields is included in the message trailer and encoded with chunked transfer coding.
	Transfer-Encoding	This field indicates that the message body has been subjected to a transformation (typically encryption) to safeguard privacy during communication.
	Upgrade	This field indicates which additional HTTP protocols the client supports and would like to use if possible by 'switching protocols'.
	Via	This field is used by proxies and gateways to indicate to the client the intermediate protocols and intermediaries between the server and the client on a response.
	Warning	If a cache returns a message which is no longer 'fresh enough' or is not a first-hand copy, it must include a warning to this effect.
Request-header	Accept	Specifies which media types are acceptable in the response to the client.
	Accept-Charset	Specifies the character set which is acceptable in the response to the client.
	Accept-Encoding	Specifies the data encoding scheme which is acceptable in the response to the client.
	Accept Language	Specifies the (human) language which is acceptable in the response to the client.
	Authorisation	This field is used by the client when necessary to identify itself to the server.
	Expect	This field indicates the server behaviour required by the client.
	From	If used, this field contains an Internet email address of the human user who generated the request.
	Host	This field indicates the host Internet address and port number of the resource being requested.
	If-Match	This field makes a request into a conditional request based upon a match of the entity tag (ETag).
	If-Modified-Since	This field makes a request into a conditional request, which will generate a response only if the resource requested has been updated or modified since the specified data and time.

(*continued overleaf*)

Table 11.5   *(continued)*

Header field type	Field-type or field name	Purpose
	If-None-Match	This field makes a request into a conditional request. By means of such a request the client can verify that none of the entities previously received (as identified by their ETags) are current.
	If-Range	In the case that the client has a partial copy of a particular resource, it can update that partial copy with a conditional request of this type.
	If-Unmodified-Since	This conditional request should only be undertaken by the server if the resource has not been modified since the specified date and time.
	Max-Forwards	This field is used in conjunction with TRACE requests and OPTIONS to limit the number of proxies and gateways allowed to forward the request.
	Proxy-Authorisation	This field includes information by which the client or user can identify itself to the proxy in the request and response chain.
	Range	This field specifies the byte offset range of a request for a partial response.
	Referrer	This field allows the client to indicate to the server, the address (i.e., URI) of a previous resources from which the Request-URI was obtained.
	TE	This transfer extension coding indicates which extension transfer codings the client is willing to accept.
	User-Agent	This field includes information about the user agent generating the request. It is used for statistical purposes and tracing.
Response-header	Accept-Ranges	This field indicates the Methods supported by the resource.
	Age	This field indicates the sender's estimate of the elapsed time since the response was generated by the origin server.
	Etag	The current value of the entity tag. The entity tag is used to compare with other entities provided by the same resource.
	Location	This field identifies a referral location (Request-URI) which should be requested to complete the transaction.
	Proxy-Authenticate	This response-header field must be included to achieve a successful subsequent request, after the initial request was refused because of the need for the client to authenticate itself to an intermediate proxy.
	Retry-After	This field is used in conjunction with the 503 response to indicate the expected duration of service unavailability.

Table 11.5 (*continued*)

Header field type	Field-type or field name	Purpose
	Server	This field includes information about the server software allowing the origin server to handle the request.
	Vary	This field indicates the request header fields which are to be used by a cache to determine whether the response is still 'fresh'.
	WWW_Authenticate	This field must be included in responses with the 401 (unauthorised) response. It comprises at least one authentication 'challenge' to be replied to in order to identify the client to the server.
Entity-header	Allow	This field lists the Methods supported by the resource.
	Content-Encoding	This field indicates the encoding Method applied to the entity- body.
	Content-Language	This field indicates the (human) language of the entity-body.
	Content-Length	This field indicates the length of the entity-body in octets.
	Content-Location	This field may be used to supply the resource location to the server.
	Content-MD5	This is a field used in conjunction with the MD5 encryption scheme when securing the privacy of communication between server and client.
	Content-Range	This indicates where within a full entity-body a particular partial response is to be found.
	Content-Type	This field indicates the media type of the entity-body.
	Expires	This field indicates the date and time after which the response should be considered 'stale'.
	Last-Modified	This field indicates the date and time at which the server believes the resource contained in the entity-body was modified.
	Extension-header	Extension headers may be added here.
Entity-body		When included, the data type of the entity or message body is revealed in the Content-Type and Content-Encoding entity-header fields.

Header field values comprise words separated by spaces (so-called *linear white space, LWS*) or special characters like commas. Comments can be included in brackets.

The version number of http is normally indicated the header in the following format: 'HTTP/1.1'. The first '1' is the *major version* number and the second '1' the *minor version* number.

A date and timestamp, expressed in GMT (Greenwich Mean Time), are included in all http messages.

URIs, URLs and URNs

The request-*URI* (*universal resource identifier*) is a combination of a *URL* (*universal resource locator*: which locates network resources) and a *URN* (*universal resource name*: which locates a particular file). An http URI has the standard format:

```
http://host [:port]/[abs_path [? query]]
```

where *host* is the relevant domain name of the http server, *port* is the TCP port number to be used for the http protocol transfer, *abs_path* is the absolute file path of the target file and query is further information related to the request. If the *port* is not stated, the default port (value = 80) is assumed. An example URI might thus be:

```
http://company.com:80/~admin/home.html
```

HTTP responses

On receipt of an *http request*, the origin server is expected to undertake a case-sensitive octet-by-octet comparison of the URI (except that host names and domain names are case-insensitive) to decide upon the requested file match.

The format of the *http response* message is similar to that of an http request message, except that it includes a Status-Line rather than a Request-Line. The Status-Line indicates the success or failure of the request (the possible replies are listed in Table 11.6).

Table 11.6   Http response messages: status-line codes

HTTP Status code type	HTTP Status code	Meaning
Informational	100	Continue (allows the client to determine if the server will accept the request before it has to send to the request message body).
	101	Switching Protocols (server is willing to switch protocols as requested by client).
Success	200	OK (a GET, HEAD, POST or TRACE request was successful).
	201	Created (a new resource has been created).
	202	Accepted (request accepted but still undergoing processing).
	203	Non-Authoritative Information (the information provided by the server is not the authoritative version but obtained from a local or third-party cached copy).
	204	No Content (the server has undertaken the request and is not returning an entity-body).
	205	Reset Content (the request was undertaken; the client should now reset the document view which triggered the request).
	206	Partial Content (the server has fulfilled a partial GET request).

Table 11.6   (*continued*)

HTTP Status code type	HTTP Status code	Meaning
Redirection	300	Multiple Choices (client may choose from a number of identified locations in the response where to direct the referred request).
	301	Moved Permanently (the URI has moved and all requests should be directed to the new permanent URI.)
	302	Found (the requested resource temporarily exists at another URI, but only on a temporary basis. Future requests can continue to use the 'old' URI.)
	303	See Other (the request should be directed to a specifically named URI).
	304	Not Modified (a conditional GET request was made but the response document has not been modified).
	305	Use Proxy (the requested resource must be accessed by means of an HTTP proxy).
	307	Temporary Redirect (the requested resource temporarily exists at another URI, but only on a temporary basis. Future requests can continue to use the 'old' URI).
Client error	400	Bad Request (the request could not be understood) due to incorrect syntax.
	401	Unauthorized (the request requires authorisation of the user).
	402	Payment Required (the resource requested must be paid for).
	403	Forbidden (access to the resource requested is forbidden).
	404	Not Found (the resource requested was not found).
	405	Method Not Allowed (the Method requested is not allowed).
	406	Not Acceptable (the responses which the server is able to generate are not permitted according to the parameters set in the request).
	407	Proxy Authentication Required (response similar to 401, but in which the authorisation must take place with the proxy).
	408	Request Timeout (the client did not respond to a server request within the acceptable waiting period determined by the server).
	409	Conflict (the request could not be undertaken as this conflicts with the current state of the resource).
	410	Gone (the requested resource is no longer available at this URI and the new URI is unknown).

(*continued overleaf*)

Table 11.6 (*continued*)

HTTP Status code type	HTTP Status code	Meaning
	411	Length Required (the request cannot be accepted without a defined Content-Length).
	412	Precondition Failed (one of the preconditions indicated in the request failed when evaluated by the server).
	413	Request Entity too Large (the requested resource or entity is larger than the server is able or willing to handle).
	414	Request-URI Too Long (the requested URI is longer than the server is able or willing to handle).
	415	Unsupported Media Type (the media format requested is not supported by the requested resource).
	416	Request Range Not Satisfiable (the request stated a Range which cannot be satisfied).
	417	Expectation Failed (an expect-request header field in the request could not be fulfilled).
Server error	500	Internal Server Error (an unexpected internal error of the server).
	501	Not implemented (the functionality requested is not supported by the server).
	502	Bad Gateway (while acting as a gateway or proxy a server received an invalid response in the response chain).
	503	Service Unavailable (due to temporary overload or maintenance of the server).
	504	Gateway Timeout (while acting as a gateway or proxy a server did not receive a valid response in the response chain within an acceptable waiting time).
	505	HTTP Version Not Supported (the HTTP version requested cannot be supported).

A typical response will return a copy of a requested file (or, in the case of multiple requested files, a *multipart message* response may be sent) to the *user agent*. In this case, the file is attached as an *entity-body* to the http response message. The coding of any attached response file is identified in the *entity header*. The standard MIME (multipurpose Internet mail extension)-*coding-type* formats[6] are used. In addition, a data compression technique may be used to reduce the size of the file for transmission. In this case, one of the following *content-encoding* tokens will also be indicated in the entity header of the http response:

- *gzip* — indicates that the content has been produced by the GNU zip compression pro-gramme as defined by RFC 1952 (Lempel-Ziv coding LZ77)

---

[6] See Chapter 12, Table 12.3.

- *compress* — indicates that the content has been produced by the UNIX file compression program (Lempel-Ziv-Welch coding — LZW)

- *deflate* — indicates the content is in the zlib format and deflate mechanism (RFCs 1950 and 1951)

- *identity* — indicates that the content has not been altered (i.e., transformed). It is in its original format.

When a general header transfer-encoding value is set, the http message is said to be *chunked*. In this case a special coding is being used (e.g., encryption) to ensure the 'safe transport' of the message across a shared network such as the Internet.

HTTP operational considerations

Before ending our brief review of the operation of the hypertext transfer protocol (http), it is worth considering a number of features which have been incorporated into it to improve its operational performance and to increase the security of data transported by it. We shall consider in turn:

- how an http server decides which response file to send;

- the better performance afforded by *persistent* TCP connections;

- the processing and response capacity of http servers;

- http access authentication; and

- the security problems associated with DNS *spoofing*.

In some cases, the http server may have the same basic document available in a number of different data formats. Thus, for example, some of the RFCs on the RFC-editor website (www.rfc-editor.org) are available in either 'text', 'pdf' (portable document format) or 'ps' (postscript) formats. Which is the 'best' file to respond with will depend upon the preference of the human user or his *user agent*. The preference can be indicated by means of *content-negotiation*, which may be either server-driven, agent driven, or a combination of the two (called *transparent negotiation*). Alternatively, the server can always respond with all available file formats and leave the user or user agent to choose the one he or she needs.

By using *persistent* TCP connections for http sessions, the performance of the session can be improved. A persistent TCP connection is not cleared after each individual request/response pair, but instead is left to 'time out'. By this means, any subsequent short-term http requests to the same http server are spared the need to wait while a new TCP connection is set up every time. This reduces the potential network congestion which might be caused by the TCP connection set-up *handshake* messages, and simultaneously greatly improves the speed of response to the http request.

Understandably, some very popular web servers (http servers) are subjected to a very large number of http requests each day, and the processing capacity of the server hardware greatly affects the speed at which responses can be generated. To increase the capacity of the http server, duplicate or *cluster* servers are sometimes used. You may have noticed that sometimes that the URI-address of a responding server sometimes appears with a prefix *www1, www2, www3* rather than just the simple 'www'. The numbers refer to duplicate http servers which are *load-sharing* the requests — answering individual requests in rotation.

A challenge response mechanism (*www_authenticate*) is included in http to allow a client to identify itself to the server, and encryption may be used to secure the content of http messages. Despite this, security of information is one of the major operational challenges associated with the use of http. Http is heavily dependent upon the domain name system (DNS) for resolving the IP addresses of http servers (in order that appropriate TCP connections can be set up to them). This alone makes http prone to DNS *spoofing* (the receipt of wrong IP address information generated by bogus DNS name servers). We shall discuss prudent security precautions in Chapter 13.

11.6   Hypertext markup language (html)

Hypertext markup language (html) is a software language used to format 'documents' intended for human viewing which include http commands to retrieve component text, image and other files access the Worldwide Web. We shall not cover html in great detail here, as to do so would cross the boundary from 'networking' into 'computing' — which is beyond the scope of this book. However, we shall briefly present the form of html commands and the manner in which http commands are incorporated into html-type files.

HTML is an application of *SGML (standard generalised markup language* — ISO 8879). It was invented for the Worldwide Web and has been in use since 1990. Initially, the aim was to cross-reference scientific papers with *hyperlinks* giving direct access to other referenced documents (e.g., appearing in the bibliography).

An html editor is a software program (like Microsoft Word) which allows the creation of html documents or the conversion of text documents into html code. An html editor uses the html coded-character set (ISO 10646 — Table 11.7). An html-tag is a symbol used in an html document to identify and delimit a *page element*'s type, format and/or structure.

An html-tag is associated with each html element on the page and appears as in the format ⟨Tag⟩. A typical html ⟨tag⟩ might be a ⟨link⟩ or ⟨A⟩ *(anchor)* element — identifying one of the two ends of a hyperlink.

A hyperlink is a relationship between two *anchors*, called the *head* and *tail* of the hyperlink. An html *user agent* allows the human user to navigate through the html document and request the activation of hyperlinks denoted by any of the following *tail anchors*: ⟨A⟩ ⟨LINK⟩ ⟨IMG⟩ ⟨INPUT⟩ ⟨ISINDEX⟩ or ⟨FORM⟩ elements.

The *head anchor* comprises a *URI (universal resource indicator)* to 'point' to a resource at the other end of the hyperlink which may be retrieved by one of a number of different retrieval protocols (http — hypertext transfer protocol — is the one used most commonly nowadays). The URI may be optionally followed by a #-sign and a sequence of further characters called the *fragment identifier.*

The head anchor is determined from the *base URI* (defined in the html document — see Table 11.8) and any hyperlink *calls*. Thus, if for example, the base URI is:

`http://company.com/sales/order.html`

and the document contains a hyperlink as follows:

`<img one="../icons/logo.gif">`

then the user agent will use the following URI in conjunction with HTTP to retrieve and show the image:

`http://company.com/icons/logo.gif`

The *message entity* transferred by means of the hyperlink and http (hypertext transfer protocol is typically a text, image, video or other Internet Media type (IMEDIA) or MIME Content

Table 11.7   HTML character set (based on iso 8859-1]

HEX CODE YX			X	0	1	2	3	4	5	6	7	8	9	A	B	C	D	E	F
			4	0	0	0	0	0	0	0	0	1	1	1	1	1	1	1	1
	BITS		3	0	0	0	0	1	1	1	1	0	0	0	0	1	1	1	1
			2	0	0	1	1	0	0	1	1	0	0	1	1	0	0	1	1
Y			1	0	1	0	1	0	1	0	1	0	1	0	1	0	1	0	1
	8 7 6 5																		
0	0 0 0 0			0	1	2	3	4	5	6	7	8	9 HT	10 LF	11	12	13 CR	14	15
1	0 0 0 1			16	17	18	19	20	21	22	23	24	25	26	27	28	29	30	31
2	0 0 1 0			32 spce	33 !	34 "	35 #	36 $	37 %	38 &	39 '	40 (	41)	42 *	43 +	44 ,	45 -	46 .	47 /
3	0 0 1 1			48 0	49 1	50 2	51 3	52 4	53 5	54 6	55 7	56 8	57 9	58 :	59 ;	60 <	61 =	62 >	63 ?
4	0 1 0 0			64 @	65 A	66 B	67 C	68 D	69 E	70 F	71 G	72 H	73 I	74 J	75 K	76 L	77 M	78 N	79 O
5	0 1 0 1			80 P	81 Q	82 R	83 S	84 T	85 U	86 V	87 W	88 X	89 Y	90 Z	91 [	92 \	93]	94 ^	95 _
6	0 1 1 0			96 `	97 a	98 b	99 c	100 d	101 e	102 f	103 g	104 h	105 i	106 j	107 k	108 l	109 m	110 n	111 o
7	0 1 1 1			112 p	113 q	114 r	115 s	116 t	117 u	118 v	119 w	120 x	121 y	122 z	123 {	124		125 }	126 ~	127
8	1 0 0 0			128	129	130	131	132	133	134	135	136	137	138	139	140	141	142	143
9	1 0 0 1			144	145	146	147	148	149	150	151	152	153	154	155	156	157	158	159
A	1 0 1 0			160 NBS	161 ¡	162 ¢	163 £	164 ¤	165 ¥	166 ¦	167 §	168 ¨	169 ©	170 ª	171 «	172 ¬	173 -	174 ®	175 ¯
B	1 0 1 1			176 °	177 ±	178 ²	179 ³	180 ´	181 µ	182 ¶	183 ·	184 ¸	185 ¹	186 ♂	187 »	188 ¼	189 ½	190 ¾	191 ¿
C	1 1 0 0			192 À	193 Á	194 Â	195 Ã	196 Ä	197 Å	198 Æ	199 Ç	200 È	201 É	202 Ê	203 Ë	204 Ì	205 Í	206 Î	207 Ï
D	1 1 0 1			208 Ð	209 Ñ	210 Ò	211 Ó	212 Ô	213 Õ	214 Ö	215 ×	216 Ø	217 Ù	218 Ú	219 Û	220 Ü	221 Ý	222 Þ	223 ß
E	1 1 1 0			224 à	225 á	226 â	227 ã	228 ä	229 å	230 æ	231 ç	232 è	233 é	234 ê	235 ë	236 ì	237 í	238 î	239 ï
F	1 1 1 1			240 ð	241 ñ	242 ò	243 ó	244 ô	245 õ	246 ö	247 ÷	248 ø	249 ù	250 ú	251 û	252 ü	253 ý	254 þ	255 ÿ

CODED CHARACTER

Key: CR = Carriage Return; HT = Horizontal Tab; LF = Line Feed; NBS = Non-Breaking Space;   shaded boxes represent unused values.

Type (MIME) file). Files in these formats are also commonly sent as Internet mail attachments, and we shall discuss them in more detail in Chapter 12.

HTML document structure and format

An html (hypertext markup language) document always commences with a *document type definition (DTD)*, or the start symbol ⟨HTML⟩. The definition is followed by other html-coded elements (each contained between start '<' and end '>' tags) comprising the *HEAD* and the *BODY* of the document. The *HEAD* contains the ⟨TITLE⟩. The rest of the document is a *tree* of html elements, e.g., headings, paragraphs, lists, forms etc. Table 11.8 lists a selection of the most commonly used html elements and their meaning.

An example short html document is illustrated in Figure 11.8. The document generates a very crude html form for inputting a sales order. The *form* inputs the variable named *product* to the URL '/sales/orders'.

Table 11.8 Hypertext markup language (html) elements (in approximate order of appearance in an html document)

HTML element name	Meaning or usage
DTD (document type definition)	e.g., ⟨DOCTYPE HTML PUBLIC "-//IETF//DTD HTML 4.0/EN"⟩
⟨HEAD⟩	The header portion of an html document.
⟨TITLE⟩	The title of an html document.
⟨BASE⟩	The base address (i.e., the reference address element). The href = notation in a hyperlink head anchor element provides the hyperlink URI relative to this base address.
⟨ISINDEX⟩	A keyword index element.
⟨LINK⟩	An html element used to represent a hyperlink head anchor. This form is typically used to provide a link in a glossary.
⟨NEXTID⟩	This element indicates the name used for a new anchor ⟨A⟩ element when editing an html document.
⟨BODY⟩	This element contains the text flow of the document, including headings, paragraphs, lists, forms etc.
⟨H1⟩ to ⟨H6⟩	These elements denote section headings within the ⟨BODY⟩.
⟨P⟩ (paragraph) ⟨LISTING⟩ ⟨PRE⟩ (pre-formatted text) ⟨XMP⟩ (example)	These are all examples of 'block-structuring elements' used for the page formatting of html documents.
⟨ADDRESS⟩	This element is intended to contain the document author's name, address and signature.
⟨BLOCKQUOTE⟩	These elements contain test quoted from another source.
⟨LI⟩ (list) ⟨OL⟩ (ordered list) ⟨UL⟩ (unordered list)	⟨OL⟩ and ⟨UL⟩ are list elements containing lists of multiple ⟨LI⟩ elements.
⟨DIR⟩ (directory)	The directory element is a sequence of short list items (up to 20 characters).
⟨MENU⟩	This is an element comprising a list of items with one line per item.
⟨DD⟩ (definition description) ⟨DL⟩ (definition list) ⟨DT⟩ (definition term)	The ⟨DL⟩ element is a list of ⟨DT⟩ (definition term) elements and their respective ⟨DD⟩ (definition description) elements.
⟨CITE⟩ (citation) ⟨CODE⟩ ⟨EM⟩ (emphasis) ⟨KBD⟩ (keyboard) ⟨SAMP⟩ (sample) ⟨STRONG⟩ (strong emphasis) ⟨VAR⟩ (variable)	These elements are all examples of phrase and text markup elements. They include ⟨EM⟩ (emphasis), ⟨STRONG⟩ (strong emphasis), ⟨SAMP⟩ (sample — a sequence of characters). ⟨VAR⟩ is placeholder for a variable element. The ⟨CITE⟩ element can indicate a book title or other citation and the ⟨CODE⟩ element marks computer code. The ⟨KBD⟩ elements indicates text typed by a user.
⟨B⟩ (bold) ⟨BR⟩ (line break) ⟨HR⟩ (horizontal rule) ⟨I⟩ (italic) ⟨TT⟩ (teletype)	These are typographical elements controlling the presentation of text: ⟨BR⟩ indicates a line break between words; ⟨HR⟩ a horizontal rule — a divider between sections of text.
⟨FORM⟩ ⟨INPUT⟩ ⟨OPTION⟩ ⟨SELECT⟩ ⟨TEXTAREA⟩	These are html elements concerned with forms and with the input of data to a hyperlink.

```
<HTML>
<HEAD><TITLE>order Form</TITLE></HEAD>
<BODY>
<FORM METHOD=GET ACTION="/sales/orders">
What is your order?
<INPUT TYPE=TEXT NAME=product SIZE=10>
<INPUT TYPE=SUBMIT VALUE="Product Required">
</FORM>
</BODY>
</HTML>
```

Figure 11.8 Example html document: an html 'form'.

Figure 11.9 The graphical user interface (GUI) of Microsoft's Internet Explorer web browser [reproduced under the advice of Waggener Edstrom on behalf of Microsoft Corporation].

Note how the start and end of each html element are labelled with the relevant element ⟨tag⟩. The start tag simply closes the element name between '<' and '>' symbols. The end tag is similar, except that an extra 'slash' (/) appears in front of the element name. Note how each html element has both start and end tags, e.g.: ⟨HTML⟩ and ⟨/HTML⟩, ⟨HEAD⟩ and ⟨/HEAD⟩ etc.

11.7 Web browsers

When viewed using a web browser software (such as Netscape *Navigator* or Microsoft *Internet Explorer*) or using an html-editor software (such as Microsoft *Word*), an html document assumes the formatted graphical-style format intended for the human user to see (rather than the tagged format of the original text/html document). This is the professional-looking graphical-format familiar to all 'surfers' of the Worldwide Web (Figure 11.9).

The address field which appears near the top of most browsers allows a human user to input http-format Request-Lines, including the relevant head anchor address of a hyperlink. Hitting the 'enter' key after this input activates the hyperlink — triggering the http user agent into action. The first action is undertaken by the DNS resolver software incorporated into the web browser software. The DNS resolver queries the domain name system (DNS) for the IP-address of the head end of the hyperlink. Once the address has been returned in a DNS response message, a TCP connection is set up to this address by the web browser software, and the hypertext transfer protocol (http) is used to locate and retrieve the requested html file. Once received, the web browser displays each of the elements of the html document.

Should the human user 'click' on any of the further hyperlinks within the document, a new retrieval process begins again involving DNS, TCP/IP and http.

11.8   Web-based applications

The popularity of the Worldwide Web (www) has grown rapidly since the early 1990s and the computing world has adapted to embrace it. Nowadays, web browser software is not simply used to navigate around the web pages and hyperlinks of the Worldwide Web, but also widely used as the *client software* for many enterprise applications. Browsers lend themselves well as the 'user front-end' for complex enterprise applications. Thus, for example, Figure 11.10 illustrates a multi-layered web-based application accessed by a human user by means of a PC and web-browser software. In all, five different servers are involved in the application: a DNS server to resolve the address of the application server, the application server itself, *a RADIUS (remote authentication dial-in user service) server and LDAP (lightweight directory access protocol)* to perform the client authentication and a database server, where the results of the process will be stored (e.g., input of a sales order).

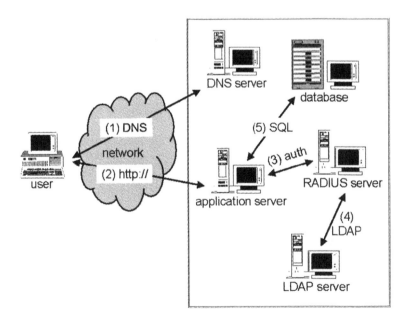

Figure 11.10   An example multi-layered web-based application.

By the use of the web browser as the main client interface, the user-end software is simplified, and the same software (the web browser) can be used for accessing multiple applications. This leads to lower costs and easier service support. The complex part of the application, the *remote procedure calls (RPCs)*, the UNIX sockets and *SQL (standard query language)* typical of 'classical' client/server applications are kept in the application server and database server—near at hand for the IT (information technology) support staff.

12

Electronic Mail (email)

Electronic mail (email) is a reliable and exceptionally fast means of message communication between human users equipped with computer terminals or personal computers. Large tracts of text, and diagrams too, can be quickly delivered across great geographical distances and either printed to a high quality paper format using local computer printing resources or subjected to further processing. With the assistance of electronic mail software, a human user can quickly formulate a message: he or she types the name of the main recipient and any other individuals who should receive a 'carbon copy', writes the 'title' or 'subject' and then the main text of the message. In addition, he/she might add complete computer files (e.g., a document, a spreadsheet or a presentation file) as attachments. Once the electronic mail (email) message is ready for sending, the sender may set a priority rating for the message, decide whether he/she requires confirmation of receipt and set a time and date for delivery (if this is not to be immediate). After this, the 'email' message is 'timestamped' and submitted to the Internet mail system. A message transfer system (MTS), based upon the simple mail transfer protocol (SMTP) sees to the electronic transport and delivery of the message, which usually arrives more-or-less instantly in the mailboxes of all the intended recipients. In this chapter we explain the principles of Internet mail: the format of messages and mail addresses; the message transfer system (MTS) and the various protocols associated with it: SMTP (simple mail transfer protocol), IMAP (Internet message access protocol) and POP (post office protocol).

12.1 A typical electronic mail

The general form of an electronic mail message is a familiar sight nowadays:

```
To: sales@company.com
Cc: purchasing@customer.org
From: consumer@customer.org

Date: Monday 15th January 2002 20:03

Subject: New Product Catalogue

Please send me a copy of your latest product catalogue and your
current price list.
```

Data Networks, IP and the Internet: Protocols, Design and Operation   Martin P. Clark
© 2003 John Wiley & Sons, Ltd   ISBN: 0-470-84856-1

```
Best Regards,
A very good customer

Attachments:     <none>
```

The above example represents the *content*-part of a typical electronic mail conforming to the *Internet message format*. The *email addresses* which appear after the *keywords* 'To:', 'Cc:' and 'From:' are *Internet mail addresses*. The 'Subject:' field is the title of the email, as typed by the message sender. This is followed by the main text of the message and any attachments.

The 'To:', 'From:' 'Date:' and 'Subject:' fields typically appear listed in the 'Inbox' and 'Outbox' of recipient and sender respectively. These fields give the electronic mail user a quick overview of the messages currently in his *mailbox*. The main text section of the message, however, usually only appears when the message is *opened*.

Should any attachments have been included with the message, they must conform to one of a set of standard datafile types, as defined by the *MIME (multipurpose Internet mail extension)* specifications (RFC 2045-9).

12.2 The benefits of electronic mail (email)

Electronic mail (email) messages are delivered extremely quickly with no possibility of getting lost or only part-delivered. There can be confirmation of receipt if required, and in some email systems even confirmation of the fact that recipients have read their messages. Broadcasting of messages is quick and easy to achieve. Editing of text and returning or forwarding the amended version can be achieved with minimal re-typing. Messages can be filed and quickly retrieved later. Messages can be posted for exactly timed delivery, and can be prioritised according to the urgency with which they need to be dealt with.

On the receipt of a new message, some email systems immediately advise 'online' users (i.e., active users at their terminals) of the message arrival with a 'beeping' noise or a short message displayed at the bottom of their terminal screen). The recipient then has the option to read the message immediately or later, depending on how they regard its priority. Otherwise, recipients simply find their incoming mail on the next occasion on which they check their mailbox (like emptying the paper mail out of a postal mailbox).

As messages are read, confirmation is returned to the sender (if required). Each recipient has the choice to reply to the message, forward it, file it, print it out, amend it or delete it.

In some systems, users are able to check their electronic mailboxes even when they are away from their normal offices — either by downloading them across the Internet using *POP3 (post office protocol version 3)* or by using a *webmail* interface to view their messages directly on the server 'mailbox'.

At the start of electronic mail during the 1980s, various different technical standards appeared, including: the *Internet mail system*, a rival system based on the *X.400 (message handling system)* recommendation of the CCITT[1] and various computer manufacturers' proprietary systems (e.g., USENET, XNS — Xerox network system etc.).

Market demand for email was so great that a widespread take-up took place during the 1990s. And because the ability to communicate with company-external parties by means of email was considered a high priority, most companies elected for the *Internet mail system* because of its widespread coverage. As a result, the alternative technologies largely died out and a *de facto* standard for email thus emerged. Internet mail is nowadays widely used for all sorts of general business communication, including memos, receiving customer orders,

---

[1] CCITT stands for International Telephone and Telegraph Consultative Committee. The name has subsequently been changed to ITU-T — International Telecommunications Union Standardization sector.

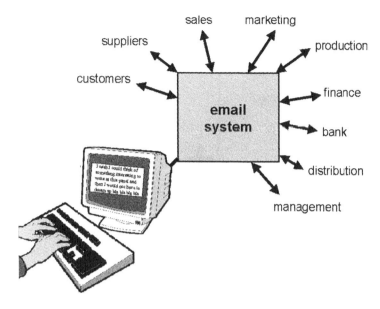

Figure 12.1 Typical usage and benefits of electronic.

laying-off orders on suppliers, etc. It is widely considered to be a reliable, accurate and quick means of communication, overcoming the barriers of geography and time zones (Figure 12.1).

Companies who have successfully introduced electronic mail have observed a beneficial change in the whole culture of how they do business. Questions and responses are dealt with more quickly and more directly than before, messages have become shorter and less formal — and tend to be typed by the managers themselves rather than by their secretaries. Workgroups composed of members in widespread locations have evolved, and it is possible to draw together new teams for previously impossible tasks.

12.3   The principles of the Internet mail transfer system (MTS)

All electronic mail (email) messages comprise the message itself (called the *content*) and an *envelope* (Figure 12.2). The envelope provides a 'label' for the message, telling the *message transfer system* (*MTS* — which is equivalent to the postal service) where to deliver the message, without the need to open it and inspect the contents.

In reality, the envelopes are simply extra data and control commands, sent in a standard format accompanying the message content. These commands control and 'steer' the message transfer system (MTS).

The Internet mail message transfer system (MTS) allows the conveyance of messages across a network on a *store-and-forward* or *store-and-retrieve* basis. Because of the capability of an email network to store messages, information can be sent at any time without interrupting the recipient from his current activities: the message is retrieved by and dealt with by the recipient when it is convenient. This is important in computing activities if the receiving machine is already busy. It is also important for human recipients who might be away from their desk. Much like a postal system, messages may be *posted* into the system at any time of day and will be delivered to the recipient's *mailbox*. The recipient is free to sort through the incoming *mail* at any time of day, and as soon as an item is accepted by the designated recipient,

content   envelope

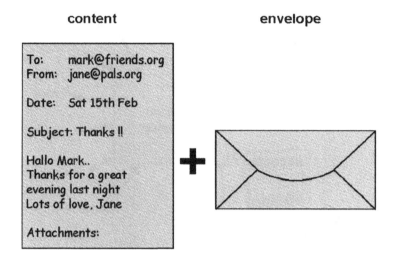

Figure 12.2   Content and envelope of an Internet email message.

automatic confirmation of receipt can be despatched back to the sender. This eliminates any possible concern about whether the recipient has seen a given item.

Figure 12.3 illustrates the elements of the Internet mail system and message transfer system (MTS). The two basic components of the Internet mail system are the *message user agent (MUA)* and the *message transfer agent (MTA)*. The message user agent (MUA) or user agent (UA) function is undertaken by email software on a personal computer (or by a combination of software on the user PC and his *email server*). The *user agent* helps the human user to compose messages in a standard form suitable for transmission and provides a 'filing cabinet' for previously received and sent messages which may be read, filed or otherwise processed.

Having prepared a message with the help of the user agent, the human user may trigger the user agent to *submit* the message to the local message transfer agent (MTA). The message is conveyed to its final destination via a number of MTAs, collectively called the message transfer system. The first MTA in the connection is typically the electronic mail server associated

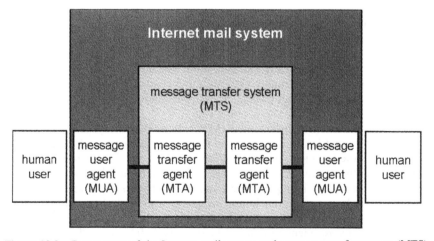

Figure 12.3   Components of the Internet mail system and message transfer system (MTS).

with the sender. Intermediate MTA devices may then be used to relay the message to a destination *postmaster* server, where the mailbox of the intended recipient is located. (This is the equivalent of a PO box at a receiving post office.)

The message transfer from MTA to MTA through the message transfer system (MTS) occurs in a step-by-step (store-and-forward) fashion until it reaches the destination mailbox, where it is stored until *retrieved*.

Users may be either *originators* or *recipients* of messages. Messages are carried between originator and recipient in a sealed electronic envelope. The envelope indicates the name of the recipient and records further information which may be necessary to cater for special delivery needs. For example, a confidential marking could be added to ensure that only the named addressee may read the information.

12.4   Operation of the Internet mail system

The content of electronic mail (email) messages is required to conform to the *Internet message format* laid out in RFC 2822 (formerly RFC 822). The format requires that the message content provided by the human user be split into a *content-header* and a *content-body*. The content-header includes a number of the standard fields regular email users will be familiar with:

- To:

- From:

- Cc: (carbon copy)

- Date:

- Subject:

The content-header fields are input by the human user (the message *originator*). They indicate the intended *recipients* of the message, the *subject* and so on. The address fields must be filled in using standard Internet mail addresses in the form 'martin.clark@company.com'. Such mail addresses conform to the standard defined by the *domain name system (DNS)*. The first part of the address (preceding the @-symbol) is the *mail username*. This identifies the mailbox of the recipient user. This mailbox is located on the *mailbox exchange (MX)* server or *postmaster* associated with the *domain*. The server is identified by the *domain name* which follows the @-symbol (in the example above, the domain name is 'company.com').

The *content-body* includes the basic text message and any attachments (in *MIME — multipurpose Internet mail extension* — format).

Once the human user has composed the content-header and content-body, the message is ready for submission to the message transfer system (MTS). At this stage, the user agent (UA) (usually the mail server of the mail originator) makes a number of copies of the message (one for each of the recipients) and prepares a separate envelope for each. Each envelope is addressed appropriately to one of the intended message recipients. (Figure 12.4). Any message encryption or other end-to-end security measures are also undertaken at this stage — in effect 'sealing' the envelope with the contents inside it.

Messages are next introduced to the message transfer system. Messages are relayed from one message transfer agent (MTA) to the next on a step-by-step basis by means of the *simple mail transfer protocol (SMTP)*. The sending MTA is called the *SMTP-sender* (also called the *SMTP client*) and the receiving MTA is the *SMTP-receiver* (also called *SMTP-server*). SMTP determines the address of the message by inspecting the *mail address* on the *envelope*.

The simple mail transfer protocol (SMTP) is an *application layer* protocol which relies on the transmission control protocol (TCP) *port* 25 and the Internet protocol (IP) for data transport.

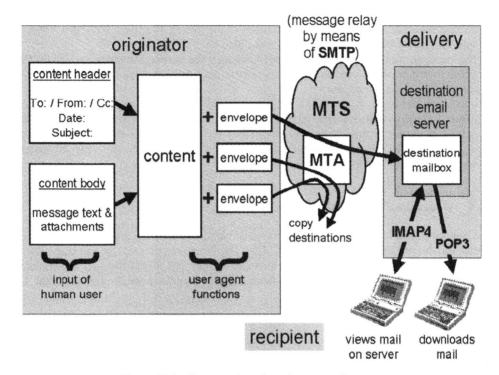

Figure 12.4   The operation of the Internet mail system.

The DNS service (as already explained in detail in Chapter 11) is used to *resolve* the Internet protocol address of the *mail exchange (MX)* server associated with the destination mail address. Once the address is known, the email message can be forwarded to the destination mailbox by means of *SMTP (simple mail transfer protocol)*. If possible, the delivery should be directly from the originating MTA to the delivery MTA (Figure 12.4). But if necessary, the message may also traverse a number of intermediary MTAs. When present, the intermediary MTAs typically perform one of three functions: a *relay MTA*, a mail *gateway MTA* or a *mail proxy* MTA.

A relay MTA may be used in the case where the originating MTA has been unable to resolve the IP address of the destination, (maybe the DNS service was temporarily unavailable to the originating MTA).

A *mail gateway* MTA may be used to convert the format of the mail message or to connect to a mail system conforming to a different technical standard (e.g., to an X.400 mail system). Alternatively, a gateway might be used to arrange for message delivery via some other type of network (e.g., fax, telex or even voicemail!).

A *mail proxy* is often present in an enterprise firewall. The proxy typically has the role of checking the content of the mail for viruses or other malicious material, before allowing the message to be transferred into an internal company network. This is a security measure called *content filtering*. The filter typically searches for 'unallowed' or 'dangerous' file types (e.g., .vba etc.), deleting or *quarantining* these if necessary.

Once the mail message has traversed the message transfer system (MTS) to the recipient's mailbox on the destination *email exchange server*, the message is ready to be picked up by the human recipient. There are two normal means of final delivery (Figure 12.4). *The IMAP4 (Internet mail access protocol version 4)* allows a user equipped with appropriate software to view the mail messages directly on the server, replying, copying, deleting or filing them without removing them from the server. The IMAP4 protocol may also be used to maintain a duplicate 'offline' mailbox (say, on a user's laptop PC), and to *synchronise* the contents of the laptop and server copies of the mailbox. Alternatively, the *post office protocol (POP3)* is a simple protocol for retrieving (i.e., downloading) all the messages from the server mailbox into the user's PC. Following a successful POP3 download, the copies of the messages left on the server are deleted.

12.5   The Internet message format

An Internet mail message comprises an envelope and the message content (Figure 12.5). The envelope comprises a series of SMTP (simple mail transfer protocol) *commands* and *replies* (defined by RFC 2821) which control the message transfer from message transfer agent (MTA) to message transfer agent (MTA) across the message transfer system (MTS). We shall discuss these commands later in the chapter.

The message content of an Internet mail message (email) is formatted according to the Internet message format (as defined in RFC 2822). The content is sub-divided into a content-header and a content-body (Figure 12.4). The content-header and content-body are transferred together by SMTP as an SMTP *protocol DATA unit (PDU)*. SMTP protocol DATA units are divided into lines of characters, each terminated with a ⟨CRLF⟩ sequence. The maximum line length is 998 characters but it is recommended that a 'normal line length' of 78 characters be used (with ⟨CRLF⟩, this is equivalent to a standard line length of 80 characters).

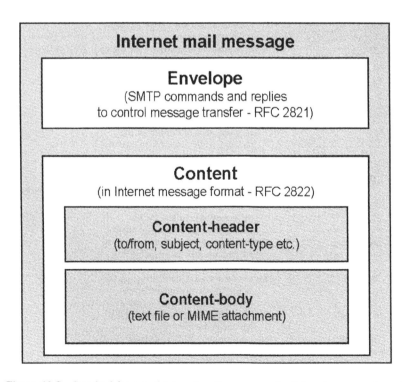

Figure 12.5   Standard format of Internet mail messages: the Internet message format.

Content-header format

Mail message content-header fields are composed of a *field-name* followed by a colon, a list of arguments (*field-body*) and a *carriage return-line feed* (CRLF), thus:

```
field-name:field-body <CRLF>
```

Table 12.1 lists the possible header fields. The 'From:' and 'Date:' fields must always be present. The other fields are optional, though most email software will check for at least one addressee ('To:', 'Cc:' or 'Bcc:') and that the 'Subject:' field has been filled in. The addressee fields ('To:', 'Cc.' and 'Bcc:') may each appear a maximum of one time, but may include a list of multiple email addresses.

Table 12.1   SMTP protocol DATA unit: header fields

Header field	Header field name	Meaning and purpose	Min number in a header	Max number in a header
Orig-date	Date:	The origination date and time.	1	1
From	From:	The mail address of the mail originator. In the case of a list of originators, the sender field must also appear.	1	1
Sender	Sender:	The mail address of the sender of the message in the case that the from field is a mailbox list.	0	1
Reply-to	Reply-To:	The mail address suggested to be used for replies.	0	1
To	To:	A destination mailbox address (or list of mailbox addresses) for the message (primary recipients)	0	1
Cc	Cc:	A destination mailbox address (or list of mailbox addresses) for the message (so-called carbon copies)	0	1
Bcc	Bcc:	A destination mailbox address (or list of mailbox addresses) to which the message is being copied (Blind copy). The message is copied to this recipient, though this line of the message will be deleted, so none of the other recipients will be aware of this copy.	0	1
Message-id	Message-ID:	An optional but recommended unique message identifier.	0	1
in-reply-to	In-Reply-To:	This field is recommended to appear in reply messages to identify the message to which the reply refers.	0	1
References	References:	A reference to the unique message-id of another message as appearing in a reply.	0	1
Subject	Subject:	An unstructured text field indicating the subject of the mail message and intended for the human recipient.	0	1
Comments	Comments:	An unstructured text field intended for the human recipient.	0	Unlimited
Keywords	Keywords:	An unstructured text field intended for the human recipient.	0	Unlimited

Table 12.1 (*continued*)

Header field	Header field name	Meaning and purpose	Min number in a header	Max number in a header
Resent-date	Resent-Date:	Resent fields have the same meaning as the 'original' fields with the same name. The 'resent' tag indicates they have been re-introduced to the mail transport system.	0	Unlimited
Resent-from	Resent-From:		0	Unlimited
Resent-sender	Resent-Sender:		0	Unlimited
Resent-to	Resent-To:		0	Unlimited
Resent-cc	Resent-Cc:		0	Unlimited
Resent-bcc	Resent-Bcc:		0	Unlimited
Resent-msg-id	Resent-Message-ID:		0	Unlimited
Trace	Return-Path: Received:	The optional trace-header fields can be used for debugging mail faults.	0	Unlimited
Optional-field		Optional fields not defined in the formal standard.	0	Unlimited

Content-body format

The content-body may include one of a number of different general types of mail message bodies:

- text message bodies in US-ASCII (this was the standard format of mail message bodies prior to MIME);

- text message bodies using character sets other than US-ASCII;

- extensible set of different non-text message body formats;

- multi-part message bodies; and/or

- body-header information about content-bodies which use character sets other than US-ASCII.

The content-body may also include one or more computer files as attachments. When attachments are included, the data file format of the attachment needs to be indicated in line with the file-type definitions specified by *MIME* (*multipurpose Internet mail extensions* — RFC 2045-9 [Nov. 1996]).

The format of an attached file in the content-body (including text-body and any attachments) is indicated in the content-header by the following additional *MIME message header fields* (see also Table 12.2):

- MIME-version header field

- Content-Type

- Content-Transfer-Encoding

- Content-ID

- Content-Description

Table 12.2   MIME (multipurpose Internet mail extensions): additional body header fields [RFC 2045]

Header field name	Meaning and purpose	Field values
MIME-Version:	The argument value of this header field indicates the MIME-version number.	the current version is '1.0' (RFC 2045-9)
Content-Type:	The argument value of this header field indicates the mail body [media] content type and subtype.	Indicates the MIME-Media type (see Table 12.3) Values are either of a discrete-type or of composite-type
Content-Transfer-Encoding:	The argument value of this header field indicates the character set used to code the message body.	Alternative values: 7bit 8bit binary quoted-printable base64 ietf-token x-token
Content-ID:	The argument value of this header field is a unique identifier to identify the MIME content in the message body.	unique-message-id
Content-Description:	A textual string follows this header-field name to provide descriptive information about the message content intended for a human recipient.	Text
Content-Duration:	The argument value of this header filed provides an indication of the content duration. It is intended for use with any timed media (i.e., audio or video) content (defined in RFC 2424).	Value of duration in seconds
Content-features:	This field provides additional information about the main message content (defined in RFC 2912).	Various parameters

Table 12.3   MIME media-types (RFC 2046)

Top-level media type	Media-subtypes	Usage or meaning
Text	text/plain; charset=iso-8859-1	A message in plain text encoded using the ISO8859-1 character set.
	text/plain; charset=us-ascii	A message in plain text encoded using the US-ASCII character set (ANSI $\times$3.4–1986).
	text/enriched	Rich text format (RFC 1896).
	text/html	A text file coded using hypertext markup language (html) (RFC 2854).
	text/directory	A text file containing directory information.
	text/parityfec	An RTP (real-time application transport protocol) format employing generic forward error correction (fec) (RFC 3009).
Image	image/jpeg	An image file in jpeg (joint photographic experts group) format.

Table 12.3 *(continued)*

Top-level media type	Media-subtypes	Usage or meaning
	image/g3fax	An image file in group3 fax format according to CCITT/ITU-T recommendation T.30 (RFC 2159).
	image/gif	An image file in gif-format (graphics interface format).
	image/t38	An image file in ITU-T facsimile format (ITU-T rec. T.38).
	image/tiff	An image file in tiff-format (tag image file format).
Audio	audio/basic	8-bit μ-law pulse code modulation.
	audio/32kadpcm	An attached file in 32 kbit/s adaptive differential pulse code modulation (ITU-T Rec G.726 and RFC 2422).
	audio/L16	An audio-format file encoded according to L16 coding (RFC 1890 and RFC 2586).
	audio/parityfec	An RTP (real-time application transport protocol) format employing generic forward error correction (fec) (RFC 3009).
Video	video/mpeg	A video file in mpeg (motion picture experts group) format.
	video/parityfec	An RTP (real-time application transport protocol) format employing generic forward error correction (fec) (RFC 3009).
Application	application/octet-stream	A file that comprises data of an arbitrary type or otherwise unspecified format.
	application/postscript	A file in the Adobe Systems PostScript format (typically used for printing).
	application/oda	A file in office data architecture (ODA) format (CCITT recommendation T.411 and RFC 21619).
	application/iso-10161-ill-1; transfer encoding...	The carried object is a BER (basic encoding rules) encoded ISO ILL (interlibrary loan) PDU (protocol data unit).
	application/ill-ddi; transfer encoding...	The carried object is a BER (basic encoding rules) encoded ISO ILL (interlibrary loan) PDU (protocol data unit).
	application/parityfec	An RTP (real-time application transport protocol) format employing generic forward error correction (fec) (RFC 3009).
	application/ISUP	The carried object is an ISUP (integrated services user part) message of signalling system number 7 (used in digital telephone networks) (RFC 3204).
	application/QSIG	The carried object is a QSIG signalling message (as used in digital private telephone networks) (RFC 3204).
	application/xhtml + xml	A file in xhtml or xml format (RFC 3236).
	application/dicom	A file in a format specified by DICOM (digital imaging and communications in medicine) (RFC 3240).
Multipart (composite-type)	multipart/mixed	The message body comprises a number of separate 'attachments' of different types, separated by 'boundaries'.

(continued overleaf)

Table 12.3   *(continued)*

Top-level media type	Media-subtypes	Usage or meaning
message (composite-type)	multipart/alternative	The message body comprises a number of versions of the same 'basic content' in different formats. The receiver should decide which one is most appropriate to use.
	multipart/digest	Intended to be used to send collections of messages.
	message/rfc822	An encapsulated message in RFC 822 format.
	message/partial	An encapsulated fragment of a message.
	message/external-body	This indicates that the body data are not included in the message but instead only referenced.
	multipart/related	This MIME-type is accompanied by additional information indicating how to unpack or process data (e.g., which program to use) (RFC 2387).
	multipart/voice-message	A file to be used in conjunction with the Voice Profile for Internet Mail (VPIM) (RFC 1911 and RFC 2423).
	multipart/signed	A file made secure for transport by the use of a digital signature (RFC 2480).
	multipart/encrypted	A file made secure for transport by encryption transformation (RFC 2480).

The file attachments to an electronic mail conform to one of the *MIME media types* as listed in Table 12.3:

12.6   Simple mail transfer protocol (SMTP)

Once the content of an electronic mail (email) message is ready for sending (i.e., both the content-header and content-body are complete as an SMTP protocol DATA unit), it is the simple mail transfer protocol (SMTP) which is used to transfer it to its destination. SMTP is defined in RFC 2821 and RFC 821. The SMTP communication model (Figure 12.6) is somewhat similar to the FTP (file transfer protocol) model (explained in Chapter 10). Like FTP and http (hypertext transfer protocol), SMTP is an application layer protocol.

The mail addresses provided as the intended recipients of an electronic mail are expected to be *fully qualified domain names (FQDN)*, though some email software may be able to infer names from partial inputs or *alias* names. Like http, SMTP relies upon the *domain name system (DNS)* for the resolution of email domain name addresses — thereby deriving the IP address of the destination *mail exchange (MX)* server.

An SMTP message transfer system (MTS) (Figure 12.3) comprises message transfer agents (MTAs) and message user agents (MUAs — also called simply user agents, UAs). An originating MUA collects mail from a user and hands it off to an MTA. In the final delivery the MTA hands off the message to the MUA (which may be an end user (i.e., a computer terminal), although nowadays it is more normal for email to be delivered to a message depository, i.e., mailbox — for awaiting pick-up by the user).

SMTP MUAs and MTAs may also undertake the following SMTP functions:

- An *SMTP originator* (a message user agent, MUA) introduces mail into the Internet mail transport system — creates copies and makes up an envelope for each of them (like an office secretary when sending the post).

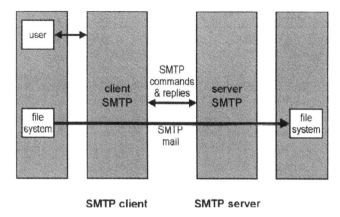

SMTP client   **SMTP server**

Figure 12.6   SMTP client/server (sender/receiver) model.

- An *SMTP delivery* agent is usually the destination mail exchange (MX) — it receives mail and stores it in a depository.

- An *SMTP relay* stores and forwards mail without opening the envelope. It may provide functionality which a simple SMTP originator may itself be incapable of. A relay might be used, for example, to use the domain name service (DNS) to resolve the destination mail address for the IP address of the destination mail exchange.

- A mail gateway is a particular type of SMTP relay station, whose function is to convert the format of the mail (e.g., from the Internet message format into the ITU-T X.400 *message handling system (MHS)* format).

- A mail *proxy* is a particular type of SMTP relay station which is often a constituent part of a company's security *firewall*. Mail proxies are usually used to vet the contents of emails. They may check the 'From:' and 'To:' addresses as well as performing content-filtering of the message content-body (to ensure it does not contain known computer *viruses* or other 'undesirable' contents).

When an *SMTP client* (which is also called an SMTP-sender: and may be either an *SMTP originator* or an SMTP relay agent) has a message to transmit, it establishes a two-way transmission channel to an *SMTP server* (also called an SMTP-receiver) on TCP (transmission control protocol) port 25 (Figure 12.6). Normally, the SMTP server selected will be the final destination (e.g., destination postoffice or mailbox), but the SMTP server may also be an intermediate relay point (such as a mail relay, mail gateway or mail proxy). Relaying takes place by means of a formal hand-off of the message from SMTP client (SMTP-sender) to SMTP server (SMTP receiver).

In the case of a message relayed across intermediate points, each SMTP server reverts to become an SMTP client after having received the message from the previous station. Only as an SMTP client can a message transfer agent (MTA) forward a message. SMTP message transfer takes place under the control of the SMTP server (SMTP receiver) but at the initiation and only following the requests of the SMTP client (SMTP sender).

SMTP protocol commands: the message 'envelope'

The envelope of an Internet mail message (Figure 12.5) is composed of a series of SMTP *commands* (Table 12.4), sent on the SMTP connection prior to the message itself. Each *SMTP*

Table 12.4 SMTP commands (the message *envelope*)

Commands (in approximate order of usage)	Syntax	Meaning	Usual response
EHLO (extended hello) [replaces HELO]	EHLO ⟨SP⟩ ⟨domain⟩ ⟨CRLF⟩	Used by the SMTP client to start an SMTP session.	250
MAIL	MAIL FROM: ⟨reverse-path⟩ [⟨SP⟩ Mail parameters] ⟨CRLF⟩	Used by the SMTP client to initiate a mail transaction.	
RCPT (recipient)	RCPT TO: Postmaster@domain [⟨SP⟩ Rcpt-parameters] ⟨CRLF⟩	Used to identify the intended recipient of the mail.	
DATA	DATA ⟨CRLF⟩	Request from an SMTP client to send data. After the 354 response, the following transmission is treated as the mail data.	354
BDAT	BDAT ⟨SP⟩ chunk-size [⟨SP⟩ end-marker] ⟨CR⟩ ⟨LF⟩	An alternative to the DATA command (if the EHLO CHUNKING extension is supported, allowing large messages to be sent in chunks of binary data. The last chunk is sent with a chunk-size of 0 to indicate the end of the data.	250
RSET (reset)	RSET ⟨CRLF⟩	This command aborts the current mail transaction.	
VRFY (verify)	VRFY ⟨SP⟩ String ⟨CRLF⟩	The SMTP client requests confirmation from the server that the argument identifies a user mailbox	
EXPN (expand)	EXPN ⟨SP⟩ String ⟨CRLF⟩	The SMTP client requests confirmation from the server that the argument identifies a mailing list and to return a list of the individual members mailbox addresses.	Mailing list
HELP	HELP [⟨SP⟩ String] ⟨CRLF⟩	The SMTP client requests that the server return helpful information to aid the human user.	
NOOP	NOOP [⟨SP⟩ String] ⟨CRLF⟩	The SMTP client requests a sign of life from the server, which should ignore the string if sent.	220 OK
QUIT	QUIT ⟨CRLF⟩	The SMTP client requires that the server (the SMTP receiver) send an OK reply and then close the transmission channel.	221 OK

command is answered with an *SMTP reply*. Replies indicate acceptance of the previous command, that additional commands are expected or that an *error* has occurred. As in the *file transfer protocol (FTP* — discussed in Chapter 10), a simple *dialog* is expected to take place: each command followed by a reply.

SMTP *session initiation* commences when an SMTP client *opens* a connection and the SMTP server *responds* with an *opening message*. Following session initiation the client sends the *EHLO (extended hello)* command with the client's identity (this is called *client initiation*). This has the effect of opening the session and clarifying any *SMTP extensions* (Table 12.5) which are supported.

The message transfer *transaction* commences with the SMTP client (SMTP sender) sending the *MAIL* command to indicate to the SMTP receiver (SMTP server) that a new mail transaction is starting [The ⟨reverse-path⟩ is the 'From:' mail address].

        MAIL FROM: <reverse-path> [<SP> <mail-parameters>] <CRLF>

If the SMTP server is prepared to accept the *MAIL*, it responds with the reply '250 OK'. The next step is for the SMTP client (sender) to indicate the destination address of the message. This it does with the *RCPT (recipient)* command. [The ⟨forward-path⟩ is a destination mail address].

        RCPT: <forward-path> [SP <rcpt-parameters>] <CRLF>

Once the message destination (i.e., the message envelope) has been sent, the SMTP may send the message content (an *SMTP protocol DATA unit*). This follows the SMTP DATA command

Table 12.5   SMTP service extensions

EHLO keyword	Service extension	Defined in	Purpose
CHUNKING	Large Message	RFC 3030	As an alternative to the SMTP command DATA, the BDAT command allows a large message to be sent in chunks of binary data.
DELIVERBY	Deliver By	RFC 2852	The MAIL FROM command is extended by the optional keyword BY to request delivery within the time given.
PIPELINING	Command Pipelining	RFC 2920	Using a single TCP send operation, the SMTP client is able to send multiple commands simultaneously. This is intended to improve SMTP performance.
SAML	Send and Mail	RFC 821 & RFC 1869	This EHLO keyword and SMTP command requires that the mail be delivered directly to the user's terminal (if the user is active), and to his mailbox.
SEND	Send	RFC 821 & RFC 1869	This EHLO keyword and SMTP command requires that the mail be delivered directly to the user's terminal (not to his mailbox). If this is not possible the 450 error reply is returned.
SOML	Send or Mail	RFC 821 & RFC 1869	This EHLO keyword and SMTP command requires that the mail be delivered directly to the user's terminal (if the user is active), otherwise to his mailbox.
TURN	Turn	RFC 821 & RFC 1869	This EHLO keyword and SMTP command allows the roles of SMTP client (SMTP-sender) and SMTP server (SMTP-receiver) to be reversed.

(Table 12.4). Once a message has been transmitted, the SMTP client may either request to terminate the SMTP session and shut down the connection or may undertake further mail transactions.

Sometimes an SMTP client may try to forward an email to a particular SMTP server in order to *verify (VRFY)* or correct a destination mailbox address. This may be necessary, for example, in the case of the use of an *alias* address (a pseudo-mailbox address). In this case it will receive an updated address as a reply. The *expand (EXPN)* command, meanwhile, is used to obtain the individual recipient mail addresses, in the case that a mail is sent to a mail distribution list.

SMTP reply codes (Table 12.6) have a very similar format to those used in FTP (file transfer protocol).

Table 12.6   SMTP reply message codes and meanings

Reply types	SMTP reply meaning
x0z	Syntax error; unrecognised or superfluous command or simply OK confirmation
x1z	Request for further information such as status or help information
x2z	Reply refers to the transmission channel
x3z	Unspecified
x4z	Unspecified
x5z	Status of the receiver mail system (i.e., SMTP server)
Positive preliminary reply (1yz)	The requested action is being initiated — wait for a further reply before issuing a new command (this type of reply is only used by extended SMTP)
Positive completion reply (2yz)	The requested action had been completed successfully
211	System status or system help ready
214	Help message
220	⟨domain⟩ service ready
221	⟨domain⟩ service closing transmission channel
250	Requested mail action OK and completed
251	User not local, will forward mail to ⟨forward-path⟩
252	Cannot verify user but will accept message and attempt to deliver it
Positive intermediate reply (3yz)	The command has been accepted but the requested action has not been commenced, while awaiting further information — which should be sent by the user
354	Start mail input; end with ⟨CRLF⟩.⟨CRLF⟩
Transient negative completion reply (4yz)	The command was not accepted and the requested action not undertaken, but the action may be requested again — if so by returning to the beginning of the command sequence
421	⟨domain⟩ service not available, closing transmission channel
450	Requested mail action not undertaken, mailbox temporarily unavailable (e.g., busy)
451	Requested action aborted due to local processing error
452 ·	Requested action not possible due to insufficient storage
Permanent negative completion reply (5yz)	The command was not accepted and the requested action not undertaken. The same request command sequence should not be repeated
500	Syntax error, command unrecognised
501	Syntax error, parameter or argument error
502	Command not implemented
503	Bad sequence of commands

Table 12.6 (*continued*)

Reply types	SMTP reply meaning
504	Command not implemented for parameter specified
550	Requested mail action not undertaken, mailbox unavailable (e.g., not found, access not allowed or command rejected for administrative policy reasons)
551	User not local, please try ⟨forward-path⟩
552	Requested mail action aborted, exceeds storage allocation
553	Requested action not undertaken, mailbox name not allowed (e.g., incorrect syntax)
554	Transaction failed (or if in response to a HELO or EHLO — 'no SMTP service here')

Table 12.7 SMTP parameter constraints

SMTP parameter	Maximum size
local-part (user name length appearing before @)	64 characters
domain (name length)	255 characters
path (⟨reverse-path⟩ or ⟨forward-path⟩)	256 characters (including @ and. characters)
command line	512 characters (including command word and ⟨CRLF⟩ but not including parameters, specifically MAIL FROM and RCPT TO parameters and lists)
reply line	512 characters (including command word and ⟨CRLF⟩)
text line	1000 characters (including ⟨CRLF⟩) [but can be increased by means of SMTP service extensions]

Table 12.7 lists the length constraints on SMTP address, command, reply and text lengths.

Once an SMTP message is received at its final destination (the destination message user agent, MUA), the presentation format of the message mail (how it appears on the screen) is not standardized. This is left to the electronic mail software developer to decide.

12.7 Internet mail access protocol (IMAP4)

The *Internet mail access protocol* (*IMAP* — also called the *interactive message access protocol*) allows a *client* to access and manipulate electronic mail messages on an (IMAP) server (i.e., a mail exchange (MX) server or electronic mail *post office*). The current version (IMAPv4rev1) is defined in RFC 2060 (Dec 1996).

Access to the mailbox on the IMAP server might allow an IMAP client simply to read and reply to his email. Alternatively, by manipulating his/her mailbox, he/she can perform administrative functions like filing and deleting old messages. A common usage of IMAP is to maintain an 'offline' copy of the mailbox — typically on a laptop computer. Thus a human user might choose to create two copies of his or her mailbox — one on the office electronic mail server and a second *synchronised* copy on a laptop computer used while on business trips.

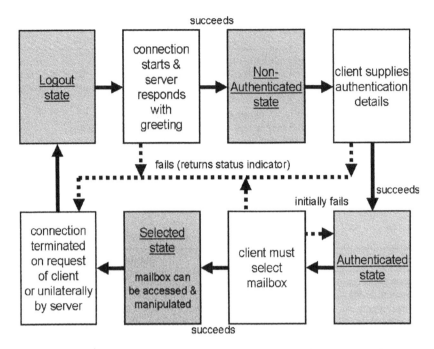

Figure 12.7   Internet mail access protocol v4 (IMAP4): IMAP server state diagram.

IMAP is an application protocol and operates in conjunction with the Internet protocol (IP) and TCP (transmission control protocol) on port 143. An IMAP server is always in one of four states (Figure 12.7):

- non-authenticated state;

- authenticated state;

- selected state; or

- logout state.

A series of IMAP commands and responses lead the IMAP server from one state to another. Starting in the logged out state, the IMAP client starts by opening a connection (TCP port 143). If successful, the IMAP server responds with a greeting, whereupon it enters the *non-authenticated state*. Now the *client* must authenticate itself. Success in doing so leads the IMAP server to progress to the *authenticated state*, whereupon the client may select a given mailbox. Finally, in the *selected state*, the client may view the contents of the mailbox, access individual messages and manipulate them (using the various commands listed in Table 12.8).

Each IMAP command (issued by the IMAP client) receives a response from the IMAP server. Three main possible server completion responses: OK, NO or BAD (i.e., protocol or syntax error). In addition, there are two further responses: PREAUTH and BYE. PREAUTH (pre-authorisation) is used as a possible greeting at the time of connection set-up. BYE indicates that the server is about to end the connection.

Table 12.8 IMAP commands from client to server

States in which command may be issued	Command	Arguments of command	Meaning and purpose	Response
Any	CAPABILITY	None	Client requests a list of capabilities supported by the IMAP server	OK BAD
	LOGOUT	None	Client request to logout	BYE
	NOOP	None	Client uses this command as a periodic check for new messages	OK BAD
Non-Authenticated	AUTHENTICATE	Authentication name	Client supplies optional authentication information	OK — authenticated NO — refused BAD
	LOGIN	⟨user-name⟩ ⟨password⟩	Client identifies himself to the server and provides password	OK login complete NO — rejected BAD
Authenticated	SELECT	⟨mailbox-name⟩	Client selects relevant mailbox	OK (after server responds with number of messages and flags) NO — failure BAD
	EXAMINE	⟨mailbox-name⟩	Client selects the mailbox as 'read-only'	Responses as 'SELECT' responses
	CREATE	⟨mailbox-name⟩	Client creates a mailbox with the specified name	OK NO BAD
	DELETE	⟨mailbox-name⟩	Client requests the permanent deletion of the mailbox with the specified name	OK NO BAD
	RENAME	⟨mailbox-name⟩ ⟨new-mailbox-name⟩	Client renames the mailbox	OK NO BAD
	SUBSCRIBE	⟨mailbox-name⟩	Client adds the specified mailbox to the list of the server's 'active' mailboxes	OK NO BAD
	UNSUBSCRIBE	⟨mailbox-name⟩	Client removes the specified mailbox from the list of the server's 'active' mailboxes	OK NO BAD
	LIST	⟨user-reference-name⟩	Client requests a list of all mailbox names available to him	OK — (after list complete) NO — failure BAD
	LSUB	⟨user-reference-name⟩	Client requests a list of all the 'active' mailbox names available to him	OK (after list complete) NO — failure BAD
	STATUS	⟨mailbox-name⟩ ⟨status-item(s)⟩	Client requests to know the status of the identified mailbox (e.g., number of messages, unseen messages)	OK (after listing status) NO BAD
	APPEND	⟨mailbox-name⟩ ⟨options⟩ ⟨message⟩	Client appends a message to the specified destination mailbox	OK — completed NO — error BAD

(continued overleaf)

Table 12.8   (*continued*)

States in which command may be issued	Command	Arguments of command	Meaning and purpose	Response
Selected	CHECK	None	Client asks whether the server is performing 'housekeeping' duties	OK—check complete BAD—command unknown
	CLOSE	None	Client removes all the \Deleted flags associated with the current mailbox and returns to authenticated state	OK—closed NO—failed—no mailbox selected BAD
	EXPUNGE	None	Client permanently removes all messages in the current mailbox marked with the Deleted flag	OK—completed NO—permission denied BAD
	SEARCH	⟨criteria⟩	Client searches through messages in current mailbox for those matching the search criteria (e.g., BEFORE, FROM, TO etc.)	OK—search complete NO—search error BAD
	FETCH	⟨message⟩ ⟨items⟩	Client requests data associated with a particular message (e.g., HEADER, BODY, ENVELOPE etc.)	OK—fetch complete NO—error BAD
	STORE	⟨message⟩ ⟨items⟩ ⟨value⟩	Client alters some element of data associated with a message	OK—completed NO—can't store BAD
	COPY	⟨message⟩ ⟨mailbox-name⟩	Client copies the message to the specified mailbox	OK—complete NO—can't copy BAD
	UID	⟨command-name⟩ ⟨arguments⟩	Used with SEARCH or FETCH commands to specify unique message identifiers	OK—complete NO—error BAD
	X		An experimental command	

Table 12.9   IMAP message system flags

System flags	Meaning
\Seen	Message has been read
\Answered	Message has been answered
\Flagged	Message is flagged for urgent or special attention
\Deleted	Message is marked as deleted for later removal by the EXPUNGE routine
\Draft	Message is still in the draft or composition stage
\Recent	Message recently arrived in the mailbox

Mailboxes on an IMAP server must have names based on original ASCII (UTF-7) characters. The messages within the mailbox are all marked with message flags (Table 12.9), which are updated during the course of an IMAP session. At the end of the session, it is normal for the client to request a mailbox 'clean up' by issuing the EXPUNGE command to permanently remove any messages at that time marked for deletion. The client then logs out (LOGOUT) and closes the TCP connection.

To protect the user's mailbox on the IMAP server from malicious manipulation, an auto logout may be undertaken by the IMAP server, should the client fail to authenticate himself fast enough, or should he/she fail to issue further commands to the server within a given timeout (when in the authenticated or selected states).

12.8   The post office protocol version 3 (POP3)

POP3 (post office protocol version 3) is intended to be used on smaller Internet nodes to allow for receipt of Internet electronic mail without the need for a full message transport system (MTS) or SMTP server. POP3 provides for the support of a message user agent (MUA), and allows a workstation to retrieve mail that a mail server is holding for it or forward all outgoing mail for onward relaying. Unlike IMAP (Internet mail access protocol), however, POP3 does not allow extensive manipulation of the mail on the server. Instead the normal mode of operation is a download of the mail, followed by a deletion of the mail copy held on the server. The POP3 protocol and procedure are defined in RFCs 1939, 2384 and 2449.

The POP3 client establishes a TCP (transmission control protocol) connection with the POP3 server (mailbox server) on TCP port 110. The POP3 server responds with a greeting. The client sends commands and the server responses to progress from an *offline state*, through an *authorisation state* and *transaction state* to the *update state*, after which the TCP connection is closed (Figure 12.8).

Mail transfer occurs in the transaction state, following successful authentication of the client to the server and successful *locking* of the user's mailbox by the server. The mailbox is *locked* prior to the mail download transaction state to ensure that no further modifications are made to the mailbox during this period. After mail has been successfully transferred to the POP3 client, the POP3 server enters the update state, during which the copies of the messages remaining in the locked mailbox are deleted. Following the update state the TCP connection

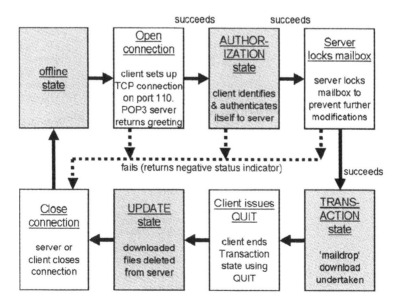

Figure 12.8   POP3 (post office protocol version 3): POP3 server state diagram.

is closed. The state progression is always offline to authorization state to transaction state to update state to offline, unless the session is aborted for some reason. If the session is *aborted*, the update state is not entered, thus preserving the entire contents of the server mailbox.

POP3 commands (Table 12.10) consist of 3 or 4-character *keywords* and relevant ASCII character arguments. The command is terminated by ⟨CRLF⟩. POP3 responses are either '+OK' or '−ERR', together with a text message intended for the human user (which may be up to 512 characters long including the ⟨CRLF⟩ which must terminate them).

Table 12.10   POP3 commands

State	Command	Meaning	Possible responses
AUTHORISATION	QUIT	Client Quitting	+OK server signing off
	USER name	Client identifies his username after receiving the POP3 server greeting	+OK name is a valid mailbox −ERR never heard of mailbox
	PASS string	Client follows up his USER command and mailbox name with a server/mailbox-specific password	+OK maildrop locked and ready −ERR invalid password −ERR unable to lock maildrop
	APOP name digest	Client indicates with a single command his mailbox name and an MD5 encryption digest string after the POP3 greeting or an unsuccessful USER/PASS	+OK maildrop locked and ready −ERR permission denied
	AUTH	Client requests an SASL authentication mechanism with POP3	+OK maildrop locked and ready −ERR permission denied
	CAPA	Client requests a list of POP3 server capabilities	+OK capabilities list follows −ERR command not implemented
TRANSACTION	STAT	Client requests maildrop listing	+OK nn mm
	LIST [msg]	Client requests scan listing of numbered message	+OK scan listing follows −ERR no such message
	RETR [msg]	Client retrieves message with number given	+OK message follows −ERR no such message
	DELE [msg]	Client requests that the numbered message be deleted (typically having received it correctly)	+OK message deleted −ERR no such message
	NOOP	Client requests a 'sign of life' from the server	+OK

Table 12.10 *(continued)*

State	Command	Meaning	Possible responses
	RSET	Client requests that any messages marked as deleted be unmarked	+OK
	QUIT	Client triggers server to enter UPDATE state	+OK
	TOP msg n	Client requests that the n lines from the top of message number msg be sent	+OK top of message follows −ERR no such message
	UIDL [msg]	Client requests the unique ID listing of message numbered msg. The UIDL is the message number and its unique ID	+OK unique-id listing follows −ERR no such message
—	CAPA	Client requests a list of POP3 server capabilities (see Table 12.11)	+OK capabilities list follows −ERR command not implemented
UPDATE	QUIT	POP3 server removes all messages marked as deleted from the mailbox	+OK −ERR some deleted messages not removed

Table 12.11 POP3 capabilities (responses to the POP3 CAPA command)

Capability	Meaning/purpose
EXPIRE	Indicates the minimum retention period of messages by the POP3 server in days.
IMPLEMENTATION	A string describing information about the POP3 server implementation.
LOGIN-DELAY	Some POP3 servers try to reduce server loads by requiring a minimum delay between successive client logins. The response indicates the minimum delay required in seconds.
PIPELINING	When supported, this capability means that the POP3 server is capable of accepting multiple commands at a time.
RESP-CODES	This capability indicates that any server responses commencing with a square bracket ([) is an extended response code.
SASL	This indicates that the AUTH command permits the use of SASL (simple authentication and security layer) type authentication.
TOP	This indicates that the TOP command is available.
UIDL	This capability indicates that the (optional) UIDL command is supported.
USER	This indicates that the USER and PASS commands are supported for authentication of the client to the POP3 server.

Table 12.11 lists the allowable POP3 capabilities, as indicated in response to the POP3 CAPA (capabilities) command.

13

Data Network Security

The boom in electronic business (ebusiness) and the use of the Internet for all kinds of business communication have left many companies open to breaches in confidentiality, industrial espionage and abuse. Such breaches can sometimes go unnoticed for long periods, and can have serious business or cost implications. But just as damaging can be the impact of distorted, corrupted, neglected, misinterpreted or spoofed information (i.e., misleading information input by 'hackers' to confuse a system). Fortunately, the great public concern about the security risks associated with data networking across the Internet has spurred a great deal of activity in the development of counter-measures. Today, a range of different protocols and other security techniques is available. This chapter describes the various levels of information protection provided by different data network security means, explaining how they work and the threats (both malicious and non-malicious) which they attempt to eliminate. We discuss simple password techniques, methods of path protection, tunnelling, firewalls, VPNs (virtual private networks), as well as digital signatures and data encryption. In developing a full security strategy for data networking, it is important to understand the risks, consider the motivations of 'hackers' and develop a pragmatic policy to counter the most likely and most threatening dangers.

13.1 The trade-off between confidentiality and interconnectivity

The man or woman who sold the first telephone must have been a brilliant salesman — for there was no-one for the first customer to talk to! On the other hand, what confidence the customer could have had that there were no eavesdroppers on his or her conversations! The simplicity of the message should be a warning to all: the more people connected to the network which you are using, the greater your risk. Justifiably, the public Internet has the reputation of being an insecure network!

As the number of connections on a network increases, users are subjected to:

- the risk of interception, 'tapping' or eavesdropping;

- greater uncertainty about who they are communicating with (have you reached the right telephone or not? which caller might be masquerading as someone else? how can I be sure that the document received has not been tampered with or altered by a third-party en route?);

Data Networks, IP and the Internet: Protocols, Design and Operation   Martin P. Clark
© 2003 John Wiley & Sons, Ltd   ISBN: 0-470-84856-1

- the risk of time-wasting mistakes (an incorrect access to a database or misinterpretation of data may lead to the corruption or deletion of substantial amounts of data);

- the nuisance of disturbance ('wrong number' calls, unsolicited calls from salesmen; worse still; 'forced entry' by computer 'hackers', or abuse of the network by third parties to gain free calls at your expense); and

- problems associated simply with the complexity or the network, or the lack of understanding by users of how services or security measures are intended to work.

Too often, much thought goes into improving the connectivity of networks, but comparatively little effort is applied to information protection. Risks creep in — often unnoticed. So what different risks and means of network protection are available?

13.2   Data network protection: the main types of threat and counter-measure

Data network security techniques protect the information conveyed across communication networks from breaches in confidentiality and from possible tampering by external third parties by any one of five basic means (as illustrated in Figure 13.1):

1. *destination access control* — ensures that only properly identified and authorised users gain access to a specific destination server or *host*.

2. *firewalls* — perform 'gateway checks' thereby controlling who and what may enter or leave the destination local network.

3. *path protection* — ensures that only properly identified and authorised users may use specific network paths.

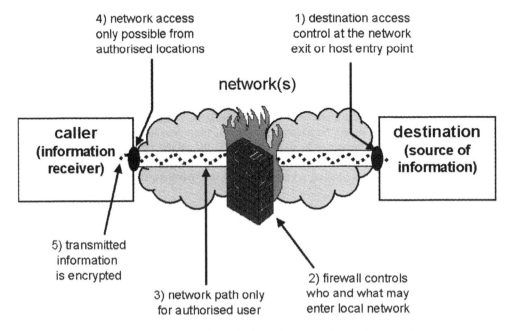

Figure 13.1   The five basic methods of data network security protection.

4. *network access control* — ensures that only properly identified and authorised users can gain access to the communications network at its entry point.

5. encryption — coding of the information on an end-to-end basis, so that only the desired sender and receiver of the information can understand it, and can tell if it has been tampered with.

In the following sections, we review real data security protocols and protection mechanisms which have been broadly categorised into the five basic types of Figure 13.1 according to their prime mode of operation. The techniques we shall cover are listed in Table 13.1. But please note that the categorisation is not an 'official' one — it is only provided as assistance in understanding the various techniques, their modes of operation and the inter-relationship of one to another. A combination of the five different protection methods will give the maximum overall security, and many modern security methods are a 'mixture' of the different basic techniques.

Table 13.1   Data networking security methods: their usage and basic method of operation

Basic methodology	Security method	Full name	Usage
Destination access control	PAP	Password authentication protocol	Used by a client (when accessing a server) to identify himself for the purpose of access authorisation.
	CHAP	Challenge handshake authentication protocol	A challenge protocol initiated by a server to demand that a user identify himself for purpose of access authorisation.
	SASL	Simple authentication and security layer	A simple authentication and security layer protocol which may be added to other protocols to improve their security.
	One-time passwords		One-time passwords are used in conjunction with password authentication protocols, but are only valid once and for a very short period of time. They are intended to prevent 're-use' of 'overheard' passwords.
Firewall methods	Stateful inspection		A firewall of this type inspects the contents of individual packets and their addresses to determine what is 'allowed'.
	ACL	Access list or access control list	A list held by a router used as a 'firewall' which defines which originating users (identified by their IP addresses) may access which servers/hosts (also identified by their IP addresses).
	NAT	Network address translation	By the use of network address translation, enterprise networks can be operated with 'private' IP-addressing schemes but still be connected to the Internet. Only the addresses of selected servers within the company network need be 'translated' into public IP addresses — thereby making them reachable by public Internet users.

(continued overleaf)

Table 13.1   *(continued)*

Basic methodology	Security method	Full name	Usage
	DMZ	Demilitarized zone	A DMZ is a 'quarantine' network comprising proxy servers and content filters used in a multi-stage firewall to protect a private enterprise IP network from security threats posed by public Internet users.
	Proxy server	—	A proxy server is an application server which forms part of a firewall. It checks the validity and acceptability of IP messages directed by a public Internet user to a server behind the firewall. Only 'allowed' messages are relayed by the proxy to the 'real' application server.
	Content filter	—	A proxy server used to check the 'acceptability' of incoming electronic mail content. In effect this is a 'virus scanner'.
Protected path	Tunnelling	—	A means of securely 'encapsulating' data in the case that a public or other 'insecure' network needs to be traversed.
	GRE	Generic routing encapsulation	A generic protocol used for tunnelling.
	PPTP	Point-to-point tunnelling protocol	A tunnelling protocol developed by Microsoft and US Robotics (now 3Com) for providing secure dial-in access to 'private' enterprise IP-networks by means of dial-up public Internet access services.
	L2TP	Layer 2 tunnelling protocol	A further development of the PPTP protocol, now standardised by IETF (Internet engineering task force).
	Mobile IP	Internet protocol mobility	A protocol intended to allow for mobility of Internet users, which is in effect a combination of user identification and tunnelling security methods.
	VPN	Virtual private network	The term VPN is used to describe any of a range of different methods by which 'private' enterprise IP networks (Intranets) can be extended using secure 'connections' across a public Internet or router network.
	CUG	Closed user group	A particular form of VPN in which only certain pre-defined access points to a public network may inter-communicate with one another.
Network access control	NAS	Network access server	A network access server is the point-of-entry to an IP-based network (e.g., the Internet) for a caller accessing the network by means of a dial-up telephone connection. Various security methods, including TACACS/RADIUS, PPTP and L2TP are typically implemented at NASs.
	CLI	Calling line identity	A network-generated identification of the user or caller. Since the identity is network-generated it is more difficult to 'spoof' or falsify an identity.
	Callback		A means of ensuring that only pre-defined network addresses may access servers. After a request by the remote user, the pre-defined network address is called back. This helps to eliminate falsified caller identities.

Table 13.1   *(continued)*

Basic methodology	Security method	Full name	Usage
	TACACS	Terminal access controller access control system	A username/password means of identifying and authenticating users when trying to gain dial-in access to an IP network.
	RADIUS	Remote authentication dial-in user service	A centralised server and database used to store the identities of users who are allowed access to a given service. Centralisation of the database makes for easier administration. The standard password authentication protocols used in association with RADIUS is TACACS, PAP and/or CHAP.
Encryption	IPsec	Security architecture for Internet protocol	A complete security architecture for end-to-end encryption or tunnelling of sensitive data.
	DES	Defense encryption standard	A highly robust and difficult to crack 'symmetrical' encryption method for coding data while in transit.
	MD	Message digest	A characteristic 'fingerprint' or 'signature' on a message which is intended to prove its authenticity.
	PKI	Public key infrastructure	A modern method of encryption based upon 'asymmetrical' encryption. A public key is used for encryption but the secret private key is needed for decryption.
	SSH	Secure shell	A modern protocol which in effect adds a 'security layer protocol' on top of TCP (transmission control protocol) and allows for secure forwarding of TCP port segments. As the name suggests, the secure shell (SSH) protocol was originally designed for secure login across a network to a remote server. SSH is described in detail in Chapter 10.
	SSL	Secure sockets layer	A security protocol (https) developed by Netscape for use in conjunction with websites with high security requirements (e.g., for transmission of financial transaction data).
	TLS	Transport layer security	A security protocol developed from SSL intended to be used as a standard protocol for transport layer security. The protocol is independent of the application layer protocol used.
	PGP	Pretty good privacy	A simple but 'pretty good' encryption methodology based on a mixture of 'symmetrical' and 'asymmetrical' encryption intended to be used to secure electronic mail content during transmission.
	S/MIME	Secure multipurpose Internet mail extensions	An extension to the MIME (multipurpose Internet mail extension) standards to allow encryption of electronic mail content during transmission.
	S-HTTP	Secure hypertext transfer protocol	An experimental protocol for securing website communications.

We shall explain the basic principles of each main technique in turn, and then give detailed examples of real security protocols and mechanisms (as listed in Table 13.1) which are based upon the basic technique.

Some of the techniques may be combined easily with one another, while others may conflict and overlap. This reflects the relative immaturity of data security standards. There is not yet a single accepted industry standard or security framework, but a range of alternative techniques, with different strengths and weaknesses.

13.3   Destination access control methods

Protection applied at the destination end is analogous to the security guard at the office door or the keep of a medieval castle — having got past the other layers of protection it is the last hope of preventing a raider from looting your prized possessions. If an intruder gets past this stage, it may be nearly impossible to control his further activities.

In highly interconnected networks, destination protection may be the only feasible means available for securing data resources which must be shared and used by different groups of people. Destination protection is undertaken by maintaining a list (at the destination) of the users allowed to access a given database or use a given server or application. From an operational viewpoint, destination access control is relatively easy to administer and maintain, since the list of 'allowed users' is associated with, and stored on, the server and linked with the application to which it relates. (Resource and list of 'allowed users' are 'all in one place'.)

Typically, companies apply access control methods at either the entry point to a private network (e.g., an IP-based *intranet*) or at the entry point to a particular server or a particular application. The user identification is usually by means of a *username* and *authorisation* is usually granted based upon successful *authentication* of the user's identity. The authentication is typically undertaken by the submission of a *password*. The identification and authentication of the user (Figure 13.2) are usually initiated by the user (the 'caller' in Figure 13.2) using a protocol like *PAP (password authentication protocol)*. Alternatively, the destination server can *challenge* the caller for identification (or for additional identification information). In this case, the authentication details are sent by means of a *challenge protocol* like *CHAP (challenge handshake authentication protocol)*.

Password control

Destination access control is based upon a list of *authorised* users who are expected to *identify* themselves by means of an *identifier* or *username* (or a combination of identifiers — e.g., the source Internet protocol address and a username). Following *identification*, the user is usually also expected to *authenticate* his identity. This is usually done by means of revealing a *shared secret* known only to the user and the server or other device he or she wishes to *access*. The shared secret is typically a simple *password* or *personal identification number (PIN)*.

Destination *access control* systems are generally based on *user authentication* by means of a simple password. The list of users allowed to access the particular network, server or application is held directly within the server or other 'entry-point' device. Perhaps the main advantage of this type of access control is that the username and password database are directly associated with the application to which they relate. Multiple or distributed databases do not need to be maintained, so it is easier to keep the list up-to-date.

As an analogy: if there is only one doorway and one guard controlling entry to your office you only need to make sure that the one guard is kept informed about any particular 'shady characters' who are to be kept out. On the other hand, if there are six doors to the building, all the security guards will need to be told. This takes more time to coordinate.

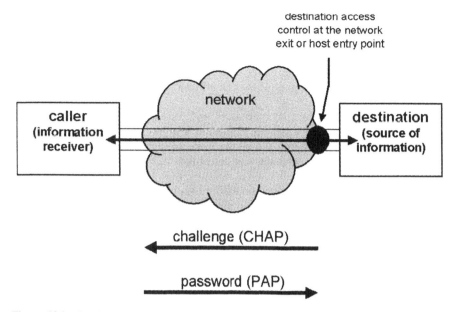

Figure 13.2   Destination access control by means of password authentication and 'challenge'.

Each doorman maintains a separate 'ring binder' with the list of names and descriptions of the 'shady characters' to be denied building access. But inevitably, none of the ring binders are the same … different information has 'gone missing' from each.

The problem of mimicked identity and 'spoofing'

The problem with simple password access control methods is that people determined to get in just keep trying different combinations until they stumble on a valid password. The computer *hacker* gains access to confidential information simply by posing as someone authorised to receive that information. The hacker tries to *spoof* a valid identity.

Aided by computers, the first computer hackers simply tried all the possible password combinations. Such password 'crackers' can determine a simple 8–10 digit number or alphabetic sequence within a matter of seconds. One possible counter-measure is to limit the number of different password attempts which may be made consecutively before the user account is barred. Bank cash teller machines, for example, typically retain the customer's card if he or she does not type in the correct authorisation code within three attempts.

Another manner in which computer hackers may learn of valid username and password combinations is simply by monitoring the communications undertaken by a particular server. The hacker records a valid session with the server by an authorised user — noting both the username and the password. Subsequently, the hacker himself logs on using the valid username and password which he or she learned about by 'eavesdropping'. To counter such 'password stealing' it is important to encrypt passwords for transmission across the network rather than sending them as 'plain text' sequences.

PAP/CHAP protocols

The PAP (password authentication protocol) and CHAP (challenge handshake authentication protocol) protocols are optional features of the *link control protocol (LCP)* of the *point-to-point protocol (PPP)*. PAP and CHAP, like the PPP protocol itself, were developed as a means of providing dial-up access to the Internet or other IP-based data networks. The point-to-point protocol (PPP) provides the link protocol necessary for transitting the first dial-up part of the connection (across a public telephone network). The link control protocol (LCP), as we saw in Chapter 8, provides for the set-up of the connection. During the connection phase, either PAP or CHAP may be used as a means of user identification and authentication to the *network access server (NAS)*.

The database listing the users (the usernames and corresponding passwords) allowed to 'exit' the telephone network and 'enter' or access the Internet (or other IP-based network) of Figure 13.3 is (in principle) stored in the network access server (NAS), though as we shall discover later in the chapter, the database may sometimes be held remotely from the NAS, at a central *RADIUS* server.

PAP (password authentication protocol — RFC 1334) is nowadays considered to be a relatively insecure method of providing access control of remote users to an IP-network, since it employs a simple 2-way handshake (consisting of an *authentication request* and an *authentication accept* (or an *authentication reject*) — Figure 13.4a). The acceptance or rejection is dependent only upon the use of a valid password.

CHAP (challenge handshake authentication protocol — RFC 1994) is considered more secure than PAP since it uses a 3-way handshake procedure (challenge/response/succeed) and communicates the password in an encrypted form (using the MD5 encryption algorithm which we shall discuss later). The use of encryption makes the task of overhearing or interception of the password (for a later log-on by a hacker) much harder (Figure 13.4b).

SASL simple authentication and security layer

The *simple authentication and security layer (SASL)* protocol (RFC 2222) is designed to provide a 'generic' means of user authentication for connection-based transport protocols. Like

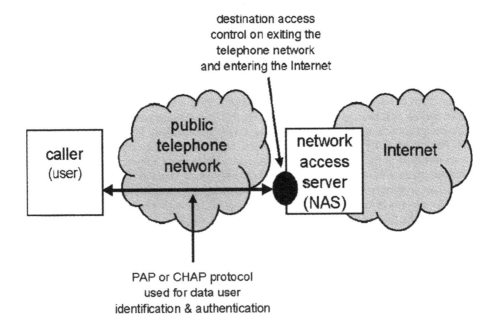

Figure 13.3 Typical use of PAP and CHAP protocols: authorising dial-up access an Internet or IP-based network.

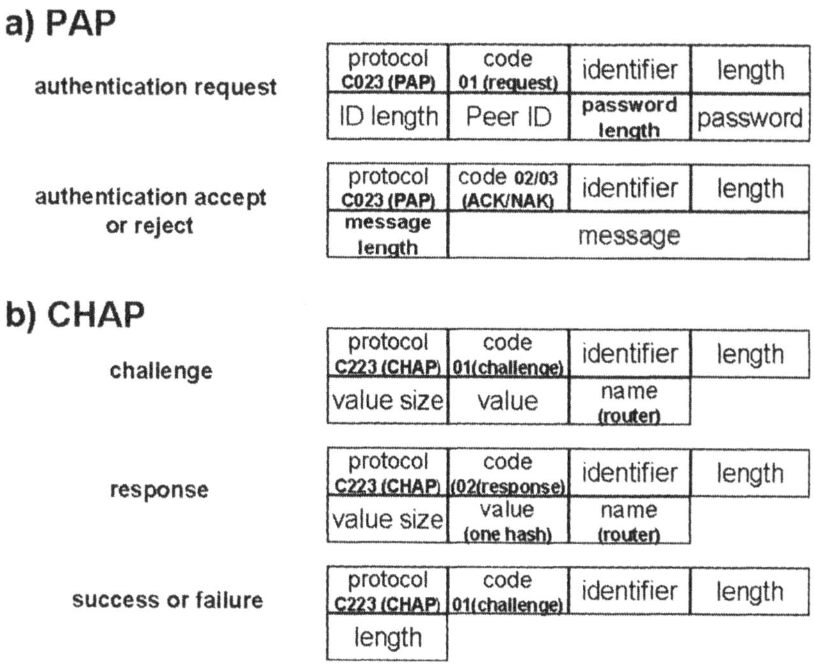

a) PAP

authentication request

protocol C023 (PAP)	code 01 (request)	identifier	length
ID length	Peer ID	password length	password

authentication accept or reject

protocol C023 (PAP)	code 02/03 (ACK/NAK)	identifier	length
message length	message		

b) CHAP

challenge

protocol C223 (CHAP)	code 01(challenge)	identifier	length
value size	value	name (router)	

response

protocol C223 (CHAP)	code (02(response)	identifier	length
value size	value (one hash)	name (router)	

success or failure

protocol C223 (CHAP)	code 01(challenge)	identifier	length
length			

Figure 13.4 PAP (password authentication protocol) and CHAP (challenge handshake authentication protocol) formats.

PAP and CHAP, it relies upon the identification and authentication of a user to a server using a password (or some other shared secret — known only to the user and the server). A number of different authentication mechanisms are supported:

- *Kerberos* — a security system for UNIX-based client/server-based computing in the 1980s which employs a distributed database for user authentication;

- *GSS API (general security service application program interface)* — an application program interface (API) intended to provide a standard interface between applications and security mechanisms. Any security mechanism using this interface can thus easily be incorporated into SASL;

- *S/Key* — a one-time password system defined in RFC 1760 and based upon the MD4 digest algorithm. (we shall discuss both one-time passwords and digest algorithms later in the chapter);

- an *external* security mechanism (such as IPsec or TLS — both of which we shall discuss later in the chapter).

In order to use SASL, a transport or application protocol includes a command for the identification and authentication of the user. If the server (the remote device to which the user is requesting access) supports the requested authentication mechanism, it initiates an *authentication protocol exchange*.

After an access and authentication request by a user (an SASL client), an SASL server may either issue a *challenge* (requesting further authentication or identification information), it may indicate *failure* (thereby denying user access) or may indicate *completion* of authentication (which leads to the next stage of the login procedure to the remote service).

Following a challenge made by the server, an SASL client may issue a *response* (thus continuing its request for access) or may instead *abort* the authentication protocol exchange.

Challenges and responses are based on binary tokens as defined by the particular authentication mechanism in use.

One-time passwords

One-time passwords are only valid for one use (i.e., one-time). Their period of validity is typically limited to a very short period of time (e.g., one minute). The very short validity period of the password and the fact that it may only be used once are intended to overcome the main weakness of password-oriented user authentication: the danger of password *re-use* by a third-party having 'overheard' or otherwise found out a user's password. Otherwise the procedures and protocol mechanisms used for submitting passwords and conducting user authentication are as for standard password authentication and challenge protocols.

S/Key one-time password system

The S/Key one-time password system defined by Bellcore in RFC 1760 generates one-time passwords of 64 bits (8 characters) in length. The client generates the password, based upon a secret *pass-phrase* which is known to both client and server, but communicated only when changed and, then, only in a secure manner. *A secure hash algorithm (SHA)* is applied to the pass-phrase a given number of times to generate the one-time S/Key password. On the subsequent occasion, the number of repeats of the secure hash algorithm (SHA) on the pass-phrase is reduced by one, thereby creating a different password, which to all intents and purposes is 'totally unrelated to the first'. Periodically, the client registers a new pass-phrase with the server — by means of secure and encrypted communication. The secure hash algorithm (SHA) used by the S/Key system is the MD4 algorithm (which we shall discuss later in the chapter).

SecurID one-time password system

One of the best-known commercially available one-time password systems is the SecurID product of RSA Security Inc. The SecurID system generates a 'random' one-time numerical password every 60 seconds. Like the S/Key system, the calculation of the value is based upon a 64-bit key (equivalent to the S/Key pass-phrase) and a security hash algorithm.

SecurID passwords are typically made known to the human user by means of an electronic 'keyfob' or 'smartcard' with a digital calculator-like display. The authentication procedure typically requires that the human user input his username, a secret PIN (personal identification number) plus the one-time securID password. Since the password values generated by SecurID appear to be unrelated to one another they are 'virtually hackerproof'. In reality, of course, anyone with the right key and algorithm could calculate the passwords. But these are much harder to discover or 'overhear' than normal passwords, since they are not communicated over the network during normal login and *authentication protocol exchanges*.

13.4   Firewalls

A *firewall* is a device used to protect a private (typically company-internal) *intranet* from intrusions by unauthorised third-parties attempting to gain access from the public Internet

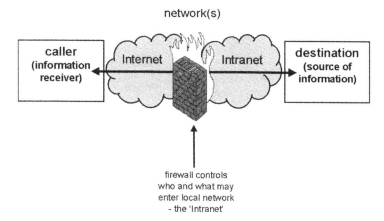

Figure 13.5   Firewalls: a means of protecting company 'Intranets' from intrusion by users of the public Internet.

(Figure 13.5). The firewall may comprise of one or a number of different devices, which together are intended to control:

- which external users (i.e., Internet users) may access the intranet;

- which servers these external users may access;

- which information may be exported from these servers; and

- what type of information may be sent into the network.

Firewalls typically comprise routers, application *proxies* and *content filters* (including *virus scanners*). Firewall routers are used to check source and destination IP addresses and to allow communication only between allowed combinations of source and destination. *Application proxies* protect the 'real' application servers by checking the communication between the external user and the server. Only 'acceptable' requests are actually relayed to and from the 'real' application server. Content filters are used to check the nature of data sent into the network. The objective is to prevent intrusion by *viruses* or other harmful data or application programs.

The operation of a firewall is usually independent of the physical network media and protocols used to connect to both the Internet and intranet sides of the firewall.

Holes may be made through firewalls. They allow unrestricted communication through the firewall. Such unrestricted communication may be desired by certain mobile users (e.g., travelling employees) who wish to be able to access the company network via the Internet from remote locations when travelling. Alternatively, selected 'trusted' external destinations (e.g., the bank, etc.) may also require holes in order to allow more 'intimate' communication than the firewall normally permits.

Holes should be avoided as far as possible. A hole creates a weakness in the firewall and can potentially be exploited by a hacker. When multiple holes are created, the administration of the firewall needs to be impeccable! Forgotten disused holes create the potential for major breaches of security! When a hole is made through the firewall, it is important to use a *path protection* or *extranet* methods such as *tunnelling* or *VPN (virtual private network)* on the Internet-side of the firewall (Figure 13.6), in order to prevent 'allcomers'

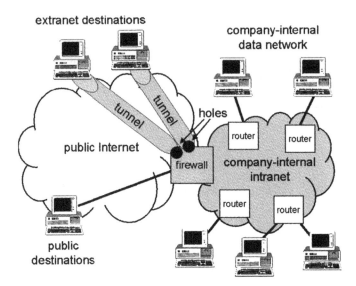

Figure 13.6   Interfacing an intranet with the Internet using a *firewall* to control access: extending the Intranet to *Extranet* destinations, reached by means of the Internet, VPN or tunnelling technology and a hole through the firewall.

accessing the intranet through the hole. Path protection, including tunnelling and VPN, we shall discuss shortly.

Intrusion and intrusion detection

Firewalls are complex devices. Sometimes they comprise special *hardware firewall* devices, sometimes pure *software firewalls* are used. But despite what the different names might suggest, all firewalls are heavily reliant upon their *software* and the up-to-date maintenance of relevant configuration data. A virus scanner that hasn't been updated with the latest virus patterns for more than 6 months can provide no protection against a rapidly spreading new virus. Similarly, a firewall not updated or maintained to cope with the latest criminal techniques used for network intrusion will not provide for security of data.

Firewalls are devices intended to prevent network *intrusion* and to detect and report any intrusion attempts. For maximum security it is important that the *counter-intrusion* software is kept up-to-date and that the *intrusion detection* records are regularly analysed — to determine the source and motivation of the network attacks being attempted by external third parties.

Firewalls vary in their capabilities and their complexity — ranging from simple single-stage firewalls to complex multiple-stage firewalls with *quarantine* or *demilitarized zones (DMZ)*. We next review the main types and capabilities of 'real' firewalls.

Single-stage firewalls

Access control list (ACL)

Routers employing *access lists* or *access control lists (ACLs)* are the simplest form of firewalls. The access control list is a list of the IP addresses which are allowed to communicate with

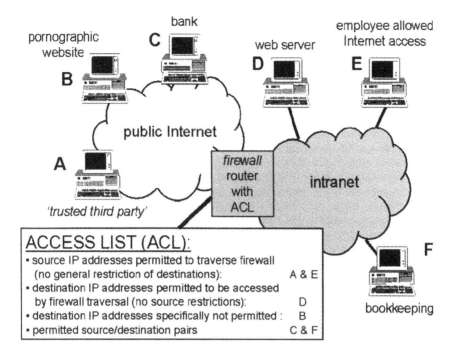

Figure 13.7   The use of a router with an access control list (ACL) as a firewall.

one another across the firewall. The access list may apply one or more of three different types of access restriction (Figure 13.7):

- Access restriction based upon destination IP address, thereby:

  — permitting only certain IP addresses 'behind' the firewall (i.e., in the intranet) to be reached from the Internet (i.e., by untrusted external parties);

  — permitting intranet users only to make 'calls' to certain public IP addresses;

- Access restriction based upon source IP address, thereby:

  — permitting only certain IP addresses 'behind' the firewall (i.e., in the intranet) to make 'calls' to external destinations in the Internet;

  — permitting only certain public IP addresses (i.e., *trusted third parties*) to make 'calls' into the intranet;

- Access restriction limited to pre-defined pairs of source and destination IP addresses, thereby:

  — permitting communication only between specified pairs of source and destination IP addresses.

Stateful inspection

Stateful inspection is an extension of the idea of an access control list (ACL), to include not only access restrictions based upon source and destination IP-addresses, but also dependent upon the protocols and services requested. Thus, for example, it might be appropriate to allow messages to traverse the firewall to and from the Intranet email server, but only when a suitable mail transfer protocol is used. By such means, the use of network management protocols (e.g., SNMP) by external parties could be prohibited.

Network address translation (NAT)

Network address translation (NAT), as we discussed in Chapter 5, allows a 'private' IP addressing scheme to be used in an intranet. Rather than having to get a public IP address allocation for all the devices connected to the intranet, a company can simply think up its own addressing scheme (for example, using the IANA ranges reserved for *private addressing* — 192.168.x.x or 10.x.x.x.) Addresses in the private address ranges allocated by IANA (Internet Assigned Numbers Authority) have only 'local significance'. In other words, they are not publicly recognised IP addresses, and will not be forwarded across the public Internet.

The main advantage of a 'private' IP-addressing scheme is the fact that intranet and other IP network operators need not obtain official public IP address range allocations. But when the 'private' network is to be connected for communication with the public Internet, network address translation (NAT) becomes necessary. Thus, if the web server of Figure 13.7 has an intranet IP address of, say 10.45.24.2, then the web server could not be reached by public Internet users. In order that the web server can be reached, a public IP address needs to be allocated to it. Let us assume that the allocated public address is 173.145.23.5. Then packets sent from Internet users to the web server will have an IP packet header with the address 173.145.23.5 indicated as the *destination IP address*. The router at the entry point to the intranet needs to *translate* this number (an act called network address translation) to the IP address of the web server as known within the intranet (i.e., 10.45.24.2). Similarly, any reply packets sent by the web server to the external destination will initially have an indicated *source IP address* (in the IP-packet header) of 10.45.24.2. This address also needs to be appropriately translated by the exit router to the publicly recognised address 173.145.23.5.

The 'security' advantage of using 'private' IP address schemes and network address translation (NAT) in conjunction with intranets which are connected to the Internet, is that only allowed *hosts* within the intranet are able to be communicated with from the 'outside' and *untrusted* world of the *Internet*. Communication between the Internet and the intranet has to be 'specifically permitted'. From an administrative point of view this has potential benefits — particularly in large networks with large numbers of users. Users have to request permission — they cannot just 'slip by'.

Multi-stage firewalls

Multi-stage firewalls are more secure than simple single stage firewalls. In a multi-stage firewall (as illustrated in Figure 13.8) a *demilitarized zone (DMZ)* exists between the 'public' *Internet* and the 'protected' *intranet*. The *demilitarized zone* is like the 'no-man's land' sometimes found along the border between warring countries. In order to cross the border between warring countries, you first have to cross the 'no man's land'. You are subjected to severe checks by each of the two countries border patrols, and you have an uncomfortable walk across the barren 'no-man's land'.

DMZ (demilitarized zone)

The *demilitarized zone (DMZ)* of a multi-stage firewall is a *quarantine* area sandwiched between *relaying* routers which connect on either side to the 'public' (and untrusted) Internet

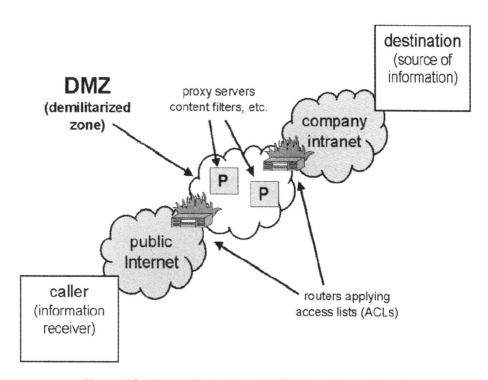

Figure 13.8   The demilitarized zone (DMZ) of a multi-stage firewall.

and the 'private' (and to be protected) intranet. Communications between the Internet and the intranet have to cross the demilitarized zone (DMZ) and are thus subject to the various communication checks which are performed within it. *Application proxy servers* and *proxy agents*, as well as *content filters, virus scanners* and *intrusion detection* devices control this area.

The two *relaying routers* typically both employ network address translation (NAT), and a different IP-address scheme is used in the DMZ than in either the Internet or the intranet. The double use of NAT prevents any possibility of 'bypassing' the DMZ. Thus a packet originating from the Internet first has to traverse a router into the DMZ. The destination IP-address indicated in the IP packet might be the public IP address of the mail server in the intranet, but the address will first be translated to the DMZ-IP-address of the mail proxy server in the DMZ. The proxy will check the packet, and then *relay* it to the DMZ-exit/intranet-entry access router, where the destination IP-address is once again translated — this time to the intranet IP address of the destination mail server. Any packets which try to bypass the mail proxy in the DMZ will have a source address not allowed by the access list (ACL) of the second router, and also will have unknown destination IP addresses.

Proxies

There are two types of *proxies* — *proxy servers* and *proxy agents*. As their names suggest, both types perform a proxy (or substitute) role — either taking on the role of a server (a proxy server does this) or of a client (a proxy agent performs this function). In the case of a multi-stage firewall, the proxy servers (and proxy agents, if used) are located on hardware in the demilitarized zone (DMZ) part of the network.

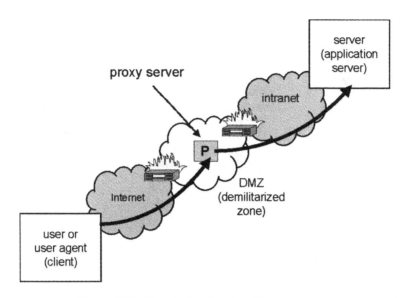

Figure 13.9   The relaying function of a proxy server.

As far as the *untrusted client* (the Internet user) is concerned, the proxy server does not exist. All IP packets are addressed and sent as if they will travel directly to the *application server* in the intranet (behind the firewall). In reality, the packets are forwarded by the firewall routers first to the proxy server in the demilitarized zone (marked (P) in Figure 13.9). The proxy server will inspect the contents of the IP packet, taking into account the application protocols, any application layer user names, the nature of the service requested and maybe even the contents of any other data enclosed in the packet. Only once the proxy server is satisfied that the incoming packet is from an 'acceptable' origin with a bona fide request, is the packet relayed on to the 'real' application server. 'Unacceptable' packets are either rejected (e.g., with an application protocol rejection or a suitable ICMP (Internet control message protocol) response) or simply ignored.

The use of proxy servers can be a valuable way of preventing *service attacks* or *denial-of-service (DOS)* attacks on web servers. Certain service attacks on *web servers* try to 'crash' the web server by overloading it with requests. The result is the denial-of-service (DOS) to other bona fide customers. By using a proxy server to *filter* the requests to the web server, the rate at which new requests are relayed to the *application server* can be controlled to prevent server overload.

Proxy agents are devices which receive messages from servers and convert them into a form which can be understood by the client software. Such proxy agents are widely used in the network management field (as we saw in Chapter 9): for example, to convert standard network management protocols such as SNMP (simple network management protocol) into 'proprietary' network management protocols used in certain manufacturers' networking equipment. Proxy agents, however, are rarely used in firewalls. Packets leaving the intranet are generally assumed not to pose a risk to the clients or other hosts that they are directed at.

SOCKSv5

The *SOCKS* protocol is intended to provide for secure client/server communication in the case that a firewall needs to be traversed. It provides a method for *authenticated firewall traversal*

(AFT). In effect, it is an alternative to the use of a proxy server, and has the benefit of not requiring application-specific software to be developed especially for the proxy server. On the other hand, it is not as secure.

SOCKS provides a generic connection-level technology for authenticating users, offering a variety of *strong authentication* methods. It also provides for application protocol *filtering* and an enhanced level of access control (though based on 'generic' rather than application-specific methods).

SOCKS allows firewall traversal (following user authentication) by means of networking proxies at the transport layer. The current and most widely used version is SOCKS version 5 (SOCKSv5). This is defined in RFC 1928. SOCKS operates on TCP port 1080. A SOCKS system comprises a SOCKS server (which operates at the application layer) and a SOCKS client (which is 'squeezed between' the application and transport layers on the remote user's machine (Figure 13.10).

The SOCKSv4 protocol allowed for connection requests (traversing the firewall), for the setting up of *proxy* circuits and for the subsequent relaying of application data (as shown in Figure 13.10). The SOCKSv5 protocol works in the same manner, except with the addition of *strong user authentication* prior to acceptance of a connection request.

A *SOCKSv5 client* attempts to establish a *SOCKSv5 connection* by sending a request, including the *method herald* (Table 13.2) to indicate which authentication method is to be applied. The SOCKSv5 server authenticates the request. An acceptance or rejection is then returned to the client, whereupon the SOCKSv5 client may send a *SOCKSv5 proxy request*. The SOCKSv5 server (usually a firewall) responds by setting up a SOCKSv5 proxy circuit. Thereafter application data can be relayed through the firewall.

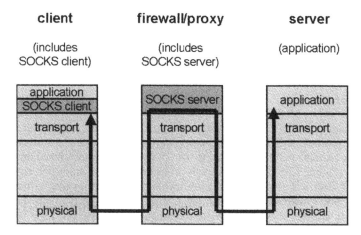

Figure 13.10   SOCKS: a security technology and protocol for secure connection-oriented client/server communication.

Table 13.2   SOCKS authentication methods

Method value	Method meaning
00	No authentication required
01	GSS API (general security service application program interface)
02	Authentication by username and password
03 to 7F	Values assigned by IANA
FF	No acceptable methods

Content filters

A content filter is a particular type of firewall proxy server, usually specially designed to filter the contents of incoming electronic mail messages bound for destinations within an intranet. In particular, the filter may be used to reject (either by deleting, ignoring or *quarantining*) incoming emails with particular 'undesirable' contents, for example:

- emails containing known computer viruses;

- *SPAM* emails (i.e., junk emails) from previously adjudged 'undesirable' email address origins;

- emails containing attached files with particular 'undesired' data formats. Some companies, for example, ban video and MPEG-files, on the basis that such files are 'rarely for business purposes'. Other file types which might be adjudged risky might include those types most likely to contain new *viruses* or *worms* (e.g., application files of .exe or .vdb types). Emails containing such file types might be *quarantined* for further attention of the firewall administrator, before being manually forwarded (if appropriate).

Virus scanners

Virus scanners are a particular type of content filter, intended to detect incoming packets (particularly emails with file attachments) which contain known computer viruses, worms or *trojan horses*. Viruses and worms generally work to disrupt computer systems by feeding them with information to confuse or corrupt it. Trojan horses, meanwhile, attempt to breach firewalls by the placement of a 'spy' or 'foreign body' behind the firewall by means of an initial 'bona fide' communication. Once the trojan horse is in place, it is considered 'trustworthy' by the firewall, but then, like a spy starts to feed secret information to the external world.

13.5   Path protection

The communication path itself is almost bound to run through public places and, in consequence, pass sources of potential eavesdropping, interception and disturbance. For highly secure communications, it is thus essential to take measures to protect the communications path from 'tapping' (Figure 13.11). The best path protection depends upon the right combination of physical and electrical telecommunication techniques, but from the serious eavesdropper there is no absolute protection. To reduce the risk of interception the path should be kept as short as possible and not used if electrical disturbances are detected upon it. For the most secure communication there is nothing better than sitting in the same room!

In the early days of telephony, individual wires were used for individual calls and thus the physical paths for all callers were separate. Using dedicated communications circuits or even laying special cables provides a fair level of data communications security — especially preventing remote Internet 'hackers' from easy access to the network. The firms who build intranets (IP-based router networks intended exclusively for the use of employees) order their 'own' point-to-point leased lines from remote sites to their computer centre to ensure that only authorised callers can access their data.

But for the determined eavesdropper the physical separation of a dedicated circuit or cable may be an advantage: it is much easier to identify a dedicated cable and *tap* into it at a manhole in the street than it is to find a single communication among a mass of others using the same medium. The latter is closer to the analogy of having to 'find a needle in a haystack'. A very determined communications intruder can surround an individual copper cable (or even

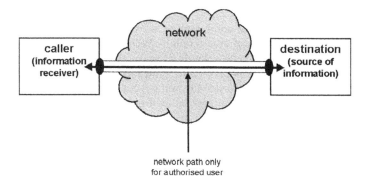

network path only
for authorised user

Figure 13.11 By protecting the path from access by third parties, the security of data in transport is increased.

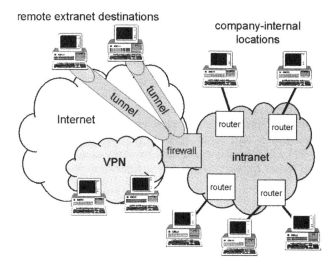

Figure 13.12 Path protection in corporate IP-based networks: intranets, extranets, tunnelling and VPN.

a glassfibre one!) with a detection device (a modified standard cable measurement device) to sense the electromagnetic signals passing along the cable, and interpret these for his own use. On the other hand, a protected path carrying an encrypted signal provides a very high level of security!

Intranets, extranets, tunnelling and VPN
The following are all networking methods intended to provide for path protection security in corporate IP-based data networks (see Figure 13.12):

- intranet;

- extranet;

- tunnelling; and

- VPN (virtual private network).

An intranet is a dedicated set of routers, cables, LANs (local area networks) and data equipment intended for the exclusive use of the employees of a company. It is a 'private' network. No unintended third parties have physical connection to the network. In the case that the intranet needs to extend beyond the bounds of a single campus, this can be achieved by means of *private circuits*. These are point-to-point circuit connections between two end-points leased from the public telecommunications carrier for exclusive use.

An *extranet* is an extension of an intranet, allowing secure access by pre-defined remote users to the data networking and processing resources of the intranet. An extranet uses the public Internet for the physical connection of the remote user and for the transport of data to and from the Intranet. The extranet is typically connected to the intranet by means of a firewall (with relevant holes made through the firewall to facilitate firewall traversal, as necessary).

The term extranet does not describe a particular technology, but instead is a generic term for secure remote access via the Internet to an intranet. An extranet may be based upon any of a number of different path protection technologies. The two we shall describe later in this chapter in more detail are tunnelling and virtual private networking (VPN).

A tunnel is a point-to-point connection of a remote user to the intranet. The most commonly used *tunnelling* technologies are: *PPTP (point-to-point tunnelling protocol); L2TP (layer 2 tunnelling protocol), IPsec, mobile IP* (also called *IP mobility*) and *GRE (general routing encapsulation)*. We shall describe each of these in more detail later in the chapter.

A VPN (virtual private network) is intended to allow the connection of a number of different locations in the 'public' Internet as if they were connected to a 'private' network (e.g., an intranet). Confusingly, sometimes the term VPN is used synonymously for the term extranet — i.e., as a generic term for remote user connection to an intranet. But VPN is also the name of a number of specific technologies which achieve the aim. One of the most popular VPN technologies is that based upon *MPLS (multiprotocol label switching)*.[1] The security of MPLS-VPN technology is considered by many companies good enough even for data transport of the 'core' Intranet inter-router connections. We shall describe it in more detail shortly.

Other important path protection considerations for network designers
In addition to the path protection measures offered by intranets, extranets, tunnelling and VPN, network designers also need to consider less-obvious causes of possible data corruption or network intrusion. The electro-magnetic radiation emitted by high speed data cables, for example, can lead to corruption of information or data loss. This may be maliciously motivated or unintended. Radio jamming is often used in wars as a means of corrupting or interrupting enemy radio communication. But even when not maliciously intended, high speed data cables or power cables, when laid alongside one another, tend to cause *crosstalk* disturbances for speech and corruption problems for data. Both are examples of *electro-magnetic interference (EMI)*. EMI has been a particular problem is recent years for companies running high speed LAN systems and for PC users who also have portable telephones. Careful cable route planning (including rigid observance of equipment operating conditions and specified maximum workable cable *lengths*) is therefore very important.

Where radio is used as the communications path (you may not know this if you order a leased line from the telephone company), interception by eavesdroppers is made relatively simple. Protection of radio (both from *radio interference* and from eavesdropping) can be achieved at least to some extent by modern radio modulation techniques. *Frequency hopping*, for example, is a modulation method in which both transmitter and receiver jump different *carrier* frequencies every few seconds. Jumping about like this reduces the possible chance

---

[1] See Chapter 7.

of prolonged interference which may be present on a particular frequency, and makes it very difficult for eavesdroppers to catch much of a conversation or data communication.

The effects of statistical multiplexing and datagram routing

The statistical multiplexing method (packet switching) used for modern data and IP (Internet protocol) communication, as we discussed in the early chapters of this book, makes it possible for many different communications to coexist on the same physical cable at the same time. On the one hand this makes it harder to perform interception through *tapping* since the electrical signal carried by the wire has to be decomposed into its constituent parts before any sense can be made of a particular communication. On the other hand, it may mean that an electrically-coded version of your data is available in the machine of someone you might like to keep it from. A message sent across a *(shared medium)* LAN, for example, may appear to go directly from one PC to another. In reality the message is broadcast to all PCs connected to the LAN and the LAN software is designed to ensure that only the intended recipient PC is activated to decode it! The secure solution to this problem is to use *virtual bridging LAN (VLAN)* technology. This requires the use of LAN *switches* as opposed to LAN *hubs* (LAN *switches* and *hubs* we explained in Chapter 4).

In theory, the *Internet protocol (IP)*, by exploiting its potential for *datagram routing*, has the means to provide for quite secure transmission of long, multiple packet messages. Theoretically, the IP forwarding of individual packets independently from one another (so-called *datagram routing*) could lead to the use of different paths being used by different packets, even though they are transported from the same source to the same destination. Sometimes this is claimed as a 'security benefit' of datagram routing and the Internet protocol (IP). In practice, however, any stream of packets passing from a single source location to a single destination location across an IP-based router network will tend to follow the same path. This is a consequence of the manner in which the routing protocols work. The routing protocols always calculate a single 'shortest' routing path and favour this single route from source to destination. This has benefits for the application protocol since the receipt of packets is 'predictable'. Packets are received in the right order and each is subjected to a similar path propagation delay. But the 'cost' of better application and network performance is lower security of data during transmission.

We continue with a review of 'real' path protection techniques in detail.

VLAN (virtual bridged LAN)

By splitting a campus or other complex local area network (LAN) into a series of smaller VLANs (virtual LANs or virtual-bridged LANs), a network administrator may derive a number of benefits.[2] Figure 13.13 illustrates how VLAN technology (defined by IEEE 802.1q and IEEE 802.1p) provides the ability to create separate *layer 2 broadcast domains* for the distribution of packets, even though the individual ports or MAC addresses are spread across different LAN switches. For some network administrators, the data security benefits are the main reasons for VLAN deployment. The security benefits of VLANs are of a 'path protection' nature:

- the local area network (LAN) may be structured and subdivided into VLANs reflecting the organisational structure of the company. This gives all departmental users equal access to shared files and other resources but on a *closed user group (CUG)* basis. Unauthorised employees of other departments (who are not members of the VLAN) do not get such easy access;

---

[2] See Chapter 4.

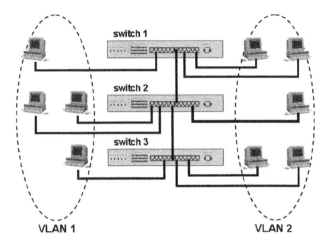

Figure 13.13 Using VLANs to segregate large campus networks and LANs into defined closed user groups.

- VLAN technology (IEEE 802.1p) allows packet carriage across the LAN or router backbone network to be prioritized, giving higher priority to certain users, as defined to belong to a particular VLAN.

The membership of a VLAN may be defined according to one of the following three schemes:

- *port-based VLANs(Class 1 VLANs)* — the individual switch ports (and all connected devices) belong to the VLAN;

- *MAC address-based VLANs (Class 2 VLANs)* — individual device MAC addresses are defined in the VLAN configuration server to belong to one or more VLANs. The advantage of basing the VLAN on MAC addresses is that no network reconfiguration is necessary when a particular device is moved from one port to another;

- *upper layer protocol (ULP)-based VLANs (Class 3 VLANs)* — define VLAN members as all endpoints (service address points) of a particular protocol type (e.g., for conveniently splitting *multiprotocols* on the same network, such as *IP, IPX, Appletalk,* etc.)

Individual switch ports, MAC addresses or protocols may be configured to belong to multiple VLANs. The configuration itself is carried out manually by the network administrator. But once the membership of the VLAN has been defined and the *VLAN identifier (VID)* has been allocated, the network automatically takes over the necessary VLAN network functions.

Figure 13.14 illustrates the VLAN tagging of layer 2 (ethernet LAN) frames (as defined by IEEE 802.1q) to incorporate the VLAN identifier (VID) and the user priority defined by IEEE 802.1p. The use of the VLAN tag in conjunction with ethernet LANs is defined by the standard IEEE 802.3ac. The *tag protocol identification (TPID)* indicates which layer 2 protocol is in use. The *canonical form identifier (CFI)* indicates the bit order of address information.

Tunnelling

Tunnelling is sometimes referred to as 'dial-type VPN (virtual private network)'. Figure 13.15 illustrates a typical application of tunnelling. A remote extranet client gains access to his

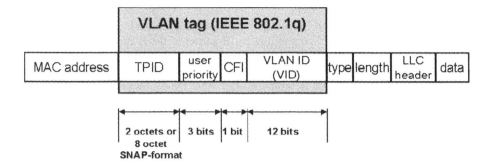

CFI   canonical form identifier
LLC   logical link control
MAC   medium access control
SNAP   subnetwork access protocol (see chapter 4)
TPID   tag protocol identification (ethernet value is 81-00)
VID   VLAN identifier
VLAN   virtual LAN or virtual-bridged LAN

Figure 13.14   VLAN tag-format (IEEE 802.1q).

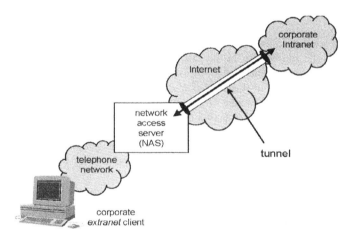

Figure 13.15   The use of *tunnelling*.

company's intranet by means of a dial-up telephone connection and the Internet. The dial-up connection is usually terminated by a *network access server (NAS)*, sometimes also referred to as a *remote access server (RAS)*. Formerly the term *terminal access controller* was also used to describe similar devices which offered 'dial-in terminal connections'. The network access server (NAS) employs a standard means of user identification, authentication and authorisation (e.g., *PAP, password authentication protocol*; *CHAP, challenge handshake authentication protocol; or TACACS, terminal access controller access control system*). Having authenticated and authorised the user, the NAS sets up a tunnel through the Internet to the remote intranet.

The tunnel is created using one of a number of alternative tunnelling protocols. The tunnelling protocol provides for the *encapsulation* of intranet packets. A tunnel makes the remote user (the 'extranet client' of Figure 13.15) appear to the intranet to be directly connected.

Thus the intranet IP packet and a layer 2 (datalink) header, if appropriate, are simply 'packed into' (i.e., *encapsulated*) within the *payload* of the tunnelling protocol to traverse the Internet. The tunnelling protocol, in turn, is carried by an IP packet of the Internet. The overall effect is that an intranet IP packet is carried encapsulated within an Internet IP packet.

Tunnelling is synonymous with encapsulation. You might ask, what makes encapsulation inherently so secure? The answer is 'nothing'. But what makes tunnelling protocols secure is the use of *encryption* for protecting the encapsulated payload of the tunnel. Thus the tunnel from the NAS to the *intranet* of Figure 13.15 is secure, because the communication is *encrypted* during this part of its path.

Although in theory, most tunnelling protocols allow for the establishment of both *incoming call* tunnels (from the NAS *incoming* to the intranet) and *outgoing call* tunnels (from the intranet to the NAS), generally the tunnel is NAS initiated (i.e., *incoming*).

Tunnels are normally connected to the relevant intranet via a firewall. The SOCKS protocol discussed earlier, or some other hole in the firewall is used to allow firewall traversal.

GRE (generic routing encapsulation)

The *generic routing encapsulation (GRE)* protocol (defined in RFC 1701) provides for a flow- and congestion-controlled 'encapsulated datagram' (i.e., tunnelling) service.

The GRE protocol defines a header for the encapsulated payload as illustrated in Figure 13.16. The complete GRE *protocol data unit* (header and encapsulated payload) itself becomes the payload of a *delivery protocol* (e.g., IP — Internet protocol). The protocol is coded as follows.

C is the *checksum-bit*. In the case that this bit is set at value '1', a *checksum field* is present in the header. R is the *routing-bit*. In the case that this bit is set at value '1', a *source routing field* is present in the header, and if the s-bit (strict source-bit) is also set at value '1', then the routing information provided in this field is a *strict source route*. K is the encryption *key-bit*. In the case that this bit is set at value '1', a *key-field* is included in the GRE header. S is the *sequence number bit*. In the case that this bit is set at value '1', a *sequence number* field is included in the header. The *sequence number* field may be used in the case that a *transport protocol* (such as TCP or UDP) is not used as the *delivery protocol*. The *sequence number* and *checksum* fields provide optional *transport layer* functionality for the GRE protocol.

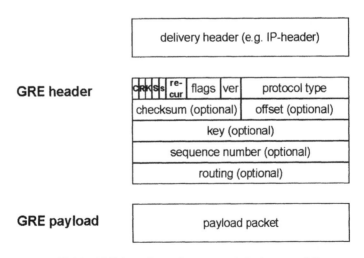

Figure 13.16 GRE (generic routing encapsulation): protocol format.

Table 13.3 General routing encapsulation (GRE) protocol types [ethernet numbers]

Ethernet number (protocol type)	Protocol
0800	Internet protocol version 4 (IPv4)
0806	Address resolution protocol (ARP)
8035	Reverse address resolution protocol (RARP)
814C	Simple network management protocol (SNMP)
86DD	Internet protocol version 6 (IPv6)
877B	TCP/IP compression
867C	IP autonomous systems
867D	Secure data
880B	Point-to-point protocol (PPP)
8847	Multiprotocol label switching (MPLS) unicast
8848	Multiprotocol label switching (MPLS) multicast

The *recursion control* (*recur*) field is a 3-bit number (value 0 to 7) which indicates how many additional encapsulations are permitted (i.e., how many 'tunnels within tunnels' are allowed). The default setting is value '0'. The *version* (ver) number (a 3-bit field) should be set to value '0'.

The protocol type field contains the protocol type of the payload packet. The allowed protocol types are so-called ethernet numbers, as listed in Table 13.3.

The optional *offset* field indicates the number of octets from the start of the routing field to the first octet of the active source route entry to be used next. The *checksum* included in the optional *checksum field* conforms to standard IP suite practice: the value is the one's complement checksum of the GRE header and the GRE payload. Finally, the key-field (if included) contains a four octet number inserted by the GRE encapsulator and used by the receiver to authenticate the source of the packet.

The GRE protocol specification does not provide any advice on the use of encryption. Instead it is a pure encapsulation protocol. If encryption is required, this can be carried out using an appropriate encryption protocol header coded within the first part of the GRE payload.

PPTP (point-to-point tunnelling protocol)

The *point-to-point tunnelling protocol* (*PPTP* — RFC 2637) is used mainly for client-initiated creation of tunnels between Microsoft PPTP clients and Microsoft servers. The protocol is designed to carry *encapsulated PPP (point-to-point protocol)* packets in an encrypted format. PPTP was developed by Microsoft, in conjunction with US Robotics (now part of 3Com). Recently the protocol has been largely superceded by *L2TP (layer 2 tunnelling protocol)*, which is in effect an improved version of PPTP.

As the name suggests, PPTP is intended to be used to create *point-to-point tunnel* connections via the Internet to corporate IP-networks (*intranets*). Such connection are sometimes also called extranet or VPN connections as we saw earlier.

A PPTP tunnel is created between a *PPTP access concentrator (PAC)* client and a *PPTP network server (PNS)* as illustrated in Figure 13.17. The tunnel may be established in either direction. Typically the PAC is a *network access server (NAS)*, or as Microsoft calls it, a *remote access server (RAS)*, as we saw in Figure 13.15. The PNS is typically a WindowsNT or Windows 2000 server.

PPTP uses separate *control* and *data* channels. The *control channel*, which is carried by means of TCP (transmission control protocol) is used for the authentication of the user. Once the user has identified himself/herself and been authenticated, he/she may request the establishment of a tunnel (i.e., a *data channel*). The data channel is carried by means of

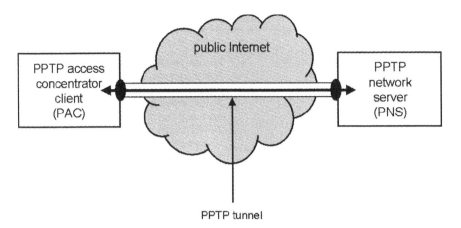

Figure 13.17 is illustrated with the following labels: "public Internet", "PPTP access concentrator client (PAC)", "PPTP network server (PNS)", "PPTP tunnel".

Figure 13.17   PPTP (point-to-point tunnelling protocol): access concentrator and network server.

the GRE (generic routing encapsulation) protocol, carried directly on IP (Internet protocol) discussed earlier. The data channel employs the source route data option of GRE — coding the GRE routing field appropriately.

Before tunnelling can commence between a PAC and a PNS, a *control connection* must be established (on TCP port 1723). The control channel may be established by either the PAC or the PNS. The control connection is maintained (i.e., 'kept-alive') until cleared by *echo messages*. The messages are designed to detect any connectivity failure — and therefore prevent the potential for re-routing the tunnel (e.g., by third parties). The PPTP control channel carries call control and management information messages, as listed in Table 13.4.

Once the control channel is established (following proper authentication), either the PAC or the PNS may request the establishment of a tunnel. The tunnel is established between a *PAC/PNS pair*. The tunnel comprises PPP (point-to-point protocol) packets carried between the PAC and the PNS encapsulated in an adapted form of the GRE packet format (illustrated in Figure 13.18). PPTP data packets are carried directly by IP (i.e., do not use a transport protocol like TCP or UDP). The key-field in the header indicates which PPP session the packet belongs to. A sliding window protocol (similar to that used in TCP, as we discussed in Chapter 7) is used for PPTP data channel *flow control*. Multiple packet acknowledgement (*cumulative acknowledgement*) is possible using a single ACK message.

L2TP (layer 2 tunnelling protocol)

The *layer 2 tunnelling protocol (L2TP)* is a protocol which facilitates the tunnelling of PPP packets across a network in a manner which is transparent to both end-users and the applications which they are using. It bears a number of similarities to PPTP (point-to-point tunnelling protocol) from which it was developed, but uses a combined control/data channel (rather than separate control and data (tunnel) channels). The control/data channel (tunnel) is carried on port 1701 by UDP (user datagram protocol) as the transport protocol. L2TP is defined by RFC 2661.

The typical use of the layer 2 tunnelling protocol (L2TP) is for the dial-up (i.e., extranet or VPN) access of remote users to an Intranet on a point-to-point (tunnel) basis (Figure 13.19). Once the tunnel has been established (by means of a control connection), a number of L2TP sessions may be established to use the tunnel simultaneously. As a layer 2-based tunnelling protocol, L2TP has been conceived to allow L2TP sessions to extend beyond the tunnel itself

Table 13.4   PPTP control channel: control and management messages

Message type	Message name	Message code	Message sent by	Message purpose
Control connection management	Start-Control-Connection-Request	1	PAC or PNS to peer	Sent by either partner to request the establishment of a PPTP control connection.
	Start-Control-Connection Reply	2	PAC or PNS to peer	Sent in response to the Start-Control-Connection-Request. A result code indicates the result of the attempt to start the PPTP control connection.
	Stop-Control-Connection-Request	3	PAC or PNS to peer	Sent to inform the peer partner that the PPTP control connection should be closed.
	Stop-Control-Connection-Reply	4	PAC or PNS to peer	Sent in response to a Stop-Control-Connection-Request.
	Echo-Request	5	PAC or PNS to peer	Sent by either partner as a 'keep-alive' message for the PPTP control connection.
	Echo-Reply	6	PAC or PNS to peer	Sent in response to an Echo-Request message.
Call Management	Outgoing-Call-Request	7	PNS to PAC	Sent by the PNS to indicate that an outbound call from PNS to PAC is to be established and provides call details.
	Outgoing-Call-Reply	8	PAC to PNS	Sent by the PAC to indicate whether the Outgoing-Call-Request attempt was successful or not.
	Incoming-Call-Request	9	PAC to PNS	Sent by the PAC to indicate that an inbound call from PAC to PNS is to be established and provides call details.
	Incoming-Call-Reply	10	PNS to PAC	Sent by the PNS to indicate whether the Incoming-Call-Request attempt was successful.
	Incoming-Call-Connected	11	PAC to PNS	Sent by the PAC in response to a received Incoming-Call-Reply message from the PNS (3-way handshake for incoming calls).
	Call-Clear-Request	12	PNS to PAC	An indication from the PNS to the PAC that a call (incoming or outgoing) is to be disconnected.
	Call-Disconnect-Notify	13	PAC to PNS	A notification from the PAC to the PNS that a call has been disconnected.
Error reporting	WAN-Error-Notify	14	PAC to PNS	Indicates an error on the wide area network which forms the end-user to PAC connection (e.g., a dial-in network)
PPP Session Control	Set-Link-Info	15	PNS to PAC	Sent by the PNS to set PPP (point-to-point protocol) options.

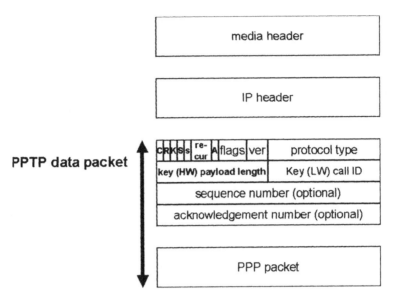

Figure 13.18 PPTP (point-to-point tunnelling protocol): protocol format.

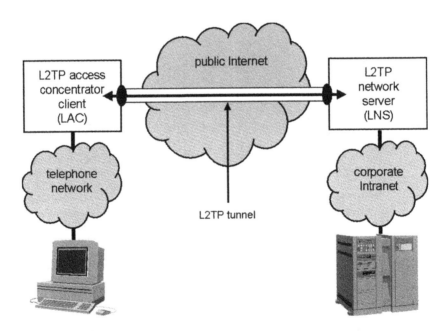

Figure 13.19 Use of L2TP (layer 2 tunnelling protocol).

(as shown in Figure 13.20). This reduces the workload of the LAC (L2TP access concentrator) in that it need not *terminate* L2TP sessions (it need only relay them). It also allows for simpler encryption of the transported data for the entire length of the extranet connection.

The layer 2 forwarding capabilities of L2TP are derived from the proprietary Cisco-developed protocol *L2F (layer 2 forwarding)*. In effect, L2TP is a 'mixture' and further development of

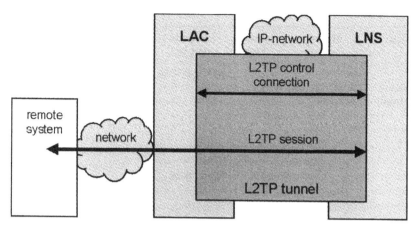

Figure 13.20   Layer 2 tunnelling protocol (L2TP): L2TP tunnels and L2TP sessions.

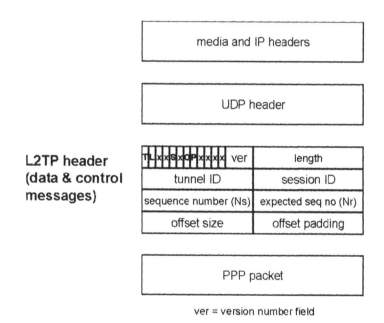

ver = version number field

Figure 13.21   Layer 2 tunnelling protocol (L2TP): protocol format.

L2F and PPTP. Its PPTP roots can be found both in the protocol header format (Figure 13.21) and in the range of control messages (Table 13.5). But be careful, there are important differences between L2TP and PPTP!

An *L2TP tunnel* comprises a *control connection* and zero or more *L2TP sessions*. The tunnel carries encapsulated PPP datagrams and control messages between the LAC and LNS. A 'virtual' PPP connection is thereby created. The tunnel is secured by means of IPsec encryption (as defined in RFC 3193). The tunnel protocol comprises both control and data messages, both with the same format (Figure 13.21).

The *T(type)* bit indicates whether the L2TP packets contain a *data message* (value '0') or a *control message* (value '1'). The *L(length)-bit* indicates (when set at value '1') that a *length field* is present. The *X-bits* are not used but are reserved for future use (they are all set at

Table 13.5   L2TP control messages

Message type	Message name	Message code	Message sent by	Message purpose
Control connection management	Start-Control-Connection-Request (SCCRQ)	1	LAC or LNS to peer	Sent by either partner to request the establishment of a L2TP tunnel and control connection between LNS and LAC.
	Start-Control-Connection-Reply (SCCRP)	2	LAC or LNS to peer	Sent in response to the Start-Control-Connection-Request to indicate that the establishment of a tunnel should continue.
	Stop-Control-Connection-Connected (SCCCN)	3	LAC or LNS to peer	Sent in reply to an SCCRP to complete tunnel establishment.
	Stop-Control-Connection-Reply (StopCCN)	4	LAC or LNS to peer	Sent by either LNS or LAC to shut down a tunnel and close the control connection.
	[reserved]	5		
	Hello (HELLO)	6		This message is a 'keep-alive' message for maintaining the tunnel.
Call Management	Outgoing-Call-Request (OCRQ)	7	LNS to LAC	Sent by the LNS to indicate that an outbound call from LNS to LAC is to be established and provides call details.
	Outgoing-Call-Reply (OCRP)	8	LAC to LNS	2nd step of a 3-way handshake for setting up outgoing calls (L2TP sessions).
	Outgoing-Call-Connected (OCCN)	9	LNS to LAC	Sent by the LNS in response to a received Outgoing-Call-Reply message from the LAC (3-way handshake for outcoming calls).
	Incoming-Call-Request (ICRQ)	10	LAC to LNS	Sent by the LAC to indicate that an inbound call from LAC to LNS is to be established and provides call details.
	Incoming-Call-Reply (ICRP)	11	LNS to LAC	2nd step of a 3-way handshake for setting up incoming calls (L2TP sessions).
	Incoming-Call-Connected (ICCN)	12	LAC to LNS	Sent by the LAC in response to a received Incoming-Call-Reply message from the LNS (3-way handshake for incoming calls).
	[reserved]	13		
	Call-Disconnect-Notify (CDN)	14	LAC to LNS	A notification from the LAC to the LNS that a call has been disconnected.
Error reporting	WAN-Error-Notify (WEN)	15	LAC to LNS	Indicates an error on the wide area network which forms the end-user to LAC connection (e.g., a dial-in network).
PPP Session Control	Set-Link-Info (SLI)	16	LNS to LAC	Sent by the LNS to set PPP (point-to-point protocol) options.

value '0'). The *S-bit*, when set to value '1', indicates that *sequence number fields* are present in the header. *Sequence numbers* used on the *control channel* but not on the *data channel*. The *Ns*-field value indicates the *sequence number* of the message in which the value appears. *Ns* values are incremented in each subsequent packet sent, starting at value '0' at the start of a session. The *Nr*-field value indicates the *sequence number* of the next expected control message. This is used to *acknowledge* received packets and thus to undertake data flow control (as we discussed in Chapter 3).

The *O(offset)-bit* indicates (when set at value '1') that an *offset field* is included in the L2TP header. The *P(priority)-bit* (when set at value '1') indicates that the message should receive preference. The *session-ID-field* contains the *session identifier*. This identifies the particular *L2TP session* within an L2TP tunnel (Figure 13.20) to which the packet belongs. It is important to note that two different *session-ID* numbers are used at either end of a session.

The entire L2TP packet including payload and L2TP header is sent within a UDP datagram (port 1701).

Tunnel establishment takes place by means of control messages (Table 13.5). The establishment commences with a *control connection* request. Tunnel *authentication* takes place during the control connection establishment by a simple CHAP-like authentication system based upon a single *shared secret*.

Once the *L2TP control connection* is established, individual L2TP sessions can be created. During an L2TP session, L2TP data packets are exchanged between the remote system and the LNS (Figure 13.20). A three-way message handshake (Table 13.5) is used to set up a session (i.e., a *call*).

Tunnels maintain a queue of control messages for the peer. The message at the front of the queue is sent when the appropriate *Nr sequence number* value is received.

The following RFCs define security methods relevant to the use of the layer 2 tunnelling protocol (L2TP):

- RFC 2661: the layer 2 tunnelling protocol

- RFC 2809: the use of RADIUS (remote authentication dial-in user service) with L2TP

- RFC 2888: secure remote access using L2TP

- RFC 3070: L2TP and frame relay

- RFC 3145: L2TP disconnect cause information

- RFC 3193: securing L2TP using IPsec encryption (this is the assumed 'default' configuration of L2TP).

L2F (layer 2 forwarding)

The layer 2 forwarding (L2F) protocol (as defined in RFC 2341) is a proprietary Cisco-developed protocol which allows for tunnelling (i.e., encapsulation) of the link layer frames and higher protocols within IP. The aim of the protocol was to decouple the location of the initial dial-up server from the point at which the protocol is terminated and network access (to the intranet) is provided. It has been largely superseded by L2TP (layer 2 tunnelling protocol).

Like L2TP, L2F uses UDP port 1701 and the Internet protocol. Alternatively, L2F packets may be carried by frame relay or ATM (asynchronous transfer mode) networks. L2F and L2TP packet formats are somewhat similar, but may be distinguished by means of the version number (ver) field value (Figure 13.21). L2F packets have their version field value set at '1'. The L2TP version value is '2'. Version '0' is GRE (generic routing encapsulation). The L2F equivalent of the L2TP network server (Figures 13.19) is called the *home gateway (HGW)*.

Mobile IP — another use of tunnelling

Mobile IP, also called *IP mobility*, is defined in RFC 3344. It was developed to provide a secure and reliable means of communication with mobile IP devices, without the mobile device (called the *mobile node*) having to change its IP *home address*. In this way, a communication partner may continue communication with the mobile node without interruption, no matter what the location of the *foreign link* (this is the name of the current physical attachment point of the mobile node). Mobile IP relies upon tunnelling.

As shown in Figure 13.22, the mobile node is connected to the Internet or IP-based network by means of a *foreign link* and a *foreign agent*. The IP address of this location is the current *care-of-address* of the mobile node. The foreign agent *registers* the care-of-address with the *home agent*. During the *registration*, the home agent must authenticate the mobile node, after which a secure tunnel is established to the foreign agent (using the care-of-address). Any packets received for the mobile node by the home agent are simply forwarded over the tunnel.

The *mobility binding* is the record held by the home agent of the current authenticated care-of-address for a given *mobile node*. The visitor list, meanwhile, is the list maintained by a foreign agent of all the mobile nodes currently *visiting* it.

The tunnelling methods intended to be used in conjunction with *mobile IP* are either *IP in IP encapsulation* (as defined in RFC 2003) or the generic routing encapsulation (GRE), discussed earlier (and defined in RFC 1701).

The main advantage of the *mobile IP* protocol is the ability of communication partners to continue to communicate with mobile IP devices without having themselves to track the current location and care-of-address (i.e., *foreign link* attachment point) of the *mobile node.* Communication can be maintained, even if the mobile node changes location during the communication session. It was intended, in particular for secure IP communication over packet radio networks.

The major security risk associated with mobile IP is the authentication of the mobile node.

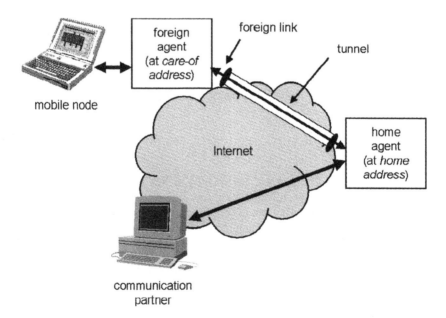

Figure 13.22   Mobile IP: network components and the operation of IP mobility.

Ascend tunnel management protocol (ATMP)

The *Ascend tunnel management protocol (ATMP)* defined in RFC 2107 laid some of the roots for mobile IP. Specifically, the protocol allows for tunnelling (like Figure 13.22) between a *foreign agent* in a *network access server (NAS)* and a *home agent (HA)*. In this way, the protocol allowed for simple mobility based upon dial-in by the mobile node. Having dialled-into the network, a mobile node may utilize a fixed *home network address* to receive communications.

Unlike Mobile IP, the ATMP mobile node is not able to roam between different foreign agents during a single communications session. Instead, a single NAS will provide the physical connection for the remote client for the duration of the session.

Virtual private network (VPN)

Virtual private networks (VPNs) interlink pre-defined end-points connected to a *public* network in a manner which guarantees private or *closed user group (CUG)* communication between the parties. As far as the end-points are concerned, they appear to be linked to one another by a *private network*.

Figure 13.23 illustrates the idea of *closed user group (CUG)* networking between end-point connections interlinked by means of a public network. To a given exit connection from the network for which a CUG has been defined, only pre-determined calling connections (as recorded within the network) are permitted to make calls. Typically a small number of connections within a CUG are permitted to call one another. Additionally, they may be able to call users outside the CUG, but these general users will not be able to callback. In effect, communication to a member of the group is closed except for the other members of the group, hence the name. CUG services are generally offered in ISDN, frame relay and ATM networks.

The critical quality about closed user groups (CUGs) or virtual private networks (VPNs) is that the members of the CUG or VPN are recorded within the network. Neither CUG nor VPN techniques rely upon identification or authentication information provided by callers. As a result they are far more secure.

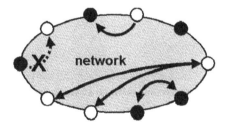

○   member connections of the closed user group (CUG) - these can call
    to all other 'white' or 'black' connections. Together, they form a
    virtual private network (VPN)

●   ordinary network connections - can only call other 'black' connections

⟷ calls possible between these endpoints in either direction

⟶ calls possible only when set up in the specified direction

✗⟶ calls not possible to be set up in this direction between these endpoints

Figure 13.23   The principle of closed user groups (CUGs) and virtual private networks (VPNs).

Because VPN links and networks provided across a public network are immune to unwanted third-party intrusion, they are nowadays used by many enterprises not only for connection of remote users to an intranet, but may also be used as a 'trunk network technology' for links between routers in the 'core part' of the intranet itself.

There are various different alternative 'secure' VPN technologies. The most popular is VPN based upon MPLS (multiprotocol label switching). As we saw in Chapter 7, MPLS relies upon the addition of a *flow label* or *tag* to a *stream* of packets between pre-defined connection end-points for fast relaying of the packets across an MPLS router network. The fact that the flow labels are determined by the network devices and not provided (as in the case of IP-header information) by the end-user devices, makes for the security of MPLS-based VPN.

The principal advantage of using virtual private network (VPN) services of public IP-backbone operators and telecommunications carriers is their low price. A VPN link can be much cheaper than a dedicated *leaseline* connection between two endpoints when relatively low volumes of data are to be carried across long distance connections. The major drawback of using carrier VPN services is the fact that the network is shared between all the other public network users. As a result, the quality of service (QOS) can be unpredictable and uncontrollable. To overcome this problem, MPLS VPN and BGP/MPLS VPN variants allow for the implementation of VPN with or without QOS guarantees. By contracting to provide a QOS-guaranteed VPN service, a public carrier may guarantee a number of different network performance parameters, including minimum guaranteed bit rate, maximum end-to-end packet delay, etc.

Other path protection considerations

In the immediately preceding sections of this chapter, we have reviewed path protection mechanisms, including tunnelling and VPN, which are intended to prevent stealing of data by third parties. These mechanisms are designed greatly to increase the security of the IP-suite communications protocols themselves. But in addition to such precautions, network designers and operators should also consider a number of other path risks. We discuss next the problems caused by *EMI (electromagnetic interference)* and the risks of radio communication links.

EMI (electro-magnetic interference)

Electromagnetic Interference has recently become a significant problem as the result of high power and high speed data communications devices (e.g., mobile telephones and office LAN systems). EMI can lead to corruption of data information and general line degradation, particularly with intermittent and unpredictable errors. Such problems may be caused by bad data network design or installation. Alternatively, they could result from the *jamming* attempts of a malicious third party.

The problem of EMI is recognised as so acute that a range of international technical conformance standards has been developed which define the acceptable electromagnetic radiation and compatibility of individual devices. In practice, the most common problems are experienced with high speed data networks (eg LANs), particularly when the cabling has not been well designed. Simple network design and operations precautions are:

- the rigid separation of telecommunications and power cabling in office buildings;

- the use of specified cable material only; and,

- the rigid observance of specified maximum cable lengths.

Radio transmission, LANs and other broadcast-type media

Broadcast-type telecommunications media, while otherwise being technically very reliable, are not well suited to high security applications. Many celebrities have discovered to their cost just how easily mobile telephones (cellular radio handsets) can be intercepted. But other broadcast telecommunication media may not be so apparent to users — satellite, LANs and radio-sections of leaselines rented from the telephone company.

Satellite transmission has proved to be one of the most reliable means of international telecommunication. Satellite media do not suffer the disturbances of cables by fishing trawlers and by sharks and achieve near 100% availability over long periods of time. But from a security standpoint, just about anyone can pick up a satellite signal. Thus satellite pay-TV channels need much more sophisticated coding equipment than do cable TV stations to prevent unauthorised viewing.

Local area networks (LANs) of interconnected PCs work by broadcasting information across themselves. So while LANs achieve a very high degree of connectivity (particularly those connected to the public Internet network), they could also present a security risk for sensitive information (unless VLANs are carefully deployed as we discussed earlier).

13.6   Network entry or access control

By controlling who has access to a network we minimise both intentional and unintentional disturbances to communication, in much the same way we might reduce the road hold-ups, hazards and hijacks by limiting the number of cars on the road.

The simplest way of limiting network access is to restrict the number of network connections. Without a connection, a third party cannot access a network and cannot cause disturbance.

Like destination access control methods, entry to a network can be protected by password or equivalent software-based means. The simplest procedures require a user to 'log on' with a recognised username, and then be able to provide a corresponding *authorisation code, password* or *personal identification number (PIN)*. Password-based network access control mechanisms have a lot in common with password-based destination access control mechanisms. Indeed, you may be wondering why it is worth making any distinction.

There are three main distinctions between network access control and destination access control:

- network access control does not even allow contact with the destination server until after successful user authentication;

- the maintenance of the permitted combinations of user and accessed applications is more challenging to maintain for network access control, because of its distributed nature; but

- the authentication of users may be more reliable (you can't claim to be in London, if you're connected to an access node in New York!).

By denying access to the network, an intruder is denied even temporary access to his target server. (In contrast, when <u>destination</u> access control is employed, the user (or intruder) is logged on to the destination server even during the user authentication phase).

The drawback of performing user authentication at the network entry point, is that this is a *distributed task*. Each user connection line and each network ingress point counts as a network entry point, and at each one, there has to be a record of which target applications each individual user is allowed to access. The distributed and duplicated nature of the database (about which users are allowed to access which applications) makes the maintenance of the network access database more difficult than maintaining a single list of allowed users in a destination server.

In the case that users can be authenticated simply by virtue of determining which network connection line (network access line) is originating a communication, the user authentication will be much more reliable. It is potentially much harder to *spoof* another persons identity when network access control security is used.

AAA (authentication, authorisation and accounting)

During the initial access to a network and the establishment of a communications path, the first network node is usually made responsible for *authentication, authorisation and accounting (AAA)*. *Authentication* is the process under which the identity of the calling user is determined and confirmed. The process of *authorisation* determines whether the user is allowed to access the given destination or use the particular service as requested. Having authenticated and authorised the user, the appropriate *accounting* procedure and rate is determined. This governs the charges which will be levied on the user for network and service use.

NAS (network access server)

A *network access server (NAS)* provides for network access control to Internet and IP-based router networks (see Figure 13.24) in the case of users employing dial-in telephone connection lines. The network access server (NAS) performs three main functions:

- physical interfacing to both the IP-network and to the telephone network (either the PSTN (public switched telephone network) or ISDN (integrated services digital network) by means of a modem pool);

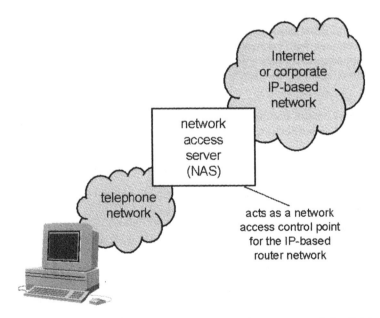

Figure 13.24   Network access server (NAS): performs network access control for dial-in users of IP networks.

- interworking of relevant data protocols, including the logical termination of PPP LCP (point-to-point protocol link control protocol) which is used on the telephone network part of the connection; and

- authentication, authorisation and accounting (AAA) of dial-in users.

The authentication of dial-in users by the NAS may use a number of different techniques. Most commonly, the PPP PAP/CHAP protocols are used (as explained earlier). Alternatively, a *calling line identity (CLI)*, as explained next, may be used.

Calling line identity (CLI)

Calling line identity (CLI) is a feature available on telex networks, on X25 *packet-switched* data networks and on modern ISDN telephone networks. The network (and not the caller himself) identifies the caller to the receiver, thus giving the receiver absolute confidence in the authenticity of the calling user's identity. This gives the receiver the opportunity to refuse the call during the connection establishment phase, if it is from an unauthorised calling location (see Figure 13.25). Call-in to a company's computer centre can thus be restricted to remote pre-authorised company locations. Password protection should additionally be applied as a safeguard against intruders (e.g., the office cleaner!) in the remote sites.

Not all systems which might appear to offer the calling line identity are reliable. Fax machines and fax messaging software, for example, often letterhead their messages with 'sent from' and 'sent to' telephone numbers. These addresses are unreliable. They are only numbers which the machine owner has programmed in himself.

Callback

A popular method of authenticating callers who access company data networks by means of dial-in connections is by the use of *callback* to pre-defined user telephone numbers. Under callback (which is an option of PPP, point-to-point protocol), the user calls up the NAS and identifies himself. The dial-up connection is then cleared, whereafter the NAS calls back the user. The user authentication is by virtue of a pre-defined callback telephone number for the user stored in the NAS. Thus, while it may be possible to *spoof* a valid user identity during

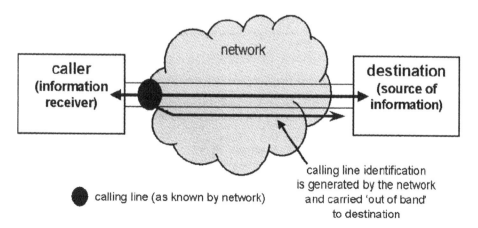

Figure 13.25   The authentication of callers using CLI (calling line identity).

the initial call to the NAS, this does not help an intruder, since the returned callback will always be to the authorised user's telephone number.

The use of *callback* has become popular in corporate networks, in order to allow employees with laptop computers to access the company data network (*intranet*) from home. It not only ensures security of access, but also minimises the 'out-of-pocket' company expenditure which the employee might otherwise incur on his private telephone bill.

Callback does not offer a 100% solution to user authentication. In the case where the calling user is able to input the telephone number on which he or she wishes to be called back, callback offers no user authentication. And do not forget the possibility of a callback call being diverted to somewhere else. Modern telephone networks give householders, for example, the chance to divert calls to their holiday cottage while on vacation. Such call diversion also provides an opportunity for criminals to intercept even callback calls!

Telnet option for TACACS (terminal access controller — access control system):RFC 927

The original TACACS (as defined by RFC 927 in 1984) is a simple password and authentication challenge protocol for the telnet protocol. It is achieved by means of an optional telnet command: *TUID (Telnet user identification)*. The telnet NVT-ASCII code value for TUID is 26 (decimal) — see Table 13.6 (and Table 10.1 of Chapter 10). The user identification takes the form of a standard ASCII character string.

This form of TACACS was used for the original dial-in access to the ARPANET/Internet telnet service. A *terminal access controller* is a device roughly equivalent to a modern NAS (network access server).

TACACS, TACACS+ and extended TACACS (RFC 1492)

TACACS (terminal access controller access control system) is one of the most commonly used AAA (authentication, authorisation, accounting) protocols used in conjunction with dial-in services to the Internet (in particular using a predecessor of PPP — *SLIP* — *serial line Internet protocol*. With the introduction of PPP and the PAP/CHAP protocols TACACS has been somewhat superseded). There are three four different and incompatible versions of TACACS:

- TACACS (original UDP encoding);

- TACACS+ (Extended TACACS);

Table 13.6 TACACS (terminal access controller access control system): TUID (telnet user identification) command

Telnet command	Command meaning
IAC WILL TUID	Telnet user proposes to or agrees to authenticate the user and send the identifying UUID.
IAC WON'T TUID	Telnet user refuses to authenticate the user.
IAC DO TUID	Telnet server proposes that the telnet user authenticate the user by sending the UUID.
IAC DON'T TUID	Telnet server refuses the user authentication.
IAC SB TUID (uuid) IAC SE	Telnet user sends the UUID (uuid) of the user (a 32-bit binary number) to authenticate the user.

- TACACS (TCP encoding);

- TACACS/Telnet option (RFC 0927) (see above).

The TACACS daemon (a software which runs on the terminal server or network access server, NAS) is in complete control of the decision about which users may access which applications/destinations at which times, and subject to which constraints and accounting charges.

The TACACS protocol operates in conjunction with either TCP (transmission control protocol) or UDP (user datagram protocol) — on port 49. Three types of connections may be established using the TACACS protocol:

- an *authentication* without a connection;

- a *login* connection; or

- a serial line Internet protocol (SLIP) connection.

Table 13.7 lists the various TACACS message requests, including the parameter values (i.e., arguments) which must accompany them.

In the case of an authentication with no connection, the TACACS client sends an *AUTH* message and the server sends a *RESPONSE* with a reply, result or reason (for rejection).

In the case of a *login connection,* the client sends a *LOGIN.* The server sends a *RESPONSE* either accepting or rejecting the *authentication.* If accepted, the client may send a *CONNECT* request to set up a connection. The server responds accordingly. Multiple connections

Table 13.7 TACACS: request messages

Tacacs request	Full meaning	Arguments	Purpose
AUTH	Authentication	Username, password, line, style	Request for a user authentication. Username is up to 128-octet ASCII string (codes 33–126). Password is also up to 1128 octets. Line is decimal integer and identifies the line number (0 = console port). Authentication *style* is a string identifying the method used for authentication
LOGIN	Log in	Username, password, line	Requests an authentication similar to AUTH, which if successful starts a LOGIN connection.
CONNECT	Connect	Username, password, line, destinationIP, destinationPort	Requests a TCP connection be established following a LOGIN.
SUPERUSER		Username, password, line	Used in conjunction with an already existing connection, to request that the SUPERUSER or ENABLE mode be entered on the terminal server.
LOGOUT	Log out	Username, password, line, reason	Request to terminate an existing connection (password may be an empty string).
SLIPON	SLIP on	Username, password, line, SLIPaddress	Asks whether the existing connection (defined by username and line) can be connected to the defined SLIP (serial line Internet protocol) address as a remote connection.
SLIPOFF	SLIP off	Username, password, line, reason	Request to terminate an already existing SLIP connection.

may be established if allowed by the TACACS daemon. When finished, the client sends a *LOGOUT* request to terminate the connection and session. The server responds accordingly.

In the case of a *SLIP connection*, the client commences with a *LOGIN* request to which the server responds. The client then sends a CONNECT message to arrange for its connection. Having received a server RESPONSE, it sends a SLIPADDR packet to register its SLIP address. Subsequently it may establish SLIP connections using SLIPON and SLIPOFF commands.

If any client requests do not receive a RESPONSE from the server, the request is simply repeated.

The UDP (user datagram protocol) version protocol packet format of both simple TACACS and extended TACACS (TACACS+) is shown in Figures 13.26a and 13.26b respectively. The field coding is as follows.

The version number is set at value '0' for simple TACACS or value '128' (80 hexadecimal) for extended TACACS (TACACS+). The type field indicates the request or response type. The values are coded as indicated in Tables 13.8 and 13.9.

The *nonce* field may be used by clients to carry an arbitrarily chosen value which will be transferred to the corresponding response, thereby indicating which request a response applies to.

The *username length* and *password length* fields indicate the length in number of ASCII characters (i.e., octets). The username and password fields themselves are coded in plain (i.e., unencrypted) 'standard' alphabetic ASCII characters (values 33–126) (see Table 2.1 in Chapter 2). Their unencrypted carriage makes the TACACS protocol susceptible to *snooping*. For this reason, among others, dial in network access servers (NAS) have moved on to the use of PPP (point-to-point protocol) and encrypted authentication using PAP or CHAP.

The UDP response field is coded with value '1' for an *accept*; or value '2' for *reject*.

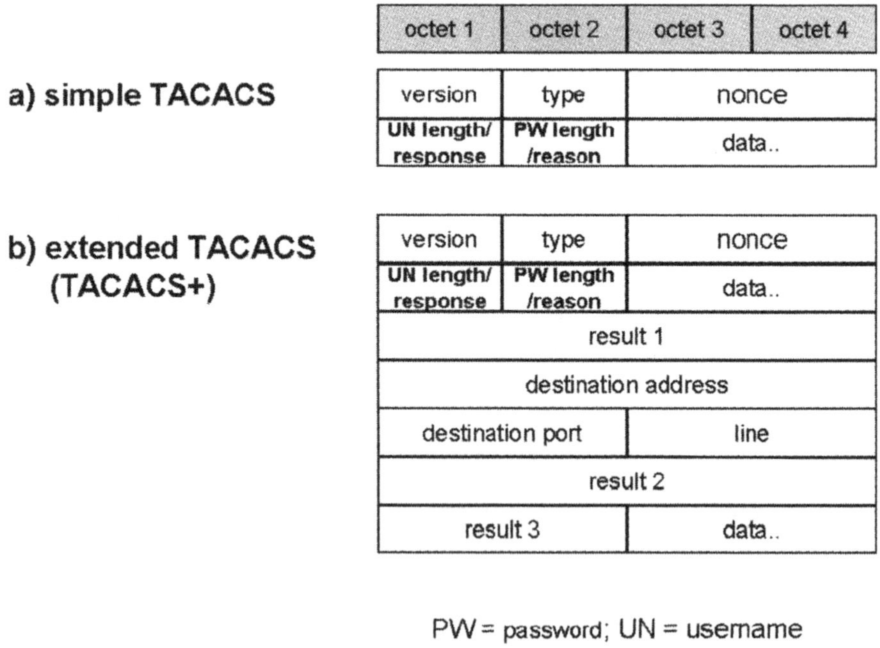

octet 1	octet 2	octet 3	octet 4

a) simple TACACS

version	type	nonce	
UN length/ response	PW length /reason	data..	

b) extended TACACS (TACACS+)

version	type	nonce	
UN length/ response	PW length /reason	data..	
result 1			
destination address			
destination port		line	
result 2			
result 3		data..	

PW = password; UN = username

Figure 13.26   TACACS and TACACS+ protocol packet formats (UDP format).

Table 13.8   TACACS and TACACS+: message type field
coding

Type field value	Request or response TYPE
1	LOGIN
2	RESPONSE (server to client only)
3	CHANGE
4	FOLLOW
5	CONNECT
6	SUPERUSER
7	LOGOUT
8	RELOAD
9	SLIPON
10	SLIPOFF
11	SLIPADDR

Table 13.9   TACACS server response reasons

Reason field value	Server message to client
0	None or 'Accepted'
1	Expiring
2	Password
3	Denied
4	Quit (user quit normally)
5	Idle timeout
6	Carrier dropped
7	Reject — too many bad passwords

The more modern, and alternative TCP (transmission control protocol) encoding of TACACS is incompatible with the historic UDP formats, though the same basic request types (Table 13.7) are used. The basic format of a TCP-format TACACS request is as follows:

```
<version> <type> <parameters> <CRLF>
<username> <CRLF>
<password>
<line>
```

The version number for the TCP-format is value '1'. The TACACS server responses have also been extended beyond the simple accept/reject alternatives of the UDP versions. These a similar '3-digit code and brief text'-format also used in FTP (file transfer protocol) and http (hypertext transfer protocol). The basic TACACS (TCP) responses are:

* `201 accepted`

* `202 accepted, password is expiring`

* `401 no response; retry`

* `501 invalid format`

* `502 access denied`

The TCP version of TACACS is most common of the TACACS forms used, since it is easier to implement than the older UDP versions. But the UDP version has the advantage of a lower protocol overhead and is compatible with older devices.

RADIUS (remote authentication dial-in user service)

Remote authentication dial-in user service (RADIUS) allows for the centralization of the user access and authorization database for Internet or IP-network dial-in users. Instead of each NAS (network access server) having to have a duplicated copy of the AAA (authentication, authorisation and accounting) database, a single, reliable and consistent database (the RADIUS database) is held at a central location. Each time any of the NASs receives a new incoming call, the username, password and user authentication data will be verified with the dial-in user by means of a standard *AAA (authentication, authorization and accounting) protocol* such as *PAP (password authentication protocol), CHAP (challenge handshake authentication protocol) or TACACS (terminal access controller access control system)*. But then, rather than checking the authorization in a local database held at the NAS, the NAS makes a *remote authentication* enquiry to the *RADIUS server* and database using the *RADIUS protocol* (Figure 13.27).

Remote authentication dial-in user service (RADIUS) is defined in RFCs 2865-9 (IPv4) and RFC 3162 (IPv6). The service is widely used in conjunction with network access servers (NASs) to provide *for* AAA (authentication, authorisation and accounting) of dial-in users to IP networks including the Internet itself. Typically the data transport protocol used by the remote dial-in user to transport data across the telephone network part of the connection is PPP (point-to-point protocol), SLIP (serial line Internet protocol — now largely superseded by PPP), telnet or the UNIX rlogin service.

Standard user authentication protocols (e.g., PAP or CHAP in the case of PPP or TACACS in the case of SLIP or telnet) are used to obtain the remote user's username and password and the NAS acts as a relaying agent between these protocols and RADIUS. Having received the authentication data, the NAS (Figure 13.27) operates as a client of the RADIUS server and uses the RADIUS protocol to submit a connection request. RADIUS servers receive connection requests. They authenticate the user data, then deliver service and accounting information to the client (i.e., the NAS) to allow the service requested by the dial-in user to be provided and accounted properly.

Transactions between the RADIUS client (i.e., the NAS) and the RADIUS server are authenticated by means of the use of a shared secret. User passwords are sent in an encrypted form to eliminate snooping. In addition, tunnelling may be used between the RADIUS client and server.

In the case that a network access server (NAS) chooses to authenticate using RADIUS, it becomes a *RADIUS client*. An *Access-Request* is submitted by the client via the network to the *RADIUS server*. In the case that no response is received within a reasonable period of time, the client may repeat the request any number of times.

A RADIUS server responds to a client's Access-Request with one of three possible replies:

- an Access-Reject is sent if any authentication or authorisation condition is not met by the supplied request data;

- an Access-Challenge may be sent to request further authentication data; or

- an Access-Accept is sent to authorise the use of the service.

Figure 13.28 illustrates the RADIUS protocol message format. Messages are carried by UDP (user datagram protocol) on port 1812. Table 13.10 lists the various message types and the message code values.

The major problem encountered with the use of RADIUS is that it can suffer degraded performance and lost data in very large-scale systems.

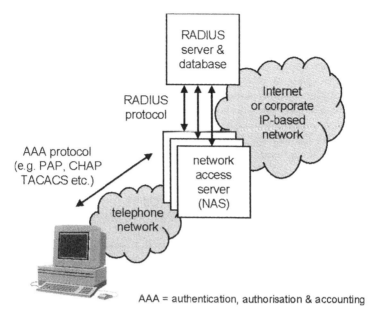

Figure 13.27 Remote authentication dial-in user service (RADIUS).

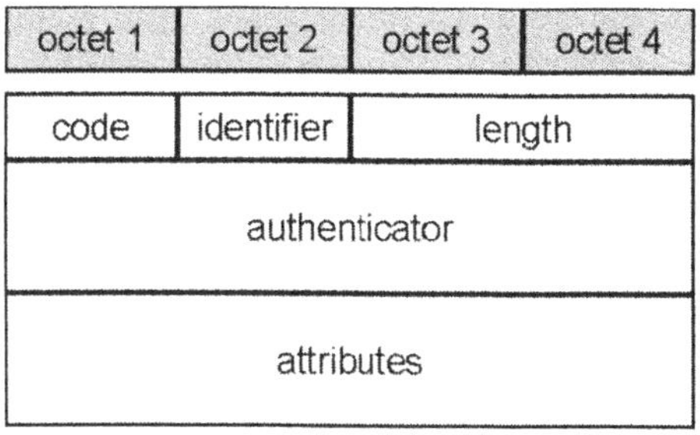

octet 1	octet 2	octet 3	octet 4
code	identifier	length	
authenticator			
attributes			

Figure 13.28 RADIUS protocol message format (UDP port 1812).

Table 13.10 RADIUS message codes

RADIUS code (decimal)	Meaning
1	Access-Request
2	Accept-Accept
3	Accept-Reject
4	Accounting-Request
5	Accounting-Response
11	Access-Challenger
12	Status-Server
13	Status-Client
255	Reserved

13.7   Encryption

Encryption prevents the eavesdropper from understanding what he or she might pick up. Encryption (sometimes called *scrambling*) is available for the protection of both speech and data information. A *cypher* or electronic *algorithm* can be used to code the information in such a way that it appears to third parties like meaningless garbage. A combination of a known *codeword* and a decoding (or *decryption*) formula are required at the receiving end to recon-vert the message into something meaningful. The most sophisticated encryption devices were developed initially for military use. They continuously change the precise codewords and/or algorithms which are being used, and employ special means to detect possible disturbances and errors.

To give maximum protection, information encryption needs to be coded as near to the source and decoded as near to the destination as possible (Figure 13.29). The encryption should be on an end-to-end basis! There is nothing to compare with speaking a language which only you and your fellow communicator understand!

Greatest protection is achieved when the data itself is also stored in an encrypted form, and not just encrypted at times when it is to be carried across telecommunications networks. Permanent encryption of the data renders it in a meaningless or inaccessible form for even the most determined computer hacker. Thus, for example, encrypted confidential information held on an executive's laptop computer can be prevented from falling into unwanted hands, should the laptop go missing.

There are two types of encryption: *symmetric* and *asymmetric* encryption. When using symmetric encryption, a *codeword* (based upon a shared secret known only to the two com-municating parties at either end of the communications path) is used to *code* or *encrypt* the data or message before its transmission. The data is then transmitted in the encrypted form. At the receiver, the *decryption* process uses the *encryption* process in reverse — employing the same codeword. The major problem with symmetric encryption is the need for the same codeword to be known by both sending and receiving parties. A secure *codeword distribution* method is paramount, if the encryption is not to be 'cracked' by third parties who steal or overhear the codeword.

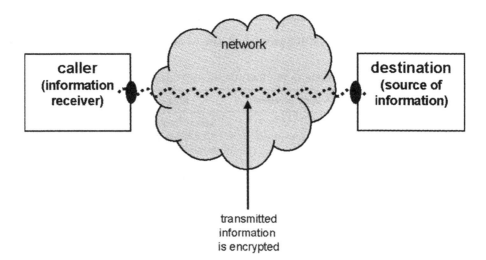

Figure 13.29   Encryption garbles a message during transmission preventing snoopers from understand-ing it.

Asymmetric encryption is nowadays considered more reliable than symmetric encryption and is becoming ever more popular. Asymmetric encryption relies upon the use of two separate *keys* for encryption and decryption. One of the keys (a codeword) is used for encryption, the other for decryption. The keys, however, are not identical. The encryption and decryption processes are not identical, but instead are asymmetric. Thus the key used for encryption is not the correct key for decryption. The advantage of asymmetric encryption is that the key used for encryption can be made relatively *public* (hence the term *public key*). The public key is used by any sender to encrypt data intended for a given destination. But only the *private key* held at the destination is able to decrypt the message. Each destination thus 'owns' a pair of keys — a public key which is made available to anyone who wishes to send encrypted data, and the matching private key retained by the destination for decryption.

Asymmetric encryption is often called *public key cryptography (PKC)*. Public keys are published in open directories — part of an infrastructure for encryption called the *public key infrastructure (PKI)*.

A second important usage of encryption techniques (apart from message content coding or encryption) is their use for the generation of *digital signatures* and *digital certificates* which are used for *STRONG* user identification, authentication and security. A digital signature or certificate is like the signature at the bottom of a letter on paper. The signature proves the authenticity of the letter and its sender. Then by putting the letter in an envelope and sealing it up, we ensure its confidentiality during transport. Sealing the letter is analogous to encrypting data.

Similar mathematical encryption techniques are used for the generation and authentication of digital signatures and certificates, as are used for data encryption. We shall review the most commonly used techniques next.

IP Security architecture and protocols (IPsec)

IPsec provides a complete protocol architecture for authenticating and securing IP (Internet protocol) packet contents. It allows for the complete IP packet payload (including transport and application protocols) to be encrypted and authenticated by means of a digital signature. The IPsec architecture is defined in RFC 2401, though the entire range of RFCs 2401–2411 relate to IPsec. The architecture defines:

- an *authentication header (AH)* and protocol used to provide an IP packet contents (either IPv4 or IPv6) with a *digital signature* or *fingerprint* to confirm authenticity;

- an *encapsulating security payload (ESP)* protocol mechanism for encrypting the contents of IP packets (either IPv4 or IPv6) — thereby ensuring their confidential transmission;

- *security associations* — a <u>simplex</u> end-to-end connection between two points that affords secure transmission of data. The end-to-end *security association* is identified by means of a *security parameter index (SPI)*, an *IP destination address* and a security protocol (i.e., AH or ESP) identifier. Two types of security association are defined: *transport mode* (i.e., end-to-end encryption) and *tunnel mode* (only part of the communications path is encrypted — that part between two intermediate *security gateways*);

- recommended algorithms for *authentication* and *encryption*; and

- *Internet key exchange (IKE)* — the secure management and distribution of encryption keys and codewords (when necessary).

IPsec thus provides for end-to-end access control, data origin authentication and encryption. It may be used to protect one or more paths between the end-points and can use different security associations (i.e., simplex paths) to support the two directions of communication. The architecture is also intended to provide support for IP multicast packets.

Internet protocol authentication header (AH) (RFC 2402)

The Internet protocol *authentication header (AH)* is part of the IPsec architecture for securing IP packet contents during transmission. The authentication header (AH) ensures the authenticity of data to the receiver of that data by the inclusion of an *integrity check value (ICV)*. In effect this is a digital signature. If the ICV value received with the message is correct, then there is a high level of assurance that the message contents have not been tampered with en route. The correctness of the ICV value is confirmed by means of an encryption algorithm and a shared secret (e.g., codeword or key) known only to the two end-parties of the communication.

The format (Figure 13.30) and coding of the *authentication header (AH)* is defined in RFC 2402. The header is intended to be inserted after the IP header and before the ESP header (Figure 13.31). The IPv4 *protocol field* and the IPv6 *next header field* thus indicate the AH protocol (protocol value = 51). The coding of the various protocol fields is as follows:

octet 1
octet 2
octet 3
octet 4
next header
payload length
reserved
security parameters index (SPI)
sequence number
authentication data (variable)

Figure 13.30 IPsec: authentication header (AH) protocol format (RFC 2402).

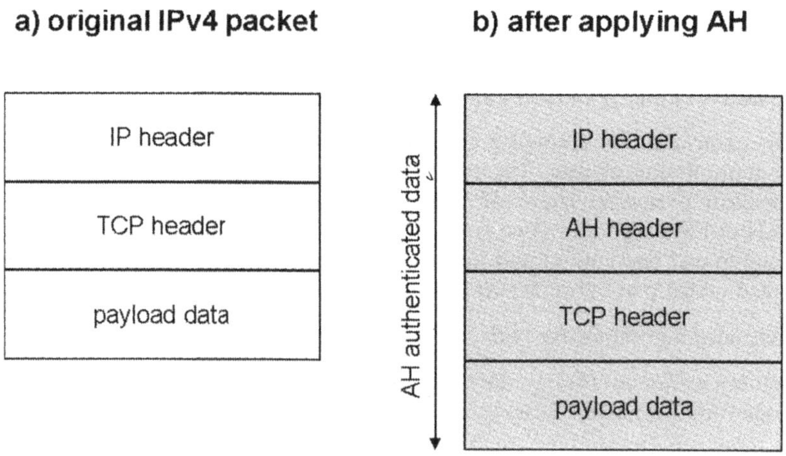

a) original IPv4 packet
b) after applying AH
IP header
TCP header
payload data
AH authenticated data
IP header
AH header
TCP header
payload data

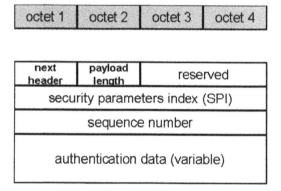

Figure 13.31 IPsec: authentication header (AH) position and scope of the authenticated data.

- The authentication header (AH) *next header* field identifies the next payload after the AH. This will typically be the *encapsulated security payload (ESP)* or a transport layer or application layer protocol, and will be identified with the standard IP protocol number (see Chapter 5–Table 5.6). In the example of Figure 13.31 the next header of the AH will be set to 'TCP' (transmission control protocol).

- The *payload length* field contains a binary number value indicating the length of the authentication header (AH) in terms of the number of 32-bit words. The *reserved* field must be set to all 0's.

- The *security parameters index (SPI)* is an arbitrary 32-bit value, which, in combination with the destination IP address and the security protocol (in this case AH), uniquely identifies the security association (SA) for this datagram. The security association is a simplex (i.e., one-way) path between the two endpoints of the secure connection. A *security association database (SAD)* held by each of the communicating parties stores information about the authentication algorithm and key or codeword information used for generating and checking the integrity check value (ICV) associated with the security association (SA).

- The *sequence number* field contains a counter which is initialized to 0 when the security association (SA) is established and incremented for each subsequent packet. The *authentication data* field contains the value of the integrity check value (ICV). The calculation of the integrity check value (ICV) is computed over the entire IP header and payload fields (Figure 13.31). Suitable security/encryption algorithms for generating the integrity check value (ICV) include symmetric codes (e.g., *DES — Defense encryption standard*) or one-way (i.e., asymmetric) *hash* functions (e.g., MD2, MD4, MD5, RC4 or SHA-1). We shall discuss the codes in more detail later.

Encapsulating security payload (ESP) (RFC 2406)

The Internet protocol *encapsulating security payload (ESP)* is part of the IPsec architecture for securing IP packet contents during transmission. The encapsulating security payload (ESP) ensures the confidentiality of the IP packet contents (i.e., the data) by coding it in an encrypted format. On receipt, the ESP has to be decrypted before sense can be made of the confidential data contents. A number of different encryption algorithms may be used for the encryption itself — either symmetric encryption used a codeword or asymmetric encryption using a public key and a private key.

The format (Figure 13.32) and coding of the encapsulating security payload (ESP) is defined in RFC 2406. The ESP is intended to be inserted after the IP header (and any authentication header, AH). The coding of the various protocol fields is as follows:

- The *security parameters index (SPI)* is the same as for the *authentication header (AH)* as described previously. The *sequence number* is a mandatory field containing a packet counter.

- The payload data is the main data payload, and contains data in a protocol format as defined by the next header field. This will typically be a transport layer or application layer protocol, and will be identified with the standard IP protocol number (see Chapter 5–Table 5.6). In the example of Figure 13.33 the next header of the AH will be set to 'TCP' (transmission control protocol).

- If the algorithm used to encrypt the payload requires *cryptographic synchronisation* (i.e., an *initialisation vector* for the encryption algorithm), then this may also be carried in the payload field (see Figure 13.34).

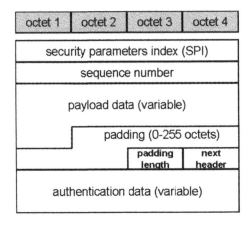

Figure 13.32   IPsec: encapsulating security payload (ESP) format of IP packet contents (RFC 2406).

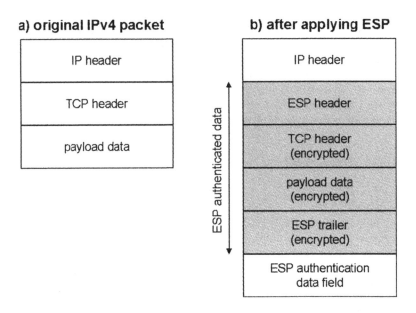

Figure 13.33   IPsec: encapsulating security payload (ESP) position and scope of the authenticated data.

- The padding field allows the payload length to be adjusted to a standard length according to the needs of the encryption algorithm.

- The authentication data field contains an integrity check value (ICV) computed over the ESP packet minus the authentication data.

IPsec conventions and terminology

IPsec may be operated either in *transport mode* (i.e., on an end-to-end *security association* basis) or in *tunnel mode* (i.e., for only that part of the path which interlinks two IPsec-capable *security gateways*).

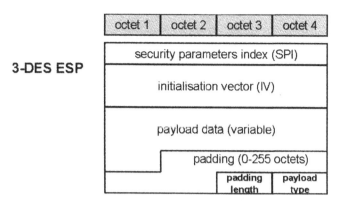

Figure 13.34 Triple DES (Defense encryption standard): use as the IPsec ESP (encapsulating security payload).

In transport mode, the higher layer protocol of IPv4 (and the next header of IPv6) is always AH (authentication header) protocol and then ESP (encapsulating security payload) and then the transport protocol. In other words, the protocols listed in order (lowest-to-highest) are: IP, AH, ESP, transport layer protocol, application layer protocol. In the IPsec tunnel mode the encrypted payload will include a second (encapsulated or tunnelled) IP header.

A *security policy database (SPD)* defines a user's IPsec security requirements and a security association database (SAD) stores information necessary for the generation and checking of authentication data and for encrypting/decrypting the encapsulated security payload.

DES *(Defense encryption standard)*

DES (Defense encryption standard) is a high grade symmetric encryption/decryption standard which uses the same *shared secret key* for both encryption and decryption. The original form of DES (US Federal information processing standard publication 46–published in 1977) uses a 64-bit key. Unfortunately, the huge increase in the processing power of computers has made it possible to 'crack' DES. In consequence, more secure 128-bit DES and triple DES (168-bit key) versions have been developed. But even despite these new developments which provide for exceptional security for most normal corporate uses, DES is no longer considered the most secure encryption algorithm.

The most commonly used version of DES is DES-CBC (cipher block chaining) mode. In this mode, the encryption is carried out progressively in blocks of data, the output of each calculation being fed back as input to the next block or iteration. This method is considered to increase significantly the computer processing power required to 'crack' the code.

In Internet data communications, the following are the most common uses of DES or its variants for data security encryption:

- *Triple DES (3-DES)* (RFC 1851) is an optional encryption method which may be used with the IPsec encapsulating security payload (ESP) to encrypt and decrypt payload data (Figure 13.34). When using triple DES (3-DES), the data to be encrypted (i.e., the *plain text*) is processed three times over — each time with a different 56-bit key ($3 \times 56 = 168$ bit total key). As well as the *shared secret key*, the receiver requires an *initialization vector (IV)* which must be included in each datagram to enable decryption.

- *DESE (PPP DES encryption protocol)* (RFC 1969) is commonly used to encrypt PPP (i.e., dial-in Internet or IP-network) connections.

- *3-DESE (PPP triple DES encryption protocol)* (RFC 2420) is a further development of DESE to incorporate the higher security offered by triple DES into PPP (point-to-point protocol).

Message digest algorithms and digital signatures

A *digital signature* (which is an encrypted form of a *message digest, message authentication code (MAC)* or *fingerprint*) is used to verify the authenticity of the origin of a message. A message digest (MD) algorithm takes a message of an arbitrary length (i.e., the data to be *signed*) and produces as an output a 128-bit or longer fingerprint or message digest of the input. The message digest is then encrypted using an asymmetric encryption algorithm and the sender's *private key* in order to generate a digital signature. The receiver uses the sender's public key (if it is available to him) to decrypt the fingerprint, and checks this value against the value calculated for the received message. If the value is correct, then the message has not been altered or tampered with during transmission.

The theory goes that, given a defined algorithm and shared secret, no two messages will have the same digest (i.e., fingerprint). It is also reckoned to be impossible to create a message to fit a pre-defined fingerprint value. In this way it is not possible without detection to copy a signature, or to 'cut it off one message' and attach it instead to a bogus message. There are four main *algorithms* widely used in Internet and IP data communications.

MD2-message digest algorithm (RFC 1319)

The length of the message to be 'signed' is adjusted by adding padding to make it a length which is an exact multiple of 16 bytes. Padding is always added (even if the message is already an exact multiple of 16 bytes). The algorithm is then applied. The calculation is quite secure (difficulty of code cracking is reckoned to be around 2^{64} (i.e., one in 10^{19} probability). The 128-bit fingerprint or message digest value is encrypted using the sender's public key and then sent with the message to authenticate its origin. The receiver first requires access to the sender's *public encryption key* to decode the fingerprint. The receiver applies the algorithm to the received message, checking that the newly calculated fingerprint matches the received decrypted value. If so, the message is authentic.

MD4-message digest algorithm (RFC 1320)

The *MD4* algorithm is similar to the MD2 algorithm. It is designed to be very fast, but is considered to be 'at the edge' of risking crypto-analytic attack. It generates a 128-bit fingerprint or message digest used for message authentication.

MD5-message digest algorithm (RFC 1321)

The *MD5* algorithm is a more robust adaptation of the MD4 algorithm, and is designed to be fast on 32-bit computer processors. The MD5 algorithm can be coded quite compactly, and does not require large substitution tables. It generates a 128-bit *fingerprint* or *message digest* used for message authentication and is probably the most popular algorithm for this purpose.

SHA-1 secure hash algorithm (RFC 2841)

The *secure hash algorithm (SHA-1)* generates a random *secret authentication key* of up to 160-bits in length to confirm the authenticity of a message. It is the standard defined for use with the *digital signature standard (DSS) public key infrastructure (PKI)* system developed by the US National Institute of Standards and Technology (NIST) and becoming very popular as an alternative to the MD5 algorithm. SHA-1 is not quite as fast as MD5, but the larger size of the fingerprint makes it more robust than either MD2, MD4 or MD5.

Public key infrastructure (PKI)

A *public key infrastructure (PKI)* is an infrastructure used for maintaining and distributing the public keys and digital certificates which play an important part in modern data security mechanisms — including message authentication and encryption. The public key infrastructure (PKI) was first internationally standardised in ITU-T recommendation X.509.

The public keys used in asymmetric encryption techniques need to be made available (on a secure and as-authorised basis) in order that secure encrypted communication can take place between a private key holder and a range of other (authorised) parties (who use the public key). Keys only work as a combination of public key/private key pairs, but while the public key is intended to be *transferable* (since it is not critical if it is 'overheard'), the private key should be stored very securely. It needs to be treated like a human's passport. It should not be stored on a computer hard-drive. A better means of storage is by means of a PKI 'smart card'.

The potential uses of a PKI (public key infrastructure) include:

- authentication of network or application users;

- email or data file encryption;

- digital signatures; and

- access controls.

A PKI (public key infrastructure) comprises a number of different components:

- communications applications that make use of certificates and seek validation of others *certificates*;

- *certificate authorities (CA)*;

- *certificate practice statements (CPS)* and associated policies;

- *registration authorities (RA)* and *commercial certificate authorities (CCAs)*;

- a certificate distribution system to store certificates, public keys and certificate management information;

- management tools and software to manage user validations, certificate renewals and revocations; and

- databases and key management software to store escrowed and archived keys.

As well as containing the public key of an organisation or individual, a certificate (which is a password-protected and encrypted data file) also includes the name and other information about the identity of the issuing party.

The certificate authority (CA) is the 'owner' of a private key. The CA publicises the associated public key (for asymmetric encryption purposes) by means of a digital certificate. The certificate authority manages and signs certificates on behalf of an individual or organisation and revokes certificates when necessary (e.g., when the public key is changed) by publishing *certificate revocation lists (CRLs)*.

Registration authorities (RAs) or commercial certificate authorities (CCA) are *trusted third-parties* responsible for storing and distributing certificates and public keys and for managing certificate distribution. They operate on behalf of the CA to validate users, to distribute certificates to authorised public key users and to maintain a list of which users have been issued certificates. In effect, they provide for a public interface between the remote user and the CA (certificate authority). It is up to the RA or CCA to prevent hackers from getting hold of certificates without proper authorisation and to prevent the generation of bogus certificates.

SSH (secure shell)

The *secure shell (SSH or SECSH)* protocol is in the stage of being an 'Internet draft' (i.e., will shortly be issued as an RFC). It adds a 'security layer protocol' on top of TCP (transmission control protocol), creating an end-to-end encrypted tunnel for the secure forwarding of segments (i.e., TCP user data 'packets') of any chosen application protocol using a TCP port. As the name suggests, the secure shell (SSH) protocol was originally designed for secure login to a remote server (operating a computing shell program in a secure manner — if you like a 'secure telnet' connection). For some purposes, SSH can be considered to be a 'lightweight version of IPsec'. SSH is described in detail in Chapter 10.

SSL (secure sockets layer)

The *secure sockets layer (SSL)* was developed by Netscape to provide privacy during communication over the Internet, particularly during http (hypertext transfer protocol) sessions. It is achieved by means of encryption: typically using either the DES (Defense encryption standard) or RC-4 encryption algorithm. When in use, it can be recognised by the mnemonic *https*. It is intended to enable http client/server applications to communicate without eavesdropping or tampering. The current version (SSL3.0) was published in November 1996, though it has been largely superseded by the *TLS (transport layer security)* protocol published in 1999 (RFC 2246).

SSL is a two-layer protocol, comprising the *SSL record protocol* and the *SSL handshake protocol*. After an initial handshake (using the SSL handshake protocol) the connection is private (by means of key encryption). The SSL record protocol is a secure tunnelling (i.e., encapsulation) protocol, similar to the tunnelling protocols described earlier in the chapter. It is used as the carriage mechanism first for the SSL handshake protocol, and subsequently for the data payload. SSL is intended to provide for connection-security which is independent of the application protocol in use.

TLS (transport layer security)

The *transport layer security (TLS)* protocol (RFC 2246 is TLS version 1.0) developed from and has superseded SSL3.0 (secure sockets layer protocol version 3). It is intended to provide end-to-end transport layer connection security for any higher layer application protocol. It provides for secure *encapsulation* (i.e., end-to-end tunnelling) of higher level protocols.

Like SSL, TLS is a two-layer protocol, comprising the *TLS record protocol* and the *TLS handshake protocol*. After an initial handshake to identify the peer communication partner (using the TLS handshake protocol) the connection is private (by means of key encryption). The *SSL record protocol* is a secure tunnelling (i.e., encapsulation) protocol based upon symmetric cryptography, and typically employing either the DES (Defense encryption standard) or the RC4 encryption algorithm. Alternatively, RFC 2712 describes how the UNIX Kerberos cipher suites may be applied to TLS. The negotiation of a shared secret (as needed for the symmetric encryption) is secure and reliable (no hacker changes are possible without being identified).

TLS relies upon TCP (transmission control protocol) to provide for a reliable end-to-end connection.

Pretty good privacy (PGP)

PGP (pretty good privacy) is a 'lightweight' protocol which uses a combination of public key and conventional symmetric encryption to provide security services for electronic mail and data files. It provides for digital signature, confidentiality and data compression. It is defined in RFC 1991.

PGP digital signature

A PGP digital signature is generated by the sender of a message to confirm his identity and the authenticity of the message. It is generated and used as follows:

- The sender creates a message, and calculates the message digest or fingerprint using a *hash code* (e.g., SHA-1). The fingerprint is then encrypted using the sender's *private key*, whereupon it is attached to the front of the message as the digital signature. Like this, it is sent to the receiver.

- The PGP receiver decrypts the digital signature using the sender's public key to regenerate the fingerprint. After this the receiver generates a new hash code for the received message and compares it to the regenerated fingerprint. If both fingerprints are the same, the message is taken to be authentic (digital signature).

PGP confidentiality

When a sender wishes to ensure confidentiality of data during transmission, the message can be encrypted using PGP as follows:

- The sender creates message and generates a random number as a *session key* for the message. The sender encrypts the entire message contents using the agreed encryption algorithm and using the session key. The session key is then encrypted using the intended recipient's public key and then is added to the message.

- The message recipient decrypts the session key by making use of the recipient's private key. With the decrypted session key, the recipient is then able to decrypt the main message contents.

Data compression and character set conversion under PGP

Data compression (if applied) is performed by a PGP sender after applying the signature to the message, but before encryption.

Once ready for transmission, a character set conversion (to *radix-64*) may be applied by PGP if necessary in order to accommodate transport networks which allow only restricted ASCII character sets.

S/MIME (Secure multipurpose internet mail extension)

S/MIME (secure multipurpose Internet mail extension) defines a methodology for securing electronic mail communications. It is based upon public key encryption and uses digital certificates and signatures to confirm that a message has not been tampered with during transmission. In addition, the signature can be used to provide for non-repudiation (in other words, the sender cannot deny having sent the message). Messages are encrypted and enclosed in a digital envelope.

S/MIME is defined in RFCs 2632-4. It is beginning to take over from the alternative methods of securing email communication: PGP (pretty good privacy) and *PEM (privacy enhanced mail)*.

Other encryption methods

Before we leave the subject of encryption, let us mention another couple of methods found in practical networks.

RC4

RC4 is an encryption/decryption algorithm supported in *cellular digital packet data (CDPD)*. It is a stream cypher, used for file encryption in the encryption of secure web sites which employ the SSL and TLS protocols.

S-http (secure hypertext transfer protocol)

The secure hypertext transfer protocol (as defined in RFC 2660) provides for secure communication between an http (hypertext transfer protocol) client and an http server. It may seem like 'competition' for the SSL or TLS protocol, but in reality S-http and SSL/TLS (https) complement one another.

13.8   Application layer interface for security protocols

The *GSS-API (general security service application program interface)* is designed as a *CAPI (communications application program interface)*, by means of which computer application software may make use of secure communications services to distributed locations. GSS-API is designed to insulate the application from having to be developed to include specifics of the underlying security mechanisms. GSS-API was published in January 2000 and is defined in RFC 2743. Like many other security protocols and encryption methods, its initial development was driven by RSA laboratories.

13.9   Other risks and threats to data security and reliable network operations

What are the main technical risks leading to potential network abuse, breaches in confidentiality or simple corruption of information? What can be done to avoid them? A proper analysis of the risks and the development of proper procedures is as important (if not more important) than the data security technology employed to protect data during transmission across a network.

Consider the motivations

Before being able to draw up a comprehensive network and data security plan, designers need to consider the risks and the motivations of 'snoopers', 'spoofers' and 'intruders'. These can be many fold:

- stealing of confidential data or business information;

- industrial espionage;

- disruption of business operations or service (e.g., as a result of a *service attack*, a *denial-of-service (DOS)* or because of service or network *jamming*);

- misleading the company into taking actions beneficial to the intruder or 'spoofer';

- misrepresentation of the company to gain information or other favours from third parties (e.g., by 'spoofing' the identity of the company);

- economic motivation (communicating across your network at your cost);

- simple 'monitoring' of who the target individual communicates with; and

- *hacking* as a 'sporting challenge'.

Different precautions will be appropriate, depending upon the overall *risk profile.*

Service attacks and denial of service (DOS)

Service attacks are attempts by malicious third parties to upset the computing operations of a given company. Such attacks are most commonly launched at website servers and ebusiness applications. As ebusiness software has become increasingly robust to fraudulent manipulation, a new form of service attack has appeared — the *denial of service (DOS)* attack. A denial of service (DOS) attack works to 'crash' the web application server by overloading it with requests. During the period of the crash, ebusiness cannot be conducted and customers get frustrated.

Early DOS attacks were simply based upon thousands or millions of requests generated by a single host and directed at the *target server*. But the use of proxy servers and traffic filters has helped to eliminate this type of attack by reducing the rate at which packets are accepted from a given source IP address. More recent evolutions of the denial of service (DOS) attack instead do not launch their attacks directly, but instead try to cause a large number of 'innocent bystander' hosts to all generate requests at once. The ingenuity of the determined criminal is unlimited!

Perhaps even more scary is that an incompetent (but authorised) user executing a poorly written software or *script* program can be almost as (if not more) damaging than a criminal hacker.

Spoofing

By the use of a stolen identity (this is called *spoofing*), a criminal may attempt to get access to information which he or she is otherwise not authorised to receive. Certain forms of spoofing rely upon a source host indicating a *spoofed* source-IP address or other higher layer source protocol address in the packets it originates. More recent forms of spoofing take advantage of the heavy reliance of modern email and web (i.e., http) communications upon the domain name system (DNS). DNS spoofing takes place by mimicking the actions of a DNS server to mislead a DNS client.

Economic motivations for network abuse

One of the most common (and obvious) motivations for network intrusions is the simple criminal desire to get something for nothing — perhaps communication at your expense.

One way in which criminals have historically tried to exploit private telecommunications networks is as a means of network transit. Such transit is possible across any network which offers both a *dial-on* and a *dial-off* capability. Thus, for example, some companies operate a reverse-charge or *freephone* telephone network *dial-on* capability to enable their executives to access their electronic mailboxes from home without expense. Some of these networks simultaneously offer a *dial-off* facility. Thus, for example, the London office of a company might call anywhere in the United States for domestic tariff, by first using the private network to reach the company's New York office, and then 'dialling-off' into the local US telephone company.

When simultaneous dial-on and dial-off is available, the outsider can make all the calls he or she wants — entirely at company expense unless the network is well enough designed to prevent simultaneous dial-on and dial-off by the same call (Figure 13.35). Simultaneous dial-on/dial-off can be a lucrative scam, and has been used many times in the past for exploiting private telephone networks for free international calls. But the same principles could easily be adapted for exploitation of a large-scale private company data network.

A possible means of prevention of criminal dial-in/dial-off activity is the rigid use of *dial-back* (or *callback*) instead of *dial-on*. Dial-back (callback) similarly reverses the charges for the caller (other than the cost of the initial set-up call), but in addition enables the company to have greater confidence that only authorised callers (i.e., known telephone numbers) are

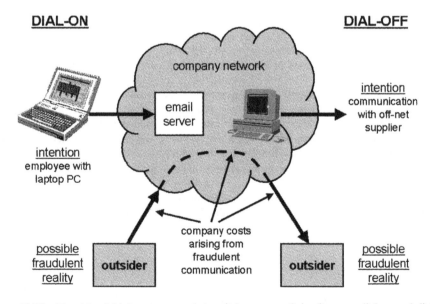

Figure 13.35 The risk of third party network transit by means of simultaneous dial-on and dial-off.

originating calls. Alternatively, the network could be configured in such a way that dial-on and dial-off are not possible on the same connection.

The confidentiality of electronic mail: DO's and DON'Ts

Electronic mail and other message switching networks offer their users a higher level of confidence that messages will be delivered correctly and completely, and usually can give confirmation of receipt. Messages are generally read by a manager himself rather than by his secretary — making them more secure — but this is not always the case. For very highly confidential information, users need to take into account the fact that a complete copy of the message is stored on the various mailbox servers along the path, and may be accessible to mail system administrators.

'Deletion' of a message from your mailbox may prevent you as a user from further accessing a message, but should not be taken to imply that the information itself has been obliterated from its storage place. A technical specialist with the right access may still be able to retrieve it.

Public telecommunication carriers in most countries are obliged by law to ensure absolute confidentiality of transmitted information and proper deletion once the transmission is completed successfully. But while this level of legal protection may be adequate for the confidentiality needs of most commercial concerns, for matters of national security it will not be.

The right confidentiality policy for communication: DO's and DONT's

Keep it simple! Minimise the number of potential weaknesses by standardising on accepted company telecommunications media (e.g., electronic mail, telephone) and building adequate safeguards around these 'normal' methods. Make sure that the company's basic security and confidentiality policy covers the particulars of acceptable telecommunications media for each classification of document or other information.

The human links in communication are usually the weakest and the most prone to 'leak' information, so ensure that adequate discipline is applied to maintaining professional care. Automate processes like the changing of computer passwords to ensure they are updated regularly.

DO: consider the possible motivations for data network intrusion, tampering or exploitation and take pragmatic steps to eliminate them.

DON'T: be careless or neglectful. You shouldn't use a fax or mobile phones for company information classified as 'confidential'. And don't assume that a firewall or other security device will work on its own without maintenance. It needs proper management, updating and administration by suitably qualified experts. Don't let the hackers get one or more steps ahead!

Carelessness

Always check addresses. I was once amazed to receive some UK-government classified SECRET documents that should have been sent to one of my namesakes!

Why even think about encrypting a fax or other message between sending and receiving machines, if either machine is to be left unattended? Do not contemplate reading it on the train or talking about it on the bus.

Computer system passwords should be changed regularly. If possible, password software should be written so that it demands a regular change of password, does not allow users to use their own names, and does not allow any previously used passwords to be re-used. Ex-employees should be denied access to computer systems and databanks by changing system passwords and by cancelling any personal user accounts.

Computer systems designed to restrict *write-access* to a limited number of authorised users are less liable to be corrupted by simple errors. Holding the company's entire customer records in a PC-based spreadsheet software leaves it very prone to unintentional corruption or deletion by occasional users of the data. Any changes to a database should first be confirmed by the user (e.g., a prompting question such as 'update database with 25 new records?' could be required to be answered with either 'confirm' or 'cancel').

System software should perform plausibility checks (as far as possible) before any old data records are changed or replaced (e.g., can a person claiming social security benefit in 2003 really have been born in 1850?).

Ensuring proper and regular back-ups of computer data help guard against corruption or loss due to viruses, intruders, technical failures or simple mistakes. Daily or weekly back-ups should be archived 'off-line'.

Simple precautions properly applied dramatically reduce the risk to most commercial concerns and sometimes even eliminate the need for complex and expensive data security technology!

Call records

For some very sensitive commercial issues, say when contemplating a company takeover, it may be important to a senior company executive that no-one should know he or she is even in contact with a particular company or advisor. Such company executives should be reminded of the increasing commonality of itemised call records from telephone companies, and similar records of data communications. Use a 'private' or 'temporary' email address.

An onus on communicators

Confidence in communication — the reliance on the safe delivery of accurate information into the right hands — depends most on the right choice of communications media by the originating party and professional care by both parties. You don't need expensive 'rocket science' data security technology if you don't even stick to the basic and obvious rules of confidentiality! Most important is the removal of temptation or opportunity, and the reduction of the potential benefit of 'snooping'.

Only the originator can be to blame if he or she sends a confidential message to an unattended fax machine. But the call receiver can also give important information away. I have been amazed to observe investigative journalists at work — simply calling up company representatives during periods of intense company activity (e.g., at a time of rumoured merger) and asking speculative questions. The journalist may know nothing to start with, but by planting direct questions as a bait to an unsuspecting manager, he or she may gain interesting information if they strike close to a current 'truth'. This is the basis of *spoofing*.

One final thought: don't get too fanatical! The determined criminal can usually find a way to 'beat the system'. One of the easiest ways is to log on as system administrator: companies often have little defence against maliciously minded current and former system administration employees!

14

Quality of Service (QOS), Network Performance and Optimisation

The maintenance of good quality for any product or service (its 'fitness for purpose') is of supreme importance to the consumer and therefore requires utmost management attention. But while it is relatively easy to test a tangible product to destruction, the objective measurement of the quality of a service is much more difficult. What often counts most is the human end-user's perception: how well did he or she think the communication went? Unfortunately, however, end-user opinion is rarely a good basis for deciding what design changes and extensions are necessary in a data network: End users may be unaware of line quality problems causing a high incidence of bit errors, due to the beneficial effects of protocol error correction and recovery. Meanwhile, the same end users may unfairly criticise the network as being slow, for problems really caused by poor application software design. The optimisation of both network quality and efficiency demands a high level of expertise and experience. The root causes of many problems are not easy to find. But this should not put off network administrators from striving to achieve the best quality of service (QOS) possible. In this chapter we set out an objective framework for measuring telecommunications network service quality. Our aim is to provide a practical framework for the continuous monitoring of network quality, describing the symptoms of typical network problems to look out for, the methods available to diagnose problems and the tools available to overcome them. We set out a structured process for network design and administration with the goal of optimum network quality and efficiency: avoiding problems as far as possible before they arise.

14.1 Framework for network performance management

Good management, in all types of industry, demands the use of simple, structured and effective monitoring tools and control procedures to maintain the efficiency of internal business processes and the quality of the output. When all is running smoothly, a minimum of effort should be required. But in order to be able quickly to correct defects or cope with abnormal circumstances, measurable means of reporting *faults* or *exceptions* and rapid procedures for

Data Networks, IP and the Internet: Protocols, Design and Operation Martin P. Clark
© 2003 John Wiley & Sons, Ltd   ISBN: 0-470-84856-1

identifying actionable tasks are required. The framework for doing so needs to be structured and comprehensive.

Unfortunately, the subject of operational service quality management has been largely neglected by the designers of the Internet. There is little formal advice and few standardised methods for network quality measurement and everyday network administration.[1] Instead, the vision of the Internet founders — of a large 'shared network' and always making 'best efforts' to carry all possible traffic (even that of third parties) — still remains the central philosophy of the Internet and many IP (Internet protocol)-based networks. The job of maintaining a defined network service quality is largely a 'black art' comprising 'empirical' methods — 'if the network appears a little slow, add a little capacity here and there, and see how things improve (or don't!)'.

With the commercialisation of the Internet, and the increasing expectations of end-users for reliable, high-speed and predictable *ebusiness* services, it seems inevitable that a formal framework[1] for network quality management will appear. In the meantime, network administrators will have to develop their own quality frameworks for IP data networking[*see Note]. Perhaps the best place to start is by considering the quality framework used by the public telephone network companies. The framework is set out in the quality standards of ITU-T (International Telecommunications Union — Standardization sector) and is based on many years experience with the public telephone service and the expectation of its users that the network shall always be available.

Table 14.1 provides a possible framework for data network quality management based upon an adapted version of the ITU-T framework intended for the public telephone service. A number of simple quality categories and 'dimensions' are presented, together with suggested performance management methods. As you will note, the dimensions cover a range of areas, some requiring more tangible monitoring measures and control procedures than others. A complete quality framework for data networks needs to consider many more factors than simply the *latency* of data traffic! It also needs to provide objective feedback for the network design process.

14.2 Quality of service (QOS) and network performance (NP)

In practice, it is difficult to separate the quality of data communications networks from the quality of the computer software applications which are running on those networks. Thus, for example, the time elapsed before data is returned from a server in response to a user's request depends not only upon the bit rate of the data transmission lines and the end-to-end packet propagation delay (the so-called latency) but also upon the protocols used by the application software and the speed of response of the server as well. The *quality of service* perceived by the human end-user thus depends upon a number of factors in addition to the *performance* of the data network.

In its recommendations on network service quality, ITU-T (International Telecommunications Union — Standardization sector) has established a clear distinction between two separate categories of performance measurement:[2]

- *quality of service (QOS)*; and

- *network performance (NP)*.

---

[1] The scope of a new area of study, setting out the 'overview and principles of Internet Traffic Engineering' was first laid out in RFC 3272 in May 2002.
[2] The ITU-T quality of service (QOS)/network performance (NP) model is not strictly ratified as applying to the Internet. Nonetheless, it presents a number of ideas which have great applicability to data network performance (NP) and the end-to-end quality of service (QOS) perceived by human users of networked computer applications. It is for this reason that we present it here.

Table 14.1 Quality framework for management of data communications network service

Category	Dimension	Tangible measurement method or control mechanism
Service ordering, provision and alteration	Availability of required services, software and features	User questionnaire
	Waiting time for a connection	Service order waiting list
	Waiting time for bit rate or other service upgrade	Change control procedure
Service availability	Geographic availability of connections of required bit rate	Number and percentage of customers not within the service area (e.g., for direct connection, secure dial-in service or some other service, etc.)
	Availability of destinations	Destinations and remote services not reachable
Quality of communication or data transfer	Access to network (connection or data transfer 'establishment phase')	Number/percentage of connections/data transfers not able to be established (e.g., destination not available/temporarily unavailable/connection limit exceeded/flow label limit exceeded, etc.)
		Percentage of misrouted packets (e.g., due to routing table errors, technology faults or network topology problems)
	Data transfer phase	Bit error ratio (BER) (a measure of line quality)
		Number/percentage of lost, discarded or resent packets (caused by poor line quality or network congestion)
		Network latency — mean and peak network propagation delays for end-to-end packet transfer (a measure of network congestion)
		Network load as percentage of capacity
		Number and percentage of lost connections or sessions (system 'hang ups')
	Connection clearance (*Disengagement*)	Correct network 'reset' should occur automatically after each communication. A high frequency of the need for manual resets signals a problem.

(continued overleaf)

Table 14.1 (*continued*)

Category	Dimension	Tangible measurement method or control mechanism
Service reliability	Service faults Service availability or non-availability	Number, frequency and duration of faults Frequency of necessary network or software upgrades to overcome bugs Total period of lost service (per user or server connection) per month or per year. Percentage availability. (A target availability of 99% allows only 3 days outage due to faults each year)
Customer service and support	Accuracy and speed of helpdesk in recording problems	Helpdesk time to respond (e.g., time to answer telephone, elapsed time before expert is despatched or calls back user)
	Technical competence of staff and speed of fault resolution	Mean-time-to-repair (MTTR) faults, resolve reported problems and clear trouble tickets Percentage of trouble tickets cleared within a given target time Percentage of trouble tickets cleared with reason 'fault not found' (a large percentage may indicate problems not properly diagnosed and resolved or user misunderstanding regarding a particular service)
	'Helpfulness' of staff	The courteousness and willingness of staff to help — measured by questionnaire
	Service documentation	Availability of documentation about the service and operational procedures
Fairness, reliability and accuracy of service charges	Probability of incorrect invoice	Percentage of disputed bills

The relationship of quality of service to network performance as defined by ITU-T is shown in Figure 14.1. According to ITU-T definition, quality of service measurements help a telecommunications service or network provider to gauge customers' perceptions of the service, while network performance parameters are direct measurements of the performance of the network, in isolation of the effects caused by human users or data terminal equipment and application software. Thus quality of service encompasses a wider domain than network performance, so that it is possible to have a poor overall quality of service even though the network performance

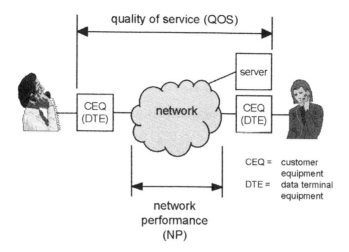

Figure 14.1   ITU-T definitions of quality of service (QOS) and network performance (NP).

may be excellent. Identifying such cases is the key to deploying the right application support staff to seek the problem rather than committing network operations staff to an attempt at network improvement without chance of success.

The measured quality of service may differ greatly from the measured network performance values in cases where the end-to-end path traverses several different networks. But while the quality of service is of utmost importance to the end-user, it cannot become the only pre-occupation of network administration staff, because it does not directly reflect the performance of their network and operations, and anyway it is very difficult to measure objectively. Quality of service is not only technically difficult to measure, it should theoretically also be measured for each individual customer and each computer application separately. Such difficulties were the reason for the development by ITU-T of the concept of network performance. Network performance can be more easily measured within the network, and provides meaningful performance targets and direct network design feedback for the technicians and network managers operating the network.

Quality of service parameters should be chosen to reflect the end-to-end communication requirements or the performance requirements of a networked computer application (as perceived by the human user). These parameters should then be correlated with one or a number of directly related network performance parameters, each NP parameter reflecting the performance of a subnetwork or other component part of the network, and therefore contributing to the end-to-end quality. In this way, both quality of service and network performance problems can be most effectively monitored and traced to their root cause.

Similar parameters may be (but are not always) used to measure both quality of service and network performance (e.g., propagation delay, *bit error ratio (BER)*, % congestion etc.). Normally the measured quality of service will be found to be lower than the measured network performance — the difference is due to the performance degradations caused by the users' data terminal equipment (DTE), software and the human user's method of use.

14.3   Quality of service (QOS), type of service (TOS) and class of service (COS)

In data protocol specifications, the term quality of service is most often used to describe the target level of communications quality and reliability which should be achieved by a given protocol (in effect: what ITU-T calls the network per *formance* of the protocol). The quality of service of a number of layer 2 and layer 3 protocols are defined in this way. Thus the target quality of service of a datacommunications protocol or network is typically defined in terms of:

- the minimum 'guaranteed' data *throughput* (i.e., bit rate);

- the maximum one-way packet propagation delay (often called the *latency*); and

- the *reliability* of the network or connection (e.g., the 'variability' of the connection quality; the fluctuation in packet delay times or the *availability* of the service).

IP-suite protocols which attempt to 'guarantee' a given network quality-of-service do so by prioritising packets. Higher priority packets are dealt with first. They are forwarded in preference to other lower priority packets and stand less chance of being discarded in the case of network congestion. By means of such prioritisation, capacity within the network can be 'reserved' for higher priority traffic during times of network congestion, thereby 'guaranteeing' the minimum bit rate available for data throughput. In parallel, the preferential forwarding of packets by intermediate routers minimises the end-to-end propagation delay or latency. The assumption is that by minimising the delay, the variation in the delay will also be minimised. But you should never forget that the <u>minimum</u> delay is not necessarily either a short delay or an acceptable delay!

Packet prioritisation schemes usually operate by means of one or more quality of service 'labels' in the packet's protocol header. A number of different 'labels' are used by different protocols as listed below. (Some of these we have discovered earlier in this book):

- *type of service (TOS) field* (used by Internet Protocol version 4 header);

- *IP precedence* field (used by Internet Protocol version 4 header);

- *T (throughput), D (delay)* and *R (reliability bits* (used by Internet Protocol version 4 header);

- *DiffServ* field as used by Internet protocol *differentiated services (DiffServ)*;

- *traffic class* field (used by Internet Protocol version 6 (IPv6) for differentiated services);

- *flow label* (used by Internet Protocol version 6 and *MPLS, multiprotocol label switching*);

- *user priority field* (defined by *IEEE 802.1p* for prioritisation of traffic between *virtual bridged LANs, VLANs*).

Type of service (TOS)

Internet protocol (IP) version 4 (IPv4), as we saw in Chapter 5, can be made to prioritise packets according to their *type-of-service (TOS)* as recorded in the IP header type of service field (Figure 14.2). The *IP precedence* value within the TOS-field is a 3-bit value indicating the priority with which routers are to deal with and forward so-labelled packets. As listed in Table 14.2, the highest priority is given to 'network control' packets, thereby ensuring that packets concerned with network reconfiguration and control have a high chance of getting through even during periods of very severe network congestion. The message, after all, might be critical to relieving the congestion. The lowest priority of packets (and most likely to be discarded at times of congestion) are 'routine' packets. Alternative, but less well defined, are packet prioritisations to achieve a target quality of service with respect to *throughput (T-bit)*, *delay (D-bit)* or *reliability (R-bit)*. The bit value settings of T-, D- and R-bits are listed in

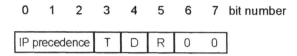

Figure 14.2 Internet protocol: type of service (TOS) field in the IP-header.

Table 14.2 IP-Precedence field determines relative priority for packet processing and forwarding

IP Precedence value decimal (binary in brackets) TOS bits 0, 1 and 2	Meaning
7 (111)	Network Control
6 (110)	Internetwork Control
5 (101)	CRITIC/ECP (critical/exceptional)
4 (100)	Flash Override
3 (011)	Flash
2 (010)	Immediate
1 (001)	Priority
0 (000)	Routine

Table 14.3 Delay, throughput and reliability — bits (D-, T- and R-bits) of the type of service (TOS) field

TOS bit	Meaning	Binary value '0'	Binary value '1'
D-bit (TOS bit 3)	Delay	Normal Delay	Low Delay
T-bit (TOS bit 4)	Throughput	Normal Throughput	High Throughput
R-bit (TOS bit 5)	Reliability	Normal Reliability	High Reliability
TOS bit 6	Reserved for future use	Always set to '0'	Not used
TOS bit 7	Reserved for future use	Always set to '0'	Not used

Table 14.3, but there is no advice in the protocol definition how the different types of packets are to be handled.

Differentiated services (DiffServ): class of service (COS) and per-hop behaviour (PHB)

A more comprehensive quality of service 'guarantee' scheme for the transport of IP (Internet protocol) packets (than TOS) is offered by *differentiated services (DiffServ)*. The DiffServ standards define an alternative use of the TOS field of IPv4 and the *traffic class* field of the IPv6 header. Like TOS, we also discussed DiffServ in detail in Chapter 5. We recap briefly.

Each packet is labelled with a *differential services codepoint (DSCP)* (Table 14.4). The DSCP reveals the *class of service (COS)* of the traffic (first three bits of DSCP) as well as the *drop priority* (the probability that the packet may have to be discarded by a router during a period of congestion on a particular network trunk). According to the value of the DSCP set in a particular IP packet (IPv4 or IPv6), the forwarding and treatment of the packet will be conducted according to one of a number of pre-defined *PHBs (per-hop behaviours)*. Four types of PHB are defined so far (see Table 14.4):

- *default PHB* (for 'routine' or 'normal' traffic);
- *assured forwarding (AF)* PHB;

Table 14.4 Differential services (DiffServ): Codepoints (DSCP) and PHB (per-hop behaviour groups)

PHB type	First three bits of DSCP			Second three bits of DSCP		Actions of PHB
Default PHB	000			000		RFC 1812
Assured forwarding (AF) PHB	001	Class 1	A l t E r n A t i V e s	010	LDP (Low Drop Priority)	RFC 2597
	010	Class 2		100	MDP (Medium Drop Priority)	
	011	Class 3		110	HDP (High Drop Priority)	
	100	Class 4				
Expedited forwarding (EF) PHB	101			110		RFC 2598
Network control traffic	11x			xxx		Highest priority

- *expedited forwarding (EF)* PHB; and
- network traffic PHB (highest priority).

The correct functioning of the *per-hop behaviour (PHB)* mechanism[3] relies upon the reservation of a certain amount or percentage of the network or trunk capacity for the carriage of the packets of the highest priority. This involves 'forward planning'. The forward planning typically takes one of two forms, either:

- a certain amount of capacity is reserved 'for all time' for high priority *classes of service*, or
- on establishment of a new data communications connection or session, the user must request the required reserved capacity. If the capacity is available, the connection is established, otherwise the new connection is rejected — and the user must wait until later (like calling back when the phone network is busy). This type of reservation is usually undertaken in IP-networks using RSVP (resource reservation protocol).[4]

Calculating how much network capacity should be provided to carry a given quantity of data *traffic* or how much capacity should be temporarily reserved to meet a given target quality of service for a particular traffic class of service (COS) is a subject in its own right. We return to this later in the chapter. In the meantime, we complete our review of packet prioritisation schemes used by common data networking protocols. . .

VLAN user priority field

In switched ethernet LANs (local area networks) it is possible to create *virtual bridged LANs (VLANs)*, as defined by IEEE 802.1q. These we discussed in Chapter 4. Two of the prime

---

[3] See Chapter 3.
[4] See Chapter 7.

reasons for establishing VLANs are to provide high quality of service and secure transmission of packets during their transport across the LAN backbone network. The quality of service is 'guaranteed' by means of a packet labelling mechanism defined by IEEE 802.1p. The user priority field of IEEE 802.1p allows packets to be prioritised into 8 different classes of service using a 3 bit field priority value used in a similar manner to the *IP precedence* value of the IPv4 header (Figure 14.2 and Table 14.2).

Quality of service control in ATM (asynchronous transfer mode) data networks

The standards defining *ATM (asynchronous transfer mode)*[5] were among the first to set out a framework for 'guaranteeing' the quality of service of data transmission across a packet-based network. Traffic- and congestion-control are achieved in ATM networks by means of two measures:

1. A negotiation process (*connection admission control, CAC*) is carried out during each call set-up (ATM networks are connection-oriented and not connectionless networks). During the negotiation, the end-user must request the network capacity required which is required. This capacity will be reserved for the duration of the call (on the basis of the *peak cell rate* required), provided the capacity is available. Otherwise the call is rejected.

2. In addition, a network congestion monitoring and relief process (*network parameter control, NPC* and *usage parameter control, UPC*) is carried out during the active data transfer phase of the connection. This discards less important cells (cell is the ATM name for a packet with a fixed length of 53 octets) at times of network congestion. In addition, *flow control* procedures are used to relieve congestion.

Each time a connection across an ATM network is requested, the user (or *requesting entity*) is required by the connection admission control (CAC) to declare the connection type needs in terms of the following parameters:

- *peak cell rate (PCR)* required;

- *sustainable cell rate (SCR)* (i.e., the minimum *persistent* cell rate required);

- *burst cell rate* (i.e., the maximum rate of cell sending over and above the SCR during very short periods);

- *quality of service*, defined in terms of the parameters: *cell delay, cell delay variation tolerance (CDVT)* and *cell loss ratio (CLR)*.

The parameters are carried in the call set up message and form part of a *traffic contract* which the network commits to at the time of connection establishment. It is a commitment to the end user or device that the network will meet the requested quality of service, provided the user complies with the conditions he or she has specified. Once the *traffic contract* is established and the connection is set up, the usage parameter control (UPC) and network parameter control (NPC) procedures take up the process of monitoring network performance and service delivery according to the contract. The main purpose of the procedures is to protect the network resources and other network users from quality of service degradations arising from unintentional or malicious violations of the negotiated traffic contracts, and take appropriate action. On detecting a violation, the UPC or NPC may elect to carry the extra cells anyway, may *re-schedule* the cells, may *discard* them or may *tag* them, by overwriting

---

[5] See Appendix 10.

the *cell loss priority (CLP)* bit, resetting the value from '0' to '1' and thereby increase the probability of cell discarding should congestion be encountered.

QOS in MPLS (multiprotocol label switching) and RSVP (resource reservation protocol) networks

Similar principles of quality assurance to those used in ATM are used in MPLS (multiprotocol label switching) and RSVP (resource reservation protocol) networks, as well as in DiffServ networks using PHB (per-hop behaviour). To date, however, the standards documentation for MPLS, RSVP and DiffServ is largely restricted to the coding of headers, flow labels and priority bit fields. There is little standardised advice on when a reservation can be allowed and how traffic contract policing should be conducted.

Obviously this will affect the <u>absolute level</u> of quality of service achieved. Currently, the correct measures to be undertaken are left for the router manufacturer or network administrator to determine for himself. (No guarantees then!)

QOS in frame relay networks

Frame relay networks,[6] like ATM networks, allow for the establishment of *traffic contracts* between the network and users regarding the quality of service 'guaranteed' for an individual

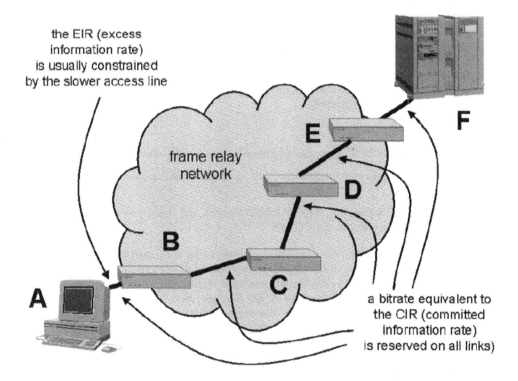

Figure 14.3   Typical frame relay connection: committed information rate (CIR) and excess information rate (EIR).

---

[6] See Appendix 9.

point-to point frame relay connection. The quality of service contract is expressed in terms of the:

- *committed information rate (CIR)*; and the

- *excess information rate (EIR)*.

Typically the traffic contract values in frame relay networks are 'pre-configured' into the network by the network administrator. Values cannot be negotiated at the time of call or session set-up. (Frame relay networks generally only support *permanent virtual circuits* and not *switched virtual circuits*). The committed information rate (CIR) is the minimum end-to-end bit rate exclusively reserved for the connection. The excess information rate (EIR) is the maximum permitted rate (over and above the CIR) at which the network is prepared to receive packets from a frame relay sender.

Network and trunk capacity has to be reserved for the CIR along the whole course of a frame relay connection (links A-B, B-C, C-D, D-E and E-F of Figure 14.3). Typically, the sum of the CIR and the EIR will equal the bit rate of the access line connecting the end user's device to the network (e.g., link A-B of Figure 14.3 or link E-F). But the EIR might also be set lower than the constraints imposed by the access line if any of the core network trunks are seriously congested. Should the sending device send frames at a rate higher than the maximum 'permitted' rate (equals CIR plus EIR), then any of the frame relay devices along the path are allowed to discard the excess frames.

14.4   Data network traffic theory: dimensioning data networks

Good network performance — fast packet propagation across the network — depends upon good network design and the provision of adequate node (e.g., router or switch) and trunk capacity to carry the *offered traffic* demand. Public telephone network operators have a long tradition of measuring the *traffic carried* by their networks. Since they operate their networks with little or no congestion, they generally assume that the *carried traffic* is equal to the offered traffic (i.e., the unsuppressed traffic demand). Calculating using the offered traffic and the target *grade of service (GOS)*[7] as inputs, it is possible to determine exactly how many telephone circuits are required between two nodes in a telephone network. Each of the trunk connections between the various nodes in the network can be *dimensioned* in this manner.

The theory underlying the mathematical analysis of telephone traffic is called *teletraffic theory*. It is based largely upon the work of a Danish scientist, A.K. Erlang, who published a teletraffic dimensioning method for *circuit-switched* networks in 1917. The method (the *Erlang lost-call formula*) is still widely used for telephone network dimensioning today. Indeed, telephone network design engineers express the volume of telephone traffic in *Erlangs*. One *Erlang* of telephone traffic between two given end-points is equivalent to one telephone circuit (on average) being permanently occupied during the whole of the *busy hour*.

ITU-T (International Telecommunications Union — Standardization sector) has published a large number of recommendations (in the E-series) which cover in detail the standardised methods for dimensioning of international telephone networks. Unfortunately, there is no similar formal advice about data- or IP-network dimensioning. This is largely left 'up to the user or network administrator', many of whom use 'empirical methods' of network dimensioning (e.g., 'keep all trunks at less than 75% loading' or simply and more crudely: 'trial and error').

---

[7] The *grade-of-service (GOS)* is the probability that a new call will be rejected by the network — due to congestion at the time of call set-up. Telephone networks are typically designed for a target grade of service (GOS) of 0.01 (i.e., 1%). At this GOS, of 100 calls offered to the network, 1 or less calls will be rejected due to network congestion.

Empirical methods used by individuals with a wide experience of real network traffic have much to recommend them, but there are also many instances where it is useful to be able to 'fall back' on a simple mathematical model to estimate likely network performance. One such model is the *Erlang call-waiting formula*, and we shall discuss this in the next few pages — to give you, the reader, a simple model to fall back on if you need to.

The Erlang call-waiting formula was originally designed to model calls made to human telephone operators, who were responsible for making telephone connections in the first half of the twentieth century. Callers had to wait until the next operator was free before being connected. The formula, though, can easily be adapted to model data packets waiting to be transmitted across a statistically multiplexed transmission line. The waiting in this case is caused by other data packets already in the 'queue' or *buffer*. The *call-waiting* formula provides a means for calculating:

- the latency (i.e., packet delay) caused by queueing buffers at intermediate nodes and limited line capacity; and

- the likely lengths of packet queues (and thus the size of queueing buffers which are required).

For a given traffic *demand* and target latency, it is thus possible to calculate the required line capacity (i.e., bit rate). Alternatively, knowing the capacity and the traffic demand, the latency can be estimated. Such 'guide calculations' are invaluable to network designers and operators in dimensioning networks efficiently to meet quality targets.

Modelling data traffic: the Erlang call-waiting formula

Devices sending information on data networks (data terminal equipment) generally do so in a packet-, frame- or cell-oriented format over a packet-, frame- or cell-switched network. The number of packets, frames or cells trying to traverse the network or a particular link at any time fluctuates from one moment to the next, as the users make their requests for information to servers, and the servers reply with responses or file transfers. The traffic profile (illustrated as the bit rate of the packets submitted to the network or link over time) typically has the form of a series of peaks, as illustrated by Figure 14.4a. Each peak corresponds with a sudden 'burst' of activity from a terminal or a server — as it submits a series of packets to the network commensurate with the needs of the associated request, response or file transfer.

At certain times, the rate of submission of packets may be so large that it exceeds the capacity of the line for a moment or two. Such instances can occur at any time and are quite common. They are caused by multiple terminals submitting traffic at the same time. The example of Figure 14.4b illustrates an example in which the line capacity (maximum line bit rate) is temporarily restricting the packet throughput. The net result is a flattening of the traffic profile. The shaded portion of the traffic peak is slightly delayed by the storage buffers involved in statistical multiplexing. Some packets are made to wait. The traffic performance of such a network can be modelled by the Erlang call-waiting dimensioning technique — to decide the required data link bit rate and buffer capacity which are required in order that a given waiting time is not exceeded.

The time during which an individual packet, frame or cell must wait in the buffer for the line to become free may be a significant proportion of the total time needed for it to traverse the network as a whole. Thus the *waiting time* in the buffer contributes significantly to the *propagation time* or latency. The propagation time (network latency) in turn will affect the apparent speed of response of a computer reacting to the typed or other commands of

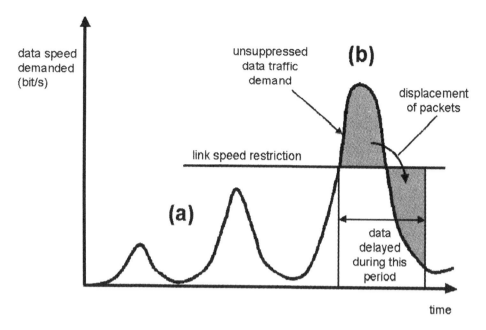

Figure 14.4 Typical fluctuation in data traffic over time.

the computer user. Most computer applications can withstand some propagation delay, though above a given threshold value further delay may be unacceptable. Thus, for example, many packet-switched data networks are designed to keep the latency (one-way propagation delay) lower than 200 ms.

The switch and network designers of a data network must ensure that the bitspeed of the line (the capacity between two points in a network) is greater than the average *offered bit rate* trying to traverse between the two points. If the line were not fast enough, a continuous build-up of data would occur in the buffer at the transmitting end, causing ever-increasing delays. In addition, the buffer must be large enough to temporarily store all waiting data. If the buffer is not big enough, then arriving packets will be lost at times when it is full. The arriving packets at these times simply *overflow* the *buffer*.

The original *Erlang call-waiting formula* (Table 14.5) is oriented to telephone *call-waiting systems* (e.g., the queueing times callers must wait before being able to speak to an operator in a *call centre*). The formula is relevant to determining how many helpdesk or call centre operators are required, given a certain volume of telephone caller traffic and a target *waiting time*. (In this sense, it may also be of direct application for data network operations or helpdesk organisations running a call centre.)

The formulae of Table 14.5 may also be easily adapted for data network dimensioning by substituting $N = 1$. This value corresponds to the assumption that there is only one line (i.e., one *server* in 'call-waiting formula language') carrying packets between any given pair of nodes in a data network. The link will have a given bit rate capacity, and the data flowing over the link could be thought of as being measured in Erlangs (the average usage of the line, assuming that permanent full usage of the single line would be equivalent to one Erlang). Thus, for example, an average bit rate of 7500 bit/s being carried on a 14 400 bit/s link is equivalent to 7500/14 400 or 0.52 Erlangs.

By setting $N = 1$ into the various formulae of Table 14.5, rearranging them and renaming some of the parameters to make them more familiar 'language of data communication', we obtain the formulae of Table 14.6.

Table 14.5   Erlang call waiting formula (original version): relevant to calculating the required number of call centre operators for a helpdesk

Parameter	Usual notation	Formula
Probability of delay (*Erlang call-waiting formula*)	C	$= \dfrac{NB.}{N - A(1 - B)}$
Average delay (held in queue)	D	$= \dfrac{Cd.}{N - A}$
Average number of waiting calls		$= \dfrac{AC.}{N - A}$
Probability of delay exceeding t seconds		$= C\,e^{-(N-A)t/d}$
Probability of j or more waiting calls		$= C(A/N)^j$
Erlang lost-call formula	B	$= \dfrac{A^N/N!.}{(1 + A + A^2/2! + A^3/3! + \cdots + A^N/N!)}$

Key: N = number of servers (e.g., human operators) processing the calls
　　j = number of calls in queue
　　B = lost call probability if there were no queue (calculated from the *Erlang lost-call formula*)
　　A = *offered traffic* in *Erlangs* (average number of simultaneous calls in busy hour)
　　d = average service time required by an active server to process a call

Table 14.6   Erlang call waiting formula: rearranged form appropriate to data network traffic modelling

Parameter	Usual notation	Formula
Probability of delay (*Erlang call-waiting formula*)	C	$= A$ (average line loading)
Average delay (packet held in buffer waiting for transmission)	D	$= \dfrac{p.}{(1/A - 1)L}$
Average number of waiting packets or frames in buffer		$= \dfrac{A.^2}{1 - A}$
Probability of buffer delay exceeding t seconds		$= A\,e^{-(1-A)tL/p}$
Probability of j or more waiting packets		$= A^{j+1}$

Key: p = data packet size in bits
　　A = *offered traffic* in *Erlangs* (average *line loading* during the busy hour)
　　L = line bit rate in bit/s

Before we move on, it is important to note the limitations of the use of the Erlang call-waiting formula. What we have not mentioned so far is that the derivation of Erlang's formulae is based upon the assumption that the traffic offered to a telecommunications network is of a statistically random nature. This assumption is not necessarily true — particularly in smaller networks dominated by a small number of computer servers which schedule or coordinate much of the traffic in the network. But in networks with a large number of users, the formulae may produce a good estimate of likely real network performance.

It is interesting to substitute a few values in the formulae of Table 14.6 to give practical insight into the operation of data networks. First, let us consider the maximum acceptable line loading for a typical IP-based network which uses 64 kbit/s trunk lines between routers. To do so, we consider the formula for calculating the 'probability of buffer delay exceeding t

seconds'. We shall substitute L = 64 000 bits/s and a packet size p of 576 bytes = 4608 bits. We then consider two cases: an 'acceptable' buffering delay of t = 200 ms and a 'maximum acceptable' buffering delay of t = 500 ms. Figures 14.5 and 14.6 illustrate the resulting plots of the relationship between the probability of exceeding the stated delay and the line loading (A). Figure 14.5 illustrates the probability of exceeding a 200 ms delay, while Figure 14.6 illustrates the probability of exceeding 500 ms.

Notice in Figure 14.6 how the probability of the buffering delay exceeding the 500 ms ('maximum acceptable delay') increases rapidly at line loadings above about 80%. At 80% line loading, this probability is only around 20%. The conclusion is that we should not expect to run a 64 kbit/s data network much above about 70%–80% line loading, if we do not want to experience 'unacceptable' delays.

What about the effect of increasing the linespeed while keeping the packet size constant, you might ask? What if we increase the linespeed to 2048 kbit/s (2 Mbit/s)? The result is shown in Figure 14.7, where we again plot the probability of exceeding the 500 ms 'maximum acceptable' buffering delay limit. Now we are able to operate the line at over 95% loading without exceeding the 500 ms delay limit more than a very tiny proportion of the time. Is this what you might expect? Of course. For a packet of a given size can be transmitted to line much faster by a high bit rate line than by a low bit rate line. So the period during the packet transmission during which other packets might have to wait is much lower. Only when a number of other packets of equal or higher priority are also already queued up will the waiting time be appreciable. But the chances of queues are also lower, since each individual packet takes a relatively small proportion of the available line capacity. As a result, higher speed lines may be loaded (measured in percentage terms) more than lower speed lines. The corollary also applies: you must be careful not to overload low speed lines. Figure 14.8 illustrates the

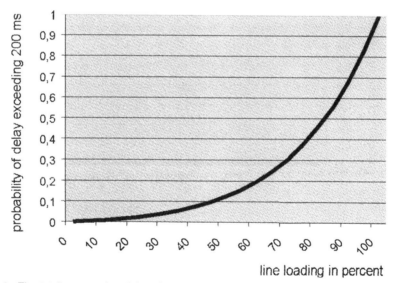

Note: The total propagation delay of a packet crossing the data link (rst bit sent to last bit received) is the sum of the delay caused by buffering plus the time required to transmit the packet to line (= p/L) plus the electrical signal propagation delay (distance/speed of propagation). In the examples of Figures 14.5 and 14.6 the packet transmission time adds 72 ms and the electrical propagation around 3–10 ms per 1000 km.

Figure 14.5 Probability of buffering delay exceeding 200 ms (for linespeed L = 64 kbit/s, packet size p = 576 octets) [Total packet propagation delay in this case around 280 ms (see Note)].

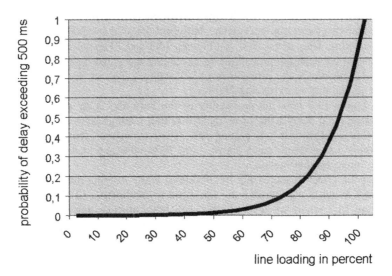

Figure 14.6 Probability of buffering delay exceeding 500 ms (for linespeed L = 64 kbit/s, packet size p = 576 octets) (Total packet propagation delay in this case around 580 ms — see Note to Figure 14.5).

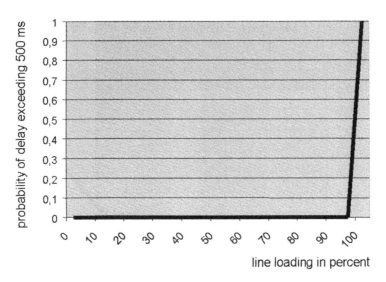

Figure 14.7 Probability of delay exceeding 500 ms (for linespeed L = 2048 kbit/s, packet size p = 576 octets).

effect of reducing the line speed to 9.6 kbit/s (a bit rate associated with older analogue modem lines). With this line speed, the probability of the buffering delay exceeding the 'maximum acceptable' of 500 ms is more than 10% when the line loading exceeds 25%. And what is the moral of the story? That using a fixed target percentage line loading (a single target value) is not a good way of managing *latency* delays caused by buffering.

Rather than using a fixed target percentage utilisation for dimensioning network links, it is better to design networks for a given packet delay performance and to use a method like

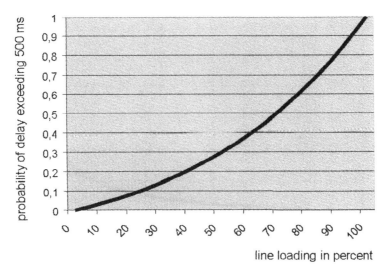

Figure 14.8   Probability of delay exceeding 500 ms (for linespeed $L = 9.6$ kbit/s, packet size $p = 576$ octets).

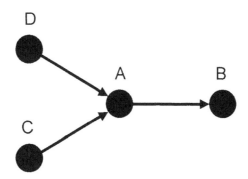

Figure 14.9   Typical data network: each link must be separately dimensioned to carry the total traffic offered to it.

the formulae of Table 14.6, inputting the value of the line loading (A), average packet size (p) and line bit rate (L) to calculate the probability of exceeding a given target 'maximum acceptable buffering delay'. If you are working to a maximum end-to-end network latency target, then you need to summate the likely delays occurring at each node along the way, as well as the electrical propagation delays and packet transmission delays.

In more complex data networks, as exemplified by Figure 14.9, the same dimensioning method may be used for the individual links of the network. Thus link D-A of Figure 14.9 can be dimensioned according to the needs resulting from the sum of D-A and D-B traffic. Link A-B is dimensioned according to the aggregate needs of D-B, C-B and A-B traffic.

Before leaving the Erlang call-waiting formula, let us also consider the practical problem of how large network packet buffers must be. (The size of the buffer in number of packets is dependent only upon the relative line loading and independent of the line speed.) We shall assume that there is a 'maximum acceptable' proportion of packets which may be lost or corrupted (due to buffer overflow) of say 0.01% (1 in 10 000). We use the fifth formula of Table 14.6 to calculate the required buffer size and derive the graph of Figure 14.10.

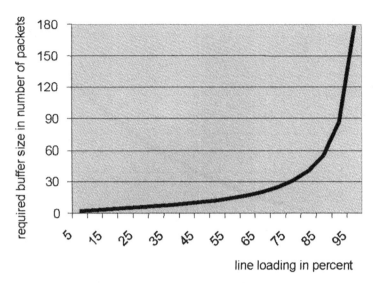

Figure 14.10   Required packet buffer size for loss or corruption of not more than 0.01% of packets.

Calculation of end-to-end latency

The end-to-end latency (i.e., one-way propagation delay of a single packet) depends upon a number of factors:

- the total of all the buffering delays caused by intermediate nodes along the path between the two end-points (these delays are dependent upon the network capacity and the level of traffic demand or congestion, as we explained in the previous section);

- the total line transmission delays (which are dependent upon the size of the packet and the bit rate of the various lines which must be traversed); and

- the propagation delay resulting from the length of the path.

Returning to the example of Figure 14.9, the total network latency experienced by packets traversing the network from point C to point B is the sum of the following components:

- the buffering delay at node C (awaiting transmission onto link C-A);

- the packet transmission delay onto link C-A (equals the packet length in bits divided by the bit rate of link C-A in bits/s);

- the buffering delay at node A (awaiting transmission onto link A-B);

- the packet transmission delay onto link A-B (equals the packet length in bits divided by the bit rate of link A-B in bits/s);

- the total length of the path from C via A to B divided by the velocity of transmission (approximately 10^5 km/s). The delay is approximately $3-10$ ms per 1000 km.

Let us assume that we are sending packets of 576 octets in size (a typical packet size for IPv4), that the first link C-A has a bit rate of 64 kbit/s and that link A-B has a bit rate of 2 Mbit/s. Let us further assume, that both links C-A and A-B are operating at 70% of their maximum load and that the total path length is 500 km. Using the second formula of Table 14.6 (for the delay, D) we are able to calculate the expected (i.e., average) end-to-end latency of the path

C-A-B (the time from the submission of the first bit of the packet at node C until the receipt of the last bit at node B) as follows:

- buffering delay at C   168 ms
- packet transmission delay at C   72 ms
- buffering delay at A   5 ms
- packet transmission delay at A   2 ms
- path length transmission delay   5 ms

Total delay: **252 ms**

252 ms total delay — this is more than 1/4 second and perhaps too long for a given application. What could we do to improve matters? Increase the bit rate of link C-A — this will reduce both the buffering delay and the packet transmission delay at C. You might like to calculate for yourself the reduction in total latency caused by a bit rate increase to 128 kbit/s and/or try out some other examples of your own!

Forecasting traffic demand

Ideally, networks should be dimensioned to carry the future forecast traffic demand, and not merely repeatedly upgraded after-the-event to overcome measured congestion. A simple forecast can be made by extrapolating the growth in measured traffic. The extrapolation can be either linear or calculated as the result of a continuous percentage growth rate.

Experienced network designers will tell you that it is not worth trying to be very precise with the forecasting — forecasts are always wrong! In practice, your forecast is only an aid to your network design decision-making and to answering questions like:

- shall I upgrade a particular 64 kbit/s line to 128 kbit/s now, or shall I wait another month?

- what is the likely network *latency* of a particular critical network application?

- shall I add a new direct trunk between two particular routers? If so, with which bit rate?

You will find that the answer to such questions are not greatly affected by quite large errors in the traffic forecast. Here are a couple of examples to illustrate this.

If you upgrade a line this month rather than next month, you may feel assured that the network latency will be much lower than the target value and users will be happy. Since the user traffic in most networks grows continuously, you can be sure that the extra capacity will be justified. And the marginal extra costs of buying the line a month early might be insignificant in comparison with the overall network budget.

As illustrated in Figure 14.11, network and performance management typically involves deciding when to add discrete capacity extensions to the network to meet the forecast demand. Thus to achieve the carriage of the forecast traffic of Figure 14.11 the stepped 'capacity' line must always be maintained above the 'demand' curve. Typically a network operator will plan that the 'steps' of the 'capacity' line do not come too close to the 'demand' curve, and always leave a 'contingency period' of time between the date of the planned capacity extension and the actual date on which it is needed.

If the optional bit rates for a new direct link between two routers are either 64 kbit/s or 2 Mbit/s, which line should I have installed? The way to answer this question is to compare the cost of the two lines with your calculated expectations of network latency. Even if an expected network latency of 252 ms using a 64 kbit/s line does not meet your 'target' of 250 ms for a

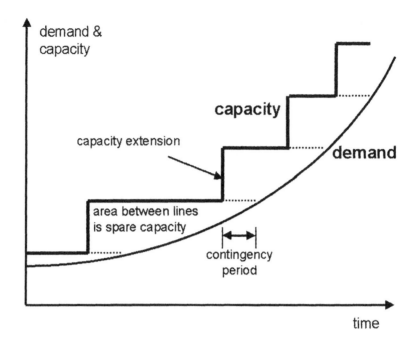

Figure 14.11   Keeping network capacity higher than forecast demand: thereby meeting users' quality
expectations.

particular application, is it really worth paying a much higher price for the 2 Mbit/s line and
the resulting improved latency, or could users 'live with' 252 ms latency after all?

Practical dimensioning of networks

Many network operators do not apply sophisticated traffic modelling techniques in support of
their network dimensioning. Instead the more pragmatic and empirical approach prevails: 'if
too many calls are being lost add some more circuits' or 'if the computer network response
time is too slow, upgrade the speed of the data transmission lines'. While such methodology
may appear crude, there is much to recommend it, since only very large networks are likely to
carry sufficient traffic to be accurately characterised by statistical methods. In addition, many
computer applications running on data networks may be designed to run at certain fixed times
(e.g., hourly update of all the sales made in the outlets of a particular shopping chain). Here
again, statistics may be of lesser value than experience.

The empirical method should, however, be used with caution. The data network is not the
only possible cause of slow-responding computer applications, and merely 'throwing more
capacity at the data network' may not provide a solution at all. You need to review the design
of the computer program itself, and the protocols being used.

Adding more capacity at one point in a network sometimes has the effect of stimulating
more traffic or of destabilising the routing of calls somewhere else in the network. This in
turn can lead to greater rather than lesser problems. Internet routing protocols which calculate
link costs based on link bit rates (such as OSPF — open shortest path first) will tend to choose
high bit rate links in preference to lower speed ones. Thus adding capacity to a single link
in isolation may (perversely) have the adverse effect of attracting more traffic to it — and

thereby making congestion worse. Meanwhile, lower speed links may become practically unused!

Ultimately, there is no better tool for dimensioning a network than a thorough knowledge of the traffic routes within it and the uses to which it is being put.

14.5   Application design factors affecting quality of service

With the objective to optimise the quality of service (QOS) of a networked application, or in the search for the cause of a particular problem, particular attention should be paid to studying the design of the computer software application and to considering the protocols which it uses. Just because an application was well suited to a slow speed line, does not mean it will communicate effectively or efficiently using a high bit rate line! Moral of the story: don't just assume that slow computer application response times are due to network capacity problems and simply add bit rate willy-nilly!

Applications which run on high bit rate lines need to be designed very carefully, if they are to benefit from the full capacity of the line. In the example of Figure 14.12, an application is shown operating in a 'conversational mode' on a high speed line. Either the protocols which have been selected, or the manner in which the application has been written, demand that requests and responses are 'ping-ponged' across the network.

The lighter-shaded diagonal bands of Figure 14.12 represent 'request' packets making their way across the network from the A-end to the B-end. The bands are relatively narrow (from top to bottom) since the request packet sizes are relatively small, and consequently are rapidly submitted onto the high bit rate line. The responses of the B-end (which are assumed to be large data messages) are the darker shaded bands, diagonally making

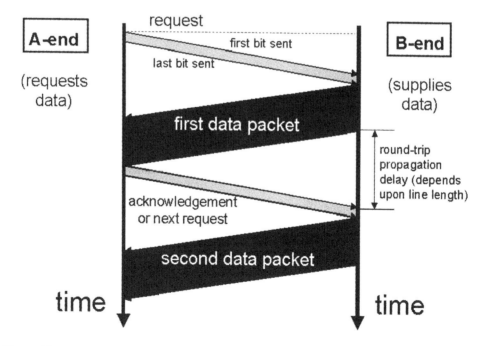

Figure 14.12   Applications which communicate in a 'conversational style' may waste the capacity of high speed lines.

their way back from the B-end to the A-end. The 'conversational' manner in which the application has been conceived to work demands a strict sequence of 'request-response-request-response-request, etc.'. Unfortunately, the 'ping-pong conversation' greatly reduces the effective capacity of the line. Consider the direction of transmission from B to A. After sending the first data packet in response to the first request, the B-end must pause sending for a period of time. The minimum duration of the sending pause is equal to the round trip propagation delay (see Figure 14.12) (assuming that the request from the A-end is a very short message).

If we assume that the line speed of Figure 14.12 is 2 Mbit/s and that each 'data packet' response of the B-end is of 576 octets in length, then the time required to transmit each packet to line is 576×8 bits/2048 kbit/s = 2.3 ms. This is the 'thickness' of the dark-shaded data packets making their way from the B-end to the A-end of Figure 14.12. The round-trip delay time, meanwhile, for a 500 km one-way path is around 10 ms. In other words, the line from B-to-A is busy for 2.3 ms (while sending a packet) but then must remain idle for at least 10 ms before the next data packet may be sent. This represents an overall efficiency of only 19% — effectively reducing the line throughput capacity to 383 kbit/s!

Based on an actual end-to-end network capacity of 2 Mbit/s and a packet size of 576 octets, the application of Figure 14.12 only achieves a maximum data throughput of 383 kbit/s! So if we want to achieve a much higher data throughput what options do we have? Increase the line bit rate perhaps? Let's try increasing the line bit rate to 34 Mbit/s: at this speed, the time required to transmit each packet to line is dramatically reduced — to only 0.14 ms — but as a result the line efficiency also drops dramatically (to 0.14 ms/10.14 ms = 1.3 %), with the net result that the absolute data throughput only increases to 454 kbit/s. Not a very good use of a 34 Mbit/s line! So what other options do we have? Answers: Increase the size of the packet (above 576 octets — if possible) or re-design the application to operate much more efficiently — managing without the 'ping-pong'.

DNS (domain name system), FTP (file transfer protocol), telnet and *http (hypertext transfer protocol)* are all somewhat 'conversational' protocols. The challenge is to integrate them into application software in such a way as to minimise the amount of conversation necessary to achieve the objective!

14.6   Network design for efficient, reliable and robust networks

Not only the design of the applications, but also the topology of a data network can have a significant impact on the quality of service (QOS) perceived by end users. The next few sections provide a few simple guidelines on network design. They are intended to aid in the design of robust and efficient networks.

Hosts with heavy traffic are best sited at major network nodal points

Hosts or servers which frequently conduct data communication with remote locations and are subjected to large volumes of data network traffic are best sited at major network nodal points. Thus, for example, in Figure 14.13, New York is the best location for a major website with large numbers of regular US national and international visitors. Why? Simply because it has the network infrastructure and capacity to deal with very large volumes of traffic.

Charleston, South Carolina, might be the location of the company headquarters and the company's computer centre, but it is simply not as accessible as New York from overseas

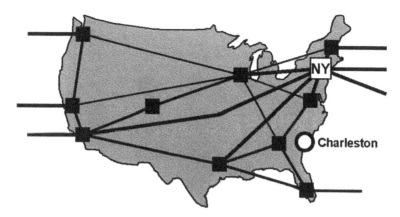

Figure 14.13 What is the best location for a web server with frequent US national and international 'visitors'?

locations. On the other hand, for a main audience of customers and Internet users in South Carolina, Charleston maybe is the perfect location!

And where exactly is the best network location in a city like New York or Charleston? The answer is.. 'as close as possible to the public network providers main location in the city'. This is the location with the best connectivity, the highest level of network redundancy and the greatest capacity. Some major public operators offer *hosting facilities* in these locations — the opportunity for corporate enterprises to locate their web servers and other network equipment in the location, and have it operated and maintained by the public network operator. The offer is worth considering purely on the grounds of the optimum network position of the server at this location.

As an analogy, consider which is the easiest place to meet at in New York City? The answer is the airport! A big office in downtown Manhattan might comfortably meet all the possible needs of participants during the meeting itself, but first they will have to make it through the traffic jams and 'gridlock' in the Taxi from the airport. Even a high-speed fibre connection to your local telecommunications network provider is like the taxi from the airport! And a second link to provide for redundant 'back-up' connection of the site is only like a second taxi! There is nothing more secure from a networking perspective than being directly located at a major node!

The siting of important computer servers at major network locations may seem like obvious advice, but it is surprising how seldom such matters are considered in network design. Many enterprise network managers imagine the company HQ to be 'at the centre of the universe'. While they may spend much effort designing an economic network considering the relative positions of their computer servers, the locations of the employees, customers and suppliers accessing them, many often neglect to consider the topology of the public telecommunications networks which will be needed to interconnect them. The form of the remote connections may differ from one case to another (e.g., leaseline, dial-up connection, VPN — virtual private network or IP backbone service), but a public operator's network is always in use!

I was once involved in an enterprise network in which the major node appeared to be 'redundantly' connected to five others (Figure 14.14a). In reality, however, all five links were carried by the same higher-order transmission system to the nearest exchange building of the public telecommunications carrier (Figure 14.14b). The network redundancy was nothing like as good as the network designer had intended!

On another occasion, I encountered a foreign exchange dealer's network with redundant international leaselines between nodes in two different capital cities. The lines left the building

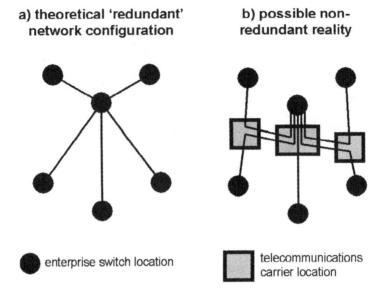

a) theoretical 'redundant' network configuration

b) possible non-redundant reality

● enterprise switch location

□ telecommunications carrier location

Figure 14.14   The possible reality of a 'redundant' set of leaselines?

on separate cables, in separate conduits to separate local exchanges, and even transmitted different countries on their way to the destination. Unfortunately, however, both lines converged at the regional switching centre of the public telecommunications carrier — and consequently often failed at the same time. A more secure location for the enterprise network node both in this and the previous example would have been at a *hosted location* within the building of the regional switching centre site.

Questions worth considering when thinking about the location of servers with heavy data traffic:

- Where are the remote users located who will access the server?

- What is the total volume of traffic?

- On which public networks will the remote users originate their traffic?

- What level of network redundancy is required?

- What are the topology and redundancy of the public telecommunications carrier's network?

- What is the traffic growth rate? Could I need further capacity on a short term basis?

- Does the public operator offer a hosting facility at a suitable location?

- Are their over-riding labour costs, expertise or security reasons for siting the server in a particular location?

- Which of the available algorithms or software design tools should I use to determine the best network topology for my network?

Fully meshed networks

By fully meshing a router network (i.e., directly connecting each router to each other router), both the packet forwarding overheads and the packet delay (or latency) can be minimised.

The load on the routing protocol is greatly reduced and many redundant paths are available to overcome single trunk failures. In such a fully-meshed configuration, each IP (Internet protocol) packet needs only to traverse two routers (the one in the originating network and the one in the destination network). The packet forwarding process (including 'looking up' the destination IP address in the routing table — a comparatively onerous task) only needs to be undertaken twice. In this way the processing effort required of the network as a whole is minimised, as is the packet delay during transmission (the latency).

Full meshing of routers can be achieved by providing direct physical (*trunk*) connections between each pair of routers, but this is an expensive way of achieving a full mesh. A cheaper alternative method, which also provides for full meshing of routers, is the use of a *frame relay, ATM (asynchronous transfer mode)* or *MPLS (multiprotocol label switching)* network in the *core* of the network — as a 'transmission medium' (Figure 14.15[8]).

Figure 14.15a illustrates a router network comprising a total of five routers. Because of the geographical elongation of the network, the operator can afford only five inter-network trunks between the routers. In consequence, packets crossing the network from router A to router E must traverse at least three IP-*hops* between routers (rather than the single *hop* needed in a fully-meshed network). Ten trunks would have been necessary for full-meshing of the routers.

Figure 14.15b shows an alternative configuration of the network in which a core network has been created using two frame relay, ATM (asynchronous transfer mode) or MPLS (multiprotocol label switching) core switches. The two core switches are labelled b' and c'. The switch b' is collocated with router B and the switch c' is collocated with router C. The trunks between the sites are connected to the 'core network' switches b' and c' rather than direct to routers B and C. Routers B and C are respectively connected to switches b' and c'. The effect is to create a 'core network' between the routers without needing to add any trunks between

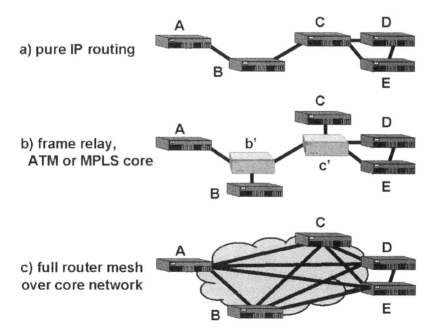

Figure 14.15   Creating a fully-meshed router network by means of a 'core transmission network'.

---

[8] See also Chapter 8, Figure 8.11.

the locations. The physical network topology appears as shown in Figure 14.15b, but a complete mesh of layer 2 *virtual connections* (i.e., frame relay, ATM or MPLS *label-switched connections*) can be created between the routers as illustrated in Figure 14.15c.

The ability to fully mesh routers in the manner illustrated in Figure 14.15 is a good reason for deploying a frame relay, ATM or MPLS network as a 'transmission core network' for an IP-router network. A similar effect can also be achieved using 'classical' telecommunications transmission technology — either PDH (plesiochronous digital hierarchy), SDH (synchronous digital hierarchy) or SONET (synchronous optical network). A further reason for the use of a different 'transmission network technology' at the core of a router network might also be the pure economics. Frame relay trunk port card hardware was in the past much cheaper than the equivalent router hardware! (A total of 7 router port cards and 7 frame relay port cards are required in the configuration of Figure 14.15b — maybe much cheaper than the 20 router port cards required for a full physical mesh router network.)

Under normal operating conditions (i.e., with no trunk failures), the fully-meshed router network of Figure 14.15b has similar advantages to a network which is fully-interconnected with separate physical links. But during periods of link failure, the router network's routing protocol has to work much harder if the 'full-mesh' is based only on *virtual connections*. A single trunk failure between router A and switch b' in Figure 14.15b will have the effect of removing all four of the *virtual* direct connections between router A and all the other routers!

The routing protocol has to detect all four link failures and try to work around them. This is harder than dealing with a single physical link failure between two routers. Not only this, but router A of Figure 14.15b can become isolated as the result of the link to switch b'. In the case of separate physical links from router A to all the other routers, a single link failure has much less impact on the network, and is dealt with more easily by the routing protocol. So in summary, the virtual full-mesh created by the 'transmission core network' is as effective as a physical full mesh in normal operation, but at times of link failure is not as robust. The question for the network designer, of course, is whether the extra cost of the full physical mesh is justified by a need for a more robust network.

Load balancing and route redundancy using parallel paths

By using multiple paths, both the capacity and the redundancy of network paths can be increased. This can be achieved in a number of ways, but requires careful network planning.

Path splitting and balancing of traffic between the different routes (called *path balancing*) can be used when more than one possible route exists between the two end-points of the communication, as for example in Figure 14.16a. In this case, the balancing of traffic between the two possible routes A-C-B and A-D-B must be carried out by careful configuration of the routing protocol. The routing protocol OSPF (open shortest path first) allows for *path balancing*, but only between paths of *equal cost*. When undertaken, path balancing causes roughly equal numbers of packets to be sent via each of the available alternative paths.

Figure 14.16b illustrates an example in which the data transport capacity between two routers has been increased over time by the addition of extra trunks. While such a 'multiple parallel link' configuration is a little more robust than the alternative 'big fat pipe' configuration of Figure 14.16c, the 'big fat pipe' is generally better. Let us consider why.

Let us assume that each of the links of Figure 14.16b has a capacity of 64 kbit/s. Then the total capacity available is $5 \times 64 = 320$ kbit/s. Let us therefore assume that the 'big fat pipe' of Figure 14.16c has a comparable total bit rate of 320 kbit/s. How do the two configurations compare in performance? Typically, the path balancing mechanism applied in the case of Figure 14.16b will direct each individual packet from router A to router B across one of the five alternative links. If we assume that each packet is 576 octets in length (a typical IPv4 packet length), then the time required to transmit the packet to line is $576 \times 8/64\,000 = 72$ ms.

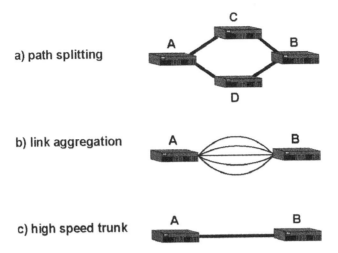

a) path splitting

b) link aggregation

c) high speed trunk

Figure 14.16   Path splitting and link aggregation.

In comparison, the time to transmit the same packet to line across the 320 kbit/s 'big fat pipe' of Figure 14.16c is only $576 \times 8/320\,000 = 14$ ms. So the 'big fat pipe' inflicts much less latency (i.e., delay) on packets during transmission.

A solution to the problem of the higher latency of the configuration of Figure 14.16b (in comparison with the configuration of Figure 14.16c) may be provided by *link aggregation*, in which the two routers A and B use special methods of *reverse multiplexing* to make the five individual links appear to be a single 320 kbit/s connection. In this way, the capacity of all five links can be used to carry each packet. But even this does not compensate for the economic benefits of the single link configuration of Figure 14.16c... Typically the price of five separate 64 kbit/s *leaselines* between two locations is approximately the same price as a 2 Mbit/s single link connection. And the cost of a single trunk port for each of the two routers A and B is likely to be cheaper than the cost of five separate port cards for each router. The configuration of Figure 14.16c genuinely offers more bit rate between the routers and better performance, for less cost!

Route redundancy

Duplicating the nodes and trunks of a network is a standard means used to eliminate major network disruptions caused by a *single point of failure*. If we consider the example of Figure 14.17a, the failures of any of the three routers or any of the trunks will isolate one part of the network from the other. The configuration of 14.17b, on the other hand, can withstand a single node or trunk failure without major impact on the overall network service. This has been achieved by a *redundant configuration* in which each of the nodes and each of the trunks is duplicated. In the particular example of Figure 14.17b a very high level of network redundancy has been implemented by the use of 'parallel' and 'cross-over' trunks between each of the routers at location A and each of the routers at location B. Ethernet switches with similar 'parallel' and 'cross-over' connections have also been included at location A to interconnect the two separate pairs of routers. This configuration is very robust even to multiple simultaneous node and trunk failures, but this has been achieved at a high cost. An alternative, cheaper, but slightly less robust redundant configuration might have used only four long distance trunks between the two locations — A and B — either the 'parallel' pair or the 'cross-over' pair.

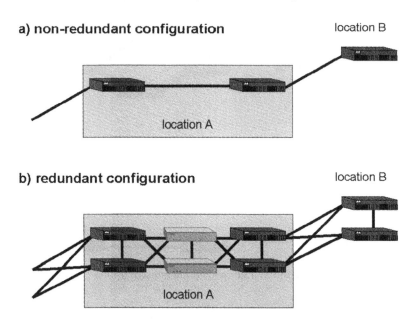

Figure 14.17 Employing a 'cross-over' topology to improve the redundancy and robustness of a router network.

Server redundancy and load-balancing

On some occasions when particular applications or servers are subjected to very heavy 'interrogation' by remote users across a network, it is useful to be able to share the data traffic destined to the *server* across a number of different hardware devices acting as if they were a single server. This is often referred to as a server *cluster* (Figure 14.18). By sharing the incoming traffic across multiple processors, a higher overall processing capacity is achieved, and the failure of a single processor hardware does not mean an interruption of all the services offered by the server cluster.

Figure 14.18   Server clusters and load balancing.

There are two main methods by which server clusters with load balancing can be realised. The first method is to purchase special 'computer cluster' hardware. To all intents and purposes such hardware appears like a single server to the network and the outside world. Such hardware may use proprietary methods for load balancing and may require optimisation of the application software. A second method is to use the *DNS (domain name system)*[9] service to load balance address resolution requests to a number of servers, variously identified by slightly different domain name prefixes: www, www2, www3, etc.

Router and gateway redundancy

Gateway redundancy protocols allow *access routers* in originating LANs (local area networks) to be duplicated as illustrated in Figure 14.19. Examples of gateway redundancy protocols are *VRRP (virtual router redundancy protocol*—RFC 2338) and the Cisco-proprietary protocol *HSRP (hot standby router protocol)*.

Both VRRP and HSRP work in the manner illustrated in Figure 14.19. Originating hosts with the originating network use the *virtual standby IP-address* as the address of their *default router* (the *default gateway* is the address of the first *hop* of an IP path when packets are sent by the host). But the virtual standby address is not the actual gateway address of either of the redundant routers A or B. Instead, each router has its own related, but different IP gateway address, and both share the virtual standby address. One of the routers is in *active mode* and the other in *standby mode*. The router in active mode operates as if it 'owned' the virtual standby address, until it fails, whereupon the standby mode router takes over. As far as the originating host is concerned, the two gateway routers appear to be a single *virtual router* with the gateway address of the virtual standby address.

VRRP or HSRP *hello* messages are sent regularly between the two routers (e.g., every 3 seconds). These messages communicate which router is in active mode and which is in standby mode. In addition, they serve to indicate to both routers that the other router is still 'alive'. A priority scheme determines which router assumes the role of the active router and

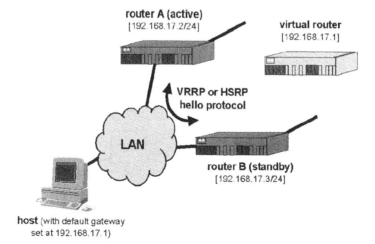

router A (active)
[192.168.17.2/24]

virtual router
[192.168.17.1]

VRRP or HSRP hello protocol

LAN

router B (standby)
[192.168.17.3/24]

host (with default gateway set at 192.168.17.1)

Figure 14.19   The operation of gateway redundancy protocols (e.g., VRRP—virtual router redundancy protocol).

---

[9] See Chapter 11.

which one shall be standby. Only the *active* router forwards IP packets outside the LAN (local area network).

The router with the *highest priority* value (as communicated by means of the hello messages) assumes the role of the active router. In fact, routers simply assume that they should be *active* unless they receive a higher priority value from one of the other routers in the local network by means of a *hello* message. A router switches over from standby mode to active mode, should the currently active router fail to send three consecutive hellos.

Both VRRP and HSRP allow two or more routers to be used in a redundant gateway configuration. Since the failure of the gateway router is one of the commonest network failures in an IP-based network, the use of such a redundant gateway configuration is important in cases where very high network availability and reliability is required.

Interconnection and peering

One of the early attractions of router networks employing the Internet protocol (IP) was their use of *routing protocols* to automatically determine *routing tables* for the forwarding of packets to all *reachable destinations*. This is a major advantage, and networks can indeed be built or attached to other networks, with little concern for how the packets will find their correct destinations. But while automatic routing protocols will always find the best available route to a given destination, this is no assurance that the end-to-end communication quality of the best route (particularly the network latency) will be acceptable to the end-users. If the network is to provide service in line with end-to-end quality targets, then there is no alternative to comprehensive network design and consideration of all possible main traffic paths through the network.

The degree to which the network is interconnected with other Internet or IP-networks has a major impact on the *reachability* of destinations and the quality of the communication. Figure 14.20 illustrates the typical dilemma of a network designer. The network designer (of the dark shaded 'network') is faced with having to decide which inter-network connections (*peer connections*) need to be made. The options under consideration are connections to the Internet service providers ISP1 and ISP2 or direct connections to the Internet exchange points IX1 and IX2. A particular overseas destination (to which a large amount of data is sent) is best reached by means of the 'overseas ISP' which is directly connected to IX2.

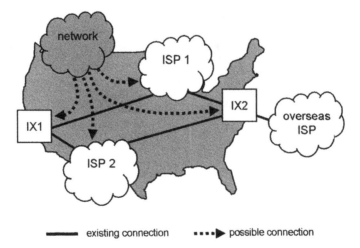

Figure 14.20   Making the right network connections impacts the reachability of destinations and the quality of service.

All possible destinations in Figure 14.20 could be reached by a single connection of the dark-shaded 'network' to either ISP1 or ISP2, and this is likely to be the lowest cost 'solution'. But when also considering the quality and performance of the network, the designer may choose to make further *peer connections*. The network designer needs to consider a number of factors in selecting the final network design:

- total network cost;

- accounting charges of ISPs, *transit* networks (transit *autonomous systems*) and *Internet exchanges (IXs)*;

- network hardware and equipment costs;

- total volume of traffic;

- network quality requirements;

- network performance and latency to frequently accessed destinations;

- maximum permissible hop count to frequently accessed destinations (this affects the network latency); and

- maximum number of transit AS (autonomous systems) in reaching a destination.

The network designer's choice of peer connections for the dark-shaded 'network' of Figure 14.20 might include any one or more of the 'possible connections'.

14.7   Network operations and performance monitoring

No matter how good your network design, unpredicted traffic demand or unexpected network failures will occasionally upset even your best-laid plans! It is critical to monitor traffic activity and network performance. Traffic demand is usually tracked as a long-term trend. Network traffic capacity planning ensures that the predicted traffic demand (including long-term growth) can be carried while simultaneously meeting prescribed communication quality targets. For the purpose of network dimensioning, the traffic demand is defined to be the maximum demand arising during the busiest hour and busiest day of a particular month. The growth in demand is tracked from one month to the next.

For the purpose of measuring traffic demand, it is normal to collect network statistical records and *post-process* them. Statistical records can be collected from network routers, switches or other nodes. Alternatively, special traffic monitoring devices (*probes* or *sniffers*) may be used to collect sample traffic data. The data may be collected either on a 'small sample' basis (e.g., a measurement made only on the assumed busiest day of a particular month) or on a 'full-time' basis. Post-processing (i.e., computer analysis after the event) of the data can be used to generate a full *traffic matrix* 'view' of the network. The traffic matrix reveals the individual sources and destinations of packets and the volumes of data sent and received by each. The sources and destinations may be analysed in terms of individual host or server addresses, but more normal is to consider the traffic flows between source and destination subnetworks (e.g., LANs). The traffic matrix (once inflated according to the predicted growth in demand) is used directly for network planning. Thus link capacity upgrades and network extensions can be planned for the upcoming months.

The traffic matrix will usually include the volume of data (i.e., number of bytes and maximum packet rate or bit rate) sent from an individual source to an individual destination. But in addition, network performance analysis also needs to consider:

- overall usage (number of bytes, Mbytes or Gbytes sent);

- maximum bit rate or packet rate demand;

- peak and mean packet size;

- individual link utilisations (i.e., percentage of capacity in use during the busiest period);

- the *top talkers* (i.e., the main sources and destinations of traffic); and

- average and maximum transaction delay.

As well as long-term monitoring of traffic demand and network quality performance, it is also essential to monitor network performance in real-time, if the service degradations caused by unpredicted peaks in traffic demand or network failures are to be minimised. There are two methods by which network failures or sudden degradations in network quality can be detected: either by means of *remote monitoring (RMON)* and equipment-reported *alarms* (as discussed in Chapter 9) or by using external monitoring equipment.

Alarms reporting failed equipment or links, *unreachable* destinations or unacceptable quality of transmission are usually sent to a network management station, where they are *filtered* and *correlated* before being presented to a human *network manager* — typically in the form of a graphical view of the network topology, with the failed equipment blinking or illuminated in red. This prompts the human manager to action.

External network monitoring equipment typically works by checking the 'heartbeat' of the network. If the 'heart' stops beating, the monitoring equipment raises the alarm. Some network administrators, for example, use *packet groper* devices to poll critical destinations every few minutes. They send a *groper (PING)* packet every few minutes to each critical destination and receive a reply in order to confirm that the destination is still *reachable* and that the latency of the network still meets the target quality level. Should the test fail, or the return packet be unduly delayed, the human network manager is alerted. Problems will typically be caused either by undue traffic demand or by a network link failure. The exact cause of the problem may require more detailed diagnosis by the human network manager (e.g., by manually PINGing the transit nodes along the route to the unreachable destination in turn).

14.8   Network management, back-up and restoration

Having located a network failure, what sort of network management control is appropriate? Network management actions can be classified into one of two categories:

- *expansive* control actions; and

- *restrictive* control actions.

The correct action to be taken in any individual circumstance needs to be considered in the light of a set of guiding principles, viz:

- use all available equipment to complete calls or deliver data packets, frames or cells;

- give priority to data packets which are most likely to reach their destination, have a high priority, and are likely to be processed immediately;

- prevent nodal (switch or router) congestion and its spread;

- give priority to connections or data packets which can be carried using only a small number of links.

In an *expansive action* the network manager makes further resources or capacity available for alleviating the congestion, whereas in a *restrictive action*, having decided that there are insufficient resources within the network as a whole to cope with the demand, the human network manager can cause attempted communications with *hard to reach* (i.e., temporarily congested) destination(s) to be rejected. It makes good sense to reject such communications close to their point of origin, since rejection of traffic early in the communication path frees as many network resources as possible, which can then be put to good use in serving communications between unaffected points of the network.

Expansive control actions

There are many examples of expansive actions. Perhaps the two most worthy of note are:

- *temporary alternative re-routing (TAR)*; and
- network *restoration* or link *back-up*.

Temporary alternative re-routing (TAR)

The use of idle capacity via third points is the basis of an expansive action called *temporary alternative re-routing (TAR)*. Re-routing is generally invoked only from computer controlled switches where routing table changes can be made easily. It involves temporarily using a different route to a particular destination. In Figure 14.21, the direct link (or maybe one of a number of direct links) between routers A and B has failed, resulting in congestion. This will change the *reachability* of destinations and the *cost* of the alternative paths to particular destinations, as calculated by the *routing protocol* (as we discussed in Chapter 6).

Some routes will thus change to temporary alternative routes (in the example of Figure 14.21, the temporary alternative route from router A to router B will be via router C). The routing tables of all the routers in the network may be changed during the period of the link failure to reflect the temporary routes which are to be used. The change will typically occur within about 5 minutes. A reversion to the direct route occurs after recovery of the failed link.

Network restoration

Network restoration is made possible by providing more plant in the network than the normal traffic load requires. During times of failure this 'spare' or *restoration* plant is used to 'stand in' for the faulty equipment, for example, a failed cable or transmission system. By restoring

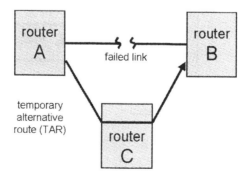

Figure 14.21   Temporary alternative routing (TAR) to overcome a link failure.

service with spare equipment, the faulty line system, switch or other equipment can be removed from service and repaired more easily.

Network restoration techniques have historically been applied to transmission links on a *1 for N* basis, i.e., 1 unit of restoration capacity for every N traffic-carrying units. The following example shows how 1 for N restoration works. Between two points of a network, A and B, a number of transmission systems are required to carry the traffic (see Figure 14.22). These are to be provided in accordance with a 1-in-4 restoration scheme. One example of how this could be met (Figure 14.22a) is with 5 systems, operated as 4 fully-loaded transmission lines plus a separate spare. Automatic *changeover* equipment is used to effect instant restoration of any of the active cables, by switching their load to the 'spare' cable should they fail.[10] An alternative but equally valid 1-in-4 configuration is to load each of the five cables at four-fifths capacity (Figure 14.22b). Should any of the cables fail, its traffic load must be restored in four parts — each of the other cables taking a quarter of the load of the failed cable.

In practice, not all cables (or network links) are of the same capacity and it is not always practicable or economic to restore cables or links exactly as shown in the examples of Figure 14.22, but the same basic principles can be applied. Another common practice used for restoration is that of *triangulation* (concatenating a number of restoration links via third points to enable full restoration). Figure 14.23 illustrates the principle of triangulation. In the simple example shown, a cable exists from node A to node B, but there is no direct restoration path. Restoration is provided instead by plant which is made available in the triangle of links A-C and C-B. These restoration links are also used individually to restore simpler cable failures, i.e., on the one-link connections such as A-C or B-C.

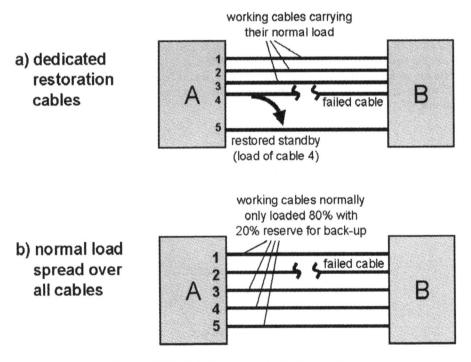

Figure 14.22   1 for N transmission link restoration.

---

[10] It is wise to keep the 'spare' link 'warm' — i.e., active — inactive plant tends not to work when called into action.

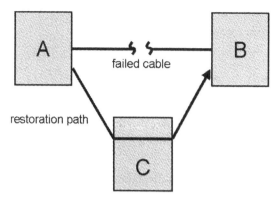

Figure 14.23   Restoration by triangulation.

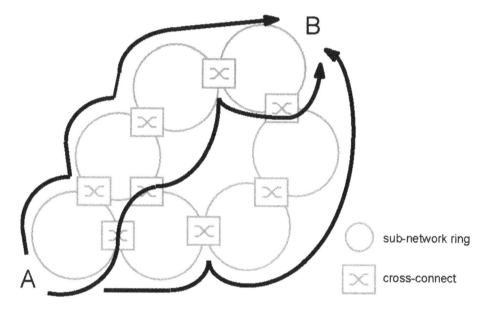

Figure 14.24   Alternative paths A-B in an SDH or SONET network made up of subnetwork rings.

Because of the scope for triangulation, restoration networks (also called *protection networks*) are often designed on a network-wide basis. This enables overall restoration network costs to be minimised without seriously affecting their resilience to problems.

Automatic restoration capabilities are built in to many modern transmission technologies (e.g., SDH — synchronous digital hierarchy and SONET — synchronous optical network). Thus in both SDH and SONET it is intended that highly resilient transmission networks should be built up from inter-meshed rings and subnetworks (Figure 14.24). Alone the use of a ring topology leads to the possibility of alternative routing around the surviving ring arc, should one side of the ring become broken due to a link failure. In addition, multiple cross-connect points between ring subnetworks further ensure a multitude of alternative paths through larger networks, as is clear from Figure 14.24. The possibilities are limited only by the capabilities of the network planner to dimension the network and topology appropriately and the ability of the *network management system* to execute the necessary path changes at times when individual links fail or come back into service.

1 : 1 link restoration by means of back-up links

In smaller networks, where 1 for N restoration might be impractical or too costly, it is common to provide for 1 : 1 restoration only of critical links in the network. Such *back-up* links for data networks are often provided by means of one of the following different types of networks or network services:

- *standby* links (links dedicated for back-up purposes, should the *main* (i.e., normal) link fail;

- *VPN (virtual private network)*;

- *dial back-up* (telephone or *ISDN—integrated services digital network*); or

- radio.

Figure 14.25 illustrates a possible network configuration for providing 1 : 1 restoration or link back-up. The 'normal connection' between routers A and B is a direct connection, dimensioned with a bit rate sufficient to carry the 'normal' traffic which flows between the two routers. Such a direct connection will generally be reliable and secure (i.e., not easy to snoop on by outsiders). Two alternative back-up links are shown. The first is a VPN (virtual private network) connection via a public Internet service provider's (ISP) network. The second is a dial-back-up connection (using either modems and the analogue telephone network or the ISDN — integrated services digital network). Either, both or neither of the back-up links may be in use at any given time.

Some routers are capable of automatically setting up the back-up or standby links when they detect that the *primary* or *main* link has failed. Otherwise external devices may be used to provide this functionality. Sometimes, the back-up link can simply be configured as a permanent part of the network topology: the VPN link of Figure 14.25, for example, could be configured as a direct link between routers A and B, but given a very high link cost weighting,

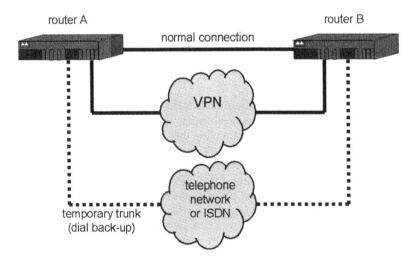

Figure 14.25 1 : 1 network restoration or link back-up.

so that the routing protocol will only select the route in preference to the 'normal' route during times of failure of the direct link.

Using a VPN (virtual private network) service (e.g., an MPLS (multiprotocol label switching) connection across a public IP-based 'backbone' or a connection across a public *frame relay* or ATM (asynchronous transfer mode) network) is a popular way of providing for network link back-up. Subscription charges must be paid for the VPN connection; usage charges for carriage of data will, however, only be incurred during the periods of temporary failure of the direct link. Most of the time, data is carried over the more secure direct link!

The use of modems and dial back-up across the analogue *public switched telephone network (PSTN)* is not common nowadays due to the very limited bit rates (typically up to 56 kbit/s) achievable by this method. Instead, *ISDN (integrated services digital network* — 'digital telephone network') is more common. Using ISDN back-up, 'dial-up' telephone connections of 64 kbit/s bit rate are provided between the two endpoints during times when the 'normal' direct link is adjudged to have 'failed'. Determining what constitutes a 'failure' of the direct link may be configurable. A 'failure' might be defined to be a 'complete loss of communication across the link' or alternatively an 'unacceptably poor quality of the direct link' may similarly be defined to be treated as a 'failure'. (Radio links, for example, are rarely completely 'lost', but radio interference may degrade the quality of communication to an unacceptable degree.)

It is important when using dial back-up that the switchover mechanism (between 'normal' and 'dial-back-up' links) is correctly configured to avoid flip-flopping between the links. If the direct link is working only intermittently, the back-up link should remain in operation all the time, and not be permanently switched on and off. Switching the connection over to a different physical connection (main to back-up, or back-up to main) is a disruptive process, requiring the lengthy processes of *link synchronisation* and data communications *session recovery*. It should be undertaken as infrequently as possible.

Dial-back-up connections with bit rates higher than 64 kbit/s can be created by means of bundling a number of individual 64 kbit/s connections and using *reverse multiplexing* (Figure 14.26). The *reverse multiplexor* splits up the bit stream comprising the 384 kbit/s connection into six separate 64 kbit/s bitstreams, which are then carried by separate dial-up

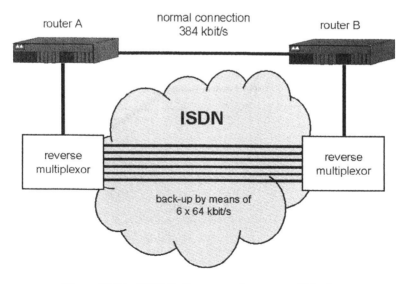

Figure 14.26   ISDN dial-back-up and reverse multiplexing.

ISDN connections to the destination. At the destination, the six separate data streams are re-assembled in the correct order to recover the original 384 kbit/s data stream.

Radio technology allows for the rapid establishment of transmission links across even the most inhospitable terrain. It can be a useful method of augmenting network capacity or backing up links which might otherwise take a long time to repair. Alternatively, radio links are sometimes built as a permanent and relatively cheap means of network back-up. A large number of different applications and users can share the same radio spectrum for network back-up purposes, since it is unlikely that all the users will require the spectrum at once!

Restrictive control actions

Unfortunately, there will always be some condition under which no further expansive action is possible. In Figure 14.27, for example, routes A-C and C-B may already be busy with their own direct traffic, or may not be large enough for the extra demand imposed by A-B traffic during the period of the failure shown. In this state, congestion on the route A-B cannot be alleviated by an alternative route via C without causing other problems. Meanwhile, attempts to reach the problematic destination (B) become a nuisance to other network users, since the extra network load they create starts to hold up traffic between otherwise unaffected end-points. When such a situation occurs, the best action is to refuse (or at least restrain) communication with the affected destination, rejecting connections or packets as near to their points of origination as possible. In the case of data networking, such call restriction may be undertaken by means of *flow control, ingress control* or *pacing.*

The principle of congestion flow control is that the traffic demand is 'diluted' or packets are 'held up' at the network node nearest their point of origin. A restricted number of packets or data frames to the affected destination (corresponding to a particular packet rate or bit rate) are allowed to pass into the network. Within the wider network, this reduces the network overload, relieves congestion of traffic to other destinations, giving a generally better chance of packet delivery across the network. There are two principal sub-variants of traffic dilution.

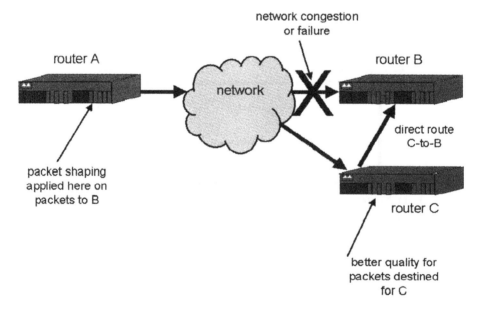

Figure 14.27 Restricting communication by congestion flow control near the point of origin.

These are by means of *1-in-N* or *1-in-T* dilution. The 1-in-N method allows every Nth packet to pass into the network. The remaining proportion of packets, (N-1)/N, may be discarded or rejected at a point near their origin, before gaining access to the main part of the network. Alternatively they may be marked for *preferential discarding* at a later (congested) point along the network path. By means of such packet dilution, the packet load on the main network (or the congested part of it) is reduced by a factor of N.

The 1-in-T method, by comparison, performs a similar traffic dilution by accepting only 1 packet every T seconds. This method provides for quite accurate allocation of a specific bit rate to a specific traffic stream.

When a very large value of N or T is used, nearly all packets will be blocked or held up at their point of origin. In effect, the destination has been 'blocked'. The action of complete blocking is quite radical, but nonetheless is sometimes necessary. This measure may be appropriate following a public disaster (earthquake, riot, major fire, etc.). Frequently in these conditions the public are given a telephone number or a website address as a point of enquiry, and inevitably there is an instant flood of enquiries, very few of which can be handled. In this instance, traffic dilution can be a useful means of increasing the likelihood of successful communication between unaffected network users.

Without moves to restrict the traffic demand on a network, the volume of successful traffic often drops as the offered traffic increases. Thus in Figure 14.28, the effective throughput of the network reduces if the traffic offered to the network exceeds value T_0. This is a common phenomenon in data networking.

A number of different IP-protocol suite congestion control methods employ restrictive control actions to protect the network against traffic overload (such as that shown in Figure 14.28). These include:

- *TCP (transmission control protocol)* flow and *congestion control*;

- *IP precedence* (Internet protocol);

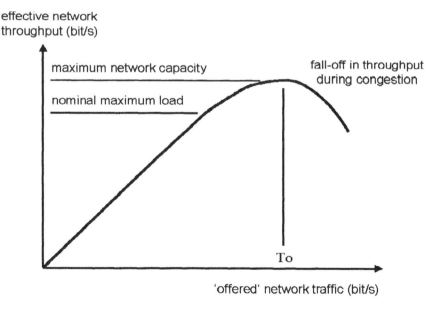

Figure 14.28 Congestion reduces the effective throughput of many networks if the traffic demand is too great.

- IP *TOS (type of service)*;

- *class of service (COS), DSCP (differentiated services codepoint)* and *PHB (per-hop behaviour)* (IP *differentiated services (DiffServ)*);

- *admission control* (as employed by *RSVP (resource reservation protocol)* and *MPLS (multiprotocol label switching)* networks);

- *quality of service (QOS)* and *user priority fields (UPF)* of protocols like IEEE 802.1p and IEEE 802.1q.

All the above are examples of automatic *congestion control* actions. An alternative, but perhaps cruder means of congestion control, is for human network managers to undertake temporary network, routing table or equipment configuration changes as they see fit.

The most widely used automatic congestion control method is the congestion window employed by the transmission control protocol (TCP). This we discussed in detail in Chapter 7. The various congestion and flow control protocols employed by TCP act to regulate the rate at which TCP datagrams may be submitted to the network at the origin (or source end) of a TCP connection. The flow rate is determined by the TCP protocol, which operates on an end-to-end basis between the two hosts which are sending and receiving data across the TCP connection across the network. The maximum allowed rate of datagram submission depends upon not only the ability of the receiving host to receive them, but also upon the delays and congestion currently being experienced by datagrams traversing the network. For completeness, we also recap briefly here the other previously discussed means of automatic congestion control based on quality-of service (QOS) methods.

IP precedence, the IP type of service (TOS) field and the *drop priority* of IP differentiated services (DiffServ) are all used simply to determine the order in which IP (Internet protocol) packets should be discarded (i.e., dropped) by a router, should an accumulation of incoming packets exceed the rate at which packets can be forwarded. The *DE (discard eligibility)* bit of frame relay and the *CLP (cell loss priority)* bit of ATM (asynchronous transfer mode) provide a similar prioritisation of which frame relay frames or ATM *cells* should be discarded first at a time of congestion. Alternatively, packets may be delayed by holding them in a buffer until the congestion subsiders. But simply dropping (or delaying) packets, frames or cells during a period of network congestion is no guarantee that the quality of service will be adequate for any of the network's users. For this purpose some kind of quality of service scheme must additionally be used in conjunction with packet, frame or cell discarding. Such schemes are defined for use with IP differentiated services (DiffServ), *admission control* (as used by RSVP/MPLS and ATM) and *virtual-bridged LAN (VLAN)* networks (IEEE 802.1q / IEEE 802.1p).

Quality of service (QOS) assurance schemes usually work by reserving link capacity, router forwarding capacity or other networks for particular users. The reservation may be on a permanent basis (i.e., configured into the network). (An example of a permanent reservation scheme is the *committed information rate (CIR)* offered in *frame relay* networks.) Alternatively, a negotiated reservation is possible with *admission control* schemes used in conjunction with connection-oriented communications networks.

When admission control is used, a request is made for the reservation of network resources at the time of *connection establishment* — in line with the bit rate, delay and other quality stipulations of the request. Should sufficient network resources not be currently available to meet the request, then the connection request is rejected, and the user must wait until a later (less congested time) before re-attempting the connection set-up. During the process of admission control, a *traffic contract* is negotiated between the network and the user requesting a connection. The contract commits the network to provide a connection meeting *the* quality of service parameters defined in the contract for the whole duration of the active phase of

Table 14.7 QOS guarantee in data networks: admission control protocols and parameters used in traffic contracts

Data network type	Admission control protocol and policing	Traffic contract (QOS parameters)
ATM (asynchronous transfer mode)	CAC (connection admission control) NPC (network parameter control) UPC (usage parameter control)	PCR (peak cell rate) SCR (sustainable cell rate) Burst cell rate CLR (cell loss ratio) CDVT (cell delay variation tolerance)
DiffServ (IP differentiated services)	RSVP (resource reservation protocol) [if used]	PHB (per-hop behaviour) DSCP (differential services codepoint)
Frame relay	Pre-configured	CIR (committed information rate) EIR (excess information rate)
MPLS (multiprotocol label switching)	RSVP (resource reservation protocol) [if used]	Bit rate Packet rate Packet size Delay

the connection. The QOS parameters typically include the bit rate, delay or latency, packet size, etc. During the active phase of communication, the network will normally monitor the connection, policing and enforcing the traffic contract as necessary. Thus if, during the duration of the connection, the network becomes subject to congestion, the network nodes will try to determine the cause of the congestion. Provided each user is only subjecting packets of a size and at a rate conforming with his traffic contract, then these packets are forwarded appropriately. Packets exceeding the traffic contract, however, may be subjected to *packet shaping*. Packet shaping actions can include:

- discarding or delaying packets;

- marking them for preferential discarding at a later point in the path;

- rejecting or fragmenting packets which exceed a certain packet size.

Examples of admission control processes and traffic contracts used by data protocols we have encountered in this book are listed in Table 14.7.

Network management systems

It is common nowadays for network management computer systems to be provided as an integral part of the subnetworks which they control. Real-time communication between network elements (e.g., routers, switches and transmission systems) and *network management systems* allow real-time network status information to be presented to the human network managers. Thus the *RMON (remote monitoring)* MIB and *the SNMP (simple network management protocol)* (as we discussed in detail in Chapter 9) allow for real-time monitoring of network performance and alerting of network failures and other alarm conditions.

As adjudged necessary by the human network manager (or automatically by the network management system software), control signals may be returned by means of SNMP to effect network configuration changes, thereby relieving congestion or overcoming network failures. For example, the network manager may choose to downgrade the handling of traffic of medium priority and temporarily to reject all communications of low priority.

Network management systems can usually be procured from network equipment manufacturers and are often sold with the equipment itself. Such 'proprietary' network management systems are usually optimised for the management of the particular *network element* — making for much easier network configuration and monitoring than the alternative 'command line interface' on the equipment console port. The drawback of such systems is that they are usually only suited to management of one type of network element. Correctly, they are termed *network element managers.*

To coordinate all the different *network elements* making up a complex network, an 'umbrella' network management system is required. Such systems are mostly the realm of specialist software-development companies (e.g., Hewlett Packard *OpenView*, Micromuse *Netcool, Syndesis*, etc.[11]). Ideally, an 'umbrella' network management system should coordinate the actions of the various network elements and subnetworks when a network failure or network congestions arises. It should be able to determine the best overall remedial action, and prevent different network element managers from undertaking contradictory actions. After all, the problem might only get worse if two parties pull in opposite directions!

14.9   Performance optimisation in practice

In practice, many data networks evolve without close management. The number of user devices and applications making use of the network, and the volume of traffic grows over time, often without close scrutiny of the implications for the network. Network dimensioning and capacity extensions are carried out on an 'empirical' basis — 'try-it-and-see'. The human network manager might monitor the link utilisation of all the links in the network on a monthly basis or even only 'as needed'. Should any of the links be found to be approaching 100% utilisation, an increase in line bit rate can be arranged. In many cases such a 'casual approach' may be entirely adequate and appropriate, but there are also occasions on which the increase of link capacity does not resolve a user's problem of poor quality. What do you do in such an instance? The answer is more detailed analysis of the network, the user and the application software. If you are not capable of this on your own, you can contract one of the specialist network analysis firms to do it for you.

A detailed performance analysis of a network requires special network monitoring (*probes*) and analysis tools. It usually starts with a basic analysis of network link performance and application transaction times (Figure 14.29). The link utilisation chart (Figure 14.29a) is the first step of analysis. If a particular link is operating near its full capacity, then the line propagation delays (and consequently the application response time) increase rapidly (as we saw in Figures 14.5 to 14.8). It may be important to consider the peak utilization (i.e., average link utilisation during the busiest one-hour or fifteen-minute period) rather than the average daily utilisation of a particular link. If a particular application is only used at a particular time of day at which time the network is likely to be heavily loaded, then the average daily link utilisation will not provide a good indication of the likely level of performance.

If particularly high network traffic demand, or rapid growth in demand is being experienced from one month to the next, it may be valuable to perform a detailed analysis of the main users of the network — i.e., the main sources and destinations of traffic. Table 14.8 illustrates a 'top talkers table' generated by some network performance analysis tools for this purpose.

Following simple analysis of the network to identify overloaded links, the next step of a detailed application performance review is likely to be a study of the average transaction delay. The example of Figure 14.29b illustrates the transaction delays of a server (i.e., application) and the network in supporting an imaginary application. There is generally a correlation in the transaction delay of the application and that caused by the network (at most times of day, the main delay is caused by the network). There are, however, two exceptional peaks of

---

[11] See Chapter 9.

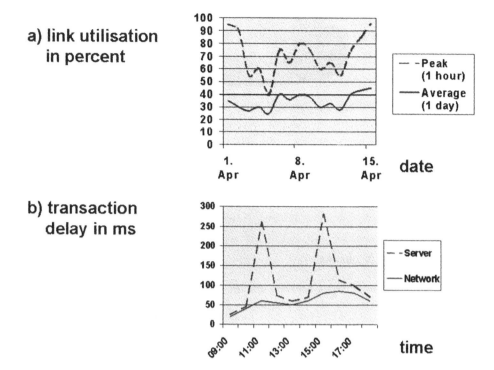

Figure 14.29 Network and application performance analysis.

Table 14.8 Top talkers (top sending hosts)

	DNS Name	IP Address	Usage (%)
1	CLARK	10.3.16.4	49
2	clark-corp	192.168.34.1	12
3	BOOKKEEPING	206.134.24.101	6
4	www.sap.com	252.234.13.38	4
5	www.company.com	178.121.101.103	4
6	www.supplier.org	23.16.1.252	4
7	DATA SERVER	10.3.16.1	4
8	SECRETARY	10.3.16.3	3
9	BOSS	10.3.16.2	3
10	www.footballscores.com	156.23.45.12	3

high transaction delays (at around 11:00 and 15:00), which cannot be accounted for by long network propagation delays. These warrant further analysis. Complaints received from users about poor performance at these times will not be resolved by merely adding further capacity to the network. The cause of the long transaction delay lies somewhere else, maybe a routine is run by the server at these times of day, or a database update is undertaken, or a particular user or application synchronises information at this time?

Some transaction delays may be explained by the packet sizes being used. A chart which plots the peak and mean packet sizes may be helpful in this case. Large packet sizes generally make for efficient usage of the network, since larger packets require relatively less packet header data (i.e., network *overhead*) for a given volume of *payload* data to be transferred. On

Figure 14.30   Packet size affects the efficiency of the network: the relative volumes of *payload* and *overhead*.

the other hand, if the sender has to wait for a large packet to be filled up before submitting a particular request for processing, then the overall time required for the transaction will accordingly be increased. So if the network needs to be made more efficient (as in Figure 14.30), increase the size of the packets but if the applications need to run faster, smaller packets may be needed!

Application and LAN checks

Studying the detailed traffic flows between specific devices within a network (Figure 14.31) reveals not only which devices are the *top talkers* but also a good deal of information about the software design of individual applications running on particular servers.

Figure 14.31a shows a typical office LAN, dominated by the office file and print server. Each client (C1 to C4–in this case simple user PCs) conducts most of its communications with the server (S). The top talker is client C4. In contrast, Figure 14.31b shows a network with multiple clients, servers and applications. In this case, the network administrator may be aware that server S1 hosts an important and highly used application, so that the heavy interaction between client C1 and server S1 is not a surprise. What maybe is a surprise to the network administrator is the heavy traffic from server S1 to server S2 and from server S2 to server S3. One of the servers might be a DNS (domain name system) server, the other a database server. Each request from client C1 may be generating related DNS and database query traffic. In this case, it would not be surprising if a shortage of capacity on the path between S2 and S3 caused poor transaction performance for client C1!

It may be pertinent in the example network of Figure 14.31b to study in detail the operation of the application and the client and server configuration settings. Are all the DNS enquiries essential, or could the number and frequency of them be reduced to improve performance? Where is the most time being used in processing transactions?

By increasing the *TTL (time to live)* of cached *DNS resource records*, it might be possible to greatly reduce the number of DNS queries which have to be made. Instead of having to query the remote DNS server each time (and thus have to wait for a DNS query to traverse the network and a response to come back), the DNS resolver can respond instantly using a *cached* copy of the resource record. If, for example, an enquiry is typically made to server S1 every 125 seconds and that the DNS route record TTL is set at 120 seconds, then most requests to server S1 will generate a DNS request to the DNS server. On the other hand, increasing the TTL to 180 seconds might mean that only every second request need generate a DNS query, thus halving the DNS traffic without significantly impacting the 'freshness'[12] of the DNS information being used. Alternatively, it may be appropriate to reduce the traffic to a remote DNS server by providing a local DNS *proxy server*.

---

[12] As we learned in Chapter 11, it is critical to reliable operation of an application, that sufficiently 'fresh' DNS route record information is used. The use of outdated information (which might arise as the result of using unduly old *cached* information) may mean that a server critical to the application cannot be contacted.

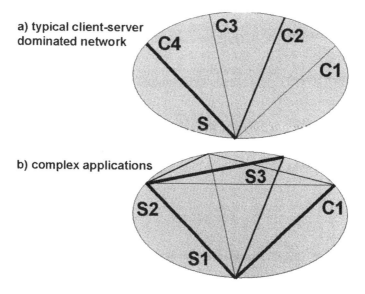

Figure 14.31 Network traffic flow analysis.

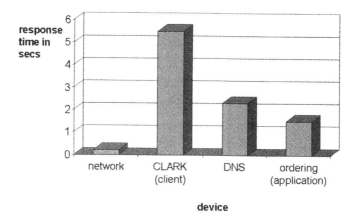

Figure 14.32 Response time components of a transaction.

An analysis of the response time components of a complete transaction (such as the example of Figure 14.32) helps to reveal 'where the most time is being used'. On the basis of the analysis of Figure 14.32, increasing the network line speed will not have an appreciable impact on the overall transaction time. Instead, client or server configuration settings might have a significant impact. In addition, a study of the individual *threads* and processes running in support of the application (e.g., Figure 14.33) may be necessary in order to localise the main cause of long transaction times.

Other service types

Certain types of services can be very sensitive to the performance of the data network on which they operate. Others, meanwhile, place such heavy demands on the network that they

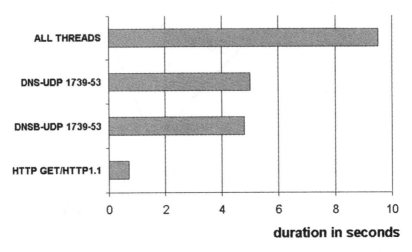

Figure 14.33 Relative durations of individual threads making up an application.

lead to significant degradation of other services. These and similar types of services need special consideration of the network designer and operator:

- some LAN server services generate huge quantities of network traffic through *service advertisements* (if the router is not properly configured to filter out such advertisements, they may inappropriately be being broadcast to the *wide area* network — and causing significant congestion);

- electronic mail (mail) can cause very heavy and 'bursty traffic', but is not time-critical;

- real-time traffic (such as video or voice-over-IP, VOIP) traffic requires very high network quality.

The right configuration settings of network services, application software, client, server and access routers are all crucial to the optimum performance of the network. Good data network design and administration are not easy. On the other hand, it is a challenging and rewarding profession!)

15

Challenges Ahead for IP

Since the early beginnings of the ARPANET in the late 1970s, the Internet protocol (IP) and the Internet itself have developed a long way, and the legacy will be long-lasting. Electronic mail (email), the Worldwide Web (www) and ebusiness are here to stay — though perhaps not in exactly the form in which we know them today! The development of the IP-suite protocols and networked applications will continue to improve the range of services available from the Internet and the manageability of the network. In this final, short chapter we discuss the five greatest challenges yet to be overcome by the Internet protocol (IP) and by users of IP-based networks.

15.1 Financing the network

The rapid growth and popularity of the Internet to date have been largely due to the easy access it has made possible to a large range of information on a cost-free basis. Once a user has paid his or her monthly Internet service access charge, the cost of 'surfing' the Internet usually comprises nothing more than a per-minute rate for the use of the telephone or other telecommunications line used to access the network. Few or no charges are made for accessing websites or other networks. Everything is 'for free'.

The network itself is actually a collection of different networks, owned by different operators but used as if it were a single large 'shared resource' and the accounting principles have historically been largely 'tit-for-tat', 'you can use my network for free provided I can use yours'. The assumption of the early Internet service providers (ISPs) was that the volume of traffic passing in the two directions between *peer* networks was the same, so that an accounting adjustment was unnecessary. Saving the bother of accounting the traffic was considered to reduce effort and costs and simultaneously to encourage usage of the network.

Over time, the number and diversity of individual service provider networks making up the Internet have grown enormously, and the networks no longer play an equal part in the transport and delivery of data packets. Internet exchanges (IXs), as we saw in Chapter 14, emerged to cater for the interconnection and peering needs of the networks, but a standard accounting system for settlement between operators for network transit or packet delivery services has not been developed. Managers at some ISPs still vigorously defend the old 'tit-for-tat' accounting ways, but this reflects in poor network quality. If you don't pay to have the packets delivered, then you can't complain about the quality of the network used for delivery! The corollary is that if you want your packets to be delivered reliably and with good quality, then you need to expect to pay commensurately.

Charging for Internet usage is bound to come. Only by charging can the continued growth and development of the network and services be financed. The more you pay, the better the

Data Networks, IP and the Internet: Protocols, Design and Operation   Martin P. Clark
© 2003 John Wiley & Sons, Ltd   ISBN: 0-470-84856-1

quality and the greater the range of services you can expect! *Internet service providers (ISPs)* are beginning to experiment with charging, but no 'industry-standard' charging structure has yet emerged. There are still a number of dilemmas to be resolved:

- How to charge fairly for the different types of services carried by the Internet? (Historically data transport was charged on a per Gbyte basis, while telephone calls were (and thus perhaps VOIP—voice over IP—should be) charged on a per-minute basis);
- What level of charging will the market accept?
- How should the accounting data necessary for charging be collected? (Not all router equipment generates accounting call records);
- Which data, and how often should accounting records be collected?[1]

15.2   Network architecture, interconnection and peering

To date, there have been no formal standards issued to advise upon the architecture and structure of large IP (Internet protocol)-based networks including the Internet. Instead, this is left to the initiative of individual network owners and operators. While the absence has clearly not hindered progress to date, it is in stark contrast to the wide range of network design and dimensioning recommendations established over many years by the International Telecommunications Union for international public telephone and telecommunications networks.

Internet exchanges (IXs) were historically founded by consortia of cooperating Internet service providers and network operators with a common interest in interconnection or *peering* within a given city. They were usually established as non-profit making 'clubs'. But with the increasing commercial importance of the Internet, we can expect their status to change. Maybe one large operator will take them all over and gain a worldwide monopoly on the Internet backbone. Or maybe world, regional or national telecommunications regulators will define how they are to be owned and maintained.

15.3   Quality of service (QOS) and network performance (NP)

There are, as yet, no formal standards defining how to measure the *quality of service (QOS)* or *network performance (NP)* of an Internet or IP-based network. Without such a common framework, it is difficult to compare the quality of networks operators by different service providers and difficult to predict the likely performance of applications which run on multiple, interconnected networks. A clear framework of quality targets is essential in establishing the goals of the network design and operations processes.

15.4   Scaling and adapting the network for higher speeds and real-time applications

Most telecommunications networks are subjected to continuous growth in traffic and ever-increasing technical demands of the services they are expected to carry. The Internet is no

---

[1] Some types of accounting 'counters' are unsuited to the volumes of data carried by modern networks. Thus, for example, a 1 Gbit/s line could carry 324 000 Gbytes in a single month. This value would overrun a 16-bit counter (which can count from 0 to 65535 before resetting). To be sure to record all the customer's volume, the network operator would have to collect call records daily. A monthly check of the counter would not be able to determine whether a counter overrun had occurred or not. But frequent transfers of accounting data across the network create additional *overhead* traffic load for the network.

exception. The traffic volumes continue to grow exponentially, and the bit rates of the user connection lines and services become ever faster.

Experience with other data network technologies and protocols suggest that the protocols and bit rates today used in the 'core' of a network, will tomorrow be migrated to the 'edge' (i.e., periphery) of the network and be replaced in the 'core' by new protocols, using even higher bit rates and faster transmission technologies. Thus, for example, the 100 Mbit/s *fast ethernet* interfaces used five years ago mainly in the *backbones* of large campus LANs (local area networks) are now commonly used as user-network interfaces (UNI) for connecting hosts and other end devices. The backbone meanwhile has migrated to the 1000 Mbit/s speed of *Gigabit ethernet*. But now, the price of Gigabit ethernet interfaces is dropping to a level, where it too can be considered for wider use as a user-network interface (UNI). The backbone will now have to migrate to even higher bit rates, to cope with the extra traffic volume that will result.

With each order-of-magnitude leap in the bit rate required, the technologies and protocols of the network backbone are severely challenged. Just because a protocol was ideally suited to a lower backbone bit rate is not any guarantee that it will work at all at a much higher bit rate! The technical, physical, electrical, optical and processing limits of network components may demand a major change in the basic operation of a protocol or even make it unviable, so requiring its replacement with a new protocol better suited to the higher bit rate.

While there is no declared upper limit (bit rate or otherwise) of the capabilities of IPv6 (Internet protocol version 6), there are bound to be constraints which will become apparent over time. The future is thus likely to bring Internet protocol versions 7, 8, 9 and so on — but other than the fact that they share the name 'IP', they may have little in common, and may be fully incompatible, with today's versions IPv4 and IPv6! Already, IPv4 and IPv6 need to be treated as different, albeit interworkable protocols.

The current trend in IP-backbone networks is for the use of *MPLS (multiprotocol label switching)* and IPv6 (Internet protocol version 6) protocols at the 'core' of the network. These protocols have been developed to be better suited to higher bit rates and the assurance of network *quality of service (QOS)*, thus making them better suited than IPv4 for the handling of real-time traffic such as video and *voice-over-IP (VOIP)*.

15.5   Network management

An unfortunate, but inevitable reality, is that the tools available for managing telecommunications networks are always somewhat behind the capabilities of the networks themselves. The tools tend to be developed to control individual elements of the network, rather than for the optimisation and control of the network as a whole. While the range and capabilities of network management tools for complete service and network management continue to increase, there is always bound to be a role for the human network specialist!!

Appendix 1

Protocol Addresses, Port Numbers, Service Access Point Identifiers (SAPIs) and Common Presentation Formats

This appendix presents for easy reference the most commonly used service access point identifiers (SAPIs), protocol addresses, protocol numbers and port numbers as used by protocol layers 1–4 to identify the hardware address, network address or nature of the next higher protocol. In addition, it presents common presentation formats, including the ASCII code (7-bit), extended ASCII (8-bit) and NVT-ASCII as well as MIME-media types.

As a general point of enquiry for the most up-to-data protocol and port number assignments, you may wish to refer also to the following website address:

www.iana.org/numbers.html

Layer 1 (physical layer): LAN MAC addresses (hardware addresses)

The *MAC (medium access control)* used in *LANs (local area networks)* uses a 48-bit *IEEE unique identifier* (also called the *MAC address* or *hardware address*). The first three bytes of the address (most significant bit values) comprise the *organisation unique identifier (OUI)* of the equipment or network interface card (NIC) manufacturer (Table A1.1). The last three bytes are a number unique to each individual piece of hardware. This value is 'burned into' the equipment at its time of manufacture. Values appear in the source and destination MAC-address fields of the MAC-header. Alternatively, a multicast address may be in use as the MAC-address (Table A1.2).

The most current list of organisation unique identifiers (OUIs) can be viewed on the IEEE website at:

http://standards.ieee.org/regauth/oui/index.shtml

Data Networks, IP and the Internet: Protocols, Design and Operation   Martin P. Clark
© 2003 John Wiley & Sons, Ltd   ISBN: 0-470-84856-1

Table A1.1   IEEE unique identifiers (48-bit) used in the MAC-address field: well known examples of the organisation unique identifier (OUI)

Organization	Organisational unique identifier (OUI) value in hexadecimal (3 bytes total)
Apple computer	08-00-07
Cisco	00-00-0C
Compaq (ex-Digital)	08-00-2B and 00-00-F8
Hewlett Packard (HP)	08-00-09
IANA (Internet Assigned Numbers Authority)	00-00-5E
IANA multicast (RFC 1054)	01-00-5E
IBM	10-00-5A
Novell	00-00-1B

LAN multicast addresses

Table A1.2   LAN-MAC (medium access control) multicast addresses

Multicast address	Meaning
01-80-C2-00-00-00	IEEE 802.1d protocol
01-80-C2-00-00-10	IEEE 802.1d All_Bridge_Management
01-00-5E-00-00-01 (224.0.0.1)	All systems on this (IP) subnet
01-00-5E-00-00-02 (224.0.0.2)	All routers on this (IP) subnet
01-00-5E-00-00-05 (224.0.0.5)	All OSPF (open shortest path first) routers
01-00-5E-00-00-06 (224.0.0.6)	All designated OSPF (open shortest path first) routers
01-00-5E-00-00-09 (224.0.0.9)	RIPv2 (routing information protocol) routers
01-00-5E-00-00-0A (224.0.0.10)	IGRP (Cisco interior gateway routing protocol) routers
01-00-5E-00-00-0D (224.0.0.13)	All PIM (protocol independent multicast) routers
01-00-5E-00-01-18 (224.0.1.24)	Microsoft WINS server autodiscovery
01-00-5E-00-01-27 (224.0.1.39)	Cisco PIM rendezvous point announcements
01-00-5E-00-01-28 (224.0.1.40)	Cisco PIM rendezvous point discovery
01-00-5E-00-01-29 (224.0.1.41)	ITU-T H.225 gatekeeper discovery
01-00-5E-00-01-4B (224.0.1.75)	SIP (session initiation protocol) ALL_SIP_Server
01-00-5E-02-7F-FE (224.2.127.254)	SAP (session announcement protocol) announcements
30-00-00-00-00-01	NetBEUI multicast

Layer 2 (datalink layer): HDLC, LLC and PPP protocol types

HDLC (higher level datalink control)

The HDLC address field comprises an 8-bit value. On a point-to-point link, only two values are assigned — one to represent the DCE (data circuit terminating equipment) and one to represent the DTE (data terminal equipment). There are two reserved addresses. These are the values 1111 1111 (all stations broadcast) and 0000 0000 (no stations)

LLC (logical link control) — datalink protocol used in LANs

In the LLC protocol, the *service access point identifier (SAPI)* in the SAP address (protocol) field indicates how the contents of the data frame are to be handled — according to Table A1.3.

Table A1.3   Logical link control (LLC) service access point identifiers (SAPIs)

Service access point (SAP) identifier (SAPI)	Protocol and address format in use
04, 08, 0C (05, 09, 0D as group identifiers)	SNA (systems network architecture)
0A	IEEE 802.10 standard for interoperable LAN/MAN security (SILS)
42	IEEE 802.1d transparent bridging
AA	IEEE 802.2 SNAP (subnetwork address protocol)
7E	X.25 over ethernet (ISO 8208)
FE	ISO network layer protocol
F0 (F1 as group identifier)	NetBEUI
E0 (E1 as group identifier)	Novell

Table A1.4   Protocol types supported by ethernet SNAP-format (*EtherType* field)

Protocol	Protocol type (PT) value in hexadecimal (2 bytes total)
Address resolution protocol (ARP)	08-06
Appletalk	80-9B
Appletalk ARP (address resolution protocol)	80-F3
DECnet maintenance operations protocol (MOP)	60-01
DECnet local area transport (LAT) protocol	60-04
DECnet routing	60-03
IBM SNA over ethernet	80-D5
IEEE 802.1Q (VLANs, virtual LANs)	81-00
Internet protocol version 4 (IP v4)	08-00
Internet protocol version 6 (IP v6)	86-DD
Novell IPX (Internetwork packet exchange) protocol	81-37
Reverse address resolution protocol (RARP)	80-35
Simple network management protocol (SNMP)	81-4C
Xerox network system (XNS)	06-00

The most common values used are the values 1111 1111 (all stations broadcast), 0000 0000 (no stations) or the value 'AA' (indicating the use of the *SNAP — subnetwork address protocol*) (as illustrated in Chapter 4–Figure 4.6).

When both the source-SAP (service access point) and destination-SAP fields of the LLC header are set to value 'AA' (10101010), then the SNAP (subnetwork address protocol) format is in use. In this format a PT (protocol type field) is included, which indicates the protocol being used at the next higher layer, e.g., as a network protocol (see Table A1.4).

The most current list of protocol types (*EtherType* field values) can be viewed on the IEEE website at:

`http://standards.ieee.org/regauth/ethertype/type-pub.html`

PPP protocol assignments

Table A1.5 lists the coding of the PPP protocol type field. This reveals the protocol in use (e.g., Internet protocol) for coding the user data held as the contents of the PPP packet.

Table A1.5   PPP protocol value field — common values

Protocol type	Protocol value range	Allocated value	Protocol
Network layer protocols	0xxx — 3xxx	0001	Padding protocol
		0003	Robust Header Compression (ROHC small-CID, context identifier: RFC 3095)
		0005	Robust Header Compression (ROHC large-CID, context identifier: RFC 3095)
		0007 — to — 001F	Reserved (transparency efficient)
		0021	Internet Protocol v4
		0023	OSI network layer
		002B	Novell IPX
		002D	Van Jacobsen compressed TCP/IP (RFC 2508)
		002F	Van Jacobsen uncompressed TCP/IP (RFC 2508)
		0031	Bridging PDU
		003D	PPP Multilink protocol (MP — RFC 1717)
		003F	NetBIOS framing
		0041	Cisco systems
		0049	Serial data transport protocol (SDTP — RFC 1963)
		004B	IBM SNA over IEEE 802.2
		004D	IBM SNA
		0057	Internet Protocol v6
		0059	PPP muxing (RFC 3153)
		0061–0069	RTP (Real-Time Transport Protocol) Internet Protocol Header Compression (IPHC — RFC 2509)
		007D	Reserved (control escape — RFC 1661)
		007F	Reserved (compression inefficient — RFC 1662)
		00CF	Reserved (PPP NLPID)
		00FB	Single link compression in multilink (RFC 1962)
		00FD	Compressed datagram (RFC 1962)
		00FF	Reserved (compression inefficient)
	02xx — 1Exx (compression inefficient)	0201	IEEE 802.1p hello protocol
		0203	IBM source routing BPDU
		0207	Cisco discovery protocol
		0281	MPLS unicast
		0283	MPLS multicast
		2063–2069	Real-time Transport Protocol (RTP) Internet Protocol Header Compression (IPHC — RFC 2509)
Low volume traffic protocols with no NCP	4xxx — 7xxx	See www.iana.org	
Network control protocols (NCPs)	8xxx — Bxxx	8001–801F	Unused
		8021	IPv4 control protocol (RFC 1332)
		8023	OSI network layer control protocol (RFC 1377)
		802B	Novell IPX control protocol (RFC 1552)
		802D	Reserved
		802F	Reserved
		8031	Bridging control protocol (BCP — RFC 2878)

Table A1.5 (*continued*)

Protocol type	Protocol value range	Allocated value	Protocol
		803D	Multilink control protocol (RFC 1717)
		803F	NetBIOS framing control protocol (RFC 2097)
		8041	Cisco systems control protocol
		8049	Serial data control protocol (SDCP)
		804B	IBM SNA over IEEE 802.2 control protocol
		804D	IBM SNA control protocol (RFC 2043)
		8057	IPv6 control protocol
		8059	PPP muxing control protocol (RFC 3153)
		807D	Unused (RFC 1661)
		80CF	Unused (RFC 1661)
		80FB	Single link compression in multilink control (RFC 1962)
		80FD	Compression control protocol (CCP — RFC 1962)
		80FF	Unused (RFC 1661)
Link control protocols (LCPs)	Cxxx — Fxxx	C021	Link control protocol (LCP)
		C023	Password authentication protocol (PAP — RFC 1334)
		C025	Link quality report
		C029	CallBack control protocol (CBCP)
		C02B	Bandwidth allocation control protocol (BACP — RFC 2125)
		C02D	Bandwidth allocation protocol (BAP — RFC 2125)
		C223	Challenge handshake authentication protocol (CHAP — RFC 1994)
		C227	Extensible authentication protocol (RFC 2284)

The most current list of protocol types (*PPP protocol numbers*) can be viewed on the IANA website at:

www.iana.org/assignments/ppp-numbers

Layer 3 (network layer): Internet addresses and protocol numbers

The Internet protocol uses both *address* format fields (source and destination network address) as well as a *protocol* type field (called *next header* field in IPv6).

Network address (IP address)

IPv4 address field

Addresses of IPv4 (Internet protocol version 4) format are 32-bit values, usually denoted as a series of decimal values separated by 'dots' thus:

d1.d2.d3.d4

where each value d1-d4 takes a numerical integer value between 0 and 255. The values d1 correspond to *class A address-ranges*. These ranges are assigned by IANA (Internet

Assigned Numbers Authority) and delegated registration authorities. Current allocations appear in Table A1.6.

The most current list of IPv4 address assignments can be viewed on the IANA website at:

www.iana.org/assignments/ipv4-address-space

IPv6 address field

Addresses of IPv6 (Internet protocol version 6) format are 128-bit values, usually denoted as a series of hexadecimal values separated by 'colons' thus:

hhhh:hhhh:hhhh:hhhh:hhhh:hhhh:hhhh:hhhh

where each value individual hexadecimal value h takes a value between 0 and A. The various different ranges of IPv6 addresses are assigned by IANA (Internet Assigned Numbers Authority) and delegated registration authorities. Current allocations appear in Table A1.7.

The most current list of IPv6 address assignments can be viewed on the IANA website at:

www.iana.org/assignments/ipv6-address-space

Table A1.6   Allocation of IPv4 class A address ranges

Decimal value (first 8 bits of IPv4 address)	Assignment
0–2	IANA — reserved
3	General Electric Company
4	BBN (Bolt, Beranek and Newman)
5	IANA — reserved
6	US Army information systems center
7	IANA — reserved
8	BBN (Bolt, Beranek and Newman)
9	IBM
10	IANA — reserved for private IP addresses
11	US Department of Defense (DoD) Intel information services
12	AT&T Bell Laboratories
13	Xerox Corporation
14	IANA — public data network
15	Hewlett Packard
16	Digital Equipment Corporation (now COMPAQ/HP)
17	Apple Computer Inc
18	MIT (Massachusetts Institute of Technology)
19	Ford Motor Company
20	CSC (Computer Sciences Corporation)
21	DDN-RVN
22	US Defense information services agency
23	IANA — reserved
24	ARIN (American Registry for Internet Numbers)
25	UK Royal Signals and Radar Establishment
26	US Defense information services agency
27	IANA — reserved
28	DSI — north
29–30	US Defense information services agency
31	IANA — reserved
32	Norsk Informasjonsteknologi

Table A1.6 (*continued*)

Decimal value (first 8 bits of IPv4 address)	Assignment
33	DLA system automation center
34	Halliburton Company
35	MERIT computer network
36	IANA — reserved (formerly Stanford University)
37	IANA — reserved
38	PSI (performance systems international)
39	IANA — reserved
40	Eli Lilly and Company
41–42	IANA — reserved
43	Japan Inet
44	Amateur Radio Digital Communications
45	Interop Show network
46	BBN (Bolt, Beranek and Newman)
47	Bell-Northern Research
48	Prudential Securities
49–50	IANA (formerly Joint Technical Command)
51	UK (Department of Social Security — DSS)
52	E.I. duPont de Nemours and Co., Inc
53	Cap Debis CCS
54	Merck and Co. Inc
55	Boeing Computer Services
56	US Postal Service
57	SITA (Société Internationale de Télécommunications Aéronautiques)
58–60	IANA — reserved
61	APNIC (Asia-Pacific Network Information Centre)
62	RIPE (Réseaux IP Européens)
63–68	ARIN (American Registry for Internet Numbers)
69–79	IANA — reserved
80–81	RIPE (Réseaux IP Européens)
82–128	IANA — reserved
129–191	Various registries
192	Various registries, multiregional
193–195	RIPE (Réseaux IP Européens)
196	Various registries
197	IANA — reserved
198	Various registries
199–200	ARIN (American Registry for Internet Numbers)
201	Central and South America
202–203	APNIC (Asia-Pacific Network Information Centre)
204–209	ARIN (American Registry for Internet Numbers)
210–211	APNIC (Asia-Pacific Network Information Centre)
212–213	RIPE (Réseaux IP Européens)
214–215	US Department of Defense (DoD)
216	ARIN (American Registry for Internet Numbers)
217	RIPE (Réseaux IP Européens)
218–220	APNIC (Asia-Pacific Network Information Centre)
221–223	IANA — reserved
224–239	IANA — multicast (see also Table A1.4)
240–255	IANA — reserved

Table A1.7   IPv6 address range allocations

IPv6 prefix	Usage
0000 0000	Reserved
0000 001	Network Service Access Point (NSAP)
0000 010	IPX
0001	Global unicast addresses (RFC 2374)
1111 1110 10	Link-local unicast addresses
1111 1110 11	Site-local unicast addresses
1111 1111	Multicast addresses

Protocol type

The *protocol* type field of the IPv4 header and the *next header* field of the IPv6 header reveal the protocol used to code the data held in the IP packet payload. Common protocol values are listed in Table A1.8. The most current list of *protocol* number assignments can be viewed on the IANA website at:

www.iana.org/assignments/protocol-numbers

Table A1.8   Protocol field values (IANA protocol numbers) and their meanings

Protocol field value (possible range 0–255)	Protocol used at next higher layer (i.e., content type of IP-data field)
0	HOPOPT — IPv6 Hop-by-Hop Option
1	ICMP — Internet Control Message Protocol
2	ÎGMP — Internet Group Management Protocol
4	ÎP in IP encapsulation (RFC 2003)
6	TCP — Transmission Control Protocol
7	CBT — Core-Based Trees
8	EGP — Exterior Gateway Protocol
9	IGP — Interior Gateway Protocol (e.g., Cisco's IGRP)
10	UDP — User Datagram Protocol
27	RDP — Reliable Data Protocol
28	IRTP — Internet Reliable Transaction Protocol
30	NETBLT — Bulk Transfer Data Protocol
35	IDPR — Inter-Domain Policy Routing Protocol
42	SDRP — Source Demand Routing Protocol
43	IPv6-Route — Routing Header for IPv6
44	IPv6-Frag — IPv6 Fragment header follows
45	IDRP — Inter-Domain Routing Protocol
46	RSVP — Resource ReSerVation Protocol
47	GRE — Generic Routing Encapsulation
48	MHRP — Mobile Host Routing Protocol
50	ESP — Encapsulation Security Payload for IPv6
51	AH — Authentication Header for IPv6
54	NARP — NBMA Address Resolution Protocol
58	IPv6-ICMP — ICMP for IPv6
59	IPv6-NoNxt — No next header — payload should be ignored
60	IPv6-Opts — Destination Options for IPv6

Table A1.8   (*continued*)

Protocol field value (possible range 0–255)	Protocol used at next higher layer (i.e., content type of IP-data field)
80	ISO-IP — ISO Internet Protocol
81	VMTP — Versatile Message Transaction Protocol
88	EIGRP — Extended Interior Gateway Routing Protocol (Cisco)
89	OSPF — Open Shortest Path First
92	MTP — Multicast Transport Protocol
95	MICP — Mobile Interworking Control Protocol
97	ETHERIP — Ethernet-within-IP encapsulation
98	ENCAP — Encapsulation header / private encryption
103	PIM — Protocol Independent Multicast
108	IP Comp — IP Payload Compression Protocol
111	Novell IPX-in-IP
112	VRRP — Virtual Router Redundancy Protocol
115	L2TP — Layer 2 Tunneling Protocol

Layer 4 (transport layer): port numbers

The service access point identifiers (SAPIs) used at the transport layer by the *transmission control protocol (TCP)* and the *user datagram protocol (UDP)* are called *port numbers* or *socket numbers* Port numbers are 16-bit binary values (decimal: 0–65 535) which are assigned by IANA according to the following broad scheme:

- *well-known ports* 0–1023

- *registered ports* 1024–49151

- *dynamic* and *private ports* 49152–65535

Commonly used port number values are listed in Table A1.9.

Table A1.9   TCP (transmission control protocol) and UDP (user datagram protocol) port numbers

UDP / TCP port number	TCP or UDP as carriage protocol	Application protocol or service
21	TCP	FTP (file transfer protocol)
22	TCP	SSH (secure shell) remote login protocol and secure forwarding protocol
23	TCP	Telnet
25	TCP	SMTP (simple mail transfer protocol)
53	TCP/UDP	DNS (domain names service)
65	TCP/UDP	TACACS (terminal access controller access control system) database service
67	UDP	BOOTP (bootstrap protocol) / DHCP (dynamic host configuration protocol) server
68	UDP	BOOTP (bootstrap protocol) / DHCP (dynamic host configuration protocol) client
69	UDP	TFTP (trivial file transfer protocol)

(*continued overleaf*)

Table A1.9   (*continued*)

UDP / TCP port number	TCP or UDP as carriage protocol	Application protocol or service
80	TCP	Worldwide web HTTP (hypertext transfer protocol)
111	UDP	Sun remote procedure call (RPC)
119	TCP	NNTP (network news transfer protocol)
123	UDP	NTP (network time protocol)
137	UDP	NetBIOS name service
138	UDP	NetBIOS datagram service
139	UDP	NetBIOS session service
161	UDP	SNMP (simple network management protocol)
162	UDP	SNMP trap
179	TCP	BGP (border gateway protocol)
194	TCP	IRC (Internet relay chat)
213	UDP	Novell IPX (internetwork packet exchange)
443	TCP	HTTPS (secure hypertext transfer protocol)
512	TCP	Rsh (BSD — Berkeley software distribution) remote shell
513	TCP	RLOGIN (remote login)
514	TCP	cmd (UNIX R commands)
520	UDP	RIP (routing information protocol)
540	TCP	UUCP (UNIX-to-UNIX copy program)
646	TCP/UDP	LDP (label distribution protocol): LDP hello uses UDP, LDP sessions use TCP
1080	TCP	SOCKS (OSI session layer security)
1645	UDP	RADIUS (remote authentication dial-in user service) authentication server
1646	UDP	RADIUS accounting server (Radacct)
1701	UDP	L2F (layer 2 forwarding)
2049	TCP/UDP	NFS (network file system)
2065	TCP	DLSw (data link switching) read port
2066	TCP	DLSw (data link switching) write port
5060	UDP	SIP (session initiation protocol)
6000-4	TCP	X-windows system display
9875	UDP	SAP (session announcement protocol)

The most current list of *protocol* number assignments can be viewed on the IANA website at:

`www.iana.org/assignments/port-numbers`

Layer 6 (presentation layer): 7-Bit ASCII, NVT-ASCII and extended ASCII (IBM PC)

It can be confusing to talk simply about the ASCII (American standard code for interchange of information) since there are many similar but different codes based on the original 7-bit ASCII code (Tables A1.10 and A1.11). In this appendix we present the original 7-bit ASCII code, as well as the 8-bit NVT-ASCII (network virtual terminal) version used by the telnet protocol and FTP (file transfer protocol). We also present the 8-bit *extended ASCII* which was developed as code page 437 for the IBM PC (personal computer). There are, however,

Table A1.10 The original 7-bit ASCII code (International alphabet IA5)

HEX CODE YX			X	0	1	2	3	4	5	6	7	8	9	A	B	C	D	E	F
			4	0	0	0	0	0	0	0	0	1	1	1	1	1	1	1	1
	BIT		3	0	0	0	0	1	1	1	1	0	0	0	0	1	1	1	1
			2	0	0	1	1	0	0	1	1	0	0	1	1	0	0	1	1
Y			1	0	1	0	1	0	1	0	1	0	1	0	1	0	1	0	1
	8	7	6	5																
0	0	0	0	0	0 NUL	1 SOH	2 STX	3 ETX	4 EOT	5 ENQ	6 ACK	7 BEL	8 BS	9 HT	10 LF	11 VT	12 FF	13 CR	14 SO	15 SI
1	0	0	0	1	16 DLE	17 DC1	18 DC2	19 DC3	20 DC4	21 NAK	22 SYN	23 ETB	24 CAN	25 EM	26 SUB	27 ESC	28 FS	29 GS	30 RS	31 US
2	0	0	1	0	32 spce	33 !	34 "	35 #	36 $	37 %	38 &	39 '	40 (| 41)	42 *	43 +	44 ,	45 -	46 .	47 /
3	0	0	1	1	48 0	49 1	50 2	51 3	52 4	53 5	54 6	55 7	56 8	57 9	58 :	59 ;	60 <	61 =	62 >	63 ?
4	0	1	0	0	64 @	65 A	66 B	67 C	68 D	69 E	70 F	71 G	72 H	73 I	74 J	75 K	76 L	77 M	78 N	79 O
5	0	1	0	1	80 P	81 Q	82 R	83 S	84 T	85 U	86 V	87 W	88 X	89 Y	90 Z	91 [	92 \	93]	94 ^	95 _
6	0	1	1	0	96 `	97 a	98 b	99 c	100 d	101 e	102 f	103 g	104 h	105 i	106 j	107 k	108 l	109 m	110 n	111 o
7	0	1	1	1	112 p	113 q	114 r	115 s	116 t	117 u	118 v	119 w	120 x	121 y	122 z	123 {	124		125 }	126 ~	127 DEL

a plethora of similar but different 'extended ASCII 8-bit codes' (e.g., that used by Microsoft Windows) which we do not present here.

The original 7-bit ASCII code

This code is presented in Table A1.10.

The explanation of the control character codes appears in Table A1.11.

Table A1.11 ASCII control characters

ASCII character	Meaning
ACK	Acknowledgement
BEL	Bell
BS	Backspace
CAN	Cancel
CR	Carriage Return
DC1	Device Control 1
DC2	Device Control 2
DC3	Device Control 3
DC4	Device Control 4
DEL	Delete
DLE	Data Link Escape
EM	End of Medium
ENQ	Enquiry
EOT	End of Transmission
ESC	Escape
ETB	End of Transmission Block

(continued overleaf)

Table A1.11   *(continued)*

ASCII character	Meaning
ETX	End of Text
FF	Form Feed
FS	File Separator
GS	Group Separator
HT	Horizontal Tab
LF	Line Feed
NAK	Negative Acknowledgement
NUL	Null
RS	Record Separator
SI	Shift In
SO	Shift Out
SOH	Start of Header
STX	Start of Text
SUB	Substitute Character
SYN	Synchronisation character
US	Unit Separator
VT	Vertical Tab

NVT (network virtual terminal) version of ASCII

The NVT-version of ASCII which is used in the telnet protocol and FTP (file transfer protocol) (see chapter 10) is based on the original 7-bit ASCII code (Table A1.10), but is coded as an 8-bit code. The 7 bits of the original code are simply prefixed with a '0' value, but a number of further control codes are included (Table A1.12).

Table A1.12   The telnet network virtual terminal (NVT) character and command set (NVT-ASCII)

Telnet control functions and signals	Code (decimal)	Code (hexa-decimal)	Function
NULL (NUL)	0	00	No operation.
Bell (BEL)	7	07	(Option) Produces an audible or visible signal without moving the print head.
Back space (BS)	8	08	(Option) Moves the printer one space towards the left margin, remaining on the current line.
Horizontal tab (HT)	9	09	(Option) Moves the printer to next horizontal tab stop (undefined is where exactly this is).
Line feed (LF)	10	0A	Moves the printer to the next line, but retaining the same horizontal position.
Vertical tab (VT)	11	0B	(Option) Moves the printer to next vertical tab stop (undefined is exactly where this is).
Form feed (FF)	12	0C	(Option) Moves the printer to the top of the next page, retaining the same horizontal position.
Carriage return (CR)	13	0D	Moves the printer to the left margin of the current line.
Alphanumeric characters and punctuation (ASCII)	32-126	20-7E	Alphanumeric text characters making up the main portion of the telnet data.
Subnegotiation end (SE)	240	F0	This signal indicates the end of subnegotiation of option parameters.
No operation (NOP)	241	F1	This signals that no operation is possible.
Data mark (DM)	242	F2	The data stream part of the SYNCH mechanism — for clearing a congested data path to the other party.

Table A1.12   (*continued*)

Telnet control functions and signals	Code (decimal)	Code (hexa-decimal)	Function
Break (BRK)	243	F3	The break command is an additional command outside the normal ASCII character set required by some types of computers to create an 'interrupt'. It is not intended to be an alternative to the IP command, but instead used only by those computer systems which require it.
Interrupt process (IP)	244	F4	This command suspends, interrupts, aborts or terminates a user process currently in operation on the remote host computer. (This is the equivalent of the 'break', 'attention' or 'escape' key.)
Abort output (AO)	245	F5	This command causes the remote host computer to jump to the end of an output process without outputting further data.
Are you there ? (AYT)	246	F6	This command causes the remote host computer to reply with a printable message that it is still 'alive'.
Erase character (EC)	247	F7	This command deletes the last character sent in the data stream currently being transmitted.
Erase line (EL)	248	F8	This command deletes all the data in the current 'line' of input.
Go ahead (GA)	249	F9	Continue...
Subnegotiation (SB)	250	FA	This signal indicates that the following codes represent the subnegotiation of telnet features.
WILL (option code)	251	FB	This signal is a request to start performing a given (indicated) option.
WON'T (option code)	252	FC	A rejection of the requested option.
DO (option code)	253	FD	An acceptance of the requested option.
DON'T (option code)	254	FE	An instruction to the remote party not to, or to stop using the indicated option.
IAC	255	FF	Data byte 255.
SYNCH	—	—	The SYNCH signal is a combination of a TCP 'urgent notification' coupled with a 'data mark' character in the data stream. SYNCH clears the data path to a remote 'timeshared' host computer. The 'urgent' notification bypasses the normal TCP flow control mechanism which is otherwise applied to telnet connection data.

8-bit extended ASCII (IBM PC code page 437)

Table A1.13 documents the extended 8-bit version of ASCII (code page 437) which is the default character set for the DOS (disk operating system) (but not quite the character set used nowadays by Microsoft Windows).

ISO 8859 character sets

ISO has standardised a number of 8-bit 'extended ASCII' character sets. The document references and character set names are listed in Table A1.14.

A useful website on the subject of character sets is:

http://czyborra.com/charsets/iso8859.html

Table A1.13 Extended 8-bit ASCII code (as developed for the IBM PC; code page 437)

HEX CODE YX	X	0	1	2	3	4	5	6	7	8	9	A	B	C	D	E	F
BITS	4	0	0	0	0	0	0	0	0	1	1	1	1	1	1	1	1
	3	0	0	0	0	1	1	1	1	0	0	0	0	1	1	1	1
	2	0	0	1	1	0	0	1	1	0	0	1	1	0	0	1	1
Y 8 7 6 5	1	0	1	0	1	0	1	0	1	0	1	0	1	0	1	0	1
0 0 0 0 0		0	1 ☺	2 ●	3 ♥	4 ♦	5 ♣	6 ♣	7 •	8 ◘	9 ○	10 ◙	11 ♂	12 ♀	13 ♪	14 ♫	15 ☼
1 0 0 0 1		16 ►	17 ◄	18 ↕	19 ‼	20 ¶	21 §	22 ▬	23 ↨	24 ↑	25 ↓	26 →	27 ←	28 ∟	29 ↔	30 ▲	31 ▼
2 0 0 1 0		32 spce	33 !	34 "	35 #	36 $	37 %	38 &	39 '	40 (	41)	42 *	43 +	44 ,	45 -	46 .	47 /
3 0 0 1 1		48 0	49 1	50 2	51 3	52 4	53 5	54 6	55 7	56 8	57 9	58 :	59 ;	60 <	61 =	62 >	63 ?
4 0 1 0 0		64 @	65 A	66 B	67 C	68 D	69 E	70 F	71 G	72 H	73 I	74 J	75 K	76 L	77 M	78 N	79 O
5 0 1 0 1		80 P	81 Q	82 R	83 S	84 T	85 U	86 V	87 W	88 X	89 Y	90 Z	91 [	92 \	93]	94 ^	95 _
6 0 1 1 0		96 `	97 a	98 b	99 c	100 d	101 e	102 f	103 g	104 h	105 i	106 j	107 k	108 l	109 m	110 n	111 o
7 0 1 1 1		112 p	113 q	114 r	115 s	116 t	117 u	118 v	119 w	120 x	121 y	122 z	123 {	124 ¦	125 }	126 ~	127 ⌂
8 1 0 0 0		128 Ç	129 ü	130 é	131 â	132 ä	133 à	134 å	135 ç	136 ê	137 ë	138 è	139 ï	140 î	141 ì	142 Ä	143 Å
9 1 0 0 1		144 É	145 æ	146 Æ	147 ô	148 ö	149 ò	150 û	151 ù	152 ÿ	153 Ö	154 Ü	155 ¢	156 £	157 ¥	158 Pts	159 ƒ
A 1 0 1 0		160 á	161 í	162 ó	163 ú	164 ñ	165 Ñ	166 ª	167 º	168 ¿	169 ⌐	170 ¬	171 ½	172 ¼	173 ¡	174 «	175 »
B 1 0 1 1		176 ░	177 ▒	178 ▓	179 │	180 ┤	181 ╡	182 ╢	183 ╖	184 ╕	185 ╣	186 ║	187 ╗	188 ╝	189 ╜	190 ╛	191 ┐
C 1 1 0 0		192 └	193 ┴	194 ┬	195 ├	196 ─	197 ┼	198 ╞	199 ╟	200 ╚	201 ╔	202 ╩	203 ╦	204 ╠	205 ═	206 ╬	207 ╧
D 1 1 0 1		208 ╨	209 ╤	210 ╥	211 ╙	212 ╘	213 ╒	214 ╓	215 ╫	216 ╪	217 ┘	218 ┌	219 █	220 ▄	221 ▌	222 ▐	223 ▀
E 1 1 1 0		224 α	225 β	226 Γ	227 π	228 Σ	229 σ	230 µ	231 τ	232 Φ	233 θ	234 Ω	235 δ	236 ∞	237 ø	238 ε	239 ∩
F 1 1 1 1		240 ≡	241 ±	242 ≥	243 ≤	244 ⌠	245 ⌡	246 ÷	247 ≈	248 °	249 •	250 ·	251 √	252 ⁿ	253 ²	254 ■	255 blnk

Table A1.14 ISO 8859 character sets

Standard defining character set	Character set name
ISO 8859-1	Latin-1 (West European)
ISO 8859-2	Latin-2 (East European)
ISO 8859-3	Latin-3 (South European)
ISO 8859-4	Latin-4 (North European)
ISO 8859-5	Cyrillic
ISO 8859-6	Arabic
ISO 8859-7	Greek
ISO 8859-8	Hebrew
ISO 8859-9	Latin-5 (Turkish)
ISO 8859-10	Latin-6 (Nordic)

Table A1.15   HTML character set (based on ISO 8859-1)

HEX CODE YX	X	0	1	2	3	4	5	6	7	8	9	A	B	C	D	E	F
	4	0	0	0	0	0	0	0	0	1	1	1	1	1	1	1	1
BITS 3	3	0	0	0	0	1	1	1	1	0	0	0	0	1	1	1	1
	2	0	0	1	1	0	0	1	1	0	0	1	1	0	0	1	1
Y (8 7 6 5)	1	0	1	0	1	0	1	0	1	0	1	0	1	0	1	0	1
0 (0 0 0 0)		0	1	2	3	4	5	6	7	8	9 HT	10 LF	11	12	13 CR	14	15
1 (0 0 0 1)		16	17	18	19	20	21	22	23	24	25	26	27	28	29	30	31
2 (0 0 1 0)		32 spce	33 !	34 "	35 #	36 $	37 %	38 &	39 '	40 (	41)	42 *	43 +	44 ,	45 -	46 .	47 /
3 (0 0 1 1)		48 0	49 1	50 2	51 3	52 4	53 5	54 6	55 7	56 8	57 9	58 :	59 ;	60 <	61 =	62 >	63 ?
4 (0 1 0 0)		64 @	65 A	66 B	67 C	68 D	69 E	70 F	71 G	72 H	73 I	74 J	75 K	76 L	77 M	78 N	79 O
5 (0 1 0 1)		80 P	81 Q	82 R	83 S	84 T	85 U	86 V	87 W	88 X	89 Y	90 Z	91 [	92 \	93]	94 ^	95 _
6 (0 1 1 0)		96 `	97 a	98 b	99 c	100 d	101 e	102 f	103 g	104 h	105 i	106 j	107 k	108 l	109 m	110 n	111 o
7 (0 1 1 1)		112 p	113 q	114 r	115 s	116 t	117 u	118 v	119 w	120 x	121 y	122 z	123 {	124 ¦	125 }	126 ~	127
8 (1 0 0 0)		128	129	130	131	132	133	134	135	136	137	138	139	140	141	142	143
9 (1 0 0 1)		144	145	146	147	148	149	150	151	152	153	154	155	156	157	158	159
A (1 0 1 0)		160 NBS	161 ¡	162 ¢	163 £	164 ¤	165 ¥	166 ¦	167 §	168 ¨	169 ©	170 ª	171 `	172 not	173 -	174 ®	175 ¯
B (1 0 1 1)		176 °	177 ±	178 ²	179 ³	180 ´	181 µ	182 ¶	183 ·	184 ¸	185 ¹	186 ♂	187 »	188 ¼	189 ½	190 ¾	191 ¿
C (1 1 0 0)		192 À	193 Á	194 Â	195 Ã	196 Ä	197 Å	198 Æ	199 Ç	200 È	201 É	202 Ê	203 Ë	204 Ì	205 Í	206 Î	207 Ï
D (1 1 0 1)		208 Ð	209 Ñ	210 Ò	211 Ó	212 Ô	213 Õ	214 Ö	215 ×	216 Ø	217 Ù	218 Ú	219 Û	220 Ü	221 Ý	222 Þ	223 ß
E (1 1 1 0)		224 à	225 á	226 â	227 ã	228 ä	229 å	230 æ	231 ç	232 è	233 é	234 ê	235 ë	236 ì	237 í	238 î	239 ï
F (1 1 1 1)		240 ð	241 ñ	242 ò	243 ó	244 ô	245 õ	246 ö	247 ÷	248 ø	249 ù	250 ú	251 û	252 ü	253 ý	254 þ	255 ÿ

Key: CR = Carriage Return; HT = Horizontal Tab; LF = Line Feed; NBS = Non-Breaking Space; shaded boxes represent unused code values.

HTML character set (based on ISO 8859-1: Latin-1)

The HTML character set is based upon the ISO 8859-1 character set, as illustrated in Table A1.15.

8-bit extended ASCII (microsoft windows latin-1 code page 1252)

Table A1.16 documents the extended 8-bit version of ASCII (code page 1252) which is the default character set for western versions of Microsoft Windows. It is based on ISO 8859-1 but is not identical.

Unicode character sets (ISO 10646)

Unicode is a 16-bit character set, laid out in ISO 10646. For a full listing and review of the current version of the code refer to the following website:

www.unicode.org

Table A1.16   Microsoft Latin-1 character set (code page 1252)

HEX CODE YX				X	0	1	2	3	4	5	6	7	8	9	A	B	C	D	E	F
	BITS			4	0	0	0	0	0	0	0	0	1	1	1	1	1	1	1	1
				3	0	0	0	0	1	1	1	1	0	0	0	0	1	1	1	1
				2	0	0	1	1	0	0	1	1	0	0	1	1	0	0	1	1
Y 8 7 6			5	1	0	1	0	1	0	1	0	1	0	1	0	1	0	1	0	1
0	0 0 0	0		0	1	2	3	4	5	6	7	8	9 HT	10 LF	11	12	13 CR	14	15
1	0 0 0	1		16	17	18	19	20	21	22	23	24	25	26	27	28	29	30	31
2	0 0 1	0		32 spce	33 !	34 "	35 #	36 $	37 %	38 &	39 '	40 (	41)	42 *	43 +	44 ,	45 -	46 .	47 /
3	0 0 1	1		48 0	49 1	50 2	51 3	52 4	53 5	54 6	55 7	56 8	57 9	58 :	59 ;	60 <	61 =	62 >	63 ?
4	0 1 0	0		64 @	65 A	66 B	67 C	68 D	69 E	70 F	71 G	72 H	73 I	74 J	75 K	76 L	77 M	78 N	79 O
5	0 1 0	1		80 P	81 Q	82 R	83 S	84 T	85 U	86 V	87 W	88 X	89 Y	90 Z	91 [	92 \	93]	94 ^	95 _
6	0 1 1	0		96 `	97 a	98 b	99 c	100 d	101 e	102 f	103 g	104 h	105 i	106 j	107 k	108 l	109 m	110 n	111 o
7	0 1 1	1		112 p	113 q	114 r	115 s	116 t	117 u	118 v	119 w	120 x	121 y	122 z	123 {	124 |	125 }	126 ~	127
8	1 0 0	0		128 €	129	130 ‚	131 ƒ	132 „	133 …	134 †	135 ‡	136 ˆ	137 ‰	138 Š	139 ‹	140 Œ	141	142 Ž	143
9	1 0 0	1		144	145 '	146 '	147 "	148 "	149 •	150 –	151 —	152 ˜	153 TM	154 š	155 ›	156 œ	157	158 ž	159 Ÿ
A	1 0 1	0		160	161 ¡	162 ¢	163 £	164 ¤	165 ¥	166 ¦	167 §	168 ¨	169 ©	170 ª	171 «	172 ¬	173 –	174 ®	175 ¯
B	1 0 1	1		176 °	177 ±	178 ²	179 ³	180 ´	181 µ	182 ¶	183 ·	184 ¸	185 ¹	186 º	187 »	188 ¼	189 ½	190 ¾	191 ¿
C	1 1 0	0		192 À	193 Á	194 Â	195 Ã	196 Ä	197 Å	198 Æ	199 Ç	200 È	201 É	202 Ê	203 Ë	204 Ì	205 Í	206 Î	207 Ï
D	1 1 0	1		208 Ð	209 Ñ	210 Ò	211 Ó	212 Ô	213 Õ	214 Ö	215 ×	216 Ø	217 Ù	218 Ú	219 Û	220 Ü	221 Ý	222 Þ	223 ß
E	1 1 1	0		224 à	225 á	226 â	227 ã	228 ä	229 å	230 æ	231 ç	232 è	233 é	234 ê	235 ë	236 ì	237 í	238 î	239 ï
F	1 1 1	1		240 ð	241 ñ	242 ò	243 ó	244 ô	245 õ	246 ö	247 ÷	248 ø	249 ù	250 ú	251 û	252 ü	253 ý	254 þ	255 ÿ

Key: CR = Carriage Return; HT = Horizontal Tab; LF = Line Feed; NBS = Non-Breaking Space; shaded boxes represent unused code values.

Layer 7 (application layer): MIME (multimedia Internet mail extensions)

Table A1.17 presents the various media types defined by MIME (multimedia Internet mail extension) standards.

Table A1.17   MIME media-types (RFC 2046)

Top-level media type	Media subtypes	Usage or meaning
Text	text/plain; charset=iso-8859-1	A message in plain text encoded using the ISO8859-1 character set.
	text/plain; charset=us-ascii	A message in plain text encoded using the US-ASCII character set (ANSI $\times$ 3.4–1986).
	text/enriched	Rich text format (RFC 1896)
	text/html	A text file coded using hypertext markup language (html) (RFC 2854).

Table A1.17   (*continued*)

Top-level media type	Media subtypes	Usage or meaning
	text/directory	A text file containing directory information.
	text/parityfec	An RTP (real-time application transport protocol) format employing generic forward error correction (fec) (RFC 3009).
Image	image/jpeg	An image file in jpeg (joint photographic experts group) format.
	image/g3fax	An image file in group3 fax format according to CCITT/ITU-T recommendation T.30 (RFC 2159).
	image/gif	An image file in gif-format (graphics interface format).
	image/t38	An image file in ITU-T facsimile format (ITU-T rec. T.38).
	image/tiff	An image file in tiff-format (tag image file format).
Audio	audio/basic	8-bit μ-law pulse code modulation.
	audio/32kadpcm	An attached file in 32 kbit/s adaptive differential pulse code modulation (ITU-T Rec G.726 and RFC 2422).
	audio/L16	An audio-format file encoded according to L16 coding (RFC 1890 and RFC 2586).
	audio/parityfec	An RTP (real-time application transport protocol) format employing generic forward error correction (fec) (RFC 3009).
Video	video/mpeg	An video file in mpeg (motion picture experts group) format.
	video/parityfec	An RTP (real-time application transport protocol) format employing generic forward error correction (fec) (RFC 3009).
Application	application/octet-stream	A file that comprises data of an arbitrary type or otherwise unspecified format.
	application/postscript	A file in the Adobe Systems PostScript format (typically used for printing).
	application/oda	A file in office data architecture (ODA) format (CCITT recommendation T.411 & RFC 2161).
	application/iso-10161-ill-1; transfer encoding...	The carried object is a BER (basic encoding rules) encoded ISO ILL (interlibrary loan) PDU (protocol data unit).
	application/ill-ddi; transfer encoding...	The carried object is a BER (basic encoding rules) encoded ISO ILL (interlibrary loan) PDU (protocol data unit).
	application/parityfec	An RTP (real-time application transport protocol) format employing generic forward error correction (fec) (RFC 3009).
	application/ISUP	The carried object is an ISUP (integrated services user part) message of signalling system number 7 (used in digital telephone networks) (RFC 3204)
	application/QSIG	The carried object is a QSIG signalling message (as used in digital private telephone networks) (RFC 3204).
	application/xhtml+xml	A file in xhtml or xml format (RFC 3236).
	application/dicom	A file in a format specified by DICOM (digital imaging and communications in medicine) (RFC 3240).
Multipart [composite-type]	multipart/mixed	The message body comprises a number of separate 'attachments' of different types, separated by 'boundaries'.

(*continued overleaf*)

Table A1.17   (*continued*)

Top-level media type	Media subtypes	Usage or meaning
	multipart/alternative	The message body comprises a number of versions of the same 'basic content' in different formats. The receiver should decide which one is most appropriate to use.
	multipart/digest	Intended to be used to send collections of messages.
Message [composite-type]	message/rfc822	An encapsulated message in RFC 822 format.
	message/partial	An encapsulated fragment of a message.
	message/external-body	This indicates that the body data are not included in the message but instead only referenced.
	multipart/related	This MIME-type is accompanied by additional information indicating how to unpack or process data (e.g., which program to use) (RFC 2387).
	multipart/voice-message	A file to be used in conjunction with the Voice Profile for Internet Mail (VPIM) (RFC 1911 & RFC 2423).
	multipart/signed	A file made secure for transport by the use of a digital signature (RFC 2480).
	multipart/encrypted	A file made secure for transport by encryption transformation (RFC 2480).

Appendix 2

Internet Top-Level Domains (TLDs) and Generic Domains

Domain	Domain usage	Domain operator/authority
.aero	aeronautical and air-transport industry	Société Internationale de Télécommunications Aéronautiques (SITA)
.biz	restricted to businesses	NeuLevel, Inc
.com	commercial organisations	VeriSign Global Registry Services
.coop	reserved for cooperative associations	Dot Cooperation LLC
.edu	reserved for higher educational institutions	Educause
.gov	reserved for government use (the top-level domain is the US government)	United States General Services Administration
.info	information domains	Afilias Limited
.int	used only for international organisations established by international government treaties	IANA. int Domain Registry
.mil	reserved exclusively for the US military	United States Department of Defense Network Information Centre
.museum	reserved for museums	Museum Domain Management Association
.name	reserved for individuals	Global name registry
.net	network organisations	VeriSign Global Registry Services
.org	organisations	VeriSign Global Registry Services

Data Networks, IP and the Internet: Protocols, Design and Operation   Martin P. Clark
© 2003 John Wiley & Sons, Ltd   ISBN: 0-470-84856-1

Appendix 3

Internet Country Code Top-Level Domains (ccTLDs — ISO 3166-1)

Code	Country
.ac	Ascension Island
.ad	Andorra
.ae	United Arab Emirates
.af	Afghanistan
.ag	Antigua and Barbuda
.ai	Anguilla
.al	Albania
.am	Armenia
.an	Netherlands Antilles
.ao	Angola
.aq	Antarctica
.ar	Argentina
.as	American Samoa
.at	Austria
.au	Australia
.aw	Aruba
.az	Azerbaijan
.ba	Bosnia and Herzegovina
.bb	Barbados
.bd	Bangladesh
.be	Belgium
.bf	Burkina Faso
.bg	Bulgaria
.bh	Bahrain
.bi	Burundi
.bj	Benin
.bm	Bermuda
.bn	Brunei Darussalam
.bo	Bolivia

Code	Country
.br	Brazil
.bs	Bahamas
.bt	Bhutan
.bv	Bouvet Island
.bw	Botswana
.by	Belarus
.bz	Belize
.ca	Canada
.cc	Cocos (Keeling) Islands
.cd	Democratic Republic of the Congo
.cf	Central African Republic
.cg	Republic of Congo
.ch	Switzerland
.ci	Cote d'Ivoire
.ck	Cook Islands
.cl	Chile
.cm	Cameroon
.cn	China
.co	Colombia
.cr	Costa Rica
.cu	Cuba
.cv	Cap Verde
.cx	Christmas Island
.cy	Cyprus
.cz	Czech Republic
.de	Germany
.dj	Djibouti
.dk	Denmark
.dm	Dominica

Code	Country	Code	Country
.dz	Algeria	.it	Italy
.ec	Ecuador	.je	Jersey
.ee	Estonia	.jm	Jamaica
.eg	Egypt	.jo	Jordan
.eh	Western Sahara	.jp	Japan
.er	Eritrea	.ke	Kenya
.es	Spain	.kg	Kyrgyzstan
.et	Ethiopia	.kh	Cambodia
.fi	Finland	.ki	Kiribati
.fj	Fiji	.km	Comoros
.fk	Falkland Islands	.kn	Saint Kitts and Nevis
.fm	Federal State of Micronesia	.kp	Democratic People's Republic of Korea
.fo	Faroe Islands		
.fr	France	.kr	Republic of Korea
.ga	Gabon	.kw	Kuwait
.gd	Grenada	.ky	Cayman Islands
.ge	Georgia	.kz	Kazakhstan
.gf	French Guiana	.la	Lao People's Democratic Republic
.gg	Guernsey	.lb	Lebanon
.gh	Ghana	.lc	Saint Lucia
.gi	Gibraltar	.li	Liechtenstein
.gl	Greenland	.lk	Sri Lanka
.gm	Gambia	.lr	Liberia
.gn	Guinea	.ls	Lesotho
.gp	Guadeloupe	.lt	Lithuania
.gq	Equatorial Guinea	.lu	Luxembourg
.gr	Greece	.lv	Latvia
.gs	South Georgia and the South Sandwich Islands	.ly	Libyan Arab Jamahiriya
		.ma	Morocco
.gt	Guatemala	.mc	Monaco
.gu	Guam	.md	Republic of Moldova
.gw	Guinea-Bissau	.mg	Madagascar
.gy	Guyana	.mh	Marshall Islands
.hk	Hong Kong	.mk	Macedonia
.hm	Heard and McDonald Islands	.ml	Mali
.hn	Honduras	.mm	Myanmar
.hr	Croatia	.mn	Mongolia
.ht	Haiti	.mo	Macau
.hu	Hungary	.mp	Northern Mariana Islands
.id	Indonesia	.mq	Martinique
.ie	Ireland	.mr	Mauritania
.il	Israel	.ms	Montserrat
.im	Isle of Man	.mt	Malta
.in	India	.mu	Mauritius
.io	British Indian Ocean Territory	.mv	Maldives
.iq	Iraq	.mw	Malawi
.ir	Iran	.mx	Mexico
.is	Iceland	.my	Malaysia

Code	Country	Code	Country
.mz	Mozambique	.sn	Senegal
.na	Namibia	.so	Somalia
.nc	New Caledonia	.sr	Suriname
.ne	Niger	.st	Sao Tome and Principe
.nf	Norfolk Island	.sv	El Salvador
.ng	Nigeria	.sy	Syrian Arab Republic
.ni	Nicaragua	.sz	Swaziland
.nl	Netherlands	.tc	Turks and Caicos Islands
.no	Norway	.td	Chad
.np	Nepal	.tf	French Southern Territories
.nr	Nauru	.tg	Togo
.nu	Niue	.th	Thailand
.nz	New Zealand	.tj	Tajikistan
.om	Oman	.tk	Tokelau
.pa	Panama	.tm	Turkmenistan
.pe	Peru	.tn	Tunisia
.pf	French Polynesia	.to	Tonga
.pg	Papua New Guinea	.tp	East Timor
.ph	Philippines	.tr	Turkey
.pk	Pakistan	.tt	Trinidad and Tobago
.pl	Poland	.tv	Tuvalu
.pm	Saint Pierre and Miquelon	.tw	Taiwan
.pn	Pitcairn Island	.tz	Tanzania
.pr	Puerto Rico	.ua	Ukraine
.ps	Palestinian Territories	.ug	Uganda
.pt	Portugal	.uk	United Kingdom
.pw	Palau	.um	US minor outlying islands
.py	Paraguay	.us	United States
.qa	Qatar	.uy	Uruguay
.re	Reunion Island	.uz	Uzbekistan
.ro	Romania	.va	Vatican state
.ru	Russian Federation	.vc	Saint Vincent and the Grenadines
.rw	Rwanda	.ve	Venezuela
.sa	Saudi Arabia	.vg	Virgin Islands (British)
.sb	Solomon Islands	.vi	Virgin Islands (USA)
.sc	Seychelles	.vn	Vietnam
.sd	Sudan	.vu	Vanuatu
.se	Sweden	.wf	Wallis and Futuna Islands
.sg	Singapore	.ws	Western Samoa
.sh	Saint Helena	.ye	Yemen
.si	Slovenia	.yt	Mayotte
.sj	Svalbard and Jan Mayen Islands	.yu	Yugoslavia
.sk	Slovak Republic	.za	South Africa
.sl	Sierra Leone	.zm	Zambia
.sm	San Marino	.zw	Zimbabwe

Appendix 4

Internet Engineering Task Force (IETF) Request for Comment (RFC) Listing

This is a listing of all the RFC documents (as issued by the Internet Engineering Task Force (IETF)) which are referred to in this book. For those who need a full and up-to-date listing, you should refer to either www.rfc-editor.org or www.faqs.org/rfcs/.

This appendix is intended to be used for tracing the subject of a given RFC document number, its date of issue and a note about whether it has subsequently been superseded by any later standards. The RFCs are listed here in number order (this is roughly the order of issue). But for looking up the relevant RFC documents related to a given service or protocol, it will be more efficient to use the Abbreviations appendix of this book, where the protocols are listed in alphabetical order and cross-related to their defining standards.

It is important to recognise that not all RFCs are technical standards. IETF nowadays classifies the various documents into *informational* (or *FYI*), *best current practice (BCP)*, *standard (STD)*, *request for comment (RFC)* and *experimental*, but in the past the documents issued with RFC numbers have served a plethora of different purposes as:

- announcements and agreements;
- background papers;
- experimental reports;
- handbooks;
- instructions to RFC authors;
- introductory and educational texts;
- invites to meetings;
- listing concurrently used protocols;
- objective definition of project teams and working groups;

Data Networks, IP and the Internet: Protocols, Design and Operation   Martin P. Clark
© 2003 John Wiley & Sons, Ltd   ISBN: 0-470-84856-1

- policy papers;

- problem description documents;

- questionnaires;

- statements of policy;

- status reports;

- studies and reports;

- tributes to contributors; and

- white papers.

RFC documents are never re-issued or updated. A new document is issued, with a new RFC number and the old RFC becomes obsolete.

The 'academic community' nature of the original RFCs is also reflected by the 'noticeboard' nature of many of the early RFCs (e.g. 'the service will be closed over the Thanksgiving weekend', etc.) and by the annual tradition of bogus proposals dated 1st April.

The authors of some of the most well-known RFCs are nowadays well-known Internet personalities:

- Tim Berners-Lee (author of HTML, HTTP and Worldwide Web RFCs);

- Vinton Cerf (author of many of the original ARPANET and protocol specifications);

- Jon Postel (RFC editor and IETF leader until his death in October 1998).

RFC number	Title	Date of issue	Remarks
0008	ARPA Network; functional specifications	05 May 69	ARPANET
0020	ASCII format for network interchange	16 Oct 69	
0761	DoD standard Transmission Control Protocol (TCP)	01 Jan 80	
0764	Telnet Protocol specification	01 Jun 80	Replaced by RFC 0854
0765	File Transfer Protocol (FTP) specification	01 Jun 80	Replaced by RFC 0959
0768	User Datagram Protocol (UDP)	28 Aug 80	
0783	TFTP (Trivial File Transfer Protocol) revision 2	01 Jun 81	Replaced by RFC 1350
0788	Simple Mail Transfer Protocol (SMTP)	01 Nov 81	Replaced by RFC 0821
0791	Internet Protocol (IP)	01 Sep 81	Standard
0792	Internet Control Message Protocol (ICMP)	01 Sep 81	Updated by RFC 0950
0793	Transmission Control Protocol (TCP)	01 Sep 81	Standard
0799	Internet name domains	01 Sep 81	
0812	NICNAME / WHOIS	01 Mar 82	Replaced by RFC 0954
0813	Window & acknowledgement strategy in TCP	01 Jul 82	
0819	Domain naming convention	01 Aug 82	
0821	Simple Mail Transfer Protocol (SMTP)	01 Aug 82	Replaced by RFC 2821

RFC number	Title	Date of issue	Remarks
0822	Standard format for ARPA Internet text messages	13 Aug 82	Replaced by RFC 2822
0826	Ethernet Address Resolution Protocol (ARP)	01 Nov 82	Standard
0854	Telnet protocol specification	01 May 83	
0877	Standard for IP datagrams over public data networks	01 Sep 83	Replaced by RFC 1356
0878	ARPANET 1822L Host Access Protocol	01 Dec 83	
0879	TCP maximum segment size and related topics	01 Nov 83	
0882	Domain names: concepts and facilities	01 Nov 83	Replaced by RFC 1034
0883	Domain names: implementation specification	01 Nov 83	Replaced by RFC 1034
0903	Reverse Address Resolution Protocol (RARP)	01 Jun 84	Standard
0904	Exterior Gateway Protocol formal specification	01 Apr 84	Updates RFCs 827, 888
0905	ISO transport protocol specification ISO DP 8073	01 Apr 84	
0926	ISO protocol: connectionless mode network service	01 Dec 84	Replaced by RFC 0994
0927	TACACS user identification Telnet option	01 Dec 84	Proposed standard
0951	Bootstrap protocol	01 Sep 85	Updated by later RFCs
0954	NICNAME/WHOIS	01 Oct 85	Draft standard
0958	Network Time Protocol (NTP)	01 Sep 85	Replaced by RFC 1059
0959	File Transfer Protocol (FTP)	01 Oct 85	
0988	Host extensions for IP multicasting	01 Jul 86	Replaced by RFC 1054
1013	X-Window systems Protocol, version 11	01 Jun 87	
1014	XDR: external data representation standard	01 Jun 87	
1034	Domain names — concepts and facilities	01 Nov 87	Updated by later RFCs
1035	Domain names — implementation and specification	01 Nov 87	Updated by later RFCs
1042	Standard for IP datagrams over IEEE 802 networks	01 Feb 88	
1054	Host extensions for IP multicasting	01 May 88	Replaced by RFC 1112
1055	Serial Line Internet Protocol (SLIP)	01 Jun 88	Standard
1057	RPC2: remote procedure call protocol version 2	01 Jun 88	
1058	Routing Information Protocol (RIP)	01 Jun 88	
1059	Network Time Protocol (NTP) version 1	01 Jul 88	
1063	IP MTU discovery options	01 Jul 88	

RFC number	Title	Date of issue	Remarks
1066	Management Information Base (MIB): TCP/IP nets	01 Aug 88	Replaced by RFC 1156
1067	Simple Network Management Protocol (SNMP)	01 Aug 88	Replaced by RFC 1098
1075	Distance Vector Multicast Routing Protocol	01 Nov 88	
1081	Post Office Protocol version 3 (POP3)	01 Nov 88	Replaced by RFC 1225
1082	Post Office Protocol version 3 (POP3): extended	01 Nov 88	
1084	BOOTP vendor information extensions	01 Dec 88	
1085	ISO presentation services on top of TCP/IP	01 Dec 88	
1098	Simple Network Management Protocol (SNMP)	April 89	Replaced by RFC 1157
1112	Host extensions for IP multicasting	01 Apr 89	
1119	Network Time Protocol version 2 (NTP2)	01 Sep 89	Replaced by RFC 1305
1122	Requirements for Internet Hosts	01 Oct 89	
1123	Requirements for Internet Hosts:Application & support	01 Oct 89	
1131	OSPF (open shortest path first) specification	01 Oct 89	
1146	TCP alternate checksum options	01 Mar 90	
1155	Structure & identification of mgmt info for TCP/IP	01 May 90	Standard
1156	MIB for TCP/IP networks	01 May 90	Standard
1157	Simple Network Management Protocol (SNMP)	01 May 90	Standard
1158	MIB-II for TCP/IP networks	01 May 90	Replaced by RFC 1213
1191	Path MTU discovery	01 Nov 90	Draft standard
1212	Concise MIB definitions	01 Mar 91	
1213	MIB-II for TCP/IP-based Internets	01 Mar 91	
1229	Extensions to the generic interface MIB	01 May 91	Updated by RFC 1239
1239	Reassignment of experimental MIBs to standards	01 Jun 91	
1247	OSPF version 2	01 Jul 91	Replaced by RFC 1583
1248	OSPF version 2 MIB	01 Jul 91	Replaced by RFC 1252
1252	OSPF version 2 MIB	01 Aug 91	Replaced by RFC 1253
1253	OSPF version 2 MIB	01 Aug 91	Replaced by RFC 1850
1256	ICMP router discovery messages	01 Sep 91	Proposed standard
1282	BSD Rlogin	Dec 91	Replaces RFC 1258
1286	Managed objects for bridges	Dec 91	Proposed standard
1305	Network Time Protocol version 3 (NTP3)	Mar 92	Draft standard
1319	MD2 message digest algorithm	Apr 92	
1320	MD4 message digest algorithm	Apr 92	

RFC number	Title	Date of issue	Remarks
1321	MD5 message digest algorithm	Apr 92	
1323	TCP extensions for high performance	May 92	Proposed standard
1331	PPP for multi-protocol datagrams over PP links	May 92	Replaced by RFC 1548
1332	PPP Internet protocol control protocol (IPCP)	May 92	Proposed standard
1333	PPP link quality monitoring	May 92	Replaced by RFC 1989
1334	PPP authentication protocols	Oct 92	Replaced by RFC 1994
1341	MIME (Multipurpose Internet Mail Extensions)	Jun 92	Replaced by RFC 1521
1350	TFTP protocol revision 2 (TFTP2)	Jul 92	Updated by RFC 1782
1351	SNMP administrative model	Jul 92	
1352	SNMP security protocols	Jul 92	
1353	Managed objects for SNMP	Jul 92	
1354	IP forwarding table MIB	Jul 92	Replaced by RFC 2096
1356	Multiprot. Interconnect: X.25 & ISDN packet-mode	Aug 92	Draft standard
1377	PPP OSI network layer control protocol (OSINLCP)	Nov 92	Proposed standard
1378	PPP Appletalk Control Protocol (ATCP)	Nov 92	Proposed standard
1389	RIP version 2 MIB extensions	Jan 93	Replaced by RFC 1724
1390	Transmission of IP and ARP over FDDI networks	Jan 93	Standard
1414	Identification MIB	Feb 93	Proposed standard
1441	Intro to Internet Network Management Framework v2	Apr 93	
1442	Structure of MIB for SNMPv2	Apr 93	Replaced by RFC 1902
1443	Textual conventions for SNMPv2	Apr 93	Replaced by RFC 1903
1444	Conformance statements for SNMPv2	Apr 93	Replaced by RFC 1904
1445	Administrative model for SNMPv2	Apr 93	
1446	Security protocols for SNMPv2	Apr 93	
1447	Party MIB for SNMPv2	Apr 93	
1448	Protocol operations for SNMPv2	Apr 93	Replaced by RFC 1905
1449	Transport mappings for SNMPv2	Apr 93	Replaced by RFC 1906
1450	MIB for SNMPv2	Apr 93	Replaced by RFC 1907
1451	Manager-to-manager MIB	Apr 93	
1452	Coexistence between v1 and v2 of Internet NMF	Apr 93	Replaced by RFC 1908
1460	Post Office Protocol version 3 (POP3)	Jun 93	Replaced by RFC 1725
1466	Guidelines for the Management of IP address space	May 93	
1492	Access Control Protocol TACACS	Jul 93	
1514	Host resources MIB	Sep 93	Replaced by RFC 2790
1534	Interoperation between DHCP and BOOTP	Oct 93	Draft standard
1552	PPP internetworking packet exch.cont.prot (IPXCP)	Dec 93	Proposed standard

RFC number	Title	Date of issue	Remarks
1567	X.500 Directory monitoring MIB	Jan 94	Replaced by RFC 2605
1570	PPP LCP extensions	Jan 94	Updated by RFC 2484
1573	Evolution of the interfaces group of MIB-II	Jan 94	Replaced by RFC 2233
1583	OSPF version 2 (OSPF2)	Mar 94	Replaced by RFC 2178
1584	Multicast extensions to OSPF (MOSPF)	Mar 94	Proposed standard
1587	The OSPF NSSA option	Mar 94	Proposed standard
1591	Domain Name System (DNS) structure& delegation	Mar 94	
1611	DNS server MIB extensions	May 94	Proposed standard
1618	PPP over ISDN	May 94	Proposed standard
1619	PPP over SONET/SDH	May 94	Replaced by RFC 2615
1626	Default IP MTU for use over ATM AAL5	May 94	Replaced by RFC 2225
1628	UPS MIB	May 94	Proposed standard
1631	IP Network Address Translator (NAT)	May 94	Replaced by RFC 3022
1633	Integrated Services in the Internet Architecture	Jun 94	
1634	Novell IPX over various WAN media (IPXWAN)	May 94	
1654	Border Gateway Protocol 4 (BGP-4)	Jul 94	Replaced by RFC 1771
1655	Application of BGP in the Internet	Jul 94	Replaced by RFC 1772
1656	BGP-4 document roadmap and implementation	Jul 94	Replaced by RFC 1773
1657	Managed objects for BGP-4 using SMIv2	Jul 94	Draft standard
1661	Point-to-Point Protocol (PPP)	Jul 94	Updated by RFC 2153
1662	PPP in HDLC-like framing	Jul 94	Standard
1663	PPP Reliable Transmission	Jul 94	Proposed standard
1692	Transport Multiplexing Protocol (TMux)	Aug 94	Proposed standard
1693	An extension to TCP: Partial Order Service	Nov 94	
1694	Managed objects for SMDS interfaces using SMIv2	Aug 94	Draft standard
1695	Managed objects for ATM management v8.0/SMIv2	Aug 94	Replaced by RFC 2515
1696	Modem management MIB using SMIv2	Aug 94	Proposed standard
1697	Rel. Database Mgm Sys (RDBMS) MIB / SMIv2	Aug 94	Proposed standard
1701	Generic Routing Encapsulation (GRE)	Oct 94	
1702	Generic Routing Encapsulation over IPv4 networks	Oct 94	
1717	PPP Multilink Protocol (MP)	Nov 94	Replaced by RFC 1900
1721	RIP version 2 Protocol Analysis	Nov 94	

RFC number	Title	Date of issue	Remarks
1722	RIP version 2 Protocol Applicability Statement	Nov 94	Standard
1723	RIP version 2–carrying additional information	Nov 94	Replaced by RFC 2453
1724	RIP version 2 MIB extension	Nov 94	Draft standard
1725	Post Office Protocol version 3 (POP3)	Nov 94	Replaced by RFC 1939
1730	Internet Message Access Protocol v4 (IMAP4)	Dec 94	Replaced by RFC 2060
1731	IMAP4 authentication mechanisms	Dec 94	Proposed standard
1732	IMAP4 compatibility with IMAP2 and IMAP2bis	Dec 94	
1733	Distributed Email models in IMAP4	Dec 94	
1734	POP3 AUTHentication command	Dec 94	Proposed standard
1735	NBMA Address Resolution Protocol (NARP)	Dec 94	
1738	Uniform Resource Locators (URL)	Dec 94	Updated by later RFCs
1752	Recommendation for the IP next generation protocol	Jan 95	Proposed standard
1755	ATM signalling support for IP over ATM	Feb 95	Proposed standard
1757	Remote Network Monitoring (RMON) MIB	Feb 95	Replaced by RFC 2819
1759	Printer MIB	Mar 95	Proposed standard
1760	The S/KEY One-Time Password system	Feb 95	
1762	PPP DECnet Phase IV control protocol (DNCP)	Mar 95	Draft standard
1766	Tags for the identification of languages	Mar 95	Replaced by RFC 3066
1771	Border Gateway Protocol 4 (BGP-4)	Mar 95	Draft standard
1772	Application of BGP4 in the Internet	Mar 95	Draft standard
1773	Experience with the BGP-4 protocol	Mar 95	
1774	BGP-4 Protocol Analysis	Mar 95	
1777	Lightweight Directory Access Protocol (LDAP)	Mar 95	Draft standard
1795	Data Link Switching (DLSw):switch/switch protocol	Apr 95	
1813	NFS (Network File System) v3 protocol spec.	Jun 95	Replaced by RFC 3010
1815	Character sets ISO-10646 and ISO-10646-J1	Jul 95	
1817	CIDR and Classful Routing	Aug 95	
1831	RPC: Remote Procedure Call protocol spec v2	Aug 95	Proposed standard
1832	XDR: External Data Representation standard	Aug 95	Draft standard
1844	Multimedia Email (MIME) user agent checklist	Aug 95	
1850	OSPF v2 MIB	Nov 95	Draft standard

RFC number	Title	Date of issue	Remarks
1851	ESP triple DES (Defence Encryption std) transform	Sep 95	
1852	IP authentication using Keyed SHA	Sep 95	Replaced by RFC 2841
1853	IP in IP Tunneling	Oct 95	
1864	The content-MD5 header field	Oct 95	Draft standard
1866	Hypertext Markup Language 2 (html 2.0)	Nov 95	Replaced by RFC 2854
1868	ARP extension—UNARP	Nov 95	
1869	SMTP service extensions	Nov 95	Replaced by RFC 2821
1878	Variable length subnet table for IPv4	Dec 95	
1881	IPv6 address allocation management	Dec 95	
1883	Internet Protocol version 6 (IPv6) specification	Dec 95	Replaced by RFC 2460
1884	IP version 6 addressing architecture	Dec 95	Replaced by RFC 2373
1885	Internet Control Message Protocol (ICMPv6)	Dec 95	Replaced by RFC 2463
1886	DNS extensions for support IP version 6	Dec 95	Proposed standard
1887	An architecture for IPv6 unicast address allocation	Dec 95	
1888	OSI NSAPs and IPv6	Aug 96	
1889	RTP Real-Time application Transport Protocol	Jan 96	Proposed standard
1890	RTP profile for audio & video conferences	Jan 96	Proposed standard
1896	Text/enriched MIME content-type	Feb 96	Replaces 1523, 1563
1901	Întroduction to community-based SNMPv2	Jan 96	
1902	Structure of Management Information for SNMPv2	Jan 96	Replaced by RFC 2578
1903	Textual conventions for SNMPv2	Jan 96	Replaced by RFC 2579
1904	Conformance statements for SNMPv2	Jan 96	Replaced by RFC 2580
1905	Protocol operations for SNMPv2	Jan 96	Draft standard
1906	Transport mappings for SNMPv2	Jan 96	Draft standard
1907	MIB for SNMPv2	Jan 96	Draft standard
1908	Coexistence between version 1 & version 2: SNMP	Jan 96	Replaced by RFC 2576
1909	Administrative infrastructure for SNMPv2	Feb 96	
1910	User-based security model for SNMPv2	Feb 96	
1911	Voice profile for Internet Mail	Feb 96	Replaced by RFC 2421–3
1918	Address Allocation for Private Internets	Feb 96	Best current practice
1928	SOCKS protocol version 5	Mar 96	Proposed standard
1950	ZLIB compressed data format specification v3.3	May 96	

RFC number	Title	Date of issue	Remarks
1951	DEFLATE compressed data format spec v1.3	May 96	
1952	GZIP file format specification v4.3	May 96	
1961	GSS-API authentication method for SOCKS V5	Jun 96	Proposed standard
1962	PPP Compression Control Protocol (CCP)	Jun 96	Updated by RFC 2153
1963	PPP serial data transport protocol (SDTP)	Aug 96	
1964	Kerberos Version 5 GSS-API mechanism	Jun 96	
1965	Autonomous System (AS) confederations for BGP	Jun 96	Replaced by RFC 3065
1966	BGP Route Reflection:alternative to full mesh IBGP	Jun 96	Updated by RFC 2796
1967	PPP LZS-DCP Compression Protocol (LZS-DCP)	Aug 96	
1968	PPP Encryption Control Protocol (ECP)	Jun 96	Proposed standard
1969	PPP DES Encryption Control Protocol (DESE)	Jun 96	Replaced yb RFC 2419
1970	Neighbour discovery for IP version 6	Aug 96	Replaced by RFC 2461
1971	Ipv6 stateless address autoconfiguration	Aug 96	Replaced by RFC 2462
1981	Path MTU discovery for IPv6	Aug 96	Proposed standard
1989	PPP link quality monitoring	Aug 96	Draft standard
1990	PPP Multilink Protocol (MP)	Aug 96	Proposed standard
1991	PGP (Pretty Good Privacy) message exchange format	Aug 96	
1992	Nimrod Routing Architecture	Aug 96	
1994	PPP Challenge Handshake Authentication P.(CHAP)	Aug 96	Updated by RFC 2484
1996	Prompt notification of zone changes (DNS NOTIFY)	Aug 96	Proposed standard
1997	BGP Communities Attribute	Aug 96	Proposed standard
2002	IP Mobility support	Oct 96	Replaced by RFC 3220
2003	Encapsulation with IP	Oct 96	Proposed standard
2004	Minimal encapsulation within IP	Oct 96	Proposed standard
2005	Applicability statement for IP mobility support	Oct 96	Proposed standard
2006	Managed objects for IP mobility support with SMIv2	Oct 96	Proposed standard
2011	SNMPv2 MIB for Internet Protocol using SMIv2	Nov 96	Updates RFC 1213
2012	SNMPv2 MIB for TCP using SMIv2	Nov 96	Updates RFC 1213
2013	SNMPv2 MIB for UDP using SMIv2	Nov 96	Updates RFC 1213
2015	MIME security with Pretty Good Privacy (PGP)	Oct 96	Updated by RFC 3156

RFC number	Title	Date of issue	Remarks
2016	Uniform Resource Agents (URAs)	Oct 96	
2018	TCP selective acknowledgement options	Oct 96	Replaces RFC 1072
2020	IEEE 802.12 interface MIB	Oct 96	Proposed standard
2021	Remote Network Monitoring(RMON) MIB2/SMIv2	Jan 97	Proposed standard
2024	Managed objects for Data Link Switching /SMIv2	Oct 96	Proposed standard
2032	RTP payload format for H.261 video streams	Oct 96	Proposed standard
2037	Entity MIB using SMIv2	Oct 96	Replaced by RFC 2737
2043	PPP SNA Control Protocol (SNACP)	Oct 96	Proposed standard
2044	UTF-8, transformation Unicode/ISO 10646	Oct 96	Replaced by RFC 2279
2045	MIME Part 1: format of Internet message bodies	Nov 96	Updated by 2184, 2231
2046	MIME Part 2: Media Types	Nov 96	Updated RFC 2646
2047	MIME Part 3: Message headers for non-ASCII text	Nov 96	Updated by 2184, 2231
2048	MIME Part 4: Registration procedures	Nov 96	Updated by RFC 3023
2049	MIME Part 5: Conformance criteria & examples	Nov 96	Replaces 1521–2, 1590
2058	Remote Authentication Dial In User Service(RADIUS)	Jan 97	Replaced by RFC 2138
2059	RADIUS accounting	Jan 97	Replaced by RFC 2139
2060	Internet Message Access Protocol (IMAP4rev1)	Dec 96	Replaces RFC 1730
2064	Traffic Flow Measurement: Meter MIB	Jan 97	Replaced by RFC 2720
2097	PPP NetBIOS Frames Control Protocol (NBFCP)	Jan 97	Proposed standard
2104	HMAC: keyed hashing for message authentication	Feb 97	
2107	Ascend Tunnel Management Protocol (ATMP)	Feb 97	
2125	PPP Bandwidth Allocation (BAP)& Control P(BACP)	Mar 97	Proposed standard
2131	Dynamic Host Configuration Protocol (DHCP)	Mar 97	Replaces RFC 1541
2135	Internet Society By-Lays	Apr 97	
2136	Dynamic updates in the DNS (DNS UPDATE)	Apr 97	Updated by RFC 3007
2137	Secure Domain Name System update	Apr 97	Updates RFC 1035
2138	Remote Authentication Dial-In User Service(RADIUS)	Apr 97	Replaced by RFC 2865
2139	RADIUS accounting	Apr 97	Replaced by RFC 2866
2140	TCP Control Block Independence	Apr 97	
2143	Encapsulating IP with SCSI (small computer sys I/F)	May 97	

RFC number	Title	Date of issue	Remarks
2144	CAST-128 Encryption Algorithm	May 97	
2147	TCP & UDP over IPv6 Jumbograms	May 97	
2153	PPP vendor extensions	May 97	Updates 1661, 1962
2159	MIME body part for FAX	Jan 98	Proposed standard
2161	MIME body part for ODA	Jan 98	
2165	Service Location Protocol	Jun 97	Updated by 2608, 2609
2189	Core Based Trees (CBT version 2) multicast routing	Sep 97	
2190	RTP payload format for H.263 video streams	Sep 97	Proposed standard
2201	Core Based Trees (CBT) Multicast routing architect.	Sep 97	
2205	Resource ReSerVation Protocol (RSVP) version 1	Sep 97	Updated by RFC 2750
2206	RSVP MIB using SMIv2	Sep 97	Proposed standard
2211	Controlled-Load Network Element Service — spec	Sep 97	Proposed standard
2212	Guaranteed Quality of Service - specification	Sep 97	Proposed standard
2213	Integrated Services MIB using SMIv2	Sep 97	Proposed standard
2222	Simple Authentication and Security Layer (SASL)	Oct 97	Updated by RFC 2444
2225	Classical IP and ARP over ATM	Apr 98	Replaces 1626, 1577
2226	IP broadcast over ATM networks	Oct 97	Proposed standard
2233	The Interfaces Group MIB using SMIv2	Nov 97	Replaced by RFC 2863
2234	Augmented BNF for syntax specifications: ABNF	Nov 97	Proposed standard
2236	Internet Group Management Protocol (IGMP) v2	Nov 97	Updates RFC 1112
2245	Anonymous SASL Mechanism	Nov 97	Proposed standard
2246	TLS Protocol version 1.0	Jan 99	Proposed standard
2248	Network Services Monitoring MIB	Jan 98	Replaced by RFC 2788
2249	Mail Monitoring MIB	Jan 98	Replaced by RFC 2789
2250	RTP payload format for MPEG1 / MPEG2 video	Jan 98	Replaces RFC 2038
2251	Lightweight Directory Access Protocol (LDAP) v3	Dec 97	Proposed standard
2252	LDAP v3: attribute syntax definitions	Dec 97	Proposed standard
2261	An architecture SNMPv3 management frameworks	Jan 98	Replaced by RFC 2271
2262	SNMP: message processing & dispatching	Jan 98	Replaced by RFC 2272
2263	SNMPv3 Applications	Jan 98	Replaced by RFC 2273
2264	User-based Security Model (USM) for SNMPv3	Jan 98	Replaced by RFC 2274
2265	View-based Access Control Model (VACM) /SNMP	Jan 98	Replaced by RFC 2275

RFC number	Title	Date of issue	Remarks
2266	Managed objects for IEEE 802.12 repeaters	Jan 98	Proposed standard
2267	Network Ingress Filtering: DOS attacks / spoofing	Jan 98	Replaced by RFC 2827
2268	RC2(r) encryption algorithm	Mar 98	
2279	UTF-8: transformation format of ISO 10646	Jan 97	Replaces RFC 2044
2284	PPP extensible authentication protocol (EAP)	Mar 98	Updated by RFC 2484
2287	Managed objects for Applications: System Level	Feb 98	Proposed standard
2311	S/MIME version 2 message specification	Mar 98	
2338	Virtual Router Redundancy Protocol (VRRP)	Apr 98	Proposed standard
2341	Cisco Layer 2 Forwarding Protocol L2F	May 98	
2344	Reverse Tunneling Mode for IP	May 98	Replaced by RFC 3024
2362	Protocol Independ. Multicast-Sparse Mode(PIM-SM)	Jun 98	Replaces RFC 2117
2373	IP version 6 addressing architecture	Jul 98	Replaces RFC 1884
2385	Protection of BGP sessions via TCP MD5 signature	Aug 98	Proposed standard
2386	Framework: QOS-based routing in the Internet	Aug 98	
2387	MIME Multipart/Related content type	Aug 98	Replaces RFC 2112
2390	Inverse Address Resolution Protocol (inARP)	Sep 98	Replaces RFC 1293
2393	IP Payload Compression Protocol (IPComp)	Dec 98	Replaced by RFC 3173
2394	IP Payload Compression using DEFLATE	Dec 98	
2395	IP Payload Compression using LZS	Dec 98	
2396	Uniform Resource Identifiers (URI): General syntax	Aug 98	Updates 1808, 1738
2401	Security Architecture for the Internet Protocol	Nov 98	Replaces RFC 1825
2402	IP Authentication Header	Nov 98	Replaces RFC 1826
2406	IP Encapsulating Security Payload (ESP)	Nov 98	Replaces RFC 1827
2414	Increasing TCP's Initial Window	Sep 98	
2417	Managed objects for ATM Multicast UNI3.0/3.1	Sep 98	Proposed standard
2419	PPP DES Encryption Protocol v2 (DESE-bis)	Sep 98	Replaces RFC 1969
2420	PPP triple-DES Encryption Protocol (3DESE)	Sep 98	Proposed standard

RFC number	Title	Date of issue	Remarks
2422	Toll Quality Voice–32 kb/s ADPCM MIME registr.	Sep 98	Replaces RFC 1911
2423	VPIM Voice Message MIME sub-type registration	Sep 98	Replaces RFC 1911
2424	Content Duration MIME Header Definition	Sep 98	Proposed standard
2428	FTP extensions for IPv6 and NATs	Sep 98	Proposed standard
2435	RTP Payload Format for JPEG-compressed video	Oct 98	Replaces RFC 2035
2437	PKCS #1: RSA cryptography spec version 2.0	Oct 98	Replaces RFC 2313
2439	BGP Route Flap Damping	Nov 98	Proposed standard
2452	IP version 6 MIB for Transmission Control Protocol	Dec 98	Proposed standard
2453	RIP version 2	Nov 98	Replaces RFC 1723
2454	IP version 6 MIB for User Datagram Protocol	Dec 98	Proposed standard
2459	Internet X.509 PKI certificate and CRL profile	Jan 99	Proposed standard
2460	Internet Protocol version 6 (IPv6) specification	Dec 98	Replaces RFC 1883
2461	Neighbour discovery for IP version 6 (IPv6)	Dec 98	Replaces RFC 1970
2462	IPv6 stateless address autoconfiguration	Dec 98	Replaces RFC 1971
2463	Internet Control Message Protocol (ICMPv6)	Dec 98	Replaces RFC 1885
2464	Transmission IPv6 packets over ethernet networks	Dec 98	Replaces RFC 1972
2465	MIB for IP version 6: textual conventions & general	Dec 98	Proposed standard
2466	MIB for IP version 6: ICMPv6 group	Dec 98	Proposed standard
2472	IP version 6 over PPP	Dec 98	Replaces RFC 2023
2473	Generic Packet Tunneling in IPv6 specification	Dec 98	Proposed standard
2474	Differentiated Service Field (DS Field) in IPv4 & v6	Dec 98	Replaces 1455, 1349
2475	Architecture for differentiated service	Dec 98	
2480	Gateways & MIME Security Multiparts	Jan 99	Proposed standard
2484	PPP LCP Internationalization Configuration Option	Jan 99	Updates RFC 2284
2493	Textual conventions for MIB modules /15 min. Intvl	Jan 99	Proposed standard
2507	IP header compression	Feb 99	Proposed standard
2508	Compressing IP/UDP/RTP headers: low speed links	Feb 99	Proposed standard
2509	IP header compression over PPP	Feb 99	Proposed standard

RFC number	Title	Date of issue	Remarks
2510	Internet X.509 PKI certificate management protocols	Mar 99	Proposed standard
2511	Internet X.509 certificate request message format	Mar 99	Proposed standard
2512	Accounting information for ATM networks	Feb 99	Proposed standard
2513	Managed objects: accounting info in CO networks	Feb 99	Proposed standard
2516	Transmitting PPP over ethernet (PPPoE)	Feb 99	
2543	SIP: Session Initiation Protocol	Mar 99	Proposed standard
2545	BGP-4 Multiprotocol exts:IPv6 Inter-domain routing	Mar 99	Proposed standard
2546	6Bone Routing Practice	Mar 99	Replaced by RFC 2772
2547	BGP/MPLS VPNs	Mar 99	
2561	Managed objects for TN3270E using SMIv2	Apr 99	Proposed standard
2564	Application Management MIB	May 99	Proposed standard
2570	Vers. 3 Internet-standard Network Mgmt Framework	Apr 99	
2571	Architecture for SNMP Management Frameworks	Apr 99	Replaces RFC 2271
2572	SNMPv3: Message processing & dispatching	Apr 99	Replaces RFC 2272
2573	SNMPv3 Applications	Apr 99	Replaces RFC 2273
2574	User-based Security Model (USM) for SNMPv3	Apr 99	Replaces RFC 2274
2575	View-based Access Control Model (VACM):SNMP	Apr 99	Replaces RFC 2275
2576	Coexistence v1, v2 and v3 of Internet Net Mgmt Fr.	Mar 00	Replaces 1908, 2089
2578	Structure of Management Information v2 (SMIv2)	Apr 99	Replaces RFC 1902
2579	SMIv2: Textual conventions	Apr 99	Replaces RFC 1903
2580	SMIv2: Conformance Statements	Apr 99	Replaces RFC 1904
2581	TCP Congestion Control	Apr 99	Replaces RFC 2001
2582	NewReno modification to TCP fast recovery algori.	Apr 99	
2586	Audio/L16 MIME content type	May 99	
2591	Managed objects for scheduling management ops.	May 99	Proposed standard
2594	Managed objects for WWW Services	May 99	Proposed standard
2597	Assured Forwarding PHB Group	Jun 99	Proposed standard
2598	An expedited Forwarding PHB	Jun 99	Proposed standard
2605	Directory Server Monitoring MIB	Jun 99	Replaces RFC 1567
2612	CAST-256 Encryption Algorithm	Jun 99	
2613	RMON MIB extensions for switched networks V1	Jun 99	Proposed standard

RFC number	Title	Date of issue	Remarks
2615	PPP over SONET/SDH	Jun 99	Replaces RFC 1619
2616	Hypertext Transfer Protocol HTTP/1.1	Jun 99	Updated by RFC 2817
2617	HTTP Authentication: Basic & Digest Access Auth.	Jun 99	Replaces RFC 2069
2618	RADIUS Authentication Client MIB	Jun 99	Proposed standard
2619	RADIUS Authentication Server MIB	Jun 99	Proposed standard
2620	RADIUS Accounting Client MIB	Jun 99	
2621	RADIUS Accounting Server MIB	Jun 99	
2630	Cryptographic Message Syntax	Jun 99	Proposed standard
2631	Diffie-Hellman Key Agreement Method	Jun 99	Proposed standard
2632	S/MIME v3 certificate handling	Jun 99	Proposed standard
2633	S/MIME v3 message specification	Jun 99	Proposed standard
2634	S/MIME: enhanced security services	Jun 99	Proposed standard
2637	Point-to-Point Tunneling Protocol (PPTP)	Jul 99	
2640	Internationalization of File Transfer Protocol (FTP)	Jul 99	Updates RFC 0959
2660	Secure HyperText Transfer Protocol	Aug 99	
2661	Layer Two Tunneling Protocol (L2TP)	Aug 99	Proposed standard
2662	Managed objects for ADSL lines	Aug 99	Proposed standard
2663	IP Network Address Translator (NAT) terminology	Aug 99	
2665	Managed objects for Ethernet-like interface types	Aug 99	Replaces RFC 2358
2668	Managed objects for IEEE 802.3 MAUs	Aug 99	Replaces RFC 2239
2675	IPv6 Jumbograms	Aug 99	Replaces RFC 2147
2677	Managed objects NBMA next hop resol. pr (NHRP)	Aug 99	Proposed standard
2712	Add Kerberos Cipher Suite TLS (Transpt. Layer.Sec)	Oct 99	Proposed standard
2743	GSS_APIv2 (Generic Security Service API)	Jan 00	Replaces RFC 2078
2808	SecurID® SASL Mechanism	Apr 00	
2819	Remote Network Monitoring (RMON) MIB	May 00	Replaces RFC 1757
2821	Simple Mail Transfer Protocol (SMTP)	Apr 01	Replaces various RFCs
2822	Internet Message Format	Apr 01	Replaces RFC 0822
2823	PPP on Simple Data Link (SDL) using SONET/SDH	May 00	
2841	IP Auth: Keyed SHA1/interleaved padding(IP-MAC)	Nov 00	Replaces RFC 1852
2852	Deliver By SMTP Service Extension	Jun 00	Updates RFC 1894
2854	Text/html Media Type	Jun 00	Replaces various RFCs
2861	TCP Congestion Window Validation	Jun 00	
2878	PPP Bridging Control Protocol (BCP)	Jul 00	Replaces RFC 1638
2888	Secure Remote Access with L2TP	Aug 00	

RFC number	Title	Date of issue	Remarks
2912	Indicating Media Features for MIME Content	Sep 00	Proposed standard
2917	Core MPLS IP VPN Architecture	Sep 00	
2920	SMTP Service extension for command pipelining	Sep 00	Replaces RFC 2197
2925	Managed objects:Remote Ping,Traceroute & Lookup	Sep 00	Proposed standard
2932	IPv4 Multicast Routing MIB	Oct 00	Proposed standard
2959	Real-Time Transport Protocol (RTP) MIB	Oct 00	Proposed standard
2962	SNMP Appl. Level Gway: Payload Address translat.	Oct 00	
2963	Rate Adaptive Shaper for Differentiated Services	Oct 00	
2988	Computing TCP's Retransmission Timer	Nov 00	Proposed standard
2989	Evaluating AAA Protocols for network access	Nov 00	
3009	Registration of parityfec MIME types	Nov 00	Proposed standard
3014	Notification Log MIB	Nov 00	Proposed standard
3029	Internet X.509 PKI: Data Validation & Certification	Feb 01	
3030	SMTP service exts: transm of large & binary MIME	Dec 00	Replaces RFC 1830
3031	Multiprotocol Label Switching (MPLS) Architecture	Jan 01	Proposed standard
3032	MPLS Label Stack Encoding	Jan 01	Proposed standard
3034	Use of Label Switching on FR networks spec	Jan 01	Proposed standard
3035	MPLS using LDP and ATM	Jan 01	Proposed standard
3036	LDP specification	Jan 01	Proposed standard
3056	Connection of IPv6 Domains via IPv4 Clouds	Feb 01	Proposed standard
3058	Use of the IDEA Encryption Algorithm in CMS	Feb 01	
3070	Layer 2 Tunneling Protocol (L2TP) over FR	Feb 01	Proposed standard
3095	RObust Header Compression (ROHC): Framework	Jul 01	Proposed standard
3145	L2TP Disconnect Cause Information	Jul 01	Proposed standard
3153	PPP Multiplexing	Aug 01	Proposed standard
3162	RADIUS and IPv6	Aug 01	Proposed standard
3165	Managed objects for Delegation of Mgmt Scripts	Aug 01	Replaces RFC 2592
3168	Addition: Explicit Congestion Notification (ECN)/IP	Sep 01	Proposed standard
3172	Address & Routing Parameter Domain (arpa)	Sep 01	Best current practice

RFC number	Title	Date of issue	Remarks
3173	IP Payload Compression Protocol (IPComp)	Sep 01	Replaces RFC 2393
3174	US (United States) Secure Hash Algorithm 1(SHA1)	Sep 01	
3175	Aggregation of RSVP for IPv4 & IPv6 reservations	Sep 01	Proposed standard
3193	Securing L2TP using IPsec	Nov 01	Updates RFC 1715
3201	Definitions managed objects:cct/interface translation	Jan 02	
3202	Definitions managed objects: frame relay service	Jan 02	
3204	MIME media types for ISUP and QSIG objects	Jan 02	
3220	IP mobility support for IPv4	Jan 02	Replaced by RFC 3344
3233	Defining the IETF	Feb 02	BCP 58
3236	The 'application xhtml+xml' media type	Jan 02	
3240	Digital imaging and comms in medicine (DICOM)	Feb 02	
3246	Expedited forwarding (EF) PHB	Mar 02	Replaces RFC 2598
3260	New terminology and clarifications for DiffServ	Apr 02	Updates 2474–5 & 2597
3272	Overview & principles: Internet traffic engineering	May 02	
3287	RMON: MIB extensions for DiffServ	Jul 02	Standards track
3344	IP mobility support for IPv4	Aug 02	Replaces RFC 3220
3345	BGP: persistent route oscillation condition	Aug 02	
3363	DNS: representing IPv6 addresses	Aug 02	
3364	DNS: support for IPv6	Aug 02	Updates 2673 & 2874
3369	Cryptographic Message Syntax (CMS)	Aug 02	Replaces 2630 & 3211
3370	CMS algorithms	Aug 02	Replaces 2630 & 3211

Appendix 5

IEEE 802 Standards for LANs and MANs

Standard and first publication date		Title, content and remarks
802	(1990)	**Main IEEE standard for local area networks (LANs) and metropolitan area networks (MANs) (re-issued as IEEE 802–2001)**
802.1	(1986)	**Higher layers and management interface standards for LANs (re-designated 802–1990)**
802.1b	(1993)	LAN/MAN management
802.1d	(1990)	Spanning tree algorithm for preventing circular loops when bridging LANs (MAC bridges)
802.1e	(1990)	System Load Control
802.1f	(1994)	Management Information definitions and procedures
802.1g	(1996)	Remote MAC bridging
802.1h	(1996)	MAC bridging
802.1i	(1992)	MAC bridges for FDDI (Fibre Distributed Data Interface) LANs
802.1j	(1997)	Managed objects for MAC bridges
802.1k	(1994)	Discovery and dynamic control of event forwarding
802.1m	(1994)	Managed object definitions and protocol implementation conformance statement (PICS)
802.1p	(1995)	Traffic class expediting & dynamic multicast filter bridges LAN quality of service (QOS)
802.1q	(1999)	Virtual bridged LANs (VLANs)
802.1r	(1998)	Generic attributes registration protocol (GARP) for LAN quality of service (QOS)
802.1s	(1998)	Multiple spanning trees
802.1t	(2001)	Mac bridges (amendment to 802.1d)
802.1u	(2001)	Virtual bridged LANs (VLANs) (amendment to 802.1q)

Data Networks, IP and the Internet: Protocols, Design and Operation   Martin P. Clark
© 2003 John Wiley & Sons, Ltd   ISBN: 0-470-84856-1

Standard and first publication date		Title, content and remarks
802.1v	(2001)	VLAN classification by protocol and port (amendment to 802.1q)
802.1w	(2001)	Rapid configuration of spanning tree
802.1x	(2001)	Port-based network access control
802.2	**(1983)**	**Logical link control (LLC)**
802.2a	(1994)	Standard flow control for bridged LANs
802.2b	(1994)	Acknowledged connectionless mode service and protocol (Type 3 operation)
802.2c	(1998)	Conformance requirements of LLC
802.2d	(1994)	LLC supplement
802.2e	(1994)	LLC bit-referencing
802.2f	(1998)	Managed objects for LLC
802.2g	(1991)	LLC type 4 high speed high performance operation [withdrawn]
802.2h	(1998)	Optional toleration of information transfer format protocol data units (IPDUs)
802.3		**Carrier Sense Multiple Access with Collision Detection (CSMA/CD) Ethernet LANs**
802.3	(1983)	10base5 (thicknet ethernet)
802.3a	(1985)	10base2 (thinnet ethernet)
802.3b	(1985)	10braod36 (ethernet based on broadband transmission with single 75 ohm cable TV cable)
802.3c	(1985)	10 Mbit/s baseband repeaters for coaxial cable ethernet LANs
802.3d	(1987)	FOIRL (Fibre Optic Inter-Repeater Link) for ethernet LANs
802.3e	(1987)	1base5 ethernet
802.3f	(1987)	1base5 ethernet LAN multipoint extension (withdrawn)
802.3h	(1990)	Layer management for ethernet LANs
802.3i	(1990)	10baseT (10 Mbit/s ethernet on two pair 100 ohm cabling category 3, 4 or 5)
802.3j	(1993)	10baseFL, 10baseFB and 10baseFP (10 Mbit/s ethernet on two optical fibres)
802.3k	(1992)	Layer management of repeaters in 10 Mbit/s baseband ethernet LANs
802.3l	(1992)	Conformance statement for the MAU (medium attachment unit) of 10baseT LANs
802.3m	(1996)	Revision of maintenance procedures for 100baseT (#2)
802.3n	(1996)	Revision of maintenance procedures for 100baseT (#3)
802.3p	(1992)	Media attachment unit (MAU) for 10 Mbit/s baseband ethernet LANs
802.3q	(1993)	Guidelines for the developments of managed objects
802.3r	(1996)	Updated version of 10base5 (thicknet ethernet LAN)
802.3s	(1996)	Revision of maintenance procedures for 100baseT (#4)
802.3t	(1996)	Support of 120 ohm cables in 10baseT simplex link ethernet LAN segment
802.3u	(1995)	100 Mbit/s fast ethernet LANs (100baseTX, 100baseFX, 100baseT4)
802.3v	(1995)	10BaseT link segments based on 150 ohm twinax cables
802.3x		Full duplex (FDX) operation of ethernet
802.3y	(1997)	100baseT2 (fast ethernet on two pairs of 100 ohm category 3,4 or 5 cable)

Standard and first publication date		Title, content and remarks
802.3z	(1998)	1000 Mbit/s Gigabit ethernet LANs (1000baseCX, 1000baseLX, 1000baseSX)
802.3aa	(1998)	Revision of maintenance procedures for 100baseT
802.3ab	(1999)	1000baseT Gigabit ethernet on four-pair Category 5 twisted pair cabling
802.3ac	(1998)	VLAN TAG switching
802.3ad	(2000)	Link aggregation
802.3ae	(2001)	10 Gigabit ethernet LANs
802.3af	(2001)	DTE (data terminal equipment) power from the medium dependent interface (MDI)
802.3ag	(2001)	Revision of maintenance procedures
802.4		**Token bus LANs**
802.5	**(1985)**	**Token ring LANs (redesignated 802.5–1998)**
802.5b	(1991)	4 Mbit/s token ring LAN over UTP (unshielded twisted pair) cable
802.5c	(1991)	Dual token ring LAN operation with wrapback
802.5f	(1987)	16 Mbit/s token ring LAN
802.5h	(1987)	LLC III on token ring LAN
802.5i	(1987)	Early token release
802.5j	(1993)	Full duplex token ring and optical fibre trunk signalling
802.5k	(1987)	Token ring media
802.5n	(1991)	4 Mbit/s and 16 Mbit/s token ring LAN over UTP (unshielded twisted pair) cable
802.5p	(1993)	Logical link control (LLC) annex: system route determination on token ring LANs
802.5t	(2000)	100 Mbit/s token ring over 2-pair cabling
802.5u	(1997)	1000 Mbit/s token ring over fibre (withdrawn)
802.5v	(2001)	1000 Mbit/s token ring (Gigabit token ring)
802.5w	(2000)	2000 version of token ring LAN (corrections to 802.5–1998)
802.6		**Dual Queue Dual Bus (switched multimegabit digital service, SMDS MAN)**
802.7		**Broadband LANs**
802.8		**Fibre optic LANs and MANs — recommended practices**
802.9		**Isochronous LANs for Integrated Services (voice and data)**
802.10		**Standard for interoperable LAN/MAN security (SILS)**
802.11		**Wireless LANs**
802.12		**100VG-AnyLAN**
802.13		**Not used**
802.14		**Cable-TV based broadband communication network (cable modem)**
802.15		**Wireless personal area network (WPAN — also called *Bluetooth*)**
802.16		**Broadband wireless access**
802.17		**Resilient packet ring**

Appendix 6

IEEE 802.11: Wireless Local Area Networks (WLANs)

The IEEE 802.11 standard specifies a *wireless local area network (WLAN)* system comprising of a *medium access control (MAC)* protocol and three alternative physical medium implementations. An IEEE 802.11 WLAN is intended to be used as a *physical layer* in conjunction with the IEEE 802.2 *logical link control (LLC)* protocol. Given that the MAC protocol is based on *CSMA/CA (collision sense multiple access with collision avoidance)*, an IEEE 802.11 WLAN is, in effect, a 'wireless ethernet' (Figure A6.1).

WLAN network architecture

A simple 802.11 WLAN comprises a number of *stations* which may operate in one of the following two configurations:

- *independent configuration (basic service set — BSS)* — in this mode, stations communicate directly with one other. There is no formal network structure and such networks are sometimes referred to as *ad hoc networks*. Ad hoc networks are relatively easy to operate, but their coverage area is limited. Such a configuration is termed a *basic service set (BSS)*. Where the BSS is not otherwise connected to an external network it is termed an *independent BSS (IBSS)*.

- *infra-structure configuration (extended service set — ESS)* — in this configuration, stations select a nearby *access point (AP)* and *associate* with it. The access point (AP) provides access to an external data network, which in IEEE 802.11-terminology is a *distribution system*. Typically most traffic within a given BSS will thus flow via the access point (AP). A number of BSSs can be grouped together to create an *extended service set (ESS)*. An ESS is intended to provide for wider WLAN coverage area — by allowing stations to *roam* from one BSS or AP area to another. This is achieved by bridging the separate BSSs across the distribution system.

Figure A6.2 illustrates the station, the basic service set (BSS), the access point (AP) and the *distribution system* of an IEEE 802.11 wireless LAN (WLAN).

Data Networks, IP and the Internet: Protocols, Design and Operation   Martin P. Clark
© 2003 John Wiley & Sons, Ltd   ISBN: 0-470-84856-1

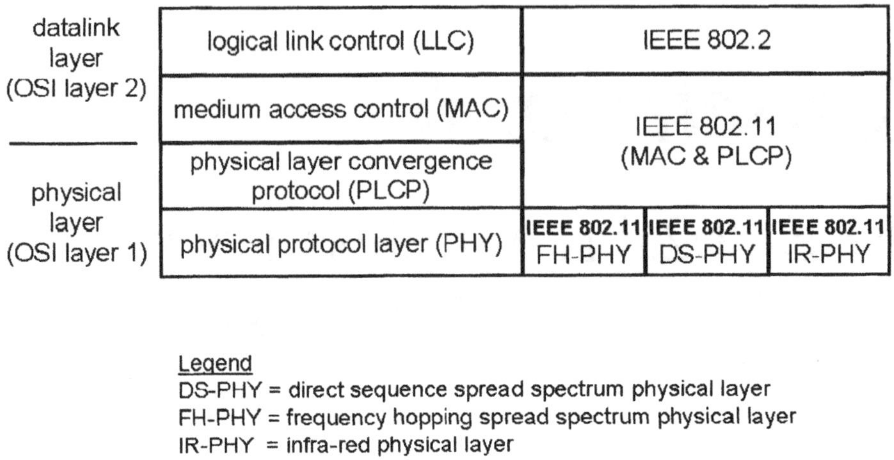

datalink layer (OSI layer 2)
physical layer (OSI layer 1)
logical link control (LLC)
IEEE 802.2
medium access control (MAC)
IEEE 802.11 (MAC & PLCP)
physical layer convergence protocol (PLCP)
physical protocol layer (PHY)
IEEE 802.11 FH-PHY
IEEE 802.11 DS-PHY
IEEE 802.11 IR-PHY
Legend
DS-PHY = direct sequence spread spectrum physical layer
FH-PHY = frequency hopping spread spectrum physical layer
IR-PHY = infra-red physical layer

Figure A6.1 IEEE 802.11 wireless local area network (WLAN): protocol stack.

server
station
distribution system (data network)
IEEE 802.11 LAN (BSS)
AP
AP
BSS (basic service set)
AP = access point
ESS (extended service set)

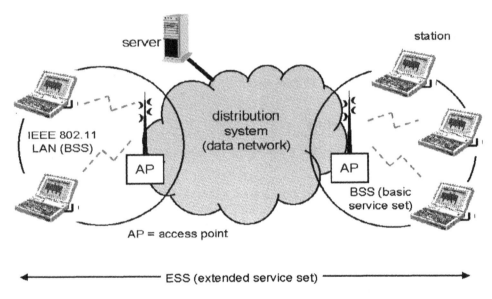

Figure A6.2 IEEE 802.11 wireless local area network (WLAN) architecture.

IEEE 802.11 Specifications

IEEE 802.11 MAC (medium access control) protocol

The MAC protocol is based upon a similar principle of *carrier sense multiple access with collision avoidance (CSMA/CA)* as ethernet LANs (IEEE 802.3). It provides for the following types of *time-bounded services, security services* and *management services*:

- authentication (station service);
- deauthentication (station service);

Table A6.1   IEEE 802.11 wireless local area network (WLAN) specifications

Standard	Contents
802.11	Wireless local area network (WLAN) MAC (medium access control) and PHY (physical layer) specification
802.11a	High speed physical layer in 5 GHz radio band
802.11b	High speed physical layer extension in the 2.4 GHz radio band
802.11c	MAC bridges
802.11d	Specification for operation in additional regulatory domains
802.11e	MAC quality-of-service (QOS) enhancements
802.11f	Multi-vendor access point (AP) interoperability via the Inter-Access-Point protocol (IAPP)
802.11g	Further higher rate data extension in the 2.4 GHz band
802.11h	Transmit power management for use in the 5 GHz band in Europe
802.11i	MAC security enhancements

- privacy (station service);

- MSDU (MAC service data unit) delivery (station service);

- association (distribution system service);

- disassociation (distribution system service);

- distribution (distribution system service);

- integration (distribution system service); and

- reassociation (distribution system service).

MAC frames may be either control frames, management frames or data frames. The format of MAC frames is as shown in Figure A6.3. The various address fields contain different addresses as indicated by the *to DS (destination station)* and *from DS (destination station)* bits (see Table A6.2).

The *power management bit* (when set to value '1') indicates that the power management protocol is in use — for saving battery life by turning off the radio transmitter when not required.

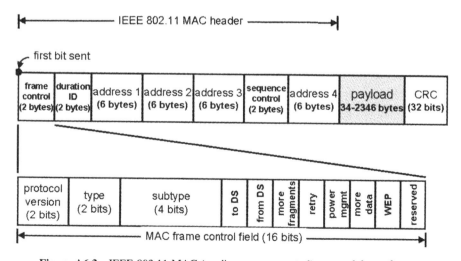

Figure A6.3   IEEE 802.11 MAC (medium access control) protocol frame format.

Table A6.2   IEEE 802.11 MAC (medium access control): use of Ds-bit and address fields

To DS bit value	From DS bit value	Address 1	Address 2	Address 3	Address 4
0	0	DA	SA	BSSID	N/A
0	1	DA	BSSID	SA	N/A
1	0	BSSID	SA	DA	N/A
1	1	RA	TA	DA	SA

Key: BSSID = basic service set identifier; DA = destination address; DS = destination stations; RA = receiver address; SA = source address; TA = transmitter address.

The *WEP* bit (when set to value '1') indicates that the *wired equivalent privacy* protocol is in use.

IEEE 802.11 Physical layer

The IEEE 802.11 standard provides three alternative physical medium implementations: two physical layer specifications for radio (operating in the 2 400-2 483.5 MHz band depending on local government regulations) and one for infrared interface (see Figure A6.1).

Frequency hopping spread spectrum radio PHYsical layer (FH-PHY)

The *frequency hopping spread spectrum radio physical layer (FH-PHY)* of IEEE 802.11 provides for either 1 Mbit/s, or (optionally) for 2 Mbit/s operation. The 1 Mbit/s version uses 2-level *Gaussian frequency shift keying (GFSK)* modulation and the 2 Mbit/s version uses 4-level GFSK. The exact number and frequency of radio channels (i.e the precise physical medium) which should be used depend upon local government radio regulations and related radio technical standards. These are listed in Table A6.3.

Figure A6.4 illustrates the *physical layer convergence protocol (PLCP)* frame format used by IEEE 802.11 FH-PHY.

The *preamble* part of the frame (which comes at the start of each frame sent) is used to *synchronise* the radio transmission. During this period, radio receivers have to 'notice' that a signal is being sent, adjust their radio frequency circuitry to the exact frequency of the signal being sent by the transmitter and adjust their signal *automatic gain control (AGC)* to ensure that the signal is amplified appropriately. (Each received signal will have a different strength,

Table A6.3   Radio operations and technical standards relevant to IEEE 802.11

Region	Radio standard	Radio frequency band of operation	Number of hopping channels	Transmitter power restriction
United States	CFR 47 parts:	2400–2483.5 MHz	75–79	1 Watt max transmit power
	15.247			4 Watt max EIRP
	15.205			
	15.209			
Europe	ETS 300 328	2400–2483.5 MHz	20–79	100 mWatt max EIRP
	ETS 300 339			
Japan	RCR STD-33A	2471–2497 MHz	10–23	

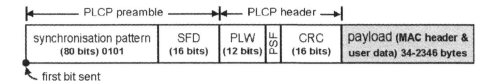

Legend
CRC   = cyclic redundancy check
MAC   = medium access control
PLCP = physical layer convergence protocol
PLW   = payload length word
PSF   = PLCP signalling field
SFD   = start of frame delimiter (always hex value 0CBD)

Figure A6.4   IEEE 802.11 frequency hopping spread spectrum radio physical layer (FH-PHY): PLCP (physical layer convergence protocol) frame format.

depending upon how far away the remote transmitter is — but there is an optimum signal level which should be sent to the *detector* circuitry.)

The *start of frame delimiter (SFD)* for FH-PHY is always set at the hexadecimal value '0CBD'.

A process called *CCA (clear channel assessment)* (one of the functions of the PLCP signalling field) performs the function of *collision detection* on behalf of the MAC layer. CCA initiates frame reception and forces back-off of transmission if the radio channel turn out to be busy.

Since radio transmission is very prone to errors (and in particular to *burst errors*[1]), data is *scrambled*[2] to reduce the problems of errors caused by interference.

A 16-bit *cyclic redundancy check (CRC)* code[3] is used as a header error check code for the *PLCP (physical layer convergence protocol)* header (i.e. the fields PLW and PSF).

Direct sequence spread spectrum radio PHYsical layer (DS-PHY)

Like the frequency hopping spread spectrum radio physical layer (FH-PHY), the *direct sequence (DS) spread spectrum (DSSS) physical layer (DS-PHY)* of IEEE 802.11 provides for either 1 Mbit/s, or (optionally) for 2 Mbit/s operation in the 2.4 GHz (2400 MHz) *ISM (industrial scientific medical)* radio band. But unlike FH-PHY, the 1 Mbit/s version of the DS-PHY uses *differential binary phase shift keying (DBPSK)* modulation and the 2 Mbit/s version uses *differential quadrature phase shift keying (DQPSK)*.

The radio *multiple access* scheme used in DS-PHY is CDMA (code division multiple access) — employing an 11 MHz chip rate and an 11-chip Barker sequence.

The physical layer convergence protocol (PLCP) and physical layer data unit frame format of DS-PHY are illustrated in Figure A6.5.

---

[1] See Chapter 8.
[2] Although *scrambling* in common parlance is synonymous with *encryption*, this is not its main purpose — and *scrambling* should not be taken to provide a suitable alternative to *encryption*.
[3] See Chapter 3.

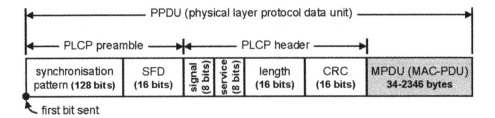

first bit sent

Legend
CRC   = cyclic redundancy check
MAC   = medium access control
PLCP = physical layer convergence protocol
PLW   = payload length word
PSF   = PLCP signalling field
SFD   = start of frame delimiter (always hex value F3A0)

Figure A6.5   IEEE 802.11 direct sequence spread spectrum radio physical layer (DS-PHY): PLCP (physical layer convergence protocol) frame format.

As with FH-PHY, the synchronisation preamble sequence of DS-PHY provides for a period during which the radio receiver can undertake signal *energy detection* (within 15 microsecs), antenna selection, frequency adjustment and signal gain settings.

The various frames have the following meanings and codings:

- the *start of frame delimiter (SFD)* for DS-PHY is always set at the hexadecimal value 'F3A0';

- the *signal* field indicates whether 1 Mbit/s DBPSK (signal = hexadecimal value '0A') or 2 Mbit/s DQPSK (signal = hexadecimal value '14') is in use;

- the *service* field (when set to hexadecimal value '00') indicates that the implementation is IEEE 802.11;

- the *length* field indicates the length of the MPDU (MAC protocol data unit) in number of bytes or octets;

- the PLCP header check provides for detection of bit errors in the signal, service and length fields. A 16-bit cyclic redundancy check (*CRC-16*) code is used and coded according to ITU-T (see Chapter 3).

Infrared PHYsical layer (IR-PHY)

The infrared PHY (IR-PHY) of IEEE 802.11 also provides for 1 Mbit/s transmission, with an option for 2 Mbit/s transmission. The 1 Mbit/s version employs *pulse position modulation* with 16 positions (*16-PPM*) and the 2 Mbit/s version uses *4-PPM (4-position pulse position modulation)*.

Appendix 7

Interfaces, Cables, Connectors and Pin-outs

This appendix illustrates the key technical specifications, connectors and cable types associated with the interfaces listed below. The information is presented in graphical and tabular format for quick reference, but without detailed explanation. Further explanation of the interfaces appears in Chapter 3.

DTE-to-DCE interfaces (for connection of data end-user devices to modems and other WAN line types)

- DTE-to-DCE interfaces of the X.21 and X.21*bis* family of interfaces
- V.24/V.28 and RS-232 interface (for DTE-to-DCE/WAN)
- V.35 and V.36 (RS-449) interfaces for DTE-to-DCE (high speed WAN lines)
- DTE/DCE and null modem cables
- X.21 interface (DTE-to-DCE for high speed WAN lines)
- G.703 interface (digital line interface used commonly for T1, T3, E1, E3, etc.)
- Copper cable types used in telephone networks

Optical fibre cable interfaces

- Table of fibre cable bitrates
- Table of fibre cable interface types
- ST and SC cable connector types

Coaxial cable interfaces

- Coaxial cable connectors: BNC, N-type, TNC
- Table of coaxial cable types used in telephone, data and ethernet networks

Data Networks, IP and the Internet: Protocols, Design and Operation   Martin P. Clark
© 2003 John Wiley & Sons, Ltd   ISBN: 0-470-84856-1

Internal office and LAN data network cabling

- Data cabling types and specifications
- Category 5 UTP (unshielded twisted pair) and STP (shielded twisted pair) cabling
- RJ-45 connector: plugs, sockets and patch panels
- RJ-45 patch cables and cabling pin-outs

Computer peripheral interfaces

- USB (universal serial bus)
- Parallel ports: DB25 and centronics
- SCSI (small computer system interface)

DTE-to-DCE interfaces X.21 and X.21bis

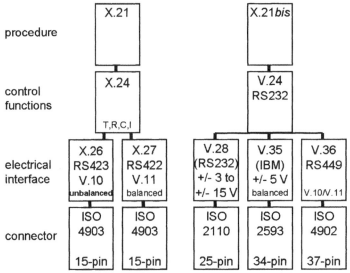

Note: You may be wondering why all the specications and standards defining DTE/DCE interfaces have such different designations. This is be cause different naming standards are used by the differentst andards publishing organisations. ISO (International Organization for Standardization) uses a simple numbering scheme (but I haven't yet worked out the logic behind the individual numbers). ITU-T (International Telecommunications Union – Standardization sector) issues *recommendations* in various *series*. The *X-series* defines interfaces and procedures intended for use ingeneral *data communications*. The *V-series* defines *data communications over the telephone network* (e.g. modems). Sometimes the same recommendation is issued with a commendation number in both series (e.g. V.10/X.26). The nomenclature RS, meanwhile, stands for *recommended standard*. This designation is used by the United States EIA/TIA (Electronic Industries Alliance or Association/Telecommunications Industries Association).

Figure A7.1   Standards, specifications and ITU-T recommendations defining DTE/DCE interfaces.

V.24/V.28 and RS-232 interface (for DTE-to-DCE)

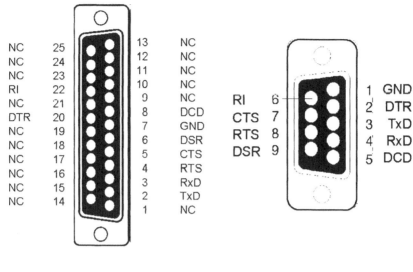

**a) DB-25 connector
(ISO 2110)**

**b) DB-9 connector
(EIA 562)**

Figure A7.2   Alternative connectors used for V.24/V.28 or RS-232 interface.

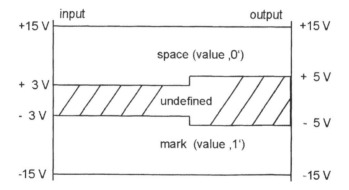

Figure A7.3   Defined output (send) and input (detect) voltages of RS-232 and V.28 circuits.

Table A7.1   DTE/DCE circuits and control signals defined by EIA RS-232/ITU-T recommendation V.24

Signal code	Signal meaning	DB-25 connector pin number (ISO 2110)	Db-9 pin number (EIA 562)	ITU-T Rec. V.24 signal name	EIA signal name	Signal meaning for binary value '1' or 'mark'
CTS	Clear-to-send	5	8	106	CB	From DCE to DTE: 'I am ready when you are'

(continued overleaf)

Table A7.1   (*continued*)

Signal code	Signal meaning	DB-25 connector pin number (ISO 2110)	Db-9 pin number (EIA 562)	ITU-T Rec. V.24 signal name	EIA signal name	Signal meaning for binary value '1' or 'mark'
DCD	Data carrier detect	8	1	109	CF	From DCE to DTE: 'I am receiving your carrier signal'
DSR	Data set ready	6(not always used)	6	107	CC	From DCE (the 'data set' to DTE): 'I am ready to send data'
DTR	Data terminal ready	20	4	108.2	CD	From DTE to DCE: 'I am ready to communicate'
GND	Ground	7	5	102	AB	Signal ground — voltage reference value
RC	Receiver clock	17(little used)	—	115	DD	Receive clock signal (from DCE to DTE)
RI	Ring indicator	22	9	125	CE	From DCE to DTE: 'I have an incoming call for you'
RTS	Request-to-send	4	7	105	CA	From DTE to DCE: 'please send my data'
RxD	Receive data	3	2	104	BB	DTE receives data from DCE on this pin
TC	Transmitter clock	15(little used)	—	114	DB	Transmit clock signal (from DCE to DTE)
TxD	Transmit data	2	3	103	BA	DTE transmits data to DCE on this pin

V.35 and V.36 (RS-449) interfaces for DTE-to-DCE (high speed WAN lines)

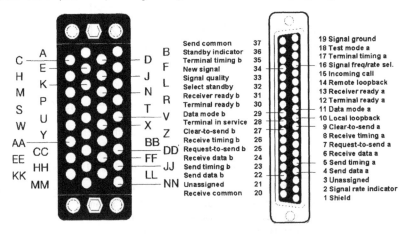

a) V.35
(connector: M/50 [ISO 2593])

b) RS-449 (V.35/V.36)
(connector: ISO 4902 [DB37])

Note: Circuits have the same meaning as ITU-T recommendation V.28 and RS-232 (see Table A7.1).

Figure A7.4   Usual connectors for V.35 and V.36 (RS-449) interfaces.

DTE-to-DCE and null modem cables

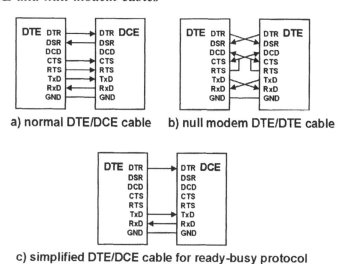

a) normal DTE/DCE cable   b) null modem DTE/DTE cable

c) simplified DTE/DCE cable for ready-busy protocol

Figure A7.5   DTE/DCE and null modem cable types.

X.21 interface (DTE-to-DCE for high speed WAN lines)

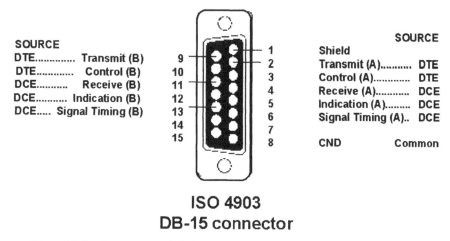

SOURCE
DTE............. Transmit (B)
DTE............. Control (B)
DCE........... Receive (B)
DCE........... Indication (B)
DCE..... Signal Timing (B)

SOURCE
Shield
Transmit (A)........... DTE
Control (A)............. DTE
Receive (A)........... DCE
Indication (A)........ DCE
Signal Timing (A).. DCE

CND   Common

ISO 4903
DB-15 connector

Figure A7.6   Connectors and pin-layout according to ITU-T recommendation X.21.

G.703 interface (digital line interface used commonly for T1, E1, T3, E3 etc.)

The G.703 interface is defined in ITU-T recommendation G.703. The interface defines the electrical characteristics of a 4-wire digital line interface. Different realisations are defined.

The line code (Figure A7.7) defined to be used on a G.703 interface may conform either to:

• AMI (alternate mark inversion — Figure A7.7a — the North American standard), or:

• HDB3 (high density bipolar 3 — Figure A7.7b — the European standard).

Table A7.2   Communication states: ITU-T recommendation X.21

State number	State name	T (transmit)	C (control)	R (receive)	I (indication)
1	Ready	1	Off	1	Off
2	Call request	0	On	1	Off
3	Proceed-to-select request (i.e., request dial)	0	On	+	Off
4	Selection signal sequence (i.e., number to be dialled)	ASCII [7 bit] (IA5)	On	+	Off
5	DTE waiting	1	On	+	Off
6	DCE waiting	1	On	SYN	Off
7	Call progress signal	1	On	ASCII [7 bit] (IA5)	Off
8	Incoming call	1	Off	Bell	Off
9	Call accepted	1	On	Bell	Off
10	Call Information (from DCE)	1	On	ASCII [7 bit] (IA5)	Off
11	Connection in progress	1	On	1	Off
12	Ready for data (i.e., to communicate)	1	On	1	On
13	Data transfer (i.e., communication)	Data	On	Data	On
13S	Send data	Data	On	1	Off
13R	Receive data	1	Off	Data	On
14	DTE controlled not ready, DCE ready	01	Off	1	Off
15	Call collision	0	On	Bell	Off
16	DTE clear request	0	Off	X (any signal)	X (any signal)
17	DCE clear confirmation	0	Off	0	Off
18	DTE ready, DCE not ready	1	Off	0	Off
19	DCE clear indication	X (any signal)	X (any signal)	0	Off
20	DTE clear indication	0	Off	0	Off
21	DCE ready	0	Off	1	Off
22	DTE uncontrolled not ready (fault), DCE not ready	0	Off	0	Off
23	DTE controlled not ready, DCE not ready	01	Off	0	Off
24	DTE uncontrolled fault, DCE ready	0	Off	1	Off
25	DTE provided information	ASCII (7 bit) (IA5)	Off	1	On

AMI is standardly used with the North American PDH (plesiochronous digital hierarchy) digital bitrates T1 (1.544 Mbit/s) and T3 (44.736 Mbit/s), while HDB3 is used with the European PDH digital bitrates E1 (2.048 Mbit/s) and E3 (34.368 Mbit/s).

The electrical interface may also take one of two alternative formats, either:

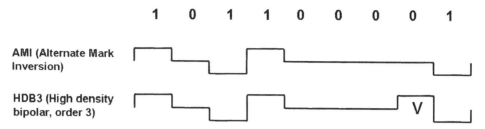

Figure A7.7   Alternative G.703 interface line codes.

Table A7.3   Copper cables as defined by the American Wired Gauge (AWG) standards

Awg	Diameter of conductor/mm	Cross-section of conductor /mm^2	Resistance in Ω/km	Weight in kg/km	Maximum current rating in Amps
12	2.0525	3.309	5.2	29.4	13.060
14	1.6277	2.081	8.3	18.5	8.2132
16	1.2908	1.309	13.2	11.6	5.1654
18	1.0237	0.823	20.9	7.3	3.2485
20	0.8118	0.518	33.3	4.6	2.0430
22	0.6438	0.326	53.0	2.9	1.2849
24	0.5106	0.205	84.2	1.8	0.8081
26	0.4049	0.129	133.9	1.1	0.5082

Table A7.4   SDH (synchronous digital hierarchy) and SONET (synchronous optical network) interfaces

North american SONET	Carried bitrate / mbit/s	SDH
VT 1.5	1.544	VC-11
VT 2.0	2.048	VC-12
VT 3.0	3.172	
VT 6.0	6.312	VC-21
	8.448	VC-22
	34.368	VC-31
	44.736	VC-32
	149.76	VC-4
STS-1 (OC-1)	51.84	
STS-3 (OC-3)	155.52	STM-1
STS-6 (OC-6)	311.04	
STS-9 (OC-9)	466.56	
STS-12 (OC-12)	622.08	STM-4
STS-18 (OC-18)	933.14	
STS-24 (OC-24)	1 244.16	
STS-36 (OC-36)	1 866.24	
STS-48 (OC-48)	2 488.32	STM-16
STS-96 (OC-96)	4 976.64	
STS-192 (OC-192)	9 953.29	STM-64

- Two coaxial cables of nominal 75 ohm (75 Ω) impedance (e.g., cable type RG58—see Table A7.6). This is the so-called *unbalanced* version of G.703, or

- Two pairs of twisted-pair copper cable of nominal 120 ohm (120 Ω) impedance (e.g., AWG22 or AWG24 of Table A7.3). This is the so-called *balanced* version of G.703.

Table A7.5   Classification of optical fibre interfaces for optical transmission lines and equipment

	Single mode fibre (SMF)		Multi-mode fibre (MMF)	
FibreCore diameter	8 μm	10 μm	50 μm	62,5 μm
Fibre mantel diameter	125 μm	125 μm	125 μm	125 μm
Glass cladding diameter	250 μm	250 μm	250 μm	250 μm
Wavelength & attenuation				
850 nm	Not used	Not used	3.0 dB/km	3.7 dB/km
1310 nm	Not used	<1.0 dB/km	2.0 dB/km	1.0 dB/km
1550 nm	0.3 dB/km	<0.5 dB/km	Not used	Not used
Relevant specifications	ITU-T rec. G.654 (loss minimized, 1550 nm only)	ITU-T rec. G.652	ITU-T rec. G.651	
	ITU-T rec. G.653 (restricted use at 1310 nm)	ITU-T rec. G.655		
Typical range	Long range > 60 km	Metropolitan < 60 km	Metropolitan and short range	Metropolitan and short range
Optical system designations	ZX	LX, LH (1300 nm-long wave)	SX (850 nm — short wave)	SX (850 nm — Short wave)
			LX, LH (1300 nm — long wave)	LX, LH (1300 nm — long wave)
	STM-S (short range)	STM-S (short range)	STM-S (short range)	STM-S (short range)
		STM-I (intra-office)	STM-I (intra-office)	STM-I (intra-office)

Figure A7.8   Typical single pair optical fibre cable showing fibres, tension members and cable sheath [reproduced courtesy of © RS Components Ltd].

The balanced version of G.703 is most commonly used in public networks in Germany, Austria, Switzerland, while the unbalanced version is most commonly used in the USA, the UK, France and other countries. Telecommunications equipment uses both types of interface, dependent upon manufacturer preference. Though there are no standardised connectors defined by G.703 for the interface, it is common to see the BNC connector (Figure A7.10) used for the unbalanced interface and RJ45 (Figure A7.13 and Table A7.8) used for the balanced version.

Table A7.6 Common coaxial cable types used in telecommunications networks

	Rg11 A/U	Rg58 C/U	Rg59 B/U	Rg62 A/U	Rg71 B/U	Rg174 A/U	Rg213/U	Rg214/U	Rg223/U
Inner conductor material	Tinned copper strands	Tinned copper strands	Copper-clad solid steel	Copper-clad solid steel	Copper-clad solid steel	Copper-clad steel strands	Bare copper strands	Silver covered solid copper	Silver covered solid copper
Inner conductor diameter in mm	1.21	0.90	0.58	0.64	0.64	0.48	2.25	2.25	0.89
Dielectric Material	Poly-ethylene	Poly-ethylene	Poly-ethylene	Poly-ethylene & air	Poly-ethylene & air	Poly-ethylene	Poly-ethylene	Poly-ethylene	Poly-ethylene
Dielectric diameter in mm	7.24	2.95	3.71	3.70	3.70	1.52	7.24	7.24	2.95
Shield Material	Copper	Tinned copper	Copper	Copper	Copper & tin	Tinned copper	Copper	Silvered copper	Silvered copper
Sheath Material	PVC	PVC	PVC	PVC	Poly-ethylene	PVC	PVC	PVC	PVC
Cable outside diameter in mm	10.3	5.0	6.1	6.1	6.2	2.5	10.3	10.8	5.4
Weight in g/m	143	43	48	57	68	12	147	187	50
Nominal impedance in Ω	75 ± 3	50 ± 2	75 ± 3	93 ± 5	93 ± 5	50 ± 2	50 ± 2	50 ± 2	50 ± 2
Nominal capacitance pF/km	68	101	68	43	43	101	101	101	101
Maximum operating voltage	5 000	1 900	2 300	750	750	1 500	5 000	5 000	1 900
Attenuation at 400 MHz dB/ 100 m	16.7	35.2	25.6	25.2	25.2	58.5	16.1	16.1	30.5

Notes:
RG cable specifications are laid out in the US government/military specification document MIL-C-17
Operating temperature: −40°C to +80°C

Figure A7.9   ST and SC connector types for optical fibres [reproduced courtesy of Black Box Corporation].

Figure A7.10   BNC connector (bayonet connector) commonly used in association with coaxial cables [reproduced courtesy of Black Box Corporation].

Copper cable types used in telephone networks

Copper cables of AWG 22 and AWG 24 are widely used as *unshielded twisted pair (UTP)* cable pairs in *outdoor* public telephone networks.

Optical fibre cable interfaces

Coaxial cable interfaces used in public telephone networks

75 Ω cable (e.g., RG58) is typically used for *unbalanced* G.703 digital line interface and other broadband connections. 50 Ω cable, meanwhile, is typically used for old-style ethernet networks. 93 Ω cable is used, for example, for ARCNET networks.

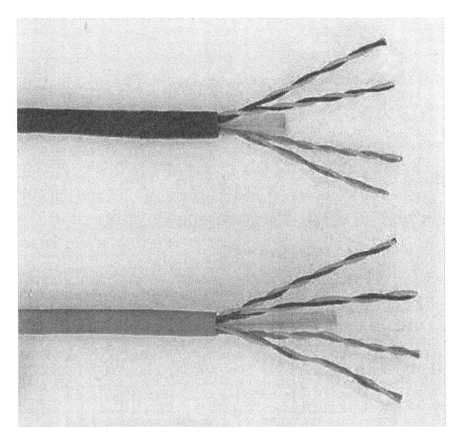

Figure A7.11   Category 5 UTP (unshielded twisted pair) cabling [reproduced courtesy of © RS Components Ltd].

Other coaxial cable types used in indoor data network cabling include the IBM twinax cable (used in particular in association with IBM AS400 computers and IBM token ring) and ethernet cabling. Table A7.7 lists standard coaxial cables used for 'classical' ethernet.

Other types of coaxial cable connector used commonly in telecommunications are the N-connector, TNC connector and micro-coaxial cable connectors. All have a similar appearance to the BNC connector, but are of different sizes.

Data cabling types and specifications

Other shielded twisted pair cables used in indoor data network cabling include the IBM cabling system cables: Type 1a, Type 6a and Type 9a. Type 1a is a general purpose cable intended for use in structured cabling systems. Type 6a is intended as a flexible patch cable.

In the *standard* or *parallel* RJ-45 patch cables, each pin of the RJ-45 plug at one end of the cable is connected to the same numbered pin in the opposite plug. Thus pin 1 connects to pin 1, pin 2 to pin 2, etc. Looking down on the RJ-45 plugs from the top, the individual wire colours appear as shown in Figure A7.15a (when coloured according to the EIA/TIA 568B standard — see Table A7.9). This is the most commonly used type of patch cable, and serves for most purposes.

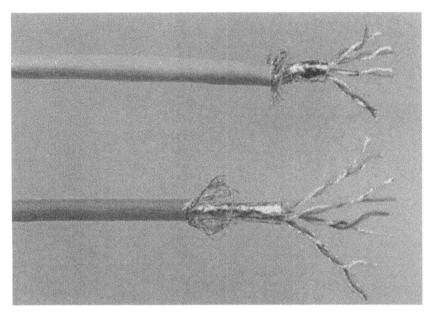

Figure A7.12   Category 5 STP (shielded twisted pair) cabling [reproduced courtesy of © RS Components Ltd].

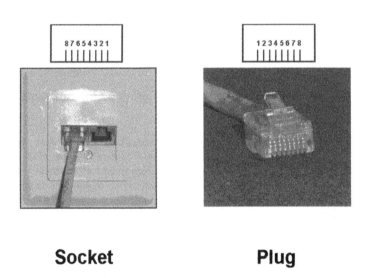

Figure A7.13   RJ-45 connector: 8-pole socket and plug and pin layout.

A *crossed* patch cable is intended to allow ethernet DTEs (e.g., PCs) to be directly cabled to one another without using a LAN hub or LAN switch. Alternatively, LAN switches or hubs can be connected directly with one another. For such connection to be possible, the ethernet transmit pair at one end of the cable (DTE transmits on pins 3 and 6) have to be 'crossed' with the receive pair at the other end (pins 1 and 2). As Figure A7.15 illustrates, this leads to a non-standard order of the cabling colours at one end of the patch cable (the left-hand end in the case of Figure A7.15b).

Table A7.7   Coaxial cables used for 'classical' style ethernet bus cabling

Cable type	Cable size	Nominal impedance	Usage
Standard ethernet bus cable (Yellow-cable)	2.2 mm inner 6.2 mm shield	50Ω	Ethernet bus cables up to 500 m length
Thinwire (Thinnet)	0.9 mm inner 2.5 mm shield	50Ω	Ethernet bus cables up to 500 m length

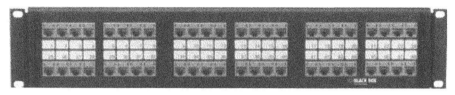

Figure A7.14   Patch panel comprising 48 RJ-45 sockets [reproduced courtesy of Black Box Corporation].

a) standard RJ45 patch cable - *parallel* cable (EIA/TIA 568B)

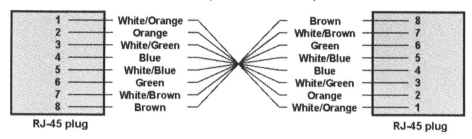

b) RJ45 *cross* cable
(Ethernet Tx and Rx 'crossed': pin 2- pin 6 and pin 1-pin 3)

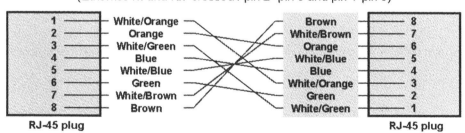

Figure A7.15   Standard and 'cross' RJ-45 patch cables.

USB (universal serial bus) interface

USB type A plugs are used *upstream* (i.e., towards the host system). USB type B plugs are to be oriented *downstream* towards the USB device. The USB standards refer to *receptacles* instead of sockets. *Receptacles* are what plugs are inserted into.

Table A7.8 Common *in-house* data cabling types and specifications (EIA/TIA 568, ISO/IEC 11801 and EN 50173)

Cabling type: Parameter:	Category 3	Category 4	Category 5 (100 Ω) [EIA/TIA 568 B-1 & B-2: April 2001]	Category 6 (100 Ω) [TIA/EIA 568 B-2-1: July 2002]	Category 7* (100 Ω)
Max attenuation at:					
1 MHz	2.6 dB/100 m	2.1 dB/100 m	2.1 dB/100 m	2.0 dB/100 m	2 dB/100 m
10 MHz	9.8 dB/100 m	7.2 dB/100 m	6.6 dB/100 m	6.0 dB/100 m	6 dB/100 m
20 MHz		10.2 dB/100 m	9.2 dB/100 m	8.5 dB/100 m	9 dB/100 m
62.5 MHz			17.1 dB/100 m	15.4 dB/100 m	16 dB/100 m
100 MHz			22.0 dB/100 m	19.8 dB/100 m	21 dB/100 m
250 MHz				32.8 dB/100 m	28 dB/100 m
600 MHz					47 dB/100 m
Min near end crosstalk (NEXT) attenuation at:					
1 MHz	41 dB	56 dB	62 dB	74 dB	
10 MHz	26 dB	41 dB	47 dB	59 dB	
20 MHz		36 dB	42 dB	55 dB	
100 MHz			32 dB	44 dB	
250 MHz				38 dB	
600 MHz					
Resistance at 10 MHz			<100 mΩ/m		68 dB

Note: *Standards still being agreed.

Table A7.9 RJ-45 patch cable pin-out configurations

	Pin 1	Pin 2	Pin 3	Pin 4	Pin 5	Pin 6	Pin 7	Pin 8
RJ-45 pin colours (EIA/TIA 568B)	White/Orange	Orange	White/Green	Blue	White/Blue	Green	White/Brown	Brown
	Pair 2←	→Pair 2	Pair 3←	Pair 1←	→Pair 1	→ Pair 3	Pair 4←	→Pair 4
Uses:								
Standard ethernet patch cable	DTE receive + (hub transmit)	DTE receive — (hub transmit)	DTE transmit + (hub receive)	Not used	Not used	DTE transmit- (hub receive)	Not used	Not used
2-wire telephone line (DTE-to-WAN)	Not used	Not used	Not used	+	—	Not used	Not used	Not used
4-wire telephone line (DTE-to-WAN)	DTE receive +	DTE receive -	Not used	DTE transmit +	DTE transmit-	Not used	Not used	Not used
V.24/V.28 or RS-232	RTS (cct 4)	DTR (cct 20)	TXD (cct 2)	GND (cct 7)	GND (cct 7)	RXD (cct 3)	DSR/CD (ccts 6/8)	CTS (cct 5)
RJ-45 pin colours (EIA/TIA 568A)	White/Green	Green	White/Orange	Blue	White/Blue	Orange	White/Brown	Brown
	Pair 3←	→ Pair 3	Pair 2←	Pair 1←	→Pair 1	→ Pair 2	Pair 4←	→ Pair 4
Older RJ45 colouring	Blue	Orange	Black	Red	Green	Yellow	Brown	Grey

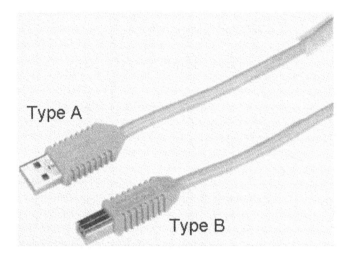

Figure A7.16   Alternative connectors (plugs) for USB interfaces [reproduced courtesy of Black Box Corporation].

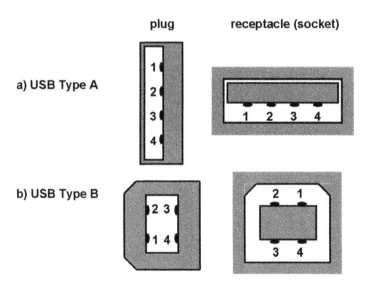

Figure A7.17   USB plugs and receptacles.

Table A7.10   USB pin layout and usage

Pin (circuit) number	Signal usage	Cabling colour	Purpose
1	V_{BUS}	Red	Power supply
2	Ground	Black	Ground
3	D+	Green	Data transmission 1
4	D−	White	Data transmission 2
Shell	Drain wire	Shield	Noise protection

USB cables comprise four conducting cables (red, black, green and white) — two for power and two are signal conductors.

More information on the USB 2.0 standard can be found at www.usb.org

Computer parallel port interfaces

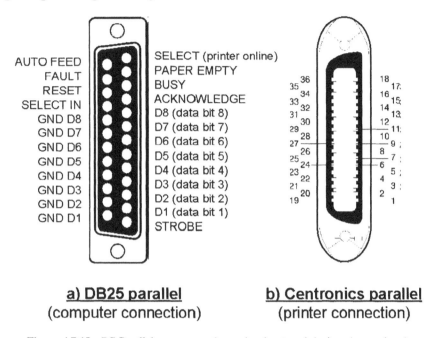

AUTO FEED
FAULT
RESET
SELECT IN
GND D8
GND D7
GND D6
GND D5
GND D4
GND D3
GND D2
GND D1

SELECT (printer online)
PAPER EMPTY
BUSY
ACKNOWLEDGE
D8 (data bit 8)
D7 (data bit 7)
D6 (data bit 6)
D5 (data bit 5)
D4 (data bit 4)
D3 (data bit 3)
D2 (data bit 2)
D1 (data bit 1)
STROBE

a) DB25 parallel
(computer connection)

b) Centronics parallel
(printer connection)

Figure A7.18   PC Parallel port connection to local external devices (e.g., printer).

SCSI (small computer system interface)

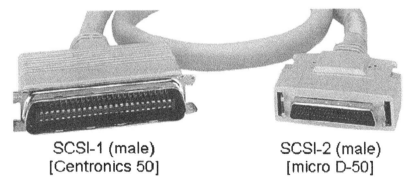

SCSI-1 (male)
[Centronics 50]

SCSI-2 (male)
[micro D-50]

Figure A7.19   Alternative connector types for SCSI interface (SCSI-1 and SCSI-2) [reproduced courtesy of Black Box Corporation].

The SCSI-2 standard is defined in ANSI X3.131–1994.

Appendix 8

X.25 Packet Switching (ITU-T Recommendation X.25)

The X.25-protocol is a network (layer 3) protocol used between peer devices at the *UNI (user-network interface)* of a public packet-switched data network (often called an *X.25 network* or *packet-switched network*).

Being a *connection-oriented* protocol, X.25 (ITU-T recommendation X.25) defines distinct procedures for *signalling* and *data transfer* during the various phases of the *call*, including:

- *call request* and *connection set-up*;

- *data transfer*, including flow control;

- *supervision*, including retransmission, interrupt, reset, restart, registration and diagnostics;

- *call clearing* (once communication is finished).

Before being able to communicate across a packet-switched-network, an X.25 DTE first has to signal its desire for a *connection* to be set up. For this purpose the DTE generates an X.25 *call request packet* which includes all the information needed by the network 'control point' to set up the connection (including *called address* (the B-end destination), *calling address* (the A-end origin of the call), features, *facilities* and network services needed for the connection and any related *call user data*. The DTE also selects the preferred *logical channel number (LCN)* it would like to use for the call, *signalling* this in the *call request* packet (Figure A8.1). This logical channel is then put in the *DTE-waiting* state, while the network node (i.e., DCE) decides what to do next.

The network address format used in the X.25 call request packet to indicate the *called* (i.e., B-end destination) and *calling* (i.e., A-end call origin) addresses conforms to the style defined by ITU-T recommendation X.121. This format comprises a 14-digit binary coded decimal (BCD) number, the first four digits of which comprise the *data network identification code (DNIC)*. The remaining digits are the *national terminal number (NTN)*.

During the *DTE-waiting* period, the network node uses its internal *routing table* to determine the best route to the indicated *called address*, and continues the call set-up procedure by negotiating with nodes further along the connection. When all the nodes have mutually

Data Networks, IP and the Internet: Protocols, Design and Operation   Martin P. Clark
© 2003 John Wiley & Sons, Ltd   ISBN: 0-470-84856-1

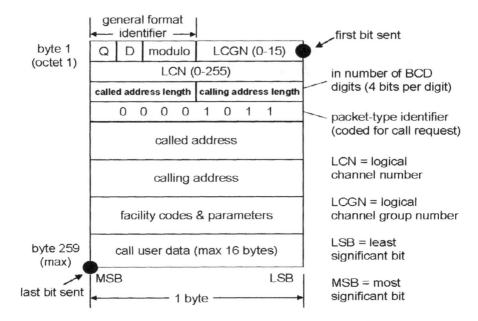

Figure A8.1   X.25 call request packet format (call set-up phase).

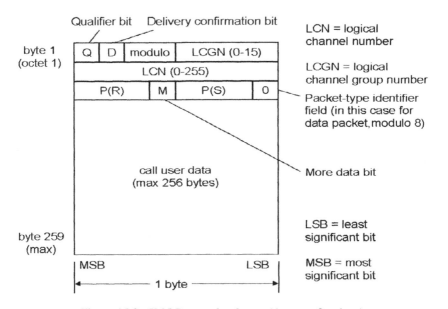

Figure A8.2   X.25 Data packet format (data transfer phase).

'agreed', the connection of the various links and nodes to the destination can be established. At this point, the DCE sends a *call accept* packet back to the DTE and *data transfer* can commence.

During data transfer, a much simpler packet can be sent (Figure A8.2), including only the *general format identifier* (bits Q, D and *modulo*), the *logical channel number (LCN)* (comprising the 4-bit *logical channel group number (LCGN)* and the 8-bit *logical channel number (LCN))* the user data and the packet type, which during data transfer contains nothing more than the send and receive packet sequence numbers, P(S) and P(R), as used for data flow control. As in HDLC, the sequence numbers may be either modulo 8 (values 0–7) or modulo 128 (values 0–127–this is called *extended packet sequence numbering*). The value of the two *modulo bits* in the first byte of the packet indicate the sequence number type. (For modulo 8, the bits are set to '01'; for modulo 128, the bit values are '10'.) The logical channel numbers comprise a total of 12 bits enabling 4096 possible separate logical channels (values: 0–4095).

In an X.25 *data packet*, up to 256 bytes of end-user data (termed *call user data* by ITU-T) may be transferred per packet. (Alternatively, you may also have noticed that a short message of up to 16 bytes of call user data could be sent as part of the call request packet.) The user data field may be coded either in a *canonical format* (i.e., sending the *least significant bit* first) or in a *non-canonical format* (*most significant bit* sent first). This contrasts with the header fields, which are always transmitted in the canonical format (least significant bit first).

The *Q-bit (qualifier bit)* is intended to be used for distinguishing user data packets from control packets. When not used, the value should be set at '0'. Together with the D-bit and the *modulo* bits, the Q-bit forms the *general format identifier*.

Figure A8.3 and Tables A8.1 and A8.2 provide an overview of the complete capabilities of X.25, including the various packet types and network facilities which can be made available. Figure A8.3 illustrates the general format of an X.25 layer 3 packet. Bits are transmitted starting at the top right-hand corner and working along each row (i.e., byte) in turn. Note that not all the fields are used in all the different packet types — the exact format depends upon the packet type (as indicated in the packet identifier field — see Table A8.1). Table A8.2 lists the network facilities made available by X.25 networks.

Typical parameter default settings used in X.25 networks

A typical public X.25 network (e.g., Deutsche Telekom's Datex-P network) might be set up with the following default settings for the packet level interface:

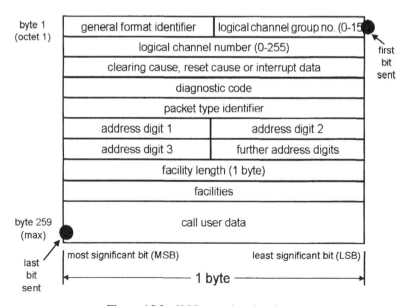

Figure A8.3   X.25 general packet format.

Table A8.1   X.25 packet type identifier coding

Packet type		Coding							
From DCE to DTE	**From DTE to DCE**	**bit 8**	**bit 7**	**Bit 6**	**bit 5**	**bit 4**	**bit 3**	**bit 2**	**bit 1**
Call set-up and clearing									
Incoming call	Call request	0	0	0	0	1	0	1	1
call connected	Call accepted	0	0	0	0	1	1	1	1
clear indication	Clear request	0	0	0	1	0	0	1	1
DCE clear confirmation	DTE clear confirmation	0	0	0	1	0	1	1	1
Data and interrupt									
DCE data	DTE data	X	X	X	X	X	X	X	X
DCE interrupt	DTE interrupt	0	0	1	0	0	0	1	1
DCE interrupt confirmation	DTE interrupt confirmation	0	0	1	0	0	1	1	1
Flow control and reset									
DCE RR (modulo 8)	DTE RR (modulo 8)	X	X	X	0	0	0	0	1
DCE RR (modulo 128)	DTE RR (modulo 128)	0	0	0	0	0	0	0	1
DCE RNR (modulo 8)	DTE RNR (modulo 8)	X	X	X	0	0	1	0	1
DCE RNR (modulo 128)	DTE RNR (modulo 128)	0	0	0	0	0	1	0	1
	DTE REJ (modulo 8)	X	X	X	0	1	0	0	1
	DTE REJ (modulo 128)	0	0	0	0	1	0	0	1
Reset indication	Reset request	0	0	0	1	1	0	1	1
DCE reset confirmation	DTE reset confirmation	0	0	0	1	1	1	1	1
Restart									
Restart indication	Restart request	1	1	1	1	1	0	1	1
DCE restart confirmation	DTE restart confirmation	1	1	1	1	1	1	1	1
Diagnostic		1	1	1	1	0	0	0	1
	Registration								
Registration request	1	1	1	1	0	0	1	1	
Registration confirmation		1	1	1	1	0	1	1	1

- layer 3 window size 2;
- packet size 128;
- window size negotiation allowed;
- packet size negotiation allowed.

Frame relay and ATM protocols — a further development of X.25

Practical experience has shown that ITU-T's X.25 *packet-switching* protocol (developed in the 1970s) does not work well at rates much above about 256 kbit/s but is very error-resistant. For this reason, ITU-T developed further, more specialised data communications protocols based on the same basic connection-oriented packet-switching protocols. *Frame relay* (ITU-T recommendations Q.922 (core aspects) and Q.933 (signalling) is suited for higher quality

Table A8.2   X.25 network facilities

X.25 facilities (optional)	Description
on-line facility registration	if subscribed to, allows user to request facility registration
extended packet sequence numbering	packets numbered modulo 128 rather than normal modulo 8
D bit modification	support of D-bit procedure
packet retransmission	allows DTE to request DCE to retransmit packets
incoming calls barred	prevents incoming calls being presented to DTE
outgoing calls barred	prevents DCE from accepting outgoing calls
one way logical channel outgoing	restricts logical channel use to originating outgoing virtual calls only
one-way logical channel incoming	restricts logical channel use to incoming virtual calls only
non-standard default packet sizes	provides for the selection of non-default packet sizes
non-standard default window sizes	provides for the selection of non-default window sizes
default throughput classes assignment	provides for the selection of default throughput classes
flow control parameter negotiation	allows the DTE to alter window and packet sizes for each virtual call
throughput class negotiation	allows throughput class (i.e., bitspeed) negotiation for each call
closed user group	restricts incoming calls to the DTE to be from other specific DTEs which are members of a closed user group
bilateral closed user group	a closed user group of only two DTEs
fast select	allows the call request packet to contain up to 128 octets of user information, thereby speeding the sending of short user messages
fast select acceptance	authorises DCE to forward fast select packets to the DTE
Reverse charging	allows calls to be requested for charging to recipients rather than callers
Reverse charging acceptance	DTE authorisation to the DCE that it is willing to accept reverse charge calls
local charging prevention	prevents the DCE from allowing virtual calls to be set up for which the local DTE will be charged
network user identification (NUI)	an authorisation procedure allowing for the dial-in to an X.25 network port across a public telephone network
charging information	DTE may request charging information
RPOA related facilities	allows the calling DTE to specify transit networks through which the call should be routed
hunt group	allows several separate DTE /DCE network connections to share the same network address. Incoming calls are routed to any free logical channel within any of the DTEs
call redirection and call deflection	allows deflection on busy or redirection on no answer
called line address modified notification	notifies the calling DTE that the called address has been modified (diverting to a new destination)
transit delay selection and indication	allows the calling DTE to specify a desired transit delay
TOA / NPi address subscription	DTE /DCE to use TOA/NPi address format

lines (BER $< 10^{-9}$) with bitrates between about 64 kbit/s and 34 Mbit/s. *ATM (asynchronous transfer mode)* (defined in ITU-T recommendations I.150 and I.361) is intended to be a multi-purpose protocol capable of carrying voice, data, video and other *multimedia* signals across the very high bitrate transmission lines making up the *broadband integrated services digital network (B-ISDN)*. Frame r'elay and ATM are explained in more detail in Chapters 8 and 14 and in Appendices 9 and 10.

Appendix 9

Frame Relay

Frame relay provides for relatively cheap *wide area* data communication. It is a protocol suitable for data transfer at rates between about 9600 bit/s and 2 Mbit/s, and has proved a viable alternative to leaselines for trunk lines in data networks.

Initially, frame relay networks only supported *PVC (permanent virtual circuit)* service between pairs of fixed end-points. *Frame relay service (FRS)* is typically used as a substitute for a 64 kbit/s leaseline, but without the full costs of a leaseline, since *statistical multiplexing* in the wide area part of the network allowed resources to be shared across a number of users and therefore costs to be saved by each of them (Figure A9.1).

Frame relay UNI (user-network interface)

In the arrangement of Figure A9.1, which is typical for a frame relay network, each of the routers is connected to the frame relay network using a single connection, typically of 64 kbit/s, employing the Frame relay *UNI (user-network interface)*. Over this single physical connection, up to 1024 (2^{10}) logical *data link* channels (usually PVCs — permanent virtual circuits) may be connected, each to a separate end destination.

These channels are available on a permanent basis, but the capacity of the wide area part of the network is only used when there are actually *frames* requiring to be *relayed* from one end of the *data link* to the other.

The *frame header* contains a numbered value called the *data link connection identifier (DLCI)* which identifies the logical connection to which the frame belongs.

As shown in Figure A9.1, it is common in frame relay networks to build fully-meshed networks of logical connections (PVCs) between individual routers (a triangle in our case). This circumvents the need for the routers themselves to act as transit nodes for inter-router traffic, and thus improves the overall performance perceived by the LAN users and can even also reduce the overall cost of the network hardware and the transmission lines needed in the *wide area network (WAN)*.[1]

---

[1] See Chapter 14.

Data Networks, IP and the Internet: Protocols, Design and Operation   Martin P. Clark
© 2003 John Wiley & Sons, Ltd   ISBN: 0-470-84856-1

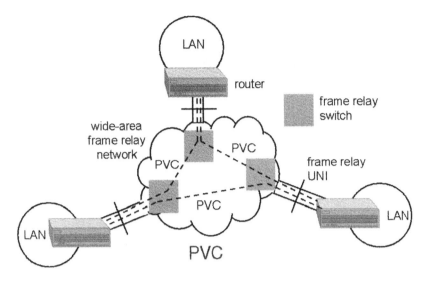

Figure A9.1   Typical use of frame relay to improve the wide area efficiency of LAN/router networks.

Frame relay SVC (switched virtual circuit) service

The main drawback of the arrangement shown in Figure A9.1 is the management effort needed to establish and maintain the large number of PVC connections within the network. Potentially, each time a link in the network fails or a new link is added, human administration work may be necessary to reconfigure some or all of the PVCs to new, more efficient paths. To get around this problem, the Europeans in particular have driven the development of an enhancement of the frame relay UNI to include the capability for *on-demand* establishment of *data links* (i.e., *switched virtual circuits, SVCs*) using a dial-up procedure. While this adds layer 3 functions to the frame relay protocol stack, these are only connection and clearing functions, which do not adversely affect the subsequent end-to-end carriage characteristics (high speed, low delay).

The main benefit of an SVC network is that individual data links need only be established when needed and may be cleared afterwards. This simplifies the network and its management, and in addition has the effect of automatically optimising the routing of connections each time they are newly established.

Congestion control in frame relay networks

The high speed of the computer devices using frame relay networks leads to the need for special measures to control network congestion, since the layer 3 protocol is almost non-existent. Figure A9.2 illustrates a case in which one of the intermediate links within the *wide area* or *backbone* part of the network is in congestion. As a result, frames are accumulating rapidly (and unabated) in the buffer immediately preceding the congested link, as they wait to be transmitted over the link. Ultimately, the buffer will overflow and frames will be lost, so affecting all the data links sharing the congested link. Worse still, once the end user devices detect the loss of information, retransmission will commence, and the load on the network will only further increase.

Since the connection from the sending device of Figure A9.2 to the network is not congested, the sending of frames would continue unabated, were it not for the *congestion notification* procedures within the frame relay protocols.

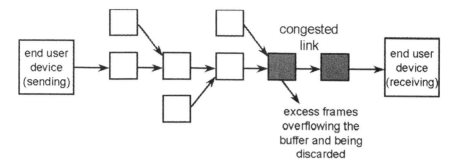

Figure A9.2   Congestion in a Frame Relay network.

At any time when a frame underway encounters congestion (the exceeding of a given wait-ing time or a given buffer queue length) then the frame header is tagged with a notification message, called the *forward explicit congestion notification (FECN)*. This notifies the receiv-ing device and the switch to which it is connected of the congestion, which may then be communicated back to the source end by means of a *backward explicit congestion notifica-tion (BECN)* message. The sending device may voluntarily respond to receipt of the BECN by reducing its transmitted output, or may be forced by the network to do so. By reducing the frame transmission rate from relevant causal sources, the congestion will ease, and the transmission rate restriction may be removed. Meanwhile, further service degradation within the network as a whole is avoided.

A further refinement of the congestion control procedure of frame relay is provided by the *committed information rate (CIR)* and *excess information rate (EIR)* parameters. The committed information rate (CIR) is the agreed minimum bitrate to be provided by the network between the two ends of the frame relay data link. The CIR is agreed at the time of configuring the connection (in the case of PVC) or at the time of connection set-up (in the case of an SVC). Provided the frame transmission rate of the sending device is at or below the CIR, then the network is not permitted to force a reduction in the frame sending rate of the sending device, and is not permitted wilfully to discard frames. However, where the sending device is exceeding the CIR at a time when the BECN message is received, then the network may first request reduction in the rate of frame transmission to the CIR. Should the reduction not be undertaken (for example, because the sending device cannot respond to the request), then the network is permitted to discard the *excess* frames.

At times of no congestion, sending devices are permitted for defined short periods of time (called the *excess burst,* or *excess burst duration, B_e*) to transmit at bitrates higher than the CIR. The maximum bitrate at which the device may send is termed the *EIR (excess information rate)*. The EIR is always greater than or equal to the CIR. It is a management decision for the network operator how high EIR and CIR may be set for a given connection, and usually these values are included in the contract or order for the user's connection (for a PVC) or negotiated at connection set-up time for an SVC. The ability to handle short bursts of high speed information (above the CIR) is what makes frame relay networks attractive to data applications requiring fast response times.

Frame relay NNI (network-network interface)

Since the switches in a frame relay network are usually supplied by a single manufacturer, it is usually not necessary to use a standardised interface between the *nodes* within the network. As a result, the individual manufacturers tended to develop extra congestion controls, network

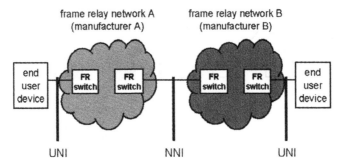

Figure A9.3 Frame relay NNI (network-network interface).

management and service features over and above those required by the frame relay standards, in order to try to improve the market value of their products. All very well. Except, of course when a given frame relay connection needs to be switched across two different networks or subnetworks, supplied by different manufacturers (as in Figure A9.3). For this a standardised interface, the *NNI (network-network interface)* is required.

While the frame relay NNI allows for interconnection of subnetworks of switches supplied by different manufacturers, and while reliable data transfer is possible, it is true that the congestion control and management capabilities of the combined network are much more restricted than the capabilities available within each of the subnetworks independently. This reflects the relative immaturity of the frame relay NNI standard.

Frame format

Figure A9.4 illustrates the format of a single frame. It consists of five basic information fields, much like the data link layer format of X.25 (i.e., HDLC). The *flag* marks the beginning of the frame, delineating it from the previous frame. The *address* field carries the *DLCI (data link connection identifier)*, the equivalent of the *logical channel number (LCN)* of X.25.[2] In addition, the address field also contains control information (*command/response*), the forward and backward explicit congestion notification (FECN and BECN), the *discard eligibility (DE)* indication and some extra fields used for *extended addressing*.

The control field contains supervision information for the connection. These are signals like *receiver ready (RR), receiver not ready (RNR)*, etc. For user *information frames*, this field indicates the length of the frame.

The *information field* contains the user information. This may be up to 65 536 bytes in length. Finally, the *frame check sequence (FCS)* is a *cyclic redundancy check (CRC)* code[3] providing for error detection.

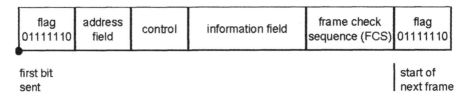

Figure A9.4 Frame format for *frame relay*.

---

[2] See Chapter 3.
[3] See Chapter 3.

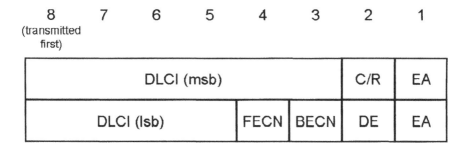

| 8 | 7 | 6 | 5 | 4 | 3 | 2 | 1 |
| (transmitted first) | | | | | | | |

| DLCI (msb) | | | | | | C/R | EA |
| DLCI (lsb) | | | | FECN | BECN | DE | EA |

BECN	= backward explicit congestion notification
C/R	= command / response bit
DE	= discard eligibility
DLCI	= data link connection identifier
EA	= extended addressing
FECN	= forward explicit congestion notification
lsb	= least significant bits
msb	= most significant bits

Figure A9.5 Address field format for frame relay.

Address field format

Figure A9.5 illustrates the two octet (i.e., two byte) *address* field used in frame relay. The main function of the address field is to carry the *data link connection identifier (DLCI)*, which identifies the end device to which the frame is to be sent (OSI layer 2 address). The DLCI is a 10-bit field,[4] allowing up to 1024 separate *virtual connections* to share the same physical connection.

Should congestion be encountered by a given frame during its transit through the network, then the affected intermediate switch will toggle the FECN (forward explicit congestion notification). This alerts the receiving device of the congestion. In response, returned frames are marked using *the BECN (backward explicit congestion notification)*. This allows the flow control procedures to be undertaken. Should congestion become so serious that frames need to be discarded, then frames marked with the *discard eligibility (DE)* set to '1' will be discarded first. The DE bit is set to '1' by the first frame relay switch (near the origin) on *excess* frames (i.e., those causing the information rate to exceed the *committed information rate (CIR)*).

ITU-T recommendations pertinent to frame relay

The following ITU-T recommendations define frame relay:

• recommendation I.233 describes the frame relay service;

• recommendation I.122 defines the framework of recommendations which specify frame relay, referring to the complete list of relevant recommendations;

• recommendation Q.922 is perhaps the most important. It defines the *core aspects* of frame relay — specifically the data link procedure (i.e., frame format, address field etc.);

---

[4] When *extended addressing* is employed, the DLCI may be extended to either a 16-bit (65 536 virtual channels) or a 23-bit field (8.3 million virtual channels).

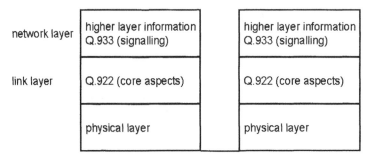

network layer	higher layer information Q.933 (signalling)	higher layer information Q.933 (signalling)
link layer	Q.922 (core aspects)	Q.922 (core aspects)
	physical layer	physical layer

Note: Q.933 is the network layer protocol used for establishing switched connections (SVCs). It is not used in the PVC (permanent virtual circuit) service.

Figure A9.6 Protocol stack for frame relay.

- recommendation I.370 defines the congestion management procedures;
- recommendation Q.933 defines the *signalling* procedures for setting up *switched virtual connections*. This recommendation is not relevant for *permanent virtual circuit (PVC)* service.

Figure A9.6 shows the layered protocol structure of frame relay.

FRAD (frame relay access device)

Frame relay has historically been offered by public telecommunications operators as a cheap alternative to a leaseline in cases where high bitrates were desirable but the number hours usage per day was relatively low. In order to access a frame relay network, a DTE (such as a router) must either incorporate a frame relay interface card, or alternatively use some other standard interface and an external conversion device. The most important of the available external conversion devices is the FRAD. *Frame relay access devices (FRADs)* provide for the conversion of *continuous bit stream oriented (CBO)* data signals into a *frame format*, much

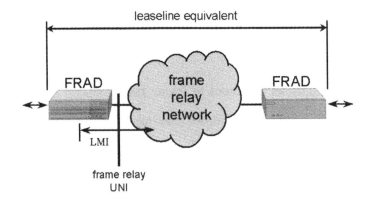

FRAD = frame relay access device
LMI = local management interface
UNI = user-network interface

Figure A9.7 A frame relay access device (FRAD).

like the way a *PAD (packet assembler/disassembler)* converts the data stream from a 'dumb' computer terminal into the packet format required by an X.25 network. A FRAD may thus be used to convert a continuous data stream from an end user device normally expecting to use a 'transparent' leaseline-like connection into a format suitable for carriage by a frame relay network. Figure A9.7 illustrates the principle.

Figure A9.7 illustrates the equivalent of a data leaseline, created using two *frame relay access devices (FRADs)* and a frame relay network. This is usually more economical than the equivalent leaseline where the CIR (committed information rate) is lower than the maximum speed of the leaseline. Usually the EIR is set to the maximum speed of the equivalent leaseline, and CIR is set somewhat above the average actual user information throughput rate. This virtually guarantees throughput of the user data (at the CIR). Bursts up to the maximum leaseline speed (the EIR) are still possible for short periods, but at more 'quiet' times other users may share the trunk resources within the network on a 'statistical' basis. This is reflected in the lower costs of frame relay service compared to 'transparent' leaseline service.

Also shown in Figure A9.7 is the *local management interface (LMI)*. This allows *status* and other *network management* information to be shared between the end terminal and the network in an SNMP-like format, so enabling overall monitoring and management of the network.[5]

---

[5] See Chapter 9.

Appendix 10

Asynchronous Transfer Mode (ATM)

ATM (asynchronous transfer mode) networks were designed to be able to support:

- usage by multiple users simultaneously;

- different *reat-time (RT)* and *non-real-time (NRT)* telecommunications needs (e.g., telephone, data transmission, LAN interconnection, videotransmission, etc.);

- each application running at different transmission speeds with differing bandwidth needs.

But these capabilities are also offered by predecessor technologies, so why bother with ATM, you might ask? What distinguished ATM from its predecessors (X.25, frame relay and IPv4) was its ability to conduct fast relaying and instant-by-instant adjustment in the allocation of the available network capacity between the various users competing for its use. The dynamic allocation of bandwidth is achieved by ATM using a newly developed technique called *cell relay switching*.

While some 'IP purists' like to proclaim the obsolescence of ATM, it remains an important data networking technology. It is used widely in public telecommunications networks as a transport technology in access networks — being used particularly in conjunction with xDSL.[1] And while many of its revolutionary features have already been 'stolen' for incorporation into IP-suite protocols (including TCP, MPLS and IPv6), it still provides quality of service guarantee measures which are unmatched by IP protocols.

The technique of *cell relay*

Cell relay is a form of statistical multiplexing similar in many ways to *packet switching*, except that the *packets* are called *cells* instead. Each of the cells is of a fixed rather than a variable size. The fixed cell size defined by ATM standards includes a 48 octet (byte) *payload* plus a 5 octet header (i.e., 53 octets in all — see Figure A10.1). The transmission line speeds

---

[1] See Chapter 8

Data Networks, IP and the Internet: Protocols, Design and Operation   Martin P. Clark
© 2003 John Wiley & Sons, Ltd   ISBN: 0-470-84856-1

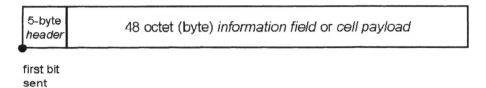

first bit
sent

Figure A10.1   ATM 53 byte cell format.

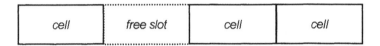

Figure A10.2   The *cells* and *slots* of cell relay.

currently foreseen to be used are either 2 Mbit/s, 12 Mbit/s, 25 Mbit/s, 34 Mbit/s, 45 Mbit/s, 52 Mbit/s, 155 Mbit/s or 622 Mbit/s. We may thus conclude that:

- the overhead is at least 5 bytes in 53 bytes, i.e., > 9 %;

- the duration of a cell is at most (i.e., at 2 Mbit/s line rate) 53×8 bits $/ 2$ Mbits^{-1} $= 0.2$ ms. (12 µs at 34 Mbits^{-1}).

Since the cell duration is relatively short, provided a priority scheme is applied to allow cells from delay-sensitive signal sources (e.g., speech, video, etc.) to have access to the next cell *slot* (see Figure A10.2) then the jitter (i.e., the variation in signal propagation delay) can be kept very low (of the order of 0.2 ms). Not zero as is possible with circuit switching, but at least low enough to give a subjectively acceptable quality for telephone listeners or video watchers. Jitter-insensitive traffic sources (e.g., datacommunication channels) can be made to wait for the allocation of the next *low priority slot*. It is the ability to limit jitter and to prioritise the use of the bandwidth in this 'near real-time' manner which particularly marks out ATM.

The jitter could be reduced still further by reducing the cell size, but this would increase the proportion of the line capacity needed for carrying the cell headers, and thus reduce the line efficiency.

The ATM cell header

The *cell header* carries information sufficient to allow the ATM network to determine to which connection (and thus to which destination port and end user) each cell should be delivered. We could draw a comparison with a postal service and imagine each of the cells to be a letter of 48 alphanumeric characters contained in an envelope on which a 5 digit postcode appears. You simply drop your letters (cells) in the right order and they come out in the same order at the other end, though maybe slightly jittered in time. Just like a postal service has numerous vans, lorries and personnel to carry different letters over different stretches and sorting offices to direct the letters along their individual paths, so an ATM network can comprise a mesh of transmission links and switches to direct individual cells by inspecting the *address* contained in the *header* (Figure A10.3).

You may think that a 5-digit postcode is rather inadequate as a means of addressing all the likely users of an ATM network, and it might be, were it not for several provisions of the ATM specifications. First, the 'digits' are whole octets (base 256) rather than decimal digits (base 10). This means that the header has the range for over 10^{12} combinations (40 bits), though

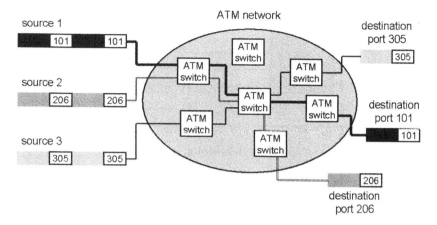

Figure A10.3 Switching in an ATM network.

only a maximum of 28 bits (270 million combinations) are ever used for addressing. Second, the addresses (correctly called *identifiers*) are only allocated to active connections.

ATM connections are allocated an identifier during call set-up, and this is re-allocated to another connection when the connection is cleared. In this way the number of different identifiers need not directly reflect the number of users connected to the network (which may be many millions), only the number of simultaneously active connections. In addition, various subregions of the network (indeed each individual interface if necessary) may use different identifier schemes, thus multiplying the available capacity, but then demanding the ability of network nodes to *translate* (i.e., amend) identifiers in the 5-octet header.

By highly efficient usage of the information carried in the header, the length of the header can be kept to a minimum. As a result, the network *overhead* is minimised.

The components of an ATM network

There are four basic types of equipment which go to make up an ATM network. These are:

- *customer equipment (CEQ)*, also called *B-TE (broadband-ISDN terminal equipment)*;

- ATM *switches*;

- ATM *cross-connects*, and;

- ATM *multiplexors*.

These elements combine together to make a network as shown in Figure A10.4. A number of standard interfaces are also defined by the ATM specifications as the basis for the connections between the various components. The most important interfaces are:

- the *user-network interface (UNI)*;

- the *network node interface (NNI)*, and;

- the *inter-network interface (INI)*.

These are also shown in Figure A10.4.

ATM *customer equipment (CEQ)* is any item of equipment capable of communicating across an ATM network. Customer equipments communicate with one another across an ATM

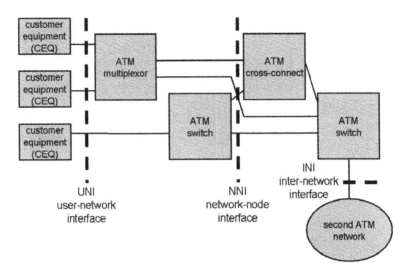

Figure A10.4   The components of an ATM network (ITU-T network reference model for ATM).

network by means of a *virtual channel (VC)*. The VC may either be set up and cleared down on a call-by-call basis similar to a telephone network, in which case the connection is a *switched virtual circuit (SVC)* or it may be a permanently dedicated connection (like a leaseline or private wire), in which case it is a *permanent virtual circuit (PVC)*.

An *ATM multiplexor* allows different *virtual channels* from different ATM UNI ports to be bundled for carriage over the same physical transmission line. This is called ATM *multiplexing*. Thus two or three customers outlying from the main exchange (Figure A10.4) could share a common line. Returning to our analogy with the postal system, the ATM multiplexor performs a similar function to a postal sack — it makes easier the task of carrying a number of different messages to the sorting station (ATM switch) by bundling a number of *virtual channels* into a single container, a *virtual path (VP)*. More about virtual channels and virtual paths later (Figures A10.7–A10.10).

An *ATM cross-connect* is a slightly more complicated device than the ATM multiplexor. It is analogous to a postal depot, where the various vanloads of mail are unloaded, the various sacks are sorted and adjusted into different van loads. At the postal depot, the individual sacks remain unopened, and at the ATM cross-connect, the virtual path contents, the individual virtual channels remain undisturbed. The ATM cross-connect appears again in Figure A10.9.

A full *ATM switch* is the most complex and powerful of the elements making up an ATM network. It is capable not only of cross-connecting virtual paths, but also of sorting and switching their contents, the virtual channels (Figure A10.10). It is the equivalent of a full postal sorting office, where sacks can either pass through unopened, or can be emptied and each letter individually re-sorted. It is the only type of ATM node device capable of interpreting and reacting upon user or network signalling for the establishment of new connections or the clearing of existing connections.

The ATM adaptation layer (AAL)

An extra functionality is added to a basic ATM network (correctly called the *ATM layer*) to accommodate the carriage of various different types of *connection-oriented* and *connectionless* network services. This functionality is contained in the *ATM adaptation layer*. The ATM

Table A10.1   Service classification of the ATM adaption layer (AAL)

Transmission characteristic	Class A	Class B	Class C	Class D
AAL Type	AAL Type 1 (AAL1)	AAL Type 2 (AAL2)	AAL Type 3/4 (AAL3/4), AAL Type 5 (AAL5)	AAL Type 3/4 (AAL3/4), AAL Type 5 (AAL5)
Timing relation between source and destination		Required	Not Required	
Bit rate	Constant		Variable	
Connection mode		Connection-Oriented		Connectionless

adaptation layer (AAL) lays out a set of rules about how the 48-byte *cell payload* can be used, and how it should be coded. The services offered by the ATM adaptation layer (AAL) are classified into 4 *classes* or *types* (the standards use both terminologies). The distinguishing parameters of the various classes are as illustrated in Table A10.1.

An example of a Class A service is *circuit emulation* (i.e., a connection service providing for 'clear channel' connections like hard-wired digital circuits). In the ATM specifications such services are referred to as *constant bit rate (CBR)* or *circuit emulation services (CES)*. Thus a constant bit rate video or speech signal would be an AAL Class A service and would use AAL1.

Variable bit rate (VBR) video and audio is an example of a class B service. Thus an audio speech signal which sends no information during silent periods is an example of a Class B VBR service and would use AAL2.

Class C and Class D cover the connection-oriented and connectionless data transfer services. Thus an X.25 packet switching service would be supported by a Class C service, and connectionless data services like electronic mail and certain types of LAN router service would be Class D. Both classes C and D use AAL types AAL3/4 or AAL5.

ATM virtual channels and virtual paths

A virtual channel extended all the way across an ATM network (ATM layer) is actually a *virtual channel connection (VCC)*. This connection is composed of a number of shorter length *virtual channel links*, which when laid end-to-end make up the VCC.

A *virtual channel link* is a part of the overall VCC, and shares the same endpoints as a *virtual path connection (VPC)* (Figure A10.5). The idea of a *virtual path (VP)* is valuable in the overall design and operation of ATM networks. As we have already explained, a virtual path has a function rather like a postal sack. In the same way that a postal sack helps to ease the handling of letters which all share a similar destination, so a virtual path helps to ease the workload of the ATM network nodes by enabling them to handle bundled groups of virtual channels. Thus a virtual path (VP) carries a number of different *virtual channel links*, which in their own separate ways may be concatenated with other virtual channel links to make VCCs.

Just like virtual channels, virtual paths can be classified into *virtual path connections (VPCs)* and *virtual path links*, where a VPC is made up by the concatenation of one or more virtual path links. A virtual path link is derived directly from a physical transmission path.

User, control and management planes

Before two *customer equipments (CEQ)* may communicate with one another (i.e., transfer information) across the so-called *user plane (U-plane)* of an ATM network, a connection must

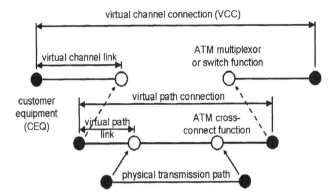

Figure A10.5   The relationship between virtual channels (VCs), virtual paths (VPs) and physical transmission paths.

first be established (ATM is a *connection-oriented* protocol!). The connection is established by means of a *control* or a *management* communication between the CEQ and the network. This communication may take one of five forms:

- *control plane communication (access)*;

- *control plane communication (network)*;

- *management plane communication type 1*;

- *management plane communication type 2*;

- *management plane communication type 3*.

A control plane communication (access) is a one conducted between *CEQ (customer equipment)* and an ATM switch. During such a communication, which uses *UNI signalling*, a connection is established or released (in the case of *SVCs, switched virtual circuits*) much like dialling a telephone number in a telephone network. Control plane communications (network) will follow, as the ATM switch communicates (using network signalling) with other nodes in the network to establish the complete network connection. Once the connection is established, the user *transfers information* (i.e., communicates) across the *user plane*.

The connection could also have been established manually by the service technicians at the network management centre (a *PVC, permanent virtual circuit*). In this case, the user uses a management plane communication type 1 from his CEQ to the NMC to request the establishment of a permanent connection. This could be carried by UNI signalling or could simply be a telephone call. The various switches and other network elements are then configured from the NMC by means of messages sent by management plane communication type 2.

Management plane communication type 3 is initiated by ATM switches which require to refer to the NMC for information, authority or other assistance in the process of connection set-up. (It may be, for example, that certain high bandwidth connections require authorisation from the NMC in order to prevent network congestion at peak times.)

How is a virtual channel connection (VCC) set up?

A UNI *signalling virtual channel* (*SVC*, but not to be confused with *SVC—switched virtual connection*) is a virtual channel or virtual path connection at a UNI dedicated specifically to *UNI signalling*. Signalling virtual channels may also exist at an NNI interface.

A signalling message sent over the *SVC (signalling VC)* might be 'set up virtual connection number one between user A and user B'. Another message might be 'clear the connection between A and B'. ATM uses a dedicated channel for signalling information (this is termed *common channel signalling*).

At the time when a CEQ requests to set up a new *SVC (switched virtual circuit)* connection across an ATM network, it must first negotiate with the network over the *UNI signalling VC*, declaring the required *peak cell rate, quality of service (QOS) class* and other parameters needed. The *connection admission control (CAC)* function at the ATM switch then decides whether sufficient resources are available to allow immediate connection. If so, the connection is set up. If not, the connection request is rejected in order to protect the quality of existing connections. (The analogy is the telephone user's receipt of busy tone when no more lines are available.) During the negotiation, virtual paths and connections between the various nodes and other equipments are allocated, and the reference numbers of these connections, the combination of *virtual path identifiers (VPIs)* and *virtual connection identifiers (VCIs)*, are confirmed over the signalling channel. These values (VPI and VCI) then appear in the *header* of any cells sent, in order to identify all those cells which relate to this connection.

Signalling virtual channels and meta signalling virtual channels

Both management and control communication in an ATM network take place via signalling virtual channels (SVCs). At NNI interfaces, SVCs are usually permanently configured between the various servers (i.e., control processors) controlling a particular B-ISDN service (e.g., *video on demand*, picture telephone service etc). But unlike narrowband ISDN, signalling virtual channels (SVC) are not normally permanently available at UNI. Instead, they are established on demand by means of a *meta signalling virtual channel (MSVC)*. This is a permanently allocated UNI signalling channel of a fixed bandwidth. It is found in the virtual path VPI = 0 and has a VCI value standard to the particular network.

By means of the meta signalling virtual channel, the end device (CEQ) can establish an SVC (signalling VC) to the ATM switch (*c-plane*) or to the network management centre (*m-plane*) for signalling communication. A *service profile identifier (SPID)* carried in the meta-signalling determines which service the user requires, and enables a signalling virtual channel to the appropriate *signalling point* server to be established.

The functionality of the device which exists at the end of a signalling virtual channel and conducts the act of signalling is called a *signalling point (SP)*. Such functionality exists in customer equipment (CEQ) and in ATM switches. A *signalling transfer point (STP)* is a switching point for the information carried in signalling virtual channels. Using a single *signalling VC* via an STP, an SP may communicate signalling messages to numerous other SPs using either *associated-mode* or *quasi-associated* mode signalling.

In the associated-mode, signalling messages are passed directly between communicating *signalling points (SPs)*. In the quasi-associated-mode, the signalling messages follow a different route from the VCCs which will subsequently be set-up. The route of the signalling messages is via a dedicated signalling data network, and transits special *signalling transfer points (STPs)*. STPs improve the efficiency and reliability of the signalling network (Figure A10.6).

Virtual channel identifiers (VCIs) and virtual path identifiers (VPIs)

As we saw in Figure A10.5, *virtual channel connections* comprise concatenated *virtual path connections*. Each is *identified* by reference numbers carried by the cell headers of active connections called *virtual channel identifiers (VCIs)* and *virtual path identifiers (VPIs)*.

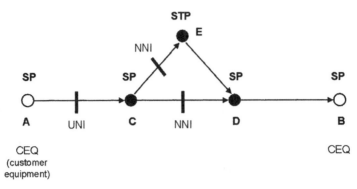

Figure A10.6 Signalling points (SPs) and signalling transfer points (STPs).

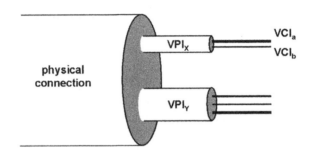

Figure A10.7 Virtual path and virtual channel identifiers.

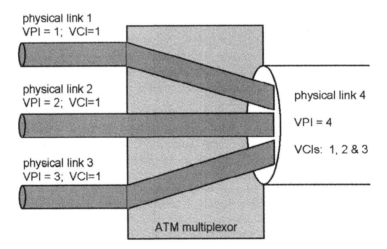

Figure A10.8 ATM multiplexor.

Figure A10.7 illustrates how a physical connection may be subdivided into a number of different virtual paths, each with a unique VPI. Each VPI in turn may be subdivided into several virtual channels, each with a separate VCI. The combination of VPI and VCI values is unique to each UNI or NNI and is sufficient to identify any active connection at the interface (i.e., on the same physical connection).

An ATM multiplexor, as we discussed earlier, allows a number of virtual channels from separate virtual paths to be combined over a single virtual path. Thus the virtual channels carried by physical links 1, 2 and 3 of Figure A10.8 are combined together into a single virtual path carried by physical link 4. In this way three separate end user devices use separate virtual channels to share a single physical connection line from ATM multiplexor to the ATM network.

An ATM cross-connect (i.e., VP cross-connect) allows rearrangement of virtual paths without disturbance of the virtual channels which they contain. Thus in Figure A10.9, the contents of incoming virtual path VPI = 1 are crossconnected to outgoing virtual path VPI = 6, the VCIs remaining unchanged. An ATM cross-connect is thus a simple form of ATM switch, but one which need only process (and *translate* (i.e., amend)) VPI values.

A full ATM switch (Figure A10.10) has the capability not only to cross-connect virtual paths, but also to switch virtual channels between different virtual paths. This requires the additional ability to process and translate VCIs held in cell headers. It is thus a relatively

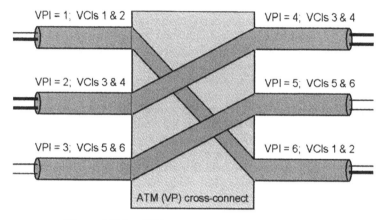

Figure A10.9   ATM cross-connect (VP crossconnect).

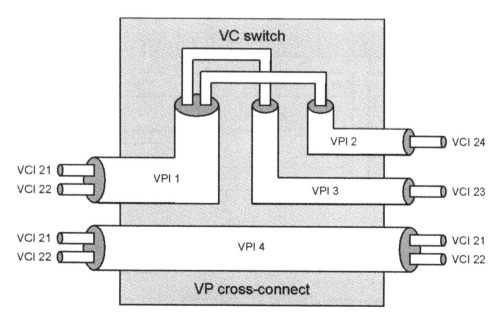

Figure A10.10   Full ATM switch.

complex device. Good performance depends upon very fast processing of both VPIs and VCIs in the cell headers. A full ATM switch is consequently a costly device.

Information content and format of the ATM cell header

The main function of the ATM cell header is to carry the VPI and VCI information which allows the active network elements to switch the cells of active connections through the network. The exact format of the cell header is as shown in Figure A10.11.

The cell header comprises 40 bits, of which 24 (UNI) or 28 (NNI) are used for the virtual path and virtual channel identifiers. Together, the VPI / VCI fields are called the *routing field*. There are four other *fields* which occupy the remainder of the header. The *PT (payload type)* field is occupied by the *payload type identifier (PTI)*. This identifies the contents of the cell (the *information field* or *payload*) as either a user data cell, a cell containing network management information, or a resource management cell. The *cell loss priority (CLP)* bit (when set to value 1 is used to identify less important cells which may be discarded first at a time of network or link congestion. The *generic flow control (GFC)* field is used to control the cell transmission between the customer equipment (CEQ) and the network (Figure A10.12). Finally, the *header error control (HEC)* field serves to detect errors in the cell header caused during cell transmission.

When there is no trunk congestion (i.e., there is no appreciable accumulation of cells waiting in the multiplexor buffer to be transmitted over the trunk) then the GFC field is set to the *uncontrolled transmission mode*. However, if there is a sudden surge of cells from all of the CEQ devices and the multiplexor experiences congestion (i.e., the filling of its cell buffers to a critical threshold level) then the GFC field is used to subject the cell flow from the various CEQ devices to *controlled transmission*. This limits the rate at which the CEQ devices may continue to send cells of one or more different types to the network.

An ATM cell is transmitted in the order of octets (i.e., octet 1 first, followed by octet 2, then octet 3, etc.). The *most significant bit (MSB)* of each octet (i.e., bit 8) is transmitted first. Thus first the header and then the payload are transmitted, MSB first.

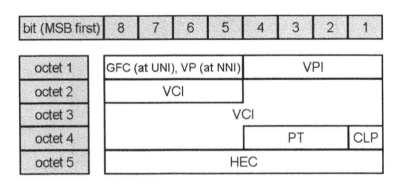

CLP	= cell loss priority	PT	= payload type
GFC	= generic flow control	VCI	= virtual channel identifier
HEC	= header error control	VPI	= virtual path identifier
MSB	= most significant bit		

Note: The GFC is used only at the UNI, at the NNI bits 5-8 of octet 1 are used as VPI.

Figure A10.11   Structure of the ATM cell header.

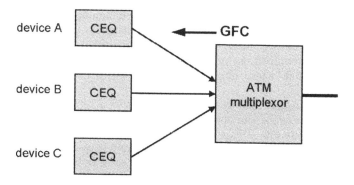

Figure A10.12   Generic flow control regulates trunk congestion at a multiplexor.

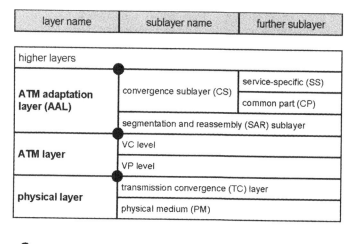

●   service access point (SAP) - an imaginary point between functional layers

Figure A10.13   The functional layers of ATM.

ATM protocol layers

Figure A10.13 illustrates the protocol layers of ATM. We shall use this in the discussion which follows to define the terminology of ATM and to explain the relationship of the various layers to one another.

The ATM transport network

The foundation of the various protocol layers (the *protocol stack* — the set of functions which together make information transfer possible) is the physical medium used for the carriage of electrical or optical signals. The *physical layer* is a specification which defines what electrical or optical signals and voltages, etc. should be used. In addition, it sets out a procedure for transferring data information across the line, providing for clocking of the bits sent and the monitoring of the equipment.

    The physical layer is divided into two sublayers. These are the *physical medium* sublayer and the *transmission convergence (TC) sublayer*. The physical medium sublayer defines the exact electrical and optical interface, the line code and the bit timing. The TC sublayer provides for framing of cells, for cell delineation, for cell rate adaption to the information carriage capacity of the line, and for operational monitoring of the various line components (*regenerator section (RS), digital section (DS)* or *transmission path (TP)*). Preferred physical media defined for use with ATM include optical fibre and coaxial cable. There is also some scope to use twisted pair cable.

It is the ATM layer which controls the transport of cells across an ATM network, setting up *virtual channel connections* and controlling the submission rate (generic flow control) of cells from user equipment. The *service* provided to the ATM layer by the physical layer is the physical transport of a valid flow of cells. This is 'delivered' at a conceptual 'point' called the physical layer *service access point (PL-SAP)*. The flow of cells is correctly called a *service data unit (SDU)* — in fact the *PL-SDU (physical layer service data unit)*).

The ATM layer controls the service provided to it by the physical layer by means of *service primitive* commands. These are standardised requests and commands exchanged between the control function within the ATM layer (in the jargon called the *ATM layer entity (ATM-LE)*) and the *physical layer entity (PL-LE)*. They allow, for example, a particular ATM-LE to request transfer of a flow of cells (service data unit). Conversely, the physical layer may wish temporarily to halt the transfer of cells to it by the ATM layer because of a problem with the physical medium.

The *transmission convergence sublayer* receives information in the form of cells provided to it by the ATM layer. This is the PL-SDU, or more specifically, the *TC-SDU*. These cells are supplemented by further information, including PL-cells (physical layer cells) and OAM cells (operations and maintenance cells). The extra information, an example of *protocol control information (PCI)* turns the TC-SDU into a *TC-PDU (protocol data unit)*. It ensures the correct transmission of information across the physical medium. The TC-PDU is passed to the physical medium sublayer, where it is called the *PM-SDU (physical medium service data unit)*.

The *operation and maintenance (OAM)* cells of ATM provide for advanced measurement of network performance without affecting live connections.

Finally, the PM-SDU is converted to the PM-PDU by addition of further *PCI* and is passed to the medium itself. The form of the PM-PDU (and thus the conversion performed by the physical medium sublayer) is dependent upon the type of medium used (e.g., electrical, optical, etc.). To accommodate a change of the physical medium, only the *physical medium sublayer* need be swapped. Other hardware and software components (e.g., corresponding to the ATM layer) can be re-used.

Together the ATM layer, the physical layer and the physical medium itself are called an *ATM transport network*. An ATM transport network is capable of conveying information in the form of cells between network end-points. However, in order that the information content carried by an ATM transport network can be correctly interpreted by the receiver, there are further higher layer protocols defined. The most important of these is the *ATM adaptation layer (AAL)*.

Capability of the ATM adaptation layer (AAL)

As the name suggests, the *ATM adaptation layer (AAL)* provides for the conversion of the *higher layer information* into a format suitable for transport by an ATM transport network. The higher layers are information, devices or functions of unspecific type which require to communicate across the `ATM network. Higher layer information carried by the ATM network may be either:

- *user information (user plane)* of one of a number of different forms (e.g., voice, data, video, etc.) as categorised by the AAL service classes (Table A10.1),

- *control information (control plane)* for setting up or clearing connections, or

- *network management information (management plane)* for monitoring and configuring network elements or for sending requests between network management staff.

Like the other layers, the AAL accepts *AAL-SDUs* from the higher layers and passes an *AAL-PDU* to the layer below it (the ATM layer), where it is known as an ATM-SDU. But unlike the ATM and physical layers a number of different alternative services can be made available to the higher layers above the AAL, thus allowing different types of information to be *adapted* for carriage across a common ATM transport network. It is the ATM adaptation layer which gives ATM networks their capability to transfer all sorts of different information types. It is split into two sublayers, *the convergence sublayer, CS* (where the alignment of the various information types into a common format takes place and division into cells occurs) and the *segmentation and reassembly sublayer, SAR* (where the cells are numbered sequentially to allow reconstruction in the right order at the receiving end).

Protocol stack when communicating via an ATM transport network

Figure A10.14 illustrates the peer-to-peer communications which take place when two user end devices communicate with one another by means of an ATM transport switch. The ATM

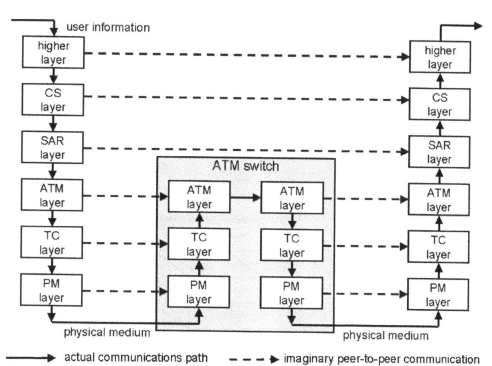

Figure A10.14 Protocol layer representation of two end devices communicating via ATM Layer switch.

switch supports only the lowest three protocol layers, and 'speaks' peer-to-peer with each of the ends, translating protocol information such as VPIs and VCIs as necessary and relaying user information. Meanwhile, at the ATM adaptation layer (AAL) and the higher layers, the two end devices communicate peer-to-peer directly over the connection established by the lower three layers. This information remains uninterpreted and passes 'transparently' through the network.

If we were to monitor the wire between either of the end devices and the ATM switch of Figure A10.14 then we would observe communication at each of the layers. What we actually observe are cells, but structured a little like a Russian doll. The smallest doll (right inside) is the information we want to carry between the users (the *higher layer information*). All the other dolls are the protocol information (PCI) — one doll for each of the lower layers, each providing a function critical to the reliable carriage correct interpretation of the message.

ATM protocol reference model (PRM)

Strictly, Figure A10.14 illustrates only the protocol stack for the *user* plane (i.e., for the transfer of information between end devices once the connection has been established. The AAL protocols used on the *control* and *management* planes will usually differ from the AAL protocols used on the user plane of the same connection, though identical protocols will be used at the ATM transport layers. This is illustrated schematically in the *ATM protocol reference model (PRM)* as shown in Figure A10.15.

In the case of the *control* and *management planes*, the network itself must interpret the higher layer information and react to it. The *control plane* AALs (for the user-network signalling in setting up a switched virtual circuit, SVC) will typically need to be suited for

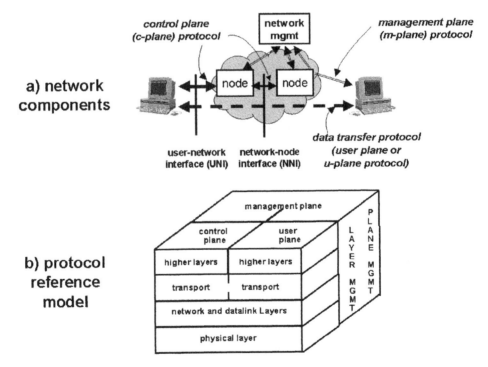

Figure A10.15   B-ISDN interfaces and protocol reference model [Figure A10.15b reproduced courtesy of ITU].

data information transfer. Similar protocols will also be necessary for the management plane. In contrast, in the case of communication across the user plane, the higher layer information may take any number of different forms (speech, data, etc.), but the individual network elements themselves (e.g., switches) may be incapable of recognising and interpreting these various forms.

In real switches and ATM end user devices, common or duplicate hardware and software may thus be used for management, control and user planes at ATM and physical layers, but distinct hardware and software will be necessary for signalling and user information transfer at the AAL and higher layers of these planes.

The different types of user, control and management services are carried by the AAL by means of *service-specific convergence services*. Examples of specific user plane services offered by the ATM adaptation layer (AAL) are:

- frame relay SSCS (service specific convergence sublayer) service;

- SMDS (switched multimegabit data service) SSCS service;

- reliable data delivery SSCS service (a packet-network like data network service);

- LAN emulation SSCS service;

- desktop quality video SSCS service;

- entertainment quality video SSCS service;

- further services still in the stage of development by ATM Forum.

There are a set of service-specific protocols used on the control plane. These are known as the *signalling AAL (SAAL)* and are defined by ITU-T recommendations Q.2110, Q.2130 and Q.2140. They operate in association with AAL5. The SAAL in turn carries either *UNI signalling* or *NNI signalling*. UNI signalling is a combination of *MTP3 (message transfer protocol layer 3)* and *DSS2 (digital subscriber signalling system version 2)*. NNI signalling instead uses *B-ISUP (broadband integrated services user part)* and *MTP3*. Figure A10.16 illustrates them.

The higher layer protocols of the control plane (DSS2 and B-ISUP/MTP) are equivalent to the DSS and ISUP/MTP signalling protocols of narrowband ISDN. These are the signalling systems used respectively between an ISDN telephone and the network exchange (UNI) and between nodes in the network (NNI). As their names suggest, DSS2 and B-ISUP are based heavily upon their narrowband ISDN counterparts, DSS1 and ISUP. The numbering plan used in B-ISDN is also based on that used in modern digital telephone and ISDN networks (ITU-T E.164).

ATM forum network reference model

The network reference model of ATM forum differs slightly from that of ITU-T (Figure A10.4), distinguishing between the *private* and *public* parts of an ATM network. Thus the model caters for the connection of private corporate and campus ATM networks to a public ATM network.

The ATM forum network defines the *public UNI* (also called the *public network interface*) and the *private UNI* (the *private local interface*). In addition, there are two types of the NNI, the *PNNI* (*private network-node interface* or *private network-network interface*) and *B-ICI* (*broadband inter-carrier interface*). At a basic level, these both use the NNI. They differ in the types of network management and administration possible over them.

The ATM forum UNI is the best specified of all interfaces available. Three versions are currently available in most ATM switches. These are the versions 3.0 (UNI v3.0), 3.1 (UNI

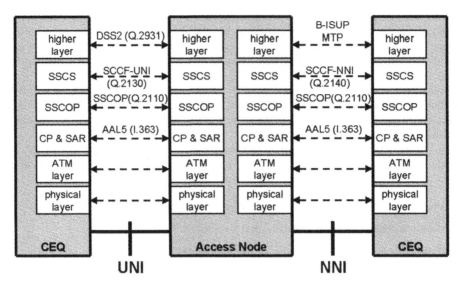

← – → communication or signalling

Key: AAL5 = AAL service type 5 ; B-ISUP = broadband integrated services user part; CP = common part (convergence sublayer); DSS2 = digital subscriber signalling system 2; MTP3 = message transfer protocol layer 3; NNI = network-node interface; SAR = segmentation and reassembly sublayer; SSCF = service specfic coordination function; SSCOP = service specfic connection-oriented protocol; SSCS = service specfic convergence sublayer; UNI = user- network interface.

Figure A10.16   ATM UNI and NNI protocols used on the control plane.

v3.1) and 4.0 (UNI v4.0). Version 3.1 incorporates the ITU-T recommendation Q.2931 (DSS2) for call set-up of *switched virtual circuits (SVCs)*. Version 3.0 is an earlier version, and in practical terms only supports *permanent virtual connections (PVCs)*. Version 4.0 meanwhile is the latest version, in which AAL types 2 and 3/4 have been eliminated by concentration on types AAL1 and AAL5. The specifications are published by Prentice Hall.

PNNI includes a number of functions for *discovery* of the network topology and for routing through it. This is well suited, for example, to a university or large campus private ATM network in which individual departments may be responsible for adding new switches and end user devices to the network on their own initiative. The PNNI *topology state* functions enable other switch nodes within the network to keep abreast of the topology and connected devices.

B-ICI defines a more secure interface than that of PNNI — reinforcing the organisational boundary between different public network operators. The B-ICI interface adds specific functions to the standard NNI to allow for *contracting traffic*, and for monitoring and management of an *interconnect* between different operators networks (e.g., in the USA the *interconnect* between *LECs (local exchange carriers)* and *IECs (inter-exchange carriers)*). The B-ICI interface supports the capabilities necessary for transmitting networks (like IEC networks), and allows for the support of *cell relay service (CRS), frame relay service (FRS)* and *SMDS (switched multimegabit data service)*. The B-ICI interface is thus an important precursor to the regulated interconnection of public ATM networks — at least in the USA.

ATM forum has also defined three *interim* interfaces for use at the PNNI, B-ICI and for management control of the network. These are respectively:

- *IISP (interim inter-switch signalling protocol* — a forerunner of the PNNI signalling protocol based on ATM forum UNI v3.1),

Table A10.2   ATM Forum-defined network management interfaces

M-Interface type (ATM forum)	End points of management interface	
M1	End-user ATM device	Private network management system
M2	Private ATM network	Private network management system
M3	Private network management system	Public network management system
M4	Public ATM network	Public network management system
M5	Public network management system	Public network management system

- *BISSI (broadband inter-switching system interface* — a forerunner of the NNI signalling protocol for use in- and between public ATM networks),

- *ILMI (interim local management interface* — a protocol for management of ATM network devices based on the Internet *simple network management protocol, SNMP)*.

ITU-T has not recognised these interim standards.

ATM forum network management model

The ATM forum network management model also differs from that of ITU-T by the strict separation of public and private networks. Confusingly, ATM forum refers to M-interfaces (Table A10.2) rather than to management plane types, and uses different numbering (ATM forum interface M1 is equivalent to ITU-T management plane type 1, but ATM forum interface M3 is not equivalent to ITU-T's management plane 3).

ATM forum's M3- and M4-interfaces are already extensively defined by ATM forum, respectively for *customer network management (CNM)* of *public ATM network service* and for network management of public ATM networks. The other interfaces are less well defined.

Glossary of Selected Terms

Term	Explanation
2-wire communication	A communication path between end points using only 2 wires. Its difficulties and limitations arise from the shared use by transmit and receive signals of the same path.
2B + D interface	An alternative term used to describe the ISDN basic date interface.
4-wire communication	A communication path between end points using 4 wires, two each for each direction of transmission, transmit and receive. 4-wire communication paths have better isolation of transmit and receive signals than 2-wire paths and are thus easier to operate.
adaptive differential pulse code modulation (ADPCM)	A highly efficient, low bit speed method of digital encoding of voice signals.
administrator	A human responsible for the operation and administration of a network.
ADSL (asymmetric digital subscriber line)	A transmission technique allowing high speed signals (e.g. 6–8 Mbit/s) to be transmitted to customer premises over existing copper lineplant. In the upstream direction typically a maximum of 384 kbit/s is available.
advanced peer to peer networking (APPN)	A communications architecture developed by the IBM company. It is used between IBM AS400 computers and between workstations and servers.
agent	In data communications, an agent is a software program which undertakes a given function on behalf of another program. The function may be delegated for security reasons. Alternatively, the main program may be unable to carry out the function itself. In this case, the agent performs a mediation or conversion function.
amplification	The boosting of a faint signal — counteracting the effects of attenuation in order to ensure good comprehension by the receiver.
amplitude	The relative strength or volume of a signal.

Term	Explanation
amplitude modulation (AM)	A technique used to enable a low bit speed or comparatively low bandwidth information signal to be carried on a higher frequency carrier signal — by adjustment of the latter's amplitude to match the former's wave shape.
analogue transmission	A method of signal transmission in which the shape of the electrical current waveform on the telecommunications transmission line is analogous to the air pressure waves of the sound or music signal it represents.
application	A software program designed to undertake a given function.
application layer	The uppermost layer (layer 7) of the open systems interconnection (OSI) model. This layer is a software function providing a program in one computer with a communication service to a 'cooperating' computer.
ARPANET	The network of the US Advanced Research Project Agency: the original name for the Internet when first set up in 1968.
AS (autonomous system)	A network under single ownership and administration — also called an administrative domain.
ASCII (American standard code for information interchange)	A binary coding scheme using 7-bit or 8-bit binary values to represent textual characters. The original version was a 7-bit code. Various 8-bit extended ASCII codes have subsequently been developed.
asynchronous data transfer	A form of data transmission in which individual text characters (typically 8-bit binary values) are always preceded by start and stop bits which serve to maintain strict delimitation of consecutive characters.
asynchronous transfer mode (ATM)	A modern technology for integrating all types of telecommunications traffic (voice, data, video, multimedia) across a single network. ATM forms the technological basis of the broadband integrated services digital network.
attenuation	The effect of signal dwindling experienced with accumulating line length or distance of radio transmission.
attenuation distortion	The various frequency components of a complex sound or data signal are often attenuated by the line to different extents. The tonal balance of the original signal is thus distorted — attenuation distortion.
audio circuits	Copper or aluminium-wired circuits being used to carry a single signal — typically a speech signal in the audio bandwidth.
balanced transmission	Telecommunications transmission across metallic conductors (i.e. wire pairs) in which both conductors play an equal role. (e.g. twisted pair 120 ohm cabling).
bandwidth	Measured in Hertz or cycles per second, the range of component frequencies making up a complex speech or data circuit. Thus speech comprises frequencies between 300 Hz and 3400 Hz, a total bandwidth of 3.1 kHz.
baseband	An analogue signal in its original form (prior to any form of multiplexing).

Term	Explanation
BCP (best current practice)	A publication produced by the Internet engineering task force which documents a given best current operational practice when administering or operating Internet protocol (IP)-based networks.
bearer service	Relating in particular to ISDN, various bit rates or bandwidths (called bearer services) may be made available between the two ends of the connection. Examples include speech bearer service, 64 kbit/s data service.
BER (bit error rate)	A measure of the quality of a digital transmission line, either quoted as a percentage, or more usually as a ratio, typically 1 error in 100 thousand or 1000 million bits carried. The lower the number of errors, the better quality the line.
binary code	A 2-state numerical code (values 'on'/'off' or 1/0) used to represent text or computer data.
B-ISDN (broadband integrated services digital network)	A powerful network capable of carrying all types of narrow and broadband telecommunications service types (e.g. voice, data, Internet, video, multimedia etc.)
bitrate or bit speed	The 'bandwidth' of a digital line or signal — governing the rate at which individual alphabetic or numeric characters may be carried by the line.
BRI (basic rate interface)	The simplest form of network access available on the ISDN (integrated services digital network). The BRI comprises 2B + D channels for carriage of signalling and user information.
bridge	A device for interconnecting LANs or LAN segments together.
broadband	A service or system supporting rates greater than 2 Mbit/s.
broadcast	A service providing unidirectional distribution to multiple receivers.
brouter	A device comprising both bridge and router functionality.
busy hour	The period of highest network usage.
busy hour traffic	A measure of the maximum user demand for network throughput.
byte	A binary digital signal value of 8 bits (decimal value 0–255).
carrier frequency	A high frequency signal used in FDM or radio transission. The information is modulated onto the carrier prior to transmission.
carrier sense multiple access with collision detection (CSMA/CD)	The transmission technique employed in ethernet LANs. Data packets are sent onto a common bus when the bus is seen to be idle. Addresses in the packets allow the correct destination device to take them from the bus.
Cat.5	A type of twisted pair cabling commonly used to provide 'structured wiring' for data communications sockets in company offices and for patch cabling. (Intended for Baud rates up to 100 MHz.)
Cat.6	A shielded foil cabling system similar to Cat.5 but intended for signal Baud rates up to 200 MHz.

Term	Explanation
Cat.7	A shielded foil cabling system similar to Cat.5 but intended for signal Baud rates up to 600 MHz.
cell	A data block of fixed length (48 byte information field and 5 byte header). Cells are transmitted across a network using ATM.
CELP (code excited linear predication)	A fast means of low bitrate digital speech compression. Because the algorithm is linear, the processing may be carried out more quickly, thus minimising signal delay.
circuit-switching	A method of switching provides for a direct connection of fixed bandwidth or bitrate (e.g. 64 kbit/s) between originating and terminating points of a communication for the entire duration of a call.
circuit transfer mode	A telecommunication transfer technique based on circuit-switching and requiring permanent allocation of bandwidth.
client	A software program usually designed to run on a human user's workstation or PC to provide a given service or application to the human user. The client software typically provides a graphical user interface (GUI) to the human user and interacts with a server (client/server architecture) to obtain data and functions supporting the application.
CMIP (common management information protocol)	The standard data protocol defined by ISO and ITU-T as part of the telecommunications management network for requesting management status information from a network element or sending control commands to it. CMIP is used at the Q3-interface.
coaxial cable	A cable used for high frequency and high bandwidth analogue and digital signal transmission. The cable comprises a single inner conductor plus a conducting outer sheath — in cross-section a dot in a circle.
codec (coder/decoder)	A device for interfacing a 4-wire digital transmission line to a 4-wire analogue. The coder codes the to-be-sent 2-wire analogue signal in a form suitable for 2-wire digital transmission. The decoder meanwhile decodes the received 2-wire digital signal.
configuration management	One of the five main network management functions defined by the ISO management model. The others are accounting, security, performance and fault management.
connection admission control	The procedure by which a network decides whether to accept a new connection.
connectionless	A type of communication across a telecommunications network in which a datagram (analogous to a postal telegram) is submitted to the network without first checking whether the intended recipient is ready to receive it (and indeed even without checking whether the addressed destination actually exists).

Term	Explanation
connection-oriented	A type of communication across a telecommunications network in which a connection is first set-up between caller and destination and the communications path is checked before data is sent.
console	Another name for a computer terminal.
console port	A connection port (typically DB-25 or DB-9) on a networking device to which a console can be connected locally (using a serial cable) for purpose of configuration of the device.
control plane	A conceptual communications connection between data switching devices for communicating control information (e.g. for establishing or clearing connections, etc.).
CPE (customer premises equipment)	The telephone or data equipment on a customer's premises — that using the public telephone or data network for connection to a remote location.
CRC (cyclic redundancy check)	A binary checksum enabling the detection and (possibly also) correction of bits within a data frame which have become corrupted during passage over a long distance connection.
crosstalk	The interference of telephone and other signals resulting from the induction of an unwanted signal from an adjacent telephone line (the crossed-wire).
data	Data is the term used to describe alphabetic and numeric information stored in and processed by computers.
data compression	Any of various techniques, of both reversible and irreversible types, designed to reduce the bit speed needed for a particular type of communication (e.g. facsimile).
datagram	The name given to an Internet protocol (IP) packet — a single block of data or small message carried across a data network. Analogous to postal telegram.
datalink layer	Layer 2 of the open systems interconnection (OSI) model. This software fuction operates on an individual link within the connection to ensure that individual bits are conveyed without error. The best-known layer 2 protocol is HDLC.
DCE (data circuit-terminating equipment)	A device which provides for connection of a computing device (data terminal equipment — DTE) to a wide area data network. A DCE (an example of which is a modem) provides an interface between the short range communications capabilities of the DTE and the modulation techniques required for long-distance communication.
DNS (domain name system)	A service and protocol which provide for resolution of website addess (e.g. www.company.com) and email domain addresses (e.g. martin.clark@company.com) into the IP addresses of the relevant servers.

Term	Explanation
domain name	The part of a web address which follows 'www.' or the part of an email address which follows the @ sign. In the case of the website address www.company.com or the email address martin.clark@company.com, the domain name is company.com.
DTE (data terminal equipment)	A computer, computer terminal, PC, server, workstation or similar device connected to a data network for purpose of data communication.
duplex (full duplex)	A mode of telecommunications transmission in which unrestricted two-way communication is possible at all times. Both parties may 'talk-at-once'.
EBCDIC	Extended binary coded decimal interchange code. An 8-bit computer code for representing alphabetic and numeric characters.
ebusiness	A term used to describe the conduct of business via the Internet.
EDI (electronic data interchange)	A term used to describe the exchange of standardised and formal business documents and transactions by electronic or data networking means. Electronic purchase orders, invoices and other financial transactions are typically carried out by means of EDI.
EFS (error free seconds)	A performance measure used to specify the quality of data lines. Higher specification lines require a higher proportion of error free seconds.
email	A shortening of the term electronic mail — the sending of letters, documents and files by means of a data network or the Internet.
EMC (electromagnetic compatibility)	EMC is the term applied to a series of specifications which define how telecommunications devices should be immune to electromagnetic disturbance.
EMI (electromagnetic interference)	Electromagnetic interference is the disturbance caused by a telecommunications device to other neighbouring devices.
encryption	The coding of data information prior to transmission in order to make the contents of the communication meaningless to third parties intent on overhearing or intercepting the message.
ethernet LAN	A local area network (LAN) conforming to the IEEE 802.3 standard. The commonest form of LAN used nowadays. There are different variants — 10baseT, 10/100baseT, fast ethernet (100baseT) and Gigabit ethernet.
extranet	A remote site connected to an intranet (internal private company data network) by means of a secure connection across the public Internet.
fading	The loss of radio signal caused by interference or weather attentuation effects.
fault management	One of the five main network management functions defined by the ISO management model. The others are accounting, security, performance and configuration management.

Term	Explanation
FDDI (fibre distributed data interface)	A type of metropolitan area network (MAN) capable of bitrates up to about 100 Mbit/s. FDDI is typically used to link LAN segments within a large office complex or on a campus site.
female connector	A plug connector or socket which comprises the 'holes' for the 'pins' of the connector rather than the 'pins' which appear in the corresponding male connector.
file transfer	The carriage of a complete data file from one location to another — transferred from a location called in jargon the ftp server (file transfer protocol server) to the ftp client.
firewall	A computer server or similar device installed at the interconnection point between a private IP-network (intranet) and the public Internet and used to 'police' the data crossing between the networks. The firewall prevents unwanted outsiders gaining access to the intranet, checks and filters incoming emails for unwanted content (e.g. viruses) and seeks to ensure that malicious attacks on website servers are not successful.
flow control	The technique used by a data communications protocol to ensure that devices receiving data across a data network are not inundated with a flood of information which they are unable to handle.
frequency modulation (FM)	A technique used to enable a low bit speed or low bandwidth information signal to be carried on a much higher frequency carrier signal — by adjustment of the latter's frequency by a small amount corresponding to the amplitude of the former.
forwarding	The process of transport of data packets across a data network. Routers inspect the address of each packet and then forward the packet along the best next hop towards the destination.
FRAD (frame relay access device)	A device which converts data from a data terminal equipment (DTE) for carriage by a frame relay network.
frame	The name given to a block or packet of data having received its layer 2 (datalink) protocol header. (If you like, a layer 2 'packet' of data.)
frame relay	A modern form of packet switching, capable of much higher bitrates than X.25-based packet-switched networks.
framing	The structuring framework required in digital line systems in order that a number of different users or channels may share the line by time division means.
FSK (frequency shift keying)	When a carrier signal is frequency modulated with a binary data signal, the effect is merely to alternate the carrier signal between two or more fixed values. This is known as frequency shift keying.

Term	Explanation
gateway	Two alternative meanings. Most commonly nowadays, the term gateway is used to describe a device which provides for protocol conversion — particularly of higher layer (e.g. application layer) protocols. A mail gateway might, for example, provide conversion between Internet mail system messages and X.400-format messages. In the early days of the Internet, the term gateway was also used to describe the nodes of a data network.
GOS (grade of service)	The proportion of calls in a circuit switched network which are lost due to network congestion. Since it is uneconomic to provide networks of enormous uncongested proportion, it is usual to design networks to a given grade of service - typically 1 in 100.
GPRS (general packet radio service)	An adaptation of the GSM mobile telephone system to incorporate packet data transmission
GSM (global system for mobilecommunication)	One of the most common types of mobile telephone network technology, and one which allows full roaming of subscribers between networks around the world. Most common in Europe.
GUI (graphical user interface)	A software program which converts a graphical-style computer screen output to or input from a human computer user into the 'hieroglyphic gibberish' that the computer understands. Thus a GUI program might convert a 'click' on a 'button' into a computer command like 'atur'.
half duplex	A mode of telecommunications transmission in which both directions of communication are possible, but in which only one party may 'speak' at any given time. There is no possible to talk at the same time. . .even if only to 'butt in'.
Hamming code	An error correction/detection code used in datacommunications.
Hayes protocol	A command language used between a computer and a modem in order that the computer can control the modem. Commands which may be issued include 'dial', 'dial number ***', 'clear', etc.
HDLC (high level data link control)	A protocol conforming with the open systems interconnection (OSI) model, layer 2, allowing for the error-free conveyance of bits over a single link of a connection.
HDSL (high bitrate digital subscriber line)	A transmission technique allowing a high speed duplex signal of 2 Mbit/s to be transmitted to customer premises over existing copper lineplant.
header	The bits within a block or cell which provide for correct delivery of the payload.

Term	Explanation
hexadecimal	A 16-state numerical code (values 0, 1, 2, 3, 4, 5, 6, 7, 8, 9, A, B, C, D, E, F). Hexadecimal numbers (typically prefixed by '0x') are often used as a 'shorthand' for binary numbers, into which they are easy to convert. Each digit of a hexadecimal number is simply replaced in turn by its four-digit binary equivalent.
host	In data communications vocabulary, a host is a computer (of any type — PC, server or mainframe) connected to a data network and running one or more applications which communicate across the data network. Often the host is synonomous with the DTE. The term DTE is more commonly used when describing physical layer and the electrical itnerface. At the software and higher protocol layers, the term host is more commonly used. In the computing world the term host is sometimes used synonomously with 'mainframe computer'.
html (hypertext markup language)	A text 'markup' language in which websites and web documents are written.
http (hypertext transfer protocol)	An application layer protocol used to create hyperlinks between documents and websites held on different physical servers within the Internet.
hub	A hardware device providing the physical interconnections and bus or ring topology necessary as the basis of a given type of local area network (e.g. ethernet, token ring, etc.)
huffman code	A code used in datacommunications for data compression. This has the benefit of reducing the overall number of bits which need to be carried across the network, so improving speed of transmission and minimising costs.
hyperlink	By means of a hyperlink documents and files stored on different physical servers connected a data network such as the Internet can be related to one another. By means of a web browser to interpret the hyperlinks, the separate documents (e.g. different text blocks and pictures) appear on the human user's computer screen to be a single document. By clicking on further hyperlinked text (typically underlined) a fast connection to a related document can be achieved.
IANA (Internet assigned numbers authority)	A governing authority of the Internet, responsible for overall control and assignment of protocol-relevant configuration information (e.g. IP protocol field values, TCP/UDP port numbers, protocol parameter value assignment, etc.).
IEEE 802	A technical committee and standards documentation covering data and wireless networking in LANs (local area networks) and MANs (metropolitan area networks).
IEEE 802.2	The IEEE standard for the layer 2 (datalink) protocol used in local area networks (LANs). The protocol is usually referred to as LLC (logical link control).

Term	Explanation
IEEE 802.3	The IEEE standard defining the physical layer of the ethernet LAN. There are a large number of different variants (see Appendix 5).
IEEE 802.5	The IEEE standard defining the physical layer of the token ring LAN.
IETF (Internet engineering task force)	The main organisation and working group responsible for defining Internet standards and issuing RFCs.
interface	A standardized mechanical, electrical and protocol format used to connect two or more pieces of equipment together.
interference	A signal impairment caused by the interaction of an unwanted adjacent signal.
Internet	A large public 'shared network' based upon the Internet protocol (IP).
InterNIC	The Internet network information centre. InterNIC provides for Internet domain name registration services.
intranet	A private (usually enterprise-internal) data network based upon similar equipment to that used in the Internet. Such networks use the Internet protocol (IP) but are either isolated from the public Internet or connected only by means of special controlled connection points (e.g. firewalls).
IP (Internet protocol)	The layer 3 (network) protocol used in Internet and router-based data networks.
IP address	The network address used to identify the intended destination of a data message sent across the Internet or some other IP-based network. IP addresses intended for use with IPv4 have the format d.d.d.d where each d is a decimal value between 0 and 255. IPv6 addresses, on the other hand, have the form hhhh:hhhh:hhhh:hhhh:hhhh:hhhh:hhhh:hhhh where each h is a hexadecimal value in the range 0 to F.
IPv4 (Internet protocol version 4)	The version of the Internet protocol defined in RFC 791 (issued in September 1981).
IPv6 (Internet protocol version 6)	The version of the Internet protocol defined in RFC 2460 (issued in December 1998). IPv6 is similar to IPv4 in its basic functioning and architecture, but was specifically designed to provide a much larger IP-addressing space and to cope with the specific needs of modern 'flow-oriented' communication (e.g. real-time voice, video and multimedia).
ISDN (integrated services digital network)	The digital telephone network.
ISO (International Organization for Standardization)	An international organisation, based in Geneva, Switzerland, responsible for coordinating and publishing international standards.

Term	Explanation
ISP (Internet service provider)	A public telecommunications provider to whom businesses and private individuals may subscribe for provision of network access to the Internet.
ITU (International Telecommunications Union)	An agency of the United Nations (UN) set up in 1934. The world's foremost regulator of radio spectrum and technical standards for public telecommunications networks
ITU-T (ITU Standardization sector)	A body within ITU responsible for the development of technical 'standards' (called 'recommendations') which are widely used as the basis for international telecommunications technology and networks.
ITU-T recommendation	A technical 'specification' of the ITU-T, however, not having official 'standards' status, and not being mandatory.
jitter	Jitter arises when the timing of the pulses of a digital signal varies slightly, so that the pulse pattern is not quite regular. The effects of jitter can accumulate over a number of regenerated links and result in received bit errors.
LAN (local area network)	A type of data network allowing the interconnection of computer devices within a relatively restricted area, typically lying along a spine or circle of length less than about 100–500 metres.
latency	The propagation delay encountered by a packet of data while traversing a network.
layered protocol	Individual protocols are generally designed to carry out a single control or data transfer function, as two peer devices communicate with one another. A number of different protocols are combined in a modular and layered fashion to undertake all the required communications functions. The combination of protocols is termed the protocol stack.
line code	The technique used when actually transmitting a digital signal onto a transmission line. The line code helps to ensure the synchronisation and correct receipt of data.
LLC (logical link control)	The layer 2 (datalink) protocol used in conjunction with local area networks (LANs).
MAC (medium access control)	A layer 1 protocol used in conjunction with a given LAN or MAN physical medium to provide a standard interface to the layer 2 protocol.
MAC address	The 'hardware address' usually 'burned into' ethernet or other LAN network interface cards (NICs) at the time of their manufacture. The address provides for the correct delivery of data frames across the LAN, and are coded in the form defined as the 48-bit IEEE unique identifier.
male connector	A plug connector or socket which comprises the 'pins' of the connector rather than the 'holes' which appear in the corresponding female connector.

Term	Explanation
MAN (metropolitan area network)	A type of data communications network suitable for interconnecting computer devices across a metropolitan area. Such networks typically have a bus or ring topology of up to about 100 km in total length.
managed object	A conceptual 'object' which may be monitored or controlled by means of a standard management protocol such as SNMP or CMIP. Managed objects are defined devices or functions, with a discrete number of possible states (e.g. on, off, transmit, receive, etc.).
management plane	A conceptual communications connection between data switching devices for communicating network management information (e.g. for configuring the network, setting up permanent connections or monitoring performance).
manager	Alternatively: a network management system or control software. Or: a human network manager or administrator.
Manchester code	The line code used in ethernet LANs.
mark	A binary digit of value '1'.
media	Plural of medium.
medium	A means by which information can be perceived, expressed, stored or transmitted.
message handling system (MHS)	A conceptual model specified by ITU-T recommendation X400, describing an electronic mail system in which message devices (e.g. facsimile and telex machines, computers, word processors, etc.) intercommunicate.
message switching	A means of data communication in which complete messages are conveyed together across the transport medium.
messaging service	An interactive service based upon store-and-forward and mailbox transfer.
MIB (management information base)	A collection of defined managed objects relating to given type of telecommunications device or interface (e.g. radio link, telephone exchange, subscriber ISDN line, etc.) and allowing for remote monitoring and configuration.
modem	A term derived from a shortening of Modulator/DEModulator. A modem provides for the carriage of digital data information over analogue transmission lines.
modulation	Modulation is a means of encoding information on to a carrier signal in order that it may be transmitted over a transmission line or radio link.
monomode fibre	An optical fibre with a narrow central core of a different refractive index from the cladding which surrounds it. The narrowness of the core allows only very few ray paths to exist — so that dispersion is minimised — a single mode of transmission prevails. Such fibres allow for reliable long distance communication.

Term	Explanation
MPEG (motion pictures experts group)	An industry grouping which has developed a number of coding standards for transmission of video signals (e.g. MPEG1, MPEG2, etc.).
MPLS (multiprotocol label switching)	A network layer protocol designed for fast forwarding of flow-oriented data streams. MPLS is likely to be deployed at the core of high performance IP backbone networks and also used widely as the basis of VPN (virtual private network) services.
multimedia service	A service in which interchanged information is a mixture of text, sound, graphics, video.
multimode fibre	An optical fibre with a relatively wide central core. The relative large dimension of the core allows many differnet ray paths to exist — so that dispersion may be a problem. Typically multimode fibres are limited in range compared with monomode fibres.
multiplexing	Multiplexing is a technique for combining a number of full duplex channels together to share the same transmission line or radio medium.
multipoint	A communication or network configuration involving more than two end points.
multiprotocol	A multiprotocol network is typically a LAN or layer 3 (e.g. IP)-based network capable of carrying multiple different types of transport, session and application protocols simultaneously.
network-node interface (NNI)	A standardised interface used between switches or subnetworks within a given (data) network. Also called 'network-network interface'
network layer	The network layer is layer 3 of the open systems interconnection (OSI) model. This layer sets up an end-to-end connection across a real network, determining which permutation of individual links to be used.
network performance (NP)	A specific term applied to performance measurement parameters which are designed to monitor only the end-to-end network connection part of a given communication (e.g. telephone call or computer application running via a data network).
NMS (network management system)	A combination of workstation hardware and application software designed specifically for the purpose of aiding human network administrators in the management of a telecommunications network.
noise	Stray, unwanted and random signals interfering with the desired signal. Common forms of noise manifest as a low level crackling noise.
optical fibre	A glass fibre of only a hair's breadth dimension, capable of carrying very high digital bit rates with very little signal attenuation.

Term	Explanation
OSI (open systems interconnection)	A conceptual model specified by the ITU-T X.200 series. The model describes the 7-layer process of communication between 'cooperating' computers. The model is the standard for open communication between computers made by different manufacturers.
OSI model	The seven layer conceptual model defining the process of communication between computer systems.
OSPF (open shortest path first)	A standardised protocol used widely in the Internet and IP networks for conveying routing information between routers.
0x	The prefix used to denote a hexadecimal number value.
packet	An information or data block carried by a network layer (layer 3) protocol and identified by a protocol header (such as that of IP or X.25).
packet assembler/disassembler (PAD)	A device enabling a character (i.e. asynchronous) terminal to be connected to a packet network.
packet-switching	A type of exchange or network which conveys a string of information from origin to destination by cutting it up into a number of packets and carrying each independently.
packet transfer mode	A telecommunication transfer technique in which information is carried in packets.
parallel data transmission	Data transmission between computer devices using multiple lead wires or ribbon cables. Whole bytes of information (8 bits) are sent simultaneously by using a separate wire for each of the individual bits of the byte.
PCM (pulse code modulation)	The method of conversion used to enable speech or some other analogue audio signal to be carried over a digital transmission path.
peer partner	Protocols used for data communications are designed in a modular and layered fashion. Each layered protocol of the protocol stack is communicated between the two peer partner devices at either end of the link, connection or other communications path.
peering	Peering is the term used to describe interconnections made between IP networks managed and operated by different organisations.
performance management	One of the five main network management functions defined by the ISO management model. The others are accounting, security, configuration and fault management.
phase shift keying (PSK)	When a binary data signal is phase modulated onto a carrier the effect is to create a transmitted carrier signal of one of a number of fixed phases. Jumping between these phases according to the incoming bit stream is known as keying.

Term	Explanation
physical layer	Layer 1 of the open systems interconnection (OSI) model. The physical layer protocol is the hardware and software in the line terminating device which converts the databits needed by the datalink layer into the electrical pulses, tones or other form.
plesiochronous digital hierarchy (PDH)	Digital transmission hierarchy in which individual line systems within a network do not run in exact synchronisation with one another. Instead free running or plesiochronous systems sometimes have to be corrected for slip and justification errors.
POP	Alternative: post office protocol — a protocol used to retrieve Internet electronic mail into a client PC from a mail server.
	Or: point-of-presence — a connection point of a network — e.g. the site of the nearest NAS (network access server) of an Internet service provider's network.
port	Alternative: a hardware connection on a networking device used to connect a line to another device.
	Or: an address (port number) used by a transport protocol (e.g. TCP or UDP) to indicate the protocol being used to code the user data field.
post office	A term sometimes used to describe an electronic mail service or message depository.
PPP (point-to-point protocol)	A protocol used particularly on dial-up access lines to the Internet to establish a datalink protocol and data frame carrying layer once the physical connection has been set up.
presentation layer	Layer 6 of the open systems interconnection (OSI) model. The presentation layer defines the manner in which the data is encoded, eg binary, ASCII, IA5, IA2, EBCDIC, facsimile, etc.
presentation medium	The means or device used to reproduce information to the user (multimedia).
PRI (primary rate interface)	The 23B + D or 30B + D ISDN interface typically used to connect digital company telephone systems to the public ISDN, but also used for back-up lines.
primary multiplex (PMUX)	A device performing the first stage of time division multiplexing. In the UK and Europe, a primary multiplexor converts 32 analogue signals to 2.048 Mbit/s time division multiplexed signal. In the US equivalent, instead 24 are multiplexed to 1.544 Mbit/s.
protocol	A protocol is a procedure, set of commands or rules by which computer devices intercommunicate. Thus a protocol is the equivalent of a human language, with punctuation and grammatical rules.

Term	Explanation
protocol stack	Individual protocols are generally designed to carry out a single control or data transfer function, as two peer devices communicate with one another. A number of different protocols are combined in a modular and layered fashion to undertake all the required communications functions. The combination of protocols is termed the protocol stack.
proxy	A device (either client or server) which undertakes a given function on behalf of another device which is either incapable of the action itself or maybe for security reasons has delegated the task.
QAM (quadrature amplitude modulation)	A complex method of modulation used in modern modems to allow very high data rates to be carried reliably and relatively error-free. QAM is a combination of phase and amplitude modulation.
quality of service (QOS)	Expressed as one of a number of different parameters (e.g. latency) which are designed to measure the performance of communications applications which operate by means of telecommunications networks.
real-time application	An application requiring the real-time transfer of live data or other signals (e.g. an audio channel or a live video link).
regenerator	A device inserted at an intermediate stage of a digital transmission line to counteract the effects of signal deterioration which occur on long lines. As far as possible the device regenerates the original bit stream.
repeater	A device inserted at an intermediate stage of an analogue transmission line to counteract the effects of signal deterioration (particularly attenuation) which occur on long lines.
RFC (request for comments)	A document issued by the Internet engineering task force (IETF) and defining technical standards, protocols or procedures suggested to be used in IP-data networks and the Internet.
RIP (routing information protocol)	A simple routing protocol used widely in small-scale router and IP networks.
RJ-11 connector	A small, usually clear plastic telephone connector or socket, approximately 10 mm wide and 7 mm high (with 6 contacts), typically used to connector telephones or modems to the analogue telephone network.
RJ-45 connector	A small, usually clear plastic data and general-purpose telecommunications connector or socket approximately 12 mm wide and 7 mm high (with 8 contacts), typically used to connect computers and other DTE to LAN switches. Also used widely to make ISDN connections.
RMON (remote monitoring)	A MIB (management information base) intended to be used by networking monitoring devices (probes) for observing the network performance of an IP-based network.

Term	Explanation
router	A switching device (i.e. node) used in the Internet or Intranet for determining the appropriate routing of a given data message through a data network to the desired destination and for forwarding the message.
routing protocol	A protocol used to share network topology and routing information between routers, in order that they may calculate the best routes to all reachable destinations and therefore generate their routing tables.
routing table	A table held in a router which relates the network address (e.g. IP-address) to the appropriate outgoing route (i.e. next hop) to be used for forwarding the associated data.
SDH (synchronous digital hierarchy)	Modern digital transmission hierarchy in which individual line systems within a network are designed to run exactly synchronously with one another. This gives major management and topology administration benefits.
SDLC (synchronous datalink control)	Synchronous data link control. IBM's equivalent of HDLC, forming the layer 2 of IBM's systems network architecture (SNA).
security management	One of the five main network management functions defined by the ISO management model. The others are accounting, configuration, performance and fault management.
segment	An information or data block carried by a transport layer (layer 4) protocol and identified by a protocol header (such as that of TCP or UDP).
serial data transmission	Data transmission between computer devices using only a single circuit path. Whole bytes of information (8 bits) are sent in a sequential pattern. Serial transmission is used across long-distance telecommunications networks (cf. parallel transmission).
server	Term applied to a computer device or software program running an application serving clients.
session	A relationship (if you like, a 'connection') between two communicating computer applications in different locations for the purpose of data transfer.
session layer	Layer 5 of the open systems interconnection (OSI) model. When established for a session of communication, the two devices at each end of the overall connection must conduct their 'conversation' in an orderly manner.
simplex	A mode of telecommunications transmission in which only one direction of communication is possible.
SNA (systems network architecture)	A standard framework of communications methods used in particular in networks of IBM computer devices.
SNMP (simple network management protocol)	A network management protocol widely used for managing corporate data networks and the Internet.

Term	Explanation
socket	A combination of an IP address and a TCP or UDP port number — the socket provides all the address details necessary to facilitate setting up communication.
SONET (synchronous optical network)	The North American technology which preceded SDH (synchronous digital hierarchy) and is based upon similar principles.
statistical multiplexing	A method of data lineplant economy in which different data communication 'conversations' share the same line by making use of each other's idle periods.
STD (standard)	A document issued by the Internet engineering task force (IETF) now declared to be a standard, and probably previously published as an RFC (request for comments) document.
storage medium	The physical means to store information or data (multimedia).
STP (shielded twisted pair)	Term applied particularly to indoor cabling of a type in which twisted copper pairs are 'wrapped' immediately inside the plastic sheathing with a foil shield. The shielding serves to reduce electromagnetic disturbances from or to the cable.
switch	A device or node used in a network to make connections.
synchronisation	The method by which the bit patterns appearing on digital line systems may be properly clocked and interpreted — allowing the beginning of particular patterns and frame formats to be correctly identified.
synchronous data transfer	Data transfer employing a strictly regular pattern, rather than (as in asynchronous operation) using start and stop bits to distinguish character patterns from idle line operation.
TCP (transmission control protocol)	The most commonly used transport protocol of the IP-suite. TCP provides for a reliable and connection-oriented transport service.
TCP/IP	The combination of transport and network protocols most commonly used for data communication across IP networks or the Internet.
telecommunications management network (TMN)	The technical architecture conceived by ITU to provide for overall coordinated management of telecommunications networks.
time division multiplex (TDM)	A technique allowing a number of different digital signals to share the same high bit rate digital transmission line by dividing it into a number of smaller component bit rates, each interleaved in time with the others.
token ring LAN	A popular type of physical infrastructure used for local area network interconnection of PCs and other computer devices within an office environment. Used particularly in networks of IBM mainframes and personal computers.
transaction	A single communications activity undertaken between two intercommunicating devices.

Term	Explanation
transport layer	Layer 4 of the open systems interconnection (OSI) model. The transport layer provides for end-to-end data relaying service across any type of data network (e.g. packet network or circuit switched network).
trunk	A circuit interconnecting network nodes. Particularly a long distance circuit.
tunneling	The encapsulation of data packets encoded in one protocol for carriage by a network using a different protocol. Often used to provide for secure remote connections to a private network (e.g. intranet) across the public Internet.
twisted pairs	Metallic conductor wires, twisted in pairs corresponding to the legs of a 2-wire circuit.
unbalanced transmission	Telecommunications transmission across metallic conductors (i.e. wire pairs) in which the two conductors do not play an equal role. (e.g. 75 ohm coaxial cabling)
UDP (user datagram protocol)	A transport protocol of the IP-suite (and alternative to TCP). UDP provides an unreliable and connectionless transport service.
UMTS (universal mobile telephone service)	The most modern type of mobile telephone network, which is designed to be capable of efficient carriage both of telephone calls and high speed data messages.
unicode	A binary coding scheme using 16-bit binary values to represent textual characters. It was developed as an extension of the ASCII code to include all possible international characters.
URL (uniform resource locator)	A fancy name for a Worldwide Web address in the form http://www.company.com
usage parameter control (UPC)	The execution of appropriate action when negotiated values of information transfer are exceeded.
user-network interface (UNI)	The interface (physical interface and protocols) between a user's terminal equipment and the network.
user plane	A conceptual end-to-end communications connection between user terminal equipments for communicating user information (i.e. the useful information content of a communications session).
UTP (unshielded twisted pair)	Term applied particularly to indoor cabling of a type in which twisted copper pairs are sheathed directly in a plastic covering without a foil shield. This is the simplest and cheapest form of indoor telephone and data network cabling.
valid cell, packet or frame	A cell in which the header is declared by the header error control process to be free of errors.
virtual connection	A means of conveying data from origin to destination without there ever being a dedicated physical path between the two. An example is a packet-switched connection.
VLAN (virtual-bridged LAN)	A means by which LAN users on different physical segments can be interconnected as if they were part of a single broadcast domain.

Term	Explanation
VOIP (voice over IP)	The practice of communicating live telephone conversation or audio conferencing signals in real-time by means of the Internet protocol (IP) or the Internet itself.
VPN (virtual private network)	As far as the user of a VPN is concerned, the network appears to be a private network — with the security and network performance benefits of a dedicated network — but is actually carried by means of a public network (e.g. the Internet). The result is a virtual private network achieved at lower cost.
(WAN) wide area network	A term applied to describe the long distance part of a network.
web browser	A software used by human users to 'surf' the Internet and view documents held in html format.
webpage	An html document held somewhere in the Worldwide Web. When viewed using a web browser, the page appears like a page of text, maybe including pictures and (underlined) hyperlinks which provide for quick connection to other related web pages.
web server	A computer and associated software connected to the Internet on which the web pages related to a given domain (e.g. all those of www.company.com) are stored. The server responds to requests from Internet 'surfers' by means of http.
website	Any machine connected to the Internet running a web server.
Worldwide Web (www)	An interconnected network of computer servers around the world, which provide for fast and easy access to online information via the Internet.

Abbreviations and Standards Quick-Reference

Because the world of data communications is punctuated by acronyms and abbreviations, it is often helpful to have a means not only of looking up the full form of an abbreviation, but also to refer to the standards documentation which defined it in the first place. After all, knowing that IPv6 stands for *Internet Protocol version 6* is all very well, but maybe you need to know the frame format of the protocol and the fields it defines. This, as you will discover in this unique Abbreviations listing is defined in the document RFC 2460.

Before using this appendix, it is valuable to read the explanation and 'key' below which will help you to make the best use of it.

As far as possible, the listing gives the abbreviated and full forms of an expression, and is then followed by a list of relevant standards documentation. The standards documents generally originate from three sources:

- Internet Engineering Task Force (IETF). These documents are labelled with *RFC* (stands for *request-for-comment*) and a number. RFCs are issued in number order and never updated. Instead, updated versions of the same document receive a new number. The text of most of the RFCs can be obtained online from www.rfc-editor.org or www.faqs.org/rfcs/.

- International Telecommunications Union — Standardization Sector (ITU-T). These documents are called *recommendations (rec.)*. They are issued with an alphabetical *series* letter and a number (e.g. X.400). These documents are sometimes updated, so that, for example there is an X.400 (1984) and another version called X.400 (1988). Most ITU-T recommendations are available online from the ITU's electronic bookshop at www.itu.org.

- International Organization for Standardization (ISO) and its related agency the International Electrotechnical Commission (IEC). International standards typically have a designation ISO or ISO/IEC and a number (e.g. ISO 8073). Most standards are available online from www.iso.org.

Since the RFCs are of central importance to modern IP-based data networking, most of the technical and standards-related RFCs are listed below, cross referenced to the relevant abbreviation. But this leads in some cases to a long list of RFCs following a particular abbreviation (e.g. MIME). To help you identify the document which defines the original term (e.g. the specification of a particular protocol), the most current version of the relevant main document has been underlined. RFCs with lower numbers than the RFC underlined are probably superseded by the underlined RFC. Meanwhile, RFCs with higher numbers are most likely to be 'extensions' or standardised 'uses' of the basic standard. Thus, for example, the many RFCs associated with MIME define a whole gamut of different content and message types which may sent as 'attachments' to Internet emails. The MIME standard meanwhile is defined by the documents RFC 2045-9.

Where long lists exist, these are sometimes punctuated with brackets enclosing either one of the terms in the key below (indicating, for example, that the relevant document defines the MIB (management information base) of a particular protocol) or with another brief explanation. If none of the listed standards documents is underlined, then the listed reference relates only to a usage or application of the particular function or protocol.

Key

IPv4	Version relevant to Internet Protocol V4
IPv6	Version relevant to Internet Protocol V4
MIB	Document defines the Management information base (MIB) for the protocol, interface or other network element, usually according to the Structure of Management Information version 2 (SMIv2)
MOs	Document defines managed objects for the protocol, interface or other network element
V	Version number
V1-V2	Covers compatibility issues between different versions of a given function or protocol

Abbreviation	Full form and references
3270	IBM 3270 (a data terminal device) (SNA)
⟨A⟩	Anchor tag (HTML)
⟨B⟩	Bold tag (HTML)
⟨BR⟩	Break tag (HTML)
⟨DD⟩	Definition Description tag (HTML)
⟨DL⟩	Definition List tag (HTML)
⟨DT⟩	Definition Term tag (HTML)
⟨HR⟩	Horizontal Rule tag (HTML)
⟨I⟩	Italic tag (HTML)
⟨LI⟩	List tag (HTML)
⟨OL⟩	Ordered List tag (HTML)
⟨P⟩	Paragraph tag (HTML)
⟨tag⟩	HTML TAG
⟨TT⟩	TeleType tag (HTML)
⟨UL⟩	Unordered List tag (HTML)
0x	prefix for number value coded in hexadecimal
10/100baseT	ethernet LAN with autosensing between 10 Mbit/s and 100 Mbit/s
1000baseT	1 Gbit/s (Gigabit) ethernet LAN over twisted pair cabling IEEE 802.3z
1000baseX	Gigabit ethernet LAN
100baseT	100 Mbit/s fast ethernet LAN over twisted pair cabling IEEE 802.3u
10baseT	10 Mbit/s ethernet LAN over twisted pair cabling IEEE 802.3i
1822L	host access protocol, ARPANET host-to-IMP interface RFC 0802, RFC 0851, RFC 0878
1VF	One voice frequency (signalling)
2VF	Two voice frequency (signalling)
2 w	2-wire
3270 protocol	Protocol used to terminal (IBM SNA)

Abbreviation	Full form and references
3DES	Triple Defense Encryption Standard RFC 1851
3DESE	PPP Triple-Defense Encryption Standard Encryption protocol RFC 2420, RFC 3217
4B/5B	4 Binary coded as 5 Binary (a block code)
4B3T	4-Binary 3-Tertiary (ISDN line code)
4D-PAM5	a block code
4-FSK	4-state (quarternary) Frequency Shift Keying (digital modulation)
4 w	4-wire
6Bone	Backbone network for IPv6 RFC 2546, RFC 2772, RFC 2921
8B/10B	4 Binary coded as 10 Binary (a block code)
8B/6T	8 Binary coded as 6 Tertiary (a block/line code)
A	Agent
A	Applicable
A	ASCII (FTP)
A	hexadecimal digit value equal to decimal '10'
A	Offered traffic in Erlangs
A/D	Analogue-to-Digital conversion
A + S	Alarm and state management
A =	Administrative domain (X.500)
AA	Administrative Authority (numbering)
AAA	Authentication, Authorisation and Accounting RFC 2903-6, RFC 2977 (Mobile IP), RFC 2989, RFC 3127, RFC 3141
AAA	Autonomous Administrative Area
AAI	Administration Authority Identifier
AAL	ATM Adaptation Layer
AAL1	AAL service type 1 (constant bit rate service)
AAL2	AAL service type 2 (variable bit rate, but time critical service)
AAL3/4	AAL service type 3/4 (class C frame relaying service or class D connectionless service)
AAL5	AAL service type 5 RFC 2364
AAL-IE	AAL Information Element
AAR	Automatic Alternative Routing
AARP	Appletalk Address Resolution Protocol
ABM	Asynchronous Balanced Mode (HDLC)
ABNF	Augmented Backus-Naur-Form RFC 2234
ABOR	ABORt (FTP)
ABR	Area Border Router
ABR	Available BitRate (ATM)
ac	Ascension Island
AC	Access Concentrator (PPPoE) RFC 2516
AC	Access control (IBM token ring LAN)
AC	Alternating current
ACAP	Application Configuration Access Protocol RFC 2244
ACB	Access method Control Block (IBM VSAM or VTAM)
ACCT	ACCounT (FTP)
ACD	Automatic Call Distribution
ACE	Access Connection Element
ACF	Access Control Field (DQDB)
ACF	Advanced Communication Function (IBM SNA)

Abbreviation	Full form and references
ACFC	Address-and-Control-Field-Compression (PPP) <u>RFC 1661</u>
ACI	Access Control Information
ACK	Acknowledgement message
ACL	ACcess List RFC 1009, RFC 1716, <u>RFC 1812</u>
ACR	Available Cell Rate
ACSE	Association Control Service Element (OSI layer 7)
ad	Andorra
AD	Administrative Domain <u>RFC 1136</u>
ADC	Advice of Duration and Charge
ADDMD	Administration Directory Domain
ADM	Add/Drop Multiplexor (SDH/SONET)
ADMD	Administration management domain (X400)
ADPCM	Adaptive Differential Pulse Code Modulation (low bitrate speech)
ADS	Active Directory Service (Microsoft Windows)
ADSL	Asymmetric Digital Subscriber Line RFC 2662 (MOs)
ADSP	Appletalk Data Stream Protocol (Appletalk)
adspec	Advertising SPECification (RSVP) <u>RFC 2205</u>
ae	United Arab Emirates
AE	Application Entity (OSI layer 7)
AEP	Appletalk Echo Protocol (Appletalk)
aero	AEROnautical and air-transport industry Internet domain
AES	Advanced Encryption Standard (Rijndael code) <u>FIPS-197</u>, RFC 3268
af	Afghanistan
AF	Assured Forwarding PHB group (differential services) <u>RFC 2597</u>
AFI	Address Format Identifier or Authority and Format Identifier (numbering)
AFP	Appletalk Filing Protocol (Appletalk)
AFT	Authenticated Firewall Traversal
ag	Antigua and Barbuda
AgentX	Agent eXtensibility protocol RFC 2257 (v1), <u>RFC 2741 (v1)</u>
AH	IP Authentication Header RFC 1826, <u>RFC 2402</u>
AHIP	ARPANET Host Interface Protocol <u>RFC 1005</u>
ai	Anguilla
AII	Active Input Interface (optical fibre)
AIM	ATM Inverse Multiplexing
AIS	Alarm Indication Signal (digital transmission systems)
al	Albania
AL	Access Link
AL	Alignment (AAL3/4 CPCS)
A-law	PCM speech encoding (European standard)
ALG	Application Level Gateway
ALLO	ALLOcate (FTP)
am	Armenia
AM	Accounting management
AM	Amplitude Modulation
AMI	Alternate Mark Inversion (line code)
AMPS	Advanced Mobile telePhone System (American cellular radio)
AMR	Adaptive Multi-Rate audio codec RFC 3267
AMR-WB	Adaptive Multi-Rate Wide Band audio codec RFC 3267
AMS	Audio visual Multimedia Services

Abbreviation	Full form and references
AMSIX	AMSterdam Internet eXchange (Internet exchange & peering point)
an	Netherlands Antilles
AN	Access Network or Node
AN	Auto-Negotiation
ANA	Article Numbering Association (UK)
ANI	Automatic Number Identification (US equivalent of CLI)
ANS	American National Standard
ANSI	American National Standards Institute www.ansi.org
anycast	host Anycasting service RFC 1546
ao	Angola
AO	Abort Output (NVT-ASCII)
AOI	Active Output Interface (optical fibre)
AOL	America OnLine
AOW	Asia-Oceania Workshop (conformance testing body)
AP	Access Point (WLAN) IEEE 802.11
AP	Acknowledgement Packet
AP	Application Process
APDU	Application Protocol Data Unit
APEX	Application EXchange RFC 3340-2
API	Application Programming Interface
APL RR	Address Prefix Lists — Resource Record RFC 3123
APNIC	Asia Pacific Network Information Centre www.apnic.net
APOP	mailbox name and message digest (POP)
APP	APPlication specific function (RTCP) RFC 1889
APPC	Advanced Program-to-Program Communication (IBM SNA LU6.2 RFC 2051 (MOs)
APPE	APPEnd (FTP)
APPL	APPLication program
APPN	Advanced Peer-to-Peer Networking (IBM) RFC 2155 (MOs), RFC 2455-7 (MOs)
APS	Automatic Protection Switching
aq	Antarctica
ar	Argentina
AR	Access Router
ARC	Alarm Reporting Capability
ARCFOUR	RC-4 encryption algorithm (Internet draft)
ARCNET	Attached Resource Computer NETwork RFC 1051, RFC 1201, RFC 2497 (IPv6)
ARE	All Routes Explorer
AREQUIPA	Application REQUested IP over ATM RFC 2170
ARIN	American Registry for Internet Numbers www.arin.net
ARL	Adjusted Ring Length
ARP	Address Resolution Protocol RFC 0826, RFC 1027
arpa	Address and Routing Parameter Area Internet domain RFC 3172
ARPA	Advanced Research Projects Agency RFC 0961
ARPANET	Advanced Research Projects Agency NETwork RFC 0008
ARQ	Automatic Repeat reQuest RFC 3366
ARR	Automatic Repeat Request
ARR	Automatic Re-Routing

Abbreviation	Full form and references
ARS	Address Resolution Server
ARS	Automatic Route Selection
as	American Samoa
AS	Autonomous System <u>RFC 1812</u>, <u>RFC 1930 (AS number allocations)</u>, RFC 1965, RFC 2270, RFC 3065, RIPE-109 (application form)
ASBR	Autonomous System Border Router
ASC	Accredited Standards Committee
ASCII	American Standard Code for Interchange of Information RFC 0020, ANSI X3.4-1968, ANSI X3.110-1983, ANSI X3.4-1986, ISO 8859, <u>ISO 14962 (1997)</u>
ASE	Application Service Element (OSI layer 7)
ASIC	Application-Specific Integrated Circuit
ASK	Amplitude Shift Keying (digital modulation)
ASM	Any Source Multicasting
ASN.1	Abstract Syntax Notation 1 <u>ISO 8824 & 5</u>, ITU-T X.208-9, ITU-T X.680-3, ITU-T X.690
ASO	Application Service Object
ASP	Appletalk Session Protocol (Appletalk)
at	Austria
AT&T	American Telephone and Telegraph company
ATCP	AppleTalk Control Protocol (PPP) <u>RFC 1378</u>
ATE	ATM Terminating Equipment
ATM	Asynchronous Transfer Mode ITU-T I.113, ITU-T I.121, ITU-T I.150, ITU-T I.211, ITU-T I.311, ITU-T I.327, <u>ITU-T I.361</u>, RFC 1695 (MOs)
ATMARP	Asynchronous Transfer Mode Address Resolution Protocol RFC 2601
ATME	Automatic Transmission, Measuring and signalling test Equipment
ATM-LE	Asynchronous Transfer Mode — Layer Entity
ATMM	ATM Management
AtmMIB	Asynchronous Transfer Mode objects (mib-2) <u>RFC 1694</u>
ATMP	Ascend Tunnel Management Protocol <u>RFC 2107</u>
ATO	Adaptive Time-Out
ATP	Appletalk Transaction Protocol (Appletalk layer 4)
ATS	Abstract Test Suite
au	Australia
AU	Administrative Unit (SDH/SONET)
AuC	Authorisation Centre
AUG	Administrative Unit Group (SDH/SONET)
AUI	Attachment Unit Interface (ethernet LAN) <u>IEEE 802.3</u>
AUTH	Authentication/Authorise command RFC 1734, RFC 2095, <u>RFC 2195</u>
AUU	ATM layer User-to-User indication
AVI	Audio Video Interleaved file format RFC 2361
AVI	AudioVisual Interactive service
AVMMS	Audio-Visual MultiMedia Service
AVP	Attribute Value Pair
aw	Aruba
AW	Administrative Weight
AWG	American Wired Gauge (cable size)
Awnd	Allowed WiNDow (TCP) <u>RFC 2581</u>

Abbreviation	Full form and references
AYT	Are You There ? (NVT-ASCII)
az	Azerbaijan
B	Erlang lost-call probability
B	hexadecimal digit value equal to decimal '11'
B	suffix for number value coded in binary
B	Block mode (FTP)
B8ZS	Bipolar 8-Zero Substitution (line code)
ba	Bosnia and Herzegovina
BA	Behaviour Aggregate (Differentiated Services) RFC 2475
BABT	British Approvals Board for Telecommunications
BACP	Bandwidth Allocation Control Protocol (PPP) RFC 2125
BAKOM	Bundesamt für Telekommunikation (Switzerland)
balanced	A type of data transmission using coaxial cable (usually 75 ohm)
Balun	BALance-to-Unbalance transformer
BAP	Bandwidth Allocation Protocol (PPP) RFC 2125
BAPT	Bundesamt für Post und Telekommunikation (Germany)
BAS	Bitrate Allocation Signal
BASize	Buffer Allocation SIZE (AAL 3/4 CPCS)
BAT	Bridge Address Table
bb	Barbados
BB	BaseBand
B-BC	Broadband bearer capability
BBN	(Richard)Bolt, (Leo) Beranek and (Robert) Newman (data switch manufacturer)
Bc	Committed burst size (frame relay)
bcc	Blind Carbon Copy (SMTP)
BCC	Bearer Channel Connection protocol
BCC	Block Check Character
BCD	Binary-Coded Decimal
BCDBS	Broadband Connectionless Data Bearer Service
B-channel	Bearer Channel
BCN	Backward Congestion Notification
BCOB	Broadband Connection-Oriented Bearer (service)
BCP	Best Current Practice
BCP	Bridging Control Protocol (PPP) RFC 1638, RFC 2878
BCS	Block Check Sequence
bd	Bangladesh
BDAT	Binary DATa (SMTP)
BDR	Back-up Designated Router (OSPF)
be	Belgium
Be	Excess burst duration (frame relay)
BECN	Backward Explicit Congestion Notification (frame relay)
BEDC	Block Error Detection Code (ATM OAM performance cell)
BEEP	Blocks Extensible Exchange Protocol RFC 3080, RFC 3081, RFC 3288
BEL	Bell (see ASCII)
BER	Basic Encoding Rules
BER	Bit Error Ratio
BERT	Bit Error Rate Tester
bf	Burkina Faso

Abbreviation	Full form and references
BFTP	Background File Transfer Program RFC 1068
bg	Bulgaria
BGMP	Border Gateway Multicast Protocol
BGP	Border Gateway Protocol RFC 1105 (v1), RFC 1163 (v2), RFC 1164, RFC 1267 (v3), RFC 1268, RFC 1269 (MOs), RFC 1364, RFC 1397, RFC 1403, RFC 1654-7 (v4), RFC 1745, RFC 1771-4 (v4), RFC 1863, RFC 1965-6, RFC 1997-8, RFC 2042, RFC 2283, RFC 2439, RFC 2545-7, RFC 2796, RFC 2842, RFC 2858, RFC 2918, RFC 3065, RFC 3107, RFC 3345
BGP-ASC	Autonomous System Confederations for BGP RFC 3065
bh	Bahrain
bi	Burundi
B-ICI	Broadband Inter-Carrier Interface (ATM)
BID	Bridge Identity IEEE 802.1d
BIND	Berkeley Internet Name daemon
BIP	Bit Interleaved Parity code (block error detection code)
BIS	Border Intermediate System
BIS	Bringing Into Service
BIS	Bump-In-the-Stack technique RFC 2767
B-ISDN	Broadband Integrated Services Digital Network ITU-T I.121
BISSI	Broadband Inter-Switching System Interface
B-ISUP	Broadband Integrated Services User Part
Bisync	IBM protocol (BSC)
bit/s	Bits per second
BIU	Basic Information Unit (IBM SNA)
biz	Business Internet domain
bj	Benin
BLER	BLock Error Rate
block	A unit of data information consisting of a header and an information field
bm	Bermuda
BML	Business Management Layer (ISO management model) ITU-T M.3000
BMPT	Bundesministerium für Post und Telekommunikation (Germany)
bn	Brunei Darussalam
BN	Bridge Number
BNC	BayoNet Connector
BNF	Backus-Naur Form
BNN	Boundary Network Node
BNS	Backus-Naur Form RFC 2205
B-NT	Network Termination for B-ISDN
B-NT1	Network Termination 1 for B-ISDN
B-NT2	Network Termination 2 for B-ISDN (multipoint configuration)
bo	Bolivia
BOC	Bell Operating Company (USA)
BOM	Beginning Of Message
BOOTP	BOOTstrap Protocol RFC 0906, RFC 0951, RFC 1048, RFC 1084, RFC 1395, RFC 1497, RFC 1532-4, RFC 1542, RFC 2132
BOP	Bit-Oriented Protocol
BPDU	Bridge Protocol Data Unit IEEE 802.1d
BPON	Broadband Passive Optical Network

Abbreviation	Full form and references
BPP	Bridge Port Pair (service routing descriptor)
BPSK	Binary Phase Shift Keying (modulation)
br	Brazil
BR	Backbone Router
BRA	Basic Rate Access (ISDN)
BRI	Basic Rate Interface (ISDN) ITU-T I.420, ITU-T I.430 (physical layer), ITU-T I.441 (datalink layer), ITU-T Q.921 (datalink layer), ITU-T I.451 (network layer), ITU-T Q.931 (network layer)
BRK	BReaK (NVT-ASCII)
BRL	Balance Return Loss
Brouter	Combination Bridge and ROUTER
bs	Bahamas
BS	BackSpace (see ASCII)
BS	Base Station (mobile telephone network)
BSC	Base Station Controller (mobile telephone network)
BSC	Binary Synchronous Communication (IBM)
BSD	Berkeley Software Distribution RFC 1258, RFC 1977, RFC 3164
BSI	British Standards Institution
BSN	Block Sequence Number
B-SP	B-ISDN signalling point
BSS	Base Station Sub-system (mobile telephone network)
BSS	Basic Service Set (WLAN) IEEE 802.11
BSSID	Basic Service Set Identifier (WLAN) IEEE 802.11
B-STP	B-ISDN signalling transfer point
bt	Bhutan
BT	British Telecommunications plc
B-TA	Terminal Adaptor for B-ISDN
Btag	Beginning TAG (AAL3/4 CPCS)
BTAM	Basic Telecommunications Access Method
BTC	empirical Bulk Transfer Capacity metrics RFC 3148
B-TE	Terminal Equipment for B-ISDN
B-TE1	B-ISDN Terminal Equipment type 1 (designed for ATM)
B-TE2	B-ISDN Terminal equipment type 2 (connected via B-TA)
BTS	Base Transmitter Station (mobile telephone network)
BTx	Bildschirmtext (Deutsche Telekom name for videotext)
BUS	Broadcast and Unknown Server
bv	Bouvet Island
BVCP	PPP Banyan Vines Control Protocol RFC 1763
BVPS	Broadband Virtual Path Services
bw	Botswana
by	Belarus
BYE	end of participation, logout or quit command
bz	Belize
C	Capacitance
C	Carriage control (FTP)
C	Characteristic
C	Compressed mode (FTP)
C	Conditional
C	Control leads ITU-T X.21

Abbreviation	Full form and references
C	hexadecimal digit value equal to decimal '12'
C/I	Carrier to Interference ratio (radio)
C/R	Command / Response
C =	Country identification (X.500)
C-4	Container-4 (149.76 Mbit/s) (SDH/SONET)
ca	Canada
CA	Certificate Authority (PKI) ITU-T X.509
CAC	Channel Access Control
CAC	Connection Admission Control (ATM)
CAD	Computer Aided Design
CAD-CAM	Computer Aided Design / Computer Aided Manufacturing
CAI	Computer assisted instruction
CALS	Continuous Acquisition and Life-cycle Support industry steering group RFC 1895
CALSCH	CALendaring and SCHeduling
CAN	CANcel (see ASCII)
CAP	Carrierless Amplitude modulation / Phase modulation
CAPA	CAPAbilities (POP)
CAPI	Communications Application Program Interface (Microsoft)
CAPI	Cryptography Application Program Interface
CARP	Cache Array Routing Protocol RFC 3040
CAS	Channel Associated Signalling (telephony)
CAS	Communication Applications Specification API (fax modem API)
CASE	Common Application Service Element
CAST	Cryptogaphic Algorithm using STrict avalanche criterion RFC 2144 (CAST-128), RFC 2612 (CAST-256), RFC 2984
CAT	Common Authentication Technology
Cat.5	Category 5 twisted pair cable EIA/TIA 568
Cat.6	Category 6 twisted pair cable EIA/TIA 568
Cat.7	Category 7 twisted pair cable EIA/TIA 568
CATNIP	Common ArchiTecture for the Internet RFC 1707
CBC	Cypher Block Chaining (DES encryption standard) RFC 1829, RFC 2040, RFC 2451
CBCP	CallBack Control Protocol (PPP)
CBDS	Connectionless Broadband Data Service
C-bit	Checksum bit (GRE) RFC 1701
CBO	Continuous Bit stream-Oriented
CBR	Constant Bit Rate (ATM)
CBT	Core Based Trees RFC 2189, RFC 2201
cc	Carbon Copy (SMTP)
cc	Cocos (Keeling) Islands
CC	CSRC count (contributing source count, RTP) RFC 1889
CCA	Clear Channel Assessment (WLAN carrier sensing function) IEEE 802.11
CCA	Commercial Certificate Authority ITU-T X.509
CCD	Charge coupled device
CCD	Contract Completion Date
C-channel	Communication channel
CCIR	International radiocommunication consultative committee (now ITU-R - the radiocommunication sector of ITU)

Abbreviation	Full form and references
CCIS	Common Channel Interoffice Signalling
CCITT	International Telephone and Telegraph Consultative Committee (now renamed ITU-T)
CCP	Compression Control Protocol (PPP) <u>RFC 1962</u>
CCR	Commitment, Concurrency and Recovery services (OSI layer 7)
CCS	Common Channel Signalling
CCS	Console Communication Service (IBM SNA)
CCSO	Computing and Communications Services Office (Univ. of Illinois) RFC 2378
ccTLDs	Country Code Top-Level Domains ISO 3166-1, <u>www.iana.org/cctld/cctld.htm</u>
CCU	Communications Control Unit
CCW	Channel Command Word (IBM ESCON)
cd	Democratic Republic of the Congo
CD	Collision Detection
CD	CountDown counter (DQDB)
CDDI	Copper Distributed Data Interface
CDLC	Cellular Data Link Control
CDMA	Code Division Multiple Access RFC 3141 (AAA)
CDN	Call-Disconnect-Notify (L2TP) <u>RFC 2661</u>
CDS	Common Directory Service
CDUP	Change to Parent Directory (FTP)
CDV	Cell Delay Variation (ATM)
CDVT	Cell Delay Variation Tolerance (ATM)
CE	Connection Element
CE	Connection Endpoint
CE	Customer Edge router (MPLS) RFC 2917
CEC	Cell Error Control (physical layer)
CEI	Connection Endpoint Identifier
CellB	Sun Microsystems video encoding RFC 2029
CELP	Code Excited Linear Predication (low bitrate audio) ITU-T G.723
CEN	Comité Européen de Normalisation
CEP	Connection EndPoint
CEPT	European Conference for Posts and Telecommunications
CEQ	Customer Equipment
CER	Cell Error Ratio (ATM)
CES	Circuit Emulation Service
CES	Connection Endpoint Suffix
cf	Central African Republic
CF	Contention Free (WLAN protocol) IEEE 802.11
CFCCH	Compact Frequency Correction burst CHannel (EDGE)
CFDP	Coherent File Distribution Protocol <u>RFC 1235</u>
CFI	Canonical Form Identifier IEEE 802.1q
CGI	Common Gateway Interface or Common Gate Interface
ch	Switzerland
chaddr	Client Hardware ADDRess (BOOTP/DHCP) RFC 0951
CHAP	CHAllenge handshake Protocol (PPP) <u>RFC 1994</u>, RFC 2433, RFC 2759, RFC 2484
CHARGEN	CHARacter GENerator protocol <u>RFC 0864</u>

Abbreviation	Full form and references
CHPID	CHannel Path IDentifier (IBM ESCON)
ci	Cote d'Ivoire
CI	Congestion Indication
CI	Company Identifier
ciaddr	Client IP ADDRess (BOOTP/DHCP) RFC 0951
CIC	Circuit Identification Code
CICS	Communication Information Control System (SNA)
CID	Communication IDentifier (IBM VTAM)
CID	Context Identifier RFC 3095
CIDR	Classless Inter-Domain Routing RFC 1517-20, RFC 1817
CIFS	Common Internet File System (Microsoft Networking)
CIM	Common Information Model RFC 3060
CIM	Computer Integrated Manufacturing
CIP	Common Indexing Protocol RFC 2651-6
CIPX	Compressing IPX headers over WAN media RFC 1553
CIR	Committed Information Rate (frame relay) ITU-T Q.922
ck	Cook Islands
cl	Chile
CL	ConnectionLess
Clav	Cell available (flow control)
CLEC	Competitive Local Exchange Carrier
CLI	Calling Line Identity
CLI	Command Line Interface
CLI	ConnectionLess Internetworking
CLIP	Calling Line Identity Presentation
CLIR	Calling Line Identity Restriction
Clk	Clock
CLLM	Consolidated Link Layer Management (FR)
CLNAP	ConnectionLess Network service Access Protocol
CLNES	Controlled-Load Network Element Service (QOS and RSVP) RFC 2211-2, RFC 2381 (ATM)
CLNP	ConnectionLess Network Protocol RFC 1526, RFC 1561, RFC 1575, ISO 8473
CLNS	ConnectionLess Network Service RFC 0926, RFC 0986, RFC 0994, RFC 1069, RFC 1162, RFC 1238 (MIB), RFC 1240 ISO 8473
CLP	Cell Loss Priority bit (ATM)
CLR	Cell Loss Ratio (ATM)
CLS	Controlled-Load network element Service (QOS and RSVP) RFC 2211-2, RFC 2381 (ATM)
CLSF	ConnectionLess Service Function
CLTP	ConnectionLess Transport Protocol
CLTS	ConnectionLess Transport Service
CLU	Control Logical Unit (IBM)
cm	Cameroon
CM	Configuration Management RFC 3139
CME	Common Management Environment
CME	Connection Management Entity
CMI	Coded Mark Inversion (line code)
CMIP	Common Management Information Protocol RFC 1095, RFC 1189

Abbreviation	Full form and references
CMIS	Common Management Information Service RFC 1095, RFC 1189
CMISE	Common Management Information Service Entity
CMOL	CMIP Over LLC (Logical Link Control)
CMOT	CMIS & CMIP over TCP/IP RFC 1095, RFC 1189
CMR	Cell Misinsertion Rate (ATM)
CMS	Content Management Software (web authoring software)
CMS	Cryptographic Message Syntax RFC 2315, RFC 2630, RFC 2797, RFC 2984, RFC 3058, RFC 3185, RFC 3211, RFC 3218, RFC 3274, RFC 3278, RFC 3369-70
CMT	Connection management mechanism (FDDI)
cn	China
CN	Congestion Notification
C-n	Container of bitrate order n (SDH)
CNAME	Canonical NAME (DNS) RFC 1035
CNLS	CoNnectionLeSs
CNRP	Common Name Resolution Protocol RFC 3367-8
co	Colombia
CO	Central Office (local public telephone exchange)
CO	Connection-Oriented
CO&M	Centralised Operation and Maintenance
COBS	Consistent Overhead Byte Stuffing
CODEC	COder/DECoder
COH	Connection OverHead
COI	Connection-Oriented Internetworking
COLT	City of London telecommunications (network operator)
com	COMmercial Internet domain
COM	Continuation Of Message
CON	CONcentrator
Conf	Confirm
CONS	Connection-Oriented Network Service
COOKIE	Internet mechanism for placing information onto a Client computer from a web server RFC 1312
coop	COOPerative associations Internet domain
COP	Character-Oriented Protocol
COPS	Common Open Policy Service (RSVP) RFC 2748, RFC 2759, RFC 2940 (MOs)
COPS	Connection-Oriented Presentation Service ISO 8823, ITU-T X.216, ITU-T X.226
COPS-PR	COPS-policy Provisioning RFC 3084
CORBA	Common Object Request Broker Architecture
COS	Class Of Service (DiffServ)
COSS	Connection-Oriented Session Service
COTS	Connection-Oriented Transport Service
CP	Common Part (ATM adaptation layer)
CP	Control Point or Control Program
CPBCCH	ComPact Broadcast Control CHannel (EDGE)
CPCCCH	ComPact Common Control CHannel (EDGE)
CPCS	Common part convergence sublayer (AAL3/4 and AAL5)
CPE	Customer Premises Equipment

Abbreviation	Full form and references
CPL	Call Processing Language RFC 2824
c-plane	Control plane (used for user-network and network-network signalling) ITU-T I.321
CPS	Certificate Practice Statements (PKI) ITU-T X.509
CPU	Central Processing Unit
cq	Republic of Congo
cr	Costa Rica
CR	Carriage Return (see ASCII)
CR	Command / Response
CR	Core Router
CRC	Cyclic Redundancy Check
CRF	Connection Related Function
CRL	Certificate Revocation List RFC 2459, RFC 3279, RFC 3280
CRLF	Carriage Return Line Feed (used as an End-Of-Line, EOL marker)
CRM	Cell Rate Margin (ATM)
CRMA	Cyclic Reservation Multiple Access
CRS	Cell Relay Service (ATM)
CRS	Certification Request Syntax RFC 2314, RFC 2986 (v1.7)
Crypto API	Simple Cryptographic Program Interface RFC 2628
CS	Circuit Switching
CS	Convergence Sublayer (ATM adaptation layer)
CS7	7-bit character set mode (pseudo-terminal mode)
CS8	8-bit character set mode (pseudo-terminal mode)
CS-CELP	Conjugate Structure — Code Excited Linear Predication (low bitrate audio) ITU-T G.729
CSCH	Compact Synchronization CHannel (EDGE)
CSMA	Carrier Sense Multiple Access
CSMA/CD	Carrier Sense Multiple Access with Collision Detection IEEE 802.3
CSR	Cell misSequenced Ratio (ATM)
CSRC	Contributing SouRCe (RTP) RFC 1889
CSU	Channel Switching Unit
Ctag	Confirmation TAG
CTD	Cell Transfer Delay (ATM)
CTP	Connection Termination Point
CTS	Clear-To-Send ITU-T V.24
CTS	Common Transport Semantics
CTS	Conformance Test Services
CTS	Cypher Text Stealing mode of RC5 RFC 2040
cu	Cuba
CU	Currently Unused bits
CUG	Closed User Group
cv	Cap Verde
CV	Code Violation
CWD	Change Working Directory (FTP)
Cwnd	Congestion WiNDow (TCP) RFC 2581
cx	Christmas Island
CX	CoaXial cable
cy	Cyprus
cz	Czech Republic

Abbreviation	Full form and references
D	Delivery confirmation bit ITU-T X.25
D	hexadecimal digit value equal to decimal '13'
DA	Destination Address
Daemon	Disk And Execution MONitor (UNIX background agent program) RFC 0361
DAL	Dedicated access line
DAMA	Demand assigned multiple access
DAP	Data Access Protocol (DECnet)
DAP	Directory Access Protocol
DARPA	Defense Advanced Research Project Agency
Das	Directory monitoring (mib-2) RFC 1567
DAS	Directory Assistance Service RFC 1202, RFC 1564, RFC 2792
DAS	Directory System Agent
DAS	Double Attached Station (FDDI)
DASD	Digital Access Storage Device
DASS	Distributed Authentication Security Service RFC 1507
DAT	Digital Audio Tape RFC 3190
DAYTIME	DAYTIME protocol RFC 0867
dB	DeciBel
DB-9	9-pin D-plug connector EIA 562
DB-25	25-pin D-plug connector ISO 2110
DB-37	37-pin D-plug connector ISO 4902
D-bit	Delay (IP type of service)
DBPSK	Differential Binary Phase Shift Keying IEEE 802.11
DC	Device Control (see ASCII)
DC	Direct Current
DC-bit	Demand Circuits (OSPF) RFC 1793
DCC	Data Country Code (X.121)
DCCS	Distributed Capability Computing System RFC 0708, RFC 0712
DCD	Data Carrier Detect ITU-T V.24
DCE	Data Circuit-terminating Equipment
DCE	Distributed Computer Environment
DCF	Data Communications Function
DCF	Distributed Coordination Function (WLAN) IEEE 802.11
D-channel	Data Channel
DCL	Digital Command Language (Digital equipment corporation)
DCLASS	DCLASS object of RSVP for admission control and QOS policy on DiffServ network RFC 2996
DCN	Data Communications Network
DCS	Digital Cellular System (mobile telephone system)
DD	Database Description packet (OSPF) RFC 2328
DDCMP	Digital Data Communications Message Protocol (DECnet)
DDDB	Distributed Directory DataBase (IBM APPN)
DDI	Direct Dialling In
DDI	Document Delivery Instructions protocol
DDL	Data Descriptive Language for shared data RFC 0242
DDP	Datagram Delivery Protocol (Appletalk layer 3)
DDS	Digital Data System
de	Germany

Abbreviation	Full form and references
DE	Directory Entries (X.500)
DE	Discard Eligibility (ATM & frame relay)
DEA	Directory Entry Attributes (X.500)
DEBUG	Network Debugging Protocol RFC 0643
DEC	Digital Equipment Corporation (now Compaq/Hewlett Packard)
DEC	Digital Equipment Corporation
DECNET	Networking protocol of the Digital equipment corporation
DEFLATE	DEFLATE compression algorithm RFC 1951, RFC 1979 (PPP), RFC 2394
DEL	DELete (see ASCII)
DELE	DELEte (FTP)
DES	Defense Encryption Standard (US Government) RFC 1851, RFC 2947-8, RFC 2952-3
DESE	PPP DES Encryption Protocol RFC 1969
DESEbis	PPP DES Encryption Protocol version 2 RFC 2419
DFI	Domain specific part Format Identifier (numbering)
DGP	Dissimilar Gateway Protocol
DGPT	Direction de la Reglementation Générale de Postes et Telecommunications (France)
DGT	Dirección General de Telecommunicaciones (Spain)
DGT	Director General of Telecommunications
DHC	Dynamic Host Configuration load balancing algorithm RFC 3074
DHCP	Dynamic Host Configuration Protocol RFC 1531, RFC 1533-4, RFC 1541, RFC 2131-2, RFC 2563, RFC 2610, RFC 2855, RFC 2937, RFC 3004, RFC 3011, RFC 3046, RFC 3118, RFC 3203 (reconfigure), RFC 3361
DIB	Directory Information Base (X.500)
DICOM	Digital Imaging and Communications In Medicine RFC 3240
DICT	DICTionary Server Protocol RFC 2229
DID	Direct Inward Dialling
DIF	Directory Interoperability Forum
DiffServ	Differentiated Services RFC 2474-5, RFC 2638, RFC 2873, RFC 2963, RFC 2983, RFC 2998, RFC 3086, RFC 3260, RFC 3270 (MPLS), RFC 3287 (RMON), RFC 3289 (MIB), RFC 3290
DIS	Draft International Standard
DISC	DISConnect (HDLC,LLC)
DISCARD	Discard Protocol RFC 0863
DISP	Directory information shadowing protocol
DIT	Directory Information Tree (X.500)
DIX	Digital Equipment Corporation / Intel /Xerox
DIXIE	Directory Interface and assistance protocol for smaller hosts RFC 1249
dj	Djibouti
dk	Denmark
DL	Data Language RFC 0515
DLCI	Data Link Connection Identifier (Frame Relay) ITU-T Q.922
DLE	Data Link Escape see ASCII
DLS	Data Link Switching RFC 1434, RFC 1795, RFC 1937, RFC 2024 (MOs), RFC 2106, RFC 2114

Abbreviation	Full form and references
DLSw	Data Link Switching RFC 1434, RFC 1795, RFC 1937, RFC 2024 (MOs), RFC 2106, RFC 2114
DlswMIB	Data Link SWitching MIB(mib-2) RFC 2024
DLU	Destination Logical Unit
DLUR	Dependent Logical Unit Requester (IBM APPN) RFC 2232 (MOs)
DLUS	Dependent Logical Unit Server (IBM APPN) RFC 2232 (MOs)
dm	Dominica
DM	Data Mark (NVT-ASCII)
DM	Delta Modulation
DM	Dense Mode protocol-independent multicasting (PIM)
DM	Disconnected Mode (HDLC, LLC)
DMA	Deferred Maintenance Alarm
DMD	Directory Management Domain
DME	Distributed Management Environment
DMI	Digital Multiplex Interface
DMO	Domain Management Organization
DMTF	Desktop Management Task Force
DMZ	DeMilitarized Zone (in multi-stage firewall)
DN	Directory Number
DN	Distinguished Name
DNA	Digital Network Architecture (DECNET)
DNCP	DECNet Control Protocol RFC 1376, RFC 1762 (ph IV)
DNIC	Data Network Identification Code ITU-T X.121
Dns	DNS MIB (mib-2) RFC 1611
DNS	Domain Name System RFC 0897, RFC 0921, RFC 0974, RFC 1034-5, RFC 1591, RFC 1611-2 (MIB), RFC 1794, RFC 1876, RFC 1886 (IPv6), RFC 1995-6, RFC 2065, RFC 2136-7, RFC 2181 RFC 2230, RFC 2535-41, RFC 2671-3, RFC 2929, RFC 3071, RFC 3197 (MIB extensions), RFC 3363-4 (IPv6)
DNS NCACHE	Negative CACHEing of DNS queries RFC 2308
DNS RR	Domain Name System Resource Record RFC 1183, RFC 1348, RFC 1637, RFC 1706, RFC 2782, RFC 2915, RFC 3123
DNS SRV	DNS RR for location of SeRVices RFC 2052, RFC 2782
DNS UPDATE	Dynamic UPDATEs in the Domain Name System RFC 2136
DNSSEC	Secure Domain Name System RFC 3007-8, RFC 3090, RFC 3225-6
DNS_ALG	DNS Application Level Gateway RFC 2694
DOCSIS	Data Over Cable Service Interface Specification RFC 2669-70 (MIB), RFC 3083 (MIB), RFC 3256
DoD	Department of Defense (USA)
DODIIS	Department of Defense Intelligence Information System (TELNET) RFC 1043
DOM	Document Object Model RFC 2803
DOP	Directory Operational and binding management Protocol
DOS	Denial Of Service RFC 2267, RFC 2827
DOS	Disk Operating System
dot12	IEEE 802.12 MIB (mib-2) RFC 2020
DPC	Destination Point Code

Abbreviation	Full form and references
DPCM	Differential Pulse Code (speech) Modulation
DPI	Distributed Program Interface (part of SNMP) RFC 1228, RFC 1592 (v2)
DPS	Distributed Programming System RFC 0708
DPSK	Differential Phase Shift Keying
DQDB	Dual Queue Dual Bus IEEE 802.6
DQPSK	Differential Quadrature Phase Shift Keying IEEE 802.11
DR	Designated Router (OSPF)
DRS	Data Reconfiguration Service RFC 0166
DS	Destination Station (IEEE 802.11)
DS	Differentiated Services (see DiffServ)
DS	Digital Section (transmission line)
DS	Direct Sequence
DS	Directory Service
DS Field	Differentiated Service Field RFC 2474, RFC 2475, RFC 2873
DS0	Digital Signal hierarchy level 0 (N. America) (64 kbit/s PDH) RFC 2494 (MOs)
DS1	Digital Signal hierarchy level 1 (N. America) (1.544 Mbit/s) RFC 1232 (MIB), RFC 1406 (MOs), RFC 2495 (MOs)
DS2	Digital Signal hierarchy level 2 (N. America) (6.312 Mbit/s) RFC 2495 (MOs)
DS3	Digital Signal hierarchy level 3 (N. America) (45 Mbit/s) RFC 1233 (MIB), RFC 1407 (MOs), RFC 2496 (MOs)
DSA	Directory Service Agent (X.500)
DSAP	Destination Service Access Point
DSCP	Differentiated Services CodePoint (DiffServ) RFC 2474
DSE	Data Switching Exchange
DSE	DSA-Specific Entry RFC 3045
DSG	Default Slot Generator (DQDB)
DSL	Digital Subscriber Line RFC 3276 (MOs)
DSLAM	Digital Subscriber Line Access Module
DSLCP	Dynamically switched Link Control Protocol RFC 1307
DSML	Directory Services Markup Language
DSN	Delivery Status Notification RFC 1894, RFC 2530, RFC 2852
DSN	Distributed Systems Network (Hewlett Packard)
DSP	Digital Signal Processing
DSP	Directory System Protocol
DSP	Domain-Specific Part RFC 0982
DS-PHY	Direct Sequence Spread Spectrum PHYsical layer protocol (WLAN) IEEE 802.11
DSR	Data Set Ready ITU-T V.24
DSS	Data Security Solutions www.datasecuritysolutions.com
DSS	Digital Signalling System ITU-T Q.931, ITU-T Q.2931
DSS1	Digital subscriber signalling system 1 (ISDN) ITU-T Q.931
DSS2	Digital subscriber signalling system 2 (ATM) ITU-T Q.2931
DSSS	Direct Sequence Spread Spectrum
DSU	Digital Service Unit or data service unit

Abbreviation	Full form and references
DTD	Data Type Definition (HTML)
DTE	Data Terminal Equipment
DTMF	Dual Tone Multi-Frequency telephone handset signalling RFC 2833
DTP	Data Transfer Process (FTP) RFC 0959
DTP	Data Transfer Protocol RFC 0163, RFC 0171, <u>RFC 0264</u>
DTR	Data Terminal Ready <u>ITU-T V.24</u>
DU	Data Unit
DUA	Directory User Agent RFC 1373, ITU-T X.500
DUAL	Diffusing Update Algorithm (Cisco routing algorithm used in EIGRP)
DUT	Device Under Test
DV	Digital Video RFC 3189, IEC 61834
DVMRP	Distance Vector Multicast Routing Protocol <u>RFC 1075</u>
DVP	Distance Vector Protocol (routing protocol)
DWDM	Dense Wave Division Multiplexing
DXC	Digital crossConnect
DXI	Data eXchange Interface
dz	Algeria
E	EBCDIC (FTP)
E	hexadecimal digit value equal to decimal '14'
E1	Digital Signal hierarchy level 1 (Europe) (2.048 Mbit/s) RFC 1406 (MOs), RFC 2495 (MOs)
E2	Digital Signal hierarchy level 2 (Europe) (8.448 Mbit/s) RFC 2495 (MOs)
E3	Digital Signal hierarchy level 3 (Europe) (34 Mbit/s) RFC 1407 (MOs), RFC 2496 (MOs)
E.164	ITU-T recommendation defining the telephone/ISDN numbering scheme RFC 2916 (DNS), ITU-T E.164
EA	Extended Address
EAB	Extended Address Bit
EA-bit	External Attribute LSA (OSPF)
EAP	Extensible Authentication Protocol (PPP) <u>RFC 2284</u>, RFC 2484, RFC 2716
EAPOL	Extensible Authentication Protocol Over LANs <u>IEEE 802.1x</u>
EBCDIC	Extended Binary-Coded Decimal Interchange Code RFC 0183
EBCN	Explicit Backward Congestion Notification (frame relay)
EBGP	Exterior Border Gateway Protocol (see BGP)
E-bit	External area flag (OSPF)
EBN	Extended Border Node RFC 2457 (MOs)
ec	Ecuador
EC	Echo Canceller
EC	Erase Character (NVT-ASCII)
EC	Error Correction
ECC	Elliptic Curve Cryptography <u>RFC 3278</u>
ECC	Embedded Control Channel (SDH/SONET)
echo	Echo Protocol <u>RFC 0862</u>

Abbreviation	Full form and references
ECHO	Enable echoing (pseudo-terminal mode)
ECHOCTL	Echo control characters as ^(Char). (pseudo-terminal mode)
ECHOE	Visually erase characters (pseudo-terminal mode)
ECHOK	Kill character discards current line (pseudo-terminal mode)
ECHOKE	Visual erase for line kill (pseudo-terminal mode)
ECHONL	Echo NL (new line) even if ECHO is off (pseudo-terminal mode)
ECMA	European Computer Manufacturers Association
ECML	Electronic Commerce Modelling Language RFC 2706 (v1), RFC 3106 (v1.1)
ECN	Explicit Congestion Notification RFC 2481, RFC 2884, RFC 3168
ECP	Encryption Control Protocol (PPP) RFC 1915, RFC 1968
ECS	Encryption Control Signal
ECSA	Exchange Carriers Standards Association (USA)
ECSD	Enhanced Circuit-Switched Data
ED	Ending Delimiter (IBM token ring LAN)
ED	Error Detection
EDC	Error Detection Codes are included (physical layer)
EDGE	Enhanced Data-rate for GSM Evolution
EDI	Electronic Data Interchange
EDIFACT	Electronic Data Interchange For Administration, Commerce and Transport
EDNS0	Extension mechanism for Domain Name System RFC 2671
edu	EDUcational Internet domain
ee	Estonia
EF	Expedited Forwarding PHB group (differential services) RFC 2598, RFC 3246-7
EFCI	Explicit Forward Congestion Indication (frame relay)
EFCN	Explicit Forward Congestion Notification (frame relay)
EFI	Errored Frame Indicator
EFI	External Functionality Interface
EFM	Ethernet in the First Mile
EFS	Error Free Seconds
EFTPOS	Electronic Funds Transfer at the Point Of Sale
eg	Egypt
EGA	Extended video Graphics Array
EGP	Exterior Gateway Protocol (replaced by BGP) RFC 0827, RFC 0888, RFC 0904
EGPRS	Enhanced GPRS (general packet radio service)
EGRP	Exterior gateway routing protocol
eh	Western Sahara
EHLO	Extended HeLlO (SMTP)
EIA	Electronic Industries Alliance or Association (USA)
EIGRP	Enhanced Interior Gateway Routing Protocol (Cisco)
EIN	Internet Engineering Note
EIR	Equipment Identity Register
EIR	Excess Information Rate (frame relay) ITU-T Q.922
EL	ELectrical (interface)
EL	Element Layer (TMN)
EL	Erase Line (NVT-ASCII)
ELAN	Emulated Local Area Network

Abbreviation	Full form and references
ELAP	EtherTalk Link Access Protocol (Appletalk layer 2 for Ethernet LAN)
EL-bit	Extended Length bit (BGP)
ELLC	Extended Logical Link Control (IBM SNA)
EM	ElectroMagnetic
EM	End of Medium (see ASCII)
email	Electronic MAIL
EMC	Electromagnetic Compatibility
EMI	ElectroMagnetic Interference
EML	Element Management Layer (ISO management model) ITU-T M.3000
EMSD	Neda Efficient Mail Submission and Delivery RFC 2524
EN	Edge Node (IBM APPN)
EN	Equipment Number
Enb	Enable
ENCAP	ENCAPsulation header
ENCAPS	scheme for IPng routing and addressing RFC 1955
encapsulation	Internet Encapsulation Protocol RFC 1241
ENQ	ENQuiry (see ASCII)
EntityMIB	Entity MIB (mib-2) RFC 2037
ENUM	Telephone Number Mapping RFC 2916, RFC 3245
EOF	End-Of-File
EOL	End-Of-Line
EOM	End Of Message
EOR	End-Of-Record
EOT	End of Transmission
EP	Emulation Program
EPOS	Electronic Point Of Sale
EPRT	Extended address PoRT (FTP)
EPSV	Extended address PaSsiVe port (FTP)
EQ	EnQuiry
er	Eritrea
ER	Explicit Route
E-R	Entity Relationship
ERA	Exterior Reachable Address
Erlang	Unit of traffic intensity measurement / telephone
ERM	Event Report Management
ERP	Error Recovery Protocol
es	Spain
ES	End System RFC 0995
ES	Errored Seconds
ESC	ESCape character (see ASCII)
ESCON	Enterprise Systems CONnection (SNA)
ESD	End-Station Density
ESI	End System Identifier (numbering)
ES-IS	End System-Intermediate System routing exchange protocol RFC 0995, ISO 4053, ISO 9542
ESMTP	Simple Mail Transfer Protocol service Extensions
ESP	Encapsulating Security Payload (IPsec) RFC 1827, RFC 1851, RFC 2405, RFC 2406, RFC 2451
ESRO	Efficient Short Remote Operations protocol RFC 2188

Abbreviation	Full form and references
ESS	Extended Service Set (WLAN) IEEE 802.11
ESSID	Extended Service Set Identifier (WLAN) IEEE 802.11
et	Ethiopia
ETACS	Extended TACS (mobile telephone system)
Etag	End TAG (AAL3/4 CPCS)
ETAG	Entity TAG (HTTP)
ETB	End of Transmission Block (see ASCII)
ETFTP	Enhanced Trivial File Transfer Protocol RFC 1986
ETHERIP	ETHERnet within IP encapsulation
EtherMIB	ETHERnet like objects MIB (mib-2) RFC 2665
ETR	Early Token Release
ETS	European Telecommunications Standard
ETSI	European Telecommunications Standards Institute
ETX	End of TeXt
ETX/ACK	flow control protocol
EU	European Union
EUI	Extended Unique Identifier www.ieee.org
EXP	EXPerimental
EXPN	EXPaNd (SMTP)
f	Frequency
f	Reference point between operating system function and workstation function (TMN)
F	File structure (FTP)
F	Final bit (HDLC,LLC)
F	hexadecimal digit value equal to decimal '15'
FAC	Flow Admission Control RFC 2205 (RSVP), RFC 2386
FADU	File access data unit (FTAM)
FANP	Toshiba Flow Attribute Notification Protocol RFC 2129
FAS	Frame Alignment Signal or sequence
FAX	Facsimile RFC 3192
F-bit	Forward unknown TLV message type indicator (MPLS) RFC 3032
FC	Frame Control (IBM token ring LAN, FDDI)
FCAPS	Fault Configuration Accounting Performance Security (ISO management model)
FCC	Federal Communications Commission (USA)
FCS	Frame Check Sequence (error detection/correction code)
FDD	Frequency Division Duplex (radio)
FDDI	Fibre Distributed Data Interface RFC 1188, RFC 1285 (MIB), RFC 1512 (MIB), RFC 2019 (IPv6)
FDMA	Frequency Division Multiple Access
FDR	Full Duplex Repeater
FDX	Full DupleX
FE	Functional Element or entity
FEAC	Far End Alarm and Control
FEBE	Far End Block Error
FEC	Forward Equivalence Class (MPLS) RFC 3031
FEC	Forward Error Correction
FECN	Forward explicit congestion notification (frame relay)
FEP	Front End Processor

Abbreviation	Full form and references
FERF	Far End Receive Failure (subsequently renamed RDI, remote defect indication)
FEXT	Far End crossTalk
FF	Fixed Filer style (RSVP) RFC 2205
FF	Form Feed (see ASCII)
FG	Frame Generator
FH	Framc Handler
FH-PHY	Frequency Hopping PHYsical layer protocol IEEE 802.11
fi	Finland
FI	Format Identifier
FIB	Forwarding Information Base RFC 3222
FID	Format Identification
FIFO	First-In-First-Out
FILO	First-In-Last-Out
FIN	FINal
FIN	FINish (TCP)
FINGER	FINGER user information protocol RFC 1288
FIPS	US Federal Information Processing Standard (published by NIST)
FITL	Fibre In The Loop
fj	Fiji
fk	Falkland Islands
FlowMIB	traffic FLOW objects (mib-2) RFC 2064
FLP	Fast Link Pulse
fm	Federal state of Micronesia
FM	Fault Management
FM	Frequency Modulation
FMBS	Frame Mode Bearer Service
FMH	Function Management Header (IBM SNA)
fo	Faroe Islands
FOIRL	Fibre Optic Inter-Repeater Link (for ethernet LAN) IEEE 802.3d
FOOBAR	FTP Operation Over Big Address Records RFC 1545, RFC 1639
FP	Format Prefix
FPE	Floating Point Exception (POSIX)
FQDN	Fully Qualified Domain Name RFC 2821
fr	France
FR	Frame Relay ITU-T Q.922, ITU-T Q.933
FRAD	Frame Relay Access Device (frame relay)
frame	A block of variable length identified by a header label
FRBS	Frame Relaying Bearer Service
FRFH	Frame Relay Frame Handler
FRMR	FRaMe Reject (HDLC, LLC)
FRR	Fast Retransmit Request (TCP) RFC 2581
FRS	Frame Relay Service RFC 2954 (MOs), RFC 3202
FRSE	Frame Relay Switching Equipment
FRTE	Frame Relay Terminal Equipment
FS	File Separator (see ASCII)
FS	Frame Status (IBM token ring LAN)
FSBS	Frame Switching Bearer Service
FSK	Frequency Shift Keying (digital modulation)

Abbreviation	Full form and references
FSS	File System Switch (UNIX)
FTAM	File Transfer Administration and Management RFC 1415
FTN	FEC (Forwarding Equivalence Class)-to-NHLFE (Next Hop Label Forwarding Entry) RFC 3031
FTP	File Transfer Protocol RFC 0114, RFC 0172, RFC 0265, RFC 0354, RFC 0542, RFC 0765, RFC 0959, RFC 1415, RFC 2228, RFC 2389, RFC 2428 (IPv6), RFC 2577, RFC 2640
FTPSRV	TENEX FTP extensions for paged files RFC 0683
FTTB	Fibre To The Building
FTTC	Fibre To The Curb
FTTH	Fibre To The Home
FTTK	Fibre To The Kerb
FUNI	Frame-based User-Network Interface (ATM) RFC 2363
FYI	For Your Information
FZA	Fast approximate arithmetic coder (Gandalf) RFC 1993
G.703	ITU-T recommendation defining a 4-wire high speed digital physical layer interface ITU-T G.703
G.704	ITU-T recommendation defining the 32 timeslot framing of a 2 Mbit/s signal ITU-T G.704
G.732	ITU-T recommendation defining the multiframe synchronisation of G.703 interfaces ITU.T G.732
ga	Gabon
GA	Go Ahead (NVT-ASCII)
GAP	Gateway Access Protocol (DECnet)
GARP	Generic Attribute Registration Protocol (VLAN) IEEE 802.1p, IEEE 802.1r
GB	GigaByte (8 000 000 000 bits)
Gbit/s	Gigabits per second (1 000 000 000 bits/second)
GBSVC	General Broadcast Signalling Virtual Channel
GCAC	Generic Connection Admission Control
GCRA	Generic Cell Rate Algorithm (ATM)
gd	Grenada
GDMO	Guidelines for the Definition of Managed Objects ITU-T X.722, ITU-T X.739, ISO 8824, ISO 8825, ISO 10164-11
ge	Georgia
GECOS	General Electric Computer Operating System
gf	French Guiana
GFC	Generic Flow Control (ATM)
GFSK	Gaussian Frequency Shift Keying (modulation) IEEE 802.11
gg	Equatorial Guinea
gg	Guernsey
GGP	Gateway to Gateway Protocol RFC 791, RFC 0823
GGSN	Gateway GPRS Support Node
gh	Ghana
gi	Gibraltar
GI	Group Identifier
giaddr	Gateway IP ADDRess (BOOTP/DHCP) RFC 0951
GIF	Graphics Interface Format
GII	Global Information Infrastructure

Abbreviation	Full form and references
GKMP	Group Key Management Protocol RFC 2093-4
gl	Greenland
GLOP	GLObal addressing Policy in 233/8 RFC 2770, RFC 3180
gm	Gambia
GME	Global Management Entity
GMII	Gigabit Medium-Independent Interface (ethernet LAN)
GMPLS	Generalised Multiprotocol Label Switching
GMRP	GARP Multicast Registration Protocol IEEE 802.1p
GMSK	Gaussian Minimum Shift Keying (radio modulation for GSM)
GMT	Greenwich Mean Time
gn	Guinea
GND	GrouND ITU-T V.24
gopher	Internet Gopher Protocol RFC 1436
GOS	Grade Of Service
GOSIP	US Government Open Systems Interconnection Profile
gov	GOVernment Internet domain
gp	Guadeloupe
GPRS	General Packet Radio Service (mobile telephone network optimised for data)
gr	Greece
GRE	Generic Routing Encapsulation RFC 1701-2, RFC 2784, RFC 2890, RFC 3147
gs	South Georgia and the South Sandwich Islands
GS	Group Separator see ASCII
GSM	Global System for Mobilecommunication
GSMP	General Switch Management Protocol (Ipsilon) RFC 1987 (v1), RFC 2297 (v2), RFC 3292 (v3), RFC 3293-5
GSN	GigaByte System Network (HIPPI-6400) RFC 2835
GSS_API	General Security Service Application Program Interface RFC 1508-9, RFC 1961, RFC 2025, RFC 2078 (v2), RFC 2478-9, RFC 2743 (v2), RFC 2744, RFC 2853
GSTN	General Switched Telephone Network (see also PSTN) RFC 2846, RFC 3191
gt	Guatemala
GTF	Generalised Trace Facility (IBM SNA)
gu	Guam
GUI	Graphical User Interface
GVRP	Generic VLAN Registration Protocol
gw	Guinea-Bissau
gy	Guyana
gzip	Compression software for UNIX and DOS RFC 1952
H	Header
H0	384 kbit/s transmission rate
H11	1536 kbit/s transmission rate
H12	1920 kbit/s transmission rate
H.261	ITU-T recommendation defining a videoconference codec standard ITU-T H.261
HAP	Host Access Protocol RFC 0802, RFC 0851, RFC 0878, RFC 0907, RFC 1221 (v2)

Abbreviation	Full form and references
HCI	Host Computer Interface
HCS	Header Check Sequence
HDB3	High Density Bipolar 3-state code (line code)
HDLC	Higher level Data Link Control
HDP	High Drop Priority (see DiffServ)
Hdr	Header
HDR	Half Duplex Repeater
HDSL	High-speed Digital Subscriber Line RFC 3276 (MOs)
HDX	Half DupleX
HEC	Header Error Control (ATM)
HELO	HELlO (SMTP)
HELP	HELP (FTP)
HEMP	High-Level Entity Management Protocol RFC 1022
HEMS	High-Level Entity Management System RFC 1021, RFC 1023, RFC 1024, RFC 1076
HFC	Hybrid Fibre Coax (cable TV)
H-FP	Host-Front end Protocol RFC 0929
HIPERLAN	HIgh PERformance (wireless) Local Area Network
HIPPI	High Performance Parallel Interface RFC 1374, RFC 2067, RFC 2834-5
hk	Hong Kong
HLF	Higher Layer Function
HLI	Higher Layer Information
hm	Heard and McDonald Islands
HMAC	Hash Message Authentication Code RFC 2085, RFC 2104, RFC 2403-4, RFC 2857
HMP	Host Monitoring Protocol RFC 0869
hn	Honduras
HOB	Head Of Bus
HOPOPT	IPv6 HOP-by-Hop OPTion header RFC 2460
Host	Host resources (mib-2) RFC 1514
Host-Host	Host-Host Protocol RFC 0018, RFC 0033, RFC 0317, RFC 0604, RFC 0695, RFC 0714, RFC 0721
HP	Hewlett Packard www.hp.com
HPR	High Performance Routing (IBM APPN)
hr	Croatia
HS	Half Session (IBM APPN)
HSRP	Cisco Hot-Standby Router Protocol (Cisco) RFC 2281
HSSI	High Speed Serial Interface EIA/TIA 612/613
ht	Haiti
HT	Horizontal Tab (see ASCII)
HTCP	HyperText Caching Protocol RFC 2756 (0.0)
HTML	HyperText Mark-up Language RFC 1866 (2.0), RFC 1980, RFC 2070, RFC 2659, ISO 8859-1 (character set)
HTTP	HyperText Transfer Protocol RFC 1945 (1.0), RFC 2068-9 (1.1), RFC 2109, RFC 2227, RFC 2296, RFC 2616 (v1.1), RFC 2617, RFC 2774, RFC 2817-8, RFC 2935-6, RFC 2964-5, RFC 3143, RFC 3229-30
HTTPS	Secure HyperText Transfer Protocol RFC 2660
hu	Hungary
HUP	Hang-Up (POSIX)

Abbreviation	Full form and references
HW	HardWare
I	Indicate leads ITU-T X.21
I	Image (FTP)
I.420	ISDN basic rate interface specification ITU-T I.420
I/G	Individual / Group identifier
I/O	Input/Output
IA2	International Alphabet number 2 (Telex)
IA5	International Alphabet number 5 (the ASCII code)
IAB	Internet Architecture Board
IANA	Internet Assigned Numbers Authority RFC 2929, www.iana.org
IANAifType	Interface types (mib-2) RFC 1573
IAP	Internet Access Provider
IAPP	Inter-Access Point Protocol (WLAN) IEEE 802.11f
IASG	Internet Address Sub Group
IBGP	Internal Border Gateway Protocol RFC 2796
I-bit	Initialise bit
IBM	International Business Machines
IBSS	Independent Base Service Set (WLAN) IEEE 802.11
IC	Interruption Control
iCalendar	Internet Calendar RFC 2445-7
ICANON	Canonicalize input lines (i.e. convert to least significant bit first) (pseudo-terminal mode)
ICCN	Incoming-Call-CoNnected (L2TP) RFC 2661
ICD	International Code Designation (address)
ICF	Information Conversion Function
ICI	Inter-Carrier Interface
ICIP	Inter-Carrier Interface Protocol
ICLS	Interoperation Controlled-Load Service RFC 2381
ICMP	Internet Control Message Protocol RFC 0760, RFC 0777, RFC 0792, RFC 1256, RFC 2521
ICMPv6	Internet Control Message Protocol for IPv6 RFC 1885, RFC 2463
ICP	Initial Connection Protocol RFC 0123, RFC 0143, RFC 0165, RFC 0215
ICP	Interconnect Control Program
ICP	Internet Cache Protocol RFC 2186 (v2), RFC 2187, RFC 3040
ICPC	Initial Connection Protocol Control RFC 0145
ICRNL	Map CR (carriage return) to NL (new line) on input (pseudo-terminal mode)
ICRP	Incoming-Call-RePly (L2TP) RFC 2661
ICRQ	Incoming-Call-ReQuest (L2TP) RFC 2661
ICS	Implementation Conformance Statement
iSCSI	Small Computer System Interface protocol over the Internet RFC 3347
ICV	Integrity Check Value (Ipsec) RFC 2402, RFC 2406
id	Indonesia
ID	Identifier
IDEA	International Data Encryptions Algorithm RFC 3058
Ident	IDENTification protocol RFC 1414
IDI	Initial Domain Identifier (numbering)
IDL	Interface Definition Language
IDMIB	Inter-Domain Management Information Base

Abbreviation	Full form and references
IDMR	Inter-Domain Multicast Routing
IDN	Integrated Digital Network
IDP	Initial Domain Part
IDPR	Inter-Domain Policy Routing protocol RFC 1477-9
IDR	Inter-Domain Routing RFC 3221
IDRP	Inter-Domain Routing Protocol (BGP-4) RFC 1745, RFC 1863
IDU	Interface Data Unit
IDUP	Independent Data Unit Protection (GSS_API) RFC 2479
ie	Ireland
IE	Information Element
IE	Internet Explorer (Microsoft)
IEC	Inter-Exchange Carrier (USA)
IEC	International Electrotechnical Commission
IEE	Institution of Electrical Engineers (UK)
IEEE	Institute of Electrical and Electronic Engineers (USA)
IEEE 802.2	LAN Logical Link Control (LLC) protocol IEEE 802.2
IEEE 802.3	IEEE standard for the ethernet LAN physical layer IEEE 802.3
IEEE 802.5	physical layer specification for token ring LAN IEEE 802.5
IEN	Internet Engineering Note
IESG	Internet Engineering Steering Group www.iesg.org/iesg.html
IETF	Internet Engineering Task Force RFC 3233, www.ietf.org
IEXTEN	Enable extensions (pseudo-terminal mode)
IFG	Inter-Frame Gap (ethernet LAN) IEEE 802.3
IfMIB	Interface types (mib-2) RFC 1573
i-frame	information frame (HDLC,LLC)
IGM	Internet Group Management
IGMP	Internet Group Management Protocol RFC 1112, RFC 2236 (v2), RFC 2933 (MIB)
IGNCR	Ignore CR (carriage return) on input (pseudo-terminal mode)
IGNPAR	The ignore parity flag. The parameter is 0 (FALSE) or 1 (TRUE) (pseudo-terminal mode)
IGP	Interior Gateway Protocol RFC 1074
IGRP	Interior Gateway Routing Protocol (Cisco)
IHL	Internet Header Length field (IP) RFC 0791
IISP	Interim Interswitch Signalling Protocol (ATM NNI)
IKE	Internet Key Exchange RFC 2409
il	Israel
ILEC	Independent Local Exchange Carrier (USA)
ILL	ILLegal instruction (POSIX)
ILL	ISO Interlibrary Loan protocol RFC 2503
ILM	Incoming Label Map (MPLS) RFC 3031
ILMI	Integrated Local Management Interface RFC 2601-3
ILMI	Interim Local Management Interface (ATM)
im	Isle of Man
IMAC	Isochronous Media Access Control

Abbreviation	Full form and references
IMAP	Interactive Mail Access Protocol or Internet message access protocol RFC 1176 (v2), RFC 1203 (v3), RFC 1730-3 (v4), <u>RFC 2060-2</u>, RFC 2086-8 (v4), RFC 2177, RFC 2180, RFC 2192-3, RFC 2221, RFC 2342, RFC 2683, RFC 2971, RFC 3348
IMAXBEL	Ring bell on input queue full (pseudo-terminal mode)
iMIP	iCalendar Message-based Interoperability Protocol <u>RFC 2447</u>
IMP	Interface Message Processor (ARPANET packet switch) RFC 0018, RFC 0660
IMP	Internet Message Processor
IMP	Internet Message Protocol <u>RFC 0759</u>
IMP/host	IMP/host Protocol RFC 0012, RFC 0209, RFC 0660, RFC 0687, RFC 0704
IMPP	Instant Messaging and Presence Protocol
IMSI	International Mobile Station Identity
IMSSI	Inter-MAN (metropolitan area network) Switching System Interface (DQDB)
in	India
IN	Intelligent Network
INAP	Intelligent network application part (SS7)
InARP	Inverse Address Resolution Protocol RFC 1293, <u>RFC 2390</u>
IND	INDicate or indication
info	INFOrmation Internet domain
INI	Inter-Network Interface
INLCR	Map NL (new line) into CR (carriage return) on input. (pseudo-terminal mode)
INN	Intermediate Network Node
INPCK	Enable checking of parity errors (pseudo-terminal mode)
int	INTernational organizations Internet domain
INT	INTerrupt process (POSIX)
INTELSAT	International Telecommunications Satellite organisation
IntSrv	Integrated Services MIB (mib-2) <u>RFC 2213</u>
INX	INternet eXchange
io	British Indian Ocean Territory
IOC	Input/Output Control
IOCP	Input/Output Channel Program (IBM ESCON)
IOP	InterOPerability testing
IOS	Internetworking Operating System (Cisco)
IOTP	Internet Open Trading Protocol RFC 2802, <u>RFC 2935</u>, RFC 3354 (v2)
IP	Intelligent Peripheral
IP	Internet Protocol (see IPv4 and IPv6)
IP	Interrupt Process (NVT-ASCII)
IP address	Internet Protocol addressing scheme RFC 1166 (IPv4), RFC 1918 (allocation private IP), RFC 2050 (IP allocation), <u>RFC 2373 (IPv6)</u>
IP address class	Internet Protocol Address Classes A, B, C, D and E RFC 1812
IP Comp	IP payload COMPression
IPComp	IP Payload Compression Protocol RFC 2393, <u>RFC 3173</u>
IPCP	Internet Protocol Control Protocol (PPP) <u>RFC 1332</u>, RFC 2290, RFC 2472 (IPv6-CP)

Abbreviation	Full form and references
IP Mobility	Internet Protocol for Mobility (mobile IP) RFC 2002, RFC 2005-6, RFC 2290, RFC 2344, RFC 2794, RFC 3012, RFC 3024-5, RFC 3115, RFC 3220, RFC 3344
IP Paging	dormant mode host alerting RFC 3132
IP Precedence	IP Precedence field (IPv4) RFC 791, RFC 2873
IP-E	IP datagrams over Ethernet networks RFC 0895
IPHC	Internet Protocol Header Compression (PPP) RFC 2509
IPI	Initial protocol identifier ISO/IEC TR 9577
IPL	Primary Link for Interactive services
IpMIB	Internet protocol MIB (mib-2) RFC 2011
IPMS	Inter-Personal Message Service ITU-T X.400
IP-MTU	Path maximum Message Transmission Unit Discovery RFC 1191
Ipng	Internet Protocol next generation (equivalent to IPv6) RFC 1454, RFC 1550, RFC 1752, RFC 1883, RFC 2460
IPO	Internet Protocol over Optical media
IPOA	Internet Protocol Over ATM RFC 1577, RFC 2225, RFC 2320 (MOs), RFC 2331
IPP	Internet Printing Protocol RFC 2565-9 (v1.0), RFC 2639 (v1.0), RFC 2910-1 (v1.1), RFC 3196 (v1.1), RFC 3239
IPPM	Internet Protocol Performance Metrics RFC 2330, RFC 2498, RFC 2679-81
IPS	Intrusion Prevention System
IPsec	Security architecture for the Internet Protocol RFC 1825, RFC 2401, RFC 2402, RFC 2406, RFC 2411, RFC 2709, RFC 3193 (L2TP)
IPv4	Internet Protocol version 4 RFC 0760, RFC 0791, RFC 1122 (requirements of hosts), RFC 1812 (requirements of routers)
IPv6	Internet Protocol version 6 RFC 1454, RFC 1550, RFC 1752, RFC 1883, RFC 2023 (PPP), RFC 2460-1, RFC 2464-76, RFC 2491-2, RFC 2497, RFC 2675, RFC 2711, RFC 3053, RFC 3056
IPv6-ADDR	ADDRessing scheme (128 bit) for Internet Protocol version 6 RFC 1881, RFC 1884, RFC 1887, RFC 1897, RFC 1955, RFC 1971, RFC 2073, RFC 2373-5, RFC 2462, RFC 2526, RFC 2732, RFC 3041, RFC 3177, RFC 3306, RFC 3307
IPv6-AUTH	AUTHentication header for IPv6 RFC 2402
IPv6-CP	PPP network Control Protocol for IPv6 RFC 2472
IPv6-DISC	neighbour DISCovery for IPv6 RFC 2461
IPv6-ESP	Encapsulating Security Payload for IPv6 RFC 2406
IPv6-Frag	Fragment header for IPv6 RFC 2460
IPv6-ICMP	Internet Control Message Protocol for IPv6 RFC 2463
IPv6-NoNxt	No Next header in IPv6 header RFC 2460
IPv6-Opts	Destination Options header in IPv6 RFC 2460
IPv6-Route	Routing header for IPv6 RFC 2460
IPv6 routers	RFCs relating to routers, hosts and transition to IPv6 RFC 2893-4, RFC 3068, RFC 3122, RFC 3142
IPv6-SA	Security Architecture for IPv6 RFC 2401
IPX	Internetwork Packet eXchange (Novell Netware)
IPXCP	Internetworking Packet eXchange (Novell) Control Protocol (PPP) RFC 1552
IPXWAN	Novell IPX over various WAN media RFC 1362, RFC 1551, RFC 1634

Abbreviation	Full form and references
iq	Iraq
IQUERY	Inverse QUERY (DNS) RFC 1035
ir	Iran
IR	Internal Router
IRA	Internal Reachable ATM address
irc	Initialise-Restart-Count RFC 1661 (PPP)
IRC	Internet Relay Chat protocol RFC 1459, RFC 2810-3
IRDP	ICMP Router Discovery Protocol RFC 1256
IRN	Intermediate Routing Node
IRP	Interior Routing Protocol (router networks)
IRP	Internal Reference Point
IR-PHY	Infra-Red PHYsical layer protocol IEEE 802.11
IRS	Intermediate Reference System
IRTF	Internet Research Task Force
IRTP	Internet Reliable Transaction Protocol RFC 0938, RFC 1663 (PPP)
is	Iceland
IS	Intermediate System RFC 0995
IS	International Standard
ISAKMP	Internet Security Association & Key Management Protocol RFC 2407-8
ISDN	Integrated Services Digital Network RFC 2127 (MIB)
ISDU	Isochronous Service Data Unit (DQDB)
ISI	In-Span Interconnection
ISI	Inter-System Interface
ISIG	Enable signals INTR, QUIT, [D]SUSP (pseudo-terminal mode)
IS-IS	Intermediate System-Intermediate System (intra-domain routing protocol) RFC 1142, RFC 1195 (Integrated IS-IS), RFC 2763, RFC 2966, RFC 2973, ISO 10589 RFC 3277, RFC 3358-9
ISM	Industrial Scientific Medical (radio band - 2400 MHz)
ISN	Initial Sequence Number (TCP)
ISO	International Organization for Standardization www.iso.org
ISO 2110	Standard 25-pin plug for V.24/V.28 DTE-to-DCE interface
ISO-10646	Universal Multiple Octet (Unicode) Character Set (ISO 10646)
ISO-10646-J-1	Universal Multiple Octet (Unicode) Character Set (ISO 10646) Japanese version RFC 1815
ISOC	Internet SOCiety RFC 2134, www.isoc.org
ISO-IP	ISO Internet Protocol
ISP	International Standardised Profile
ISP	Internet Service Provider RFC 3013
ISP	Internet Stream Protocol RFC 1190, RFC 1819, RFC 1946
ISP	Interswitch Signalling Protocol (ATM NNI)
ISPBX	ISDN Private Branch eXchange
ISR	Intermediate Session Routing (IBM APPN)
ISSI	Inter-Switching System Interface
ISSN	International Serial Standard Number RFC 3044
ISTRIP	Strip 8th bit off characters (pseudo-terminal mode)
ISUP	Integrated Services digital network User Part (SS7)
it	Italy
IT	Information Technology

Abbreviation	Full form and references
IT	Information Type
ITC	Independent Telephone Company
iTIP	iCalendar Transport-Independent interoperability Protocol RFC 2446
ITOT	ISO Transport service On top of TCP RFC 2126
ITT	Invitation To Tender
ITU	International Telecommunications Union
ITU-R	International Telecommunications Union Radiocommunication sector (former CCIR)
ITU-T	International Telecommunications Union Standardization sector (former CCITT)
IUCLC	Translate uppercase characters to lowercase (pseudo-terminal mode)
IUT	Implementation under test
IVD	Integrated Voice Data
IW	Initial Window (TCP)
IWF	InterWorking Function
IWP	Inter-Working Point
IWU	InterWorking Unit
IX	Internet eXchange
IXANY	Any character will cause a restart after stop (pseudo-terminal mode)
IXC	IntereXchange Carrier
IXOFF	Enable input flow control (pseudo-terminal mode)
IXON	Enable output flow control (pseudo-terminal mode)
J-bit	Justification bit (of line code)
je	Jersey
jm	Jamaica
jo	Jordan
jp	Japan
JPEG	Joint Photographic Experts Group (compressed picture standard) RFC 2035, RFC 2435
JunOS	Juniper Operating System (Juniper Networks)
kB	kiloByte (8000 bits)
K-bit	Justification bit (of line code)
K-bit	Key bit (GRE) RFC 1701
kbit/s	kilobits per second (1000 bits/second)
kbps	kilobits per second
ke	Kenya
KEA	Key Exchange Algorithm RFC 2528, RFC 2773, RFC 2876
KEX	Key Exchange
kg	Kyrgyzstan
kh	Cambodia
ki	Kiribati
km	Comoros
km	kilometre
kn	Saint Kitts and Nevis
kp	Democratic People's Republic of Korea
kr	Republic of Korea
kw	Kuwait
ky	Cayman Islands
kz	Kazakhstan

Abbreviation	Full form and references
L	Local byte size (FTP)
L2F	Layer 2 Forwarding protocol (Cisco) RFC 2341
L2TP	Layer 2 Tunneling Protocol RFC 2661, RFC 2809, RFC 2888, RFC 3070, RFC 3145, RFC 3193 (IPsec), RFC 3301 (ATM), RFC 3355 (AAL5), RFC 3371 (MIB)
L2TP-FR	Layer 2 Tunneling Protocol over Frame Relay RFC 3070
la	Lao People's Democratic Republic
LA	Link Acknowledgement
LA	Link Adaptation
LAC	L2TP Access Concentrator Client (L2TP) RFC 2661
LAN	Local Area Network
LANE	LAN Emulation
LANG	LANGuage (FTP)
LAP	Link Access Procedure
LAPB	Link Access Procedure Balanced ITU-T X.25
LAPD	Link Access Procedure (ISDN D-channel) ITU-T Q.931
LAPM	Link Access Procedure Modem
LASER	Light Amplification by Stimulated Emission of Radiation
LATA	Local Access and Transport Area (USA)
lb	Lebanon
L-bit	Length bit (L2TP) RFC 2661
lc	Saint Lucia
LCGN	Logical Channel Group Number ITU-T X.25
LCN	Logical Channel Number (statistical multiplexing) ITU-T X.25
LCP	Link Control Protocol (PPP) RFC 1570, RFC 1661, RFC 2484
LCT	Last Compliance Time or last conformance time (leaky bucket algorithm)
LD	LAN Destination
LD	Loop disconnect (telephone signalling)
LDAP	Lightweight Directory Access Protocol RFC 1487, RFC 1558, RFC 1777, RFC 1798, RFC 1959, RFC 2164, RFC 2251-6 (v3), RFC 2307, RFC 2559 (v2), RFC 2589 (v3), RFC 2596, RFC 2649, RFC 2657, RFC 2696 RFC 2739, RFC 2820, RFC 2829-31(v3), RFC 2891, RFC 2926-7, RFC 3062, RFC 3112 RFC 3296
LDAP API	LDAP Application Program Interface RFC 1823
LD-CELP	Low Delay Code Excited Linear Prediction (low bitrate speech)
LDDB	Local Directory DataBase
LDIF	LDAP Interchange Format RFC 2849
LDP	Label Distribution Protocol (ATM, MPLS) RFC 3031, RFC 3032, RFC 3035-8, RFC 3212-4 (CR-LDP), RFC 3215 (state machine)
LDP	Listserv Distribute Protocol RFC 1429
LDP	Loader Debugger Protocol RFC 0909
LDP	Low Drop Priority (see DiffServ)
LDUP	LDAP Duplication/replication Update Protocol
LE	LAN Emulation
LE	Layer Entity
LE	Local Exchange
LEC	LAN Emulation Client
LEC	Local Exchange Carrier (USA)
LED	Light Emitting Diode

Abbreviation	Full form and references
LEN	Low Entry Networking
Level 0	(Network Standard) Graphics Protocol: Level 0 RFC 0199, RFC 0292, RFC 0336, RFC 0493, RFC 0553
LF	Largest Framesize
LF	Line Feed (see ASCII)
LFA	Cabletron Lightweight Flow Admission protocol RFC 2124
LFA	Loss of Frame Alignment
LFC	Local Function Capabilities
LFSID	Local Form Session IDentifier (IBM APPN logical channel identifier)
LGN	Logical Group Node
LH	Link Header
LH	Long Wave
li	Liechtenstein
LI	Length Indicator
LI	Link Identifier
LIB	Label Information Base (LDP — Label Distribution Protocol) RFC 3036
LIC	Line Interface Coupler
LIFO	Last In First Out (queueing technique)
LINX	London INternet eXchange (Internet exchange & peering point)
LIPKEY	Low Infrastructure public KEY using SPKM RFC 2847
LIST	LIST (FTP)
Listserv	List Server (automatic mailing list distribution system) RFC 1429
LIVT	Link Integrity Verification Test
lk	Sri Lanka
LLA	Logical Layered Architecture
LLAP	LocalTalk Link Access Protocol (Appletalk layer 2)
LLC	Logical Link Control (LAN) IEEE 802.2
LLC	Lower Layer Compatibility (ISDN)
LLI	Lower Layer Information
LMCS	Local Multipoint Communications System (broadband wireless local loop)
LMDRP	Lost Message Detection and Recovery Protocol RFC 0663
LMDS	Local Multipoint Distribution Service (broadband wireless local loop)
LME	Layer Management Entity
LMI	Local Management Interface
LMTP	Local Mail Transfer Protocol RFC 2033
LN	Link attentioN
LNA	Link attentioN Acknowledgement
LNS	L2TP Network Server (L2TP) RFC 2661
LOC	Loss Of Cell delineation
LOF	Loss Of Frame
LOP	Loss Of Pointer
LPD	Line Printer Daemon RFC 2569
LPDU	Logical link control Protocol Data Unit
LPR	Line PRinter
LQC	Link Quality Control
lr	Liberia
ls	Lesotho
LSA	Link State Advertisement (OSPF) RFC 2328

Abbreviation	Full form and references
LSAck	Link State ACKnowledgement (OSPF)
LSB	Least Significant Bit
LSD	Link State Database (OSPF)
LSNAT	Load-Sharing using IP Network Address Translation RFC 2391
LSP	Label-Switched Path (MPLS) RFC 3031, RFC 3212
LSP	Link State routing Protocol
LSR	Label-Switched Router (MPLS)
LSR	Link State Request (OSPF) RFC 2328
LSRR	Loose Source and Route Record
LSU	Link State Update (OSPF) RFC 2328
LSUB	List of available usernames (IMAP)
lt	Lithuania
LT	Line Termination
LTE	Line Terminating Equipment
LTH	LengTH field
LTU	Line Terminating Unit
lu	Luxembourg
LU	Logical Unit (IBM SNA)
LUNI	LAN-User-Network-Interface
lv	Latvia
LW	Loss Window (TCP) RFC 2581
LX	Long Wave
ly	Libyan Arab Jamahiriya
LZS	Lempel-Ziv-Stac compression protocol RFC 2395
LZS-DCP	Lempel-Ziv-Stac Data Compression Protocol (PPP) RFC 1967, RFC 1974
LZW	Lempel-Ziv-Welch data compression coding
M	Manager
M	Mandatory
M1-interface	Management interface between ATM CEQ and private NMS (ATM Forum)
M2-interface	Management interface between private ATM network and private NMS (ATM Forum)
M3-interface	Management interface between private NMS and public NMS (ATM Forum)
M4-interface	Management interface between public ATM network and public NMS (ATM Forum)
M5-interface	Management interface between two different public NMSs (ATM Forum)
ma	Morocco
MA	Medium Adaptor
MAC	Medium Access Control IEEE 802.3 (ethernet), IEEE 802.5 (token ring)
M-ACTION	(CMIP) request an object to perform an action
MADCAP	Multicast Address Dynamic Client Allocation Protocol RFC 2730, RFC 2907
MAE	MAnaged Ethernet (Internet exchange & peering point)
MAIL	Internet Message Format RFC 0822, RFC 2822
MAN	Metropolitan Area Network
MANET	Mobile Ad-hoc NETworking RFC 2501
MAP	Manufacturing Automation Protocol
MAP	Mobile Application Part (SS7)

Abbreviation	Full form and references
MAPOS	Multiple Access Protocol on SONET / SDH RFC 2171-6, RFC 3186 (tunneling mode)
MARC	MAchine Readable Cataloguing records RFC 2220
MARS	Multicast Address Resolution Server RFC 2149, RFC 2269, RFC 2443, RFC 2602
MASC	Multicast Address Set-Claim protocol RFC 2365, RFC 2909
MAU	Media Access Unit (Token ring LAN)
MAU	Medium Attachment Unit (ethernet LAN) IEEE 802.3l
MaxCR	MAXimum Cell Tate
MaXIM-11	Mapping between X.400 / Internet email and Mail-11 RFC 2162
MB	MegaByte (8 000 000 bits)
Mb/s	Megabits per second
M-bit	Marker bit (RTP) RFC 1889
M-bit	More packets follow
Mbit/s	Megabits per second (1 000 000 bits/second)
MBONE	Multicast backBONE
MBP	Mail Box Protocol RFC 0196, RFC 0221 (v2), RFC 0278
Mbps	Megabits per second
MBS	Maximum Burst Size
MBZ	Must Be Zero
mc	Monaco
M-CANCEL-GET	(CMIP) cancels previous M-GET command
MC-bit	MultiCast extensions for OSPF (MOSPF) RFC 1584
MCDV	Maximum Cell Delay Variation (ATM)
MCF	Message Communication Function
MCGAM	MIXER Conformant Global Address Mapping RFC 2163
MCI	Microwave Communications Inc (USA)
MCLR	Maximum Cell Loss Ratio (ATM)
MCNS	Multimedia Cable Network System partners Ltd (developing DOCSIS) RFC 2670 (MIB)
MCP	MAC Convergence Protocol (DQDB)
MCP	Multilink Control Protocol (PPP) RFC 1717
MCR	Maximum Cell Rate (ATM)
M-CREATE	(CMIP) creates objects
MCS	Modulation and Coding Scheme
MCS	Multicast Server RFC 2022
MCTD	Maximum Cell Transfer Delay (ATM)
MCU	Multipoint Control Unit
md	Republic of Moldova
MD	Management Domain
MD	Mediation Device
MD	Message Digest algorithm RFC 1186 (MD4), RFC 1319 (MD2), RFC 1320 (MD4), RFC 1321 (MD5), RFC 1544 (MD5), RFC 1864 (MD5), RFC 2082 (MD5)
MD2	Message Digest 2 algorithm RFC 1319
MD4	Message Digest 4 algorithm RFC 1320
MD5	Message Digest 5 algorithm RFC 1321, RFC 2385 (TCP signature)

Abbreviation	Full form and references
M-DELETE	(CMIP) deletes objects
MDI	Medium-Dependent Interface
MDL	Medium-Dependent Layer
MdmMIB	dial-up MoDeM objects (mib-2) RFC 1696
MDN	Message Disposition Notification RFC 2530
MDP	Medium Drop Priority (see DiffServ)
MDU	Management Data Unit
Megaco	Media Gateway Control Protocol RFC 2705 (v1), RFC 2805, RFC 2897, RFC 2885-6, RFC 3015 (v1.0), RFC 3064 RFC 3149, ITU-T H.248
MEMS	MicroElectroMechanical System
M-EVENT-REPORT	(CMIP) allows a network resource to announce the occurrence of an event
MF	Management Function (TMN)
MF	Mapping Function
MF	Mediation Function
MF	More Fragments bit (IPv4 protocol)
mg	Madagascar
MGCP	Media Gateway Control Protocol RFC 2705 (v1), RFC 2805, RFC 2897, RFC 2885-6, RFC 3015 (v1.0), RFC 3064 RFC 3149, ITU-T H.248
M-GET	(CMIP) read value of an attribute
mh	Marshall Islands
MH	Message Handling
MHRP	Mobile Host Routing Protocol
MHS	Message Handling System ITU-T X.400
MHTML	Multipart (multimedia) HyperText Markup Language RFC 2112, RFC 2387
MIB	Management Information Base RFC 1065, RFC 1212, RFC 1213, RFC 1214, RFC 2578-80 (v2)
mib-2	Management Information Base for network management of TCP/IP internets RFC 1156, RFC 1213
MIB-II	Management Information Base for network management of TCP/IP internets RFC 1156, RFC 1213
MIC	Medium Interface Connector (jack, plug or socket)
MICP	Mobile Internet Control Protocol
MID	Message IDentifier (DQDB)
MID	Multiplexing IDentification (ATM)
midcom	Middlebox communication RFC 3303-4
MIF	Management Information File
MII	Medium-Independent Interface
mil	MILitary Internet domain
MIL-STD	MILitary STanDard
MIM	Management Information Model
MIME	Multipurpose Internet Mail Extensions RFC 1341, RFC 1521-4, RFC 1556, RFC 1563, RFC 1740-1, RFC 1820, RFC 1844, RFC 1847-8, RFC 1872, RFC 1896, RFC 2017, RFC 2045-9, RFC 2077, RFC 2110, RFC 2112, RFC 2159-61, RFC 2184, RFC 2231, RFC 2387, RFC 2422-6, RFC 2480, RFC 2503, RFC 2557, RFC 2586, RFC 2912-3, RFC 2927, RFC 3009, RFC 3204, RFC 3236, RFC 3240

Abbreviation	Full form and references
MipMIB	Mobile IP MIB (mib-2) RFC 2006
MIS	Management Information System
MIS	Mobile Interface Server
MISTY1	Mitsubishi SecuriTY block cipher encryption algorithm RFC 2994
MIXER	MIME Internet X.400 Enhanced Relay RFC 2156
mk	Macedonia
MKD	MaKe Directory (FTP)
ml	Mali
ML	Maximum Length
MLDP	Multicast Listener Discovery Protocol RFC 3019 (MIB)
MLID	Multiple Link Interface Driver (Novell Netware)
MLP	MultiLink Procedure
MLT	MultiLevel Transmission (line code)
MLTNET	MuLti-TelNet subsystem RFC 0339
mm	Myanmar
mm	millimetre
MM	Message Mode (AAL3/4 or AAL5)
MM	MultiMode fibre
MMF	Multi-Mode Fibre ITU-T G.651, ITU-T G.657
MMI	Man-Machine Interface
MMS	Multimedia Message Service
mn	Mongolia
MNP	Microcom Networking Protocol
MNT	MaNagemenT
mo	Macau
MO	Managed Object (MIB)
mobile IP	IP mobility (see IP mobility)
MOCS	Management Object Conformance Statement
MODE	transfer MODE (FTP)
modem	MODulator / DEModulator
MOSPF	Multicast extensions for OSPF (open shortest path first) RFC 1584
MOTIS	Message Oriented Text Interchange System ISO/IEC 10021
mp	Northern Mariana Islands
MP	Measurement Point
MP	Multilink Protocol (PPP) RFC 1717, RFC 1990, RFC 2686
MP+	Ascend (Lucent) Multilink Protocol Plus RFC 1934
MP3	MPEG Layer 3 (audio compression format) RFC 3119
MPEG	Motion Picture Experts Group (video encoding standard) RFC 2038, RFC 2343, RFC 3003, RFC 3016
MPH	Management Physical Header (physical layer primitive)
m-plane	Management plane (used for network management messages) ITU-T I.321
MPLS	Multi-Protocol Label Switching RFC 2547 (BGP-MPLS VPNs), RFC 2702 (traffic engineering), RFC 2917 (MPLS-VPN), RFC 3031 (MPLS architecture), RFC 3032 (label coding), RFC 3034 (MPLS-FR), RFC 3035 (MPLS-ATM), RFC 3036 (MPLS-LDP), RFC 3270 (DiffServ) RFC 3346, RFC 3353
MPLS-ATM	MPLS over ATM (Asynchronous Transfer Mode) RFC 3035
MPLS-BGP	MPLS using the Border Gateway Protocol RFC 2547

Abbreviation	Full form and references
MPLS-FR	MPLS over Frame Relay RFC 3034
MPLS-SHIM	SHIM header for MultiProtocol Label Switching RFC 3032
MPLS-VPN	Virtual Private Networks by means of MPLS RFC 2917
MPOA	Multi-Protocol encapsulation Over ATM RFC 1483, RFC 2684
MPOFR	Multi-Protocol Over Frame Relay RFC 1294, RFC 1490, RFC 2427
MPP	Message Posting Protocol RFC 1204
MPP	Multi-Protocol Package (Appletalk)
MPPC	Microsoft Point-to-Point Compression protocol RFC 2118
MPPE	Microsoft Point-to-Point Encryption RFC 3078-9
MPT	Ministero delle Poste e Telecommunicazioni (Italy)
mq	Martinique
mr	Mauritania
MRRU	Maximum-Receive-Reconstructed Unit (PPP multilink protocol) RFC 1717
MRU	Maximum Receive Unit (PPP) RFC 1661
ms	Montserrat
MS	Management Services
MS	Mobile Station
MS	Multiplex Section (SDH)
MSAP	MAC Service Access Point
MSB	Most Significant Bit
MS-bit	MaSter bit (as opposed to slave — OSPF synchronisation)
MSC	Mobile Switching Centre (mobile telephone network)
MSDP	Multicast Source Discovery Protocol
MSDTP	Message Services Data Transmission Protocol RFC 0713
MSDU	MAC Service Data Unit
M-SET	(CMIP) add, remove or replace command
MSF	Management Systems Framework
msg	MeSsaGe
MSL	Maximum Segment Lifetime (TCP) RFC 0793
MSN	Message Sequence Number
MSN	Microsoft Network (Internet Service Provider)
MSN	Multiple Systems Networking
MSOH	Multiplex Section OverHead (SDH)
MSP	Management Service Provider (TMN)
MSP	Message Security Protocol
MSP	Message Send Protocol RFC 1159, RFC 1312 (v2)
MSS	MAN Switching System (SMDS)
MSS	Maximum Segment Size (TCP) RFC 0793, RFC 0879, RFC 1122
MST	Multiplex Section Termination
MSVC	Meta Signalling Virtual Channel (ATM)
mt	Malta
Mta	Mail monitoring (mib-2) RFC 2249
MTA	Message Transfer Agent ITU-T X.400
MTAE	Message Transfer Agent Entity ITU-T X.400
MTBF	Mean Time Between Failures
MTL	Message Transfer Layer ITU-T X.400
MTP	Mail Transfer Protocol (replaced by SMTP) RFC 0772, RFC 0780
MTP	Message Transfer Part (SS7)

Abbreviation	Full form and references
MTP	Message Transmission Protocol <u>RFC 0680</u>
MTP	Multicast Transport Protocol <u>RFC 1301</u>
MTS	Message Transfer Service or message transfer system ITU-T X.400
MTTR	Mean Time To Repair
MTU	Maximum Transmission Unit RFC 791 (IPv4), RFC 877 (X.25), RFC 894 (ethernet V2), RFC 1042 IEEE 802.2), RFC 1063 (path discovery options), <u>RFC 1191 (PMTU discovery)</u>, RFC 1209 (SMDS), RFC 1356 (X.25/ISDN), RFC 1390 (FDDI), RFC 1626 (ATM AAL5), RFC 1661 (PPP), RFC 1981 (IPv6), RFC 2225 (IP over ATM), RFC 2923 (TCP problems)
MTU	Message Transfer Unit
MTU	Message Transmission Unit
mu	Mauritius
MUA	Message User Agent
MUF	Maximum Usable Frequency
Mu-law	PCM speech encoding (North American and Japanese standard)
Multicast	over IP and ATM RFC 2365, RFC 2366
museum	MUSEUMs Internet domain
MUX	MUltipleXor
mv	Maldives
MVI	Major Vector Identifier (Token ring MAC)
MVL	Major Vector Length (Token ring MAC)
MVS	Multiple Virtual Storage (IBM operating system)
mw	Malawi
mx	Mexico
my	Malaysia
mz	Mozambique
MZAP	Multicast Zone Announcement Protocol <u>RFC 2776</u>
N	Non-print (FTP)
na	Namibia
NA	Network Adaptor (An interworking function between B-ISDN and narrowband ISDN)
N/A	Not Applicable
NAE	Network Address Extension
NAI	Network Access Identifier <u>RFC 2486</u>
NAK	Negative AcKnowledgement (rejection) RFC 1106
name	Internet domain for individuals
Name/Finger	Name/Finger Protocol <u>RFC 0742</u>
NAPTR	Naming Authority PoinTeR (DNS RR) <u>RFC 2915</u>
NARP	Non-broadcast multiple access Address Resolution Protocol <u>RFC 1735</u>
NAS	Network Access Server RFC 2881-2 (next gen)
NASA	National Aeronautical and Space Administration
NAT	Network Address Translation RFC 1631, RFC 2663, RFC 2993, <u>RFC 3022</u>, RFC 3027, RFC 3235
NAT-PT	Network Address Translator - Protocol Translation RFC 2766, RFC 3152
NAU	Network Addressable Unit (SNA)
NAV	Net Allocation Vector IEEE 802.11
NBFCP	PPP NetBIOS Frames Control Protocol <u>RFC 2097</u>

Abbreviation	Full form and references
NBMA	Non-Broadcast Multiple Access network RFC 1735, RFC 2332, RFC 2491, RFC 2677 (MOs)
NBNS	NetBIOS over TCP/IP Name Server
NBP	Name Binding Protocol (Appletalk)
NBS	Non-Breaking Space
nc	New Caledonia
NCOP	Network Code Of Practice
NCP	Netware Control Protocol (Novell Netware)
NCP	Network Control Program (IBM SNA)
NCP	Network Control Protocol (PPP) RFC 1332 (IPCP), RFC 1377 (OSINLCP), RFC 1378 (ATCP), RFC 1552 (IPXCP), RFC 1638 (BCP), RFC 1661, RFC 1762 (DNCP), RFC 1962 (CCP)
NCP	Network Control Protocol or Network Control Program RFC 0055, RFC 0060, RFC 0215
NCS	Network Computing System (UNIX)
NDF	New Data Flag
NDIS	Network Driver Interface Specification (Microsoft Windows)
NDS	Novell Directory Service RFC 2241
ne	Niger
NE	Network Element
NECP	Network Element Control Protocol RFC 3040
NEF	Network Element Function
NEL	Network Element Layer (ISO management model) ITU-T M.3000
NEM	Network Element Manager
net	NETwork organisations Internet domain
NET	Norme Européene de Télécommunication
netascii	the NVT (Network Virtual Terminal) ASCII character set
NetBEUI	NetBIOS Extended User Interface
NetBIOS	NETwork Basic Input/Output System
NETBLT	NETwork BLock Transfer (bulk transfer protocol) RFC 0969, RFC 0998
NETCRT	NETwork ChaRacTer display protocol RFC 0205
NETRIS	NETwork Remote Input Service (replaced by NETRJS) RFC 0088
NETRJS	NETwork Remote Job entry Service RFC 0189, RFC 0325, RFC 0338, RFC 0599, RFC 0740
NETRJT	NETwork Remote Job service for TIPS RFC 0283
NewsML	Multimedia News extensible Markup Language (XML) RFC 3085
NEXT	Near End crossTalk
nf	Norfolk Island
NFB	Number of Fragment Blocks RFC 0791
NFILE	New FILE (file access protocol) RFC 1037
NFS	Network File System RFC 1094, RFC 1813 (v2), RFC 2623, RFC 2624 (v4), RFC 3010 (v4)
ng	Nigeria
NGL	Network Graphics Loader RFC 0186
NHC	Next Hop Client RFC 2520, RFC 2583
NHFLE	Next Hop Label Forwarding Entry (MPLS) RFC 3031
NHRP	Next Hop Resolution Protocol RFC 2332-3, RFC 2335-6, RFC 2520, RFC 2603, RFC 2677 (MOs), RFC 2735, RFC 2991

Abbreviation	Full form and references
ni	Nicaragua
NIC	Network Information Centre
NIC	Network Interface Card
NICNAME	Network Information Centre NAME database service RFC 0812, RFC 0954, RFC 1834
NIS	Network Information Service
N-ISDN	Narrowband ISDN
NIST	US National Institute of Standards and Technology
NIU	Network Interface Unit
nl	Netherlands
NL	New Line (FTP)
NLA	Network Layer Address RFC 2450, RFC 2921 (IPv6)
NLA	Next-Level Aggregation identifier (IPv6 address) RFC 2373
NLM	Network Lock Manager (Novel Netware)
NLPID	Network Layer Protocol IDentifier ISO/IEC TR 9577
NLRI	Network Layer Reachability Information (BGP)
NLST	Name LiST (FTP)
nm	nanometre
NM	Network Management
NMB	Number of Monitored Blocks (physical layer)
NMC	Network Management Centre
NMF	Network Management Forum
NMF	Network Management Framework
NML	Network Management Layer (ISO management model) ITU-T M.3000
NMS	Network Management System
NN	National Number
NN	Network Node (IBM APPN)
NNI	Network-Network Interface (or Network-Node Interface)
NNTP	Network News Transfer Protocol RFC 0977, RFC 2980
no	Norway
NOC	Network Operations Centre
NOFLSH	Do not flush after interrupt (pseudo-terminal mode)
NOOP	No Operation (FTP)
NOP	No Operation (NVT-ASCII)
np	Nepal
NP	Network Performance
NPA	Number Plan of America
N/P-bit	Not-So-Stubby-Area LSA (OSPF) RFC 1587
NPC	Network Parameter Control (ATM)
NPDU	Network Protocol Data Unit
NPMA	Non Pre-emptive priority Multiple Access
NPSI	NCP-Packet-Switching-Interface (IBM SNA/X.25)
nr	Nauru
N(R)	Receive sequence Number (HDLC)
NREN	National Research and Education Network
NRM	Network Resource Management

Abbreviation	Full form and references
NRT	Non-Real Time
NRZ	Non-Return-to-Zero (line code)
NRZI	Non Return-to-Zero Inverted (line code)
N(S)	Send sequence Number (HDLC)
NSAP	Network Service Access Point RFC 0982, RFC 1237, RFC 1348, RFC 1629, RFC 1637, RFC 1706, RFC 1888 (IPv6)
NSDU	Network Service Data Unit
NSF	National Science Foundation
NSF	Network Search Function (IBM APPN)
NSFnet	National Science Foundation NETwork RFC 1074
NSN	National Significant Number ITU-T E.164
NSP	Network Services Protocol (DECnet)
NSR	Non-Source Routed
NSS	Network and Switching Sub-system
NSS	Network Security Services
NSSA	Not-So-Stubby-Area (OSPF)
NT	Network Termination
NT1	Network Terminating equipment type 1 (ISDN)
NT2	Network Terminating equipment type 2 (ISDN)
NTBBA	Network Terminator BroadBand Access
NTE	network Terminating Equipment
NTN	National Terminal Number ITU-T X.121
NTP	Network Time Protocol RFC 0958, RFC 1059, RFC 1119 (v2), RFC 1165, RFC 1305 (v3), RFC 1708
NTU	Network Termination Unit
nu	Niue
NUI	Network User Identification ITU-T X.25
NUL	NULl (see ASCII)
NULL encryption	encryption algorithm RFC 2410
NULL service	Null service type RFC 2997
NVFS	Network Virtual File System (FTP)
NVP	Network Voice Protocol RFC 0741
NVT	Network Virtual Terminal (see telnet)
NVT-ASCII	Network Virtual Terminal-ASCII character set (see telnet)
NWG	Network Working Group
nz	New Zealand
O	Optional
OAF	Origination Address Field
OAKLEY	generic key exchange and determination protocol RFC 2412
OAM	Operations And Maintenance or Operations, Administration and Maintenance
OAM&P	Operations, Administration, Maintenance and Provisioning
OAMC	Operations, Administration and Maintenance Centre (TMN)
O-bit	Offset bit (L2TP) RFC 2661

Abbreviation	Full form and references
O-bit	Opaque option (OSPF) RFC 2370
O-bit	Optional bit (BGP)
OC	Optical Carrier (SONET)
OC-1	Optical carrier (SONET) OC-1 has bitrate of 51.84 Mbit/s
OC-12	622 Mbit/s SONET digital line system (SONET)
OC-192	10 Gbit/s SONET digital line system (SONET)
OC-3	155 Mbit/s SONET digital line system (SONET)
OC-48	2.5 Gbit/s SONET digital line system (SONET)
OCC	Other Common Carrier
OCCN	Outgoing-Call-CoNnected (L2TP) RFC 2661
OCD	Out of Cell Delineation
OC-n	Optical carrier level-n (SONET)
OCRNL	Translate carriage return to newline (output) (pseudo-terminal mode)
OCRP	Outgoing-Call-RePly (L2TP) RFC 2661
OCRQ	Outgoing-Call-ReQuest (L2TP) RFC 2661
OCS	Object Conformance Statement
OCSP	Online Certificate Status Protocol RFC 2560
ODA	Office Document Architecture
ODETTE	Organisation for Data Exchange by Tele-Transmission in Europe file transfer protocol RFC 2204
ODI	Open Datalink Interface (Novell Netware)
ODMR	On-Demand Mail Relay RFC 2645
ODP	Open Distributed Processing
ODP	Originator Detection Pattern
official standards	Internet Official Protocol Standards STD 1
OFN	Optical Fibre Node
OfTel	Office of Telecommunications (UK)
OH	OverHead
OID	Object Identifier
OLCUC	Convert lowercase to uppercase (pseudo-terminal mode)
OLRT	OnLine Real-Time
OLS	On-Line System (UCSB) RFC 0217
OLTU	Optical Line Terminating Unit
om	Oman
OM	Object Management
OMAP	Operations and Maintenance Application Part (SS7)
OMC	Operations and Maintenance Centre
OMG	Object Management Group
ONC	Open Network Computing RFC 2695 (ONC RPC)
ONLCR	Map NL (new line) to CR-NL (carriage return new line) (pseudo-terminal mode)
ONLRET	New line (NL) performs a carriage return (output) (pseudo-terminal mode)
ONOCR	Translate newline to carriage return-newline (output) (pseudo-terminal mode)
ONP	Open Network Provision (European Union)
OO	Object Oriented
OOF	Out Of Frame

Abbreviation	Full form and references
OOK	On-Off Keying (digital modulation)
OPAL	OPtical Access Line
OPC	Originating Point Code (SS7)
opcode	OPerations CODE
OpenPGP	Open standard Pretty Good Privacy RFC 3156
OPOST	Enable output processing (pseudo-terminal mode)
O/R	Originator / Recipient (X.400) RFC 1836, RFC 2294
ORB	Object Request Broker
org	ORGanisations Internet domain
ORS	Optical Repeater Station
OS	Operating System
OSF	Open Software Foundation
OSF	Operations Systems Function
OSH	Optical Splitter Head
OSI	Open Systems Interconnection ISO 7498
OSIE	OSI Environment
OSINLCP	OSI Network Layer Control Protocol (PPP) RFC 1377
OSIRM	OSI Reference Model
OSPF	Open Shortest Path First RFC 1131, RFC 1247 (v2), RFC 1248 (MIB), RFC 1252 (MIB), RFC 1253 (MIB), RFC 1364, RFC 1370, RFC 1403, RFC 1583-7 (v2), RFC 1745, RFC 1765, RFC 1793, RFC 1850 (MIB), RFC 2154, RFC 2178 (v2), RFC 2328-9 (v2), RFC 2370, RFC 2676, RFC 2740 (IPv6), RFC 2844, RFC 3137
OSS	Operations and support sub-system
OTA	Over-The-Air
OTAPA	Over-The-Air Service Parameter Administration of mobile stations RFC 2604, RFC 2636
OTASP	Over-The-Air Service Provisioning of mobile stations RFC 2604, RFC 2636
OUI	Organizational Unique Identifier
OWF	Optimum Working Frequency
P	Padding field bit
P	Page structure (FTP)
P	Performance
P	Poll Bit
P/F	Poll / Final bit
P1-interface	X.400 interface for relaying between post offices
P3-interface	X.400 interface between client and post office (submission/delivery)
pa	Panama
PA	PreArbitrated segment (DQDB)
PABX	Private Automatic Branch eXchange (nowadays synonymous with PBX)
PAC	PPTP Access concentrator Client (PPTP) RFC 2367
PAC	Proxy Auto Configuration RFC 3040
PACCH	Packet Associated Control CHannel (EDGE)
PAD	Packet Assembler / Disassembler RFC 1053, ITU-T X.3
PAD	PADding
PADI	PPPoE Active Discovery Initiation message RFC 2516
PADO	PPPoE Active Discovery Offer message RFC 2516
PADR	PPPoE Active Discovery Request message RFC 2516

Abbreviation	Full form and references
PADS	PPPoE Active Discovery Session confirmation message RFC 2516
PADT	PPPoE Active Discovery Termination message RFC 2516
PAF	PreArbitrated frame (DQDB)
PAM	Pulse Amplitude Modulation
PAMA	Pre-Assigned Multiple Access
PANS	Pretty Amazing New Stuff
PAP	Password Authentication Protocol RFC 1334
PARC	Palo Alto Research Centre (Xerox Corporation)
PARENB	Parity enable (pseudo-terminal mode)
PARMRK	Mark parity and framing errors (pseudo-terminal mode)
PARODD	Odd parity (otherwise even parity). (pseudo-terminal mode)
PAS	Port Address Support
PASS	PASSword (FTP)
PASTE	Provider Architecture for differentiated Services & Traffic Engineering RFC 2430
PASV	PASsiVe port (FTP)
Path	PATH message (RSVP) RFC 2205
PathErr	PATH message ERRor (RSVP) RFC 2205
PathTear	PATH TEARdown message (RSVP downstream) RFC 2205
P-bit	Partial bit (BGP)
PBX	Private Branch eXchange (an office telephone system)
PC	Personal Computer
PC	Priority Control
PCCP	PPP Connection Control Protocol RFC 1915
PCF	Point Coordination Function (WLAN) ÎEEE 802.11
PCI	Programming Communication Interface (like an API)
PCI	Protocol Control Information
PCIM	Policy Core Information Model (extension to CIM) RFC 3060 (v1)
PCM	Pulse Code Modulation ITU-T G.711
PCMAIL	distributed MAIL system for Personal Computers RFC 0984, RFC 0993, RFC 1056
PCN	Personal Communications Network (mobile telephone system)
PCR	Peak Cell Rate (ATM)
PCS	Personal Communications System (mobile telephone system)
PCS	Physical Coding Sublayer
PDA	Personal Digital Assistant (handheld computer)
PDCH	Packet Data physical CHannel (GPRS)
PDF	Portable Document Format RFC 2346
PDH	Plesiochronous Digital Hierarchy ITU-T G.701-4
PDN	Public Data Network
PDO	Packet Data Only
PDTCH	Packet Data Traffic CHannel (GPRS)
PDU	Packet Data Unit
PDU	Protocol Data Unit
pe	Peru
PE	PolyEthylene
PENDIN	Retype pending input (pseudo-terminal mode)
pf	French Polynesia
PF	Presentation Function

Abbreviation	Full form and references
PFC	Protocol Field Compression (PPP) RFC 1661
PF-KEY	(packet forwarding) key management API derived from BSD PF-ROUTE RFC 2367
PFR	Portable Font Resource RFC 3073
pg	Papua New Guinea
PG	Peer Group
PGL	Peer Group Leader
PGM	Pragmatic General Multicast RFC 3208
PGP	Password Generator Protocol RFC 0972
PGP	Pretty Good Privacy (MIME security) RFC 1991, RFC 2015, RFC 2440
ph	Philippines
PH	Packet Handler
PH	Primitive used by layer 2 to control the PHysical layer
PHB	Per-Hop Behaviour (DiffServ) RFC 1812 (default PHB), RFC 2474, RFC 2475, RFC 2597-8, RFC 2836, RFC 3140
Ph-SAP	Physical layer Service Access Point
PHY	PHYsical layer or PHYsical layer protocol
PICS	Protocol Implementation Conformance Statement
PID	Protocol Identifier
PIM	Protocol Independent Multicast RFC 2362 (sparse mode), RFC 2934 (MIB)
PIM-DM	Protocol Independent Multicast - Dense Mode
PiMF	Pairs in Metal Foil (cable type)
PIM-SM	Protocol Independent Multicast — Sparse Mode RFC 2117, RFC 2337, RFC 2362
PIM-SSM	Protocol Independent Multicast - Source-Specific Multicast
PIN	Personal Identification Number
PIN	Personal Internet Name RFC 3043
PING	Packet Internet Groper (ICMP echo or timestamp request/reply) RFC 0792 (ICMP), RFC 2463 (ICMPv6), RFC 2925 (MOs)
PingMIB	Packet Internet Groper MIB (mib-2) RFC 2925
PINT	PSTN / Internet inter NeTworking RFC 2458, RFC 2848, RFC 3055
PIPE	communications PIPE (POSIX) (an interface between process and terminal device)
PIU	Path Information Unit (IBM SNA)
pixel	Picture Element
PIXIT	Protocol Implementation eXtra Information for Testing
pk	Pakistan
PKCS	Public Key Crypto-System RFC 2313-5, RFC 2437, RFC 2898, RFC 2985-6
PKI	Public Key Infrastructure RFC 2459, RFC 2510-1, RFC 2527-8, RFC 2559-60, RFC 2585, RFC 2587, RFC 3029, RFC 3039, RFC 3161, RFC 3280, ITU-T X.509
pl	Poland
PL	Pad Length (DQDB)
PL	Physical Layer
PLCP	Physical Layer Convergence Protocol
PLK	Primary LinK station
PLL	Phased Locked Loop

Abbreviation	Full form and references
PL-LE	Physical Layer - Layer Entity
PLMTS	Public Land Mobile Telephone System
PL-OAM	Physical Layer — Operation And Maintenance cell
PL-OU	Physical Layer Overhead Unit
PLP	Packet Layer Protocol
PLS	Physical Layer Signalling (ethernet LAN) IEEE 802.3
PLS	Primary Link Station
PLU	Primary Logical Unit (IBM SNA)
PLW	Payload Length Word IEEE 802.11
pm	Saint Pierre and Miquelon
PM	Performance Management
PM	Physical Medium (sublayer)
PMA	Physical Medium Attachment (ethernet LAN) IEEE 802.3
PMD	Physical layer Medium Dependent
PMP	Point-to-MultiPoint (radio system)
PMTU	Path Maximum Transmission Unit (see also MTU) RFC 1063, RFC 1191, RFC 1981 (IPv6)
PMUX	Primary MultipleXor
pn	Pitcairn Island
PNA	Private Network Administrator (MPLS) RFC 2917
PNG	Portable Network Graphics RFC 2083
PNNI	Private Network-Node Interface or Private Network-Network Interface (ATM)
PNS	PPTP (Point-to-Point Tunneling Protocol) Network Server RFC 2367
POH	Path OverHead (SDH)
POI	Path Overhead Identifier (DQDB)
POI	Point-Of-Interconnection
PolSK	POLarity Shift Keying
PON	Passive Optical Network
POP	Point Of Presence
POP	Post Office Protocol RFC 0918, RFC 0937 (v2), RFC 1081 (v3), RFC 1082, RFC 1225 (v3), RFC 1460 (v3), RFC 1725 (v3), RFC 1939 (v3), RFC 1957, RFC 2384, RFC 2449
PORT	FTP PORT (FTP)
POS	Packet-over SDH/SONET
POS	Point Of Sale (system)
POSIX	Portable Operating System Interface (UNIX)
POTS	Plain Old Telephone Service
PP	Point-to-Point link
PPD	Packet Processing Delay
PPDU	Presentation Protocol Data Unit (OSI layer 6)
PPM	Pulse Position Modulation
PPP	Point-to-Point Protocol RFC 1134, RFC 1171, RFC 1172, RFC 1220, RFC 1331-4, RFC 1471-4 (MOs), RFC 1548-9, RFC 1661-3, RFC 1973-9, RFC 2023 (IPv6), RFC 2153, RFC 2472 (IPv6), RFC 2686-7, RFC 3153, RFC 3186 (tunneling)
PPP multilink	PPP multilink protocol (MP) RFC 1990
PPP/AAL5	PPP over AAL5 (ATM adaptation layer 5) RFC 2364
PPP/FR	PPP in Frame Relay RFC 1973

Abbreviation	Full form and references
PPP/HDLC	PPP in HDLC-like framing RFC 1662
PPP-LINK	PPP LINK quality monitoring RFC 1989
PPPoE	Point-to-Point Protocol (PPP) Over Ethernet RFC 2516
PPPoISDN	PPP over ISDN RFC 1618
PPPoSONET	PPP over SONET/SDH RFC 1619, RFC 2615, RFC 2823, RFC 3255
PPPoX.25	PPP over X.25 RFC 1598
PPS	Pulse-Per-Second API (for UNIX-like operating systems) RFC 2783
PPTP	Point-to-Point Tunneling Protocol RFC 2637
pr	Puerto Rico
PRA	Primary Rate Access (ISDN)
PRBS	Pseudo Random Binary Sequence
PRDMD	PRivate Directory Management Domain
PRI	Primary Rate Interface (ISDN) ITU-T I.421, ITU-T I.430 (physical layer), ITU-T I.441 (datalink layer), ITU-T Q.921 (datalink layer), ITU-T I.451 (network layer), ITU-T Q.931 (network layer)
PRIM	PResence and Instant Messaging protocol
PrintMIB	PRINTer objects (mib-2) RFC 1759
PRM	Protocol Reference Model ITU-T I.321 (ATM)
PRMA	Packet Reservation Multiple Access
PRMD	PRivate Management Domain (X400)
PRNG	Pseudo-Random Number Generator
ProtocolID	PROTOCOL Identifier (IP protocol number)
Proxy-PAR	Proxy (minimal version of) PNNI Augmented Routing RFC 2843-4
PRR	Pulse Repetition Rate
PRS	Primary Reference Source (timing clock)
PRT	Pulse Repetition Time
Prty	Parity
ps	Palestinian Territories
PSAP	Physical Service Access Point (OSI)
PSDN	Public Switched Data Network
PSE	Packet Switched Exchange
PSF	PLCP (Physical Layer Convergence Protocol) Signalling Field IEEE 802.11
PSH	PuSH (TCP)
PSI	Packet Switched Interface
PSK	Phase Shift Keying (digital modulation)
PSN	Packet Switch Node (originally IMP: Interface Message Processor)
PSN	PL-OAM Sequence Number (ATM)
PSTN	Public Switched Telephone Network
pt	Portugal
PT	Payload Type or protocol type
PTCCH	Packet Timing advance Control CHannel (EDGE)
PTE	Path Terminating Equipment
PTI	Payload Type Identifier
PTM	Path Trace Management
PTO	Public Telecommunications Operator
PTP	Point-To-Point (radio system)
PTR	PoinTeR
PTS	Proceed To Send (signal)

Abbreviation	Full form and references
PTT	Post, Telegraph and Telephone (company)
pty	Pseudo Terminal
PU	Physical Unit (IBM SNA)
PV	Parameter Value
PVC	Permanent Virtual Circuit
PVC	Poly Vinyl Chloride (plastic used for cable sheathing)
PVCC	Permanent Virtual Channel Connection
PVPC	Permanent Virtual Path Connection
pw	Palau
PW	PassWord
PWD	Print Working Directory (FTP)
PWM	Pulse Width Modulation
py	Paraguay
Q	Qualifier bit ITU-T X.25
Q3	Interface at q3 reference point (TMN)
qa	Qatar
QA	Q-interface adaptor (TMN)
QA	Quality Assurance
QA	Queued Arbitrated (DQDB)
QAF	Q-Adaptor Function (TMN)
QAF	Queued Arbitrated Function (DQDB)
QAM	Quadrature Amplitude Modulation (digital modulation)
QCLASS	Query CLASS (DNS) RFC 1035
qdu	Quantisation Distortion Unit
QNAME	Query NAME (DNS) RFC 1035
QOS	Quality Of Service
QOS	Quality-Of-Service based routing RFC 2386, RFC 2676, RFC 2990, IEEE 802.2p
QPSK	Quarternary Phase Shift Keying (digital modulation)
QPSX	Queued Packet and Synchronous switch
Q-SIG	Q-SIGnalling (inter-PBX ISDN signalling system)
QTYPE	Query TYPE (DNS) RFC 1035
QUIT	logout (FTP)
QUOTE	QUOTE of the day protocol RFC 0865
Qx	Proprietary interface between mediation function and network element (TMN)
R	Receive leads ITU-T X.21
R	Requirement
R	Resistance
R	Record structure (FTP)
R reference point	Reference point at the ATM UNI between a B-TA and a B-TE2 (i.e. an X- or V-series interface)
RA	Receiver Address IEEE 802.11
RA	Registration Authority (PKI) ITU-T X.509
RAC	Registration Authority Committee (IEEE)
RADIUS	Remote Authentication Dial-In User Service RFC 2058-9, RFC 2138-9, RFC 2548, RFC 2618-21 (MIBs), RFC 2809, RFC 2865-9 RFC 2882, RFC 3162 (IPv6)
RAI	Remote Alarm Indication

Abbreviation	Full form and references
RAM	Random Access Memory
RAP	Internet Route Access Protocol RFC 1476
RARP	Reverse Address Resolution Protocol RFC 0903, RFC 1931
RAS	Registration/Admission/Status protocol ITU-T H.225
RAS	Remote Access Server (Microsoft terminology for NAS) RFC 2367
RATP	Reliable Asynchronous Transfer Protocol RFC 0916, RFC 0935
RBB	Residential BroadBand
R-bit	Reliability (IP type of service)
R-bit	Routing bit (GRE) RFC 1701
RBOC	Regional Bell Operating Company (USA)
RC	Receiver Clock ITU-T V.24
RC	Routing Control
RC2	Raw Cipher 2 encryption algorithm RFC 2268, RFC 3217
RC4	Raw Cipher 4 encryption algorithm (proprietary encryption algorithm of RSA)
RC5	Raw Cipher 5 encryption algorithm RFC 2040
RCA	Receive-Configure-Acknowledgement RFC 1661 (PPP)
RCN	Receive-Configure-Negative-acknowledgement (rejection) RFC 1661 (PPP)
RCPT	ReCiPienT (SMTP)
RCR	Receive-Configure-Request RFC 1661 (PPP)
Rcwd	ReCeiver WinDow (TCP) RFC 2581
RD	Routing Descriptor
RD	Routing Domain (numbering)
RDA	Remote Database Access
RDBMS	Relational DataBase Management System RFC 1697 (MIB)
RDI	Remote Defect Indication (formerly called FERF, far end receive failure)
RdmsMIB	Relational Database objects (mib-2) RFC 1697
RDN	Relative Distinguished Name
RDP	Reliable Data Protocol RFC 0908, RFC 1151 (v2)
re	Reunion Island
RE	Reference Equivalent
reconnection	Reconnection Protocol RFC 0426
Ref	Reference
REIN	REINitialise (FTP)
REJ	REJect
REQ	REQuest
RES	REServed
RESP	RESPonse
REST	RESTart (FTP)
Resv	RESerVation request (RSVP) RFC 2205
REsvConf	RESerVation CONFirmation (RSVP) RFC 2205
ResvErr	RESerVation message ERRor (RSVP) RFC 2205
ResvTear	RESerVation TEARdown message (RSVP upstream) RFC 2205
RETR	RETRieve (FTP)
REX	Remote EXecution service (UNIX)
RF	Radio Frequency
RFC	Request For Comment (IETF specification document) www.rfc-editor.org
RFI	Radio Frequency Interference

Abbreviation	Full form and references
RFP	Request For Proposal
RFQ	Request For Quotation
RFS	Ready For Service
RFS	Remote File Sharing or remote file system (SUN UNIX)
RG	ReGenerator
RI	Ring In
RI	Ring Indicator ITU-T V.24
RI	Routing Information
RID	Router Identifier (OSPF)
RID	Routing IDentifier
RIF	Routing Information File (source route bridging)
RII	Routing Information Indicator
RIP	Routing Information Protocol RFC 1058 (v1), RFC 1387-9, RFC 1581-2, RFC 1721-4 (v2), RFC 1923, RFC 2082, RFC 2091-2, RFC 2453 (v2)
RIPE	Réseaux IP Européens (European IP Coordination Centre) RFC 1181, www.ripe.net
RIPng	Routing Information Protocol next generation (for IPv6) RFC 2080-1 (IPv6)
RISC	Reduced Instruction Set Computer
RJE	Remote Job Entry RFC 0105, RFC 0217
RJOR	Remote Job Output Routine RFC 0105, RFC 0217, RFC 0360, RFC 0407, RFC 0725
RL	Return Loss
RLC	Radio Link Control
rlogin	Remote login (BSD UNIX) RFC 1258
RLP	Resource Location Protocol RFC 0887
RM	Reference Model
RM	Resource Management
RM	Routine Maintenance
RMCP	Remote Mail Checking Protocol RFC 1339
RMD	ReMove Directory (FTP)
RMON	Remote network MONitoring RFC 1271, RFC 1757, RFC 2021, RFC 2074, RFC 2613, RFC 2819, RFC 2895-6, RFC 3144, RFC 3273, RFC 3287
RMT	Reliable Multicast Transport building blocks RFC 3048, RFC 3269
RNFR	ReName FRom (FTP)
RNR	Receiver Not Ready
RNTO	ReName TO (FTP)
ro	Romania
RO	Read Only
RO	Ring Out
ROA	Recognised Operating Agency
ROHC	RObust Header Compression RFC 3095-6j RFC 3241-3
ROM	Read-Only Memory
RORE	Refresh Overhead Reduction Extensions (RSVP) RFC 2961
ROSE	Remote Operations Service Element (OSI layer 7)
routed	Route-Dee daemon of UNIX Berkeley System Distribution
router alert	IP Router Alert Option RFC 2113
RP	Rendezvous Point (sparse mode multicasting)

Abbreviation	Full form and references
RPC	Remote Procedure Call RFC 0674 (v2)j RFC 1050j RFC 1057 (v2)j RFC 1831 (v2)j RFC 1833j RFC 2695 (ONC RPC)
RPOA	Recognised Private Operating Agency (telecommunications operator recognised by ITU)
RPR	Resilient Packet Ring
RPS	Radio Protection Switching
RPS	Routing Policy System RFC 2769
RPSL	Routing Policy Specification Language RFC 2280j RFC 2622
RPSL	Routing Policy System Replication RFC 2769
RQ	ReQuest counter (DQDB)
RR	Receiver Ready
RR	Resource Record (DNS) RFC 1183, RFC 1637
RR	Route Reflector (BGP)
RR	Receiver Report (RTCP) RFC 1889
RRE	Receiving Reference Equivalent
RRP	Registry Registration Protocol RFC 2832 (v1.1.0)
RRQ	Read ReQuest (TFTP)
RRR	Radio Relay Regenerator
RRT	Radio Relay Terminal
Rs	Sustainable cell rate
RS	Reception Status (AAL3/4 SAR-SDU)
RS	Reconciliation Sublayer
RS	Regenerator Section (transmission)
RS	Record Separator (see ASCII)
RS	Repeater Station
RS-232	EIA recommended standard 232 (DTE-to-DCE interface) EIA RS-232
RS-422	EIA recommended standard 422 (DTE-to-DCE interface) EIA RS-422
RS-449	EIA recommended standard 449 (DTE-to-DCE interface) EIA RS-449
RSA	Rivest-Shamir-Adleman public key encryption algorithm RFC 2313, RFC 2437, RFC 2792, RFC 3110
RSA	RSA security Inc. www.rsasecurity.com
RSE	Real system environment
RSET	ReSET (SMTP)
Rsh	Remote SHell (Berkeley Software Distribution UNIX)
RSOH	Regenerator Section OverHead (SDH)
Rsp	ReSPonse
RSRVD	ReSeRVeD
RSS	Route Selection Service
RST	ReSeT
RSU	Remote Switching Unit
RSVD	ReSerVeD
Rsvp	ReSerVation Protocol MIB (mib-2) RFC 2206
RSVP	Resource reSerVation Protocol RFC 1633 (integrated services), RFC 2205 (architecture), RFC 2206 (MIB), RFC 2207 (IPsec), RFC 2208 (RSVP applicability), RFC 2209 (message rules), RFC 2210 (RSVP with integrated services), RFC 2211 (CLNES), RFC 2212 (QOS spec), RFC 2379-82 (RSVP over ATM), RFC 2745-53 (extensions), RFC 2814 (RSVP on IEEE 802), RFC 2872, RFC 3097 (authentication), RFC 3175 (RSVP aggregation IPv4/IPv6), RFC 3181-2, RFC 3209 (TE- traffic engineering), RFC 3210

Abbreviation	Full form and references
RT	Real Time
RT	Routing Type
RTA	Receive-Terminate-Acknowledgement RFC 1661 (PPP)
RTCP	Real-time application Transport Control Protocol
RTFM	Remote Traffic Flow Measurement RFC 2064, RFC 2720-4
RTO	Retransmission TimeOut (TCP)
RTP	Real-time application Transport Protocol <u>RFC 1889-90</u>, RFC 2862, RFC 2959 (MIB), RFC 3016, RFC 3119, RFC 3189-90, RFC 3267
RTP	Reliable Transport Protocol <u>RFC 3208</u>
RTR	Receive-Terminate-Request RFC 1661 (PPP)
RTS	Request-To-Send <u>ITU-T V.24</u>
RTS	Residual Time Stamp
RTSE	Reliable Transfer Service Element (OSI layer 7)
RTSP	Real-Time Streaming Protocol <u>RFC 2326</u>
RTT	Round Trip Time (TCP)
RTTVAR	Round Trip Time VARiance (TCP) <u>RFC 1323</u>, RFC 2988
ru	Russian Federation
RU	Remote Unit
RUC	Receive-Unknown-Code RFC 1661 (PPP)
RUIP	finger Remote User Information Protocol RFC 1196, <u>RFC 1288</u>
rw	Rwanda
RW	Read Write
Rwhois	Referral Whois protocol RFC 1714, <u>RFC 2167 (v1.5)</u>
RWP	Remote Write Protocol <u>RFC 1756</u>
Rx	Receive
RxD	Receive Data <u>ITU-T V.24</u>
RXJ	Receive-Code-Reject RFC 1661 (PPP)
RXR	Receive-Echo-Request RFC 1661 (PPP)
RZ	Return-to-Zero (line code)
S	Stream mode (FTP)
S&F	Store and Forward
sa	Saudi Arabia
SA	Security Association (IPsec) RFC 2402, RFC 2406
SA	Source MAC Address
SAA	Services And Applications
SAA	Systems Applications Architecture (IBM architecture)
SAAL	Signalling ATM Adaptation Layer ITU-T Q.2110, ITU-T Q.2130, ITU-T Q.2140
SAB	Subnetwork Access Boundary
SABM	Set Asynchronous Balanced Mode (HDLC)
SABME	Set Asynchronous Balanced ModE (HDLC)
SACK	Selective ACKnowledgement (TCP) <u>RFC 2018</u>
SAD	Security Association Database (IPsec) RFC 2402, RFC 2406
samba	initiative for the extension of the Microsoft Networking Server Message Block
SAML	Security Assertion Markup Language
SAML	Send And MaiL (SMTP)
SAP	Service Access Point
SAP	Service Advertising Protocol (Novell Netware)

Abbreviation	Full form and references
SAP	Session Announcement Protocol RFC 2974
SAPI	Service Access Point Identifier
SAR	Security Alarm Reporting
SAR	Segmentation And Reassembly
SAS	Serial Attached SCSI
SAS	Single Attached Station (FDDI)
SASL	Simple Authentication and Security Layer RFC 2222, RFC 2245, RFC 2444, RFC 2808, RFC 2831, RFC 3163, ISO/IEC 9798-3
SAT	Security Audit Trail
SAT	Source Address Table
SATA	Serial Advanced Technology Attachment
SATF	Shared Access Transport Facility
sb	Solomon Islands
SB	Reference point at the User-network interface (UNI) between terminal equipment (B-TE) and B-NT2
SB	SuBnegotiation (NVT-ASCII)
s-bit	Strict-source-bit (GRE) RFC 1701
S-bit	bottom of Stack bit (MPLS) RFC 3032
S-bit	Sequence number bit RFC 1701 (GRE), RFC 2661 (L2TP)
SBM	Subnet Bandwidth Manager RFC 2814
SBS	Selective Broadcast Signalling
SBSVC	Selective Broadcast Signalling Virtual Channel
sc	Seychelles
SC	optical fibre connector type
SC	Session Connector (IBM APPN)
sca	Send-Configure-Acknowledgement RFC 1661 (PPP)
SCCCN	Start-Control-Connection-CoNnected (L2TP) RFC 2661
SCCRP	Start-Control-Connection-RePly(L2TP) RFC 2661
SCCRQ	Start-Control-Connection-ReQuest (L2TP) RFC 2661
SCE	System Control Element (IBM SNA)
scj	Send-Code-Reject RFC 1661 (PPP)
scn	Send-Configure-Negative-acknowledgement (rejection) RFC 1661 (PPP)
scr	Send-Configure-Request RFC 1661 (PPP)
SCR	Static Conformance Requirement
SCR	Sustainable Cell Rate (ATM)
SCSI	Small Computer System Interface
SCSP	Server Cache Synchronization Protocol RFC 2334-5, RFC 2443
SCTP	Stream Control Transmission Protocol RFC 2960, RFC 3257 (applicability statement), RFC 3286
sd	Sudan
SD	Starting Delimiter (IBM token ring LAN)
SDCP	Serial Data Control Protocol RFC 1963
SDDI	Shielded Distributed Data Interface
SDE	Submission and Delivery Entity (X.400)
SDES	Source DEScription items (RTCP) RFC 1889
SDF	Service Data Function
SDH	Synchronous Digital Hierarchy ITU-T G.709, RFC 2558 (MOs)
SDL	Simple Data Link RFC 2823
SDL	Specification and Description Language

Abbreviation	Full form and references
SDLC	Synchronous Data Link Control (IBM SNA)
SDP	Session Description Protocol RFC 2327, RFC 2543, RFC 2848, RFC 3108, RFC 3264, RFC 3266
SDR	Session Detail Record (Cisco term for accounting record)
SDRP	Source Demand Routing Protocol RFC 1940
SDT	Structured Data Transfer (AAL1)
SDTP	Serial Data Transport Protocol RFC 1963
SDU	Service Data Unit
SDXF	Structured Data eXchange Format RFC 3072
se	Sweden
SE	Shared Explicit style (RSVP) RFC 2205
SE	Subnegotiation End (NVT-ASCII)
SECB	Severely Errored Cell Block
SECBR	Severely Errored Cell Block Ratio
SEC_GSS	SECurity_General Security Service protocol (RPC) RFC 2203
SECSH	SECure Shell (see also SSH) (Internet draft)
SecureFast	Cabletron's SecureFast VLAN operational model RFC 2643
SecurID	A proprietary one-time password system (RSA Security Inc)
SEGV	SEGment Violation (POSIX)
SEL	SELector (address)
ser	Send-Echo-Reply RFC 1661 (PPP)
SERIAL_	Serial Number Arithmetic RFC 1982
SES	Severely Errored Seconds
SFD	Start of Frame Delimiter (HDLC)
s-frame	supervisory frame (HDLC, LLC)
SFT	System Fault Tolerance (Novell Netware)
sg	Singapore
SGML	Standard Generalized Markup Language RFC 1874
SGMP	Simple Gateway Monitoring Protocol RFC 1028
SGSN	Serving GPRS Support Node
sh	Saint Helena
SHA	Secure Hash Algorithm RFC 1852, RFC 2841, RFC 3110, RFC 3174
SHDSL	Single-pair Highspeed Digital Subscriber Line RFC 3276 (MOs)
SHTTP	Secure HyperTest Transfer Protocol RFC 2660
si	Slovenia
SI	Shift In (see ASCII)
siaddr	Server IP ADDRess (BOOTP/DHCP) RFC 0951
Sieve	a mail filtering language RFC 3028
SIFT	Sender Initiated File Transfer RFC 1440
SIG	SIGnature RFC 2931, RFC 3125-6
SIIT	Stateless IP/ICMP Translation algorithm RFC 2765
SIM	System Identification Module
SIP	Session Initiation Protocol RFC 2543, RFC 2848, RFC 2976, RFC 3050, RFC 3087, RFC 3261-5, RFC 3351
SIP	Subscriber Interface Protocol (SMDS)
SIPP	Simple Internet Protocol Plus RFC 1710
SIR	Sustained Information Rate (SMDS)
SITA	Société Internationale de Transports Aeronautiques

Abbreviation	Full form and references
SITE	SITE parameters (FTP)
sj	Svalbard and Jan Mayen Islands
sk	Slovak Republic
S/Key	Secure Key one-time password system RFC 1760
SKIP	Sun SKIP Firewall RFC 2356
SKIPJACK	an encryption algorithm RFC 2773, RFC 2876
SKMP	Photuris Session Key Management Protocol RFC 2522-3
sl	Sierra Leone
SLA	Service Level Agreement RFC 2475, RFC 2917
SLA	Site-Level Aggregation identifier (IPv6 address) RFC 2373
SLE	SubLayer Entity
SLI	Set-Link-Info (L2TP) RFC 2661
SLIP	Serial Line Internet Protocol RFC 1055
SLP	Service Location Protocol RFC 2165, RFC 2608 (v2), RFC 2610, RFC 2926, RFC 3059, RFC 3082, RFC 3111 (IPv6), RFC 3224
SLP	Single Link Procedure
sm	San Marino
SM	Scheduling Management
SM	Security Management
SM	Session Manager
SM	Single Mode (monomode) fibre
SM	Sparse Mode protocol-independent multicasting (PIM) RFC 2362
SM	Streaming Mode (AAL3/4 or AAL5)
SMAE	System Management Application Entity
SMASE	System Management Application Service Element
SMB	Server Message Block (Microsoft Networking)
SMDS	Switched Multimegabit Digital Service RFC 1209, RFC 1694 (MOs)
SME	Society of Manufacturing Engineering (USA)
SMF	Single Mode Fibre (monomode fibre) ITU-T G.652, ITU-T G.653, ITU-T G.654, ITU-T G.655, ITU-T G.657
SMF	System Management Function
SMF	Single Mode (or monomode) Fibre
SMFS	Simple Minded File System RFC 0122, RFC 0217, RFC 0399, RFC 0431
SMI	Structure of Management Information RFC 1155 (v1), RFC 1212, RFC 2578-80 (v2), RFC 3216 (SMIng)
S/MIME	Secure MIME RFC 2311-2 (v2), RFC 2632-4 (v3), RFC 2785, RFC 3156, RFC 3183
SMK	Shared Management Knowledge
SML	Service Management Layer (ISO management model) ITU-T M.3000
SMNT	Structured MouNT (FTP)
SMON	Switched network MONitoring RFC 2613
SMP	Service Management Point
SMS	Short Message Service (GSM)
SMT	Station ManagemenT (FDDI)
SMTP	Simple Mail Transfer Protocol RFC 0788, RFC 0821-2, RFC 0913, RFC 0974, RFC 1425-7, RFC 1651-3, RFC 1854, RFC 1869-70, RFC 1891-4, RFC 1985, RFC 2034, RFC 2197, RFC 2442, RFC 2487, RFC 2554, RFC 2821-2, RFC 2852, RFC 2920, RFC 3030, RFC 3207

Abbreviation	Full form and references
SMX	Script MIB eXtensibility protocol RFC 2593 (v1.0)
sn	Senegal
SN	Sequence Number
SN	Subarea Node (IBM)
SNA	Systems Network Architecture (IBM)
SNACP	PPP SNA Control Protocol RFC 2043
sname	Server NAME (BOOTP/DHCP) RFC 0951
SNAP	Subnetwork Access Protocol IEEE 802.2
SNAP	System Network Architecture Protocol (IBM)
SNI	SNA Network Interconnection (IBM SNA)
SNI	Subscriber Network Interface
SNMP extensions	Simple Network Management Protocol extensions and uses RFC 1187, RFC 1215, RFC 1270, RFC 1283, RFC 1303, RFC1418-20, RFC 1451 (MIB), RFC 2742 (MOs), RFC 2962
snmpMIB	MIB for SNMPv2 RFC 1907
SNMPv1	Simple Network Management Protocol version 1 RFC 1067, RFC 1098, RFC 1157, RFC 1351-3
SNMPv2	Simple Network Management Protocol version 2 RFC 1441-50, RFC 1901 (SNMPv2c), RFC 1902-4, RFC 1905 (protocol operation), RFC 1906-8, RFC 1909-10 (SNMPv2u)
SNMPv3	Simple Network Management Protocol version 3 RFC 2261-5, RFC 2271-5, RFC 2570, RFC 2671-6
SNMPv3-v2-v1	Simple Network Management Protocol coexistence of different versions RFC 1452 (v2-v1), RFC 1908 (v2-v1), RFC 2089 (v2-v1), RFC 2576 (v3-v2-v1)
SNP	Sequence Number Protection (ATM AAL1)
SNPA	Sub-Network Point of Attachment
SNPP	Simple Network Paging Protocol RFC 1568, RFC 1645 (v2), RFC 1861 (v3)
SNQP	Simple Nomenclator Query Protocol RFC 2259
SNTP	Simple Network Time Protocol RFC 1361, RFC 1769, RFC 2030 (v4)
so	Somalia
SO	Shift Out (see ASCII)
SOAP	Simple Object Access Protocol in BEEP (Blocks Extensible Exchange Protocol) RFC 3288
SOC	Start Of Cell
socket	synonym for a port and application program interface (API) to TCP/IP RFC 0147, RFC 0349, RFC 2133 (IPv6), RFC 2292 (IPv6), RFC 2553 (IPv6), RFC 3152
SOCKS	OSI session layer security technology RFC 1928-9 (v5), RFC 1961, RFC 3089
SOH	Section OverHead (SDH)
SOH	Start Of Header (see ASCII)
SOIF	Summary Object Interchange Format RFC 2655-6
SOML	Send Or MaiL (SMTP)
SONET	Synchronous Optical NETwork RFC 2558 (MOs)
SP	Service Provider

Abbreviation	Full form and references
SP	Signalling Point
SP	Space
SPAM	junk email
SPC	Signalling Point Code
SPD	Security Policy Database (IPsec) RFC 2402, RFC 2406
SPE	Synchronous Payload Envelope (SONET)
SPED	Service Provider Edge Device RFC 2917
SPF	Shortest Path First algorithm (Dijkstra algorithm) RFC 2386
SPI	Security Parameters Index (IPsec) RFC 2402, RFC 2406
SPID	Service Profile IDentifier
SPIRIT	Service Providers' Integrated Requirements for Information Technology
SPIRITS	Services in the PSTN/IN Requesting InTernet Services RFC 2995, RFC 3136, RFC 3298
SPKI	Simple Public Key Infrastructure RFC 2692-3
SPKM	Simple Public Key GSS_API Mechanism RFC 2025, RFC 2847
SPL	Service Provider Link
SPM	FDDI-to-SONET Physical layer Mapping standard (FDDI)
SPN	Subscriber Premises Network
SPNA	Service Provider Network Administrator RFC 2917
SPPI	Structure of Policy Provisioning Information RFC 3159
SPT	Shortest Path Tree
SPX	Sequenced Packet eXchange (Novell Netware)
SQL	Standard Query Language
sr	Suriname
SR	Sender Report (RTCP) RFC 1889
SR	Source Routing
SRB	Source Route Bridging RFC 1525 (MIB)
SRE	Sending Reference Equivalent
SREJ	Select REJect frame
SRF	Specifically Routed Frame
SRI	Stanford Research Institute
SRL	Simple Ruleset Language RFC 2723
SRP	Spatial Re-use Protocol
SRT	Source Routing Transparent
srTCM	Single Rate Three Colour Marker RFC 2697
SRTS	Synchronous Residual Time Stamp (ATM AAL1)
SRTT	Smoothed Round-Trip Time (TCP) RFC 1323, RFC 2988
SS	Service Specific
SS	Start/Stop transmission
SS7	Signalling System number 7
SSAP	Source Service Access Point
SSCF	Service Specific Coordination Function (ATM)
SSCOP	Service Specific Connection-Oriented Protocol (ATM)
SSCP	System Services Control Point (IBM SNA)
SSCS	Service Specific Convergence Sublayer (ATM)
SSH	Secure Shell protocol or program (see also SECSH) (Internet draft)
SSH-ARCH	Secure Shell protocol ARCHitecture (Internet draft)

Abbreviation	Full form and references
SSH-CONNECT	Secure Shell CONNECTion protocol (Internet draft)
ssh-connection	Secure Shell CONNECTION service (SSH-TRANS)
SSH-TRANS	Secure Shell TRANSport layer protocol (Internet draft)
ssh-userauth	Secure Shell USER AUTHentication service (SSH-TRANS)
SSH-USER-AUTH	Secure Shell USER AUTHentication protocol (Internet draft)
SSI	Service Specific Information
SSL	Secure Sockets Layer
SSM	Single Segment Message
SSM	Source-Specific Multicast
SSRC	Synchronization SouRCe (RTP) RFC 1889
SSRR	Strict Source and Route Record
SSTHRESH	Slow Start THRESHold (TCP) RFC 2581
st	Sao Tome and Principe
ST	Internet Stream Protocol RFC 1190, RFC 1819
ST	optical fibre connector type
ST	Segment Type (AAL3/4 SAR)
ST	Signalling Terminal
S/T	S/T interface (ISDN basic rate interface passive bus)
ST2+	Internet Stream Protocol version 2+ RFC 1819, RFC 1946, RFC 2383
sta	Send-Terminate-Acknowledgement RFC 1661 (PPP)
STACK	STart ACKnowledgement message (DECnet)
STAT	maildrop listing (POP)
STAT	STATus (FTP)
STD	Standard
STD1	Internet Standard number 1 — List of current Internet Official Protocol Standards STD1
STDA	StreetTalk Directory Assistance
STDL	Structured Transaction Definition Language
STE	Section Terminating Equipment (SONET)
STE	Signalling TErminal
STE	Spanning Tree Explorer
STM	STation Management
STM	Synchronous Transfer Mode
STM	Synchronous Transport Module (SDH)
STM-1	Synchronous Transport Module of SDH with bitrate of 155 Mbit/s
STM-4	Synchronous Transport Module of SDH with bitrate of 622 Mbit/s
STM-16	Synchronous Transport Module of SDH with bitrate of 2.5 Gbit/s
STM-64	Synchronous Transport Module of SDH with bitrate of 10 Gbit/s
STM-n	Synchronous Transport Module-n
StopCCN	Stop-Control-CoNnection-reply (L2TP) RFC 2661
STOR	STORe (FTP)
STOU	STOre Unique (FTP)
STP	Shielded Twisted Pair (cable) EIA/TIA 568
STP	Signalling Transfer Point
STP	Spanning Tree Protocol IEEE 802.1d, IEEE 802.1s (multiple STPs), IEEE 802.1 w (rapid)

Abbreviation	Full form and references
str	Send-Terminate-Request RFC 1661 (PPP)
STRU	file STRUcture (FTP)
STS	Synchronous Transport System (SONET)
STS-1	Synchronous Transport System (SONET) with bitrate of 51.84 Mbit/s
STS-3	Synchronous Transport System (SONET) with bitrate of 155.52 Mbit/s
STS-12	Synchronous Transport System (SONET) with bitrate of 622 Mbit/s
STS-48	Synchronous Transport System (SONET) with bitrate of 2.5 Gbit/s
STS-192	Synchronous Transport System (SONET) with bitrate of 10 Gbit/s
STS-n	Synchronous Transport Signal level-n (SONET)
STX	Start of TeXt
SUB	SUBstitute character (see ASCII)
subnetting	Internet Protocol subnetting procedure RFC 950
SUPDUP	A virtual software display telnet protocol RFC 0734, RFC 0736, RFC 0746, RFC 0749
SUT	System Under Test
sv	El Salvador
SVC	Signalling Virtual Channel
SVC	Switched Virtual Circuit or switched virtual channel RFC 3031
SVI	SubVector Identifier (Token ring MAC)
SVL	SubVector Length (Token ring MAC)
SVP	SubVector Parameters (Token ring MAC)
SVP	Switched Virtual Path RFC 3031
SWS	Silly Window Syndrome (TCP) RFC 0813
SX	Short Wave
sy	Syrian Arab Republic
SYN	SYNchronization character or byte
Syslog	System log RFC 3164, RFC 3195
SYST	SYSTem (FTP)
sz	Swaziland
T	Telnet format effector (FTP)
T	Trailer
T	Transmit leads ITU-T X.21
T1	North American TDM hierarchy (1.5 Mbit/s PDH)
T3	North American TDM hierarchy (45 Mbit/s PDH)
TA	Terminal Adaptor
TA	Transmitter MAC Address IEEE 802.11
TACACS	Terminal Access Controller Access Control System (Telnet) RFC 0927, RFC 1492
Tag-switching	Cisco Systems Tag-Switching Architecture RFC 2105
TAPI	Telephone Application Programming Interface (Microsoft/Intel telephony)
TAR	Temporary Alternative Routing
TAT	Theoretical Arrival Time (leaky bucket algorithm)
TAT	TransAtlantic Telephone cable
TB	Reference point at the User-network interface (UNI) on the user side of the B-NT1
TB	Transparent Bridging
TBF	Temporary Block Flow

Abbreviation	Full form and references
T-bit	Throughput (IP type of service)
T-bit	Transitive bit (BGP)
T-bit	Type and quality of service (OSPF) RFC 2676
T-bit	Type bit (L2TP) <u>RFC 2661</u>
tc	Turks and Caicos Islands
Tc	Committed burst duration (frame relay)
TC	Transaction Capabilities (SS7)
TC	Transmitter Clock <u>ITU-T V.24</u>
TC	Transmission Convergence sublayer (ATM)
TCA	Traffic Conditioning Agreement (DiffServ) <u>RFC 2475</u>
TCAM	TeleCommunication Access Method (IBM)
TCAP	Transaction Capabilities Application Part (SS7)
TCE	Transmission Connection Element
TCF	Transparent Computing Facility
TCN	Topology Change Notification
TCP	Transmission Control Protocol RFC 0675, RFC 0761, <u>RFC 0793</u>, <u>RFC 1122</u>, RFC 2012 (MIB), RFC 2452 (MIB IPv6)
TCP extensions	Transmission Control Protocol extensions and uses RFC 0813 (window), RFC 0879 (max segment), RFC 1072 (long path delay), RFC 1078 (port MUX), RFC 1185 (high speed paths), RFC 1323 (high perf.), RFC 2001 (slow start), RFC 2018 (sel ack), RFC 2140 (ctrl blk), RFC 2147 (over IPv6 jumbograms), RFC 2414 (init. Window), RFC 2488 (satellite), RFC 2581 (congestion), RFC 2582 (fast recovery), RFC 2861 (cong. Window), RFC 2988 (retrans. Timer), RFC 3042 (ltd Xmit), RFC 3360
TCP/IP	Transmission Control Protocol/Internet protocol
TcpMIB	Transmission Control Protocol MIB (mib-2) <u>RFC 2011</u>
td	Chad
TDD	Time Division Duplex (radio)
TDM	Time Division Multiplexing
TDMA	Time Division Multiple Access
TE	Terminal Equipment
TE	Transfer Extension (HTTP)
TE	Traffic Engineering RFC 3209 (RSVP), <u>RFC 3272</u>
TELNET	TErminaL-to-NETwork interface and program for accessing remote computers RFC 0097, RFC 0137, RFC 0158, RFC 0215, RFC 0318, RFC 0495, RFC 0562, RFC 0587, RFC 0764, <u>RFC 0854-5</u>, RFC 1408-12, RFC 1416, RFC 2941-53
TERM	software TERMinal signal (POSIX)
tf	French Southern Territories
TFA	Tiny Fragment Attack <u>RFC 3128</u>
TFTP	Trivial File Transfer Protocol RFC 0783 (v1), <u>RFC 1350 (v2)</u>, <u>RFC 1782-5</u>, RFC 2090, RFC 2347-9
tg	Togo
TG	Transmission Group (IBM)
th	Thailand
TH	Transmission Header (IBM SNA)
THT	Token Holding Time

Abbreviation	Full form and references
TIB	Task Information Base
TIC	Token ring Interface Coupler
TIFF	Tag Image File Format RFC 2302, RFC 2306
TIG	Topology Information Group
TIM	Traffic Indication Map IEEE 802.11
TIME	TIME protocol RFC 0868
TIP	Transaction Internet Protocol RFC 2371-2 (v3)
TIPS	Terminal Internet Message Processor Service RFC 0283
tj	Tajikistan
tk	Tokelau
TKEY RR	Transaction KEY Resource Record RFC 2930
TLA	Top-Level Aggregation identifier (IPv6 address) RFC 2373
TLA	Transport Layer Address RFC 2450, RFC 2921 (IPv6), RFC 2928 (IPv6)
TLAP	TokenTalk Link Access Protocol (Appletalk layer 2 for token ring LAN)
tld	This-Layer-Down RFC 1661 (PPP)
TLD	Top-Level Domain
tlf	This-Layer-Finished RFC 1661 (PPP)
TLI	Transport Layer Interface ISO 8072
Tlr	TraiLeR
tls	This-Layer-Started RFC 1661 (PPP)
TLS	Transport Layer Security RFC 2246 (v1.0), RFC 2487, RFC 2595, RFC 2712, RFC 2716, RFC 2817-8, RFC 2830
tlu	This-Layer-Up RFC 1661 (PPP)
TLV	Type-Length-Value RFC 2460
tm	Turkmenistan
TMN	Telecommunications Management Network ITU-T M.3010
TMux	Transport Multiplexing protocol RFC 1692
tn	Tunisia
TN3270E	Telnet Protocol (IBM 3270E) RFC 1041, RFC 1647, RFC 2355, RFC 2562, RFC 3049
TNS	Transit Network Selection
TNVIP	TelNet Visual Information Projection RFC 1921
to	Tonga
TO	TimeOut RFC 1661 (PPP)
TOA	Type Of Address
TOP	Technical and Office Protocol
TOS	Type of Service field (IP) RFC 0791, RFC 1349, RFC 2474, RFC 2873
TOSTOP	Stop background jobs from output (pseudo-terminal mode)
tp	East Timor
TP	Termination Point
TP	ThroughPut
TP	Transaction Processing
TP	Transmission Path
TP	Transport Protocol RFC 0892, RFC 0905, RFC 0983, ISO 8073
TP	Twisted Pair (cable)
TP1	Transport Protocol class 1 ISO 8073
TP2	Transport Protocol class 2 ISO 8073

Abbreviation	Full form and references
TP3	Transport Protocol class 3 ISO 8073
TP4	Transport Protocol class 4 ISO 8073
TPE	Transmission Path Endpoint
TPI	Transport Provider Interface
TPID	Tag Protocol Identification (VLAN)
TPM	ToPology database Manager
TPON	Telephony Passive Optical Network
tr	Turkey
TR	Technical References
TR	Technical Report
TR	Token Ring (LAN)
TR	Trouble Resolution
TRCC	Total Received Cell Count (ATM)
TRIP	Telephony Routing over IP RFC 3219
trTCM	Two Rate Three Colour Marker RFC 2698
TS	Time Slot (PDH transmission)
TS	TimeStamp
TS16	Timeslot 16 (of 2 Mbit/s E1 transmission system)
TSB	Technical Systems Bulletin
TSF	Traffic Synchronization Function IEEE 802.11
TSI	Time Slot Interchange
TSIG	Transaction SIGnature RFC 2845
TSO	Time Sharing Option (IBM operating system)
TSOPT	TimeStamp OPTion (TCP) RFC 1323
TSP	Time Stamp Protocol RFC 0781, RFC 3161
tspec	Traffic SPECification (RSVP) RFC 2205
TSSI	Time Slot Sequence Integrity
TSTP	TimeSTamP (ATM)
TSWTCM	Time Sliding Window Three Colour Marker RFC 2859
tt	Trinidad and Tobago
TT	Trouble Ticketing
TTC	Telecommunication Technology Committee (Japan)
TTL	Time-To-Live (field — in seconds) RFC 791
TTP	Trail Termination Point
TTR	Time To Repair
TTRT	Target Token Rotation Time
TTY	TeleTYpe
TTY_OP_END	Indicates end of options (pseudo-terminal mode)
TTY_OP_ISPEED	Specifies the terminal input baud rate in bits per second (pseudo-terminal mode)
TTY_OP_OSPEED	Specifies the terminal output baud rate in bits per second (pseudo-terminal mode)
TU	Tributary Unit (SDH/SONET)
TUBA	TCP and UDP with Bigger Addresses RFC 1347, RFC 1526, RFC 1561
TUG	Tributary Unit Group (SDH)
TUID	Telnet User Identification RFC 0927
TU-n	Tributary Unit of order n (SDH)
TURN	SMTP keyword — reverses roles of SMTP server and client
tv	TeleVision Internet domain

Abbreviation	Full form and references
tv	Tuvalu
tw	Taiwan
Tx	Transmit
TxD	Transmit Data ITU-T V.24
tz	Tanzania
U	User option
ua	Ukraine
UA	Unnumbered Acknowledgement (HDLC, LLC)
UA	User Agent
UAE	User Agent Entity (X.400)
UAL	User Adaptation Layer RFC 3057, ITU-T Q.921
UAL	User Agent Layer (X.400)
UAP	open User group Authorization Protocol (DHCP) RFC 2485
UART	Universal Asynchronous Receiver/Transmitter
UAS	UnAvailable Seconds
UB	Reference point at the UNI on the network side of B-NT1
U-bit	Unrecognised TLV type (MPLS) RFC 3032
UBR	Unspecified Bit Rate (ATM)
UCLA	University of California, Los Angeles
UCS	Universal Multiple Octet (Unicode) Character Set ISO-10646
UCSB	University of California, Santa Barbara
UDDI	Universal Discovery, Description and Integration directory www.uddi.org
UDF	User DeFined
UDLR	UniDirectional Link Routing
UDP	User Datagram Protocol RFC 0768, RFC 2013 (MIB), RFC 2147 (IPv6), RFC 2454 (MIB IPv6)
UdpMIB	User Datagram Protocol MIB (mib-2) RFC 2013
UDT	Unstructured Data Transferred
u-frame	unnumbered frame (HDLC,LLC)
UFT	Unsolicited File Transfer RFC 1440
ug	Uganda
UI	Unnumbered Information (LLC)
UIDL	Unique IDentifier Listing (POP)
UIDPLUS	Using IMAP4 to support Disconnected operation PLUS RFC 2359
U-interface	2 wires only ISDN BRI interface
uk	United Kingdom
U/L	Universal/Local identifier
um	US minor outlying islands
UME	UNI Management Entity
UMTS	Universal Mobile Telephone Service
UNARP	Unavailable station announcement update to ARP RFC 1868
UNI	User-Network Interface
Unicode	a 16-bit universal code for alphanumeric character representation ISO 10646, www.unicode.org
uniq	UNIQue
UNIX	Universal Operating system for computers developed initially by AT&T
UNRP	User-to-Network Reference Points
UPC	Usage Parameter Control (ATM)
UPF	User Priority Field IEEE 802.1p

Abbreviation	Full form and references
u-plane	User plane (used for transferring user data) ITU-T I.321
UPS	Uninterruptible Power Supply RFC 1628 (MIB)
UpsMIB	Uninterruptible Power Supplies MIB (mib-2) RFC 1628
UPT	Universal Personal Telecommunication
URA	Uniform Resource Agent RFC 2016
URG	URGent (TCP)
URI	Uniform Resource Identifier RFC 1630, RFC 2168, RFC 2396, RFC 2483, RFC 2838, RFC 3305
URL	Uniform Resource Locator RFC 1738, RFC 1808, RFC 1959-60, RFC 2056, RFC 2111, RFC 2368-9, RFC 2384, RFC 2392, RFC 2397, RFC 2806, RFC 3305
URN	Uniform Resource Name RFC 2141, RFC 2169, RFC 2276, RFC 2483, RFC 2611, RFC 2648, RFC 3001, RFC 3061, RFC 3120-1, RFC 3187-8, RFC 3305
us	United States
US	Unit Separator (see ASCII)
US	United States of America
USA	United States of America
USART	Universal Synchronous/Asynchronous Receiver/Transmitter
USB	Universal Serial Bus
USENET	USENET Newsgroup Organization RFC 1036
USER	USERname (FTP)
USERS	Active USERS RFC 0866
USF	Uplink State Flag
USM	User-based Security Model (SNMP v3) RFC 2264, RFC 2274, RFC 2574
USR1	USeR-defined signal 1 (POSIX)
USR2	USeR-defined signal 2 (POSIX)
UT	Universal Time
UTC	Coordinated Universal Time
UTF	UCS Transformation Format (see also unicode) RFC 2044 (UTF8), RFC 2152 (UTF7), RFC 2279 (UTF8), RFC 2781 (UTF-16)
UTOPIA	Universal Test and Operations Physical Interface for ATM (standard ATM/PHY layer interface)
UTP	Unshielded Twisted Pair (cable) EIA/TIA 568
UUCP	Unified User Control Protocol RFC 0976
UUCP	UNIX-to-UNIX Copy Program
UULP	Unified User Level Protocol RFC 0451, RFC 0666
UUP	User-User Protocol RFC 0091
UUS	User-to-User Signalling
uy	Uruguay
uz	Uzbekistan
V	Version number
V	Violation
V.10	ITU-T recommendation for physical layer DTE/DCE data interface ITU-T V.10
V.11	ITU-T recommendation for physical layer DTE/DCE data interface ITU-T V.11
V.24	ITU-T recommendation V.24 (DTE-to-DCE interface) ITU-T V.24
V.32	ITU-T recommendation for inter-modem modulation up to 9600 bit/s ITU-T V.32

Abbreviation	Full form and references
V.34	ITU-T recommendation for inter-modem modulation up to 28 800 bit/s ITU-T V.34
V.35	ITU-T recommendation V.35 (highspeed DTE-to-DCE interface) ITU-T V.35
V.36	ITU-T recommendation V.36 (highspeed DTE-to-DCE interface) ITU-T V.36
V.90	ITU-T recommendation V.90 (high speed modem) ITU-T V.90
V+D	Voice and Data
va	Vatican state
VACM	View-based Access Control Model (SNMPv3) RFC 2265, RFC 2275, RFC 2575
VAN	Value Added Network
VANS	Value Added Network Services
VarBind	VARiable BINDing (SNMP)
VAS	Value Added Service
VASP	Value Added Service Provider
VAX	Virtual Address eXtended (Digital equipment corporation)
VBR	Variable Bit Rate (ATM)
VBR-NRT	Variable BitRate (Non Real Time)
VBR-RT	Variable BitRate (Real Time)
vc	Saint Vincent and the Grenadines
VC	Virtual Call
VC	Virtual Channel, virtual connection or virtual circuit RFC 3031
VC	Virtual Container (SDH)
VC-4	Virtual Container of SDH with 149.76 Mbit/s transmission rate
VC-11	Virtual Container of SDH with 1544 kbit/s transmission rate
VC-12	Virtual Container of SDH with 2048 kbit/s transmission rate
VC-21	Virtual Container of SDH with 6312 kbit/s transmission rate
VC-22	Virtual Container of SDH with 8448 kbit/s transmission rate
VC-31	Virtual Container of SDH with 34.368 Mbit/s transmission rate
VC-32	Virtual Container of SDH with 44.736 Mbit/s transmission rate
VC CEPF	Virtual Connection Connection End Point Functions
VC CPF	Virtual Connection Connecting Point Functions
vCard	visiting Card file of type .vcf RFC 2739
VCC	Virtual Channel Connection (ATM)
VCCE	Virtual Channel Connection Endpoint
VCDIFF	also Vcdiff—a compressed data format based on the Vdelta algorithm RFC 3284
vcf	visiting card file RFC 2739
VCI	Virtual Channel Identifier (ATM) ITU-T I.361
VCI	Virtual Circuit Identifier RFC 3031
VCID	Virtual Connection Identifier (ATM) RFC 3038
VCL	Virtual Channel Link
VC-n	Virtual Container of order n (SDH)
VDISCARD	Toggles the flushing of terminal output (pseudo-terminal mode)
VDSUSP	Another suspend character (pseudo-terminal mode)
VDU	Video Display Unit
ve	Venezuela
VEMMI	VErsatile MultiMedia Interface RFC 2122

Abbreviation	Full form and references
VENUS	Very Extensive Non-Unicast Service RFC 2191
VEOF	End-of-file character (pseudo-terminal mode)
VEOL	End-of-line character in addition to (carriage return line feed) (pseudo-terminal mode)
VEOL2	Additional end-of-line character (pseudo-terminal mode)
ver	version number
VERASE	Erase the character to left of the cursor (pseudo-terminal mode)
VF	Variance Factor
VF	Voice Frequency
VFIP	Voice File Interchange Protocol RFC 0978
VFLUSH	Character to flush output (pseudo-terminal mode)
VFS	Virtual File System
vg	Virgin Islands (British)
VGA	Video Graphics Array
vi	Virgin Islands (USA)
VID	VLAN Identifier IEEE 802.1p
VINTR	Interrupt character (pseudo-terminal mode)
VISCII	VIetnamese Standard Code for Interchange of Information RFC 1456
VKILL	Kill the current input line (pseudo-terminal mode)
VLAN	Virtual-bridged Local Area Network or Virtual Local Area Network IEEE 802.1q IEEE 802.1u
VLAN	Virtual bridged Local Area Network RFC 2643, RFC 2674 (MOs), RFC 3069, IEEE 802.1q
VlanHello	Cabletron's VlanHello protocol RFC 2641
VLNEXT	Enter the next character typed, even if it is a special character (pseudo-terminal mode)
VLS	Cabletron Virtual LAN Link State Protocol RFC 2642
VLSM	Variable Length Subnet Mask RFC 1219, RFC 1517-8, RFC 1519 (CIDR)
VM	Virtual Machine (IBM)
VMA	VPN (Virtual Private Network) Multicast Address RFC 2917
VMTP	Versatile Message Transaction Protocol RFC 1045
vn	Vietnam
VNS	Nortel Virtual Networking Switching RFC 2340
VoD	Video On Demand
VOIP	Voice Over Internet Protocol ITU-T H.323
VP	Virtual Path (ATM) RFC 3031
VP CEPF	Virtual Path Connection End Point Functions
VP CPF	Virtual Path Connecting Point Functions
VPC	Virtual path connection (ATM)
VPCE	Virtual Path Connection Endpoint
VPI	Virtual Path Identifier (ATM) ITU-T I.361
VPI	Virtual Path Identifier RFC 3031
VPIM	Voice Profile for Internet Mail RFC 1911, RFC 2421 (v2), RFC 2423
VPL	Virtual Path Link
VPN	Virtual Private Network RFC 2547 (BGP), RFC 2735, RFC 2764 (IP), RFC 2917 (MPLS)

Abbreviation	Full form and references
VPNI	Virtual Private Networks Identifier RFC 2685, RFC 2917
VPNID	Virtual Private Networks IDentifier RFC 2917
VPT	Virtual Path Terminator
VP-XC	Virtual Path crossConnect
VQUIT	quit character (pseudo-terminal mode)
VR	Virtual Route (IBM SNA)
VR	Virtual Router RFC 2917
VRC	Virtual Router Console RFC 2917
VREPRINT	Reprints the current input line (pseudo-terminal mode)
VRFY	VeRiFY (SMTP)
VRRP	Virtual Router Redundancy Protocol RFC 2338, RFC 2787 (MOs)
VSTART	Continues paused output (normally control-Q) (pseudo-terminal mode)
VSTATUS	Prints system status line (load, command, pid, etc.) (pseudo-terminal mode)
VSTOP	Pauses output (normally control-S) (pseudo-terminal mode)
VSUSP	Suspends the current program (pseudo-terminal mode)
VSWTCH	Switch to a different shell layer (pseudo-terminal mode)
VT	Virtual Terminal RFC 0782
VT	Virtual Tributary (SONET)
VT	Vertical Tab (see ASCII)
VT1.5	Virtual Tributary of SONET with 1544 kbit/s transmission rate
VT100	ASCII based Virtual Terminal developed by the Digital equipment corporation
VT-100	Virtual Terminal-100 (a terminal emulation device conforming to DECNET)
VT2.0	Virtual Tributary of SONET with 2048 kbit/s transmission rate
VT3.0	Virtual Tributary of SONET with 3152 kbit/s transmission rate
VT6.0	Virtual Tributary of SONET with 6312 kbit/s transmission rate
VTAM	Virtual Telecommunications Access Method (IBM SNA)
VTE	Virtual Terminal Environment
VTOA	Voice Over ATM
vu	Vanuatu
VWERASE	Erases a word left of cursor (pseudo-terminal mode)
WAE	Wireless Application Environment (WAP) www.wapforum.org
WAL2	(line code)
WAN	Wide Area Network
WAP	Wireless Application Protocol www.wapforum.org
WARC	World Administrative Radio Council
WAVE	wave file of type. wav RFC 2361
WCCP	Web Cache Control Protocol RFC 3040
WDM	Wavelength Division Multiplexing
WDP	Wireless Datagram Protocol (WAP) www.wapforum.org
WDS	Wireless Distribution System (WLAN) IEEE 802.11
WEBDAV	HTTP extension for Distributed Authoring and Versioning RFC 2518, RFC 3253
WEBI	WEB Intermediaries

Abbreviation	Full form and references
WebNFS	Web Network File System RFC 2054-5, RFC 2755
WEN	WAN-Error-Notify (L2TP) RFC 2661
WEP	Wireless Equivalent Privacy (WLAN) IEEE 802.11
wf	Wallis and Futuna Islands
WF	Wildcard Filter style (RSVP) RFC 2205
WG	Working Group
WHOIS	A command associated with the IETF NICNAME service RFC 0812, RFC 0954, RFC 1834, RFC 1835 (Whois + +), RFC 1913-4 (Whois + +), RFC 2957-8
WINS	Windows Internet Name Server
WLAN	Wireless Local Area Network IEEE 802.11
WLL	Wireless Local Loop
WML	Wireless Markup Language www.wapforum.org
WPAD	Web Proxy Auto-discovery Protocol RFC 3040
WPKI	WAP Public Key Infrastructure www.wapforum.org
WP-TCP	Wireless Profiled Transport Control Protocol www.wapforum.org
WRQ	Write ReQuest (TFTP)
ws	Western Samoa
WS	Window Size
WS	WorkStation
WSF	WorkStation Function
WSOPT	Window Scale OPTion (TCP) RFC1323
WSP	Wireless Session Protocol www.wapforum.org
WSP	Wireless Session Protocol (WAP)
WTA	Wireless Telephony Application www.wapforum.org
WTAI	Wireless Telephony Application Interface www.wapforum.org
WTLS	Wireless Transport Layer Security protocol (WAP) www.wapforum.org
WTP	Wireless Transaction Protocol (WAP) www.wapforum.org
W/U	Wanted to Unwanted signal ratio
WWW	World Wide Web RFC 1630, RFC 2594 (MOs)
X	eXtension bit
X.3	packet assembler/disassembler specification ITU-T X.3
X.21	ITU-T recommendation X.21 (highspeed DTE-to-DCE interface) ITU-T X.21
X.25	packet network UNI ITU-T X.25
X.28	interface between a DTE and a PAD (packet network) ITU-T X.28
X.29	procedure for synchronous/asynchronous communication across a packet network ITU-T X.29
X.75	packet network NNI ITU-T X.75
X.121	ITU-T recommendation defining public data network numbering scheme ITU-T X.121
X.400	ITU-T recommendation defining the message handling service (MHS) ITU.T X.400
X.500	ITU-T recommendation defining the directory service ITU-T X.500
X.509	ITU-T recommendation defining public key infrastructure (PKI) for network security ITU-T X.509
X11	X-windows protocol version 11 www.x.org

Abbreviation	Full form and references
XACML	eXtensible Access Control Markup Language
XCASE	Enable input and output of uppercase characters by preceding their lowercase equivalents with '\'. (pseudo-terminal mode)
XDR	eXternal Data Representation RFC 1014, RFC 1832
xDSL	generic name for Digital Subscriber Line systems
XID	eXchange Identification or transaction Identification
XML	eXtensible Markup Language RFC 2376, RFC 2807, RFC 3017, RFC 3023, RFC 3075, RFC 3076, RFC 3120, RFC 3275
XMP	X Manager Protocol (a mapping function — common handler between CMIS/CMIP)
XNS	Xerox Network System
XNSCP	FTP Recipient Scheme selection RFC 1764
X-OFF	Transmit 'off' at DTE/DCE interface - data may not be sent
X-ON	Transmit 'on' at DTE/DCE interface - data may be sent
XON/XOFF	flow control protocol
XOT	X.25 Over TCP/IP
XPG	X/Open Portability Guides
XRCP	PPP XNS IDP Control Protocol RFC 0743
XRSQ	FTP ReCiPient specification RFC 0743
XSEN	FTP SENd to terminal RFC 0737
XTERM	X-windows TERMinal emulator www.x.org
XTI	X/Open Transport Interface
XTP	Xpress Transfer Protocol
X-Windows	uniX-Windows — a graphical user interface for UNIX similar to Windows RFC 1013 (v11)
ye	Yemen
yiaddr	Your IP ADDRess (BOOTP/DHCP) RFC 0951
YP	Yellow Pages (SUN)
yt	Mayotte
yu	Yugoslavia
za	South Africa
ZBTSI	Zero Byte Time Slot Interchange (line code)
ZIP	Zone Information Protocol (Appletalk)
ZLB	Zero-Length Body
ZLIB	compressed data format RFC 1950 (v3.3)
zm	Zambia
zrc	Zero-Restart-Count RFC 1661 (PPP)
zw	Zimbabwe

Bibliography

Useful websites

Subject area	Organisation	Website
Internet standards and RFCs	Internet Society /IETF/ RFC editor	www.rfc-editor.org
	Internet Archives	www.faqs.org/rfcs/
Administration and design of the Internet	The Internet Society	www.isoc.org
	Internet Architecture Board (IAB)	www.iab.org
	Internet Engineering Steering Group (IESG)	www.ietf.org/iesg.html
	Internet Engineering Task Force (IETF)	www.ietf.org
	Internet Research Task Force (IRTF)	www.irtf.org
Internet addressing, domain names and protocol parameter assignment	Internet Assigned Numbers Authority (IANA)	www.iana.org
	Internet Network Information Centre	www.internic.org or www.internic.net
	Asia-Pacific Network Information Centre (APNIC)	www.apnic.net
	American Registry for Internet Numbers (ARIN)	www.arin.net
	Réseaux IP Européens (RIPE — European IP Network Coordination Centre)	www.ripe.net

Subject area	Organisation	Website
International telecommunications standards	International Telecommunications Union	www.itu.int
	ITU-T recommendations	www.itu.int/ITU-T/publications/index.html
	Institution of Electrical and Electronics Engineers (IEEE)	www.ieee.org
	International Organization for Standardization (ISO)	www.iso.org
	International Electrotechnical Commission	www.iec.ch
Protocols	General protocols	www.protocols.com
	ATM (asynchronous transfer mode) Forum	www.atmforum.com
	Frame Relay Forum	www.frforum.com
	IEEE 802 standards committee	http://grouper.ieee.org/groups/802/
	IPv6 forum	www.ipv6forum.com
	MPLS (multiprotocol label switching) Forum	www.mplsforum.org
	UMTS (universal mobile telephone service — also called 3GSM)	www.gsmworld.com
		www.3gpp.org
		www.umts-forum.com
	VOIP (voice over IP)	www.iptelephony.org
	WAP (wireless access protocol) Forum	www.wapforum.org
	W3C (world wide web consortium)	www.w3.org
Coding	ASCII and other codes	www.jimprice.com/jim-asc.htm
		http://czyborra.com/charsets/
		http://czyborra.com/charsets/iso8859.html
	Unicode Inc.	www.unicode.org
Cryptography	RSA Security	www.rsasecurity.com
	Data Security Solutions	www.datasecuritysolutions.com
	Counterpane Internet Security Inc. (CTO is Bruce Schneier)	www.counterpane.com
Router and network equipment manufacturers	Cisco Systems	www.cisco.com

Subject area	Organisation	Website
	Extreme Networks	www.extremenetworks.com
	Foundry Networks	www.foundrynet.com
	Juniper Networks	www.juniper.net
	Lucent Technologies	www.lucent.com
	Northern Telecom (Nortel)	www.nortelnetworks.com
	3 Com	www.3com.com
Network management software manufacturers	Micromuse Netcool	www.micromuse.com
	HP Openview	www.openview.hp.com
	Remedy Corporation	www.remedy.com
	Syndesis NetProvision	www.syndesis.com
Cabling and connectors	EIA/TIA (Electronics Industries Alliance/ Telecommunications Industry Association) standards	www.tiaonline.org
	Review of structured cabling & cable types	www.kwhw.co.uk/structured.htm
		www.dennislundin.com/stuntcom munications/australia.datacablenet. com/dctpages/cablingstandards.html
	AMP	www.amp.com
	Belden Inc	www.belden.com
	Black Box Corporation	www.blackbox.com
	Pirelli	www.pirelli.com/cables/telecom.htm
	RS Components Ltd.	www.rs-components.com
	Tyco Electronics	www.tycoelectronics.com
	USB Implementors Forum Inc	www.usb.org
Computer equipment manufacturers	Hewlett Packard/COMPAQ	www.hp.com
	IBM	www.ibm.com
	Sun Microsystems	www.sun.com
Web browser software	Microsoft Internet Explorer	www.microsoft.com/windows/ie/ default.htm
	Nescape Navigator	www.netscape.com
Internet service providers (ISPs)	AOL (America Online)	www.aol.com
	AT&T Internet Services	www.att.net
	British Telecom openworld	www.btopenworld.com
	Cable & Wireless	www.cwplc.com
	Deutsche Telekom T-Online	www.t-online.de
	France Telecom wanadoo	www.wanadoo.fr
	Sprint Corporation	www.sprint.com
	Worldcom UUNET	www.worldcom.com/uunet/

General books on data communications, protocols and encryption

Book title	Author	Publication date & edition	Publisher
Data and Computer Communications	William Stallings	2000/6th edition	Prentice Hall
Computer Networks	Andrew Tanenbaum	1996/4th edition	Prentice Hall
Internetworking with TCP/IP Volume I	Douglas Comer (Purdue University)	2000/4th edition	Prentice Hall
Interconnections: Bridges, Routers, Switches and Internetworking Protocols	Radia Perlman (Sun Microsystems)	2000	Addison-Wesley
Netzwerkprotokolle in Cisco-Netzwerken (detailed coverage of protocol formats but written in German)	Andreas Aurand	2000	Addison-Wesley
Emerging Communications Technologies	Uyless D. Black	1997/2nd edition	Prentice Hall
TCP/IP	Behrouz Forouzon	1999	McGraw-Hill
An Engineering Approach to Computer Networks	Srinivasan Keshav	1997	Addison-Wesley
Data Communications, Computer Networks and Open Systems	Fred Halsall	1996/4th edition	Addison-Wesley
Computer Networks: a Top-down Approach	Jim Kurose and Keith Ross	2000	Addison-Wesley Longman
TCP/IP Illustrated Volume 1	Richard Stevens	1994	Addison-Wesley
Computer Networks: A Systems Approach	Larry L. Peterson and Bruce S. Davie	1999/2nd edition	Morgan Kaufmann
Data Communications, Computer Networks, and Open Systems	Fred Halsall	1996/4th edition	Addison-Wesley
Designing TCP/IP Networks	Geoff Bennet	1994	Van Nostrand Reinhold

Book title	Author	Publication date & edition	Publisher
Cisco CCIE: preparation book texts	QUE		Baer Wolf
Applied Cryptography: Protocols, Algorithms, and Source Code in C	Bruce Schneier	1995/2nd edition	John Wiley & Sons

General reference

Newton's Telecom Dictionary (Harry Newton) 16th edition (2000) Telecom Books/CMP Media Inc.

Index

Printed and bound by CPI Group (UK) Ltd, Croydon, CR0 4YY

Printed and bound by CPI Group (UK) Ltd, Croydon, CR0 4YY

27/10/2024

14580295-0004